Formula Weights

AgBr	187.78	$K_2Cr_2O_7$	294.19
AgCl	143.32	$K_3Fe(CN)_6$	329.26
Ag_2CrO_4	331.73	$K_4Fe(CN)_6$	368.38
AgI	234.77	$KHC_8H_4O_4$ (phthalate)	204.23
$AgNO_3$	169.87	$KH(IO_3)_2$	389.92
AgSCN	165.95	K_2HPO_4	174.18
Al_2O_3	101.96	KH_2PO_4	136.09
$Al_2(SO_4)_3$	342.14	$KHSO_4$	136.17
As_2O_3	197.85	KI	166.01
B_2O_3	69.62	KIO_3	214.00
$BaCO_3$	197.35	KIO_4	230.00
$BaCl_2 \cdot 2H_2O$	244.28	$KMnO_4$	158.04
$BaCrO_4$	253.33	KNO_3	101.11
$Ba(IO_3)_2$	487.14	KOH	56.11
$Ba(OH)_2$	171.36	KSCN	97.18
$BaSO_4$	233.40	K_2SO_4	174.27
Bi_2O_3	466.0	$La(IO_3)_3$	663.62
CO_2	44.01	$Mg(C_9H_6ON)_2$	312.59
$CaCO_3$	100.09	$MgCO_3$	84.32
CaC_2O_4	128.10	$MgNH_4PO_4$	137.35
CaF_2	78.08	MgO	40.31
CaO	56.08	$Mg_2P_2O_7$	222.57
$CaSO_4$	136.14	$MgSO_4$	120.37
$Ce(HSO_4)_4$	528.4	MnO_2	86.94
CeO_2	172.12	Mn_2O_3	157.88
$Ce(SO_4)_2$	332.25	Mn_3O_4	228.81
$(NH_4)_2Ce(NO_3)_6$	548.23	$Na_2B_4O_7 \cdot 10H_2O$	381.37
$(NH_4)_4Ce(SO_4)_4 \cdot 2H_2O$	632.6	NaBr	102.90
Cr_2O_3	151.99	$NaC_2H_3O_2$	82.03
CuO	79.54	$Na_2C_2O_4$	134.00
Cu_2O	143.08	NaCl	58.44
$CuSO_4$	159.60	NaCN	49.01
$Fe(NH_4)_2(SO_4)_2 \cdot 6H_2O$	392.14	Na_2CO_3	105.99
FeO	71.85	$NaHCO_3$	84.01
Fe_2O_3	159.69	$Na_2H_2EDTA \cdot 2H_2O$	372.2
Fe_3O_4	231.54	Na_2O_2	77.98
HBr	80.92	NaOH	40.00
$HC_2H_3O_2$ (acetic acid)	60.05	NaSCN	81.07
$HC_7H_5O_2$ (benzoic acid)	122.12	Na_2SO_4	142.04
HCl	36.46	$Na_2S_2O_3 \cdot 5H_2O$	248.18
$HClO_4$	100.46	NH_4Cl	53.49
$H_2C_2O_4 \cdot 2H_2O$	126.07	$(NH_4)_2C_2O_4 \cdot H_2O$	142.11
H_5IO_6	227.94	NH_4NO_3	80.04
HNO_3	63.01	$(NH_4)_2SO_4$	132.14
H_2O	18.015	$(NH_4)_2S_2O_8$	228.18
H_2O_2	34.01	NH_4VO_3	116.98
H_3PO_4	98.00	$Ni(C_4H_6O_2N_2)_2$	286.91
H_2S	34.08	$PbCrO_4$	323.18
H_2SO_3	82.08	PbO	223.19
H_2SO_4	98.08	PbO_2	239.19
HgO	216.59	$PbSO_4$	303.25
Hg_2Cl_2	472.09	P_2O_5	141.94
$HgCl_2$	271.50	Sb_2S_3	339.69
KBr	119.01	SiO_2	60.08
$KBrO_3$	167.01	$SnCl_2$	189.60
KCl	74.56	SnO_2	150.69
$KClO_3$	122.55	SO_2	64.06
KCN	65.12	SO_3	80.06
K_2CrO_4	194.20	$Zn_2P_2O_7$	304.68

FUNDAMENTALS OF ANALYTICAL CHEMISTRY

THIRD EDITION

DOUGLAS A. SKOOG
Stanford University

DONALD M. WEST
San Jose State University

HOLT, RINEHART AND WINSTON   New York   Chicago   San Francisco   Atlanta
Dallas   Montreal   Toronto   London   Sydney

Copyright © 1963, 1969, 1976 by Holt, Rinehart and Winston
All rights reserved

Library of Congress Cataloging in Publication Data

Skoog, Douglas Arvid, 1918–
    Fundamentals of analytical chemistry.

    Includes bibliographical references and index.
    1. Chemistry, Analytic.   I. Title.
QD75.2.S55   1976   543   75-25554
ISBN 0-03-089495-6

Printed in the United States of America
    8 9   032   9 8 7 6

PREFACE

This edition of *Fundamentals of Analytical Chemistry* contains several innovations. Chapter 3 is a new chapter which has been added to review equilibrium concepts in the same way that Chapter 2 reviews stoichiometric relationships. All specific laboratory directions are collected in a single chapter (31). Chapters concerned with the evaluation of data, potentiometry, spectroscopy, and separations have been revised and expanded. New problems have been prepared; the answers to nearly half the problems in the text are given at the back of the book. A solutions manual is also available. To make room for new material, discussions concerned with gravimetric analysis have been substantially condensed and the chapter on conductometric titration deleted.

Although it is tempting to emphasize changes, it is also important to note that the principal aims of this edition are the same as those of earlier editions. We remain convinced that a thorough grounding in theory and—insofar as it can be considered in an elementary text—an appreciation for the entire analytical process are of prime importance. The student with a clear understanding of the chemical background for an analysis is most likely to identify the experimental details that are crucial to success and to distinguish them from those which require little or no attention.

With the exception of the changes noted at the outset, we have retained the format of the earlier editions. Although gravimetric analysis is covered before volumetric methods, the order of presentation can be reversed if the instructor so chooses. Sufficient material concerned with instrumental methods for completion of an analysis is presented to allow coverage of those topics that are deemed to be most important; of particular importance for a one-semester course are Chapter 17, in which ion-selective membrane electrodes are discussed, and Chapters 23 and 24, which are concerned with methods based on the absorption of electromagnetic radiation.

Many colleagues have generously shared suggestions for improvement of the presentation. As before, we are particularly grateful to Professor Alfred Armstrong of The College of William and Mary who read the entire manuscript and contributed many incisive criticisms. We also acknowledge with gratitude the comments of M. R. Bacon, University of Nevada; R. G. Bates, University of Florida; D. G. Berge, University of Wisconsin; E. J. Billo, Boston College; R. L. Birke, University of South Florida; G. L. Blackmer, Texas Technological University; J. Q. Chambers, University of Tennessee; J. A. Cox, Southern Illinois University; V. P. Guinn, University of California/Irvine; T. P. Hadjiioannou, University of Illinois/Urbana-Champaign; T. J. Haupert, California State University/Sacramento; J. W. Knoeck, North Dakota State University; R. C. Legendre, University of Southern Alabama; D. E. Leyden, University of Georgia; C. K. Mann, Florida State University; H. B. Mark, Jr., University of Cincinnati; R. S. Mitchell, Arkansas State University; J. H. Nelson, University of Nevada; R. T. O'Donnell, State University of New York/Oswego; G. K. Pagenkopf, Montana State University; F. W. Plankey, University of Pittsburgh; D. S. Polcyn, University of Wisconsin; M. W. Rowe, Texas A&M University; J. T. Stock, University of Connecticut/Storrs; W. E. Swartz, Jr., University of South Florida; P. J. Taylor, Wright State University; C. E. Wilson, Indiana/Purdue University; D. C. Young, Oakland University; S. T. Zenchelsky, Rutgers University.

December, 1975
Stanford, California
San Jose, California

D.A.S.
D.M.W.

CONTENTS

FUNDAMENTALS OF
ANALYTICAL CHEMISTRY

INTRODUCTION

Analytical chemistry deals with methods for the identification of one or more of the components in a sample of matter and the determination of the relative amounts of each. The identification process is called a *qualitative analysis* while the determination of amount is termed a *quantitative analysis*. In this text, we will deal largely with the latter.

The results of a quantitative analysis are expressed in such relative terms as the percent of the *analyte* (the substance being determined) in the sample, the parts of analyte per thousand, per million, or even per billion parts of sample, the grams or milliliters of analyte per liter of sample, the pounds of analyte per ton of sample, or the mole fraction of the analyte in the sample.

Applications of Quantitative Analysis

The results of chemical analyses have widespread practical applications. To cite just a few examples of the way in which quantitative data influence the life of modern man, consider the following. Information as to the parts per million of hydrocarbons, nitrogen oxides, and carbon monoxide in exhaust gases serves as a means of defining the quality of smog-control devices for the automobile.

Determination of the concentration of ionized calcium in blood serum is an important method for the diagnosis of hyperparathyroidism in human patients. Quantitative data on nitrogen in breakfast cereals and other foods are directly related to their nutritional qualities. Periodic quantitative analyses during the production of a steel permit the manufacture of a product having a desired strength, hardness, ductility, or corrosion resistance. The continuous analysis for mercaptans in the household gas supply assures the presence of an odorant which warns of dangerous leaks in the gas-distribution system. The analysis of soils for phosphorus, nitrogen, sulfur, and moisture throughout the growing season makes it possible for the farmer to tailor fertilization and irrigation schedules to meet plant needs most efficiently; significant reductions in costs for fertilizer and water as well as increases in yield result.

In addition to practical applications of the types just considered, quantitative analytical data are at the heart of research activity in chemistry, biochemistry, biology, geology, and the other sciences. In support of this assertion, consider the following few examples. Much of what the chemist knows about the mechanisms by which chemical reactions occur has been learned from kinetic studies in which the rate of disappearance of reactants or appearance of products is followed by quantitative analyses for these reactants or products. Recognition that the conduction of nerve signals in animals and the contraction or relaxation of muscles involves the transport of sodium and potassium ions across membranes was made possible by quantitative measurements of these species on the two sides of such membranes. Studies concerned with the mechanism by which oxygen and carbon dioxide are transported in blood have required methods for continually monitoring the concentration of these and other compounds within a living organism. The understanding of the behavior of semiconductor devices has required development of methods for the quantitative determination of impurities in pure silicon and germanium in the range of 1×10^{-6} to 1×10^{-10} %. Recognition that the amounts of various minor elemental constituents in obsidian samples permit identification and location of their sources has enabled archeologists to trace prehistoric trade routes for tools and weapons fashioned from these materials. In some instances, analytical data on the composition of surface soils have permitted geologists to detect the presence of major ore bodies at considerable depths. Quantitative analyses of minute samples from works of art have provided historians with important clues as to the materials and technologies employed by artists of the past as well as an important tool for the detection of art forgeries.

For the typical research worker in chemistry, biochemistry, and some of the biological sciences, the acquisition of quantitative information represents a significant fraction of his laboratory efforts. Analytical procedures, then, are among the important tools employed by such a scientist in pursuit of his research goals. The development of an understanding of the basis of the quantitative analytical process and the competence and confidence to perform analyses is therefore a prerequisite for research in many of these fields. The role of analytical chemistry in the education of chemists and biochemists can be viewed as being analogous to that of calculus and matrix algebra for those aspiring toward a career in theoretical physics or to the role of Greek or other ancient languages in the education of the scholar of classics.

Performance of Quantitative Analysis

The results of a typical quantitative analysis are based upon two measurements (or sometimes two series of measurements), one of which is related to the quantity of sample taken and the second to the quantity of analyte in that sample. Examples of the quantities measured include weight, volume, light intensity, absorption of radiation, fluorescent intensity, and quantity of electricity. It is important to recognize, however, that these measurements are but a part of the typical quantitative analysis. Indeed, some of the preliminary steps are as important and often more difficult and time-consuming as the measurements themselves.

For the most part, the early chapters of this text are devoted to the final measurement steps, and the other aspects of an analysis are not treated in detail until near the end of the book. Thus, to lend perspective, it is useful at the outset to identify the several steps that make up the analytical process and to indicate their importance.

SAMPLING

To produce meaningful results, an analysis must be performed on a sample whose composition faithfully reflects that of the bulk of material from which it was taken. Where the bulk is large and inhomogeneous, great effort is required to procure a representative sample. Consider, for example, a railroad car containing 25 tons of silver ore. Buyer and seller must come to agreement regarding the value of the shipment, based primarily upon its silver content. The ore itself is inherently heterogeneous, consisting of lumps of varying size and of varying silver content. The actual assay of this shipment will be performed upon a sample that weighs perhaps 1 g; its composition must be representative of the 25 tons (or approximately 22,700,000 g) of ore in the shipment. It is clear that the selection of a small sample for this analysis cannot be a simple one-step operation; in short, a systematic preliminary manipulation of the bulk of material will be required before it becomes possible to select 1 g and have any confidence that its composition is typical of the nearly 23,000,000 g from which it was taken.

The sampling problem often is not so formidable as that outlined in the preceding paragraph. Still, the chemist cannot afford to proceed with an analysis until he has convinced himself that the fraction of the material with which he plans to work is truly representative of the whole.

PREPARATION OF THE LABORATORY SAMPLE FOR ANALYSIS

Often, solid materials must be ground to reduce particle size and then thoroughly mixed to ensure homogeneity. In addition, removal of adsorbed moisture is often required for solid samples. Adsorption or desorption of water causes the percentage composition of a substance to depend upon the humidity of its surroundings at the time of the analysis. To avoid the problems arising from such variations, it is common practice to base the analysis on a dry sample.

MEASUREMENT OF THE SAMPLE

Quantitative analytical results are usually reported in relative terms; that is, in some way that expresses the quantity of the desired component present per unit weight or volume of sample. It is therefore necessary to know the weight or volume of the sample upon which the analysis is performed.

SOLUTION OF THE SAMPLE

Most, but certainly not all, analyses are performed on solutions of the sample. Ideally the solvent should dissolve the entire sample (not just the analyte) rapidly and under sufficiently mild conditions that loss of the analyte cannot occur. Unfortunately, such solvents do not exist for many or perhaps most materials that are of interest to the scientist. Commonly he deals with chemically intractable substances such as an ore sample, a high-molecular-weight polymer, or a piece of animal tissue. Converting the analyte in such material into a soluble form is often a formidable and time-consuming task.

SEPARATION OF INTERFERING SUBSTANCES

Few, if any, chemical or physical properties of importance in analysis are unique to a single chemical species; instead, the reactions used and the properties measured are characteristic of a number of elements or compounds. This lack of truly specific reactions and properties adds greatly to the difficulties faced by the chemist when undertaking an analysis; it means that a scheme must be devised for isolating the species of interest from all others present in the original material that produce an effect upon the final measurement. Compounds or elements that prevent the direct measurement of the species being determined are called *interferences*; their separation prior to the final measurement constitutes an important step in most analyses. No hard and fast rules can be given for the elimination of interferences; this problem is often the most demanding aspect of the analysis.

THE COMPLETION OF THE ANALYSIS

All preliminary steps in an analysis are undertaken to make the final measurement a true gauge of the quantity of the species being determined.

The chapters that follow contain descriptions of many types of final measurement, along with discussions of the chemical principles upon which such measurements are based.

Choice of Methods for an Analysis

The chemist or scientist who has need for analytical data usually finds himself faced with an array of methods which could be used to provide the desired information. His choice among these will be based on such considerations as speed, convenience, accuracy, availability of equipment, number of analyses, amount of sample that can be sacrificed, and concentration range of the analyte. The success or failure of an analysis is often critically dependent upon the proper selection of method. Unfortunately, there are no generally applicable rules that can be applied; the choice of method is thus a matter of judgment. Such judgments are difficult, and the ability to make them well comes only with experience.

This text presents many of the most common unit operations associated with chemical analyses and includes a variety of methods for the final measurement of analytes. Both theory and practical detail are treated. Mastery of this material will permit the student to perform many useful analyses and will provide him with background from which he can develop the judgment needed for the prudent choice of an analytical method.

2

A REVIEW OF SOME ELEMENTARY CONCEPTS

Most quantitative analytical measurements are performed on solutions of the sample. The study of analytical chemistry, therefore, makes use of solution concepts with which the student should have considerable familiarity. It is the purpose of this chapter, and the one that follows, to review the most important of these concepts.

The Chemical Composition of Solutions

Both aqueous and organic solvents find widespread use in chemical analysis. Nonpolar solvents, such as hydrocarbons and halogenated hydrocarbons, are employed when the analyte itself is nonpolar. Organic solvents, such as alcohols, ketones, and ethers, which are intermediate in polarity and which form hydrogen bonds with solutes, are considerably more useful than their less polar counterparts because they dissolve a larger variety of both organic and inorganic species. Aqueous solvents, including solutions of the common inorganic acids and bases, are perhaps the most widely used of all for analytical purposes. Our discussion will therefore focus on the behavior of solutes in water; reactions in nonaqueous polar media will be considered in less detail.

ELECTROLYTES

Electrolytes are solutes which ionize in a solvent to produce an electrically conducting medium. *Strong electrolytes* ionize completely whereas *weak electrolytes* are only partially ionized in the solvent. Table 2-1 summarizes the common strong and weak electrolytes in aqueous media.

TABLE 2-1   Classification of Electrolytes

Strong Electrolytes	Weak Electrolytes
1. The inorganic acids HNO_3, $HClO_4$, H_2SO_4,[a] HCl, HI, HBr, $HClO_3$, $HBrO_3$	1. Many inorganic acids such as H_2CO_3, H_3BO_3, H_3PO_4, H_2S, H_2SO_3
2. Alkali and alkaline-earth hydroxides	2. Most organic acids
3. Most salts	3. Ammonia and most organic bases
	4. Halides, cyanides, and thiocyanates of Hg, Zn, and Cd

[a] H_2SO_4 is completely dissociated into HSO_4^- and H_3O^+ ions, and for this reason is classified as a strong electrolyte. However, it should be noted that the HSO_4^- ion is a weak electrolyte, being only partially dissociated.

THE SELF-IONIZATION OF SOLVENTS

Many common solvents are weak electrolytes which react with themselves to form ions (this process is termed *autoprotolysis*). Some examples are

$$2H_2O \rightleftarrows H_3O^+ + OH^-$$
$$2CH_3OH \rightleftarrows CH_3OH_2^+ + CH_3O^-$$
$$2HCOOH \rightleftarrows HCOOH_2^+ + HCOO^-$$
$$2NH_3 \rightleftarrows NH_4^+ + NH_2^-$$

The positive ion formed by the autoprotolysis of water is called the *hydronium* ion, the proton being bonded to the parent molecule via a coordinate covalent bond involving one of the unshared electron pairs of the oxygen. Higher hydrates such as $H_5O_2^+$ and $H_7O_3^+$ are also present, but their stabilities are significantly less than that of H_3O^+. No unhydrated ions appear to exist in aqueous solutions.

To emphasize the high stability of the singly hydrated proton, many chemists use the notation H_3O^+ in writing equations for reactions in which the proton is a participant. As a matter of convenience, others use H^+ to symbolize the proton, whatever its actual degree of hydration may be; this notation possesses the advantage of simplifying the writing of equations that require the proton for balance.

ACIDS AND BASES IN VARIOUS SOLVENTS

Historically speaking, the classification of substances as acids or bases was founded upon several characteristic properties that these compounds impart to an aqueous solution. Typical properties include the red and blue colors that are associated with the reaction of acids and bases with litmus, the sharp taste of a dilute acid solution, the bitter taste and slippery feel of a basic solution, and the formation of a salt by interactions of an acid with a base.

In the late nineteenth century, Arrhenius proposed a more sophisticated classification. He defined acids as hydrogen-containing substances that dissociate into hydrogen ions and anions when dissolved in water and bases as compounds containing hydroxyl groups that give hydroxide ions and cations upon the same treatment. This fruitful proposal allowed a quantitative treatment of the idea of acidity and basicity. Thus, the relative strengths of acids and bases could be compared by measuring the degree of dissociation in aqueous solution. In addition, the theory provided a foundation for the mathematical treatment of the equilibria that are established when acids and bases react with one another.

A serious limitation to the Arrhenius theory is its failure to recognize the role played by the solvent in the dissociation process. It remained for Brønsted and Lowry to propose independently in 1923 a more generalized concept of acids and bases. In the Brønsted-Lowry view, an acid is any substance that is capable of donating a proton; a base is any substance that can accept a proton. The loss of a proton by an acid gives rise to an entity that is a potential proton acceptor and thus a base; it is called the *conjugate base* of the parent acid. The reaction between an acid and water is a typical example.

$$H_2O + acid \rightleftarrows conjugate\ base + H_3O^+$$

It is important to recognize that the acidic character of a substance will be observed only in the presence of a proton acceptor; similarly, basic behavior requires the presence of a proton donor.

Neutralization, in the Brønsted-Lowry sense, can be expressed as

$$acid_1 + base_2 \rightleftarrows base_1 + acid_2$$

This process will be spontaneous in the direction that favors production of the weaker acid and base. The dissolving of many solutes can be regarded as neutralizations, with the solvent acting as either a proton donor or acceptor. Thus,

$acid_1$	+ $base_2$ $\rightleftarrows$	$base_1$	+ $acid_2$
HCl	+ H_2O $\rightleftarrows$	Cl^-	+ H_3O^+
$HC_2H_3O_2$	+ H_2O $\rightleftarrows$	$C_2H_3O_2^-$	+ H_3O^+
$Al(H_2O)_6^{3+}$	+ H_2O $\rightleftarrows$	$AlOH(H_2O)_5^{2+}$	+ H_3O^+
$H_2PO_4^-$	+ H_2O $\rightleftarrows$	HPO_4^{2-}	+ H_3O^+
NH_4^+	+ H_2O $\rightleftarrows$	NH_3	+ H_3O^+
H_2O	+ NH_3 $\rightleftarrows$	OH^-	+ NH_4^+
H_2O	+ CO_3^{2-} $\rightleftarrows$	OH^-	+ HCO_3^-

Note that acids can be anionic, cationic, or electrically neutral. It is also seen that water acts as a proton acceptor (a base) with respect to the first five solutes and as a proton donor or acid with respect to the last two; solvents that possess both acidic and basic properties are called *amphiprotic*.

Acids (and bases) differ in the extent to which they react with solvents. To illustrate, the reaction between hydrochloric acid and water is essentially complete; this solute is thus classed as a strong acid in the solvent water. Acetic acid and ammonium ion react with water to a lesser degree, with the result that these substances are progressively weaker acids.

The extent of reaction between a solute acid (or base) and a solvent is also critically dependent upon the tendency of the latter to donate or accept protons. Thus, for example, perchloric, hydrochloric, and hydrobromic acids are all classed as strong acids in water. If glacial acetic acid,[1] a poorer proton acceptor, is used *as the solvent* instead, only perchloric acid undergoes complete dissociation and remains a strong acid; the process can be expressed by the equation

$$\underset{\text{acid}_1}{HClO_4} + \underset{\text{base}_2}{HC_2H_3O_2} \rightleftarrows \underset{\text{base}_1}{ClO_4^-} + \underset{\text{acid}_2}{H_2C_2H_3O_2^+}$$

Because they undergo only partial dissociation, hydrochloric and hydrobromic acids are weak acids in glacial acetic acid.

A consequence of the Brønsted theory is that the most effective proton donors (that is, the strongest acids) give rise, upon loss of their protons, to the least effective proton acceptors (the weakest conjugate bases). Referring again to the reaction of hydrochloric acid, acetic acid, and ammonium ion with water, it follows that chloride ion is the weakest base, acetate ion is next, and ammonia is the strongest.

The *general solvent theory* includes not only species that qualify as acids or bases in the Brønsted-Lowry sense but also extends the concept of acid-base behavior to solvents that do not necessarily contain protons. The theory is particularly useful with respect to these latter systems. Here an acid is any substance that either contains, or reacts with the solvent to produce, the cation that is characteristic of the solvent; similarly, a base is any substance that directly or indirectly yields the anion characteristic of the solvent.

A still more general view of acids and bases was proposed by Lewis, who defined an acid as an electron-pair acceptor and a base as an electron-pair donor. His concepts go further toward freeing acid-base behavior from the involvement of protons and greatly increase the number of processes that can be considered as acid-base reactions as well. The Lewis ideas are useful in explaining organic reaction mechanisms but are too broad for useful applications in analytical chemistry.

We shall employ the Brønsted view where we are concerned with acid-base equilibria and shall employ the symbol H_3O^+ in this context. On the other hand, we shall employ the more convenient symbol H^+ to represent the proton where chemical weight relationships and balanced equations are of principal concern.

---

[1] Pure or 100% acetic acid is commonly called glacial acetic acid because it forms icelike crystals at $16.7\,°C$; small quantities of water lower the freezing point drastically.

Chemical Units of Weight and Concentration

In the laboratory, the mass of a substance is ordinarily determined in such metric units as the kilogram (kg), the gram (g), the milligram (mg), the microgram (μg), the nanogram (ng), or the picogram (pg).[2] For chemical calculations, however, it is more convenient to employ mass units that express the weight relationship or *stoichiometry* among reacting species in terms of small whole numbers. The gram formula weight, the gram molecular weight, and the gram equivalent weight are employed in analytical work for this reason. These terms are often shortened to the formula weight, the molecular weight, and the equivalent weight.

CHEMICAL FORMULAS, FORMULA WEIGHTS, AND MOLECULAR WEIGHTS

An empirical formula expresses the simplest combination of atoms in a substance. It also serves as the chemical formula unless experimental evidence exists to indicate that the fundamental aggregate is actually some multiple of the empirical formula. For example, the chemical formula for hydrogen is H_2 because the gas exists as diatomic molecules under ordinary conditions. In contrast, Ne serves adequately to describe the composition of neon, which is observed to be monatomic.

The entity expressed by the chemical formula may or may not actually exist. For example, no evidence has been found for sodium chloride molecules, as such, in the solid state or in aqueous solution. Rather, this substance consists of sodium ions and chloride ions, no one of which can be shown to be in simple combination with any other single ion. Nevertheless, the formula NaCl is convenient for stoichiometric accounting and is so used. It is also necessary to note that the chemical formula is frequently that of the principal species only. Thus, for example, water in the liquid state contains small amounts of such entities as H_3O^+, OH^-, H_4O_2 (and undoubtedly others), in addition to H_2O. Here the chemical formula of H_2O is that for the predominant species and is perfectly satisfactory for chemical accounting; it is, however, only an approximation of the actual composition of the real substance.

The *gram formula weight* (gfw) is the summation of atomic weights, in grams, of all the atoms in the chemical formula of a substance. Thus, the gram formula weight for H_2 is 2.016 (2 $\times$ 1.008) g; for NaCl it is 58.44 (35.45 + 22.99) g. The definition for the gram formula weight carries with it no inference concerning the existence or nonexistence of the substance for which it has been calculated.

We shall employ the term *gram molecular weight* (gmw) rather than gram formula weight when we are concerned with a real chemical species. Thus, the gram molecular weight of H_2 is its gram formula weight 2.016 g. If we are dealing with the substance NaCl in water, we will not assign to it a gram molecular weight because this species is not found in aqueous media. It is perfectly proper to assign gram molecular weights to Na^+ (22.99 g) and Cl^- (35.45 g) since these are real chemical entities (strictly, these should be called gram ionic

---

[2] The relationship among these units is g $= 10^3$ mg $= 10^6\ \mu$g $= 10^9$ ng $= 10^{12}$ pg $= 10^{-3}$ kg.

weights rather than gram molecular weights, although this terminology is seldom encountered).

One molecular weight of a species contains 6.02×10^{23} particles of that species; this quantity is frequently referred to as the *mole*. In a similar way, the formula weight represents 6.02×10^{23} units of the substance, whether real or not, represented by the chemical formula.

Example. A 25.0-g sample of H_2 contains

$$\frac{25.0 \text{ g}}{2.016 \text{ g/mole}} = 12.4 \text{ moles of } H_2$$

$$12.4 \text{ moles} \times \frac{6.02 \times 10^{23} \text{ molecules}}{\text{mole}} = 7.47 \times 10^{24} \text{ molecules } H_2$$

The same weight of NaCl contains

$$\frac{25.0 \text{ g}}{58.44 \text{ g/fw}} = 0.428 \text{ fw NaCl}$$

which corresponds to 0.428 mole Na^+ and 0.428 mole Cl^-

Let us further distinguish between the formula weight and the molecular weight by considering exactly one formula weight of water which, by definition, weighs 18.015 g. Such a quantity contains slightly less than one mole of H_2O, owing to the existence of H_3O^+, OH^-, H_4O_2, and such other species as may be present.

Laboratory quantities are frequently more conveniently expressed in terms of *milliformula weights* (mfw) or *millimoles* (mmole); these represent, respectively, $\frac{1}{1000}$ of the quantities defined previously.

CONCENTRATION OF SOLUTIONS

Several methods are employed to describe the concentrations of solutions.

Formality or Formal Concentration. The *formality*, *F*, gives the *number* of formula weights of a substance contained in one liter of *solution*. The term also expresses the number of milliformula weights per milliliter of solution.

Example. Exactly 4.57 g of $BaCl_2 \cdot 2H_2O$ (gfw = 244) were dissolved in water and diluted to exactly 250 ml in a volumetric flask. What is the formal concentration of $BaCl_2 \cdot 2H_2O$ and Cl^-?

The number of milliformula weights of the compound taken is

$$\text{no. mfw } BaCl_2 \cdot 2H_2O = \frac{4.57 \text{ g } BaCl_2 \cdot 2H_2O}{0.244 \text{ g } BaCl_2 \cdot 2H_2O/\text{mfw}} = 18.73$$

$$F = \frac{\text{no. mfw } BaCl_2 \cdot 2H_2O}{\text{ml}} = \frac{18.73 \text{ mfw}}{250 \text{ ml}} = 0.0749 \text{ mfw/ml}$$

$$\text{no. mfw } Cl^- = 2 \times \text{no. mfw } BaCl_2 \cdot 2H_2O$$

Therefore,

$$F_{Cl^-} = 2 \times F_{BaCl_2 \cdot 2H_2O} = 2 \times 0.0749 = 0.1498 \text{ mfw/ml}$$

Molarity or Molar Concentration. The *molarity, M,* expresses the number of molecular weights or moles of a solute per liter of solution or the number of millimoles per milliliter. As shown in the following examples, the formal and the molar concentration of a solution may in some cases be identical; in others, they will be quite different.

Example.   Calculate the formal and the molar concentrations of the constituents in (a) a solution that contains 2.30 g of ethanol, C_2H_5OH (gfw = 46.1), in 3.50 liters of aqueous solution; (b) a solution that contains 285 mg trichloroacetic acid, Cl_3CCOOH (gfw = 163), in 10.0 ml of aqueous solution (assume that the acid is 73% ionized in the aqueous medium).

(a) $$F = \frac{2.30 \text{ g } C_2H_5OH}{46.1 \text{ g } C_2H_5OH/fw} \times \frac{1}{3.50 \text{ liters}} = 0.0143 \text{ fw/liter}$$

In an aqueous solution of ethanol, the only solute species present in any significant quantity is C_2H_5OH. Therefore,

$$M = \frac{2.30 \text{ g } C_2H_5OH}{46.1 \text{ g } C_2H_5OH/\text{mole}} \times \frac{1}{3.50 \text{ liters}} = 0.0143 \text{ mole/liter}$$

(b) Employing HA as the symbol for Cl_3CCOOH, we write

$$F = \frac{285 \text{ mg HA}}{163 \text{ mg HA/mfw}} \times \frac{1}{10 \text{ ml}} = 0.175 \text{ mfw/ml}$$

Because all but 27% of the Cl_3CCOOH is dissociated as H_3O^+ and Cl_3CCOO^-, the molar concentration of Cl_3CCOOH is given by

$$M_{HA} = \frac{(285 \times 0.27) \text{ mg HA}}{163 \text{ mg HA/mmole}} \times \frac{1}{10.0 \text{ ml}} = 0.0472 \text{ mmole/ml}$$

The molarity of H_3O^+ as well as Cl_3CCOO^- is equal to the formal concentration of the acid minus the concentration of undissociated acid; that is,

$$M_{H_3O^+} = M_{A^-} = 0.175 - 0.0472 = 0.128 \text{ mmole/ml}$$

It is important to appreciate that the practice of restricting the terms "mole" and "molarity" to a real species and its solution is not universally followed. Thus, many chemists employ the terms "formal concentration" and "molar concentration" interchangeably. The solution just considered is then described as having an analytical or total acid concentration of 0.175 *M.* When specifying the concentration of the undissociated acid, statements such as "0.047 *M* in the undissociated acid" or "a species concentration of 0.047-*M* acid" are employed.

The foregoing example reveals that quantitative information concerning the fate of a solute is needed before the molar concentration of its solution can be specified. In contrast, the formal concentration can be established from the specifications for preparation of the solution and the formula weight of the solute.

Example.   Describe the preparation of 2.00 liters of 0.100-*F* Na_2CO_3 from the pure solid.

$$\text{no. fw } Na_2CO_3 \text{ needed} = 0.100 \frac{\text{fw } Na_2CO_3}{\text{liter}} \times 2.00 \text{ liters} = 0.200$$

$$\text{g } Na_2CO_3 = 0.200 \text{ fw} \times 106 \frac{\text{g } Na_2CO_3}{\text{fw}} = 21.2$$

Therefore, dissolve 21.2 g of the Na_2CO_3 in water, and dilute to exactly 2.00 liters.

Example. Describe the preparation of 2.00 liters of 0.100-M Na^+ from pure Na_2CO_3.

$$\text{no. fw } Na_2CO_3 \text{ needed} = 0.100 \frac{\text{mole } Na^+}{\text{liter}} \times 2.00 \text{ liters} \times \frac{1 \text{ fw } Na_2CO_3}{2 \text{ moles } Na^+}$$

$$= 0.100$$

$$\text{g } Na_2CO_3 = 0.100 \text{ fw} \times 106 \frac{\text{g } Na_2CO_3}{\text{fw}} = 10.6$$

Dissolve 10.6 g Na_2CO_3 in water and dilute to 2.00 liters.

Normality or Normal Concentration. This specialized method for expressing concentration is based upon the number of equivalents of solute which are contained in a liter of solution. Normality and equivalent weight are defined in Chapter 7.

Titer. Titer defines concentration in terms of the weight of some species with which a unit volume of the solution reacts. The applications of titer are considered in Chapter 7.

Parts per Million; Parts per Billion. For very dilute solutions, it is convenient to express concentrations in terms of parts per million.

$$\text{ppm} = \frac{\text{weight of solute}}{\text{weight of solution}} \times 10^6$$

For even more dilute solutions 1×10^9 rather than 1×10^6 is employed in the foregoing equation; the results are then given as parts per billion (ppb). The term parts per thousand (ppt) also finds use. If the solvent is water and the quantity of solute is so small that the density of the solution is essentially 1.00 g/ml,

$$\text{ppm} \cong \frac{\text{mg solute}}{\text{liter solution}}$$

Example. What is the molarity of K^+ in a solution that contains 63.3 ppm of $K_4Fe(CN)_6$? We shall assume that the density of the solution is 1.00 g/ml. Therefore,

$$63.3 \text{ ppm } K_4Fe(CN)_6 \cong 63.3 \text{ mg } K_4Fe(CN)_6/\text{liter}$$

$$M = \frac{63.3 \text{ mg } K_4Fe(CN)_6}{\text{liter}} \times 10^{-3} \frac{\text{g}}{\text{mg}} \times \frac{1}{368 \text{ g } K_4Fe(CN)_6/\text{fw}} \times \frac{4 \text{ moles } K^+}{1 \text{ fw } K_4Fe(CN)_6}$$

$$= 6.88 \times 10^{-4}$$

p-Functions. In some circumstances it is convenient to express the concentration of an ion in terms of the negative logarithm of its molar concentration. The most common p-function is

$$\text{pH} = -\log [H_3O^+]$$

The term in brackets is the molar hydronium ion concentration. Sample calculations of p-functions are found in Chapter 7.

Density and Specific Gravity. The *density* of a substance measures its mass per unit volume whereas the *specific gravity* of a material is the ratio of its mass to that of an equal volume of water at 4°C. In the metric system, density has units of kg/liter or g/ml. Specific gravity, on the other hand, is unitless and is thus not tied to any particular system of units; for this reason it is widely used in describing items of commerce. Because water at 4°C has a density of exactly 1.00 g/ml, and because we shall be employing the metric system throughout the text, density and specific gravity are, from a practical standpoint, used interchangeably.

Percentage Concentration. Chemists frequently express concentrations in terms of percentage. This practice, unfortunately, can be a source for ambiguity because of the many ways the percentage composition of a solution can be expressed. Common methods include:

$$\text{weight percent (w/w)} = \frac{\text{wt of solute}}{\text{wt of soln}} \times 100$$

$$\text{volume percent (v/v)} = \frac{\text{volume of solute}}{\text{volume of soln}} \times 100$$

$$\text{weight-volume percent (w/v)} = \frac{\text{wt of solute, g}}{\text{volume of soln, ml}} \times 100$$

It should be noted that the denominator in each of these expressions refers to the solution rather than to the solvent. Moreover, the first two expressions do not depend on the units employed (provided, of course, that there is consistency between numerator and denominator), whereas units must be defined for the third. Of the three expressions, only weight percentage has the virtue of being temperature-independent.

Weight percent is frequently used to express the concentration of commercial aqueous reagents; thus, nitric acid is sold as a 70% solution, which means that the reagent contains 70 g of HNO_3 per 100 g of solution.

Weight-volume percent is often employed to indicate the composition of dilute aqueous solutions of solid reagents; thus, a 5% aqueous silver nitrate usually refers to a solution that is prepared by dissolving 5 g of silver nitrate in sufficient water to give 100 ml of solution.

To avoid uncertainty, it is necessary to specify explicitly the type of percentage composition that is being used. If this information is lacking, the user is forced to decide intuitively which of the several types is involved.

Example. Describe the preparation of 250 ml of 6.0-F NH_3 from the concentrated reagent. The label on the commercial NH_3 bottle states that its specific gravity is 0.90 and that it contains 27% NH_3.

Generally the percentages employed in describing commercial reagents are weight-weight. Therefore,

$$\text{g NH}_3 \text{ required} = 250 \text{ ml} \times \frac{6.0 \text{ mfw NH}_3}{\text{ml}} \times \frac{0.017 \text{ g}}{\text{mfw NH}_3} = 25.5$$

$$\frac{\text{g NH}_3}{\text{ml concd soln}} = \frac{0.90 \text{ g soln}}{\text{ml soln}} \times \frac{27 \text{ g NH}_3}{100 \text{ g soln}} = 0.243$$

$$\text{ml concd soln} = \frac{25.5 \text{ g NH}_3}{0.243 \text{ g NH}_3/\text{ml concd soln}} = 105$$

Dilute 105 ml of the concentrated reagent to a volume of about 250 ml.

Solution-Diluent Volume Ratios. The composition of a dilute solution is sometimes specified in terms of the volume of a more concentrated reagent and the volume of solvent to be used in diluting it. The volume of the former is separated from that of the latter by a colon. Thus, a 1 : 4 HCl solution contains four volumes of water for each volume of concentrated hydrochloric acid taken. This method of notation is frequently ambiguous in that the concentration of the original reagent solution is not always obvious to the reader; the use of formal concentrations is greatly to be preferred.

Stoichiometric Relationships

A balanced chemical equation is a statement of the combining ratios (in formula weights) that exist between reacting substances and their products. Thus, the equation[3]

$$2\text{NaCl(aq)} + \text{Pb(NO}_3)_2\text{(aq)} = \text{PbCl}_2\text{(s)} + 2\text{NaNO}_3\text{(aq)}$$

indicates that two formula weights of sodium chloride combine in aqueous solution with one formula weight of lead nitrate to produce one formula weight of solid lead chloride and two of aqueous sodium nitrate.[4]

Experimental measurements are never obtained directly in terms of formula weight; instead they have units such as grams, milligrams, liters, or milliliters. The chemist, interested in relating his raw data to the weight of some other compound, must first transform his figures into units of gram formula weights, taking into account the stoichiometry of the process; reconversion to ordinary metric units of weight then follows. These transformations, which are of fundamental importance to analytical chemistry, are summarized in Tables 2-2 and 2-3. The combination of two or more of these definitions will permit the solving

---

[3] Here it is advantageous to depict the reaction in terms of chemical compounds. If we wish to focus on reactive species, the net ionic representation is preferable:

$$2\text{Cl}^-\text{(aq)} + \text{Pb}^{2+}\text{(aq)} = \text{PbCl}_2\text{(s)}$$

[4] Chemists frequently include information as to the physical state of substances in equations (as we have done here). Thus, (g), (l), (s), and (aq) refer to gaseous, liquid, solid, and aqueous solution states, respectively.

TABLE 2-2   Expression of Weight in Chemical Units

Chemical Unit	Weight of Unit in Grams Given by	Method of Conversion from Metric Units to Chemical Units
Formula weight (fw)	gfw	$\text{no. fw} = \dfrac{\text{gram of substance}}{\text{gfw}}$
Milliformula weight (mfw)	$\dfrac{\text{gfw}}{1000}$	$\text{no. mfw} = \dfrac{\text{gram of substance}}{\text{gfw}/1000}$
Mole	gmw	$\text{no. mole} = \dfrac{\text{gram of species}}{\text{gmw}}$
Millimole	$\dfrac{\text{gmw}}{1000}$	$\text{no. mmole} = \dfrac{\text{gram of species}}{\text{gmw}/1000}$
Equivalent (eq)	eq wt	$\text{no. eq} = \dfrac{\text{gram of substance}}{\text{eq wt}}$
Milliequivalent (meq)	$\dfrac{\text{eq wt}}{1000}$	$\text{no. meq} = \dfrac{\text{gram of substance}}{\text{eq wt}/1000}$

TABLE 2-3   Expression of Concentration in Chemical Units

Chemical Term for Concentration	Method of Calculation from Chemical Units of Weight	Method of Calculation from Metric Units of Weight
Formality, F	$F = \dfrac{\text{no. fw}}{\text{liter of soln}}$	$F = \dfrac{\text{gram solute}}{\text{liter of soln} \times \text{gfw}}$
	$= \dfrac{\text{no. mfw}}{\text{ml of soln}}$	$= \dfrac{\text{gram solute}}{\text{ml of soln} \times \text{gfw}/1000}$
Molarity, M	$M = \dfrac{\text{no. mole}}{\text{liter of soln}}$	$M = \dfrac{\text{gram solute}}{\text{liter of soln} \times \text{gmw}}$
	$= \dfrac{\text{no. mmole}}{\text{ml of soln}}$	$= \dfrac{\text{gram solute}}{\text{ml of soln} \times \text{gmw}/1000}$
Normality, N	$N = \dfrac{\text{no. eq}}{\text{liter of soln}}$	$N = \dfrac{\text{gram solute}}{\text{liter of soln} \times \text{eq wt}}$
	$= \dfrac{\text{no. meq}}{\text{ml of soln}}$	$= \dfrac{\text{gram solute}}{\text{ml of soln} \times \text{eq wt}/1000}$

of any stoichiometric problem; the ability to manipulate them must be cultivated. In setting up equations, it is helpful to supply units to all quantities that possess dimensions; the best proof that a correct relationship has been generated is agreement between the units that appear on either side of the equal sign.

Example. Calculate the weight in grams w of $AgNO_3$ (gfw = 170) required to convert 2.33 g of Na_2CO_3 (gfw = 106) to Ag_2CO_3 (gfw = 276).

$$\text{no. fw } AgNO_3 \text{ required} = 2 \times \text{no. fw } Na_2CO_3$$

From Table 2-2 we substitute

$$\frac{w}{170 \text{ g } AgNO_3/fw} = \frac{2 \text{ fw } AgNO_3}{\text{fw } Na_2CO_3} \times \frac{2.33 \text{ g } Na_2CO_3}{106 \text{ g } Na_2CO_3/fw}.$$

$$w = \frac{170 \times 2 \times 2.33}{106} = 7.47 \text{ g } AgNO_3$$

Example. How many milliliters of 0.0669-F $AgNO_3$ will be needed to convert 0.348 g of pure Na_2CO_3 to Ag_2CO_3?

Because the volume V in milliliters is desired, it will be more convenient to base our calculations on milliformula weights than on formula weights. Thus,

$$\text{no. mfw } AgNO_3 = \frac{2 \text{ mfw } AgNO_3}{\text{mfw } Na_2CO_3} \times \text{no. mfw } Na_2CO_3$$

$$\frac{0.0669 \text{ mfw } AgNO_3}{\text{ml } AgNO_3} \times V = \frac{2 \text{ mfw } AgNO_3}{\text{mfw } Na_2CO_3} \times \frac{0.348 \text{ g } Na_2CO_3}{0.106 \text{ g } Na_2CO_3/\text{mfw } Na_2CO_3}$$

$$V = 2 \times \frac{0.348 \text{ g}}{0.106 \text{ g/mfw}} \times \frac{1}{0.0669 \text{ mfw/ml}}$$

$$= 98.1 \text{ ml } AgNO_3$$

Example. What weight of Ag_2CO_3 is formed upon mixing 25.0 ml of 0.200-F $AgNO_3$ with 50.0 ml of 0.800-F Na_2CO_3?

We must first determine which of the reactants is present in the stoichiometrically lesser amount since this species will limit the amount of product.

$$\text{no. mfw } AgNO_3 = 25.0 \text{ ml} \times 0.200 \text{ mfw/ml} = 5.00$$
$$\text{no. mfw } Na_2CO_3 = 50.0 \text{ ml} \times 0.0800 \text{ mfw/ml} = 4.00$$
$$\text{no. mfw } AgNO_3 \text{ required} = 2 \times \text{no. mfw } Na_2CO_3$$

Thus, the reaction is limited by the number of milliformula weights of $AgNO_3$, and

$$\text{no. mfw } Ag_2CO_3 = \frac{5.00}{2} = 2.50 \text{ mfw}$$

$$w = 2.50 \text{ mfw} \times \frac{0.276 \text{ g } Ag_2CO_3}{\text{mfw}} = 0.690 \text{ g}$$

PROBLEMS

*1. How many formula weights and how many milliformula weights are contained in
    (a) 27.3 g of Mn_3O_4?
    (b) 163 μg of BF_3?

* Answers to problems or parts of problems marked with an asterisk are located at the end of the book.

      (c) 6.92 liters of 0.0400-F $Na_2B_4O_7$?

      (d) 10.0 ml of 2.00 × 10^{-3} F $HgCl_2$?

      (e) 10.0 ml of an aqueous solution containing 143 ppm SO_2?

2. How many formula weights and how many milliformula weights are contained in

      (a) 23.4 mg of $Mg_2P_2O_7$?

      (b) 100 g of dry ice (CO_2)?

      (c) 1.00 lb of NaCl?

      (d) 7.50 liters of 0.0525-F carbonic acid?

      (e) 0.500 ml of a 0.0300-F solution of ascorbic acid (gfw = 176)?

*3. How many grams are contained in

      (a) 2.00 moles of CO_2?

      (b) 1.84 mfw of benzene (gfw = 78.1)?

      (c) 40.0 fw of NaOH?

      (d) 6.24 ml of 0.121-F sucrose (gfw = 342.3)?

      (e) 3.33 liters of 12.2-F HCl?

4. How many grams are contained in

      (a) 0.842 mole of Br^-?

      (b) 7.35 mfw of Na_2SO_4?

      (c) 135 ml of 0.200-F $K_2Cr_2O_7$?

      (d) 2.50 liters of 0.600-F acetone (gfw = 58.1) in cyclohexane?

      (e) 1.00 ml of 2.00 × 10^{-6} F KCl?

*5. A solution was prepared by dissolving exactly 2.42 g of $MgCl_2$ in water and diluting to 2.00 liters. Calculate

      (a) the formal concentration of $MgCl_2$.

      (b) the molar concentration of Cl^-.

      (c) the weight-volume percent of $MgCl_2$.

      (d) the weight-weight percent of $MgCl_2$ if the density of the solution is 1.01 g/ml.

      (e) the number of millimoles of Mg^{2+} in 25.0 ml of the solution.

6. A solution was prepared by dissolving 1.68 g of $K_4Fe(CN)_6$ in water and diluting to exactly 500 ml. Calculate

      (a) the formal concentration of $K_4Fe(CN)_6$.

      (b) the molar concentration of K^+ assuming complete dissociation.

      (c) the weight-volume percent of $K_4Fe(CN)_6$.

      (d) the weight-weight percent of $K_4Fe(CN)_6$ if the density of the solution is 1.008 g/ml.

      (e) the number of moles $Fe(CN)_6^{4-}$ in 16.0 ml of the solution.

*7. The average elemental composition of sea water is as follows:

Element	Concn, ppm	Element	Concn, ppm
Cl	1.90 × 10^4	Ca	4.00 × 10^2
Na	1.06 × 10^4	K	3.80 × 10^2
Mg	1.27 × 10^3	Br	6.51 × 10^1
S	8.84 × 10^2	C (inorganic)	2.80 × 10^1
		Sr	1.30 × 10^1

Calculate the formal concentration of each element if the average density of sea water is 1.024 g/ml.

8. The average concentrations of the common ions in human blood serum are as follows:

Ion	Concn, mg/100 ml	Ion	Concn, mg/100 ml
Na^+	335	HCO_3^-	164
K^+	19	Cl^-	370
Ca^{2+}	10	PO_4^{3-}	10
Mg^{2+}	2.7	SO_4^{2-}	19

Calculate the molar concentration of each constituent.

*9. Calculate the formal concentration of a solution that is 25.0% in H_2SO_4 (w/w) and has a specific gravity of 1.19.

10. Calculate the formal concentration of a 12.0% solution (w/w) of $CuSO_4$ which has a specific gravity of 1.13.

*11. The average concentration of silica (SiO_2) in the world's rivers is 15 ppm. What is the corresponding formal concentration, assuming the density of water is 1.00 g/ml?

12. The average concentration of Ca^{2+} in the world's rivers is 55 ppm. What is the corresponding molar concentration, assuming the density of the water is 1.00 g/ml?

*13. Describe the preparation of
(a) 200 ml of a 10% (w/v) aqueous glucose solution.
(b) 200 g of a 10% (w/w) aqueous glucose solution.
(c) 200 ml of a 10% (v/v) aqueous ethanol solution.

14. Describe the preparation of
(a) 500 ml of a solution that is 1.0% (w/v) I_2 in ethanol.
(b) 500 g of a solution that is 1.0% (w/w) I_2 in ethanol.
(c) 500 ml of a 1.0% (v/v) aqueous ethanol solution.

*15. Describe the preparation of
(a) 525 ml of 0.400-F $BaCl_2$ from solid $BaCl_2 \cdot 2H_2O$.
(b) 2.30 liters of 0.200-M K^+ from solid K_2SO_4.
(c) 100 ml of 0.100-F $AgNO_3$ from a solution that was 0.441 F in the salt.
(d) 500 ml of a 1.00% solution (w/v) of $KMnO_4$ from a 1.21-F solution of $KMnO_4$.
(e) 3.00 liters of a solution containing 50 ppm of K^+ from a solution that was 2.00×10^{-3} F in K_2SO_4.

16. Describe the preparation of
(a) 500 ml of a 1.00×10^{-3} F solution of iodine in CCl_4 from pure I_2.
(b) 3.00 liters of 0.150-M Cl^- from solid $BaCl_2 \cdot 2H_2O$.
(c) 1500 ml of 0.100-F H_2SO_4 from 6.10-F H_2SO_4.
(d) 250 ml of 1.00% (w/v) $AgNO_3$ from a 0.800-F solution of $AgNO_3$.
(e) 6.00 liters of a solution containing 10.0 ppm NH_3 from a 0.116-F solution of NH_3.

*17. Describe the preparation of
(a) 750 ml of 0.172-F $K_2Cr_2O_7$ from the solid.
(b) 50.0 liters of a solution that is 0.100 F in Na_2SO_4 from solid Na_2SO_4.
(c) 2.00 liters of a solution that is 0.0150 M in Na^+ from solid NaCl.
(d) 20.0 liters of a solution that is 0.202 M in Na^+ from a 2.42-F solution of Na_2SO_4.
(e) 2.00 liters of a solution that is 0.150 M in Na^+ from solid Na_2SO_4.

18. Describe the preparation of
(a) 750 ml of a solution that is 0.0150 M in $K_4Fe(CN)_6$ from solid $K_4Fe(CN)_6$.
(b) 3.00 liters of a solution that is 0.0150 M in K^+ from solid $K_4Fe(CN)_6$.

(c) 300 ml of a solution that is 0.0150 M in K^+ from solid $K_3Fe(CN)_6$.

(d) 5.00 liters of a solution that is 0.0150 M in K^+ from a 0.700-F solution of K_2SO_4.

(e) 16.0 liters of a solution that is 0.0200 F in $BaCl_2$ from pure $BaCl_2 \cdot 2H_2O$.

*19. Concentrated HNO_3 has a specific gravity of 1.42 and is 69% HNO_3 (w/w).

(a) How many grams of HNO_3 are contained in 500 ml of this reagent?

(b) Describe the preparation of 800 ml of a solution having a concentration of about 0.20 F in HNO_3 from the concentrated reagent.

20. Concentrated HCl has a specific gravity of 1.185 and is 36.5% (w/w) in HCl.

(a) How many milliliters of HCl gas (measured at S.T.P.) are contained in 1.00 liter of the reagent?

(b) Describe how 1.50 liters of approximately 0.30-F HCl should be prepared from the concentrated reagent.

*21. Describe the preparation of 400 ml of 6.0-F H_3PO_4 from the commercial reagent which is 85% (w/w) in H_3PO_4 and has a specific gravity of 1.69.

22. Describe the preparation of 200 ml of 3.0-F H_2SO_4 from the concentrated reagent which is 95% (w/w) H_2SO_4 and has a specific gravity of 1.84.

*23. Lanthanum ion reacts with iodate ion to form a slightly soluble precipitate having the formula $La(IO_3)_3$ and a gram formula weight of 664.

(a) How many grams of KIO_3 are required to react completely with 2.15 g of $La(NO_3)_3$ (gfw $= 325$)?

(b) What weight of KIO_3 will react with 2.00 fw of $La(NO_3)_3$?

(c) How many grams of $La(NO_3)_3$ are required to react completely with 2.15 g of KIO_3?

(d) How many grams of $La(IO_3)_3$ are formed when 2.00 g of $La(NO_3)_3$ are mixed with 3.00 g of KIO_3?

(e) How many grams of $La(IO_3)_2$ are formed when 20.0 ml of 0.100-F $NaIO_3$ are mixed with 12.0 ml of 0.0200-F $La(NO_3)_3$?

24. Silver ion reacts with arsenate ion to give the sparingly soluble Ag_3AsO_4 (gfw $= 463$).

(a) What weight of $AgNO_3$ will react with 6.00 fw of Na_3AsO_4 (gfw $= 208$)?

(b) What weight of Ag_3AsO_4 can be formed from 2 fw of Na_3AsO_4?

(c) What weight of Ag_3AsO_4 will form when 2.15 g of $AgNO_3$ are mixed with 6.50 g of Na_3AsO_4?

(d) What weight of Ag_3AsO_4 will form when 2.00 liters of 0.303-F $AgNO_3$ are mixed with 1.50 liters of 0.212-F Na_3AsO_4?

(e) What volume of 0.600-F $AgNO_3$ is needed to react completely with 5.64 g of Na_3AsO_4?

*25. How much of the substance in the second column is needed to react completely with the indicated amount of substance in the first column?

(a) 18.0 mfw H_2SO_4 (a) g KOH

(b) 5.00 mfw H_2SO_4 (b) ml of 0.150-F KOH

(c) 6.00 g $Ba(OH)_2$ (c) ml of 0.200-F HCl

(d) 23.0 ml of 0.292-F $BaCl_2$ (d) ml of 0.100-F $AgNO_3$

(e) 10 ml HCl, sp gr $= 1.12$,   (e) ml of 0.0666-F $Ba(OH)_2$
% HCl (w/w) $= 24.0$

26. How much of the substance in the second column is needed to react completely with the indicated amount of substance in the first column?

(a) 11.2 mfw $Ba(OH)_2$ (a) g HCl

(b) 11.2 mfw $Ba(OH)_2$ (b) ml 0.200-F HCl

(c) 1.41 g $Pb(NO_3)_2$ (c) g KCl

(d) 1.20 ml $HClO_4$, sp gr $= 1.60$,   (d) g $Ca(OH)_2$
% $HClO_4$ (w/w) $= 70.0$
(e) 14.0 ml of 0.700-F H_2SO_4 (e) ml of 0.300-F KOH

*27. How many grams of $BaCl_2 \cdot 2H_2O$ are required to precipitate all of the $SO_4{}^{2-}$ (as $BaSO_4$) in 2.00 liters of 0.160-F H_2SO_4?

28. How many grams of silver nitrate are required to precipitate all the chromium from 200 ml of a solution that is 0.0800 F in $K_2Cr_2O_7$?

$$K_2Cr_2O_7 + 4Ag^+ + H_2O \rightarrow 2Ag_2CrO_4 + 2H^+ + 2K^+$$

*29. When calcium oxalate is ignited at 500°C, the following reaction occurs:

$$CaC_2O_4 \rightarrow CaCO_3 + CO$$

(a) What weight of $CaCO_3$ is formed by igniting 1.22 g of CaC_2O_4?
(b) What would be the weight loss upon ignition of 20.0 g of CaC_2O_4?

30. When $Ca(HCO_3)_2$ is ignited, the following reaction occurs:

$$Ca(HCO_3)_2 \rightarrow CaCO_3 + H_2O + CO_2$$

(a) How many grams of CO_2 are formed from ignition of 1.00 g of $Ca(HCO_3)_2$?
(b) What loss in weight accompanies ignition of 4.24 g of $Ca(HCO_3)_2$?

3

A REVIEW OF SIMPLE EQUILIBRIUM-CONSTANT CALCULATIONS

Most reactions which are useful for chemical analysis proceed rapidly to a state of *chemical equilibrium* in which reactants and products exist in constant and predictable ratios. A knowledge of these ratios, under various experimental conditions, often permits the chemist to decide whether or not a reaction is suitable for analytical purposes and to choose conditions which will minimize the error associated with an analysis.

Equilibrium-constant expressions are algebraic equations that relate the concentrations of reactants and products in a chemical reaction to one another by means of a numerical quantity called an *equilibrium constant*. A chemist must know how to derive useful information from equilibrium constants; it is the purpose of this chapter to review the calculations that are applicable to systems in which only a single equilibrium predominates. Methods for treating complex systems involving several equilibria will be dealt with in later chapters.

The Equilibrium State

For purposes of discussion, consider the equilibrium

$$2Fe^{3+} + 3I^- \rightleftarrows 2Fe^{2+} + I_3^- \tag{3-1}$$

The rate of this reaction and the extent to which it proceeds to the right can be readily judged by observing the orange-red color imparted to the solution by the triiodide ion (at low concentrations, the other three participants in the reaction are essentially colorless). If, for example, 2 mfw of iron(III) are added to a liter of solution containing 3 mfw of potassium iodide, color appears instantaneously; within a second or less the color intensity becomes constant with time, showing that the triiodide concentration has become invariant.

A solution of identical color intensity (and hence triiodide concentration) can be produced by adding 2 mfw of iron(II) to a liter of solution containing 1 mfw of triiodide ion. Here, an immediate decrease in color is observed as a result of the reaction

$$2Fe^{2+} + I_3^- \rightleftarrows 2Fe^{3+} + 3I^- \tag{3-2}$$

Many other combinations of the four reactants could be employed to yield solutions indistinguishable from the two just described. The only requirement would be that for each liter of solution, the quantities of added species bear the following relationships to one another:

$$\text{no. mfw } Fe^{2+} = 2 \times \text{no. mfw } I_3^-$$

$$\text{no. mfw } Fe^{3+} = \tfrac{2}{3} \times \text{no. mfw } I^-$$

$$\text{no. mfw } Fe^{3+} + \text{no. mfw } Fe^{2+} = 2.00$$

For example, a solution identical to those just described could be synthesized by adding 0.8 mfw of iron(II), 1.2 mfw iron(III), 0.4 mfw of triiodide ion, and 1.8 mfw of potassium iodide to a liter of solution.

The foregoing examples illustrate that the concentration relationships at chemical equilibrium (that is, the *position of equilibrium*) are independent of the route by which the equilibrium state is achieved. On the other hand, it is readily shown that these relationships are altered by the application of stress to the system—for example, by changes in temperature, in pressure (if one of the reactants or products is a gas), or in the total concentration of one of the reactants. These effects can be predicted qualitatively from the *principle of Le Châtelier*, which states that the position of chemical equilibrium will always shift in a direction that counteracts the effect of an applied stress. Thus, an increase in temperature will alter the concentration relationships in the direction that tends to absorb heat; an increase in pressure favors those participants that occupy the smaller total volume. Of particular importance in an analysis is the effect of introducing an additional amount of one of the species to the reaction mixture; here, the resulting stress is relieved by a shift in equilibrium in a direction that partially consumes the added substance. Thus, for the equilibrium we have been considering, addition of iron(III) would cause an increase in color as more triiodide ion and iron(II) are formed; addition of iron(II), on the other hand, would have a reverse effect. An equilibrium shift brought about by changing the amount of one of the participants is called a *mass-action effect*.

If it were possible to examine a system at the molecular level, it would be found that interactions among the species present continue unabated even after equilibrium is achieved. The observed constant concentration relationship is thus the consequence of an equality in the rates of the forward and reverse reactions; that is, chemical equilibrium is a dynamic state.

Equilibrium-Constant Expressions

The influence of concentration (or pressure, if the species are gaseous) on the position of a chemical equilibrium is conveniently described in quantitative terms by means of an equilibrium-constant expression. These expressions are readily derived from thermodynamic theory; they are of great practical importance because they permit the chemist to predict the direction and the completeness of a chemical reaction. It is important to note, however, that equilibrium-constant expressions yield no information concerning reaction rates.

FORMULATION OF EQUILIBRIUM-CONSTANT EXPRESSIONS

The equilibrium-constant expression for a reaction can take one of two forms— an exact form (called the thermodynamic expression) or an approximate form that applies exactly under a very limited set of conditions only. Because it is much easier to use, the approximate form is often employed, even though errors may be introduced as a consequence.

Assumptions Inherent in Deriving Approximate Equilibrium Constants. In an approximate equilibrium-constant expression it is assumed that concentration is the sole factor that governs the influence of any ion (or molecule) upon the condition of equilibrium—that is, that each species acts independently of its neighbors. In the reaction symbolized by Equation 3-1, for example, the influence of each iron(III) ion upon the equilibrium state is assumed to be the same in very dilute solutions (where the ion is most likely surrounded by neutral water molecules) as it is in very concentrated solutions (where other charged species are in close proximity). A solution (or a gas) in which the ions or molecules act independently of one another is termed *ideal* or *perfect*. Truly ideal solutions and ideal behavior are seldom encountered in the laboratory.

Approximate Equilibrium-Constant Expressions. Consider the generalized equation for a chemical equilibrium

$$mM + nN \rightleftarrows pP + qQ \qquad (3\text{-}3)$$

where the capital letters represent the formulas of participating chemical species and the italic letters are the small integers required to balance the equation. Thus, the equation states that m moles of M react with n moles of N to form p moles of P and q moles of Q. The equilibrium-constant expression for this reaction is

$$K = \frac{[P]^p[Q]^q}{[M]^m[N]^n} \qquad (3\text{-}4)$$

where the letters in brackets represent the molar concentrations of dissolved solutes or partial pressures (in atmospheres) if the reacting substances are gases.

The letter K in Equation 3-4 is a temperature-dependent, numerical constant called the *equilibrium constant*. By convention, the concentrations of the products *as written* are always placed in the numerator and the reactant concentrations in the denominator. Note also that each concentration is raised to a power that is identical to the integer that accompanies the formula of that species in the balanced equation describing the equilibrium.[1]

Equation 3-4 can be readily derived from thermodynamic concepts by assuming that the participants in a reaction exhibit ideal behavior.[2]

Thermodynamic Equilibrium-Constant Expressions. Equation 3-4 is useful for determining the approximate equilibrium composition of a dilute solution. More exact results require the use of thermodynamic equilibrium-constant expressions, which are considered in Chapter 5. The approximate form is easier to employ and suffices to provide the chemist with adequate information for many purposes.

Common Types of Equilibrium-Constant Expressions

In this section we shall consider the common types of equilibria encountered in analyses and describe some of the characteristics of the corresponding equilibrium-constant expressions.

DISSOCIATION OF WATER

Aqueous solutions always contain small amounts of hydronium and hydroxide ions as a consequence of the dissociation reaction

$$2H_2O \rightleftarrows H_3O^+ + OH^- \qquad (3\text{-}5)$$

An equilibrium constant for this reaction can be formulated as shown in Equation 3-4; that is,

$$K = \frac{[H_3O^+][OH^-]}{[H_2O]^2}$$

---

[1] Derivation of equilibrium-constant expressions from theory reveals that the bracketed terms are in fact *concentration ratios* rather than absolute concentrations. The denominator in each of these ratios is the concentration of the species in its so-called *standard state*. The standard state is a one-molar solution for a solute. For pure elements or compounds, it is their state (solid, liquid, or gas) at 25°C and one atmosphere pressure. Thus, if P in Equation 3-4 is a solute,

$$[P] = \frac{\text{concn of P, } \cancel{\text{mole/liter}}}{1.00 \ \cancel{\text{mole/liter}}}$$

where the denominator is the concentration of solute in the standard state. An important consequence of the relative nature of [P] is that this quantity, as well as the equilibrium constant, is unitless. Strictly speaking then, [P] must be multiplied by 1.00 mole/liter to obtain units of concentration. Ordinarily, this fine distinction is of no importance and [P] is treated as an absolute concentration.

[2] See L. K. Nash, *Elements of Classical and Statistical Thermodynamics*, Book 1, pp. 124–130. Reading, Mass.: Addison-Wesley Publishing Company, 1970.

In dilute aqueous solutions, the concentration of water is large compared with the concentration of solutes and can be considered to be invariant. For example, if 0.1 mole of gaseous hydrogen chloride is bubbled into a liter of water [1000 g/ (18.0 g/fw) = 55.6 fw H_2O], the equilibrium shown by Equation 3-5 is shifted to the left, and the number of formula weights of water increases from 55.6 to 55.7; thus, unless the amount of hydrochloric acid is unusually large, the concentration of water will not change significantly. That is, [H_2O] in Equation 3-5 can, in most instances, be taken as constant; therefore, we may write

$$K[H_2O]^2 = K_w = [H_3O^+][OH^-] \tag{3-6}$$

where the new constant K_w is given the special name, the *ion-product constant for water.*

At 25°C, the ion-product constant for water has a numerical value of 1.01×10^{-14} (for convenience we shall normally use the approximation $K_w \cong 1.00 \times 10^{-14}$). Table 3-1 shows the dependence of this constant upon temperature.

TABLE 3-1   Variation of K_w with Temperature

Temperature °C	K_w
0	0.114×10^{-14}
25	1.01×10^{-14}
50	5.47×10^{-14}
100	49×10^{-14}

The ion-product constant for water permits the ready calculation of the hydronium or hydroxide ion concentration of aqueous solutions.

Example. Calculate the hydronium and hydroxide ion concentration of pure water at 25°C and at 100°C.

Because OH^- and H_3O^+ are formed from the dissociation of water only, their concentrations must be equal. That is,

$$[H_3O^+] = [OH^-]$$

Substitution into Equation 3-6 gives

$$[H_3O^+]^2 = [OH^-]^2 = K_w$$
$$[H_3O^+] = [OH^-] = \sqrt{K_w}$$

At 25°C,

$$[H_3O^+] = [OH^-] = \sqrt{1.00 \times 10^{-14}} = 1.00 \times 10^{-7}$$

At 100°C,

$$[H_3O^+] = [OH^-] = \sqrt{49 \times 10^{-14}} = 7.0 \times 10^{-7}$$

Example. Calculate the hydronium and hydroxide ion concentrations in 0.200-F aqueous NaOH.

Sodium hydroxide is a strong electrolyte and its contribution to the hydroxide ion concentration of the solution will be 0.200 mole per liter. Hydroxide ions and hydronium ions are also formed *in equal amounts* from the dissociation of water. Therefore, we may write

$$[OH^-] = 0.200 + [H_3O^+]$$

where $[H_3O^+]$ accounts for the hydroxide ions contributed by the solvent. The concentration of OH^- from the water will be small when compared with 0.200; therefore,

$$[OH^-] \cong 0.200$$

We can then employ Equation 3-6 to calculate the hydronium ion concentration

$$[H_3O^+] = \frac{1.00 \times 10^{-14}}{0.200} = 5.00 \times 10^{-14}$$

Note that the approximation

$$[OH^-] = 0.200 + 5.00 \times 10^{-14} \cong 0.200$$

will cause no significant error.

EQUILIBRIUM INVOLVING SLIGHTLY SOLUBLE IONIC SOLIDS

When an aqueous solution is saturated with a sparingly soluble salt, one or more equilibria will be established. With silver chloride, for example,

$$AgCl(s) \rightleftarrows AgCl(aq)$$

$$AgCl(aq) \rightleftarrows Ag^+ + Cl^-$$

where AgCl(aq) represents the undissociated solute. The equilibrium constant for the first reaction can be written as

$$K = \frac{[AgCl]_{aq}}{[AgCl]_s} \tag{3-7}$$

The term $[AgCl]_s$, which is the concentration of AgCl *in the solid phase*, is a constant because the concentration of a compound in a pure solid is invariant. (From footnote 1, p. 25, the magnitude of this constant is 1.00.) Thus, Equation 3-7 becomes

$$K[AgCl]_s = K_1 = [AgCl]_{aq} \tag{3-8}$$

We see from Equation 3-8 that the concentration of AgCl in a saturated solution is constant at any given temperature.

The equilibrium constant for the second reaction is given by

$$K_2 = \frac{[Ag^+][Cl^-]}{[AgCl]_{aq}} \tag{3-9}$$

It is convenient to multiply Equation 3-8 by 3-9 and thereby eliminate $[AgCl]_{aq}$.

$$K_1 K_2 = K_{sp} = [Ag^+][Cl^-] \tag{3-10}$$

The product of these two constants, K_{sp}, is called the *solubility-product constant*. The solubility-product constant is thus the equilibrium constant for the overall reaction

$$AgCl(s) \rightleftarrows Ag^+ + Cl^-$$

It is of *utmost importance* to keep in mind that Equations 3-8 and 3-10 apply *only to saturated solutions* that are in contact with an excess of undissolved solid.

Most sparingly soluble inorganic salts are essentially completely dissociated in aqueous solutions; that is, the concentration of the undissociated solute is very nearly zero with respect to the concentration of the ions. Under these circumstances the overall reaction adequately describes the dissolution equilibrium; thus, only solubility-product constants are to be found in the literature for most inorganic solids (see Appendix 3). For example, there is no evidence to suggest that saturated solutions of barium iodate contain undissociated $Ba(IO_3)_2$ or $BaIO_3^-$. The only significant equilibrium, therefore, is

$$Ba(IO_3)_2(s) \rightleftarrows Ba^{2+} + 2IO_3^-$$

and the constant associated with this equilibrium is the only one that needs to be considered; that is,

$$[Ba^{2+}][IO_3^-]^2 = K_{sp}$$

The examples which follow demonstrate some typical uses of solubility-product expressions. Other applications will be considered in Chapters 5 and 8.

Example. How many grams of barium iodate can be dissolved in 500 ml of water at 25°C? The solubility-product constant for $Ba(IO_3)_2$ is 1.57×10^{-9} (Appendix 3). Thus,

$$Ba(IO_3)_2(s) \rightleftarrows Ba^{2+} + 2IO_3^-$$

and

$$[Ba^{2+}][IO_3^-]^2 = 1.57 \times 10^{-9} = K_{sp}$$

It is seen from the equation describing the equilibrium that one mole of Ba^{2+} is formed for each formula weight of dissolved $Ba(IO_3)_2$. Thus,

$$\text{formal solubility of } Ba(IO_3)_2 = [Ba^{2+}]$$

We also see from the equation that the iodate concentration is twice that for barium ion. That is,

$$[IO_3^-] = 2[Ba^{2+}]$$

Substituting the latter into the equilibrium-constant expression gives

$$[Ba^{2+}](2[Ba^{2+}])^2 = 1.57 \times 10^{-9}$$

or

$$[Ba^{2+}] = \left(\frac{1.57 \times 10^{-9}}{4}\right)^{1/3} = 7.3 \times 10^{-4}$$

and

$$\text{solubility} = 7.3 \times 10^{-4}\,F$$

To obtain the solubility of $Ba(IO_3)_2$ in grams per 500 ml, we write

$$\text{solubility} = 7.3 \times 10^{-4} \frac{\text{mfw } Ba(IO_3)_2}{\text{ml}} \times 500 \text{ ml} \times 0.487 \frac{g}{\text{mfw } Ba(IO_3)_2}$$

$$= \frac{0.178 \text{ g}}{500 \text{ ml}}$$

where 0.487 is the milliformula weight of $Ba(IO_3)_2$.

The common-ion effect predicted from the Le Châtelier principle is demonstrated by the following examples.

Example. Calculate the formal solubility of $Ba(IO_3)_2$ in a solution that is $0.0200\,F$ in $Ba(NO_3)_2$.

In this example the solubility is not directly related to $[Ba^{2+}]$ but can be described in terms of $[IO_3^-]$; that is,

$$\text{solubility of } Ba(IO_3)_2 = \tfrac{1}{2}[IO_3^-]$$

Here barium ions arise from two sources—namely, $Ba(NO_3)_2$ and $Ba(IO_3)_2$. The contribution from the former is $0.0200\,M$ while that from the latter is equal to the formal solubility or $\tfrac{1}{2}[IO_3^-]$. Thus,

$$[Ba^{2+}] = 0.0200 + \tfrac{1}{2}[IO_3^-]$$

Substitution of these quantities into the solubility-product expression yields

$$(0.0200 + \tfrac{1}{2}[IO_3^-])[IO_3^-]^2 = 1.57 \times 10^{-9}$$

Since the exact solution for $[IO_3^-]$ will involve a cubic equation, it is worthwhile to seek an approximation that will simplify the algebra. The small numerical value for K_{sp} suggests that the solubility of $Ba(IO_3)_2$ is not large; therefore, it is reasonable to suppose that the barium ion concentration derived from the solubility of $Ba(IO_3)_2$ is small with respect to that from the $Ba(NO_3)_2$. That is, $\tfrac{1}{2}[IO_3^-] \ll 0.0200$ and

$$0.0200 + \tfrac{1}{2}[IO_3^-] \cong 0.0200$$

The original equation then simplifies to

$$0.0200[IO_3^-]^2 = 1.57 \times 10^{-9}$$

$$[IO_3^-] = 2.80 \times 10^{-4}$$

The assumption that

$$(0.0200 + \tfrac{1}{2} \times 2.80 \times 10^{-4}) \cong 0.0200$$

does not appear to cause serious error because the second term is only about 0.7% of 0.0200. Ordinarily we shall consider an assumption of this type to be satisfactory if the discrepancy is less than 5 to 10%. Therefore,

$$\text{solubility of } Ba(IO_3)_2 = \tfrac{1}{2}[IO_3^-] = \tfrac{1}{2} \times 2.80 \times 10^{-4} = 1.40 \times 10^{-4}\,F$$

If we compare this result with the solubility of barium iodate in pure water, we see that the presence of a small concentration of the common ion has lowered the formal solubility of $Ba(IO_3)_2$ by a factor of about 5.

Example. Calculate the solubility of $Ba(IO_3)_2$ in the solution that results when 200 ml of 0.0100-F $Ba(NO_3)_2$ are mixed with 100 ml of 0.100-F $NaIO_3$.

We must first establish whether either reactant will be present in excess. The amounts available are

$$\text{no. mfw } Ba^{2+} = 200 \text{ ml} \times 0.0100 \text{ mfw/ml} = 2.00$$

$$\text{no. mfw } IO_3^- = 100 \text{ ml} \times 0.100 \text{ mfw/ml} = 10.00$$

If formation of $Ba(IO_3)_2$ is complete,

$$\text{excess } IO_3^- = 10.0 - 2(2.00) = 6.00 \text{ mfw}$$

Thus,

$$F_{IO_3^-} = \frac{6.00 \text{ mfw}}{300 \text{ ml}} = 0.0200$$

As in the first example,

$$\text{formal solubility of } Ba(IO_3)_2 = [Ba^{2+}]$$

Here, however,

$$[IO_3^-] = 0.0200 + 2[Ba^{2+}]$$

where $2[Ba^{2+}]$ represents the contribution of the solubility of the precipitate to the iodate concentration. We can obtain a provisional answer after making the assumption that

$$[IO_3^-] \cong 0.0200$$

Thus,

$$\text{solubility of } Ba(IO_3)_2 = [Ba^{2+}] = \frac{K_{sp}}{[IO_3^-]^2}$$

$$= \frac{1.57 \times 10^{-9}}{(0.0200)^2} = 3.9 \times 10^{-6}$$

The approximation used in this calculation is seen to have been reasonable.

Note that the results from the last two examples demonstrate that the presence of excess iodate is more effective in decreasing the solubility of $Ba(IO_3)_2$ than is an equal excess of barium ions.

DISSOCIATION OF WEAK ACIDS AND BASES

When a weak acid or base is dissolved in water, partial dissociation occurs. Thus, for nitrous acid we may write

$$HNO_2 + H_2O \rightleftarrows H_3O^+ + NO_2^- \qquad K_a = \frac{[H_3O^+][NO_2^-]}{[HNO_2]}$$

where K_a is the *acid-dissociation constant* for nitrous acid. In an analogous way, the *basic-dissociation constant* for ammonia is given by

$$NH_3 + H_2O \rightleftarrows NH_4^+ + OH^- \qquad K_b = \frac{[NH_4^+][OH^-]}{[NH_3]}$$

Note that a concentration term for water ($[H_2O]$) does not appear in either equation; as with the ion-product constant for water, the solvent concentration is assumed to be large and constant and is thus incorporated in the equilibrium constants K_a and K_b.

Relationship between Dissociation Constants for Conjugate Acid-Base Pairs. Consider the dissociation-constant expression for ammonia and its conjugate acid, ammonium ion. Here we may write

$$NH_3 + H_2O \rightleftarrows NH_4^+ + OH^- \qquad K_b = \frac{[NH_4^+][OH^-]}{[NH_3]}$$

and

$$NH_4^+ + H_2O \rightleftarrows NH_3 + H_3O^+ \qquad K_a = \frac{[NH_3][H_3O^+]}{[NH_4^+]}$$

Multiplication of the two equilibrium-constant expressions together gives

$$K_a K_b = \frac{[\cancel{NH_3}][H_3O^+]}{[\cancel{NH_4^+}]} \times \frac{[\cancel{NH_4^+}][OH^-]}{[\cancel{NH_3}]} = [H_3O^+][OH^-]$$

But

$$[H_3O^+][OH^-] = K_w$$

Therefore,

$$K_a K_b = K_w \qquad (3\text{-}11)$$

This relationship is general for all conjugate acid-base pairs. Most tables of dissociation constants do not list both the acid- and the base-dissociation constants for conjugate pairs, since it is so easy to calculate one from the other by Equation 3-11.

Example.   What is K_b for the reaction

$$CN^- + H_2O \rightleftarrows HCN + OH^-$$

Examination of Appendix 5 (dissociation constants for bases) reveals no entry for CN^-. In Appendix 4, however, K_a for HCN is found to have a value of 2.1×10^{-9}. Thus,

$$K_b = \frac{[HCN][OH^-]}{[CN^-]} = \frac{K_w}{K_{HCN}}$$

$$K_b = \frac{1.00 \times 10^{-14}}{2.1 \times 10^{-9}} = 4.8 \times 10^{-6}$$

Applications of Acid-Dissociation Constants. Dissociation constants for many common weak acids are tabulated in Appendix 4. These constants find wide application for the calculation of the hydronium or hydroxide ion concentrations of solutions of weak acids or bases.

For example, the equilibrium established when the weak acid HA is dissolved in water may be written as

$$HA + H_2O \rightleftarrows H_3O^+ + A^- \qquad K_a = \frac{[H_3O^+][A^-]}{[HA]}$$

In addition, hydronium ions result from the equilibrium

$$2H_2O \rightleftarrows H_3O^+ + OH^-$$

Ordinarily, the hydronium ions produced from the first reaction will suppress the dissociation of water to such an extent that the concentration of hydronium and hydroxide ions from this source can be considered to be negligible. Under these circumstances, we see from the acid dissociation equilibrium that one A^- ion is produced with each H_3O^+ ion; that is,

$$[H_3O^+] \cong [A^-] \tag{3-12}$$

Furthermore, the sum of the molar concentrations of the weak acid and its conjugate base must equal the formal concentration of the acid since the solution contains no other species that contributes A^-. Thus,

$$F_{HA} = [A^-] + [HA] \tag{3-13}$$

Substituting Equation 3-12 into 3-13 and rearranging gives

$$[HA] = F_{HA} - [H_3O^+] \tag{3-14}$$

When $[A^-]$ and $[HA]$ are replaced by Equations 3-12 and 3-14, the acid-dissociation expression becomes

$$\frac{[H_3O^+]^2}{F_{HA} - [H_3O^+]} = K_a \tag{3-15}$$

Equation 3-15 can be rearranged to the form

$$[H_3O^+]^2 + K_a[H_3O^+] - K_aF_{HA} = 0$$

The solution to this quadratic equation is

$$[H_3O^+] = \frac{-K_a + \sqrt{(K_a)^2 + 4K_aF_{HA}}}{2} \tag{3-16}$$

It is frequently possible to simplify Equation 3-15 by assuming that $[H_3O^+]$ is much smaller than F_{HA}. Thus, when $[H_3O^+] \ll F_{HA}$, the equation can be rearranged to yield

$$[H_3O^+] = \sqrt{K_aF_{HA}} \tag{3-17}$$

The magnitude of the error introduced by this assumption will increase as the concentration of acid becomes smaller and the dissociation constant of the acid becomes larger. This statement is supported by the data in Table 3-2. Note that the error introduced by the assumption is about 0.5% when the ratio F_{HA}/K_a is 10^4. The error increases to about 1.6% when the ratio is 10^3, to about 5% when it is 10^2, and to about 17% when it is 10. Figure 3-1 illustrates the effect graphically. It is noteworthy that when the ratio is one or smaller, the hydronium ion concentration from the approximate solution becomes equal to or greater than the formality of the acid itself—clearly the approximation leads to meaningless results under these circumstances.

In general, it is good practice to make the simplifying assumption and obtain a trial value for $[H_3O^+]$ that may be compared with F_{HA} in Equation 3-15. If the trial value alters $[HA]$ by an amount smaller than the allowable error in the calculation (5 to 10% in most cases in this text), the solution may be considered satisfactory. Otherwise, the quadratic equation must be solved to give a more exact value for $[H_3O^+]$.

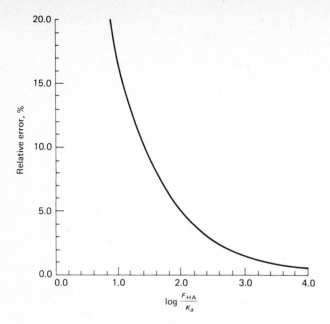

Figure 3-1 Relative Error Resulting from Assumption That $F_{HA} - [H_3O^+] \cong F_{HA}$ in Equation 3-15.

TABLE 3-2   Errors Introduced by Assuming H_3O^+ Concentration Small Relative to F_{HA} in Equation 3-15

Value of K_a	Value of F_{HA}	Value for $[H_3O^+]$ Using Assumption	Value for $[H_3O^+]$ by More Exact Equation	Percent Error
1.00×10^{-2}	1.00×10^{-3}	3.16×10^{-3}	0.92×10^{-3}	244
	1.00×10^{-2}	1.00×10^{-2}	0.62×10^{-2}	61
	1.00×10^{-1}	3.16×10^{-2}	2.70×10^{-2}	17
1.00×10^{-4}	1.00×10^{-4}	1.00×10^{-4}	0.62×10^{-4}	61
	1.00×10^{-3}	3.16×10^{-4}	2.70×10^{-4}	17
	1.00×10^{-2}	1.00×10^{-3}	0.95×10^{-3}	5.3
	1.00×10^{-1}	3.16×10^{-3}	3.11×10^{-3}	1.6
1.00×10^{-6}	1.00×10^{-5}	3.16×10^{-6}	2.70×10^{-6}	17
	1.00×10^{-4}	1.00×10^{-5}	0.95×10^{-5}	5.3
	1.00×10^{-3}	3.16×10^{-5}	3.11×10^{-5}	1.6
	1.00×10^{-2}	1.00×10^{-4}	9.95×10^{-5}	0.5
	1.00×10^{-1}	3.16×10^{-4}	3.16×10^{-4}	0.0

Example. Calculate the hydronium ion concentration of an aqueous 0.120-F nitrous acid solution. The principal equilibrium in this solution is

$$HNO_2 + H_2O \rightleftarrows H_3O^+ + NO_2^-$$

In Appendix 4 we find

$$K_a = \frac{[H_3O^+][NO_2^-]}{[HNO_2]} = 5.1 \times 10^{-4}$$

Thus,

$$[H_3O^+] = [NO_2^-]$$
$$[HNO_2^-] = 0.120 - [H_3O^+]$$

and

$$\frac{[H_3O^+]^2}{0.120 - [H_3O^+]} = 5.1 \times 10^{-4}$$

If we now assume $[H_3O^+] \ll 0.120$, we find

$$[H_3O^+] = \sqrt{0.120 \times 5.1 \times 10^{-4}} = 7.8 \times 10^{-3}$$

We must now examine our assumption that $(0.0120 - 0.0078) \cong 0.012$; a difference of about 7% is involved. The relative error in $[H_3O^+]$ will be smaller than this figure, however, as can be seen by calculating $\log F_{HA}/K_a = 2.4$; referring to Figure 3-1, we see that an error of about 3% results. If a more accurate figure was needed, solution of the quadratic equation would yield a value of 7.6×10^{-3} M.

Example. Calculate the hydronium ion concentration of a solution which is 2.0×10^{-4} F in aniline hydrochloride $C_6H_5NH_3Cl$. In aqueous solution, dissociation of this compound to Cl^- and $C_6H_5NH_3^+$ is complete. The weak acid $C_6H_5NH_3^+$ dissociates as follows:

$$C_6H_5NH_3^+ + H_2O \rightleftarrows C_6H_5NH_2 + H_3O^+ \qquad K_a = \frac{[H_3O^+][C_6H_5NH_2]}{[C_6H_5NH_3^+]}$$

Inspection of Appendix 4 reveals no entry for $C_6H_5NH_3^+$, but Appendix 5 gives a basic-dissociation constant for aniline, $C_6H_5NH_2$. That is,

$$C_6H_5NH_2 + H_2O \rightleftarrows C_6H_5NH_3^+ + OH^- \qquad K_b = 3.94 \times 10^{-10}$$

Thus, Equation 3-11 is used to obtain a value for K_a:

$$K_a = \frac{1.00 \times 10^{-14}}{3.94 \times 10^{-10}} = 2.54 \times 10^{-5}$$

Proceeding as in the previous example,

$$[H_3O^+] = [C_6H_5NH_2]$$
$$[C_6H_5NH_3] = 2.0 \times 10^{-4} - [H_3O^+]$$

Let us now assume that $[H_3O^+] \ll 2.0 \times 10^{-4}$ and substitute the simplified value for $[C_6H_5NH_3]$ into the dissociation-constant expression

$$\frac{[H_3O^+]^2}{2.0 \times 10^{-4}} = 2.54 \times 10^{-5}$$

$$[H_3O^+] = 7.1 \times 10^{-5} \text{ mole/liter}$$

Comparison of 7.1×10^{-5} with 2.0×10^{-4} suggests that a significant error exists in the value for $[H_3O^+]$ (using Figure 3-1, we find that the error is greater than 20%). Thus, unless only a crude approximation is needed, it is necessary to use the more exact expression

$$\frac{[H_3O^+]^2}{2.0 \times 10^{-4} - [H_3O^+]} = 2.54 \times 10^{-5}$$

which rearranges to

$$[H_3O^+]^2 + 2.54 \times 10^{-5}[H_3O^+] - 5.08 \times 10^{-9} = 0$$

and

$$[H_3O^+] = \frac{-2.54 \times 10^{-5} + \sqrt{(2.54 \times 10^{-5})^2 + 4 \times 5.08 \times 10^{-9}}}{2}$$

$$[H_3O^+] = 6.0 \times 10^{-5}$$

Application of Dissociation Constants for Weak Bases. The techniques discussed in previous sections are readily adapted to the calculation of the hydroxide ion concentration in solutions of weak bases.

Aqueous ammonia is basic by virtue of the reaction

$$NH_3 + H_2O \rightleftarrows NH_4^+ + OH^-$$

Here the predominant species has been clearly demonstrated to be NH_3. Nevertheless, such solutions are sometimes called ammonium hydroxide, the terminology being vestigial from the time when the substance NH_4OH rather than NH_3 was believed to be the undissociated form of the base. Application of the mass law to this equilibrium yields the expression

$$K_b = \frac{[NH_4^+][OH^-]}{[NH_3]}$$

The magnitude of this constant is independent of the formula used in the denominator, whether it is the more correct NH_3 or the historical NH_4OH.

Example. Calculate the hydronium ion concentration of a 0.075-F NH_3 solution. The predominant equilibrium in this solution is

$$NH_3 + H_2O \rightleftarrows NH_4^+ + OH^-$$

From the table of basic-dissociation constants (Appendix 5)

$$\frac{[NH_4^+][OH^-]}{[NH_3]} = 1.76 \times 10^{-5} = K_b$$

The equation for the equilibrium indicates that

$$[NH_4^+] = [OH^-]$$

and

$$[NH_4^+] + [NH_3] = F_{NH_3} = 0.075$$

If we substitute $[OH^-]$ for $[NH_4^+]$ and rearrange the second of these equations, we find that

$$[NH_3] = 0.075 - [OH^-]$$

Substituting these quantities into the dissociation-constant expression yields

$$\frac{[OH^-]^2}{7.5 \times 10^{-2} - [OH^-]} = 1.76 \times 10^{-5}$$

If we assume that $[OH^-] \ll 7.5 \times 10^{-2}$, the equation then simplifies to

$$[OH^-]^2 \cong 7.5 \times 10^{-2} \times 1.76 \times 10^{-5}$$

and

$$[OH^-] = 1.15 \times 10^{-3}$$

When we compare $[OH^-]$ with 7.5×10^{-2}, we see that the error in $[OH^-]$ will be less than 2%. A better value for $[OH^-]$ could be obtained by solving the quadratic equation.

To obtain $[H_3O^+]$, we write

$$[H_3O^+] = \frac{1.00 \times 10^{-14}}{1.15 \times 10^{-3}} = 8.7 \times 10^{-12}$$

Example.   Calculate the hydroxide ion concentration of a 0.010-*F* sodium hypochlorite solution.

The equilibrium between OCl^- and water is

$$OCl^- + H_2O \rightleftarrows HOCl + OH^-$$

for which we can write

$$K_b = \frac{[HOCl][OH^-]}{[OCl^-]}$$

Appendix 5 does not contain a value for K_b; an examination of Appendix 4, however, shows that the acid-dissociation constant of HOCl has a value of 3.0×10^{-8}. Therefore, employing Equation 3-11, we write

$$K_b = \frac{K_w}{K_a} = \frac{1.00 \times 10^{-14}}{3.0 \times 10^{-8}} = 3.3 \times 10^{-7}$$

Proceeding as in the previous example,

$$[OH^-] = [HOCl]$$
$$[OCl^-] + [HOCl] = 0.010$$

or

$$[OCl^-] = 0.010 - [OH^-] \cong 0.010$$

That is, we assume $[OH^-] \ll 0.010$. Substitution into the equilibrium-constant expression gives

$$\frac{[OH^-]^2}{0.010} = 3.3 \times 10^{-7}$$

$$[OH^-] = 5.7 \times 10^{-5}$$

The error resulting from the approximation is clearly small.

COMPLEX FORMATION

An analytically important class of reactions involves the formation of soluble complex ions. Two examples are

$$Fe^{3+} + SCN^- \rightleftarrows Fe(SCN)^{2+} \qquad K_f = \frac{[Fe(SCN)^{2+}]}{[Fe^{3+}][SCN^-]}$$

$$Zn(OH)_2(s) + 2OH^- \rightleftarrows Zn(OH)_4^{2-} \qquad K_f = \frac{[Zn(OH)_4^{2-}]}{[OH^-]^2}$$

where K_f is called the *formation constant* for the complex.[3] Note that the second constant applies only to a solution that is kept saturated by the presence of an excess of the sparingly soluble zinc hydroxide. Note also that no concentration term for zinc hydroxide appears in the formation-constant expression because its concentration in the solid is invariant and is included in the constant K_f. In this regard, the treatment is analogous to that described for solubility-product expressions.

Formation constants for numerous common complex ions appear in Appendix 6. An application involving the use of formation constants is given in the following example.

Example. The average person can see the red color imparted by $FeSCN^{2+}$ to an aqueous solution if the concentration of the complex is $6 \times 10^{-6}\,M$ or greater. What minimum concentration of KSCN would be required to make it possible to detect 1 ppm of iron(III) in a natural water sample?

Appendix 6 contains the following:

$$Fe^{3+} + SCN^- \rightleftarrows FeSCN^{2+} \qquad K_f = 1.4 \times 10^2$$

or

$$\frac{[FeSCN^{2+}]}{[Fe^{3+}][SCN^-]} = 1.4 \times 10^2$$

To convert the minimum detectable concentration of iron(III) to a formal concentration, we recall (p. 13) that 1 ppm corresponds to 1 mg/liter. Thus,

$$F_{Fe^{3+}} = \frac{1\ mg}{liter} \times \frac{1}{1000\ mg/g} \times \frac{1}{55.8\ g/fw}$$

$$= 1.8 \times 10^{-5}\ fw/liter$$

The detection limit for the complex is stated to be

$$[FeSCN^{2+}] = 6 \times 10^{-6}$$

At equilibrium, the iron(III) will be in two forms, Fe^{3+} and $FeSCN^{2+}$. Thus,

$$F_{Fe^{3+}} = 1.8 \times 10^{-5} = [Fe^{3+}] + [FeSCN^{2+}]$$
$$[Fe^{3+}] = 1.8 \times 10^{-5} - [FeSCN^{2+}] = 1.8 \times 10^{-5} - 6 \times 10^{-6}$$
$$= 1.2 \times 10^{-5}$$

---

[3] Less commonly, equilibria involving complex ions are described in terms of dissociation reactions and *instability constants*; the latter are the reciprocals of formation constants. For example,

$$Fe(SCN)^{2+} \rightleftarrows Fe^{3+} + SCN^- \qquad K_{inst} = \frac{1}{K_f} = \frac{[Fe^{3+}][SCN^-]}{[FeSCN^{2+}]}$$

We now substitute these concentrations into the formation-constant expression in order to determine the SCN^- concentration that will be required. That is,

$$\frac{6 \times 10^{-6}}{1.2 \times 10^{-5}[SCN^-]} = 1.4 \times 10^2$$

$$[SCN^-] = 0.0036$$

Thus, if the test solution is made 0.0036 F or greater in KSCN, a detectable amount of $FeSCN^{2+}$ will form, provided total iron(III) concentration is 1 ppm or greater.

OXIDATION-REDUCTION

Equilibrium constants for oxidation-reduction reactions can be formulated in the usual way. For example,

$$6Fe^{2+} + Cr_2O_7^{2-} + 14H_3O^+ \rightleftarrows 6Fe^{3+} + 2Cr^{3+} + 21H_2O$$

$$K = \frac{[Fe^{3+}]^6[Cr^{3+}]^2}{[Fe^{2+}]^6[Cr_2O_7^{2-}][H_3O^+]^{14}}$$

Note that, as in earlier examples, no term for the concentration of water is needed.

Tables of equilibrium constants for oxidation-reduction reactions are not generally available because these constants are readily derived from more fundamental constants called *standard electrode potentials*. Derivations of this kind are treated in detail in Chapter 14.

DISTRIBUTION OF A SOLUTE BETWEEN TWO IMMISCIBLE LIQUIDS

Another useful type of equilibrium-constant expression involves the distribution of a solute between two liquid phases. For example, if an aqueous solution of iodine is shaken with an immiscible organic solvent such as hexane or chloroform, a portion of the solute is extracted into the organic layer. Ultimately, an equilibrium is established between the two phases which can be described by

$$I_2(aq) \rightleftarrows I_2(org) \qquad K = \frac{[I_2]_{org}}{[I_2]_{aq}}$$

The equilibrium constant for this reaction, K, is often called a *distribution* or *partition coefficient*.

As will be seen in Chapter 29, distribution equilibria are of vital importance in understanding many separation processes.

STEPWISE EQUILIBRIA

Many weak electrolytes associate or dissociate in a stepwise manner, and equilibrium constants can be written for each step. For example, when ammonia is added to a solution containing silver ions, at least two equilibria involving the two species are established.

$$Ag^+ + NH_3 \rightleftarrows AgNH_3^+ \qquad K_1 = \frac{[AgNH_3^+]}{[Ag^+][NH_3]}$$

$$AgNH_3^+ + NH_3 \rightleftarrows Ag(NH_3)_2^+ \qquad K_2 = \frac{[Ag(NH_3)_2^+]}{[AgNH_3^+][NH_3]}$$

Here K_1 and K_2 are stepwise formation constants for the two complexes. The two constants can be multiplied together to give an overall formation constant for the reaction

$$Ag^+ + 2NH_3 \rightleftarrows Ag(NH_3)_2^+ \qquad \beta_2 = K_1K_2 = \frac{[Ag(NH_3)_2^+]}{[Ag^+][NH_3]^2}$$

Overall formation constants of this type are often symbolized as β_n, where n corresponds to the number of moles of complexing species that combine with one mole of cation. Thus, for example,

$$Cd^{2+} + 3CN^- \rightleftarrows Cd(CN)_3^- \qquad \beta_3 = K_1K_2K_3 = \frac{[Cd(CN)_3^-]}{[Cd^{2+}][CN^-]^3}$$

$$Cd^{2+} + 4CN^- \rightleftarrows Cd(CN)_4^{2-} \qquad \beta_4 = K_1K_2K_3K_4 = \frac{[Cd(CN)_4^{2-}]}{[Cd^{2+}][CN^-]^4}$$

where K_1, K_2, K_3, and K_4 are the corresponding stepwise constants.

Many common acids and bases dissociate in a stepwise manner. For example, the following equilibria exist in an aqueous solution of phosphoric acid:

$$H_3PO_4 + H_2O \rightleftarrows H_2PO_4^- + H_3O^+ \qquad K_1 = \frac{[H_3O^+][H_2PO_4^-]}{[H_3PO_4]}$$
$$= 7.11 \times 10^{-3}$$

$$H_2PO_4^- + H_2O \rightleftarrows HPO_4^{2-} + H_3O^+ \qquad K_2 = \frac{[H_3O^+][HPO_4^{2-}]}{[H_2PO_4^-]}$$
$$= 6.34 \times 10^{-8}$$

$$HPO_4^{2-} + H_2O \rightleftarrows PO_4^{3-} + H_3O^+ \qquad K_3 = \frac{[H_3O^+][PO_4^{3-}]}{[HPO_4^{2-}]}$$
$$= 4.2 \times 10^{-13}$$

As illustrated by H_3PO_4, numerical values of K_n for an acid or a base become smaller with each successive dissociation step. Equilibria of this type are considered in Chapter 10.

PROBLEMS

1. Generate solubility-product expressions for
   *(a) TlCl. (d) $Cu(OH)_2$.
   *(b) $BaSO_4$. (e) $Pr(OH)_3$.
   *(c) Ag_2SO_4. (f) $Th(OH)_4$.
   *(g) $MgNH_4PO_4$. (Products are Mg^{2+}, NH_4^+, and PO_4^{3-}.)
   (h) $Pb_2NaOH(CO_3)_2$. (Products are Pb^{2+}, Na^+, OH^-, and CO_3^{2-}.)

2. For each of the compounds in Problem 1, write an equation relating its formal solubility s to its solubility-product constant K_{sp}.

3. Write the equilibrium-constant expressions, and give the numerical values for each for the following reactions:
   *(a) The basic dissociation of ethylamine, $C_2H_5NH_2$.
   (b) The acidic dissociation of hydrogen cyanide, HCN.
   *(c) The acidic dissociation of pyridine hydrochloride, C_5H_5NHCl.
   (d) The basic dissociation of NaCN.
   *(e) The formation of $AgCl_2^-$ from AgCl.
   (f) The formation of CuI_2^- from CuI.
   *(g) The formation of $Cd(NH_3)_2^{2+}$ from NH_3 and Cd^{2+}.
   (h) The formation of AlF_6^{3-} from Al^{3+} and F^-.
   *(i) The dissociation of H_3AsO_4 to H_3O^+ and AsO_4^{3-}.
   (j) The reaction of CO_3^{2-} with H_2O to give H_2CO_3 and OH^-.

4. Calculate the solubility-product constant for each of the following substances given that the formal concentrations of their saturated solutions are as indicated.
   *(a) AgSeCN (2.0×10^{-8} fw/liter; products are Ag^+ and $SeCN^-$).
   (b) $RaSO_4$ (6.6×10^{-6} fw/liter).
   *(c) $Pb(BrO_3)_2$ (1.7×10^{-1} fw/liter).
   (d) $PbBr_2$ (2.1×10^{-2} fw/liter).
   *(e) BiOCl (at $[H_3O^+] = 1.00 \times 10^{-6}$ mole/liter, solubility $= 8.4 \times 10^{-11}$ fw/liter; $BiOCl + 2H^+ \rightleftarrows Bi^{3+} + Cl^- + H_2O$).
   (f) Hg_2O (at $[H_3O^+] = 1.00 \times 10^{-3}$ mole/liter, solubility $= 1.0 \times 10^{-24}$ fw/liter; $Hg_2O + H_2O \rightleftarrows Hg_2^{2+} + 2OH^-$).
   *(g) $Ce(IO_3)_3$ (1.9×10^{-3} fw/liter).
   (h) BiI_3 (1.3×10^{-5} fw/liter).

5. Use the solubility information in parentheses to calculate the solubility product for
   *(a) $Ag_2SeO_4(s) \rightleftarrows 2Ag^+ + SeO_4^{2-}$ (0.086 g/100 ml).
   (b) $K_2SiF_6(s) \rightleftarrows 2K^+ + SiF_6^{2-}$ (1.32 g/liter).
   *(c) $Sr(IO_3)_2(s) \rightleftarrows Sr^{2+} + 2IO_3^-$ (0.48 g/250 ml).
   (d) $PbOHCl(s) \rightleftarrows Pb^{2+} + OH^- + Cl^-$ (3.5 mg/500 ml).

*6. Calculate the weight in grams of PbI_2 that will dissolve in 100 ml of
   (a) H_2O.
   (b) 2.00×10^{-2} F KI.
   (c) 2.00×10^{-2} F $Pb(NO_3)_2$.

7. Calculate the weight of $La(IO_3)_3$ that will dissolve in 100 ml of
   (a) H_2O.
   (b) 2.00×10^{-2} F $NaIO_3$.
   (c) 2.00×10^{-2} F $La(NO_3)_3$.

*8. The solubility product for Tl_2CrO_4 is 9.8×10^{-13}. What CrO_4^{2-} concentration is required to
   (a) initiate precipitation of Tl_2CrO_4 from a solution that is 2.12×10^{-3} M in Tl^+?
   (b) lower the concentration of Tl^+ in a solution to 1.00×10^{-6} M?

9. What hydroxide concentration is required to
   (a) initiate precipitation of Fe^{3+} from a 1.00×10^{-3} F solution of $Fe_2(SO_4)_3$?
   (b) lower the Fe^{3+} concentration in the foregoing solution to 1.00×10^{-9} F?

10. What is the hydroxide ion concentration of a saturated solution of
   *(a) $Mg(OH)_2$?
   (b) $Zn(OH)_2$?

*11. The solubility-product constant for $Ce(IO_3)_3$ has a value of 3.2×10^{-10}. What will be the Ce^{3+} concentration in a solution prepared by mixing 50.0 ml of 0.0500-F Ce^{3+} with

(a) 50.0 ml of water?
(b) 50.0 ml of 0.150-F IO_3^-?
(c) 50.0 ml of 0.300-F IO_3^-?
(d) 50.0 ml of 0.050-F IO_3^-?

12. The solubility-product constant for K_2PtCl_6 is 1.1×10^{-5}. ($K_2PtCl_6 \rightleftarrows 2K^+ + PtCl_6^{2-}$) What is the K^+ concentration of a solution prepared by mixing 50.0 ml of 0.400-F KCl with

(a) 50.0 ml of 0.100-M $PtCl_6^{2-}$?
(b) 50.0 ml of 0.200-M $PtCl_6^{2-}$?
(c) 50.0 ml of 0.400-M $PtCl_6^{2-}$?

13. At 25°C, what are the molar H_3O^+ and OH^- concentrations of

*(a) 0.0200-F HOCl?
(b) 0.0800-F propanoic acid?
*(c) 0.200-F methylamine?
*(g) 0.100-F hydroxylamine hydrochloride?
(h) 0.0500-F ethanolamine hydrochloride?

(d) 0.100-F trimethylamine
*(e) 0.120-F NaOCl?
(f) 0.0860-F CH_3COONa?

14. What is the hydronium concentration of water at 0°C?

15. At 25°C, what is the hydronium ion concentration of

*(a) 0.100-F chloroacetic acid?
*(b) 0.100-F sodium chloroacetate?
(c) 0.0100-F methylamine?
(d) 0.0100-F methylamine hydrochloride?
*(e) 1.00×10^{-3} F aniline hydrochloride?
(f) 0.200-F HIO_3?

16. Calculate the hydronium ion concentration of

*(a) a 0.0100-F solution of NH_4NO_3.
(b) a 0.0100-F solution of $NaNO_2$.

*17. Mercury(II) forms a soluble, neutral complex with SCN^- having the formula $Hg(SCN)_2$. The formation constant of the complex has a value of 1.8×10^{17}. Calculate the Hg^{2+} and SCN^- concentrations in a solution prepared by mixing 10 ml of 0.0200-F Hg^{2+} with

(a) 10 ml of 0.0200-F SCN^-.
(b) 10 ml of 0.0400-F SCN^-.
(c) 10 ml of 0.0600-F SCN^-.

*18. Calculate the molar concentration of CuI_2^- in a solution that is saturated with CuI and has the following formal concentration of KI, given that the formation constant for the reaction $CuI(s) + I^- \rightleftarrows CuI_2^-$ is 8×10^{-4}:

(a) 0.00100.
(b) 0.0100.
(c) 0.100.

19. The formation constant for $Pb(OH)_3^-$ is

$$Pb(OH)_2(s) + OH^- \rightleftarrows Pb(OH)_3^- \qquad K_f = 5 \times 10^{-2}$$

Calculate the concentration of $Pb(OH)_3^-$ in a saturated $Pb(OH)_2$ solution that is

(a) 0.00100 in NaOH.
(b) 0.100-F in NaOH.

4

THE EVALUATION OF THE RELIABILITY OF ANALYTICAL DATA

Every physical measurement is subject to a degree of uncertainty that, at best, can be decreased only to an acceptable level. The determination of the magnitude of this uncertainty is often a difficult task that requires additional effort, ingenuity, and good judgment on the part of the scientist. Nevertheless, it is a task that cannot be neglected because an analysis of totally unknown reliability is worthless to a scientist. On the other hand, a result which may not necessarily be highly accurate can be of signal importance provided that the limits of probable error affecting it can be set with a high degree of certainty. Unfortunately, there exists no simple, generally applicable means by which the quality of an experimental result can be assessed with absolute certainty. Consequently, it is not uncommon for the scientist to expend as much effort in evaluating his data as he does in acquiring them. This effort can involve a study of the literature to profit from the experience of others, the calibration of equipment, additional experiments specifically designed to provide clues as to possible errors, and statistical analysis of the data. It should be recognized, however, that none of these measures is infallible. Ultimately, the scientist can only make a *judgment* as to the probable accuracy of a measurement; with experience, judgments of this kind tend to become harsher and less optimistic.

A direct relationship exists between the accuracy of an analytical measurement and the time and effort expended in its acquisition. A tenfold increase in reliability may require hours, days, or perhaps weeks of added labor. As a first step, then, the experienced scientist establishes how reliable the results of an analysis must be because this consideration will determine the amount of time and effort that must be invested in the analysis. Careful thought at the outset of an investigation often provides major savings in time and effort. *It cannot be too strongly emphasized that a scientist cannot afford to waste time in the indiscriminate pursuit of the ultimate in accuracy when such is not needed.*

In this chapter we consider the types of errors encountered in analyses, methods for their recognition, and techniques for estimating and reporting their magnitude.

Definition of Terms

The chemist generally repeats the analysis of a given sample two to five times. The individual results for such a set of replicate measurements will seldom be exactly the same; it thus becomes necessary to select a central "best" value for the set. Intuitively, the added effort of replication can be justified in two ways. First, it is plausible that the central value of the set should be more reliable than any of the individual results; second, the variations among the results ought to provide some measure of reliability in the "best" value that is chosen.

Either of two quantities, the *mean* and the *median*, may serve as the central value for a set of measurements.

THE MEAN AND MEDIAN

The *mean, arithmetic mean,* and *average* ($\bar{x}$) are synonymous terms for the numerical value obtained by dividing the sum of a set of replicate measurements by the number of individual results in the set.

The *median* of a set is that result about which all others are equally distributed, half being numerically greater and half being numerically smaller. If the set consists of an odd number of measurements, selection of the median may be made directly; for a set containing an even number of measurements, the average of the central pair is taken.

Example. Calculate the mean and median for 10.06, 10.20, 10.08, 10.10.

$$\text{mean} = \bar{x} = \frac{10.06 + 10.20 + 10.08 + 10.10}{4} = 10.11$$

Since the set contains an even number of measurements, the median is the average of the middle pair:

$$\text{median} = \frac{10.08 + 10.10}{2} = 10.09$$

Ideally, the mean and median should be numerically identical; more often than not, however, this condition is not realized, particularly when the number of measurements in the set is small.

PRECISION

The term *precision* is used to describe the reproducibility of results. It can be defined as the agreement between the numerical values of two or more measurements that have been made *in an identical fashion*. Several methods exist for expressing the precision of data.

Absolute Methods for Expressing Precision. The *deviation from the mean* $(x_i - \bar{x})$ is a common method for describing precision and is simply the numerical difference, *without regard to sign*, between an experimental value and the mean for the set of data that includes the value. To illustrate, suppose that a chloride analysis yielded the following results:

Sample	Percent Chloride	Deviation from Mean $\|x_i - \bar{x}\|$	Deviation from Median
x_1	24.39	0.077	0.03
x_2	24.19	0.123	0.17
x_3	24.36	0.047	0.00
	3⟌72.94	3⟌0.247	3⟌0.20

$$\bar{x} = 24.313 = 24.31 \qquad \text{Avg} = 0.082 = 0.08 \qquad \text{Avg} = 0.067 = 0.07$$
$$w = x_{max} - x_{min} = 24.39 - 24.19 = 0.20$$

The mean value for the data is 24.31%; the deviation of the second result from the mean is 0.12%. The average deviation from the mean is 0.08%. Note that we have carried three figures to the right of the decimal point in the calculation for the average deviation even though the individual data are given to only two. The mean and the average deviation from the mean are then rounded to the proper number of figures *after the computation is complete*. This practice is generally worthwhile because it tends to minimize rounding errors.

Precision can also be reported in terms of *deviation from the median*. In the preceding example, deviations from 24.36 would be recorded, as shown in the last column of the table.

The *spread* or *range* (*w*) in a set of data is the numerical difference between the highest and lowest result and is also a measure of precision. In the previous example, the spread would be 0.20% chloride.

Two other measures of precision are the *standard deviation* and the *variance*. These terms will be defined in a later section of this chapter.

Relative Precision. We have thus far considered the expression of precision in absolute terms. It is often more convenient, however, to indicate the precision relative to the mean (or the median) in terms of percentages or as parts per thousand. For example, for sample x_1,

$$\text{relative deviation from mean} = \frac{0.077 \times 100}{24.31} = 0.32 \cong 0.3\%$$

Similarly, the average deviation of the set from the median can be expressed as

$$\text{relative average deviation from median} = \frac{0.067 \times 1000}{24.36} \cong 3 \text{ ppt}$$

ACCURACY

The term *accuracy* denotes the nearness of a measurement to its accepted value and is expressed in terms of *error*. Note the fundamental difference between this term and precision. Accuracy involves a comparison with respect to a true or accepted value; in contrast, precision compares a result with other measurements made in the same way.

The accuracy of a measurement is often described in terms of *absolute error*, which can be defined as

$$E = x_i - x_t \tag{4-1}$$

The absolute error E is the difference between the observed value x_i and the accepted value x_t. The accepted value may itself be subject to considerable uncertainty. As a consequence, it is frequently difficult to arrive at a realistic estimate for the error of a measurement.

Returning to the previous example, suppose that the accepted value for the percentage of chloride in the sample is 24.36%. The absolute error of the mean is thus $24.31 - 24.36 = -0.05\%$ chloride; here we ordinarily retain the sign of the error to indicate whether the result is high or low.

Often a more useful quantity than the absolute error is the *relative error* expressed as a percentage or in parts per thousand of the accepted value. Thus, for the chloride analysis we have been considering,

$$\text{relative error} = -\frac{0.05 \times 100}{24.36} = -0.21 = -0.2\%$$

$$\text{relative error} = -\frac{0.05 \times 1000}{24.36} = -2.1 = -2 \text{ ppt}$$

PRECISION AND ACCURACY OF EXPERIMENTAL DATA

The precision of a measurement is readily determined by replicate experiments performed under identical conditions. Unfortunately, an estimate of the accuracy is not equally available because this quantity requires sure knowledge of the very information that is being sought, namely, the true value. It is tempting to ascribe a direct relationship between precision and accuracy. The danger of this approach is illustrated in Figure 4-1 which summarizes the analysis for nitrogen in two pure compounds by four analysts. The dots give the absolute errors of replicate measurements for each sample and each analyst. Note that analyst 1 obtained relatively high precision and high accuracy. Analyst 2, on the other hand, had poor precision but good accuracy. The results from analyst 3 are of a kind that is by no means uncommon; here the precision is excellent, but a

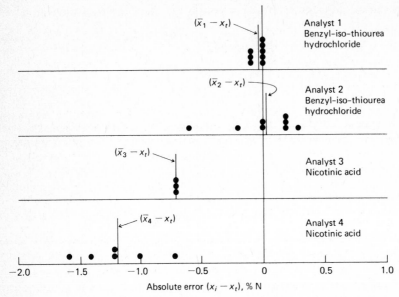

Figure 4-1   Absolute Errors in Nitrogen Analyses by a Micro Kjeldahl Procedure. Each vertical line labeled $(\bar{x}_i - x_t)$ is the absolute deviation of the mean of the set from the true value. [Data taken from C. O. Willits and C. L. Ogg, *J. Ass. Offic. Anal. Chem.*, **32**, 561 (1949).]

significant error exists in the numerical average of the data. The scientist also encounters a situation similar to that recorded by analyst 4 where both precision and accuracy are poor.

The behavior illustrated by Figure 4-1 can be rationalized by assuming that experimental measurements are afflicted by two general types of uncertainty and that the effect of one of these types is not revealed by the precision of the measurements.

Classification of Errors

The uncertainties that arise in a chemical analysis, and that are responsible for the behavior illustrated in Figure 4-1, may be classified into two broad categories, depending upon their origin. *Determinate errors* are those that have a definite value which can (in principle, if not in practice) be measured and accounted for. *Indeterminate errors* result from extending a system of measurement to its maximum. These errors cannot be positively identified and do not have a definite measurable value; instead, they fluctuate in a random manner. The scatter of individual data about the *mean* values shown in Figure 4-1 is the direct result of indeterminate-type errors. The difference between the mean and the true values reported by analysts 3 and 4 $(\bar{x}_i - x_t)$, on the other hand, is the result of one or more determinate errors.

TYPES OF DETERMINATE ERRORS

It is not possible to list all conceivable sources of determinate error; we can, however, recognize that they have their origin in the *personal* errors of the experimenter, the *instrumental* errors of his measuring devices, the errors that repose in the *method* of analysis he employs, or any combination of these.

Personal Errors. These errors are the result of the ignorance, carelessness, prejudices, or physical limitations of the experimenter. For example, they may arise from the use of an improper technique in transferring a sample, from disregard of temperature corrections for a measuring device, from over- or underwashing a precipitate, or from transposing numbers when recording an experimental measurement.

A common personal error due to a physical limitation is found in the partially color-blind person who has great difficulty in discerning color changes that are important to an analysis.

An important source of error that must be continuously guarded against is personal *bias* or prejudice. There is a natural tendency for the experimenter to estimate scale readings in a direction that improves the precision in a set of results or that causes the results to fall closer to a preconceived notion of the true value for the measurement. Determinate errors of this type can be avoided by recognition of this common human frailty and by the conscious use of objectivity in making observations.

Instrumental Errors. Instrumental errors are attributable to imperfections in the tools with which the analyst works or to the effects of environmental factors upon these tools. For example, volumetric equipment such as burets, pipets, and volumetric flasks frequently deliver or contain volumes slightly different from those indicated by their graduations, particularly when they are employed at temperatures which differ significantly from the temperature at which they were calibrated. Calibration at the proper temperature will obviously eliminate this type of determinate error.

Method Errors. Determinate errors are often introduced from nonideal chemical behavior of the reagents and reactions upon which the analysis is based. Such sources of nonideality include the slowness of some reactions, the incompleteness of others, the lack of stability of some species, the nonspecificity of most reagents, and the possible occurrence of side reactions which interfere with the measurement process. For example, in a gravimetric analysis the chemist is confronted with the problem of isolating the element to be determined as a solid of the greatest possible purity. If he fails to wash it sufficiently, the precipitate will be contaminated with foreign substances and have a spuriously high weight. On the other hand, sufficient washing to remove these contaminants may cause weighable quantities to be lost owing to the solubility of the precipitate; here, a negative determinate error will result. In either case, the accuracy of the procedure is limited by the method error associated with the analysis.

A method error frequently encountered in volumetric analysis is due to the volume of reagent, in excess of theoretical, which is consumed by an indicator to cause a color change that signifies completion of the reaction. As a consequence, the ultimate accuracy of the analysis is limited by the very phenomenon that makes the determination possible.

Yet another type of method error is illustrated in Figure 4-1. The Kjeldahl analysis for nitrogen in organic compounds is based upon oxidation of the sample with concentrated sulfuric acid; the nitrogen ordinarily is converted to ammonium sulfate. Compounds that contain a pyridine ring, such as nicotinic acid, may not be completely destroyed under the conditions of these analyses. The negative error in the results by analysts 3 and 4 is in all probability the result of this resistance to oxidation.

Errors inherent in a method are probably the most serious of determinate errors since they are the most likely to remain undetected.

EFFECT OF DETERMINATE ERROR UPON THE RESULTS OF AN ANALYSIS

Determinate errors generally fall into either of two categories, *constant* or *proportional*. The magnitude of a constant error is independent of the size of the quantity measured. On the other hand, proportional errors depend in absolute magnitude upon, and increase or decrease in proportion to, the size of the sample taken for analysis.

Constant Errors. For any given analysis, a constant error will become more serious as the size of the quantity measured decreases. This problem is illustrated by the solubility losses that attend the washing of a precipitate.

Example. Suppose that directions call for washing of the precipitate with 200 ml of water, and that 0.50 mg is lost in this volume of wash liquid. If 500 mg of precipitate are involved, the relative error due to solubility loss will be $-(0.50 \times 100/500) = -0.1\%$. Loss of the same quantity from 50 mg of precipitate will result in a relative error of -1.0%.

The amount of reagent required to bring about the color change in a volumetric analysis is another example of constant error. This volume, usually small, remains the same regardless of the total volume of reagent required. Again, the relative error will be more serious as the total volume decreases. Clearly, one way of minimizing the effect of constant error is to use as large a sample as is consistent with the method at hand.

Proportional Errors. Interfering contaminants in the sample, if not eliminated in some manner, will cause an error of the proportional variety. For example, a method widely employed for the analysis of copper involves reaction of the copper(II) ion with potassium iodide; the quantity of iodine produced in the reaction is then measured. Iron(III), if present, will also liberate iodine from potassium iodide. Unless steps are taken to prevent this interference, the analysis will yield erroneously high results for the percentage of copper since the iodine produced will be a measure of the sum of the copper and iron in the sample. The magnitude of this error is fixed by the *fraction* of iron contamination and will produce the same relative effect regardless of the size of sample taken for analysis. If the sample size is doubled, for example, the amount of iodine liberated by both the copper and the iron contaminant will also be doubled. While the *absolute* error will also undergo a twofold increase, the *relative* error will remain unchanged.

Effects of Determinate Error

Determinate errors are often large. Nevertheless, their detection is made difficult because no single procedure exists that will provide infallible information as to the presence or absence of uncertainties of this kind.

DETECTION OF INSTRUMENTAL AND PERSONAL ERRORS

Instrumental errors are usually found and corrected by calibration procedures. Indeed, periodic recalibration of equipment is always desirable since the behavior of most instruments changes with time owing to wear, corrosion, or mistreatment.

Most personal errors can be minimized by care and self-discipline. Thus, most scientists develop the habit of always rechecking instrument readings, notebook entries, and calculations. Errors that result from a physical handicap can usually be avoided by proper choice of method, provided, of course, that the handicap is recognized.

DETECTION OF METHOD ERRORS

Method errors are particularly difficult to detect; in addition, compensation for their effects is often formidable. Identification of systematic errors of this type may take one or more of the courses described in the following paragraphs.

Analysis of Standard Samples. A method may be tested for determinate error by analysis of synthetic samples whose overall composition is known and closely approximates that of the material for which the analysis is intended. Great care must go into the preparation of these standard samples to ensure that the concentration of the constituent to be determined is known with a high degree of certainty. Unfortunately, the preparation of a sample whose composition truly resembles that of a complex natural substance is often difficult if not impossible. Moreover, the problem is compounded by the requirement that the exact concentration of a particular constituent be known as a result of the method of preparation. These problems are frequently so imposing as to prevent the use of this approach.

The National Bureau of Standards has available for purchase many common substances that have been carefully analyzed for one or more constituents. These standard materials are valuable for the testing of analytical procedures for accuracy.[1]

Independent Analysis. Parallel analysis of a sample by a method of established reliability that is independent of the one under investigation is of particular value where samples of known purity are not available. In general, the independent method should not resemble the one under study in order to minimize the possibility that some common factor in the sample will have an equal effect on both methods.

Blank Determinations. Constant errors affecting physical measurements can be frequently evaluated with a blank determination in which all steps of the analysis are performed in the absence of a sample. The result is then applied as a correction to the actual measurement. Blank determinations are of particular value in exposing errors that are due to the introduction of interfering contaminants from reagents and vessels employed in an analysis.

Variation in Sample Size. The presence of a constant error will be revealed by using a range of sample sizes for the analysis of a particular material. For example, the hypothetical results in Table 4-1 assume various sample sizes for

TABLE 4-1   Effect of a Constant Error of 2 mg on the Analysis of a Silver Alloy

Weight of Sample, g	Weight of Silver Found, g	Silver Found, %
0.2000	0.0378	18.90
0.5000	0.0979	19.58
1.0000	0.1981	19.81
2.0000	0.3982	19.91
5.0000	0.9980	19.96

an alloy that contains exactly 20.00% silver. In each of these analyses a constant error of 2.0 mg causes low values for the percentage of silver. The effect of this error becomes less as the sample size is increased; indeed, the data appear to be approaching a constant value in the larger samples, as may be seen in Figure 4-2.

By way of contrast, variation in sample size is ineffective for the detection of proportional error. Figure 4-2 also summarizes results from the analysis of various samples of the same alloy by a method containing a proportional error of 5 ppt; these data also appear in Table 4-2. The plot is a straight line; clearly, without knowledge of the true percentage of silver in the alloy, the existence of such an error would go unnoticed.

[1] See the current edition of NBS Circular 552 for a description of available samples and their prices.

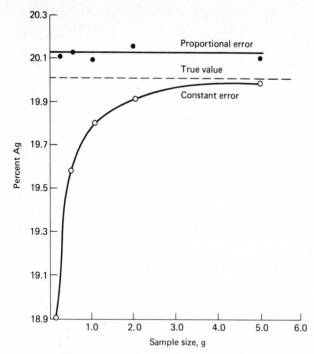

Figure 4-2 Effect of Constant and Proportional Errors upon the Results of a Silver Analysis (see Tables 4-1 and 4-2).

TABLE 4-2 Effect of a Proportional Error of 5 Parts per Thousand on the Analysis of a Silver Alloy

Weight of Sample, g	Weight of Silver Found, g	Silver Found, %
0.2000	0.0402	20.10
0.5000	0.1006	20.12
1.0000	0.2009	20.09
2.0000	0.4021	20.11
5.0000	1.0051	20.10

Effects of Indeterminate Error

As suggested by its name, indeterminate error arises from uncertainties in a measurement that are unknown and cannot be controlled by the scientist. The effect of such uncertainties is to produce a scatter of results for replicate measurements such as those for the four sets of data shown in Figure 4-1.

Table 4-3 illustrates the effect of indeterminate error upon the relatively

TABLE 4-3   Replicate Measurements from the Calibration of a 10-ml Pipet

Trial	Vol Water Delivered, ml	Trial	Vol Water Delivered, ml	Trial	Vol Water Delivered, ml
1	9.990	9	9.988	17	9.978
2	9.986	10	9.976	18	9.980
3	9.973	11	9.980	19	9.976
4	9.983	12	9.973	20	9.986
5	9.980	13	9.970[b]	21	9.986
6	9.988	14	9.988	22	9.983
7	9.993[a]	15	9.980	23	9.978
8	9.970[b]	16	9.986	24	9.988

Mean volume = 9.9816 = 9.982 ml
Average deviation from mean = 0.0054 ml
Spread = 9.993 − 9.970 = 0.023 ml
Standard deviation = 0.0065 ml

[a] Maximum value
[b] Minimum value

simple process of calibrating a pipet. The procedure involves determining the weight of water (to the nearest milligram) delivered by the pipet. The temperature of the water must be measured in order to establish its density. The experimental weight can then be converted to the volume delivered by the pipet.

The data in Table 4-3 are typical of those that might be obtained by an experienced and competent worker using a calibrated analytical balance with a sensitivity of 1 mg (corresponding roughly to 0.001 ml), with every effort being made to recognize and eliminate determinate errors. Even so, the average deviation from the mean of the 24 measurements is ± 0.0054 ml, and the spread is 0.023 ml. This dispersion among the data is the direct consequence of indeterminate error.

Variations among replicate results such as those in Table 4-3 can be rationalized by assuming that any measurement is affected by numerous small and individually undetectable uncertainties caused by uncontrolled variables in the experiment. The cumulative effect of such uncertainties will be likewise variable. Ordinarily, they tend to cancel one another and thus exert a minimal effect. Occasionally, however, they can act in concert to produce a relatively large positive or negative error. Sources for uncertainties in the calibration of a pipet might include visual judgments of the liquid level with respect to the etch mark on the pipet, the mercury level in the thermometer, and the position of an indicator with respect to a scale in the balance. Other sources include variation in the drainage time, the angle of the pipet as it drains, and temperature change resulting from the way the pipet is handled. Undoubtedly many uncertainties exist in addition to the ones cited. It is clear that many small and uncontrolled variables accompany even as simple an experiment as a pipet calibration. Although we are unable to detect the influence of any one of these uncertainties, their cumulative effect is an indeterminate error that accounts for the scatter of data about the mean.

THE DISTRIBUTION OF DATA FROM REPLICATE MEASUREMENTS

In contrast to determinate errors, indeterminate errors cannot be eliminated from measurements. Further, the scientist cannot ignore their existence on the basis of their small size. For example, it would probably be safe to assume that the average value of the 24 measurements in Table 4-3 is closer to the true volume delivered by the pipet than any of the individual data. Suppose, however, that only a duplicate calibration had been performed, and that by chance these measurements had corresponded to trials 1 and 7; the average of these two values, 9.992, differs by 0.010 ml from the mean of the 24 measurements. Note, also, that the average deviation of these two measurements *from their own mean* is only ±0.0015 ml. On the basis of this small average deviation from the mean, an overly optimistic idea of the magnitude of the indeterminate error might arise. Serious consequences would result if the user of the pipet needed to deliver volumes known, let us say, to the nearest ±0.002 ml. Here his failure to recognize the true magnitude of the indeterminate error would create a totally false sense of security with respect to the performance of the pipet. As a matter of fact, it can be shown that if this pipet were employed for 1000 measurements, 2 to 3 of the transfers would, with high probability, differ from the mean of 9.982 ml by as much as 0.02 ml, and more than 100 would differ by 0.01 ml or greater, despite every precaution on the part of the user.

In order to develop a qualitative grasp of the way small uncertainties affect the outcome of replicate measurements, let us first consider an imaginary situation in which just four such uncertainties are the cause of indeterminate error. We shall specify that each of these uncertainties has an equal probability of occurring and can affect the final result in only one of two ways, namely, to cause it to be in error by plus or minus a fixed amount, U. Further, we shall stipulate that the magnitude of U is the same for each of the four uncertainties.

Table 4-4 shows all of the possible ways the four uncertainties can combine to give the indicated indeterminate errors. We note that there is only one way in which the maximum positive error of 4U can arise, compared with four combinations that lead to a positive error of 2U, and six combinations that lead to zero error. A similar relationship exists for negative indeterminate errors. This ratio of 6 : 4 : 1 is a measure of the probability for an error of each size; if we made sufficient measurements, a frequency distribution of errors such as that shown in Figure 4-3a would be expected. Figure 4-3b shows the distribution for 10 equal-sized uncertainties. Again we see that the most frequent occurrence is zero error, while the maximum error of 10U would occur only occasionally (about once in 500 measurements).

If the foregoing arguments are extended to a very large number of uncertainties of smaller and smaller size, it can be demonstrated that the continuous distribution curve shown in Figure 4-3c will result. This bell-shaped curve is called a *Gaussian* or *normal error curve*.[2] Its properties include (1) a maximum frequency in occurrence of zero indeterminate error, (2) a symmetry

---

[2] In deriving the Gaussian curve, it is not necessary to assume, as we have done, that the individual uncertainties have identical magnitudes.

TABLE 4-4   Possible Ways Four Equal-sized Uncertainties
U_1, U_2, U_3, and U_4 Can Combine

Combinations of Uncertainties	Magnitude of Indeterminate Error	Relative Frequency of Error
$+U_1 +U_2 +U_3 +U_4$	$+4U$	1
$-U_1 +U_2 +U_3 +U_4$ $+U_1 -U_2 +U_3 +U_4$ $+U_1 +U_2 -U_3 +U_4$ $+U_1 +U_2 +U_3 -U_4$	$+2U$	4
$-U_1 -U_2 +U_3 +U_4$ $+U_1 +U_2 -U_3 -U_4$ $+U_1 -U_2 +U_3 -U_4$ $-U_1 +U_2 -U_3 +U_4$ $-U_1 +U_2 +U_3 -U_4$ $+U_1 -U_2 -U_3 +U_4$	0	6
$+U_1 -U_2 -U_3 -U_4$ $-U_1 +U_2 -U_3 -U_4$ $-U_1 -U_2 +U_3 -U_4$ $-U_1 -U_2 -U_3 +U_4$	$-2U$	4
$-U_1 -U_2 -U_3 -U_4$	$-4U$	1

about this maximum indicating that negative and positive errors occur with equal frequency, and (3) an exponential decrease in frequency as the magnitude of the error increases. Thus, a small indeterminate error will occur much more often than a very large one.

Numerous *empirical* observations have shown that the indeterminate errors arising in chemical analyses most commonly distribute themselves in a manner which approaches a Gaussian distribution. For example, if the deviations from the mean of hundreds of repetitive pH measurements on a single sample were plotted against the frequency of occurrence of each deviation, a curve approximating that shown in Figure 4-3d would ordinarily result.

The frequent experimental observation of Gaussian behavior lends credibility to the idea that the indeterminate error observed in analytical measurements can be traced to the accumulation of a large number of small, independent, and uncontrolled uncertainties. Equally important, the Gaussian distribution of most analytical data permits the use of statistical techniques for estimating the limits of indeterminate error from the precision of such data.

CLASSICAL STATISTICS

Statistics permits a mathematical description of random processes such as the effects of indeterminate error on the results of a chemical analysis. It is important to realize, however, that the techniques of classical statistics apply exactly to an *infinite* number of observations only. When these techniques are applied to the two to five replicate analyses that the chemist can afford to make, conclusions as to the probable indeterminate error can be seriously in error and misleadingly

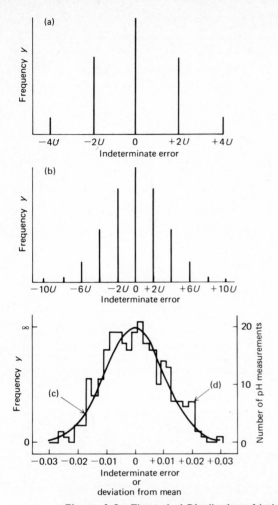

Figure 4-3 Theoretical Distribution of Indeterminate Error Arising from (a) 4 Uncertainties, (b) 10 Uncertainties, (c) a Very Large Number of Uncertainties. Curve (c) shows the normal error or Gaussian distribution. Curve (d) is an experimental distribution curve which might be obtained by plotting the deviations from the mean for about 250 replicate pH measurements against the number of times each deviation was observed.

optimistic. In these circumstances, modification of the classical techniques is necessary. Before considering these practical modifications, however, it is worthwhile to describe briefly some important relationships of classical statistics.

Properties of the Normal Error Curve. The upper two curves of Figure 4-4 are normal error curves for two different analytical methods. The topmost curve represents data from the more precise of the two methods inasmuch as the results are distributed more closely about the central value.

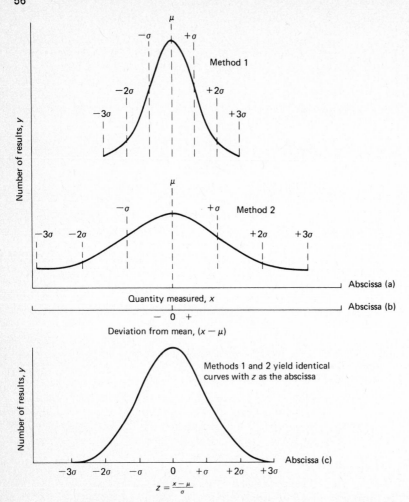

Figure 4-4 Normal Error Curves. Curve for the measurement of the same quantity by two methods. Method 1 is more reliable; thus σ is smaller. Note three types of abscissa: (a) measured quantity x with maximum at μ, (b) deviation from mean with maximum at 0, and (c) z from Equation 4-2. Abscissa (c) reduces the two curves to a single one.

As shown in Figure 4-4, normal error curves can be plotted in three different ways. In each, the ordinate is the frequency of occurrence y for each value of the abscissa. The observed values x of the measurement are plotted as abscissa (a); here the central value is the mean, which is symbolized by μ. Abscissa (b) consists of individual deviations from the mean, $x - \mu$; here the most frequently occurring deviation has a value of zero. We shall consider the third type of plot, shown by abscissa (c), presently.

It is important to emphasize that the curves under discussion are idealized because they represent the theoretical distribution of experimental results to be expected as the number of analyses involved approaches infinity. For a physically realizable set of results, a discontinuous distribution such as shown in Figure 4-3d would be more likely. Classical statistics is based on curves such as those shown in Figure 4-4 rather than on curves such as 4-3d.

The distribution data in Figure 4-4 can be described mathematically in terms of just three parameters as shown by the expression

$$y = \frac{e^{-(x-\mu)^2/2\sigma^2}}{\sigma\sqrt{2\pi}} = \frac{e^{-z^2/2}}{\sigma\sqrt{2\pi}} \tag{4-2}$$

In this equation x represents values of individual measurements, and μ is the arithmetic mean for an infinite number of such measurements. The quantity $(x - \mu)$ is thus the deviation from the mean; y is the frequency of occurrence for each value of $(x - \mu)$. The symbol π has its usual meaning, and e is the base for Napierian logarithms, 2.718.... The parameter σ is called the *standard deviation* and is a constant that has a unique value for any set containing a large number of measurements. The breadth of the normal error curve is directly related to σ.

The exponential in Equation 4-2 can be simplified by introducing the variable

$$z = \frac{x - \mu}{\sigma} \tag{4-3}$$

which then gives the deviation from the mean in units of standard deviations. As demonstrated by abscissa (c) in Figure 4-4, the substitution of z produces a single curve for all values of σ.

The Standard Deviation. Equation 4-2 indicates that a unique distribution curve exists for each value of the standard deviation. Regardless of the size of σ, however, it can be shown that 68.3% of the area beneath the curve lies within one standard deviation ($\pm 1\sigma$) of the mean, μ. Thus, 68.3% of the values lie within these boundaries. Approximately 95.5% of all values will be equal to or smaller than $\pm 2\sigma$; 99.7% will be equal to or smaller than $\pm 3\sigma$. Values of $(x - \mu)$ corresponding to $\pm 1\sigma$, $\pm 2\sigma$, and $\pm 3\sigma$ are indicated by vertical lines in the upper curves of Figure 4-4. For the bottom curve, the units for z shown on the abscissa are $\pm \sigma$.

These properties of the normal error curve are useful because they permit statements to be made about the probable magnitude of the indeterminate error of a given measurement *provided the standard deviation of the method of measurement is known.* Thus, if σ were available, one could say that the chances are 68.3 out of 100 that the indeterminate error associated with any given single measurement is smaller than $\pm 1\sigma$, that the chances are 95.5 out of 100 that the error is less than $\pm 2\sigma$, and so forth. Clearly, the standard deviation for a method of measurement is a useful parameter in estimating and reporting the probable size of indeterminate errors.

For a very large set of data, the standard deviation is given by

$$\sigma = \sqrt{\frac{\sum\limits_{i=1}^{N} (x_i - \mu)^2}{N}}$$

(4-4)

Here the sum of the squares of the individual deviations from the mean $(x_i - \mu)$ is divided by the total number of measurements in the set, N. Extraction of the square root of this quotient gives σ.

Another precision term widely employed by statisticians is the *variance* which is equal to σ^2. Most experimental scientists prefer to employ σ rather than σ^2 because the units of the standard deviation are the same as those of the quantity measured.

APPLICATIONS OF STATISTICS TO SMALL SETS OF DATA

It has been found that direct application of classical statistics to a small number of replicate measurements (2 to 20 results) often leads to false conclusions regarding the probable magnitude of indeterminate error. Fortunately, modifications of the relationships have been developed so that valid statements can be made about the random error associated with just two or three values.

Equations 4-2 and 4-4 are not directly applicable to a small number of replicate measurements because μ, the mean value of an infinitely large number of measurements (and the true value in the absence of determinate error), is always unknown. In its stead, we are forced to employ $\bar{x}$, the mean of a small number of measurements. More often than not, $\bar{x}$ will differ somewhat from μ. This difference is, of course, the result of the indeterminate error whose probable magnitude we are trying to assess. It is important to note that *any error in $\bar{x}$ causes a corresponding error in σ* (Equation 4-4). Thus, with a small number of data, not only is the mean $\bar{x}$ likely to differ from μ, but *equally important, the estimate of the standard deviation also may be misleading*. That is, we have *two* uncertainties to cope with, the one residing in the mean and the other in the standard deviation.

Uncertainties in the Estimation of σ. A decrease in the number of replicates in a set of data has two effects on the standard deviation. First, the number of very high and very low values for σ increases; that is, the reproducibility of σ becomes poorer. Second, the standard deviation as a measure of precision develops a negative *bias*. This bias is manifested by a greater frequency of low than high values for σ and a decrease in the average value of σ as the number of replicates becomes smaller.

The negative bias in σ for small sets of data is attributable to the fact that both a mean and a standard deviation must be extracted from the same small set. It can be shown that this bias can be largely eliminated by substituting the *number of degrees of freedom* $(N - 1)$ for N in Equation 4-4. That is, we define the standard deviation for a small number of measurements as

$$s = \sqrt{\frac{\sum_{i=1}^{N} (x_i - \bar{x})^2}{N - 1}} \qquad (4\text{-}5)$$

Note that Equation 4-5 differs from Equation 4-4 in two regards. First, the denominator is now $(N - 1)$. Second, $\bar{x}$, the measured mean for the small set, replaces the true but unknown mean μ. To emphasize that the resulting standard deviation is but an approximation of the true value, it is common practice to symbolize it with the letter s rather than σ.

Example.   Calculate the standard deviation s for a subset consisting of the first five values in Table 4-3.

x_i	$\lvert x_i - \bar{x} \rvert$	$(x_i - \bar{x})^2$
9.990	7.6×10^{-3}	57.8×10^{-6}
9.986	3.6×10^{-3}	13.0×10^{-6}
9.973	9.4×10^{-3}	88.4×10^{-6}
9.983	0.6×10^{-3}	0.4×10^{-6}
9.980	2.4×10^{-3}	5.8×10^{-6}

$5\lfloor 49.912$
$9.9824 = \bar{x}$

$\sum_{i=1}^{5} (x_i - \bar{x})^2 = 165.4 \times 10^{-6}$

From Equation 4-4

$$s = \sqrt{\frac{165.4 \times 10^{-6}}{5 - 1}} = 6.4 \times 10^{-3} = \pm 0.006$$

Note that the data should not be rounded until the end.

The rationale for the use of $(N - 1)$ in Equation 4-5 is as follows. When μ is unknown, we employ the set of replicate data to obtain two quantities, namely, $\bar{x}$ and s. The need to establish the mean $\bar{x}$ from the data removes one degree of freedom. That is, if their signs are retained, the individual deviations from $\bar{x}$ must total zero; once $(N - 1)$ deviations have been established, the final one is necessarily known as well. Thus, only $(N - 1)$ deviations provide independent measures of the precision for the set.

Estimation of s from w. For a small number of replicate results (up to 15), it is also possible to estimate s from the spread w of the data by means of the relationship

$$s = \frac{w}{d} \qquad (4\text{-}6)$$

where d is a statistical factor, which is dependent upon the number of measurements (see Table 4-5). Equation 4-6 is simpler to use but gives a somewhat less reliable estimate of s than does Equation 4-5.

It is interesting to note that the values of d in Table 4-5 are roughly equal

TABLE 4-5 **Factors for Calculating Standard Deviation *s* from Spread *w* Employing *s* = *w*/*d***

Number of Samples, *N*	*d*	Number of Samples, *N*	*d*	Number of Samples, *N*	*d*
2	1.128	7	2.704	12	3.258
3	1.693	8	2.847	13	3.336
4	2.059	9	2.970	14	3.407
5	2.326	10	3.078	15	3.472
6	2.534	11	3.173		

to the square root of *N*. Thus, an equally rough approximation of *s* is given by

$$s \cong \frac{w}{\sqrt{N}}$$

(4-7)

where *N* is the number of measurements.

THE USES OF STATISTICS

Experimentalists employ statistical calculations to sharpen their judgments concerning the effects of indeterminate error. Some of these include:

1. The interval around the mean of a set within which the true mean can be expected to fall with a certain probability.
2. The number of times a measurement should be replicated in order that the experimental mean will fall, with a certain probability, within a predetermined interval around the true mean.
3. Whether or not an outlying value in a set of replicate results should be retained or rejected in calculating a mean for the set.
4. The probability that two samples analyzed by the same method are significantly different in composition; that is, whether or not a difference in experimental results is likely to be the consequence of indeterminate error or of a real composition difference.
5. The probability that a difference in precision exists between two methods of analysis.

In the section that follows, we shall examine the first four of these applications.

CONFIDENCE INTERVALS

The true mean value (μ) of a measurement is a constant that must always remain unknown. With the aid of statistical theory, however, limits may be set about the experimentally determined mean ($\bar{x}$) within which we might expect to find the true mean with a given degree of probability; the limits obtained in this manner are called *confidence limits*. The interval defined by these limits is known as the *confidence interval*.

Some of the properties of the confidence interval are worthy of mention. For a given set of data, the size of the interval depends in part upon the odds for correctness desired. Clearly, for a prediction to be absolutely correct, we would have to choose an interval about the mean large enough to include all conceivable values that x_i might take. Such an interval, of course, has no predictive value. On the other hand, the interval does not need to be this large if we are willing to accept the probability of being correct 99 times in 100; it can be even smaller if 95% correctness is acceptable. In short, as the probability for making a correct prediction becomes less favorable, the interval included by the confidence limits becomes smaller.

The confidence interval, which is derived from the standard deviation s for the method of measurement, depends in magnitude upon the certainty with which s is known. Often the chemist will have reason to believe that his experimental value for s is an excellent approximation of σ. In other situations, however, a considerable uncertainty may exist in s. Under these circumstances, the confidence interval will necessarily be larger.

Methods for Obtaining a Good Approximation of σ. Fluctuations in the calculated value for s decrease as the number of measurements N in Equation 4-5 increases; in fact, it is proper to assume that s and σ are, for all practical purposes, identical when N is greater than about 20. This behavior makes it feasible for the chemist to obtain a good approximation of s when the method of measurement is not too time-consuming and when an adequate amount of sample is available. For example, if in the course of an investigation the pH of numerous solutions was to be measured, it might prove worthwhile to evaluate s in a series of preliminary experiments. Here the measurement is simple, requiring only that a pair of rinsed and dried electrodes be immersed in the test solution; the potential across the electrodes serves as a measure of the pH. To determine s, 20 to 30 portions of a solution of fixed pH could be measured, following exactly all steps of the procedure. Normally, it would be safe to assume that the indeterminate error in this test would be the same as those in subsequent measurements and that the value of s calculated by means of Equation 4-5 would be a valid and accurate measure of the theoretical σ.

For analyses that are time-consuming, the foregoing procedure is not ordinarily practical. Here, however, precision data from a series of samples can often be pooled to provide a value of s which is superior to values for the individual subsets. Again, one must assume the same sources of indeterminate error among the samples. This assumption is usually valid provided the samples are similar in composition and each has been analyzed identically. To obtain a pooled estimate of s, deviations from the mean for each subset are squared; the squares for all of the subsets are then summed and divided by an appropriate number of degrees of freedom. The pooled s is obtained by extracting the square root of the quotient. One degree of freedom is lost for each subset. Thus, the number of degrees of freedom for the pooled s is equal to the total number of measurements minus the number of subsets. An example of this calculation follows.

Example. The mercury in samples of seven fish taken from Lake Erie was determined by a method based upon absorption of radiation by elemental mercury. The results are given below. Calculate a standard deviation for the method, based upon the pooled precision data.

Sample Number	Number of Replications	Results, Hg Content, ppm	Mean, ppm Hg	Sum of Squares of Deviations from Mean
1	3	1.80, 1.58, 1.64	1.673	0.0259
2	4	0.96, 0.98, 1.02, 1.10	1.015	0.0115
3	2	3.13, 3.35	3.240	0.0242
4	6	2.06, 1.93, 2.12, 2.16, 1.89, 1.95	2.018	0.0611
5	4	0.57, 0.58, 0.64, 0.49	0.570	0.0114
6	5	2.35, 2.44, 2.70, 2.48, 2.44	2.482	0.0685
7	4	1.11, 1.15, 1.22, 1.04	1.130	0.0170
Number of measurements	28		Sum of squares =	0.2196

The value in column 5 for sample 1 is calculated as follows:

(x_i)	$|(x_i - \bar{x}_1)|$	$(x_i - \bar{x}_1)^2$
1.80	0.127	0.0161
1.58	0.093	0.0087
1.64	0.033	0.0011
$3\overline{)5.02}$		Sum of squares = 0.0259
$\bar{x}_1 = 1.673$		

The other data in column 5 were obtained similarly. Then

$$s = \sqrt{\frac{0.0259 + 0.0115 + 0.0242 + 0.0611 + 0.0114 + 0.0685 + 0.0170}{28 - 7}}$$

$$= 0.10 \text{ ppm Hg}$$

If the number of degrees of freedom is greater than 20, the estimate of s can be considered to be a good approximation of σ.

Confidence Interval When s Is a Good Approximation of σ. As indicated earlier (p. 57), the breadth of the normal error curve is determined by σ. It is also related to z in Equations 4-2 and 4-3. The area under the normal error curve *relative* to the total area can be calculated for any desired value of z by means of Equation 4-2. This ratio (usually expressed as a percent) is called the *confidence level* and measures the probability for the absolute deviation $(x - \mu)$ being less than $z\sigma$. Thus, the area under the curve encompassed by $z = \pm 1.96\sigma$ corresponds to 95% of the total area. Here the confidence level is 95%, and we may

TABLE 4-6   Confidence Levels for Various Values of z

Confidence Level, %	z
50	±0.67
68	±1.00
80	±1.29
90	±1.64
95	±1.96
96	±2.00
99	±2.58
99.7	±3.00
99.9	±3.29

state that for a large number of measurements, the calculated $(x - \mu)$ will be equal to or less than $\pm 1.96\sigma$ in 95 cases out of 100. Table 4-6 lists confidence intervals for various values of z.

The confidence limit for a single measurement can be obtained by rearranging Equation 4-3 and remembering that z can be either plus or minus in value. Thus,

$$\text{confidence limit for } \mu = x \pm z\sigma \qquad (4\text{-}8)$$

The following example shows how Equation 4-8 can be employed.

Example.   Calculate the 50% and the 95% confidence limits for the first entry (1.80 ppm Hg) in the example on page 62.

Here we calculated that $s = 0.10$ ppm Hg and had sufficient data to assume $s \cong \sigma$. From Table 4-6 we see that $z = \pm 0.67$ and ± 1.96 for the two confidence levels. Thus, from Equation 4-8,

$$50\% \text{ confidence limit for } \mu = 1.80 \pm 0.67 \times 0.10$$
$$= 1.80 \pm 0.07$$

$$95\% \text{ confidence limit for } \mu = 1.80 \pm 1.96 \times 0.10$$
$$= 1.80 \pm 0.20$$

The chances are 50 in 100 that μ, the true mean (and in the absence of determinate error the true value), will be in the interval between 1.73 and 1.87 ppm Hg; there is a 95% chance that it will be in the interval between 1.60 and 2.00 ppm Hg.

Equation 4-8 applies to the result of a single measurement. It can be shown that the confidence interval is decreased by $\sqrt{N}$ for the average of N replicate measurements. Thus, a more general form of Equation 4-8 is

$$\text{confidence limit for } \mu = \bar{x} \pm \frac{z\sigma}{\sqrt{N}} \qquad (4\text{-}9)$$

Example.   Calculate the 50% and the 95% confidence limits for the mean value (1.67 ppm Hg) for sample 1 in the example on page 62. Again, $s \cong \sigma = 0.10$.

For the three measurements

$$50\% \text{ confidence limit} = 1.67 \pm \frac{0.67 \times 0.10}{\sqrt{3}} = 1.67 \pm 0.04$$

$$95\% \text{ confidence limit} = 1.67 \pm \frac{1.96 \times 0.10}{\sqrt{3}} = 1.67 \pm 0.11$$

Thus, the chances are 50 in 100 that the true mean will lie in the interval of 1.63 to 1.71 ppm Hg and 95 in 100 that it will be between 1.56 and 1.78 ppm.

Example.   Calculate the number of replicate measurements needed to decrease the 95% confidence interval for the calibration of a 10-ml pipet to 0.005 ml, assuming that a procedure similar to the one for obtaining the data in Table 4-3 has been followed.

The standard deviation for the measurement is 0.0065 ml. Since s is based on 24 values, we may assume $s \cong \sigma = 0.0065$.

The confidence interval is given by

$$\text{confidence interval} = \pm \frac{z\sigma}{\sqrt{N}}$$

$$0.005 \text{ ml} = \pm \frac{1.96 \times 0.0065}{\sqrt{N}}$$

$$N = 6.5$$

Thus, by employing the mean of seven measurements we would have a somewhat better than 95% chance of knowing the true mean volume delivered by the pipet to ± 0.005 ml.

A consideration of Equation 4-9 indicates that the confidence interval for an analysis can be halved by employing the mean of four measurements. Sixteen measurements would be required to narrow the limit by another factor of 2. It is apparent that a point of diminishing return is rapidly reached in acquiring additional data. Thus, the chemist ordinarily takes advantage of the relatively large gain afforded by averaging two to four measurements. On the other hand, he can seldom afford the time required for further increases in confidence.

Confidence Limits When σ Is Unknown. The chemist frequently must make use of analytical methods with which he has no previous experience. Furthermore, limitations in time or amount of available sample preclude an accurate estimation of σ. Here a single set of replicate measurements must provide not only a mean value but also a precision estimate. As we have indicated earlier, s, calculated from a limited set of data, may be subject to considerable uncertainty; thus, the confidence limits will be broader under these circumstances.

To account for the potential variability of s, use is made of the parameter t, which is defined as

$$t = \frac{x - \mu}{s} \tag{4-10}$$

TABLE 4-7   Values of t for Various Levels of Probability

Degrees of Freedom	Factor for Confidence Interval, %				
	80	90	95	99	99.9
1	3.08	6.31	12.7	63.7	637
2	1.89	2.92	4.30	9.92	31.6
3	1.64	2.35	3.18	5.84	12.9
4	1.53	2.13	2.78	4.60	8.60
5	1.48	2.02	2.57	4.03	6.86
6	1.44	1.94	2.45	3.71	5.96
7	1.42	1.90	2.36	3.50	5.40
8	1.40	1.86	2.31	3.36	5.04
9	1.38	1.83	2.26	3.25	4.78
10	1.37	1.81	2.23	3.17	4.59
11	1.36	1.80	2.20	3.11	4.44
12	1.36	1.78	2.18	3.06	4.32
13	1.35	1.77	2.16	3.01	4.22
14	1.34	1.76	2.14	2.98	4.14
∞	1.29	1.64	1.96	2.58	3.29

In contrast to z in Equation 4-3, t is dependent not only on the desired confidence level but also upon the number of degrees of freedom available in the calculation of s. Table 4-7 provides values for t under various circumstances. Note that as the number of degrees of freedom becomes infinite, the values of t become equal to those for z shown in Table 4-6.

The confidence limit can be derived from t by an equation analogous to Equation 4-9. That is,

$$\text{confidence limit for } \mu = \bar{x} \pm \frac{ts}{\sqrt{N}} \qquad (4\text{-}11)$$

Example.   A chemist obtained the following data for the alcohol content in a sample of blood: percent ethanol $= 0.084$, 0.089, and 0.079. Calculate the 95% confidence limit for the mean assuming (1) no additional knowledge about the precision of the method and (2) that on the basis of previous experiences, $s \cong \sigma = 0.005\%$ ethanol.

(1) $\bar{x} = \dfrac{(0.084 + 0.089 + 0.079)}{3} = 0.0840$

$s = \sqrt{\dfrac{(0.00)^2 + (0.0050)^2 + (0.0050)^2}{3 - 1}} = 0.0050$

Table 4-7 indicates that for 2 degrees of freedom and 95% confidence, $t = \pm 4.30$. Thus,

$$95\% \text{ confidence limit} = 0.084 \pm \frac{4.3 \times 0.0050}{\sqrt{3}}$$

$$= 0.084 \pm 0.012$$

(2) Since a good value of σ is available,

$$95\% \text{ confidence limit} = 0.084 \pm \frac{z\sigma}{\sqrt{N}}$$

$$= 0.084 \pm \frac{1.96 \times 0.0050}{\sqrt{3}}$$

$$= 0.084 \pm 0.006$$

Note that sure knowledge of σ reduced the confidence interval by half.

STATISTICAL AIDS TO HYPOTHESIS TESTING

Much of scientific and engineering endeavor is based upon hypothesis testing. Thus, in order to explain an observation, a hypothetical model is advanced which then serves as a basis for experimental testing to determine its validity. If the results from these experiments do not support it, the model is rejected and a new one is sought. If, on the other hand, there is agreement between the experimental results and what would be expected from the properties of the hypothetical model, then the hypothesis can serve as the basis for further experiments. When the hypothesis is supported by sufficient experimental data, it becomes recognized as a useful theory until such time as data are developed which refute it.

Seldom will experimental results agree exactly with those predicted by a theoretical model. As a consequence, the scientist and engineer must often exercise a judgment as to whether a numerical difference is a real one that requires a rejection of the hypothesis or the result of the inevitable indeterminate error associated with the measurements. Certain statistical tests are useful in sharpening these judgments.

In approaching a test of this kind, a *null hypothesis* is employed, which assumes that the numerical quantities being compared are, in fact, the same. The probability of the observed differences appearing as a result of indeterminate error is then computed from statistical theory. Usually, if the observed difference is as large or larger than the difference that would occur 5 times in 100 (the 5% probability level), the null hypothesis is considered questionable and the difference is judged to be significant. Other probability levels such as 1 in 100 or 10 in 100 may also be adopted, depending upon the certainty desired in making the judgment.

The kinds of testing that chemists use most often include the comparison of means, $\bar{x}_1$ and $\bar{x}_2$, from two sets of analyses, the mean from an analysis $\bar{x}_1$ and what is believed to be the true value μ, the standard deviations, s_1 and s_2 or σ_1 and σ_2 from two sets of measurements, and the standard deviation s of a small set of data with the standard deviation σ of a larger set of measurements. The sections that follow consider some of the methods for dealing with these comparisons.

Comparison of an Experimental Mean with a True Value. A common way of testing for determinate errors is to employ the method for the analysis of a sample whose composition is accurately known (see p. 49). In all probability the experimental mean $\bar{x}$ will differ from the true value μ; the judgment must then be made whether this difference is the consequence of the indeterminate error of the measurement or of the presence of a determinate error in the method.

The statistical treatment for this type of problem involves comparing the difference $(\bar{x} - \mu)$ with the difference that would normally be expected as the result of indeterminate error. If the observed difference is less than that which is computed for a chosen probability level, the null hypothesis that $\bar{x}$ and μ are not different is confirmed; the conclusion can then be drawn that the experiment revealed no significant determinate error. On the other hand, if $(\bar{x} - \mu)$ is significantly larger than the expected or critical value, it may be assumed that the difference is real and that a determinate error exists.

The critical value for the rejection of the null hypothesis can be obtained by rewriting Equation 4-11 in the form

$$\bar{x} - \mu = \pm \frac{ts}{\sqrt{N}} \tag{4-12}$$

where N is the number of replicate measurements employed in the test. If a good estimation of σ is available, Equation 4-12 can be modified by replacing t and s with z and σ, respectively.

Example. A new procedure for the rapid analysis of sulfur in kerosenes was tested by the analysis of a sample which was known from its method of preparation to contain 0.123% S. The results obtained were: % S = 0.112, 0.118, 0.115, and 0.119. Do the data indicate the presence of a negative determinate error in the new method?

$$\bar{x} = \frac{0.112 + 0.118 + 0.115 + 0.119}{4} = 0.1160$$

$$\bar{x} - \mu = 0.116 - 0.123 = -0.007$$

$$s = \sqrt{\frac{(0.0040)^2 + (0.0020)^2 + (0.0010)^2 + (0.0030)^2}{4 - 1}} = 0.0032$$

From Table 4-7, we find that at the 95% confidence level that t has a value of 3.18 for 3 degrees of freedom. Thus,

$$\frac{ts}{\sqrt{N}} = \frac{3.18 \times 0.0032}{\sqrt{3}} = \pm 0.0059$$

But

$$\bar{x} - \mu = -0.007$$

Five times out of 100, an experimental mean can be expected to deviate by ± 0.0059 or more. Thus, if we conclude that -0.007 is a significant difference, we will, on the average, be right 95 times and wrong 5 times out of 100 judgments.

Comparison of Two Experimental Means. A chemist will frequently employ analytical data in an effort to establish whether two materials are different or identical. Here, the judgment must be made as to whether a difference in analytical results is the consequence of indeterminate errors in the two measurements or if it represents a real difference. To illustrate, let us assume that the N_1 replicate analyses were made on material 1 and N_2 analyses on material 2. Applying Equation 4-11, we may write

$$\mu_1 = \bar{x}_1 \pm \frac{ts}{\sqrt{N_1}}$$

and

$$\mu_2 = \bar{x}_2 \pm \frac{ts}{\sqrt{N_2}}$$

where $\bar{x}_1$ and $\bar{x}_2$ are the two experimental means. In order to establish the existence or absence of a real difference between $\bar{x}_1$ and $\bar{x}_2$, we make the null hypothesis that μ_1 and μ_2 are identical. Then, equating the two expressions gives

$$\bar{x}_1 \pm \frac{ts}{\sqrt{N_1}} = \bar{x}_2 \pm \frac{ts}{\sqrt{N_2}}$$

This equation is readily rearranged to give

$$\bar{x}_1 - \bar{x}_2 = \pm ts \sqrt{\frac{N_1 + N_2}{N_1 N_2}} \tag{4-13}$$

The numerical value for the term on the right is computed employing t for the particular confidence level desired. (The number of degrees of freedom for finding t will be $N_1 + N_2 - 2$.) If the experimental difference, $\bar{x}_1 - \bar{x}_2$, is smaller than the computed value, the null hypothesis is confirmed, and no significant difference between the means can be assumed. On the other hand, an experimental difference greater than the value computed from t indicates the existence of a significant difference.

If a good estimate of σ is available, Equation 4-13 can be modified by insertion of z and σ for t and s.

Example. The composition of a flake of paint found on the clothes of a hit-and-run victim was compared with that of paint from the car suspected of causing the accident. Do the following data for the spectroscopic analysis for Ti in the paints suggest a difference in composition between the two materials? From previous experience the standard deviation for the analysis is known to be 0.35% Ti; that is, $s \rightarrow \sigma = 0.35\%$ Ti.

Paint from clothes   % Ti = 4.0, 4.6
Paint from car   % Ti = 4.5, 5.3, 5.5, 5.0, 4.9

$$\bar{x}_1 = \frac{4.6 + 4.0}{2} = 4.30$$

$$\bar{x}_2 = \frac{4.5 + 5.3 + 5.5 + 5.0 + 4.9}{5} = 5.04$$

$$\bar{x}_1 - \bar{x}_2 = 4.30 - 5.04 = -0.74\% \text{ Ti}$$

Modifying Equation 4-13 to take into account our knowledge that $s \rightarrow \sigma$, we calculate for the 95 and 99% confidence levels

$$\pm z\sigma \sqrt{\frac{N_1 + N_2}{N_1 N_2}} = \pm 1.96 \times 0.35 \sqrt{\frac{2 + 5}{2 \times 5}} = \pm 0.57$$

$$= \pm 2.58 \times 0.35 \sqrt{\frac{2 + 5}{2 \times 5}} = \pm 0.76$$

We see that only 5 out of 100 data should differ by 0.57% Ti or greater and only 1 out of 100 should differ by as much as 0.76% Ti. Thus, it seems reasonably probable that the observed difference of -0.74% does not arise from indeterminate error but in fact is caused by a real difference between the two paint samples. Hence, we would conclude the suspected vehicle was not involved in the accident.

Example.   Two barrels of wine were analyzed for their alcohol content in order to determine whether they were from different sources. The average content of the first barrel was established, on the basis of 6 analyses, to be 12.61% ethanol. Four analyses of the second barrel gave a mean of 12.53% alcohol. The 10 analyses yielded a pooled value of $s = 0.070\%$. Is a difference between the wines indicated by the data?

Here we employ Equation 4-13, using t for 8 degrees of freedom $(10 - 2)$. At the 95% confidence level

$$\pm ts \sqrt{\frac{N_1 + N_2}{N_1 N_2}} = 2.31 \times 0.070 \sqrt{\frac{6 + 4}{6 \times 4}} = 0.10\%$$

The observed difference is

$$\bar{x}_1 - \bar{x}_2 = 12.61 - 12.53 = 0.08\%$$

As often as 5 times in 100, indeterminate error will be responsible for a difference as great as 0.10%. At this confidence level, then, no difference has been established.

In the last example, it was found that no significant difference existed at the 95% probability level. It should be noted that this statement is not equivalent to saying that $\bar{x}_1$ is equal to $\bar{x}_2$; nor do the tests prove that the wines come from the same source. Indeed, it is conceivable that one could have been a red and the other a white. To establish with a reasonable probability that the two wines were derived from the same source would require extensive testing of other characteristics such as taste, odor, refractive index, acetic acid concentration, sugar content, and trace element content. If, for all of these tests and others, no significant differences were revealed, then it might be possible to judge the two as having a common genesis. In contrast, the finding of one significant difference among these would clearly show that the two were different. Thus, the establishment of a significant difference by a single test is much more revealing than the establishment of an absence of difference.

REJECTION OF DATA

When a set of data contains an outlying result that appears to differ excessively from the average (or the median), the decision must be made to retain or to reject it. The choice of criterion for the rejection of a suspected result has its

TABLE 4-8[a]   Critical Values for Rejection Quotient Q

Number of Observations	Q_{crit} (90% confidence) Reject if $Q_{exp} >$
2	—
3	0.94
4	0.76
5	0.64
6	0.56
7	0.51
8	0.47
9	0.44
10	0.41

[a] Reproduced from R. B. Dean and W. J. Dixon, *Anal. Chem.*, **23**, 636 (1951). By permission of the American Chemical Society.

perils. If we set a stringent criterion that makes difficult the rejection of a questionable measurement, we run the risk of retaining results that are spurious and have an inordinate effect on the average of the data. On the other hand, if we set lenient limits on precision, and make easy the rejection of a result, we are likely to discard measurements that rightfully belong in the set; we thus introduce a bias to the data. It is an unfortunate fact that no universal rule can be invoked to settle the question of retention or rejection.

Of the numerous statistical criteria suggested to aid in deciding whether to retain or reject a measurement, the Q test[3] is to be preferred. Here the difference between the questionable result and its nearest neighbor is divided by the spread of the entire set. The resulting ratio, Q, is then compared with rejection values that are critical for a particular degree of confidence. Table 4-8 provides critical values of Q at the 90% confidence level.

Example. The analysis of a calcite sample yielded CaO percentages of 55.95, 56.00, 56.04, 56.08, and 56.23, respectively. The last value appears anomalous; should it be retained or rejected?

The difference between 56.23 and 56.08 is 0.15%. The spread (56.23 − 55.95) is 0.28%. Thus,

$$Q_{exp} = \frac{0.15}{0.28} = 0.54$$

For five measurements, Q_{crit} is 0.64. Since 0.54 < 0.64, retention is indicated.

Notwithstanding its superiority over other criteria, the Q test must be used with good judgment as well. For example, there will be situations in which the dispersion associated with the bulk of a set will be fortuitously small, and the indiscriminate application of the Q test will result in rejection of a value that actually should be retained; indeed, in a three-number set containing a pair of identical values, the experimental value for Q becomes indeterminately large.

[3] R. B. Dean and W. J. Dixon, *Anal. Chem.*, **23**, 636 (1951).

On the other hand, it has been pointed out[4] that the magnitudes of rejection quotients for small sets are likely to cause the retention of erroneous data.

The blind application of statistical tests to the decision for retention or rejection of a suspect measurement in a small set of data is not likely to be much more fruitful than an arbitrary decision; indeed, the application of good judgment based upon an estimate of the precision to be expected may be a more sound approach, particularly if this estimate is based upon wide experience with the analytical method being employed. In the end, however, the only entirely valid reason for rejecting an experimental result from a small set is the sure knowledge that a mistake has been made in its acquisition. Lacking this knowledge, *a cautious approach to the rejection of data is desirable*.

In summary, a number of recommendations suggest themselves for the treatment of a small set of results that contains a suspect value.

1. Reexamine carefully all data relating to the suspected result to see if a gross error has affected its value. A properly kept laboratory notebook containing careful notations of all observations is essential if this recommendation is to be helpful.
2. If possible, estimate the precision that can be reasonably expected from the procedure to be sure that the outlying result actually is questionable.
3. Repeat the analysis if sufficient sample and time are available. Agreement of the newly acquired data with those that appear to be valid will lend weight to the notion that the outlying result should be rejected. Furthermore, the questionable result will have a smaller effect on the mean of the larger set of data if its retention is still indicated.
4. If more data cannot be secured, apply the Q test to the existing set to see if the doubtful result should be retained or be rejected on these grounds.
5. If the Q test indicates retention, give consideration to reporting the median of the set rather than the mean. The median has the great virtue of allowing inclusion of all data in a set without undue influence from an outlying value. Moreover, it has been demonstrated that the median of a normally distributed set containing three measurements is more likely to provide a reliable estimate of the correct value than will the mean of the set after the outlying value has been arbitrarily discarded.[5]

Propagation of Errors in Computation

The scientist must frequently estimate the error in a result that has been computed from two or more data, each of which has an error associated with it. The way in which the individual errors accumulate depends upon the arithmetic relationship between the term containing the error and the quantity being computed. In addition, the effect of determinate errors on a computed result differs from the effect of indeterminate errors.

ACCUMULATION OF DETERMINATE ERRORS

The way in which determinate errors accumulate in a sum or difference varies from that for a product or quotient.

[4] R. B. Dean and W. J. Dixon, *Anal. Chem.*, **23**, 636 (1951).
[5] National Bureau of Standards, *Technical News Bulletin* (July 1949); *J. Chem. Educ.*, **26**, 673 (1949).

Errors in a Sum or Difference. Let us consider the relationship

$$y = a + b - c$$

where a, b, and c are the values for three measurable quantities. If Δa, Δb, and Δc are the absolute determinate errors associated with the measurement of these quantities, the actual measurements are $(a + \Delta a)$, $(b + \Delta b)$, and $(c + \Delta c)$. The resulting error in y, then, is Δy and

$$y + \Delta y = (a + \Delta a) + (b + \Delta b) - (c + \Delta c)$$

The error in the computed result can be obtained by subtracting the first equation from the second. That is,

$$\Delta y = \Delta a + \Delta b - \Delta c \tag{4-14}$$

It is seen that *for addition or subtraction, the absolute error for the sum or difference is determined by the absolute error of the numbers forming the sum or difference.*

Example. Calculate the error in the result of the following calculation:

$$
\begin{array}{r}
+0.50 \ (+0.02) \\
+4.10 \ (-0.03) \\
-1.97 \ (-0.05) \\
\hline
2.63
\end{array}
$$

where the numbers in parentheses are the absolute determinate errors. The absolute error of the summation is

$$\Delta y = 0.02 + (-0.03) - (-0.05) = +0.04$$

Errors in a Product or Quotient. Let us first consider the product

$$y = a \times b$$

We will again assume determinate errors of Δa and Δb which result in the error Δy. Thus,

$$y + \Delta y = (a + \Delta a)(b + \Delta b)$$
$$= ab + a\,\Delta b + b\,\Delta a + \Delta a\,\Delta b$$

Subtraction of the first equation from the last gives

$$\Delta y = b\,\Delta a + a\,\Delta b + \Delta a\,\Delta b$$

We now divide this equation by the first, which yields

$$\frac{\Delta y}{y} = \frac{\Delta a}{a} + \frac{\Delta b}{b} + \frac{\Delta a\,\Delta b}{ab}$$

The third term on the right-hand side of the equation will generally be much smaller than the other two inasmuch as the numerator is the product of two small terms and the denominator the product of two much larger ones. Therefore, when $\Delta a\,\Delta b/ab \ll (\Delta a/a + \Delta b/b)$,

$$\frac{\Delta y}{y} = \frac{\Delta a}{a} + \frac{\Delta b}{b}$$

Note that the three terms correspond to *relative* determinate errors rather than absolute as was the case for a sum or a difference.

An analogous relationship can be derived for the error in a quotient. Thus,

$$y = \frac{a}{b}$$

Then

$$y + \Delta y = \frac{(a + \Delta a)}{(b + \Delta b)}$$

It is convenient to rewrite these two equations as

$$yb = a$$

$$yb + b\,\Delta y + y\,\Delta b + \Delta y\,\Delta b = a + \Delta a$$

The two equations can be combined to give

$$b\,\Delta y + y\,\Delta b + \Delta y\,\Delta b = \Delta a$$

Dividing by $yb = a$ and rearranging yields

$$\frac{\Delta y}{y} = \frac{\Delta a}{a} - \frac{\Delta b}{b} - \frac{\Delta y\,\Delta b}{yb}$$

Here again we may ordinarily assume that $\Delta y\,\Delta b/yb \ll (\Delta a/a - \Delta b/b)$ and

$$\frac{\Delta y}{y} = \frac{\Delta a}{a} - \frac{\Delta b}{b}$$

For the more general case

$$y = \frac{ab}{c}$$

it can be shown by the same type of argument that

$$\frac{\Delta y}{y} = \frac{\Delta a}{a} + \frac{\Delta b}{b} - \frac{\Delta c}{c} \tag{4-15}$$

Thus, for multiplication or division, the relative error of the product or quotient is determined by the relative errors of the numbers forming the computed result.

Example.   Compute the error in the result of the following calculation, where the numbers in parentheses are absolute determinate errors:

$$y = \frac{4.10(-0.02) \times 0.0050(+0.0001)}{1.97(-0.04)} = 0.010406$$

Here we must base the calculation on *relative* errors. Thus,

$$\frac{\Delta y}{y} = \frac{-0.02}{4.10} + \frac{0.0001}{0.0050} - \frac{-0.04}{1.97}$$

$$= -0.0049 + 0.020 + 0.020 = 0.035$$

To obtain the absolute error Δy in y we write

$$\Delta y = 0.035 \times y = 0.035 \times 0.010406 = 0.0004$$

and

$$y = 0.0104(+0.0004)$$

ACCUMULATION OF INDETERMINATE ERRORS

As we have noted earlier, the absolute or the relative standard deviation serves as the most convenient measure of indeterminate errors of an experimental result. In contrast to a determinate error, however, no sign can be attached to a standard deviation, there being an equal probability of its being positive and negative. This fact leads to a range of possible standard deviations for a computed result. For example, consider the summation

$$
\begin{array}{r}
+0.50 \ (\pm 0.02) \\
+4.10 \ (\pm 0.03) \\
-1.97 \ (\pm 0.05) \\
\hline
2.63
\end{array}
$$

Here, the numbers in parentheses are standard deviations. Note that if the first two uncertainties happened to be positive and the third negative, the standard deviation of the result would be

$$s_y = +0.02 + 0.03 - (-0.05) = 0.10$$

On the other hand, under fortuitous circumstances, the accumulated uncertainty could be zero. Thus, if all three uncertainties were positive

$$s_y = +0.02 + 0.03 - (+0.05) = 0.00$$

Neither circumstance is as probable as a combination leading to an uncertainty between these two extremes.

Statistical theory demonstrates that the best or most probable value for the *absolute* standard deviation s_y of a sum or difference is given by

$$s_y = \sqrt{s_a{}^2 + s_b{}^2 + s_c{}^2 \ldots} \tag{4-16}$$

where $s_a, s_b, s_c, \ldots$ are the *absolute* standard deviations of the number making up the sum or difference. Note that the absolute *variance* of the result is the sum of the individual absolute variances. Thus, Equation 4-16 can also be written as

$$s_y{}^2 = s_a{}^2 + s_b{}^2 + s_c{}^2 \ldots$$

For multiplication and division, the *relative* standard deviations are combined in a similar way. Thus, to obtain the standard deviation for y in the relationship

$$y = \frac{ab}{c}$$

we can write

$$\frac{s_y}{y} = \sqrt{\left(\frac{s_a}{a}\right)^2 + \left(\frac{s_b}{b}\right)^2 + \left(\frac{s_c}{c}\right)^2} \tag{4-17}$$

Example. Calculate the standard deviation of the result of the summation shown in the example considered earlier in this section. Since the computation involves a sum, absolute standard deviations are combined, and we write

$$s_y = \sqrt{(\pm 0.02)^2 + (\pm 0.03)^2 + (\pm 0.05)^2}$$
$$= \pm 0.06$$

Therefore,

$$y = 2.63(\pm 0.06)$$

Here the probable uncertainty is significantly smaller than the maximum $(0.02 + 0.03 + 0.05 = 0.10)$ but larger than the minimum $(0.02 + 0.03 - 0.05 = 0.00)$.

Example.   Compute the probable standard deviation for the result of the computation

$$y = \frac{(4.10 \pm 0.02)(0.0050 \pm 0.0001)}{1.97 \pm 0.04} = 0.01041 \pm \,?$$

Here we must deal in relative standard deviations. Thus,

$$(s_y)_r = \sqrt{\left(\frac{\pm 0.02}{4.10}\right)^2 + \left(\frac{\pm 0.0001}{0.0050}\right)^2 + \left(\frac{\pm 0.04}{1.97}\right)^2}$$

$$= \sqrt{(\pm 0.0049)^2 + (\pm 0.020)^2 + (\pm 0.020)^2} = 0.029$$

The *absolute* standard deviation of y is given by

$$s_y = y \times (s_y)_r = 0.0104 \times 0.029 = 0.0003$$

Therefore,

$$y = 0.0104(\pm 0.0003)$$

Example.   Compute the standard deviation for the result of the computation

$$y = \frac{[14.3(\pm 0.2) - 11.6(\pm 0.2)] \times 50.0(\pm 0.1)}{42.3(\pm 0.4)} = 3.191 \pm \,?$$

We must first calculate the standard deviation for the difference in the numerator

$$s_1 = \sqrt{(\pm 0.2)^2 + (\pm 0.2)^2} = \pm 0.28$$

The equation is then rewritten as

$$y = \frac{2.7(\pm 0.28) \times 50.0(\pm 0.1)}{42.3(\pm 0.4)} = 3.191$$

The relative deviation of the quotient is then computed

$$(s_y)_r = \sqrt{\left(\frac{\pm 0.28}{2.7}\right)^2 + \left(\frac{\pm 0.1}{50.0}\right)^2 + \left(\frac{\pm 0.4}{42.3}\right)^2}$$

$$= 0.10$$

The absolute standard deviation of the result is

$$s_y = 3.191 \times 0.10 = 0.32$$

and the answer is

$$y = 3.2(\pm 0.3)$$

It is of interest to note the amplification of error that resulted from the subtraction step in the numerator.

PROPAGATION OF ERRORS IN EXPONENTIAL CALCULATIONS

To show how errors are propagated when the power or the root of an experimental result a is to be calculated, we write

$$y = a^x$$

where x is the power or the root and contains no uncertainty. The derivative of this expression is

$$dy = xa^{(x-1)} da$$

Dividing by the original expression gives

$$\frac{dy}{y} = \frac{xa^{(x-1)} da}{a^x}$$

But

$$\frac{a^{(x-1)}}{a^x} = \frac{1}{a}$$

Therefore,

$$\frac{dy}{y} = x \frac{da}{a}$$

or for finite increments

$$\frac{\Delta y}{y} = x \frac{\Delta a}{a} \tag{4-18}$$

Here Δy is the absolute error in y that results from the error Δa in a. Clearly, the *relative error* $\Delta y/y$ of the computed result is simply the *relative error* of the experimental number $\Delta a/a$ multiplied by the exponent x. For example, the relative error in the square of a number is twice that for the number itself, whereas the relative error of the cube root of a number is simply one-third that of the number.

It is also important to note that the propagation of an indeterminate error in raising a number of a power is treated differently from a multiplication because the possibility of errors canceling one another does not exist. Recall that the relative standard deviation in the product $a \times b$ will lie between the sum and the difference of the standard deviations of the two numbers, and a probable value can be calculated by taking the square root of the sum of the squares of the uncertainty. This technique is not applicable to the product of a single measurement $a \times a$; here the signs are necessarily identical since the numbers are identical. Thus, the relative uncertainty in a^2 must be twice that of a. Equation 4-18 also applies to indeterminate errors as well; here Δy and Δa are replaced by s_y and s_a.

Example. The standard deviation in measuring the diameter d of a sphere is ± 0.2 cm. What is the standard deviation in its calculated volume V if $d = 10.0$ cm?

$$V = \frac{4}{3} \pi \left(\frac{10.0}{2}\right)^3 = 523.6 \text{ cm}^3$$

Here we may write

$$(s_y)_r = \frac{s_V}{V} = 3 \frac{s_d}{d}$$

$$= 3 \times \frac{0.2}{10} = 0.06$$

The absolute standard deviation in V is then

$$s_y = 523.6 \times 0.06 = 31$$

Thus,

$$V = 524(\pm31) \text{ cm}^3$$

Example. The solubility product K_{sp} for the silver salt AgX is $4.0(\pm0.4) \times 10^{-8}$. What is the uncertainty associated with the calculated solubility of AgX in water? Here (see p. 28)

$$\text{solubility} = (4.0 \times 10^{-8})^{1/2} = 2.0 \times 10^{-4}$$

$$(s_x)_r = \frac{0.4 \times 10^{-8}}{4.0 \times 10^{-8}}$$

$$(s_y)_r = \frac{1}{2} \times \frac{0.4}{4.0} = 0.05$$

$$s_y = 2.0 \times 10^{-4} \times 0.05 = 0.1 \times 10^{-4}$$

and

$$\text{solubility} = 2.0(\pm0.1) \times 10^{-4}$$

PROPAGATION OF ERROR IN LOGARITHM AND ANTILOGARITHM CALCULATIONS

To show how errors are propagated when logarithms or antilogarithms are computed we take the derivative of the expression

$$y = \log a = 0.434 \ln a$$

where ln symbolizes the natural logarithm. Thus,

$$dy = 0.434 \frac{da}{a}$$

Conversion to finite increments gives

$$\Delta y = 0.434 \frac{\Delta a}{a} \tag{4-19}$$

Note that the *absolute* uncertainty in y is determined by the *relative* uncertainty in a and conversely. In employing Equation 4-19, the relative standard deviation can, as before, be substituted for $\Delta a/a$ and the absolute standard deviation in y for Δy.

Example. Calculate the absolute standard deviations of the results of the following computations. The absolute standard deviation for each quantity is given in parentheses.

(a) $y = \log [2.00(\pm0.02) \times 10^{-3}] = -2.6990 \pm ?$

(b) $a = \text{antilog} [1.200(\pm0.003)] = 15.849 \pm ?$

(c) $a = \text{antilog} [45.4(\pm0.3)] = 2.5119 \times 10^{45} \pm ?$

(a) Referring to Equation 4-19, we see that we must multiply the *relative* standard deviation by 0.434. That is,

$$\Delta y = \pm0.434 \times \frac{0.02 \times 10^{-3}}{2.00 \times 10^{-3}} = \pm0.004 = s_y$$

Thus,

$$\log [2.00(\pm 0.002) \times 10^{-3}] = -2.699 \pm 0.004$$

(b) Rearranging Equation 4-19 and replacing Δa and Δy with the corresponding standard deviations

$$\frac{s_a}{a} = \frac{s_y}{0.434} = \frac{\pm 0.003}{0.434} = \pm 0.0069$$

$$s_a = \pm 0.0069 \times a = \pm 0.0069 \times 15.849 = 0.11$$

Thus,

$$\text{antilog } [1.200(\pm 0.002)] = 15.8 \pm 0.1$$

(c)

$$\frac{s_a}{a} = \frac{\pm 0.3}{0.434} = 0.69$$

$$s_a = 0.69 \times a = 0.69 \times 2.5119 \times 10^{45}$$
$$= 1.7 \times 10^{45}$$

Thus,

$$\text{antilog } [45.4(\pm 0.3)] = 2.5(\pm 1.7) \times 10^{45}$$

Note that a large absolute error is associated with an antilogarithm of a number with few digits beyond the decimal point. The large uncertainty here arises from the fact that the numbers to the left of the decimal (the characteristic) serve only to locate the decimal point. In the last example, the large error in the anti-logarithm results from the relatively large uncertainty in the *mantissa* of the number (that is, 0.4 ± 0.3).

Significant Figure Convention

In reporting a measurement, the experimenter should include not only what he considers to be its best value, be it a mean or a median, but also his estimate of its uncertainty. The latter is preferably reported as the standard deviation of the result; the deviation from the mean, the deviation from the median, or the spread may sometimes be encountered because these precision indicators are easier to calculate. Common practice also dictates that an experimental result should be rounded off so that it contains only the digits known with certainty plus the first uncertain one. This practice is called the *significant figure convention*.

For example, the average of the experimental quantities 61.60, 61.46, 61.55, and 61.61 is 61.555. The standard deviation of the sum is ± 0.069. Clearly, the number in the second decimal place is subject to uncertainty. Such being the case, all numbers in succeeding decimal places are without meaning, and we are forced to round the average value accordingly. The question of taking 61.55 or 61.56 must be considered, 61.555 being equally spaced between them. A good guide to follow when rounding a 5 is always to round to the nearest even number; in this way, any tendency to round in a set direction is eliminated, since there is an equal likelihood that the nearest even number will be the higher or the lower in any given situation. Thus, we could report the foregoing results as

61.56 $\pm$ 0.07. On the other hand, if we had reason to doubt that ± 0.07 was a valid estimate of the precision, we might choose to present the result as 61.6 $\pm$ 0.1.

Often the significant figure convention is used in lieu of a specific estimate of the precision of a result. Thus, by simply reporting 61.6 in this example, we would be saying in effect that we believe the first 6 and the 1 are certain digits but the value of the second 6 is in doubt. The disadvantage of this technique is obvious; all the reader can discern is the range of the uncertainty—here, greater than ± 0.05 and smaller than ± 0.5.

In employing the significant figure convention, it is important to appreciate that the zero not only functions as a number but also serves to locate decimal points in very small and very large numbers. A case in point is Avogadro's number. The first three digits, 6, 0, and 2, are known with certainty; the next is uncertain but is probably 3. Since the digits that follow the 6023 are not known, we substitute 19 zeros after the digit 3 to place the decimal point. Here the zeros indicate the order of magnitude of the number only and have no other meaning. It is clear that a distinction must be made between those figures that have physical significance (that is, *significant figures*) and those that are either unknown or meaningless owing to the inadequacies of measurement.

Zeros bounded by digits only on the left may or may not be significant. Thus, the mass of a 20-mg weight that carries no correction (to a tenth of a milligram) is known to three significant figures, 20.0 mg; when this is expressed as 0.0200 g, the number of significant figures does not change. If, on the other hand, we wish to express the volume of a 2-liter beaker as 2000 ml, the latter number will contain only one significant figure. The zeros simply indicate the order of magnitude. It can, of course, happen that the beaker in question has been found by experiment to contain 2.0 liters; here the zero following the decimal point is significant and implies that the volume is known to at least ± 0.5 liter and might be known to ± 0.05 liter. If this volume were to be expressed in milliliters, the zero following the 2 would still be significant, but the other two zeros would not. The use of exponential notation eliminates the difficulty. We could thus indicate the volume as 2.0×10^3 ml.

A certain amount of care is required in determining the number of significant figures to carry in the result of an arithmetic combination of two or more numbers. For addition and subtraction, the number of significant figures can be seen by visual inspection. For example,

$$3.4 + 0.02 + 1.31 = 4.7$$

Clearly, the second decimal place cannot be significant because an uncertainty in the first decimal place is introduced by the 3.4.

When data are being multiplied or divided, it is frequently assumed that the number of significant figures of the result is equal to that of the component quantity that contains the least number of significant figures. For example,

$$\frac{24 \times 0.452}{100.0} = 0.108 = 0.11$$

Here the 24 has two significant figures and the result has therefore been rounded to agree. Unfortunately, the rule does not always apply well. Thus, the uncertainty in 24 in this calculation could be as small as 0.5 or as large as 5. The uncertainty in the quotient for these two limits is

Assumed Absolute Uncertainty in 24	Relative Uncertainty	Absolute Uncertainty in 0.108	Round to
> 0.5	0.5/24 = 0.02	0.108 × 0.02 = 0.002	0.108
< 5	5/24 = 0.2	0.108 × 0.2 = 0.02	0.11

Calculations such as this demonstrate the dilemma faced by the scientist who has only the significant figure convention to use as a guide to indeterminate errors.

Particular care is needed in rounding logarithms and antilogarithms to a proper number of significant figures. Consider again the first example on page 77; proper rounding of log 2.00×10^{-3} will yield

$$\log 2.00 \times 10^{-3} = -2.699$$

Here the mathematical manipulation appears to give a gain in the number of significant figures. In fact, however, the initial 3 in the result only indicates the location of the decimal point in the original number; information about 2.00 is contained in the three digits in 0.699. Thus, there is agreement in number of significant figures between this part of the result and the original.

As was shown by example (c) on page 78, an apparent reduction by one in number of significant figures appears to accompany the computation of an antilog. The reason for this apparent anomaly again lies in the function of the characteristic of a logarithm.

PROBLEMS

*1. In order to test an analytical method for calcium analysis, a solution was prepared by dissolving a weighed quantity of pure $CaCO_3$ in HCl. After adding other components to simulate a solution of a typical sample, the mixture was diluted to exactly 500 ml. The method was then applied to several 50.0-ml aliquots of the solution, each of which contained exactly 400 mg of Ca; the following results were obtained:

Sample No.	Ca Found, mg	Sample No.	Ca Found, mg
1	398	4	392
2	396	5	393
3	398	6	401

For the set of data calculate: (a) the mean, (b) the median, (c) the precision in terms of the spread, (d) the precision in terms of the average absolute and relative deviation from the mean, (e) the absolute and relative error of sample 1, (f) the absolute and relative error of the mean.

2. To test a method for the analysis of atmospheric SO_2, a standard sample was prepared by diluting measured quantities of SO_2 with appropriate volumes of air. Several 100-ml aliquots, with an SO_2 concentration of 9.8 ppm, were analyzed by the procedure. The results were

Sample No.	SO_2, ppm	Sample No.	SO_2, ppm
1	9.9	5	8.8
2	9.1	6	8.8
3	9.2	7	9.0
4	10.0		

For the set of data calculate (a) the mean, (b) the median, (c) the precision in terms of the spread, (d) the precision in terms of the average absolute and relative deviation (in ppt) from the mean, (e) the absolute and relative error of sample 1, (f) the absolute and relative error of the mean.

*3. A standard sample known to contain 1.31% H_2O was analyzed by student A who reported 1.28, 1.26, and 1.29% H_2O. Student B analyzed another standard having a composition of 8.67% H_2O; his results were 8.48, 8.55, and 8.53% H_2O. Compare (a) the absolute and relative deviations from the means of the two sets of data and (b) the absolute and relative errors associated with the mean of the two sets.

4. A standard sample containing 10.3% acetone yielded the following results: 10.2, 9.9, 10.3% acetone. Another standard containing 0.40% acetone upon analysis gave 0.38, 0.34, and 0.35% acetone. (a) Compare the precision of the two analyses in relative and absolute deviations from the means. (b) Compare the errors associated with the two analyses in absolute and relative terms.

*5. A method for the analysis of Br in organic compounds was found to have a constant error of -0.20 mg of Br associated with it. Calculate the relative error (in ppt) for the results of an analysis of a sample containing about 10% Br if the following sample sizes were taken: (a) 10 mg, (b) 50 mg, (c) 100 mg, (d) 500 mg, (e) 1000 mg.

6. It was found that a solubility loss of 2.5 mg was associated with a gravimetric method for the determination of Se. Calculate the relative error in parts per thousand in an analysis of a sample that contained about 16% Se if the original sample weighed: (a) 1.00 g, (b) 0.500 g, (c) 0.250 g, (d) 0.100 g.

*7. Calculate the absolute and relative standard deviations for the set of data in Problem 1 employing (a) the deviations from the mean of the data and (b) the spread of the data.

8. Calculate the absolute and relative standard deviations for the set of data in Problem 2 employing (a) the deviations from the mean of the data and (b) the spread of the data.

*9. Compare the absolute and relative standard deviations for the two sets of data in Problem 3 basing the calculations on (a) the deviations from the mean and (b) the spread of the data.

10. Compare the absolute and relative standard deviations for the two sets of data in Problem 4 basing the calculations on (a) the deviations from the mean and (b) the spread of the data.

*11. Nine samples of illicit heroin preparations were analyzed in duplicate by a gas chromatographic technique. Pool the data to establish an absolute standard deviation for the procedure.

Sample No.	% Heroin	Sample No.	% Heroin	Sample No.	% Heroin
1	2.24, 2.27	4	11.9, 12.6	7	14.4, 14.8
2	8.4, 8.7	5	4.3, 4.2	8	21.9, 21.1
3	7.6, 7.5	6	1.07, 1.02	9	8.8, 8.4

12. The following data were obtained for the analysis of sulfur in fuel oils by atomic absorption spectroscopy. Pool the data to obtain the standard deviation of the procedure.

Sample No.	% S (w/w)
1	0.50, 0.46, 0.52
2	1.96, 1.92, 2.01
3	2.65, 2.67, 2.71
4	1.25, 1.31, 1.26

*13. Calculate a pooled estimate of s from the following spectrophotometric analysis for NTA (nitrilotriacetic acid) in Ohio River water.

Sample No.	ppb NTA
1	13, 17. 14, 9
2	38, 37, 38
3	25, 29, 23, 29, 26

14. The following data were obtained for the concentration of Ca in bovine serum by an isotopic dilution technique. Pool the data to obtain a value of s for the procedure.

Sample No.	meq Ca/liter
1	3.569, 3.573, 3.569
2	4.294, 4.293
3	5.015, 5.032, 5.023, 5.020

*15. An atomic absorption method for the determination of Fe in used jet engine oil was found, from pooling 30 triplicate analyses, to have a standard deviation $s \rightarrow \sigma = 2.4$ μg/ml. Calculate the 80 and 95% confidence interval for the result, 18.5 μg Fe/ml, if it was based upon (a) a single analysis, (b) a mean of two analyses, (c) the mean of four analyses.

16. The method described in Problem 15 was found to yield a pooled standard deviation for Cu of $s \rightarrow \sigma = 0.62$ μg Cu/ml. The analysis of an oil from a reciprocating aircraft engine showed a Cu content of 8.53 μg Cu/ml. Calculate the 90 and 99% confidence interval if the result was based upon (a) a single analysis, (b) a mean of four analyses, (c) a mean of 16 analyses.

*17. How many replicate measurements would be needed to decrease the 95 and 99% confidence limits for the analysis described in Problem 15 to ± 2.0 µg Fe/ml?

18. How many replicate measurements would be necessary to decrease the 95 and 99% confidence limits for the analysis described in Problem 16 to ± 0.5 µg Cu/ml?

*19. A chemist obtained the following data in a triplicate analysis of an insecticide preparation for its percent lindane: 7.47, 6.98, 7.27. Calculate the 90% confidence interval for the mean of the three data, assuming that
(a) the only information about the precision of the method was the precision for the three data.
(b) on the basis of long experience with the method he believed that $s \rightarrow \sigma = 0.28\%$ lindane.

20. An analyst obtained the following data in a duplicate analysis of air samples: ppm $SO_2 = 10.8, 9.2$. Calculate the 95% confidence interval and the 80% confidence interval for the mean of the data, assuming that
(a) the only information about the precision of the method was the precision of these two results.
(b) based on extensive past experience with the method, $s \rightarrow \sigma = 1.1$ ppm SO_2.

*21. A chemist obtained the following results for the determination of sulfur in a contaminated kerosene sample: 0.724, 0.693, 0.755% S. Calculate the 95% confidence limit for the mean of this analysis.

22. Four replicate fluoride analyses on a sample of well water yielded the following data: 0.89, 0.96, 0.87, 0.94 ppm F^-. What are the 95 and 99% confidence limits for the mean of this analysis?

*23. A volumetric calcium analysis on triplicate samples of the blood serum of a patient believed to be suffering from a hyperparathyroid condition produced the following data: meq Ca/liter $= 3.15, 3.25, 3.26$. What is the 95% confidence interval for the mean of the data, assuming
(a) no prior information about the precision of the analysis?
(b) $s \rightarrow \sigma = 0.05$ meq Ca/liter?

*24. A physiologist, interested in the role of K^+ ion in the transmission of nerve signals, developed a potentiometric analysis for the ion in serums. In order to evaluate the precision of the method, he pooled the data on several analyses performed over several weeks on samples that contained from 3 to 6 meq K^+/liter.

Sample No.	Mean K^+ Concn Found, meq/liter	Number of Measurements	Deviation of Individual Results from the Mean, meq K^+/liter
1	4.63	5	0.12, 0.0, 0.22, 0.10, 0.00
2	5.02	2	0.13, 0.13
3	4.01	4	0.31, 0.10, 0.11, 0.10
4	6.26	9	0.03, 0.13, 0.09, 0.20, 0.27, 0.03, 0.15, 0.07, 0.12
5	3.97	5	0.12, 0.08, 0.01, 0.04. 0.17

(a) What is the standard deviation for each of the sets of data?
(b) What is the standard deviation for the method obtained by pooling the data?
(c) Would the value of s obtained in (b) be a good approximation of σ for the method?

(d) For sample 4, calculate the 95% confidence interval of the mean, first employing the sample standard deviation obtained in (a) and then the pooled standard deviation from (b).

(e) Repeat the calculation in (d) for sample 2.

25. An analytical chemist was interested in evaluating the indeterminate error in a gravimetric method for the determination of the hormone progesterone in oral tablets. Repeated use of the method yielded the following data:

Sample No.	Number of Replicate Analyses	Mean Percent Progesterone	Individual Deviations from the Mean
1	5	3.66	0.04, 0.01, 0.03, 0.05, 0.05
2	3	3.45	0.06, 0.03, 0.03
3	8	3.55	0.00, 0.07, 0.04, 0.03, 0.02, 0.02, 0.01, 0.01
4	2	3.86	0.03, 0.03
5	3	3.12	0.01, 0.00, 0.01
6	6	3.97	0.06, 0.00, 0.04, 0.01, 0.01, 0.02

(a) Calculate the standard deviation for each set of data.

(b) Calculate a standard deviation for the method by pooling the data from the six samples.

(c) Would the value of s found in (b) be expected to be a good approximation of σ?

(d) For sample 3, calculate the 95% confidence interval, first employing the s obtained in (a) and then the pooled s from (b).

(e) Repeat the calculations in (d) for sample 4.

26. The performance of a new photometer was tested by making 50 replicate measurements of the percent transmission of a solution. The standard deviation of these data was 0.15% T. How many replicate readings of the instrument should be taken for each subsequent measurement if the instrumental error associated with the mean is to be kept below

*(a) $\pm 0.2\%$ T with 99% certainty?

(b) $\pm 0.1\%$ T with 99% certainty?

*(c) $\pm 0.2\%$ T with 95% certainty?

(d) $\pm 0.1\%$ T with 95% certainty?

*27. What are the 90 and 95% confidence limits for a single measurement by the instrument described in Problem 26?

28. A standard method for the determination of tetraethyl lead (TEL) in gasoline is reported to have a standard deviation of 0.020 ml TEL per gallon. If $s \rightarrow \sigma = 0.020$, how many replicate analyses should be made in order for the mean for the analysis of a sample to be within

*(a) ± 0.03 ml/gal of the true mean 99% of the time?

(b) ± 0.03 ml/gal of the true mean 95% of the time?

(c) ± 0.015 ml/gal of the true mean 90% of the time?

29. What are the 99.9, the 99, and the 80% confidence intervals for an analysis by the method described in Problem 28 based upon

(a) the result of a single measurement?

(b) the mean of triplicate measurements?

*30. A spark-source mass spectrometric method for the determination of various elements in steel was tested by analyzing several National Bureau of Standards samples. The results from three of the analyses are given below. Assume that the NBS analyses are correct and determine whether or not a determinate error in any of the analyses is indicated at the 95% confidence level.

	Element	Number of Analyses	Mean, % (w/w)	Relative Standard Deviation, ppt	NBS Result, % (w/w)
(a)	V	8	0.090	57	0.10
(b)	Ni	5	0.36	38	0.39
(c)	Cu	7	0.55	42	0.47

31. A spectrophotometric method for the determination of boron in animal tissue was tested by adding known amounts of B as a borate-mannitol complex to samples of rat livers; the increase in B concentration was then determined. The mean result from eight replicate analyses showed an increase in B concentration of 1.490 µg/g with a standard deviation of 0.064 µg/g. The samples contained 1.60 µg/g of added B. Is a negative determinate error indicated at the 95% confidence level?

*32. A titrimetric method for the determination of Ca in limestone was tested by analysis of a National Bureau of Standards limestone containing 30.15% CaO. The mean result of four analyses was 30.26% CaO with a standard deviation of 0.085% CaO. By pooling data from several analyses it was established that $s \rightarrow \sigma = 0.094\%$ CaO.
    (a) Do the data indicate the presence of a determinate error at the 95% confidence level?
    (b) Repeat the calculation assuming that the pooled value for σ was not available.

33. By pooling data from a number of samples, it was found that the precision of a method for the determination of P in phosphate rocks was $(s)_r \rightarrow \sigma_r = 0.57\%$ relative. To test the method for determinate error, a standard sample known to contain 44.71% P_2O_5 was analyzed. The mean of five analyses was 44.93% P_2O_5.
    (a) Is a determinate error indicated at the 95% confidence level?
    (b) Repeat the calculation assuming that the pooled value for σ was unavailable and that $(s)_r = 0.57\%$ for the five analyses.

34. Determine whether a significant difference exists (at the 95% confidence level) between the mean of the results shown below and the known value.

	Known Value	Mean Value	Standard Deviation	Number of Measurements
*(a)	1.26% Cu	1.32% Cu	$s \rightarrow \sigma = 0.04\%$ Cu	4
*(b)	5.0 ppm CH_3OH	4.4 ppm CH_3OH	$s = 0.070$ ppm CH_3OH	5
(c)	55.5% Fe	55.2% Fe	$s = 0.2\%$ Fe	3
(d)	pH = 10.07	pH = 10.04	$s \rightarrow \sigma = 0.02$	5

35. In order to test the quality of the work of a commercial laboratory, duplicate analyses of a purified benzoic acid (% C = 68.74 and % H = 4.953) sample were requested. It is assumed that the relative standard deviation of the method used should be $s_r \rightarrow \sigma_r = 4$ ppt for carbon and 6 ppt for hydrogen. The means of the reported results were % C = 68.5 and % H = 4.88. At the 95% confidence level, is there any indication of determinate error in either of the analyses?

*36. Is there a significant difference (at the 95% confidence level) between the experimental means, $\bar{x}_1$ and $\bar{x}_2$, in the following cases? A pooled standard deviation is indicated by s; a known standard deviation is shown by σ.

	$\bar{x}_1$	$\bar{x}_2$	N_1	N_2	Standard Deviation
(a)	14.1	14.5	2	3	$s = 0.2$
(b)	14.1	14.5	6	8	$s = 0.2$
(c)	14.1	14.5	2	3	$\sigma = 0.2$

37. Is there a significant difference (at the 95% confidence level) between the experimental means, $\bar{x}_1$ and $\bar{x}_2$, in the following cases? A pooled standard deviation is indicated by s; a known standard deviation is shown by σ.

	$\bar{x}_1$	$\bar{x}_2$	N_1	N_2	Standard Deviation
(a)	1.078	1.063	2	2	$s = 0.010$
(b)	1.078	1.063	9	7	$s = 0.010$
(c)	1.078	1.063	2	2	$\sigma = 0.010$

*38. A prosecuting attorney in a criminal case used, as his principal evidence, the presence of small fragments of glass found imbedded in the coat of the accused that were alleged to be identical in composition to a rare Belgian stained glass window broken during the crime. The averages of triplicate analyses for five elements are shown below. On the basis of these data, does the defendant have grounds for claiming reasonable doubt as to the identity of the two materials and thus doubt as to his guilt? Employ the 99% confidence level as a criterion for doubt.

	Concentration, ppm		Standard Deviation
Element	From Clothes	From Window	$s \rightarrow \sigma$
As	1290	1090	95
Co	0.45	0.60	0.17
La	3.92	3.61	0.09
Sb	2.75	1.50	1.46
Th	0.61	0.81	0.08

39. The homogeneity of a standard chloride sample was tested by analyzing portions of the material from the top and the bottom of the container, with the following results.

% Chloride

Top	Bottom
26.32	26.28
26.33	26.25
26.38	26.38
26.39	

(a) Is inhomogeneity indicated at the 95% confidence level?
(b) Repeat the calculation, assuming that $s \rightarrow \sigma = 0.03\%$ Cl.

*40. How many significant figures are there in
    (a) 0.0607?   (d) 7357027?
    (b) 9966?   (e) 9004?
    (c) 0.0003644?   (f) 31814?

41. How many significant figures are there in
    (a) 0.008614?   (d) 83.96?
    (b) 684.5?   (e) 12.30?
    (c) 5.647?   (f) 4.175×10^{-6}?

*42. Estimate the absolute standard deviation for the results of the following calculations. Round the result to the proper number of significant figures. The numbers shown in parentheses are absolute standard deviations.
    (a) $y = 6.75(\pm 0.03) + 0.843(\pm 0.001) - 7.021(\pm 0.001) = 0.572$
    (b) $y = 19.97(\pm 0.04) + 0.0030(\pm 0.0001) + 1.29(\pm 0.08) = 21.263$
    (c) $y = 67.1(\pm 0.3) \times 1.03(\pm 0.02) \times 10^{-17} = 6.9113 \times 10^{-16}$

    (d) $y = 243(\pm 1) \times \dfrac{760(\pm 2)}{1.006(\pm 0.006)} = 183578.5$

    (e) $y = \dfrac{143(\pm 6) - 64(\pm 3)}{1249(\pm 1) + 77(\pm 8)} = 5.9578 \times 10^{-2}$

    (f) $y = \dfrac{1.97(\pm 0.01)}{243(\pm 3)} = 8.106996 \times 10^{-3}$

    (g) $y = [9.6(\pm 0.2)]^3 = 884.736$
    (h) $y = [1.03(\pm 0.04) \times 10^{-16}]^{1/3} = 4.6875 \times 10^{-6}$

43. Estimate the absolute standard deviation for the results of the following calculations. Round the result to include only significant figures. The numbers shown in parentheses are absolute standard deviations.
    (a) $y = -1.02(\pm 0.02) \times 10^{-7} - 3.54(\pm 0.2) \times 10^{-8} = -1.374 \times 10^{-7}$
    (b) $y = 100.20(\pm 0.08) - 99.62(\pm 0.06) + 0.200(\pm 0.004) = 0.780$
    (c) $y = 0.0010(\pm 0.0005) \times 18.10(\pm 0.02) \times 200(\pm 1) = 3.62$
    (d) $y = [33.33(\pm 0.03)]^3 = 37025.927$

    (e) $y = \dfrac{1.73(\pm 0.03) \times 10^{-14}}{1.63(\pm 0.04) \times 10^{-16}} = 106.1349693$

    (f) $y = \dfrac{100(\pm 1)}{2(\pm 1)} = 50$

    (g) $y = \dfrac{1.43(\pm 0.02) \times 10^{-2} - 4.76(\pm 0.06) \times 10^{-3}}{24.3(\pm 0.7) + 8.06(\pm 0.08)} = 2.948 \times 10^{-4}$

    (h) $y = [17.2(\pm 0.6)]^{1/4} = 2.036489$

*44. Estimate the absolute standard deviation for the results of the following calculations. Round the result to include only significant figures. The numbers shown in parentheses are absolute standard deviations.
    (a) $y = \log [1.73(\pm 0.030)] = 0.238046$
    (b) $y = \log [0.0432(\pm 0.004)] = -1.36452$
    (c) $y = \log [6.02(\pm 0.02) \times 10^{23}] = 23.77960$
    (d) $y = \text{antilog} [-3.47(\pm 0.05)] = 3.38844 \times 10^{-4}$
    (e) $y = \text{antilog} [5.7(\pm 0.5)] = 5.01187 \times 10^5$
    (f) $y = \text{antilog} [0.99(\pm 0.05)] = 9.77237$

45. Estimate the absolute standard deviation for the results of the following calculations. Round the result to include only significant figures. The numbers shown in parentheses are absolute standard deviations.
   (a) $y = \log [6.54(\pm 0.06)] = 0.8155777$
   (b) $y = \log [0.0022(\pm 0.0001)] = -2.657577$
   (c) $y = \log [96494(\pm 2)] = 4.984500$
   (d) $y = \text{antilog} [-6.02(\pm 0.02)] = 9.549926 \times 10^{-7}$
   (e) $y = \text{antilog} [9.83(\pm 0.07)] = 6.7608297 \times 10^{9}$
   (f) $y = \text{antilog} [0.863(\pm 0.008)] = 7.294575$

*46. Apply the Q test to the accompanying sets as a guide in determining whether the outlying result should be retained or rejected.
   (a) 41.37, 41.61, 41.84, 41.70
   (b) 7.300, 7.284, 7.388, 7.292

47. Apply the Q test to the accompanying sets as a guide in determining whether the outlying result should be retained or rejected.
   (a) 85.10, 84.62, 84.70
   (b) 85.10, 84.62, 84.65, 84.70

chapter

5

THE SOLUBILITY
OF PRECIPITATES

Reactions that yield substances of limited solubility find wide application in three important analytical processes, namely: (1) the separation of an analyte as a precipitate from soluble substances that interfere with its ultimate measurement; (2) gravimetric analysis, in which a precipitate is formed whose weight is chemically related to the amount of analyte; and (3) volumetric analysis, based on determining the volume of a standard reagent required to precipitate the analyte essentially completely. The success of each of these applications requires that the solid produced have a relatively low solubility, be reasonably pure, and have a suitable particle size. In this chapter we consider the variables which influence the first of these three physical properties.

Examples of how a solubility-product constant can be employed to calculate the solubility of an ionic precipitate in water and in the presence of a common ion have been discussed in Chapter 3. The student should be thoroughly familiar with these principles before undertaking study of this chapter. Here we will be concerned with the way such variables as pH, concentration of complexing agents, and concentration of electrolytes affect the solubility of precipitates.

Effect of Competing Equilibria on the Solubility of Precipitates

The solubility of a precipitate increases in the presence of ionic or molecular species that form soluble compounds or complexes with the ions derived from the precipitate. For example, the solubility of calcium fluoride is larger in acidic than neutral solutions because of the tendency for fluoride ions to react with hydronium ions. As a consequence, two equilibria are established when a solution is saturated with calcium fluoride:

$$CaF_2(s) \rightleftarrows Ca^{2+} + 2F^-$$
$$+$$
$$2H_3O^+$$
$$\Updownarrow$$
$$2HF + 2H_2O$$

From the Le Châtelier principle, it is evident that addition of acid causes an increase in hydrogen fluoride concentration. The consequent decrease in fluoride ion concentration is partially offset, however, by a shift of the first equilibrium to the right; a net increase in solubility results.

Another example of a solubility increase in the presence of a reactive species is shown by the following equations:

$$AgBr(s) \rightleftarrows Ag^+ + Br^-$$
$$+$$
$$2NH_3$$
$$\Updownarrow$$
$$Ag(NH_3)_2{}^+$$

Here ammonia molecules tend to decrease the silver ion concentration of the solution. A shift to the right of the solubility equilibrium and an increase in solubility of the silver bromide is the consequence.

TREATMENT OF MULTIPLE EQUILIBRIA

We shall frequently be faced with the problem of calculating the concentrations of constituents in a solution in which two or more competing equilibria are simultaneously established; the dissolution of calcium fluoride or silver bromide is an example. As a more general case, consider the sparingly soluble AB, which dissolves to give A and B ions:

$$AB(s) \rightleftarrows A + B$$
$$+ \quad +$$
$$C \quad D$$
$$\Updownarrow \quad \Updownarrow$$
$$AC \quad BD$$

If A and B react with species C and D to form the soluble AC and BD, introduction of either C or D into the solution will cause a shift in the solubility equilibrium in the direction that increases the solubility of AB.

Determination of the solubility of AB in a system such as this requires knowledge of the formal concentrations of the added C and D, as well as equilibrium constants for all three equilibria. Generally, several algebraic expressions are needed to describe completely the concentration relationships in such a solution, and the solubility calculation requires the solution of multiple simultaneous equations; frequently the solution of these algebraic equations is more formidable than the task of setting them up.

One point that should be constantly borne in mind in treating multiple equilibria is that *the validity and form of a particular equilibrium-constant expression is in no way affected by the existence of additional competing equilibria in the solution*. Thus, in the present example, the solubility-product expression for AB describes the relationship between the equilibrium concentrations of A and B regardless of whether or not C and D are present in solution. That is, at constant temperature the product [A][B] is a constant, provided only that some solid AB is present. To be sure, the *amount* of AB that dissolves is greater in the presence of C or D; the increase, however, is not because the ion product [A][B] has changed but rather because some of the precipitate has been converted to AC or BD.

In the following paragraphs we present a systematic approach by which any problem involving several equilibria can be attacked. The approach will then be demonstrated by several examples involving solubility equilibria.

SYSTEMATIC METHOD FOR SOLVING PROBLEMS
INVOLVING SEVERAL EQUILIBRIA

1. Write chemical equations for all the reactions that appear to have any bearing on the problem.
2. State in terms of equilibrium concentrations what is being sought in the problem.
3. Write equilibrium-constant expressions for all of the equilibria shown in step 1; find numerical values for the constants from appropriate tables.
4. Write mass-balance equations for the system. These are algebraic expressions relating the equilibrium concentrations of the various species to one another and to the formal concentrations of the substances present in the solution; they are derived by taking into account the way the solution was prepared.
5. Write a charge-balance equation. In any solution the concentrations of the cations and anions must be such that the solution is electrically neutral. The charge-balance equation expresses this relationship.[1]

---

[1] As a simple example of a charge-balance equation, consider a solution prepared by dissolving NaCl in water. Such a solution has a net charge of zero, although it contains both positive and negative ions. This neutrality is a direct consequence of the relationship:

$$[Na^+] + [H_3O^+] = [Cl^-] + [OH^-]$$

Thus, the solution is neutral by virtue of the fact that the sum of the concentrations of the charged positive species is equal to the sum of the concentrations of the negative species. Now consider an aqueous solution of $MgCl_2$. Here we must write

$$2[Mg^{2+}] + [H_3O^+] = [Cl^-] + [OH^-]$$

It is necessary to multiply the magnesium ion concentration by 2 in order to account for the two units of charge contributed by this ion; that is, charge balance is preserved because the chloride ion concentration is *twice* the magnesium ion concentration ($[Cl^-] = 2[Mg^{2+}]$). The concentration of a triply charged species, if present, would have to be multiplied by 3 for the same reason. Thus, for a solution containing $Al_2(SO_4)_3$, $MgCl_2$, and water, the charge-balance equation would be

$$3[Al^{3+}] + 2[Mg^{2+}] + [H_3O^+] = 2[SO_4^{2-}] + [HSO_4^-] + [Cl^-] + [OH^-]$$

6. Count the number of unknown quantities in the equations developed in steps 3, 4, and 5, and compare with the number of independent equations. If the number of equations is equal to the number of unknown concentrations, the problem can be solved exactly by suitable algebraic manipulation. If there are fewer equations than unknowns, attempt to derive additional independent equations. If this cannot be done, it must be concluded that an exact solution to the problem is not possible; it may, however, be possible to arrive at an approximate solution.

7. Make suitable approximations to simplify the algebra or to reduce the number of unknowns so that the problem can be solved.

8. Solve the algebraic equations for the equilibrium concentrations that are necessary to give the answer as defined in step 2.

9. With the equilibrium concentrations obtained in step 8, check the approximations made in step 7 to be sure of their validity.

Step 6 in this scheme is particularly significant because it indicates whether an exact solution for the problem is theoretically feasible. If the number of independent equations is as great as the number of unknowns, the problem becomes purely algebraic, involving a solution to several simultaneous equations. On the other hand, if the number of equations is fewer than the number of unknowns, a search for other equations, or for approximations that will reduce the number of unknowns, is essential. The student should never waste time in seeking a solution to a complex equilibrium problem without first establishing that sufficient data are available.

The Effect of pH on Solubility

The solubilities of many precipitates of importance in quantitative analysis are affected by the hydronium ion concentration of the solvent. Precipitates that exhibit this behavior contain an anion with significant basic properties, a cation with significant acidic properties, or both. Calcium fluoride, discussed earlier, contains the basic fluoride anion which tends to react with hydronium ions to give hydrogen fluoride. Thus, the solubility of calcium fluoride increases with acidity. An example of a compound with an acidic cation is bismuth iodide. When water is saturated with BiI_3, the following equilibria are established:

$$BiI_3(s) \rightleftarrows Bi^{3+} + 3I^-$$

$$Bi^{3+} + H_2O \rightleftarrows BiOH^{2+} + H_3O^+$$

Note that the solubility of bismuth iodide, in contrast to calcium fluoride, decreases with increasing acidity.

SOLUBILITY CALCULATIONS WHEN THE HYDRONIUM ION CONCENTRATION IS FIXED AND KNOWN

It is often desirable to perform an analytical precipitation at a fixed and predetermined hydronium ion concentration. The following example illustrates the calculation of solubility under these circumstances.

Example. Calculate the formal solubility of CaC_2O_4 in a solution maintained at a hydronium ion concentration of 1.00×10^{-4} M.

Step 1. *Chemical equations.*

$$CaC_2O_4(s) \rightleftarrows Ca^{2+} + C_2O_4{}^{2-} \tag{5-1}$$

Since oxalic acid is a weak acid, oxalate ions will react in part with hydronium ions which were added to maintain the specified hydronium ion concentration.

$$C_2O_4{}^{2-} + H_3O^+ \rightleftarrows HC_2O_4{}^- + H_2O \tag{5-2}$$

$$HC_2O_4{}^- + H_3O^+ \rightleftarrows H_2C_2O_4 + H_2O \tag{5-3}$$

Step 2. *Definition of the unknown.* What is sought? We wish to know the solubility of CaC_2O_4 in formula weights per liter. Since CaC_2O_4 is ionic, its formal solubility will be equal to the molar concentration of calcium ion; it will also be equal to the sum of the equilibrium concentrations of the oxalate species. That is,

$$\text{solubility} = [Ca^{2+}]$$
$$= [C_2O_4{}^{2-}] + [HC_2O_4{}^-] + [H_2C_2O_4]$$

Thus, if we can calculate either of these quantities, we shall have obtained a solution to the problem.

Step 3. *Equilibrium-constant expressions.*

$$K_{sp} = [Ca^{2+}][C_2O_4{}^{2-}] = 2.3 \times 10^{-9} \tag{5-4}$$

Equation 5-2 is simply the reverse of the dissociation reaction for $HC_2O_4{}^-$. We can thus use the value of K_2 for oxalic acid to provide a relationship between $[C_2O_4{}^{2-}]$ and $[HC_2O_4{}^-]$.

$$K_2 = \frac{[H_3O^+][C_2O_4{}^{2-}]}{[HC_2O_4{}^-]} = 5.42 \times 10^{-5} \tag{5-5}$$

Similarly, for Equation 5-3,

$$K_1 = \frac{[H_3O^+][HC_2O_4{}^-]}{[H_2C_2O_4]} = 5.36 \times 10^{-2} \tag{5-6}$$

Step 4. *Mass-balance equations.* Since the only source of Ca^{2+} and the various oxalate species is the dissolved CaC_2O_4, it follows that

$$[Ca^{2+}] = [C_2O_4{}^{2-}] + [HC_2O_4{}^-] + [H_2C_2O_4] \tag{5-7}$$

Furthermore, it is given that at equilibrium,

$$[H_3O^+] = 1.0 \times 10^{-4} \tag{5-8}$$

Step 5. *Charge-balance equations.* A charge-balance equation cannot be written for this system because an amount of some unknown acid HX has been added to maintain $[H_3O^+]$ at 1.0×10^{-4}; an equation based on the electrical neutrality of the solution would require inclusion of the concentration of the anion $[X^-]$ associated with the unknown acid. As it turns out, an equation containing this additional unknown term is not needed.

Step 6. *Comparison of equations and unknowns.* We have four unknowns—namely, $[Ca^{2+}]$, $[C_2O_4{}^{2-}]$, $[HC_2O_4{}^-]$, and $[H_2C_2O_4]$. We also have four independent algebraic relationships: Equations 5-4, 5-5, 5-6, and 5-7. Therefore, an exact solution is possible, and the problem has now become one of algebra.

Step 7. Approximations. Since we have sufficient data, an exact solution to the problem can be obtained.

Step 8. Solution of the equations. A convenient way to effect a solution is to make suitable substitutions into Equation 5-7 and thereby establish a relationship between $[Ca^{2+}]$ and $[C_2O_4^{2-}]$. We must first derive expressions for $[HC_2O_4^-]$ and $[H_2C_2O_4]$ in terms of $[C_2O_4^{2-}]$. Substitution of 1.00×10^{-4} for $[H_3O^+]$ in Equation 5-5 yields

$$\frac{(1.00 \times 10^{-4})[C_2O_4^{2-}]}{[HC_2O_4^-]} = 5.42 \times 10^{-5}$$

Thus,

$$[HC_2O_4^-] = \frac{1.00 \times 10^{-4}}{5.42 \times 10^{-5}} [C_2O_4^{2-}] = 1.84[C_2O_4^{2-}]$$

Upon substituting this relationship and the hydronium ion concentration into Equation 5-6, we obtain

$$\frac{1.00 \times 10^{-4} \times 1.84[C_2O_4^{2-}]}{[H_2C_2O_4]} = 5.36 \times 10^{-2}$$

Thus,

$$[H_2C_2O_4] = \frac{1.00 \times 10^{-4} \times 1.84[C_2O_4^{2-}]}{5.36 \times 10^{-2}} = 0.0034[C_2O_4^{2-}]$$

These values for $[H_2C_2O_4]$ and $[HC_2O_4^-]$ are substituted into Equation 5-7 to give

$$[Ca^{2+}] = [C_2O_4^{2-}] + 1.84[C_2O_4^{2-}] + 0.0034[C_2O_4^{2-}]$$
$$= 2.84[C_2O_4^{2-}]$$

or

$$[C_2O_4^{2-}] = \frac{[Ca^{2+}]}{2.84}$$

Substitution for $[C_2O_4^{2-}]$ in Equation 5-4 gives

$$[Ca^{2+}] \frac{[Ca^{2+}]}{2.84} = 2.3 \times 10^{-9}$$
$$[Ca^{2+}]^2 = 6.53 \times 10^{-9}$$
$$[Ca^{2+}] = 8.1 \times 10^{-5}$$

Thus, from step 2 we conclude:

solubility of $CaC_2O_4 = 8.1 \times 10^{-5}$ fw/liter

SOLUBILITY CALCULATIONS WHERE THE HYDRONIUM ION CONCENTRATION IS VARIABLE

Solutes containing basic anions (such as calcium oxalate) or acidic cations (such as bismuth iodide) contribute to the hydronium ion concentration of their aqueous solutions. Thus, if it is not held constant by means of an independent equilibrium, the hydronium ion concentration becomes dependent upon the extent to which such solutes dissolve. For example, a solution saturated with calcium oxalate becomes basic as a consequence of the reactions

$$CaC_2O_4(s) \rightleftarrows Ca^{2+} + C_2O_4^{2-}$$
$$C_2O_4^{2-} + H_2O \rightleftarrows HC_2O_4^- + OH^-$$
$$HC_2O_4^- + H_2O \rightleftarrows H_2C_2O_4 + OH^-$$

TABLE 5-1   Calculated Solubility of MA from Various Assumed Values of K_{sp} and K_b

Solubility Product Assumed for MA	Dissociation Constant Assumed for HA, K_{HA}	Basic Constant for A$^-$, $K_b = K_w/K_{HA}$	Calculated Solubility of MA, fw/liter	Calculated Solubility of MA Neglecting Reaction of A$^-$ with Water, fw/liter
1.0×10^{-10}	1.0×10^{-6}	1.0×10^{-8}	1.02×10^{-5}	1.0×10^{-5}
	1.0×10^{-8}	1.0×10^{-6}	1.2×10^{-5}	1.0×10^{-5}
	1.0×10^{-10}	1.0×10^{-4}	2.4×10^{-5}	1.0×10^{-5}
	1.0×10^{-12}	1.0×10^{-2}	10×10^{-5}	1.0×10^{-5}
1.0×10^{-20}	1.0×10^{-6}	1.0×10^{-8}	1.05×10^{-10}	1.0×10^{-10}
	1.0×10^{-8}	1.0×10^{-6}	3.3×10^{-10}	1.0×10^{-10}
	1.0×10^{-10}	1.0×10^{-4}	32×10^{-10}	1.0×10^{-10}
	1.0×10^{-12}	1.0×10^{-2}	290×10^{-10}	1.0×10^{-10}

In contrast to the example just considered, the hydroxide ion concentration now becomes an unknown, and an additional algebraic equation must therefore be developed.

In most instances, the reaction of a precipitate with water cannot be neglected without introducing an error in the calculation. As shown by the data in Table 5-1, the magnitude of the error depends upon the solubility of the precipitate as well as the basic dissociation constant of the anion. The solubilities of the hypothetical precipitate MA, shown in column 4, were obtained by taking into account the reaction of A$^-$ with water. Column 5 gives the calculated results when the basic properties of A$^-$ are neglected; here the solubility is simply the square root of the solubility product. Two solubility products, 1.0×10^{-10} and 1.0×10^{-20}, have been assumed for these calculations as well as several values for the basic dissociation constant of A$^-$. It is apparent that neglect of the reaction of the anions with water leads to a negative error, which becomes more pronounced both as the solubility of the precipitate increases (larger K_{sp}) and as the conjugate base becomes stronger.

It is not difficult to write the algebraic relationships needed to calculate the solubility of such a precipitate; solution of the equations, however, is tedious. Fortunately, it is ordinarily possible to invoke one of two simplifying assumptions to decrease the algebraic labor. The first assumption is applicable to moderately soluble compounds containing an anion that reacts extensively with water. Here it is assumed that sufficient hydroxide ions are formed to permit neglect of the hydronium ion concentration in calculations. Another way of stating this assumption is to say that the hydroxide ion concentration of the solution is determined exclusively by the reaction of the anion with water and that the contribution of hydroxide ions from the dissociation of water itself is negligible by comparison.

The second type of assumption is applicable to precipitates of very low solubility, particularly those containing an anion that does not react extensively with water. In such a system we may often assume that solution of the precipitate does not significantly change the hydronium or hydroxide ion concentration of the solution and that these concentrations remain essentially 10^{-7} mole/liter. The solubility calculation then reduces to the type considered in the preceding example.

An example of each of these types of calculations follows.

Example. Calculate the solubility of $PbCO_3$ in water. The equilibria that bear on the problem are

$$PbCO_3(s) \rightleftarrows Pb^{2+} + CO_3^{2-} \tag{5-9}$$

$$CO_3^{2-} + H_2O \rightleftarrows HCO_3^- + OH^- \tag{5-10}$$

$$HCO_3^- + H_2O \rightleftarrows H_2CO_3 + OH^- \tag{5-11}$$

$$2H_2O \rightleftarrows H_3O^+ + OH^- \tag{5-12}$$

The solubility of $PbCO_3$ can be expressed as follows:

$$\text{solubility} = [Pb^{2+}]$$
$$= [CO_3^{2-}] + [HCO_3^-] + [H_2CO_3]$$

The equilibrium-constant expressions are

$$[Pb^{2+}][CO_3^{2-}] = K_{sp} = 3.3 \times 10^{-14} \tag{5-13}$$

$$\frac{[HCO_3^-][OH^-]}{[CO_3^{2-}]} = \frac{K_w}{K_2} = \frac{1.00 \times 10^{-14}}{4.7 \times 10^{-11}} = 2.13 \times 10^{-4} \tag{5-14}[2]$$

$$\frac{[H_2CO_3][OH^-]}{[HCO_3^-]} = \frac{K_w}{K_1} = \frac{1.00 \times 10^{-14}}{4.45 \times 10^{-7}} = 2.25 \times 10^{-8} \tag{5-15}[2]$$

and

$$[H_3O^+][OH^-] = 1.00 \times 10^{-14} \tag{5-16}$$

The mass-balance expression is

$$[Pb^{2+}] = [CO_3^{2-}] + [HCO_3^-] + [H_2CO_3] \tag{5-17}$$

For this system, an expression based on the electrical neutrality of the solution can also be written

$$2[Pb^{2+}] + [H_3O^+] = 2[CO_3^{2-}] + [HCO_3^-] + [OH^-] \tag{5-18}$$

We now have six equations and also six unknowns—namely, $[Pb^{2+}]$, $[CO_3^{2-}]$, $[HCO_3^-]$, $[H_2CO_3]$, $[OH^-]$, and $[H_3O^+]$. An exact solution can be obtained. However, the algebra required would be very involved. It thus becomes necessary to seek an easier solution through the use of certain approximations. In this example we are concerned with a fairly soluble precipitate containing an anion that reacts extensively with water; these conclusions are based on the magnitude of the equilibrium constants for Equations 5-13, 5-14, and 5-15. Thus, we can reasonably expect the hydroxide ion concentration of the solution to be significantly raised as the precipitate dissolves and that the hydronium ion concentration will be consequently lowered. It seems

---

[2] These constants have numerical values equal to K_w/K_a, where K_a is the dissociation constant of the conjugate acid (p. 31). In the first dissociation step, HCO_3^- is the product; the constant associated with this process is thus numerically equal to K_w/K_2, where K_2 is the second dissociation constant for H_2CO_3. The second dissociation step gives H_2CO_3; therefore, the constant for this reaction is K_w/K_1.

probable, then, that $[H_3O^+] \ll 2[Pb^{2+}]$ in Equation 5-18 and that the former quantity can be neglected without serious error. A second assumption might be that the equilibrium involving formation of H_2CO_3 (Equation 5-11) is relatively unimportant compared with that in which HCO_3^- is the product (Equation 5-10); therefore, the concentration of H_2CO_3 is much smaller than that for HCO_3^-. This assumption appears reasonable inasmuch as the value of K_w/K_1 is only $\frac{1}{10,000}$ of that for K_w/K_2 (see Equations 5-14 and 5-15).

If $[H_3O^+]$ is indeed smaller than any of the terms in Equation 5-18 and can therefore be neglected, and if

$$[HCO_3^-] \gg [H_2CO_3]$$

Equation 5-17 then simplifies to

$$[Pb^{2+}] = [CO_3^{2-}] + [HCO_3^-] \tag{5-19}$$

and Equation 5-18 becomes

$$2[Pb^{2+}] = 2[CO_3^{2-}] + [HCO_3^-] + [OH^-] \tag{5-20}$$

Furthermore, Equations 5-15 and 5-16 are no longer needed. Thus, we have reduced the number of equations and unknowns to four.

If we multiply Equation 5-19 by 2 and subtract this from Equation 5-20, we obtain

$$0 = [OH^-] - [HCO_3^-]$$

or

$$[OH^-] = [HCO_3^-] \tag{5-21}$$

Substitution of $[HCO_3^-]$ for $[OH^-]$ in Equation 5-14 gives

$$\frac{[HCO_3^-]^2}{[CO_3^{2-}]} = \frac{K_w}{K_2}$$

$$[HCO_3^-] = \sqrt{\frac{K_w}{K_2}[CO_3^{2-}]}$$

This expression permits elimination of $[HCO_3^-]$ from Equation 5-19:

$$[Pb^{2+}] = [CO_3^{2-}] + \sqrt{\frac{K_w}{K_2}[CO_3^{2-}]} \tag{5-22}$$

From Equation 5-13 we have

$$[CO_3^{2-}] = \frac{K_{sp}}{[Pb^{2+}]}$$

Substituting for $[CO_3^{2-}]$ in Equation 5-22 yields

$$[Pb^{2+}] = \frac{K_{sp}}{[Pb^{2+}]} + \sqrt{\frac{K_w K_{sp}}{K_2[Pb^{2+}]}}$$

It is convenient to multiply through by $[Pb^{2+}]$; upon rearranging terms,

$$[Pb^{2+}]^2 - \sqrt{\frac{K_w}{K_2}K_{sp}[Pb^{2+}]} - K_{sp} = 0$$

Finally, after numerical values have been supplied for the constants, we obtain

$$[Pb^{2+}]^2 - 2.65 \times 10^{-9}[Pb^{2+}]^{1/2} - 3.3 \times 10^{-14} = 0$$

This equation is readily solved by systematic approximations. If, for example, we let $[Pb^{2+}] = 0$, the left side of the equation has a value of -3.3×10^{-14}. On the other hand, when $[Pb^{2+}] = 1 \times 10^{-5}$, the equation yields a value of 9×10^{-11}. That is,

$$(1 \times 10^{-5})^2 - (2.65 \times 10^{-9})(1 \times 10^{-5})^{1/2} - (3.3 \times 10^{-14}) = 9 \times 10^{-11}$$

Substituting $[Pb^{2+}] = 1 \times 10^{-6}$ gives

$$(1 \times 10^{-6})^2 - (2.65 \times 10^{-9})(1 \times 10^{-6})^{1/2} - (3.3 \times 10^{-14}) = -2 \times 10^{-12}$$

We see then that $[Pb^{2+}]$ must lie between 1×10^{-6} and 1×10^{-5}. The trial value $[Pb^{2+}] = 5 \times 10^{-6}$ yields 1.9×10^{-11}. The positive sign here indicates that the assumed value is too high. Further approximations reveal that

$$[Pb^{2+}] = 1.9 \times 10^{-6}$$

$$\text{solubility of } PbCO_3 = 1.9 \times 10^{-6} \text{ fw/liter}$$

To check the two assumptions that were made, we must calculate the concentrations of most of the other ions in the solution. We can evaluate $[CO_3^{2-}]$ from Equation 5-13:

$$[CO_3^{2-}] = \frac{3.3 \times 10^{-14}}{1.9 \times 10^{-6}} = 1.7 \times 10^{-8}$$

From Equation 5-19

$$[HCO_3^-] = 1.9 \times 10^{-6} - 1.7 \times 10^{-8}$$
$$\cong 1.9 \times 10^{-6}$$

From Equation 5-21

$$[OH^-] = [HCO_3^-] = 1.9 \times 10^{-6}$$

From Equation 5-15

$$\frac{[H_2CO_3][1.9 \times 10^{-6}]}{[1.9 \times 10^{-6}]} = 2.25 \times 10^{-8}$$

$$[H_2CO_3] = 2.2 \times 10^{-8}$$

Finally from Equation 5-16

$$[H_3O^+] = \frac{1.00 \times 10^{-14}}{1.9 \times 10^{-6}} = 5.3 \times 10^{-9}$$

We see that the assumptions should not lead to large errors; $[H_2CO_3]$ is approximately $\frac{1}{90}$ of $[HCO_3^-]$, and $[H_3O^+]$ is clearly much smaller than the sum of the concentrations of the two carbonate species in Equation 5-18.

Finally, failure to take account of the basic reaction of CO_3^{2-} would have yielded a solubility of 1.8×10^{-7}, which is only about one-tenth the value yielded by the more rigorous method.

Example. Calculate the solubility of silver sulfide in pure water; the pertinent equilibria are:

$$Ag_2S(s) \rightleftarrows 2Ag^+ + S^{2-} \tag{5-23}$$
$$S^{2-} + H_2O \rightleftarrows HS^- + OH^- \tag{5-24}$$
$$HS^- + H_2O \rightleftarrows H_2S + OH^- \tag{5-25}$$
$$2H_2O \rightleftarrows H_3O^+ + OH^- \tag{5-26}$$

The solubility can be expressed as follows:

$$\text{solubility} = \tfrac{1}{2}[Ag^+] = [S^{2-}] + [HS^-] + [H_2S]$$

Equilibrium-constant expressions are:

$$[Ag^+]^2[S^{2-}] = 6 \times 10^{-50} \tag{5-27}$$

$$\frac{[HS^-][OH^-]}{[S^{2-}]} = \frac{K_w}{K_2} = \frac{1.0 \times 10^{-14}}{1.2 \times 10^{-15}} = 8.3 \tag{5-28}$$

$$\frac{[H_2S][OH^-]}{[HS^-]} = \frac{K_w}{K_1} = \frac{1.0 \times 10^{-14}}{5.7 \times 10^{-8}} = 1.8 \times 10^{-7} \tag{5-29}$$

Mass- and charge-balance expressions are:

$$\tfrac{1}{2}[Ag^+] = [S^{2-}] + [HS^-] + [H_2S] \tag{5-30}$$

$$[Ag^+] + [H_3O^+] = 2[S^{2-}] + [HS^-] + [OH^-] \tag{5-31}$$

The solubility product for Ag_2S is very small; it seems probable that little alteration in the hydroxide ion concentration of the solution will occur as the precipitate dissolves. Therefore, let us assume that at equilibrium,

$$[OH^-] \cong [H_3O^+] = 1.0 \times 10^{-7}$$

This assumption will be correct, provided

$$[Ag^+] \ll [H_3O^+] \quad \text{and} \quad (2[S^{2-}] + [HS^-]) \ll [OH^-]$$

in Equation 5-31.

Substitution of 1.0×10^{-7} for $[OH^-]$ in Equations 5-28 and 5-29 gives

$$\frac{[HS^-]}{[S^{2-}]} = \frac{1.0 \times 10^{-14}}{1.2 \times 10^{-15} \times 10^{-7}} = 8.3 \times 10^7$$

$$\frac{[H_2S]}{[HS^-]} = \frac{1.0 \times 10^{-14}}{5.7 \times 10^{-8} \times 10^{-7}} = 1.8$$

When these relationships are substituted into Equation 5-30, we obtain

$$\tfrac{1}{2}[Ag^+] = [S^{2-}] + 8.3 \times 10^7[S^{2-}] + 14.9 \times 10^7[S^{2-}]$$

$$[S^{2-}] = 2.16 \times 10^{-9}[Ag^+]$$

Substituting this relationship into the solubility-product expression gives

$$2.16 \times 10^{-9}[Ag^+]^3 = 6 \times 10^{-50}$$

$$[Ag^+] = 3.0 \times 10^{-14}$$

$$\text{solubility} = \tfrac{1}{2}[Ag^+] = 1.5 \times 10^{-14} \text{ fw/liter}$$

The assumption that $[Ag^+]$ is much smaller than $[H_3O^+]$ is clearly valid. We can readily calculate a value for $(2[S^{2-}] + [HS^-])$ and confirm that this sum is likewise much smaller than $[OH^-]$. Therefore, we conclude that the assumptions made were reasonable and that the approximate solution obtained is satisfactory.

SOLUBILITY OF METAL HYDROXIDES IN WATER

In determining the solubility of metal hydroxides, two equilibria may have to be considered. For example, with the divalent metal ion M^{2+}, these are

$$M(OH)_2(s) \rightleftarrows M^{2+} + 2OH^-$$

$$2H_2O \rightleftarrows H_3O^+ + OH^-$$

Three algebraic equations are readily derived for this system, namely,

$$[M^{2+}][OH^-]^2 = K_{sp} \tag{5-32}$$

$$[H_3O^+][OH^-] = K_w \tag{5-33}$$

and from charge-balance considerations

$$2[M^{2+}] + [H_3O^+] = [OH^-] \tag{5-34}$$

If the hydroxide is reasonably soluble, the hydronium ion concentration will be small, and Equation 5-34 becomes

$$2[M^{2+}] \cong [OH^-]$$

Substitution of this expression into (5-32) and rearrangement gives

$$[M^{2+}] = \left(\frac{K_{sp}}{4}\right)^{1/3} = \text{solubility} \tag{5-35}$$

On the other hand, if the solubility of $M(OH)_2$ is very low, the situation is encountered in which $2[M^{2+}]$ is much smaller than $[H_3O^+]$. Equation 5-34 then becomes

$$[H_3O^+] \cong [OH^-] = 1.00 \times 10^{-7}$$

Again, substitution into (5-32) and rearrangement yields

$$[M^{2+}] = \frac{K_{sp}}{[OH^-]^2} = \frac{K_{sp}}{1.00 \times 10^{-14}} = \text{solubility} \tag{5-36}$$

Example. Calculate the solubility of $Fe(OH)_3$ in water.

As a hypothesis, let us assume that the charge-balance expression simplifies to

$$3[Fe^{3+}] + [H_3O^+] \cong 3[Fe^{3+}] = [OH^-]$$

Substitution for $[OH^-]$ into the solubility-product expression gives

$$[Fe^{3+}](3[Fe^{3+}])^3 = 4 \times 10^{-38}$$

$$[Fe^{3+}] = \left(\frac{4 \times 10^{-38}}{27}\right)^{1/4} = 2 \times 10^{-10}$$

and

$$\text{solubility} = 2 \times 10^{-10} \text{ fw/liter}$$

We have assumed, however, that

$$[OH^-] \cong 3[Fe^{3+}] = 3 \times 2 \times 10^{-10} = 6 \times 10^{-10}$$

which means that

$$[H_3O^+] = \frac{1.00 \times 10^{-14}}{6 \times 10^{-10}} = 1.7 \times 10^{-5}$$

Clearly $[H_3O^+]$ is not much smaller than $3[Fe^{3+}]$; indeed, the reverse appears to be the case. That is,

$$3[Fe^{3+}] \ll [H_3O^+]$$

and the charge-balance equation reduces to

$$[H_3O^+] = [OH^-] = 1.00 \times 10^{-7}$$

Substitution for $[OH^-]$ in the solubility-product expression yields

$$[Fe^{3+}] = \frac{4 \times 10^{-38}}{(1.00 \times 10^{-7})^3} = 4 \times 10^{-17}$$

$$\text{solubility} = 4 \times 10^{-17} \frac{\text{fw Fe(OH)}_3}{\text{liter}}$$

The assumption that $3[Fe^{3+}] \ll [H_3O^+]$ is clearly valid. Note the very large error in the first calculation where the faulty assumption was employed.

TABLE 5-2 Relative Errors Associated with Approximate Calculations for the Solubility of a Precipitate $M(OH)_2$

Assumed K_{sp}	Solubility Calculated without Approximations	Solubility Calculated with Equation 5-35	Percent Error with Equation 5-35	Solubility Calculated with Equation 5-36	Percent Error with Equation 5-36
1.00×10^{-18}	6.3×10^{-7}	6.3×10^{-7}	0	1.00×10^{-4}	1.6×10^4
1.00×10^{-20}	1.24×10^{-7}	1.36×10^{-7}	9.7	1.00×10^{-6}	7.1×10^2
1.00×10^{-22}	8.4×10^{-9}	2.92×10^{-8}	2.5×10^2	1.00×10^{-8}	1.9×10^1
1.00×10^{-24}	1.00×10^{-10}	6.3×10^{-9}	6.2×10^3	1.00×10^{-10}	0.00
1.00×10^{-26}	1.00×10^{-12}	1.36×10^{-9}	1.4×10^5	1.00×10^{-12}	0.00

From the foregoing example, it is apparent that solubility calculations for metal hydroxides are analogous to those for compounds containing a weak conjugate base in the sense that they frequently can be made simpler by the proper choice of one of two assumptions. It is to be expected that a range of solubility products exists for which neither assumption is valid and for which Equations 5-32, 5-33, and 5-34 must be solved for all three variables. Table 5-2 shows the extent of this range for precipitates of the type $M(OH)_2$.

COMPLEX ION FORMATION AND SOLUBILITY

The solubility of a precipitate may be greatly altered in the presence of some species that forms a soluble complex with the anion or cation of the precipitate. For example, the precipitation of aluminum with base is never complete in the presence of fluoride ion, even though aluminum hydroxide has an extremely low solubility; the fluoride complexes of aluminum(III) are sufficiently stable to prevent quantitative removal of the cation from solution. The equilibria involved can be represented by

$$Al(OH)_3(s) \rightleftarrows Al^{3+} + 3OH^-$$
$$+$$
$$6F^-$$
$$\rightleftarrows$$
$$AlF_6^{3-}$$

Fluoride ions thus compete successfully with hydroxide ions for aluminum(III); as the fluoride concentration is increased, more and more of the precipitate is dissolved and converted to fluoroaluminate ions.

Quantitative Treatment of the Effect of Complex Formation on the Solubility of Precipitates. The solubility of a precipitate in the presence of a complexing reagent can be calculated, provided the equilibrium constant for the complex-formation reaction is known. The techniques used are similar to those discussed in the preceding section.

Example. Find the solubility of AgBr in a solution that is 0.10 F in NH_3.
Equilibria:

$$AgBr(s) \rightleftarrows Ag^+ + Br^-$$
$$Ag^+ + NH_3 \rightleftarrows AgNH_3^+$$
$$AgNH_3^+ + NH_3 \rightleftarrows Ag(NH_3)_2^+$$
$$NH_3 + H_2O \rightleftarrows NH_4^+ + OH^-$$

Definition of unknown:

$$\text{solubility of AgBr} = [Br^-]$$
$$= [Ag^+] + [AgNH_3^+] + [Ag(NH_3)_2^+]$$

Equilibrium constants:

$$[Ag^+][Br^-] = K_{sp} = 5.2 \times 10^{-13} \tag{5-37}$$

$$\frac{[AgNH_3^+]}{[Ag^+][NH_3]} = K_1 = 2.0 \times 10^3 \tag{5-38}$$

$$\frac{[Ag(NH_3)_2^+]}{[AgNH_3^+][NH_3]} = K_2 = 6.9 \times 10^3 \tag{5-39}$$

$$\frac{[NH_4^+][OH^-]}{[NH_3]} = K_b = 1.76 \times 10^{-5} \tag{5-40}$$

Mass-balance expressions:

$$[Br^-] = [Ag^+] + [AgNH_3^+] + [Ag(NH_3)_2^+] \tag{5-41}$$

Since the NH_3 concentration was initially 0.10, we may also write

$$0.10 = [NH_3] + [AgNH_3^+] + 2[Ag(NH_3)_2^+] + [NH_4^+] \tag{5-42}$$

Furthermore, the reaction of NH_3 with water produces one OH^- for each NH_4^+. Thus,

$$[OH^-] \cong [NH_4^+] \tag{5-43}$$

Charge-balance equation:

$$[NH_4^+] + [Ag^+] + [AgNH_3^+] + [Ag(NH_3)_2^+] = [Br^-] + [OH^-] \tag{5-44}[3]$$

Close examination of these eight equations reveals that there are only seven independent expressions, since Equation 5-44 is the sum of Equations 5-43 and 5-41. There are only seven unknowns, however, so a solution is possible.

---

[3] We have neglected the $[H_3O^+]$, since its concentration will certainly be negligible in a 0.10-F solution of NH_3.

Approximations: (a) $[NH_4^+]$ is much smaller than the other terms in Equation 5-42. This assumption seems reasonable in light of the rather small numerical value of the dissociation constant for NH_3 (Equation 5-40).

(b) $[Ag(NH_3)_2^+] \gg [AgNH_3^+]$ and $[Ag^+]$. An examination of the constants for Equations 5-38 and 5-39 suggests that this assumption is reasonable except for very dilute solutions of NH_3.

Application of these approximations leads to the simplified equations

$$[Br^-] \cong [Ag(NH_3)_2^+] \qquad (5-45)$$

$$[NH_3] \cong 0.10 - 2[Ag(NH_3)_2^+] \qquad (5-46)$$

Upon substituting Equation 5-45 into Equation 5-46, we obtain

$$[NH_3] = 0.10 - 2[Br^-] \qquad (5-47)$$

Let us now multiply Equations 5-38 and 5-39 together to give

$$\frac{[Ag(NH_3)_2^+]}{[Ag^+][NH_3]^2} = K_1K_2 = 1.38 \times 10^7 \qquad (5-48)$$

When Equations 5-47 and 5-45 are introduced into this equation we obtain

$$\frac{[Br^-]}{[Ag^+](0.1 - 2[Br^-])^2} = 1.38 \times 10^7$$

Now, replacing the $[Ag^+]$ in this equation by the equivalent quantity from Equation 5-37, we have

$$\frac{[Br^-]}{(5.2 \times 10^{-13}/[Br^-])(0.1 - 2[Br^-])^2} = 1.38 \times 10^7$$

or

$$\frac{[Br^-]^2}{(0.1 - 2[Br^-])^2} = 7.2 \times 10^{-6}$$

This expression can be rearranged to yield the quadratic equation

$$[Br^-]^2 + 2.88 \times 10^{-6}[Br^-] - 7.2 \times 10^{-8} = 0$$

Thus,

$$[Br^-] = 2.7 \times 10^{-4}$$

$$\text{solubility} = \frac{2.7 \times 10^{-4} \text{ fw AgBr}}{\text{liter}}$$

A check of the assumptions will show that they were valid.

Complex Formation Involving a Common Ion of the Precipitate. Many precipitates tend to react with one of their constituent ions to form soluble complexes. For example, silver chloride forms chloro complexes believed to be of the composition $AgCl_2^-$, $AgCl_3^{2-}$, and so on. Such reactions cause increases in solubility at high concentrations of the common ion. This effect is illustrated by Figure 5-1, where the experimentally determined solubility of silver chloride is plotted against the logarithm of the potassium chloride concentration in the solution. At chloride concentrations less than $10^{-3} F$, the experimental solubilities do not differ greatly from those calculated with the solubility product for

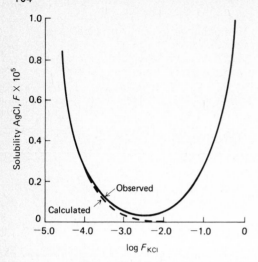

Figure 5-1 The Solubility of Silver Chloride in Potassium Chloride Solutions. The dashed curve is calculated from K_{sp}; the solid curve represents experimental values obtained by A. Pinkus and A. M. Timmermans, *Bull. Soc. Chim. Belges,* **46,** 46–73 (1937).

silver chloride. At higher chloride ion concentrations, however, the calculated solubilities approach zero while the measured values rise precipitously; in about 0.3-*F* potassium chloride, the solubility of silver chloride is the same as in pure water, and in 1-*F* solution it is approximately eight times this figure. If complete information were available regarding the composition of the complexes and their formation constants, a quantitative description of these effects should be possible (see Problem 19 in this chapter).

Solubility increases in the presence of large excesses of a common ion are by no means rare. Of particular interest are the amphoteric hydroxides such as those of aluminum and zinc, which form slightly soluble precipitates upon treatment with base; these redissolve in the presence of excess hydroxide ions to give the complex aluminate and zincate ions. For aluminum, the equilibria can be represented as

$$Al^{3+} + 3OH^- \rightleftarrows Al(OH)_3(s)$$
$$Al(OH)_3(s) + OH^- \rightleftarrows Al(OH)_4^-$$

As with silver chloride, the solubilities of aluminum hydroxide and zinc hydroxide pass through minima and then increase rapidly with increasing concentrations of the common ion. The hydroxide ion concentration corresponding to the minimum solubility can be calculated readily if the equilibrium constants for the reactions are known.

Example. At what [OH$^-$] concentration is the solubility of Zn(OH)$_2$ a minimum? Calculate the solubility at this minimum.
Equilibria:

$$Zn(OH)_2(s) \rightleftarrows Zn^{2+} + 2OH^-$$
$$Zn(OH)_2(s) + 2OH^- \rightleftarrows Zn(OH)_4^{2-}$$

Let s be the formal solubility of $Zn(OH)_2$. Then

$$s = [Zn^{2+}] + [Zn(OH)_4{}^{2-}] \tag{5-49}$$

Equilibrium constants:

$$K_{sp} = [Zn^{2+}][OH^-]^2 = 1.2 \times 10^{-17} \tag{5-50}$$

$$K_f = \frac{[Zn(OH)_4{}^{2-}]}{[OH^-]^2} = 0.13 \tag{5-51}$$

Substitution of Equations 5-50 and 5-51 into 5-49 yields

$$s = \frac{K_{sp}}{[OH^-]^2} + K_f[OH^-]^2 \tag{5-52}$$

The minimum solubility can be obtained by differentiating Equation 5-52 and setting the derivative of s with respect to $[OH^-]$ equal to zero. Thus,

$$\frac{ds}{d[OH^-]} = -\frac{2K_{sp}}{[OH^-]^3} + 2K_f[OH^-]$$

When $ds/d[OH^-] = 0$,

$$\frac{2K_{sp}}{[OH^-]^3} = 2K_f[OH^-]$$

or

$$[OH^-] = \left(\frac{2K_{sp}}{2K_f}\right)^{1/4} = \left(\frac{1.2 \times 10^{-17}}{0.13}\right)^{1/4}$$

$$= 9.8 \times 10^{-5}$$

The minimum solubility is obtained by substituting the calculated hydroxide ion concentration into Equation 5-52.

$$\text{minimum } s = \frac{1.2 \times 10^{-17}}{(9.8 \times 10^{-5})^2} + 0.13(9.8 \times 10^{-5})^2$$

$$= \frac{2.5 \times 10^{-9} \text{ fw } Zn(OH)_2}{\text{liter}}$$

SEPARATION OF IONS BY CONTROL OF THE CONCENTRATION OF THE PRECIPITATION REAGENT

When two ions react with a third to form precipitates of different solubilities, the less soluble species will precipitate at a lower reagent concentration. If the solubilities are sufficiently different, quantitative removal of the first ion from solution may be achieved without precipitation of the second. Such separations require careful control of the concentration of the precipitating reagent at some suitable, predetermined level. A number of important analytical separations, notably those employing sulfide ion, hydroxide ion, and organic reagents, are based on this method.

Calculation of the Feasibility of Separations. An important application of solubility-product calculations involves determining the feasibility and the optimum conditions for separations based on the control of reagent concentration. The following problem illustrates such an application.

Example.   Is it theoretically possible to separate Fe^{3+} and Mg^{2+} quantitatively from each other by differential precipitation with OH^- from a solution that is $0.10\,F$ in each cation? If the separation is possible, what range of OH^- concentrations is permissible? Solubility-product constants for the two hydroxides are

$$[Fe^{3+}][OH^-]^3 = 4 \times 10^{-38}$$
$$[Mg^{2+}][OH^-]^2 = 1.8 \times 10^{-11}$$

The K_{sp} for $Fe(OH)_3$ is so much smaller than that for $Mg(OH)_2$ as to suggest that the former will precipitate at a lower OH^- concentration.

We can answer the questions posed in this problem by (1) calculating the OH^- concentration required to effect the quantitative precipitation of Fe^{3+} from this solution and by (2) determining the OH^- concentration at which $Mg(OH)_2$ will just begin to precipitate. If (1) is smaller than (2), a separation is feasible, and the range of OH^- concentrations to be used will be defined by the values obtained for (1) and (2).

In order to determine (1), we must first decide what constitutes a quantitative removal of Fe^{3+} from the solution. Under no condition can every iron(III) ion be precipitated; we must, therefore, arbitrarily set some limit below which, for all practical purposes, the further presence of this ion can be neglected. When its concentration has been reduced to $10^{-6}\,M$, only $\frac{1}{100,000}$ of the original quantity of Fe^{3+} will remain in the solution; for most purposes, removal of all but this fraction of an ion can be considered a quantitative separation.

We can readily calculate the OH^- concentration in equilibrium with $1.0 \times 10^{-6}\,M$ Fe^{3+} by substituting directly into the solubility-product expression:

$$(1.0 \times 10^{-6})[OH^-]^3 = 4 \times 10^{-38}$$
$$[OH^-] = 3.4 \times 10^{-11}$$

Thus, if we maintain the OH^- concentration at 3.4×10^{-11} mole/liter, the Fe^{3+} concentration will be lowered to 1.0×10^{-6} mole/liter. It is of interest to note that quantitative precipitation of $Fe(OH)_3$ is achieved in a distinctly acidic solution.

We must now consider question (2)—that is, what is the maximum OH^- concentration that can exist in solution without causing formation of $Mg(OH)_2$? Precipitation cannot occur until the Mg^{2+} concentration multiplied by the square of the OH^- concentration exceeds the solubility product, 1.8×10^{-11}. By substituting 0.1 (the molar Mg^{2+} concentration of the solution) into the solubility-product expression, we can calculate the *maximum* OH^- concentration that can be tolerated without formation of $Mg(OH)_2$:

$$0.10[OH^-]^2 = 1.8 \times 10^{-11}$$
$$[OH^-] = 1.3 \times 10^{-5}$$

When the OH^- concentration exceeds this level, the solution will be supersaturated with respect to $Mg(OH)_2$, and precipitation can begin.

From these calculations we conclude that quantitative separation of $Fe(OH)_3$ can be expected if the OH^- concentration is greater than 3.4×10^{-11} mole/liter, and that $Mg(OH)_2$ will not precipitate until a concentration of 1.3×10^{-5} mole/liter is reached. Therefore, it should be possible to separate Fe^{3+} from Mg^{2+} by maintaining the OH^- concentration between these levels.

Sulfide Separations. A number of important methods for the separation of metallic ions involve controlling the concentration of the precipitating anion by regulating the hydronium ion concentration of the solution. Such methods are particularly attractive because of the relative ease with which the hydronium ion concentration may be maintained at some predetermined level by the use of a suitable buffer.[4] Perhaps the best known of these methods makes use of hydrogen sulfide as the precipitating reagent. Hydrogen sulfide is a weak acid, dissociating as follows:

$$H_2S + H_2O \rightleftarrows H_3O^+ + HS^- \qquad K_1 = \frac{[H_3O^+][HS^-]}{[H_2S]} = 5.7 \times 10^{-8}$$

$$HS^- + H_2O \rightleftarrows H_3O^+ + S^{2-} \qquad K_2 = \frac{[H_3O^+][S^{2-}]}{[HS^-]} = 1.2 \times 10^{-15}$$

These equations may be combined to give an expression for the overall dissociation of hydrogen sulfide into sulfide ion:

$$H_2S + 2H_2O \rightleftarrows 2H_3O^+ + S^{2-} \qquad K_1K_2 = \frac{[H_3O^+]^2[S^{2-}]}{[H_2S]} = 6.8 \times 10^{-23}$$

The constant for this reaction is simply the product of K_1 and K_2.

In sulfide separations, the solutions are ordinarily kept saturated by continuously bubbling hydrogen sulfide through them until precipitation is complete; thus, the formal concentration of the reagent is essentially constant throughout the precipitation. Since it is such a weak acid, the actual molar concentration of hydrogen sulfide will correspond closely to its solubility in water, which is about 0.1 F. For practical purposes, then, we may assume that throughout any sulfide precipitation

$$[H_2S] \cong 0.10 \text{ mole/liter}$$

Substituting this value into the dissociation-constant expression, we obtain

$$\frac{[H_3O^+]^2[S^{2-}]}{0.10} = 6.8 \times 10^{-23}$$

$$[S^{2-}] = \frac{6.8 \times 10^{-24}}{[H_3O^+]^2}$$

Thus, the molar concentration of the sulfide ion varies inversely as the square of the hydronium ion concentration of the solution. This relationship is useful for calculating the optimum conditions for the separation of cations by sulfide precipitation.

Example. Find the conditions under which Pb^{2+} and Tl^+ can be separated quantitatively by H_2S precipitation from a solution that is 0.1 F in each cation.

The equilibrium constants for the two important reactions are:

$$PbS(s) \rightleftarrows Pb^{2+} + S^{2-} \qquad [Pb^{2+}][S^{2-}] = 7 \times 10^{-28}$$

$$Tl_2S(s) \rightleftarrows 2Tl^+ + S^{2-} \qquad [Tl^+]^2[S^{2-}] = 1 \times 10^{-22}$$

---

[4] The preparation and properties of buffer solutions are considered in Chapter 9. An important property of a buffer is that it maintains the hydronium ion concentration at an approximately fixed and predetermined level.

PbS will precipitate at a lower S^{2-} concentration than the Tl_2S. Assuming again that lowering the Pb^{2+} concentration to $10^{-6}\ M$ or less constitutes quantitative removal, and substituting this value into the solubility-product expression, we can evaluate the required sulfide ion concentration

$$10^{-6}[S^{2-}] = 7 \times 10^{-28}$$
$$[S^{2-}] = 7 \times 10^{-22}$$

This value should then be compared with the S^{2-} concentration needed to initiate precipitation of Tl_2S from a 0.1-F solution:

$$(0.1)^2[S^{2-}] = 1 \times 10^{-22}$$
$$[S^{2-}] = 1 \times 10^{-20}$$

Thus, to achieve a separation the S^{2-} concentration should be kept between 7×10^{-22} and 1×10^{-20} mole/liter. Now we must compute the H_3O^+ concentrations necessary to hold the S^{2-} concentration within these confines. Using the relationship derived previously,

$$[S^{2-}] = \frac{6.8 \times 10^{-24}}{[H_3O^+]^2}$$

and substituting the two limiting values for S^{2-} concentration, we obtain

$$[H_3O^+]^2 = \frac{6.8 \times 10^{-24}}{7 \times 10^{-22}} = 0.97 \times 10^{-2}$$

$$[H_3O^+] = 0.098 \cong 0.1$$

and

$$[H_3O^+]^2 = \frac{6.8 \times 10^{-24}}{1 \times 10^{-20}}$$

$$[H_3O^+] = 0.026 \cong 0.03$$

By maintaining the H_3O^+ concentration between 0.03 and 0.1 M, it should, in theory, be possible to separate PbS without precipitation of Tl_2S. From the practical standpoint, on the other hand, it is doubtful that the acidity could be controlled closely enough to give a very satisfactory separation.

EFFECT OF ELECTROLYTE CONCENTRATION ON SOLUBILITY

It is found experimentally that precipitates are generally more soluble in an electrolyte solution than in water provided, of course, that the electrolyte contains no ions in common with the precipitate. The data plotted in Figure 5-2 demonstrate the magnitude of this effect for three precipitates. A twofold increase in the solubility of barium sulfate is observed when the potassium nitrate concentration of the solvent is increased from 0 to 0.02 F. The same change in electrolyte concentration increases the solubility of barium iodate by a factor of only 1.25 and of silver chloride by 1.20.

The effect of an electrolyte stems from the electrostatic attraction between the foreign ions and the ions of opposite charge in the precipitate. Such interactions cause a shift in the position of the solubility equilibrium. It is important to realize that this effect is not peculiar to solubility equilibria but is observed

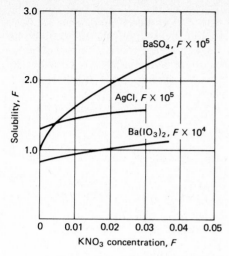

Figure 5-2 Effect of Electrolyte Concentration on the Solubility of Some Salts.

with all other types as well. For example, the data in Table 5-3 show that the degree of dissociation of acetic acid increases significantly in the presence of sodium chloride. These experimental dissociation constants were obtained by measuring the equilibrium concentrations of hydronium and acetate ions in solutions containing the indicated salt concentrations. Again, the obvious shift in equilibrium can be attributed to the attraction of the ions of the electrolyte for the charged hydronium and acetate ions.

From data such as these, one must conclude that the equilibrium law, as we have presented it, is a *limiting law* in the sense that it applies exactly only to very dilute solutions in which the electrolyte concentration is insignificant (that is, to ideal solutions only; see p. 24). We must now consider a more rigorous form of the law which applies to nonideal solutions.

TABLE 5-3 **Dissociation Constants for Acetic Acid in Solutions of Sodium Chloride at 25°C[a]**

Concentration of NaCl, F	Apparent K'_a
0.00	1.75×10^{-5}
0.02	2.29×10^{-5}
0.11	2.85×10^{-5}
0.51	3.31×10^{-5}
1.01	3.16×10^{-5}

[a] From H. S. Harned and C. F. Hickey, *J. Amer. Chem. Soc.* **59**, 1289 (1937). With permission of the American Chemical Society.

Some Empirical Observations. Extensive studies concerned with the influence of electrolyte concentration upon chemical equilibrium have led to a number of important generalizations. One is that the magnitude of the effect is highly dependent upon the charges of the species involved in the equilibrium. Where all are neutral particles, little variation in the equilibrium constant is observed. On the other hand, the effects become greater as the charges on the reactants or products increase. Thus, for example, of the two equilibria

$$AgCl(s) \rightleftarrows Ag^+ + Cl^-$$
$$BaSO_4(s) \rightleftarrows Ba^{2+} + SO_4^{2-}$$

the second is shifted farther to the right in the presence of moderate amounts of potassium nitrate than is the first (see Figure 5-2).

A second important generality is that over a considerable electrolyte concentration range, the effects are essentially independent of the kind of electrolyte and dependent only upon a concentration parameter of the solution, called the *ionic strength*. This quantity is defined by the equation

$$\text{ionic strength} = \mu = \tfrac{1}{2}(m_1Z_1^2 + m_2Z_2^2 + m_3Z_3^2 + \cdots) \qquad (5\text{-}53)$$

where $m_1, m_2, m_3, \ldots,$ represent the molar concentrations of the various ions in the solution, and $Z_1, Z_2, Z_3, \ldots,$ are their respective charges.

Example. What is the ionic strength of a 0.1-F solution of KNO_3? Of a 0.1-F solution of Na_2SO_4?

For the KNO_3 solution, m_{K^+} and $m_{NO_3^-}$ are 0.1, and

$$\mu = \tfrac{1}{2}(0.1 \times 1^2 + 0.1 \times 1^2) = 0.1$$

For the Na_2SO_4 solution, $m_{Na^+} = 0.2$ and $m_{SO_4^{2-}} = 0.1$. Therefore,

$$\mu = \tfrac{1}{2}(0.2 \times 1^2 + 0.1 \times 2^2) = 0.3$$

Example. What is the ionic strength of a solution that is both 0.05 F in KNO_3 and 0.1 F in Na_2SO_4?

$$\mu = \tfrac{1}{2}(0.05 \times 1^2 + 0.05 \times 1^2 + 0.2 \times 1^2 + 0.1 \times 2^2)$$
$$= 0.35$$

From these examples it is apparent that the ionic strength of a strong electrolyte solution consisting solely of singly charged ions is identical with the total formal salt concentration. If the species carry multiple charges, however, the ionic strength is greater than the formal concentration.

For solutions with ionic strengths of 0.1 or less, it is found that the electrolyte effect is independent of the kind of ions and dependent only upon the ionic strength. Thus, the degree of dissociation of acetic acid is the same in the presence of sodium chloride, potassium nitrate, or barium iodide, provided the concentrations of these species are such that the ionic strength is fixed. It should be noted that this independence with respect to electrolyte species disappears at high ionic strength.

Activity and Activity Coefficients. In order to describe the effect of ionic strength on equilibria in quantitative terms, chemists use a concentration parameter called the *activity*, which is defined as follows:

$$a_A = [A] f_A \qquad (5\text{-}54)$$

where a_A is the activity of the species A, $[A]$ is its molar concentration, and f_A is a dimensionless quantity called the *activity coefficient*. The activity coefficient (and thus the activity) of A varies with ionic strength such that employment of a_A instead of $[A]$ in an equilibrium-constant expression frees the numerical value of the constant from dependence on the ionic strength. To illustrate, for the dissociation of acetic acid, we write

$$K_a = \frac{a_{H_3O^+} \cdot a_{OAc^-}}{a_{HOAc}} = \frac{[H_3O^+][OAc^-]}{[HOAc]} \times \frac{f_{H_3O^+} \cdot f_{OAc^-}}{f_{HOAc}}$$

where $f_{H_3O^+}$, f_{OAc^-}, and f_{HOAc} vary with ionic strength to keep K_a numerically constant over a wide range of ionic strengths (in contrast to the *apparent* K_a' shown in Table 5-3).

Properties of Activity Coefficients. Activity coefficients have the following properties:

1. The activity coefficient of a species can be thought of as a measure of the effectiveness with which that species influences an equilibrium in which it is a participant. In very dilute solutions, where the ionic strength is minimal, this effectiveness becomes constant, and the activity coefficient acquires a value of unity. Under such circumstances, the activity and molar concentrations become identical. As the ionic strength increases, however, an ion loses some of its effectiveness, and its activity coefficient decreases. We may summarize this behavior in terms of Equation 5-54. At moderate ionic strengths, $f_A < 1$; as the solution approaches infinite dilution, however, $f_A \rightarrow 1$ and thus $a_A \rightarrow [A]$.

    At high ionic strengths, the activity coefficients for some species increase and may even become greater than one. Interpretation of the behavior of solutions in this region is difficult; we shall confine most of our discussion to regions of low or moderate ionic strengths (that is, where $\mu < 0.1$).

    The variation of typical activity coefficients as a function of ionic strength is shown in Figure 5-3.

2. In solutions that are not too concentrated, the activity coefficient for a given species is independent of the specific nature of the electrolyte and dependent only upon the ionic strength.

3. For a given ionic strength, the activity coefficient of an ion departs farther from unity as the charge carried by the species increases. This effect is shown in Figure 5-3. The activity coefficient of an uncharged molecule is approximately one regardless of ionic strength.

4. For ions of the same charge, activity coefficients are approximately the same at any given ionic strength. The small variations that do exist can be correlated with the effective diameter of the hydrated ions.

5. The activity coefficient of a given ion describes its effective behavior in all equilibria in which it participates. For example, at a given ionic strength, a single

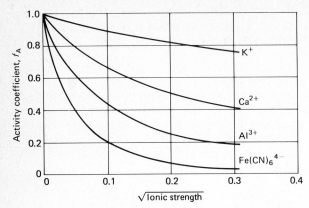

Figure 5-3   Effect of Ionic Strength on Activity Coefficients.

activity coefficient for cyanide ion describes the influence of that species upon any of the following equilibria:

$$HCN + H_2O \rightleftarrows H_3O^+ + CN^-$$

$$AgCN(s) \rightleftarrows Ag^+ + CN^-$$

$$Ni(CN)_4{}^{2-} \rightleftarrows Ni^{2+} + 4CN^-$$

Evaluation of Activity Coefficients. In 1923 P. Debye and E. Hückel derived the following theoretical expression, which permits the calculation of activity coefficients of ions.[5]

$$-\log f_A = \frac{0.5085 Z_A{}^2 \sqrt{\mu}}{1 + 0.3281 a_A \sqrt{\mu}} \tag{5-55}$$

where

f_A = activity coefficient of the species A

Z_A = charge on the species A

μ = ionic strength of the solution

a_A = the effective diameter of the hydrated ion Å in angström units (1 angström unit = 10^{-8} cm)

The constants 0.5085 and 0.3281 are applicable to solutions at 25°C; other values must be employed at different temperatures.

Unfortunately, considerable uncertainty exists regarding the magnitude of a_A in Equation 5-55. Its value appears to be approximately 3 Å for most singly charged ions so that, for these species, the denominator of the Debye-Hückel equation reduces to approximately $(1 + \sqrt{\mu})$. For ions with higher charge, a_A may become as large as 10 Å. It should be noted that the second term of the denominator becomes small with respect to the first when the ionic strength is less than 0.01; under these circumstances, uncertainties in a_A are of little significance in calculating activity coefficients.

[5] P. Debye and E. Hückel, *Physik. Z.,* **24,** 185 (1923).

TABLE 5-4   Activity Coefficient for Ions at 25°C[a]

Ion	a_A Effective Diameter, A	Activity Coefficient at Indicated Ionic Strengths				
		0.001	0.005	0.01	0.05	0.1
H_3O^+	9	0.967	0.933	0.914	0.86	0.83
Li^+, $C_6H_5COO^-$	6	0.965	0.929	0.907	0.84	0.80
Na^+, IO_3^-, HSO_3^-, HCO_3^-, $H_2PO_4^-$, $H_2AsO_4^-$, OAc^-	4–4.5	0.964	0.928	0.902	0.82	0.78
OH^-, F^-, SCN^-, HS^-, ClO_3^-, ClO_4^-, BrO_3^-. IO_4^-, MnO_4^-	3.5	0.964	0.926	0.900	0.81	0.76
K^+, Cl^-, Br^-, I^-, CN^-, NO_2^-, NO_3^-, $HCOO^-$	3	0.964	0.925	0.899	0.80	0.76
Rb^+, Cs^+, Tl^+, Ag^+, NH_4^+	2.5	0.964	0.924	0.898	0.80	0.75
Mg^{2+}, Be^{2+}	8	0.872	0.755	0.69	0.52	0.45
Ca^{2+}, Cu^{2+}, Zn^{2+}, Sn^{2+}, Mn^{2+}, Fe^{2+}, Ni^{2+}, Co^{2+}, Phthalate^{2-}	6	0.870	0.749	0.675	0.48	0.40
Sr^{2+}, Ba^{2+}, Cd^{2+}, Hg^{2+}, S^{2-}	5	0.868	0.744	0.67	0.46	0.38
Pb^{2+}, CO_3^{2-}, SO_3^{2-}, $C_2O_4^{2-}$	4.5	0.868	0.742	0.665	0.46	0.37
Hg_2^{2+}, SO_4^{2-}, $S_2O_3^{2-}$, CrO_4^{2-}, HPO_4^{2-}	4.0	0.867	0.740	0.660	0.44	0.36
Al^{3+}, Fe^{3+}, Cr^{3+}, La^{3+}, Ce^{3+}	9	0.738	0.54	0.44	0.24	0.18
PO_4^{3-}, $Fe(CN)_6^{3-}$	4	0.725	0.50	0.40	0.16	0.095
Th^{4+}, Zr^{4+}, Ce^{4+}, Sn^{4+}	11	0.588	0.35	0.255	0.10	0.065
$Fe(CN)_6^{4-}$	5	0.57	0.31	0.20	0.048	0.021

[a] From J. Kielland, *J. Amer. Chem. Soc.*, **59**, 1675 (1937). Reproduced with permission.

Kielland[6] has calculated values of a_A for numerous ions from a variety of experimental data. His "best values" for effective diameters are given in Table 5-4. Also presented are activity coefficients calculated from Equation 5-55, using these values for the size parameter.

Experimental verification of individual activity coefficients such as those shown in Table 5-4 is, unfortunately, impossible; all experimental methods give only a mean activity coefficient for the positively and negatively charged ions in a solution.[7] It should be pointed out, however, that *mean* activity coefficients

[6] J. Kielland, *J. Amer. Chem. Soc.*, **59**, 1675 (1937).
[7] The mean activity of the electrolyte A_mB_n is defined as follows:

$$f_\pm = \text{mean activity coefficient} = (f_A{}^m \cdot f_B{}^n)^{1/(m+n)}$$

The mean activity coefficient can be measured in any of several ways, but it is impossible experimentally to resolve this term into the individual activity coefficients for f_A and f_B. For example, if A_mB_n is a precipitate, we can write

$$K_{sp} = [A]^m[B]^n \cdot f_A{}^m \cdot f_B{}^n = [A]^m[B]^n \cdot f_\pm{}^{(m+n)}$$

By measuring the solubility of A_mB_n in a solution in which the electrolyte concentration approaches zero (that is, where f_A and $f_B \to 1$), we could obtain K_{sp}. A second solubility measurement at some ionic strength μ_1 would give values for [A] and [B]. These data would then permit the calculation of $f_A{}^m \cdot f_B{}^n = f_\pm{}^{(m+n)}$ for ionic strength μ_1. It is important to understand that there are insufficient experimental data to permit the calculation of the *individual* quantities f_A and f_B, and that there appears to be no additional experimental information that would permit evaluation of these quantities. This situation is general; the *experimental* determination of individual activity coefficients appears to be impossible.

calculated from the data in Table 5-4 agree satisfactorily with the experimental values.

The Debye-Hückel relationship is generally considered to give good agreement with experiment up to $\mu \sim 0.01$. Beyond this value, however, the equation fails and experimentally determined mean activity coefficients must be employed.

Solubility Calculations Employing Activity Coefficients. The use of activities rather than molar concentrations in equilibrium-constant calculations yields more accurate information. Unless otherwise specified, values for K_{sp} found in tables are generally constants based upon activities (activity-based constants are sometimes called the *thermodynamic constants*). Thus, for the precipitate $A_m B_n$, we may write

$$K_{sp} = a_A{}^m \cdot a_B{}^n = [A]^m[B]^n \cdot f_A{}^m \cdot f_B{}^n$$

or

$$[A]^m[B]^n = \frac{K_{sp}}{f_A{}^m \cdot f_B{}^n}$$

$$= K'_{sp}$$

where the bracketed terms are *molar concentrations* of A and B. By dividing the thermodynamic constant, K_{sp}, by the product of the activity coefficients for A and B (or the mean activity coefficient), one obtains a *concentration constant* K'_{sp} that is applicable to a solution of a particular ionic strength. This constant can then be employed in the equilibrium calculations that were discussed earlier. The following example will demonstrate the procedure.

Example. Use activities to calculate the solubility of $Ba(IO_3)_2$ in a 0.033-F solution of $Mg(IO_3)_2$. The thermodynamic solubility product for $Ba(IO_3)_2$ has a value of 1.57×10^{-9}.

At the outset we may write

$$[Ba^{2+}][IO_3^-]^2 = \frac{1.57 \times 10^{-9}}{f_{Ba^{2+}} \; f_{IO_3^-}^2} = K'_{sp}$$

We must next calculate activity coefficients for Ba^{2+} and IO_3^- ions from the ionic strength of the solution. Thus,

$$\mu = \tfrac{1}{2}[m_{Mg^{2+}} \times (2)^2 + m_{IO_3^-} \times (1)^2]$$

$$= \tfrac{1}{2}(0.033 \times 4 + 0.066 \times 1) = 0.099 \cong 0.1$$

In calculating μ, we have assumed that the Ba^{2+} and IO_3^- ions from the precipitate do not affect the ionic strength of the solution significantly. This simplification seems justified, considering the low solubility of barium iodate. In situations where it is not possible to make the assumption, the concentrations of the two ions can be approximated by an ordinary calculation, assuming activities and concentrations to be identical. These concentrations can then be introduced to give a better value for μ.

Turning now to Table 5-4, we find that at an ionic strength of 0.1,

$$f_{Ba^{2+}} = 0.38 \qquad f_{IO_3^-} = 0.78$$

If the calculated ionic strength did not match that of one of the columns in the table, $f_{Ba^{2+}}$ and $f_{IO_3^-}$ could be obtained from Equation 5-55.

We may now write

$$\frac{1.57 \times 10^{-9}}{(0.38)(0.78)^2} = 6.8 \times 10^{-9} = K'_{sp}$$

$$[Ba^{2+}][IO_3^-]^2 = 6.8 \times 10^{-9}$$

Proceeding now as for an ordinary solubility calculation (Chapter 3),

$$\text{solubility} = s = [Ba^{2+}]$$
$$[IO_3^-] \cong 0.066$$
$$s(0.066)^2 = 6.8 \times 10^{-9}$$
$$s = 1.56 \times 10^{-6} \text{ fw/liter}$$

It is of interest to note that the calculated solubility, neglecting the effects of ionic strength, is 3.60×10^{-7} fw/liter.

Omission of Activity Coefficients in Equilibrium Calculations. We shall ordinarily neglect activity coefficients and simply use molar concentrations in applications of the equilibrium law. This recourse simplifies the calculations and greatly reduces the amount of data needed. For most purposes, the errors introduced by the assumption of unity for the activity coefficient will not be large enough to lead to false conclusions. It should be apparent from the preceding example, however, that disregard of activity coefficients may introduce a significant numerical error in calculations of this kind; relative errors of 100% or greater are not uncommon.

The student should be alert to the conditions under which the approximation of concentration for activity is likely to lead to the largest errors. Significant discrepancies will occur when the ionic strength is large (0.01 or larger) or when the ions involved have multiple charges (see Table 5-4). With dilute solutions (ionic strength < 0.01) of nonelectrolytes or of singly charged ions, the use of concentrations in a mass-law calculation often provides reasonably accurate results.

It is also important to note that the decrease in solubility resulting from the presence of an ion common to the precipitate is in part counteracted by the concomitant increase in solubility arising from the larger electrolyte concentration associated with presence of the salt containing the common ion. This effect is illustrated by the sample calculation just completed.

ADDITIONAL VARIABLES THAT AFFECT THE SOLUBILITY OF PRECIPITATES

Temperature. Heat is absorbed as most solids dissolve. Therefore, the solubility of precipitates generally increases with rising temperatures; correspondingly, solubility-product constants for most sparingly soluble compounds become larger at high temperatures.

TABLE 5-5   Solubility of Calcium Sulfate in Aqueous Ethyl Alcohol Solution[a]

Concentration of Ethyl Alcohol (weight percent)	Solubility of $CaSO_4$, g $CaSO_4$/100 g solvent
0	0.208
6.2	0.100
13.6	0.044
23.8	0.014
33.0	0.0052
41.0	0.0029

[a] From T. Yamamoto, *Bull. Inst. Phys. Chem. Res.* (Tokyo) **9**, 352 (1930); W. C. Linke, *Seidell Solubilities of Inorganic and Metal-Organic Compounds*, 4th ed., vol. I, p. 685. Washington, D.C.: American Chemical Society, 1958. With permission.

Solvent Composition. The solubility of most inorganic substances is markedly less in mixtures of water and organic solvents than in pure water. The data for calcium sulfate in Table 5-5 are typical of this effect.

Rate of Precipitate Formation. It is important to stress that no conclusions can be drawn about the rate of a reaction from the magnitude of its equilibrium constant. Many reactions with favorable equilibrium constants approach equilibrium at an imperceptible rate.

Precipitation reactions are often slow, several minutes or even several hours being required for the attainment of equilibrium. Occasionally the chemist can take advantage of a slow rate to accomplish separations that would not be feasible if equilibrium were approached rapidly. For example, calcium can be separated from magnesium by precipitation as the oxalate, despite the fact that the latter ion also forms an oxalate of comparable solubility. The separation is possible because equilibrium for magnesium oxalate formation is approached at a much slower rate than that for calcium oxalate formation; if the calcium oxalate is filtered shortly after precipitation, a solid that is essentially free of contamination by magnesium is obtained. If, on the other hand, the precipitate remains in contact with the liquid, it will be contaminated.

PROBLEMS

*1.   The solubility products for a series of iodides are:

$$TlI \qquad K_{sp} = 6.5 \times 10^{-8}$$
$$AgI \qquad K_{sp} = 8.3 \times 10^{-17}$$
$$PbI_2 \qquad K_{sp} = 7.1 \times 10^{-9}$$
$$BiI_3 \qquad K_{sp} = 8.1 \times 10^{-19}$$

List the four compounds in order of decreasing formal solubility in
(a) water.
(b) 0.10-*F* NaI.
(c) 0.10-*F* solution of the solute cation.

2.   The solubility products for a series of iodates are:

$$AgIO_3 \qquad K_{sp} = 3.0 \times 10^{-8}$$
$$Sr(IO_3)_2 \qquad K_{sp} = 3.3 \times 10^{-7}$$
$$La(IO_3)_3 \qquad K_{sp} = 6.2 \times 10^{-12}$$
$$Ce(IO_3)_4 \qquad K_{sp} = 4.7 \times 10^{-17}$$

List the four compounds in order of decreasing formal solubility in
(a) water.
(b) 0.10-F $NaIO_3$.
(c) 0.10-F solution of the solute cation.

*3. Solubility products for $Hf(OH)_4$ and $UO_2(OH)_2$ have values of 4.0×10^{-26} and 1.1×10^{-21}, respectively. The solubility equilibrium for $UO_2(OH)_2$ is

$$UO_2(OH)_2(s) \rightleftarrows UO_2^{2+} + 2OH^-$$

If NaOH were added to a solution that was 0.100 M each in UO_2^{2+} and in Hf^{4+},
(a) which ion would begin to precipitate first?
(b) what OH^- concentration would be required to reduce the ion in (a) to a concentration of 1.0×10^{-6} M?
(c) would a quantitative separation of the two ions by control of OH^- be feasible (employ 1×10^{-6} M as a criterion for quantitative separation)? If so, what range of OH^- concentration would permit the separation?

4. Employing 1.0×10^{-6} M as the criterion for quantitative removal of an ion, determine whether it would be feasible in principle to perform the following separations if the initial concentration of each species is 0.100. If a separation is feasible, specify the conditions that must be maintained.
(a) Ag^+ from Pb^{2+} employing Br^- (K_{sp} for $PbBr_2 = 3.9 \times 10^{-5}$)
*(b) Cl^- from I^- employing Cu^+
(c) Bi^{3+} from Ag^+ employing I^- (K_{sp} for $BiI_3 = 8.1 \times 10^{-19}$)
*(d) Ti^{3+} from TiO^{2+} employing OH^- (K_{sp} for $Ti(OH)_3 = 1 \times 10^{-40}$; for $TiO(OH)_2 = 1 \times 10^{-29}$)
(e) Ba^{2+} from Ag^+ employing SO_4^{2-} (K_{sp} for $Ag_2SO_4 = 1.6 \times 10^{-5}$)
*(f) Fe^{3+} from Cu^{2+} employing OH^-

5. Determine which of the following separations is feasible by controlling the hydronium ion concentration of a saturated H_2S solution. Assume that the initial concentration of each ion is 0.100 M and that lowering of a concentration to 1.0×10^{-6} M constitutes quantitative removal. If a separation is possible, specify the range of H_3O^+ concentrations that could be employed.
*(a) Fe^{2+} and Cd^{2+}
(b) Cu^{2+} and Zn^{2+}
*(c) La^{3+} and Mn^{2+} (K_{sp} for $La_2S_3 = 2.0 \times 10^{-13}$)
(d) Ce^{3+} and Fe^{2+} (K_{sp} for $Ce_2S_3 = 6.0 \times 10^{-11}$)
*(e) Cd^{2+} and Zn^{2+}
(f) Cd^{2+} and Tl^+

*6. For a solution that is 0.050 F in KNO_3, calculate K'_{sp} for
(a) AgCl.
(b) $Cu(OH)_2$.
(c) $BaSO_4$.
(d) $Al(OH)_3$.

7. For a solution that is 0.0333 F in $Ba(NO_3)_2$, calculate K'_{sp} for
(a) TlCl.
(b) $PbCO_3$.
(c) Ag_2S.
(d) $Fe(OH)_3$.

*8. Calculate the solubilities of the following compounds in a 0.0333-F solution of $BaCl_2$, first employing activities and second, neglecting them:
(a) AgSCN.
(b) PbI_2.

      (c) $BaSO_4$.

      (d) $Cd_2Fe(CN)_6$. $[Cd_2Fe(CN)_6(s) \rightleftarrows 2Cd^{2+} + Fe(CN)_6^{4-};$
$$K_{sp} = 3.2 \times 10^{-17}]$$

9. Calculate the solubilities of the following compounds in a 0.0167-F solution of $Ba(OH)_2$, first employing activities and second, neglecting them:

    (a) $AgIO_3$.

    (b) $Mg(OH)_2$.

    (c) $BaSO_4$.

    (d) $Fe(OH)_3$.

*10. Calculate the solubility of $BaSO_4$ in a solution that is (a) neutral and (b) 0.100 F in HCl. ($HSO_4^- + H_2O \rightleftarrows H_3O^+ + SO_4^{2-}$; $K_2 = 1.2 \times 10^{-2}$)

*11. Calculate the solubility of silver oxalate, $Ag_2C_2O_4$, in a solution in which the H_3O^+ concentration is maintained at

    (a) 1.0×10^{-6} M.

    (b) 1.0×10^{-4} M.

    (c) 1.0×10^{-2} M.

12. Calculate the solubility of $Pb_3(AsO_4)_2$ ($K_{sp} = 4.1 \times 10^{-39}$) in a solution that is maintained at a molar H_3O^+ concentration of (assume that Pb^{2+} does not react with H_2O)

    (a) 1.0×10^{-3}.

    (b) 1.0×10^{-6}.

    (c) 1.0×10^{-9}.

13. Calculate the solubility of Ag_3PO_4 ($K_{sp} = 1.3 \times 10^{-20}$) in a solution that is maintained at a molar H_3O^+ concentration of (assume that Ag^+ does not react with H_2O)

    (a) 1.0×10^{-5}.

    (b) 1.0×10^{-8}.

    (c) 1.0×10^{-10}.

14. Calculate the solubility of $MgNH_4PO_4$ ($MgNH_4PO_4(s) \rightleftarrows Mg^{2+} + NH_4^+ + PO_4^{3-}$) in a solution in which the H_3O^+ concentration is maintained at

    *(a) 1.0×10^{-10} M.

    (b) 1.0×10^{-6} M.

15. Calculate the solubility of the following sulfides in water, first, ignoring the reaction of S^{2-} as a base and, second, considering the basic properties of S^{2-} (assume that the cations do not react with H_2O):

    *(a) CuS.

    (b) PbS.

    *(c) MnS.

    (d) FeS.

*16. Calculate the solubility of $BaCO_3$ in water, first, ignoring the basic properties of CO_3^{2-} and, second, considering the basic properties of the solute anion.

17. Calculate the solubility of Ag_2SO_3 in water ($K_{sp} = 1.5 \times 10^{-14}$), first, ignoring the basic properties of SO_3^{2-} and, second, considering the basic properties of the solute anion.

18. Calculate the solubility of the following bases in water:

    *(a) $Th(OH)_4$.   ($K_{sp} = 4 \times 10^{-45}$)

    (b) $Al(OH)_3$.

    *(c) $Pb(OH)_2$.

    (d) $Mn(OH)_2$.

*19. The equilibrium constants for the reactions of AgCl with Cl^- are

$$AgCl(s) + Cl^- \rightarrow AgCl_2^- \qquad K_1 = \frac{[AgCl_2^-]}{[Cl^-]} = 2.0 \times 10^{-5}$$

$$AgCl_2^- + Cl^- \rightleftarrows AgCl_3^{2-} \qquad K_2 = \frac{[AgCl_3^{2-}]}{[AgCl_2^-][Cl^-]} = 1$$

Calculate the solubility of AgCl in a solution having a formal concentration of NaCl of
(a) 2.0.
(b) 0.50.
(c) 5.0×10^{-2}.
(d) 5.0×10^{-4}.

20. The equilibrium constant for formation of $CuCl_2^-$ is given by

$$Cu^+ + 2Cl^- \rightleftarrows CuCl_2^- \qquad K = \frac{[CuCl_2^-]}{[Cu^+][Cl^-]^2} = 8.7 \times 10^4$$

What is the solubility of CuCl in solutions having the following formal NaCl concentrations:
(a) 1.0
(b) 1.0×10^{-1}
(c) 1.0×10^{-2}
(d) 1.0×10^{-3}
(e) 1.0×10^{-4}

*21. The equilibrium constant for the formation of $Al(OH)_4^-$ is given by

$$Al(OH)_3(s) + OH^- \rightleftarrows Al(OH)_4^- \qquad K = 10$$

How many milliliters of 1.0-F NaOH would be required to completely redissolve 1.00 g of $Al(OH)_3$ suspended in 100.0 ml of water?

22. What concentration of OH^- must be maintained in order to dissolve 0.200 g of $Pb(OH)_2$ in 200 ml of solution?

$$Pb(OH)_2(s) + OH^- \rightleftarrows Pb(OH)_3^- \qquad K = 5.0 \times 10^{-2}$$

*23. Formation constants for the reaction Ag^+ with $S_2O_3^{2-}$ are

$$Ag^+ + S_2O_3^{2-} \rightleftarrows AgS_2O_3^- \qquad K_1 = 6.6 \times 10^8$$

$$AgS_2O_3^- + S_2O_3^{2-} \rightleftarrows Ag(S_2O_3)_2^{3-} \qquad K_2 = 4.4 \times 10^3$$

Calculate the solubility of AgI in 0.200-F $Na_2S_2O_3$ (assume that $S_2O_3^{2-}$ does not combine with H_3O^+).

GRAVIMETRIC ANALYSIS

A gravimetric analysis is based upon the measurement of the weight of a substance of known composition that is chemically related to the analyte. Two types of gravimetric methods exist. In *precipitation methods*, the species to be determined is caused to react chemically with a reagent to yield a product of limited solubility; after filtration and other suitable treatment, the solid residue of known chemical composition is weighed. In *volatilization methods*, the substance to be determined is separated as a gas from the remainder of the sample; here the analysis is based upon the weight of the volatilized substance or upon the weight of the nonvolatile residue. We shall be concerned principally with precipitation methods because these are more frequently encountered than methods involving volatilization.

Calculation of Results from Gravimetric Data

A gravimetric analysis requires two experimental measurements: specifically, the weight of sample taken and the weight of a product of known composition derived from the sample. Ordinarily these data are converted to a percentage of analyte by a simple mathematical manipulation.

If A is the analyte, we may write

$$\% A = \frac{\text{weight of A}}{\text{weight of sample}} \times 100 \qquad (6\text{-}1)$$

Usually the weight of A is not measured directly. Instead, the species that is actually isolated and weighed either contains A or can be chemically related to A. In either case, a *gravimetric factor* is needed to convert the weight of the precipitate to the corresponding weight of A. The properties of this factor are conveniently demonstrated with examples.

Example. How many grams of Cl are contained in a precipitate of AgCl that weighs 0.204 g? From the formula for AgCl, we know that

$$\text{no. fw AgCl} = \text{no. fw Cl}$$

Since

$$\text{no. fw AgCl} = \frac{0.204}{\text{gfw AgCl}} = \text{no. fw Cl}$$

and also since

$$\text{wt Cl} = \text{no. fw Cl} \times \text{gfw Cl}$$

then

$$\text{wt Cl} = 0.204 \times \frac{\text{gfw Cl}}{\text{gfw AgCl}} = 0.204 \times \frac{35\ 45}{143.3}$$

$$= 0.204 \times 0.2474 = 0.0505 \text{ g}$$

Example. To what weight of $AlCl_3$ would 0.204 g of AgCl correspond? We know that each $AlCl_3$ yields three AgCl. Therefore,

$$\text{no. fw } AlCl_3 = \frac{1}{3} \text{ no. fw AgCl} = \frac{1}{3} \times \frac{0.204}{\text{gfw AgCl}}$$

By the preceding arguments

$$\text{wt } AlCl_3 = 0.204 \times \frac{\text{gfw } AlCl_3}{3 \times \text{gfw AgCl}} = 0.204 \times \frac{133.3}{3 \times 143.3}$$

$$= 0.204 \times 0.310 = 0.0633 \text{ g}$$

Note how these two calculations resemble each other. In both the weight of one substance is given by the product involving the known weight of some other substance and a ratio that contains their respective gram formula weights. This ratio is the gravimetric factor. In the second example it was necessary to multiply the gram formula weight of silver chloride by 3 in order to balance the number of chlorides that appear in the numerator and denominator of the gravimetric factor.

Example. What weight of Fe_2O_3 can be obtained from 1.63 g of Fe_3O_4? What is the gravimetric factor for this conversion?

Here it is necessary to assume that all Fe in the Fe_3O_4 is transformed into Fe_2O_3 and ample oxygen is available to accomplish this change. That is,

$$2Fe_3O_4 + [O] = 3Fe_2O_3$$

We see from this equation that $\frac{3}{2}$ fw of Fe_2O_3 are obtained from 1 fw of Fe_3O_4. Thus, the number of formula weights of Fe_2O_3 is greater than the number of formula weights of Fe_3O_4 by a factor of $\frac{3}{2}$, or

$$\text{no. fw } Fe_2O_3 = \frac{3}{2} \times \text{no. fw } Fe_3O_4$$

$$\frac{\text{wt } Fe_2O_3}{\text{gfw } Fe_2O_3} = \frac{3}{2} \times \frac{\text{wt } Fe_3O_4}{\text{gfw } Fe_3O_4}$$

and

$$\text{wt } Fe_2O_3 = \frac{3}{2} \times \frac{\text{wt } Fe_3O_4}{\text{gfw } Fe_3O_4} \times \text{gfw } Fe_2O_3$$

Thus, upon rearranging,

$$\text{wt } Fe_2O_3 = \text{wt } Fe_3O_4 \times \frac{3 \times \text{gfw } Fe_2O_3}{2 \times \text{gfw } Fe_3O_4}$$

substitution of numerical values gives

$$\text{wt } Fe_2O_3 = 1.63 \times \frac{3 \times 159.7}{2 \times 231.5} = 1.687 = 1.69 \text{ g}$$

In this example,

$$\text{gravimetric factor} = \frac{3 \times \text{gfw } Fe_2O_3}{2 \times \text{gfw } Fe_3O_4} = 1.035$$

A general definition of the gravimetric factor is:

$$\text{gravimetric factor} = \frac{a}{b} \times \frac{\text{gfw of the substance sought}}{\text{gfw of the substance weighed}}$$

where a and b are small integers that take such values as are necessary to make the number of formula weights in the numerator and denominator *chemically equivalent*.

Equation 6-1 can now be converted to the more useful form

$$\% A = \frac{\text{wt ppt} \times \left(\dfrac{a \times \text{gfw A}}{b \times \text{gfw ppt}}\right) \times 100}{\text{wt sample}}$$

Additional examples of gravimetric factors are given in Table 6-1. Most chemical handbooks contain tabulations of these factors and their logarithms.

In all of the gravimetric factors considered thus far, chemical equivalence between numerator and denominator has been established by simply balancing the number of atoms of an element (other than oxygen) that is common to both. Occasionally this approach will be inadequate. Consider, for example, an indirect analysis for the iron in a sample of iron(III) sulfate that involves precipitation and weighing of barium sulfate. Here the gravimetric factor will contain no element common to numerator and denominator, and we must seek further for the means of establishing the chemical equivalence between these quantities. We note that

$$2 \text{ gfw Fe} \equiv 1 \text{ gfw } Fe_2(SO_4)_3 \equiv 3 \text{ gfw } SO_4 \equiv 3 \text{ gfw } BaSO_4$$

TABLE 6-1   Typical Gravimetric Factors

Species Sought	Species Weighed	Gravimetric Factor
In	In_2O_3	$\dfrac{2 \times gfw\ In}{gfw\ In_2O_3}$
HgO	$Hg_5(IO_6)_2$	$\dfrac{5 \times gfw\ HgO}{gfw\ Hg_5(IO_6)_2}$
I	$Hg_5(IO_6)_2$	$\dfrac{2 \times gfw\ I}{gfw\ Hg_5(IO_6)_2}$
K_3PO_4	K_2PtCl_6	$\dfrac{2 \times gfw\ K_3PO_4}{3 \times gfw\ K_2PtCl_6}$

The gravimetric factor for calculating the percent Fe will be

$$\text{gravimetric factor} = \frac{2 \times gfw\ Fe}{3 \times gfw\ BaSO_4}$$

Thus, even though the species in the gravimetric factor are not directly related by a common element, we can establish their equivalence through knowledge of the stoichiometry between them.

The examples that follow illustrate the use of the gravimetric factor in the calculation of the results from analyses.

Example.   A 0.703-g sample of a commercial detergent was ignited at a red heat to destroy the organic matter. The residue was then taken up in hot HCl which converted the P to H_3PO_4. The phosphate was precipitated as $MgNH_4PO_4 \cdot 6H_2O$ by addition of Mg^{2+} followed by aqueous NH_3. After being filtered and washed, the precipitate was converted to $Mg_2P_2O_7$ by ignition at 1000°C. This residue weighed 0.432 g. Calculate the percent P in the sample.

$$\% P = \frac{0.432 \times \dfrac{2 \times gfw\ P}{gfw\ Mg_2P_2O_7} \times 100}{0.703}$$

$$= \frac{0.432 \times 0.2783 \times 100}{0.703} = 17.1$$

Example.   At elevated temperatures sodium oxalate is converted to sodium carbonate:

$$Na_2C_2O_4 \rightarrow Na_2CO_3 + CO$$

Ignition of a 1.3906-g sample of impure sodium oxalate yielded a residue weighing 1.1436 g. Calculate the percentage purity of the sample.

Here it must be assumed that the difference between the initial and final weights represents the carbon monoxide evolved during the ignition; it is this weight loss that forms the basis for the analysis. From the equation for the process, we see that

$$\text{no. fw CO} = \text{no. fw } Na_2C_2O_4$$

Thus,

$$\% \, Na_2C_2O_4 = \frac{wt \, CO \times \dfrac{gfw \, Na_2C_2O_4}{gfw \, CO}}{wt \, sample} \times 100$$

$$= \frac{(1.3906 - 1.1436) \times 4.784 \times 100}{1.3906} = 84.97$$

Example. A 0.2795-g sample of an insecticide containing only lindane ($C_6H_6Cl_6$; gfw = 290.8) and DDT ($C_{14}H_9Cl_5$; gfw = 354.5) was burned in a stream of oxygen in a quartz tube. The products (CO_2, H_2O, and HCl) were passed through a solution of $NaHCO_3$. After acidification, the chloride in this solution yielded 0.7161 g of AgCl. Calculate the percent lindane and DDT in the sample.

Here there are two unknowns, and we must therefore develop two independent equations that can be solved simultaneously.

One useful equation is

$$wt \, C_6H_6Cl_6 + wt \, C_{14}H_9Cl_5 = 0.2795 \, g$$

A second equation is

$$wt \, AgCl \, from \, C_6H_6Cl_6 + wt \, AgCl \, from \, C_{14}H_9Cl_5 = 0.7161 \, g$$

After inserting the appropriate gravimetric factors, the second equation becomes

$$wt \, C_6H_6Cl_6 \times \frac{6 \times gfw \, AgCl}{gfw \, C_6H_6Cl_6} + wt \, C_{14}H_9Cl_5 \times \frac{5 \times gfw \, AgCl}{gfw \, C_{14}H_9Cl_5} = 0.7161 \, g$$

or

$$wt \, C_6H_6Cl_6 \times 2.957 + wt \, C_{14}H_9Cl_5 \times 2.021 = 0.7161$$

Substituting the wt $C_{14}H_9Cl_5$ from the first equation gives

$$2.957 \, wt \, C_6H_6Cl_6 + 2.021(0.2795 - wt \, C_6H_6Cl_6) = 0.7161$$

Thus,

$$wt \, C_6H_6Cl_6 = 0.1616 \, g$$

and

$$\% \, C_6H_6Cl_6 = \frac{0.1616}{0.2795} \times 100 = 57.82$$

$$\% \, C_{14}H_9Cl_5 = 100 - 57.82 = 42.18$$

Properties of Precipitates and Precipitating Reagents

The ideal precipitating reagent for a gravimetric analysis would react specifically with the analyte to produce a solid that would (1) have a sufficiently low solubility so that losses from that source would be negligible, (2) be readily filtered and washed free of contaminants, and (3) be unreactive and of known composition after drying or, if necessary, ignition. Few precipitates or reagents possess all these desirable properties; thus, the chemist frequently finds it necessary to perform analyses using a product or a reaction that is far from ideal.

The variables that influence the solubility of precipitates were discussed in Chapter 5; we must now consider what can be done to achieve a pure and easily filtered solid.

FILTERABILITY AND PURITY OF PRECIPITATES

Both the ease of filtration and the ease of purification are influenced by the particle size of the solid phase. The relationship between particle size and ease of filtration is straightforward, coarse precipitates being readily retained by porous media which permit rapid filtration. Finely divided precipitates require dense filters; low filtration rates result. The effect of particle size upon purity of a precipitate is more complex. More often than not, a decrease in soluble contaminants is found to accompany an increase in particle size.

In considering the purity of precipitates we shall use the term *coprecipitation*, which describes those processes by which *normally soluble* components of a solution are carried down during the formation of a precipitate. The student should clearly understand that contamination of a precipitate by a second substance whose solubility product has been exceeded *does not constitute coprecipitation*.

Factors That Determine the Particle Size of Precipitates. Enormous variation is observed in the particle size of precipitates, depending upon their chemical composition and the conditions leading to their formation. At the one extreme are *colloidal suspensions*, the individual particles of which are so small as to be invisible to the naked eye (10^{-6} to 10^{-4} mm in diameter). These particles show no tendency to settle out from solution, nor are they retained upon common filtering media. At the other extreme are particles with dimensions on the order of several tenths of a millimeter. The temporary dispersion of such particles in the liquid phase is called a *crystalline suspension*. The particles of a crystalline suspension tend to settle out rapidly and are readily filtered.

No sharp discontinuities in physical properties occur as the dimensions of the particles comprising the solid phase increase from colloidal to those typical of crystals. Indeed, some precipitates possess characteristics intermediate between these defined extremes. The majority of precipitates, however, are easily recognizable as being predominately colloidal or predominately crystalline. Thus, while imperfect, this classification can be usefully applied to most solid phases.

Although the phenomenon of precipitation has long attracted the attention of chemists, fundamental information regarding the mechanism of the process remains incompletely understood. It is certain, however, that the particle size of the solid that forms is influenced in part by such experimental variables as the temperature, the solubility of the precipitate in the medium in which it is being formed, reactant concentrations, and the rate at which reagents are mixed. The effect of these variables can be accounted for, at least qualitatively, by assuming that the particle size is related to a single property of the system called its *relative supersaturation*,[1] where

$$\text{relative supersaturation} = \frac{Q - S}{S} \qquad (6\text{-}2)$$

[1] The name of P. P. von Weimarn is associated with the concept of relative supersaturation and its effect upon particle size. An account of von Weimarn's work is to be found in *Chem Rev.*, **2**, 217 (1925). The von Weimarn viewpoint adequately suggests the general conditions that will lead to a satisfactory particle size for a precipitate. Other theories are superior in accounting for details of the precipitation process. See, for example, A. E. Nielsen, *The Kinetics of Precipitation.* New York: The Macmillan Company, 1964.

Here Q is the concentration of the solute at any instant, and S is its equilibrium solubility.

During formation of a sparingly soluble precipitate, each addition of precipitating reagent presumably causes the solution to be momentarily supersaturated (that is, $Q > S$). Under most circumstances, this unstable condition is relieved, usually after a brief period, by precipitate formation. Experimental evidence suggests, however, that the particle size of the resulting precipitate varies inversely with the average degree of relative supersaturation that exists after each addition of reagent. Thus, when $(Q - S)/S$ is large, the precipitate tends to be colloidal; when this parameter is low on the average, a crystalline solid results.

Mechanics of Precipitate Formation. The effect of relative supersaturation on particle size can be rationalized by postulating two precipitation processes, *nucleation* and *particle growth*. The particle size of a freshly formed precipitate is governed by the extent to which one of these steps predominates over the other.

Nucleation is a process whereby some minimum number of ions or molecules (perhaps as few as four or five) unite to form a stable second phase. Further precipitation can occur either by formation of additional nuclei or by deposition of solid on the nuclei that are already present. If the former predominates, a precipitate containing a large number of small particles results; if growth predominates, a smaller number of larger particles will be produced.

The rate of nucleation is believed to increase exponentially with relative supersaturation, whereas the rate of particle growth bears an approximately linear relationship to this parameter. That is,

$$\text{rate of nucleation} = k_1 \left(\frac{Q - S}{S} \right)^n$$

where n is thought to be about 4, and

$$\text{rate of growth} = k_2 \left(\frac{Q - S}{S} \right)$$

Normally, k_2 is greater than k_1; thus, at low relative supersaturations, growth predominates. When the supersaturation is great, the exponential nature of nucleation may cause this process to occur to the near exclusion of particle growth. These effects are illustrated in Figure 6-1.

Experimental Control of Particle Size. Experimental variables that minimize supersaturation and thus lead to crystalline precipitates include elevated temperatures (to increase S), dilute solutions (to minimize Q), and slow addition of the precipitating agent with good stirring (also to lower the average value of Q).

The particle size of precipitates with solubilities that are pH-dependent can often be enhanced by increasing S during precipitation. For example, large, easily filtered crystals of calcium oxalate can be obtained by forming the bulk of the precipitate in a somewhat acidic environment in which the salt is moderately soluble. The precipitation is then completed by slowly adding aqueous

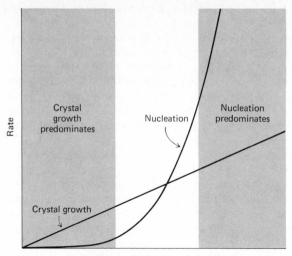

Figure 6-1 Effect of Relative Supersaturation on Precipitation Processes.

ammonia until the pH is sufficiently high for quantitative removal of the calcium oxalate; the additional precipitate produced during this step forms on the solid.

A crystalline solid is much easier to manipulate than a colloidal suspension. For this reason, particle growth is usually to be preferred over further nucleation during the formation of a precipitate. If, however, the solubility S of a precipitate is very small, it is essentially impossible to avoid a momentarily large relative supersaturation as solutions are mixed; as a consequence, colloidal suspensions often cannot be avoided. For example, under conditions feasible for an analysis, the hydrous oxides of iron(III), aluminum, and chromium(III) and the sulfides of most heavy metal ions can be formed only as colloids because of their very low solubilities. The same is true for the halide precipitates of silver ion.[2]

COLLOIDAL PRECIPITATES

Individual colloidal particles are so small that they are not retained on ordinary filtering media; furthermore, Brownian motion prevents their settling out of solution under the influence of gravity. Fortunately, however, the individual particles of most colloids can be caused to coagulate or agglomerate to give a filterable, noncrystalline mass that rapidly settles out from a solution.

Coagulation of Colloids. Three experimental measures hasten the coagulation process, namely, heating, stirring, and adding an electrolyte to the medium. To understand the effectiveness of these measures, we need to account for the stability of a colloidal suspension.

[2] Silver chloride illustrates that the relative supersaturation concept is imperfect. It ordinarily forms as a colloid; yet its solubility is not significantly different from certain other compounds such as $BaSO_4$ which generally form as crystals.

The individual particles in a typical colloid bear either a positive or a negative charge as a consequence of *adsorption* of cations or anions on their surfaces. The presence of this charge is readily demonstrated experimentally by observing the migration of the particles under the influence of an electric field.

Adsorption of ions upon an ionic solid has, as its origin, the normal bonding forces that are responsible for crystal growth. Thus, a silver ion at the surface of a silver chloride particle has a partially unsatisfied bonding capacity by virtue of its surface location. Negative ions are attracted to this site by the same forces that hold chloride ions in the silver chloride lattice. Chloride ions on the surface exert an analogous attraction for cations in the solvent.

The nature and magnitude of the charge on particles of a colloidal suspension depend, in a complex way, on a number of variables. For the colloidal suspensions of interest in analysis, however, the species adsorbed, and thus the charge on the particles, can be readily predicted from the empirical observation that lattice ions are generally more strongly adsorbed than any others. Thus, a silver chloride particle will be positively charged in a solution containing an excess of silver ions, owing to the preferential adsorption of those ions. It will have a negative charge in the presence of excess chloride ion for the same reason. It is of interest to note that the silver chloride particles formed in a gravimetric chloride analysis initially carry a negative charge but become positive as an excess of the precipitating agent is added.

The extent of adsorption increases rapidly with increases in concentration of the adsorbed ion. Ultimately, however, the surface of each particle becomes saturated; under these circumstances, further increases in concentration have little or no effect.

Figure 6-2 depicts schematically a colloidal silver chloride particle in a solution containing an excess of silver ions. Attached directly to the solid surface are silver ions in the *primary adsorption layer*. Surrounding the charged particle is a *region of solution* called the *counter-ion layer*, within which there exists an excess of negative ions sufficient to balance the charge of the adsorbed positive ions on the particle surface. The counter-ion layer forms as the result of electrostatic forces.

Considered together, the primarily adsorbed ions and their counter ions in solution constitute an *electrical double layer* which imparts a degree of stability to a colloidal suspension. These layers cause a colloidal particle to sheer away as it approaches another; the forces that would otherwise permit cohesion between the particles are insufficient to overcome the repulsions due to the electrical double layer. In order to coagulate a colloid, then, these repulsive forces must be minimized.

The effect of the charged double layer on stabilization of a colloid is readily seen in the precipitation of chloride ion with silver ion. With initial addition of silver nitrate, silver chloride is formed in an environment having a high chloride ion concentration. The negative charge for silver chloride particles is therefore high; the volume of the positive counter-ion layer surrounding each particle must also be relatively large to contain enough positive ions (hydronium or sodium ions, for example) to neutralize the negative charge of the particles. Coagulation under these circumstances does not occur. As more silver ions

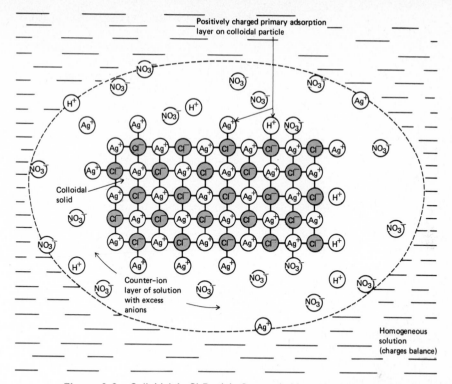

Figure 6-2 Colloidal AgCl Particle Suspended in a Solution of AgNO₃.

are added, the charge per particle decreases because the chloride ion concentration is decreased, and the number of particles is increased; the repulsive effect of the double layer is thus decreased. As chemical equivalence is approached, a sudden appearance of the coagulated colloid is observed. Here the number of adsorbed chloride ions per particle becomes small, and the double layer shrinks to a point where individual particles can approach one another closely enough to permit agglomeration. It is of interest that the agglomeration process can be reversed by addition of a large excess of silver ions; here, of course, the charge of the double layer is reversed with the counter-ion layer being negative.

Coagulation is often brought about by a short period of heating, especially if accompanied by stirring. The increased temperature reduces adsorption and thus the net charge on a particle; in addition, the particles acquire kinetic energies sufficient to overcome the barrier to close approach.

An even more effective method of coagulation is to increase the electrolyte concentration of the solution by addition of a suitable ionic compound. Under these circumstances, the volume of solution that contains enough ions of opposite charge to neutralize the charge on the particle is lessened. Thus, the introduction of an electrolyte has the effect of shrinking the counter-ion layer, with the result that the surface charge on the particles is more completely

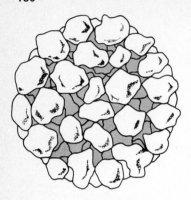

Figure 6-3   Coagulated Colloidal Particles.

neutralized. With their effective charge decreased, the particles can approach one another more closely.

Coprecipitation in Coagulated Colloids. Adsorption is the principal type of coprecipitation that affects coagulated colloids; other types are encountered with crystalline solids.

A coagulated colloid consists of irregularly arranged particles which form a loosely packed, porous mass. Within this mass, large internal surface areas remain in contact with the solvent phase (see Figure 6-3). Adhering to these surfaces will be most of the primarily adsorbed ions which were on the uncoagulated particles. Even though the counter-ion layer surrounding the original colloidal particle is properly considered to be part of the solution, it must be recognized that sufficient counter ions to impart electrical neutrality must accompany the particle (in the film of liquid surrounding the particle) through the processes of coagulation and filtration. *The net effect of surface adsorption is the carrying down of an otherwise soluble species as a surface contaminant.*

Peptization of Colloids. Peptization refers to the process whereby a coagulated colloid reverts to its original dispersed state. Peptization frequently occurs when pure water is used to wash such a precipitate. Washing is not particularly effective in dislodging adsorbed contaminants; it does tend, however, to remove the electrolyte responsible for coagulation from the internal liquid in contact with the solid. As the electrolyte is removed, the counter-ion layers increase again in volume. The repulsive forces responsible for the colloidal state are thus reestablished, and particles detach themselves from the coagulated mass. The washings become cloudy as the freshly dispersed particles pass through the filter.

The chemist is thus faced with a dilemma in handling coagulated colloids. He would like to free the precipitate of contaminants, but in so doing he runs the risk of losses from peptization. This problem is commonly resolved by washing the agglomerated colloid with a solution containing a volatile electrolyte which can subsequently be removed from the solid by heating. For example, silver chloride precipitates are ordinarily washed with dilute nitric acid. The washed precipitate is heavily contaminated by the acid, but no harm results since the nitric acid is removed when the precipitate is dried at 110°C.

Practical Treatment of Colloidal Precipitates. In general, colloids are precipitated from hot, stirred solutions to which sufficient electrolyte has been added to assure coagulation. With many colloids, the physical characteristics of the coagulated mass may be improved by permitting the solid to stand in contact with the hot solution from which it was formed. During this process, which is known as *digestion*, weakly bound water appears to be lost from the precipitate; the result is a denser mass that is easier to filter.

A dilute solution of a volatile electrolyte is used to wash the filtered precipitate. Washing does not appreciably affect primarily adsorbed ions because the attraction between these species and the solid is too strong. Some exchange may occur between the existing counter ions and one of the ions in the wash solution. Under any circumstances, it must be expected that the precipitate will still be contaminated to some degree, even after extensive washing. The error introduced into the analysis from this source can range from 1 or 2 ppt (as in the coprecipitation of silver nitrate on silver chloride) to an intolerable level (as in the coprecipitation of heavy metal hydroxides upon the hydrous oxides of trivalent iron or aluminum).

A drastic way to minimize the effects of adsorption is *reprecipitation*. Here the filtered solid is redissolved and again precipitated. The first precipitate normally carries down only a small fraction of the total contaminant present in the original solvent. Thus, the solution containing the redissolved precipitate will have a significantly lower contaminant concentration than the original. When precipitation is again carried out, less adsorption is to be expected. Reprecipitation adds substantially to the time required for an analysis; nevertheless, it is a near necessity for such precipitates as the hydrous oxides of iron(III) and aluminum, which possess extraordinary tendencies to adsorb the hydroxides of heavy metal cations such as zinc, cadmium, and manganese.

CRYSTALLINE PRECIPITATES

In general, crystalline precipitates are more easily handled than coagulated colloids. Here the size of individual particles can be varied to a degree. As a consequence, the physical properties and purity of the solid are determined by experimental variables over which the chemist has a measure of control.

Methods of Improving Particle Size and Filterability. Generally, the particle size of crystalline solids can be improved by keeping the relative supersaturation low during the period in which the precipitate is formed. From Equation 6-2, it is apparent that minimizing Q or maximizing S, or both, will accomplish this purpose.

The use of dilute solutions and the slow addition of precipitating agent with good mixing tends to reduce the momentary local supersaturation in the solution. Ordinarily, S can be increased by precipitating from hot solution. Quite noticeable gains in particle size can be obtained with these simple measures.

Purity of Crystalline Precipitates. The specific surface area[3] of crystalline precipitates is relatively small; consequently, coprecipitation by direct adsorption is negligible. However, other forms of coprecipitation, which involve incorporation of contamination within the interior of crystals, may cause serious errors.

Two types of coprecipitation, *inclusion* and *occlusion*, are associated with crystalline precipitates. The two differ in the manner in which the contaminant is distributed throughout the interior of the solid. Included impurities are randomly distributed, in the form of individual ions or molecules, throughout the crystal. Occlusion, on the other hand, involves a nonhomogeneous distribution of impurities, consisting of numerous ions or molecules of the contaminant, within imperfections in the crystal lattice.

Occlusion occurs when whole droplets of solution containing impurities are trapped and surrounded by a rapidly growing crystal. Because the contaminants are located within the crystal, washing does little to decrease their amount. A lower precipitation rate may significantly lessen the extent of occlusion by providing time for the impurities to escape before they become entrapped. Digestion of the precipitate for as long as several hours is even more effective in eliminating contamination by occlusion.

Digestion of Crystalline Precipitates. The heating of crystalline precipitates (without stirring) for some time after formation frequently yields a product with improved purity and filterability. The improvement in purity undoubtedly results from the solution and recrystallization that occur continuously and at an enhanced rate at elevated temperatures. During these processes, many pockets of imperfection become exposed to the solution; the contaminant is thus able to escape from the solid, and more perfect crystals result.

Solution and recrystallization during digestion are probably responsible for the improvement in filterability as well. Bridging between adjacent particles occurs to yield larger crystalline aggregates which are more easily filtered. The fact that little improvement in filtering characteristics is obtained if the mixture is stirred during digestion tends to confirm this view.

DIRECTION OF COPRECIPITATION ERRORS

Coprecipitated impurities may cause the results of an analysis to be either too high or too low. If the contaminant is not a compound of the ion being determined, positive errors will always result. Thus, a positive error will be observed when colloidal silver chloride adsorbs silver nitrate during a chloride analysis. On the other hand, where the contaminant contains the ion being determined, either positive or negative errors may be observed. In the determination of barium ions by precipitation as barium sulfate, for example, occlusion of barium salts

---

[3] The specific surface is defined as the area exposed by a unit weight of solid; it ordinarily is expressed in terms of cm^2/g.

occurs. If the occluded contaminant is barium nitrate, a positive error will be observed, since this compound has a greater formula weight than the barium sulfate that would have formed had no coprecipitation occurred. If barium chloride were the contaminant, however, a negative error would arise because its formula weight is less than that of the sulfate salt.

PRECIPITATION FROM HOMOGENEOUS SOLUTION

In precipitation from homogeneous solution, the precipitating agent is chemically generated in the solution at a sufficiently slow rate so that the relative supersaturation is always low. Local reagent excesses do not occur because the precipitating agent appears homogeneously throughout the entire solution. In general, homogeneously formed precipitates, both colloidal and crystalline, are better suited for analysis than precipitates formed by direct addition of a reagent.

Urea is often employed for the homogeneous generation of hydroxide ion. The reaction can be expressed by the equation

$$(H_2N)_2CO + 3H_2O = CO_2 + 2NH_4^+ + 2OH^-$$

This reaction proceeds slowly at temperatures just below boiling; typically one to two hours are needed to produce sufficient reagent to complete a precipitation. The method is particularly valuable for the precipitation of hydrous oxides or basic salts. For example, the hydrous oxides of iron(III) and aluminum, when formed by direct addition of base, are bulky, gelatinous masses that are heavily contaminated and difficult to filter. In contrast, when produced by the homogeneous generation of hydroxide ion, these same products are dense, readily filtered, and have considerably higher purity.

Homogeneous precipitation of crystalline precipitates also results in marked increases in crystal size; enhanced purity often accompanies this crystal growth.

Representative analyses based upon precipitation by homogeneously generated reagents are given in Table 6-2.

DRYING AND IGNITION OF PRECIPITATES

After filtration, a gravimetric precipitate is heated until its weight becomes constant. Heating serves the purpose of removing the solvent and volatile electrolytes carried down with the precipitate; in addition, this treatment may induce chemical decomposition to give a product of known composition.

The temperature required to produce a suitable product varies from precipitate to precipitate. Figure 6-4 shows the loss of weight as a function of temperature for several common analytical precipitates. These data were obtained with an automatic thermobalance,[4] an instrument which measures the weight of a substance continuously as its temperature is increased at a constant rate in a furnace. The heating of three precipitates—silver chloride, barium sulfate, and aluminum oxide—serves simply to remove water and perhaps

[4] For descriptions of thermobalances, see C. Duval, *Inorganic Thermogravimetric Analysis*, 2d ed. New York: Elsevier Publishing Company, 1962.

TABLE 6-2 Methods for the Homogeneous Generation of Precipitants

Precipitant	Reagent	Generation Reaction	Elements Precipitated
OH^-	Urea	$(NH_2)_2CO + 3H_2O = CO_2 + 2NH_4^+ + 2OH^-$	Al, Ga, Th, Bi, Fe, Sn
PO_4^{3-}	Trimethyl phosphate	$(CH_3O)_3PO + 3H_2O = 3CH_3OH + H_3PO_4$	Zr, Hf
$C_2O_4^{2-}$	Ethyl oxalate	$(C_2H_5)_2C_2O_4 + 2H_2O = 2C_2H_5OH + H_2C_2O_4$	Mg, Zn, Ca
SO_4^{2-}	Dimethyl sulfate	$(CH_3O)_2SO_2 + 2H_2O = 2CH_3OH + SO_4^{2-} + 2H_3O^+$	Ba, Ca, Sr, Pb
CO_3^{2-}	Trichloroacetic acid	$Cl_3CCOOH + 2OH^- = CHCl_3 + CO_3^{2-} + H_2O$	La, Ba, Ra
S^{2-}	Thioacetamide	$CH_3\overset{S}{\overset{\|}{C}}NH_2 + H_2O = CH_3\overset{O}{\overset{\|}{C}}\!-\!NH_2 + H_2S$	Sb, Mo, Cu, Cd
8-Hydroxyquinoline	8-Acetoxyquinoline	$CH_3C(=O)\!-\!O\text{-quinolinyl} + H_2O = CH_3C(=O)\!-\!OH + HO\text{-quinolinyl}$	Al, U, Mg, Zn

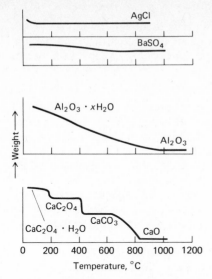

Figure 6-4 Effect of Temperature on Precipitate Weights.

volatile electrolytes carried down during the precipitation process. Note the wide range in temperature required to produce an anhydrous precipitate of constant weight. Thus, moisture is completely removed from silver chloride at a temperature above 110 to 120°C; for aluminum oxide, dehydration is not complete until a temperature greater than 1000°C is achieved. It is of interest to note that aluminum oxide, formed homogeneously with urea, can be completely dehydrated at a temperature of about 650°C.

The thermal curve for calcium oxalate is seen to be considerably more complex than the others shown in Figure 6-4. At temperatures below about 135°C, unbound water is removed to give the monohydrate, $CaC_2O_4 \cdot H_2O$; a temperature of about 225°C converts this compound to the anhydrous oxalate. The abrupt change in weight at about 450°C signals the decomposition of calcium oxalate to calcium carbonate and carbon monoxide. The final step in the curve depicts the conversion of the carbonate to calcium oxide and carbon dioxide. It is evident that the weighing form employed for a calcium oxalate precipitate will depend upon the ignition conditions.

A Critique of the Gravimetric Method

Some chemists are inclined to discount the present-day value of gravimetric methods on the grounds that they are inefficient and obsolete. We, on the other hand, believe that the gravimetric approach to an analytical problem—like all others—has strengths and weaknesses, and that ample situations exist where it represents the best possible choice for the resolution of an analytical problem.

TIME REQUIRED FOR A GRAVIMETRIC ANALYSIS

In all analytical methods, a physical quantity M is ultimately measured which varies with the amount or the concentration C_A of the analyte A in the original sample. Ideally, but by no means always, the functional relationship between C_A and M is linear; that is,

$$C_A = kM$$

Most analytical procedures require the determination of k by a standardization process or by an empirical calibration in which M is measured for one or more standards of known C_A. Two notable exceptions to this generality are the gravimetric method and the coulometric method (Chapter 20). With these procedures, k can be calculated directly from known, well-established physical constants; the need for calibration or standardization is thus avoided.

For a gravimetric method, k is simply the gravimetric factor which can be calculated with a table of atomic weights. As we shall show, this direct knowledge of k sometimes makes a gravimetric procedure the most efficient one for solving an analytical problem.

In considering the time required to perform an analysis, it is necessary to differentiate between elapsed time and operator time. The first refers to the clock hours or minutes between the start of the analysis and the report of the result. Operator time, on the other hand, represents the actual time the chemist or technician must spend in performing the various manipulations required to complete the analysis and calculate the result. A characteristic of the gravimetric method is that the difference between elapsed time and operator time is large, when compared with other methods, because the most time-consuming steps in the analysis do not require the constant attention of the analyst. For example, drying of crucibles, evaporation of solutions, digestion of precipitates, and ignition of products frequently take several hours to complete but require, at the most, a few minutes of the analyst's time, thus leaving him free to perform other tasks.

If methods for a given analysis are compared on the basis of operator time, the gravimetric approach often proves to be the most efficient, particularly where only one or two samples are to be analyzed, because no time is expended in calibration or standardization (that is, in the evaluation of k). On the other hand, as the number of samples to be analyzed increases, the time required for calibration in a nongravimetric procedure becomes smaller *on a per sample basis* and often becomes negligible when eight or ten samples are involved (assuming, of course, that a single calibration will suffice for all of the samples). With a large number of samples, then, the time required for precipitating, filtering, washing, and weighing may become larger than the equivalent operations in a nongravimetric procedure. Often, but not always, a gravimetric approach becomes less advantageous when a large number of samples is involved.

SENSITIVITY OF THE GRAVIMETRIC METHOD

In contrast to many analytical methods, a gravimetric analysis is seldom, if ever, limited in sensitivity (or accuracy) by the tool employed for measurement. Thus, with a suitable type of balance it is perfectly feasible to obtain the weight of a few micrograms of material to within a few parts per thousand of its true value; for larger masses, the weighing uncertainty can be decreased to a few parts per million.

The sensitivity of a gravimetric analysis is more likely to be limited by the difficulties encountered in attempting to separate a small amount of precipitate from a relatively large volume of solution that contains high concentrations of other constituents from the sample. In some situations solubility losses may be troublesome; in others, the precipitation rate may be so slow as to make the process impractical. For colloidal precipitates, coagulation often becomes difficult or impossible when the amounts of material are low. Finally, mechanical losses of a significant fraction of the solid become inevitable during filtration of a minute amount of solid suspended in a large volume of solution.

Because of these problems, the chemist is wise to discard the idea of employing a gravimetric analysis for a constituent if its concentration is likely to be below 0.1%; the method is best applied where the concentration is greater than 1%.

ACCURACY OF GRAVIMETRIC METHODS

It is difficult to generalize about the accuracy of gravimetric methods because each is subject in varying degree to uncertainties from such sources as solubility, coprecipitation, and variations in the chemical composition of the final product. Each of these factors is dependent upon the composition of the sample. For example, the iron content of a sample containing no other heavy metal ions is readily determined gravimetrically to within a few parts per thousand. On the other hand, in the presence of divalent cations such as zinc, nickel, and copper, errors of several percent must be expected unless an inordinate amount of time is expended in overcoming the effects of coprecipitation by these ions. Solubility losses are also affected by sample composition. With multicomponent samples, the likelihood of complex formation between the analyte and a constituent of the sample is enhanced; in addition, purification of the precipitate may require a larger volume of wash water and a concomitant increase in loss by solubility.

For simple samples containing more than 1% of analyte, the accuracy of gravimetric analysis is seldom surpassed by other methods; here errors may often be decreased to a part or two in 1000. With increasing sample complexity, larger errors are inevitable unless a great deal of time is expended in circumventing them. Here the accuracy of the gravimetric method may be no better than, and sometimes poorer than, other analytical methods.

SPECIFICITY OF GRAVIMETRIC METHODS

With a few notable exceptions, gravimetric reagents are not very specific; instead, virtually all are *selective* in the sense that they tend to form precipitates with groups of ions. Each ion within any group will then interfere with the analysis of any other ion in the group unless a preliminary separation is performed. In general, gravimetric procedures are less specific than some of the other methods that will be considered in later chapters.

APPLICABILITY OF GRAVIMETRIC METHODS

Gravimetric methods have been developed for most, if not all, inorganic anions and cations as well as for neutral species such as water, sulfur dioxide, carbon dioxide, and iodine. A variety of organic substances can also be readily determined gravimetrically. Examples include lactose in milk products, salicylates in drug preparations, phenolphthalein in laxatives, nicotine in pesticides, cholesterol in cereals, and benzaldehyde in almond extracts. Indeed, gravimetric methods are among the most widely applicable of all chemical analyses.

Applications of the Gravimetric Method

INORGANIC PRECIPITATING AGENTS

Table 6-3 lists the common inorganic precipitating agents. These reagents typically cause formation of a slightly soluble salt or a hydrous oxide. The

TABLE 6-3 Some Inorganic Precipitating Agents[a]

Precipitating Agent	Element Precipitated[b]
$NH_3(aq)$	**Be** (BeO), **Al** (Al_2O_3), **Sc** (Sc_2O_3), Cr (Cr_2O_3),* **Fe** (Fe_2O_3), Ga (Ga_2O_3), Zr (ZrO_2), **In** (In_2O_3), Sn (SnO_2), U (U_3O_8)
H_2S	Cu (CuO),* **Zn** (ZnO, or $ZnSO_4$), **Ge** (GeO_2), As ($\underline{As_2O_3}$, or As_2O_5), Mo (MoO_3), Sn (SnO_2),* Sb ($\underline{Sb_2O_3}$, or Sb_2O_5), Bi (Bi_2S_3)
$(NH_4)_2S$	Hg ($\underline{HgS}$), Co (Co_3O_4)
$(NH_4)_2HPO_4$	**Mg** ($Mg_2P_2O_7$), Al ($AlPO_4$), Mn ($Mn_2P_2O_7$), Zn ($Zn_2P_2O_7$), Zr ($Zr_2P_2O_7$), Cd ($Cd_2P_2O_7$), Bi ($BiPO_4$)
H_2SO_4	Li, Mn, **Sr, Cd, Pb, Ba** (all as sulfates)
H_2PtCl_6	K (K_2PtCl_6, or Pt), Rb ($\underline{Rb_2PtCl_6}$), Cs ($\underline{Cs_2PtCl_6}$)
$H_2C_2O_4$	Ca (CaO), Sr (SrO), **Th** (ThO_2)
$(NH_4)_2MoO_4$	Cd ($CdMoO_4$),* Pb ($\underline{PbMoO_4}$)
HCl	**Ag** ($AgCl$), Hg (Hg_2Cl_2), Na (as NaCl from butyl alcohol), Si (SiO_2)
$AgNO_3$	**Cl** ($AgCl$), Br ($\underline{AgBr}$), I($\underline{AgI}$)
$(NH_4)_2CO_3$	**Bi** (Bi_2O_3)
NH_4SCN	Cu ($Cu_2(SCN)_2$)
$NaHCO_3$	Ru, Os, Ir (pp'ted as hydrous oxides, reduced with H_2 to metallic state)
HNO_3	Sn (SnO_2)
H_5IO_6	Hg ($Hg_5(IO_6)_2$)
NaCl, $Pb(NO_3)_2$	F ($PbClF$)
$BaCl_2$	$SO_4{}^{2-}$ ($BaSO_4$)
$MgCl_2$, NH_4Cl	$PO_4{}^{3-}$ ($Mg_2P_2O_7$)

[a] From W. F. Hillebrand, G. E. F. Lundell, H. A. Bright, and J. I. Hoffman, *Applied Inorganic Analysis*. New York: John Wiley & Sons, Inc., 1953.

[b] Boldface type indicates that gravimetric analysis is the preferred method for the element or ion. Weighed form is indicated in parentheses. An asterisk indicates that gravimetric method is seldom used. Underscored entry indicates the most reliable gravimetric method.

weighing form is either the salt itself or else an oxide. The lack of specificity of most inorganic reagents is clear from the many entries in the table.

Detailed procedures for the gravimetric determination of several inorganic species are given in Experiments 1–5, Chapter 31.

REDUCING REAGENTS

Table 6-4 lists several reagents that convert the analyte to its elemental form for weighing.

TABLE 6-4   Some Reducing Reagents Employed in Gravimetric Methods

Reducing Agent	Analyte
SO_2	Se, Au
$SO_2 + H_2NOH$	Te
H_2NOH	Se
$H_2C_2O_4$	Au
H_2	Re, Ir
HCOOH	Pt
$NaNO_2$	Au
$TiCl_2$	Rh
$SnCl_2$	Hg
Electrolytic reduction	Co, Ni, Cu, Zn, Ag, In, Sn, Sb, Cd, Re, Bi

ORGANIC PRECIPITATING AGENTS

A number of organic reagents have been developed for the gravimetric analysis of inorganic species. In general, these reagents tend to be more selective in their reactions than many of the inorganic reagents listed in Table 6-3.

Two types of organic reagents are encountered. One forms slightly soluble nonionic complexes called *coordination compounds*. The other forms products in which the bonding between the inorganic species and the reagent is largely ionic.

Organic reagents which yield sparingly soluble coordination compounds typically contain at least two functional groups, each of which is capable of bonding with the cation by donation of a pair of electrons. The functional groups are located in the molecule in such a way that a five- or six-membered ring results from reaction. Coordination compounds which form complexes of this type are called *chelating agents*; their products with a cation are termed *chelates*.

Neutral coordination compounds are relatively nonpolar; as a consequence, their solubilities are low in water but high in organic liquids. Chelates usually possess low densities and are often intensely colored. Because they are not wetted by water, coordination compounds are readily freed of moisture at low temperatures. At the same time, however, their hydrophobic nature endows these precipitates with the annoying tendency to creep up the sides of the filtering medium during the washing operation; physical loss of solid may result unless care is taken. Three examples of coordination reagents are considered here.

8-Hydroxyquinoline. Approximately two dozen cations form sparingly soluble coordination compounds with 8-hydroxyquinoline, which is also known as *oxine.*

Typical of these is the product with magnesium

The solubilities of metal oxinates vary widely from cation to cation and, moreover, are pH dependent because proton formation always accompanies the chelation reaction. Therefore, by control of pH, a considerable degree of selectivity can be imparted to 8-hydroxyquinoline.

α-Nitroso-β-naphthol. This was one of the first selective organic reagents discovered (1885); its structure is

The reagent reacts with cobalt(II) to give a neutral cobalt(III) chelate having the structure CoA_3, where A^- is the conjugate base of the reagent. Note that formation of the product involves both oxidation and precipitation of the cobalt by the reagent; the precipitate is contaminated by reduction products of the reagent as a consequence. Therefore, it is common practice to ignite the chelate in oxygen to produce Co_3O_4; alternatively, the ignition is performed in a hydrogen atmosphere to produce the element as the weighed form.

The most important application of *α*-nitroso-*β*-naphthol has been for the determination of cobalt in the presence of nickel. Other ions that precipitate with the reagent include bismuth(III), chromium(III), mercury(II), tin(IV), titanium(III), tungsten(VI), uranium(VI), and vanadium(V).

Dimethylglyoxime. An organic precipitating agent of unparalleled specificity is dimethylglyoxime.

$$CH_3-C-C-CH_3$$

(with N, N and OH, OH groups)

Its coordination compound with palladium is the only one that is sparingly soluble in acid solution. Similarly, only the nickel compound precipitates from a weakly alkaline environment. Nickel dimethylglyoxime is bright red and has the structure

This precipitate is so bulky that only small amounts of nickel can be handled conveniently; it also has an exasperating tendency to creep as it is filtered and washed. The solid is readily dried at 110°C and has the composition indicated by its formula. A procedure for the determination of nickel in a steel employing dimethylglyoxime is found in Experiment 6, Chapter 31.

Sodium Tetraphenylboron. Sodium tetraphenylboron, $(C_6H_5)_4B^-Na^+$, is an important example of organic precipitating reagents that form saltlike precipitates. In cold mineral acid solutions, it is a near-specific precipitating agent for potassium ion and for ammonium ion. The precipitates are stoichiometric, corresponding to the potassium or the ammonium salt, as the case may be; they are amenable to vacuum filtration and can be brought to constant weight at 105 to 120°C. Only mercury(II), rubidium, and cesium interfere and must be removed by prior treatment.

Benzidine. Another salt-forming reagent is benzidine.

$$H_2N- \bigcirc - \bigcirc -NH_2$$

Benzidine precipitates sulfate from a slightly acidic medium as $C_{12}H_{12}N_2 \cdot H_2SO_4$. The solubility of this precipitate increases rapidly with temperature and also with the acidity of the environment; both variables must be carefully controlled. Instead of being weighed as a gravimetric precipitate, benzidine sulfate may be titrated with a standard solution of sodium hydroxide. Yet another method for completion of the analysis calls for titration of the benzidine with a standard solution of permanganate. The methods succeed in the presence of copper, cobalt, nickel, zinc, manganese(II), iron(II), chromium(III), and aluminum ions. Benzidine is well suited to the rapid, routine analysis of sulfate.

GRAVIMETRIC ORGANIC FUNCTIONAL GROUP ANALYSIS

Several reagents have been developed which react selectively with certain organic functional groups, and thus permit the determination of most compounds containing these groups. A list of gravimetric functional group reagents is given in Table 6-5. Many of these reactions can also be used for volumetric and spectrophotometric methods. For the occasional analysis, the gravimetric procedure will often be the method of choice since no calibration or standardization (p. 136) is required.

GRAVIMETRIC METHODS FOR SPECIFIC ORGANIC COMPOUNDS

Typical examples of methods developed for specific organic compounds are described in the paragraphs that follow.

Determination of Salicylic Acid. Salicylic acid can be determined by dissolving the compound in a sodium carbonate solution and digesting after addition of iodine solution. The reaction is

The yellow tetraiodophenylenequinone is filtered, dried, and weighed. Acetylsalicylic acid (aspirin) can be determined in the same way after hydrolysis to give salicylic acid.

Determination of Nicotine. Nicotine in pesticides or tobacco products can be determined, after steam distillation from the sample, by precipitation with silicotungstic acid. After filtration and washing, the residue is ignited to give $SiO_2 \cdot 12WO_3$ which is weighed.

Determination of Phenolphthalein. A standard method for the determination of phenolphthalein in laxative preparations involves separation of the compound

TABLE 6-5 Gravimetric Methods for Organic Functional Groups

Functional Group	Basis for Method	Reaction and Product Weighed[a]
Carbonyl	Weight of precipitate with 2,4-dinitrophenylhydrazine	$RCHO + H_2N\underline{NHC_6H_3(NO_2)_2} \rightarrow R\text{—}CH\text{=}\underline{NNHC_6H_3(NO_2)_2}(s) + H_2O$ (RCOR' reacts similarly)
Aromatic carbonyl	Weight of CO_2 formed at 230°C in quinoline. CO_2 distilled, absorbed, and weighed.	$ArCHO \xrightarrow[CuCO_3]{230°C} Ar + \underline{CO_2}(g)$
Methoxyl and ethoxyl	Weight of AgI formed after distillation and decomposition of CH_3I or C_2H_5I	$\left.\begin{array}{l} ROCH_3 \;\; + HI \rightarrow ROH \;\;\; + CH_3I \\ RCOOCH_3 + HI \rightarrow RCOOH + CH_3I \\ ROC_2H_5 \;\; + HI \rightarrow ROH \;\;\; + C_2H_5I \end{array}\right\}\; CH_3I + Ag^+ + H_2O \rightarrow \underline{AgI}(s) + CH_3OH$
Aromatic nitro	Weight loss of Sn	$RNO_2 + \frac{3}{2}\underline{Sn}(s) + 6H^+ \rightarrow RNH_2 + \frac{3}{2}Sn^{4+} + 2H_2O$
Azo	Weight loss of Cu	$RN\text{=}NR' + 2\underline{Cu}(s) + 4H^+ \rightarrow RNH_2 + R'NH_2 + 2Cu^{2+}$
Phosphate	Weight of Ba salt	$\begin{array}{c} O \\ \| \\ ROP(OH)_2 \end{array} + Ba^{2+} \rightarrow \begin{array}{c} O \\ \| \\ \underline{ROPO_2Ba}(s) \end{array} + 2H^+$
Sulfonic acid	Weight $BaSO_4$ after reduction with HNO_2	$RNHSO_3H + HNO_2 + Ba^{2+} \rightarrow ROH + \underline{BaSO_4}(s) + N_2 + 2H^+$
Sulfinic acid	Weight Fe_2O_3 after ignition of Fe^{3+} sulfinate	$3ROSOH + Fe^{3+} \rightarrow (ROSO)_3Fe + 3H^+$ $(ROSO)_3Fe \xrightarrow{O_2} CO_2 + H_2O + SO_2 + \underline{Fe_2O_3}(s)$

[a] The substance weighed is underlined.

by alcohol extraction followed by evaporation to dryness. The phenolphthalein is then dissolved in dilute alkali and precipitated as the tetraiodo compound by addition of iodine solution. The product can be dried to constant weight at 110°C.

VOLATILIZATION PROCEDURES

The two most common gravimetric analyses based on volatilization are for water and for carbon dioxide.

Water is quantitatively eliminated from many inorganic samples by ignition. In the direct determination, it is collected on any of several solid desiccants, and its mass is determined from the gain in weight of the desiccant. An apparatus for this procedure is shown in Chapter 27.

Less satisfactory is the indirect method based upon the loss in weight suffered by the sample as the result of ignition. Here it must be assumed that water is the only component that has been volatilized. This assumption is frequently unjustified; ignition of many samples results in decomposition and a change in weight, irrespective of the presence of water.

Carbonates are ordinarily decomposed by acids to give carbon dioxide, which is readily evolved from solution by heat. As in the direct analysis for water, the weight of carbon dioxide is established from the increase in the weight of a solid absorbent. Ascarite,[5] which consists of sodium hydroxide on asbestos, serves to retain the carbon dioxide by the reaction

$$2NaOH + CO_2 \rightarrow Na_2CO_3 + H_2O$$

The absorption tube must also contain a desiccant to prevent loss of the evolved water.

Sulfides and sulfites can also be determined by volatilization. Here hydrogen sulfide or sulfur dioxide is evolved from the sample after treatment with acid and collected in a suitable absorbent.

PROBLEMS

*1. Use chemical symbols to express the gravimetric factor for each of the following:

Sought	Weighed		Sought	Weighed
(a) $Na_2B_4O_7$	KBF_4		(d) SiF_4	$SiO_2 \cdot 12MoO_3$
(b) Nb	Nb_2O_5		(e) Mo	$SiO_2 \cdot 12MoO_3$
(c) Nb_2O_3	Nb_2O_5		(f) P	$Zn_2P_2O_7$

2. Use chemical symbols to express the gravimetric factor for each of the following:

Sought	Weighed		Sought	Weighed
(a) Tb	Tb_4O_7		(d) $C_2H_6S_2$	$BaSO_4$
(b) $C_6H_6Cl_6$	$AgCl$		(e) Rh_2SO_4	$RhB(C_6H_5)_4$
(c) Co_3O_4	$Co_2P_2O_7$		(f) Hg_2Cl_2	$Hg_3(AsO_4)_2$

[5] Arthur H. Thomas Co., Philadelphia, Pa.

*3. Use chemical symbols to express the gravimetric factor if the percentage of $Fe_3Al_2Si_3O_{12}$ is sought and the weighed form is:
(a) Fe_2O_3.
(b) SiO_2.
(c) Al_2O_3.
(d) $Al(C_9H_6ON)_3$.

4. Use chemical symbols to express the gravimetric factor if the percentage of $Co_3As_2O_8 \cdot 8H_2O$ is sought and the weighed form is:
(a) As_2S_3.
(b) $MgNH_4AsO_4$.
(c) $Mg_2As_2O_7$.
(d) H_2O.
(e) Co_3O_4.
(f) $CoSO_4$

*5. Use chemical symbols to generate the gravimetric factor required to express the results of an analysis in terms of percent $CuCrO_4 \cdot 2CuO \cdot 2H_2O$ if the weighing form is:
(a) Cu.
(b) $Ag_2Cr_2O_7$.
(c) $PbCrO_4$.
(d) H_2O.
(e) $Cu_2(SCN)_2$.
(f) CuO.

*6. Which of the weighing forms in Problem 5 will yield the greatest weight of product from the same quantity of $CuCrO_4 \cdot 2CuO \cdot 2H_2O$? Which weighing form will provide the smallest weight of product?

7. What weight of Ag_2SO_4 can be obtained from 0.420 g of
(a) K_2SO_4?
(b) $AgNO_3$?
(c) $Na[Ag(CN)_2]$?
(d) $Ag[Ag(CN)_2]$?

*8. A 9.75-g pesticide sample yielded 0.186 g of TlI after treatment with an excess of KI. Calculate the percentage of Tl_2SO_4 in the sample.

9. Calculate the percentage of copper in an ore specimen if a 0.864-g sample yielded 0.397 g of $Cu_2(SCN)_2$.

*10. The sulfur in three saccharine $(C_7H_5NO_3S)$ tablets that weighed a total of 0.118 g was oxidized to sulfate and then precipitated as $BaSO_4$. Calculate the average weight of saccharine in these tablets if 0.150 g of $BaSO_4$ was recovered.

11. The iodide ion in a 4.13-g sample that also contained chloride was converted to iodate by treatment with an excess of bromine

$$3H_2O + 3Br_2 + I^- \rightarrow 6Br^- + IO_3^- + 6H^+$$

The unused bromine was eliminated by boiling. An excess of barium chloride was then added; 0.0810 g of $Ba(IO_3)_2$ was subsequently recovered. Calculate the percentage of potassium iodide originally present in the sample.

*12. What volume of a 1% (w/v) dimethylglyoxime $(C_4H_8N_2O_2)$ solution should be used to precipitate the Ni^{2+} in a solution containing 100 mg of $NiCl_2$, assuming that a 5.0% excess of the reagent is necessary to assure quantitative removal of the cation?

13. What volume of a 4% (w/v) solution of 8-hydroxyquinoline (C_9H_7NO) should be used to precipitate Al^{3+} from a solution containing 80 mg of $Al_2(SO_4)_3$, assuming that a 10.0% excess of the reagent is necessary to assure quantitative removal of the cation? (The formal ratio of reagent to Al^{3+} in the precipitate is 3:1.)

*14. If a 10.0% reagent excess is needed to reduce solubility losses, what volume of the following solutions will be needed to precipitate the sulfate from 1.00-g samples that contain 20.0% $Al_2(SO_4)_3$:
(a) a solution that is 0.120 F in $BaCl_2$?
(b) a solution that is 3.25% (w/v) in $BaCl_2 \cdot 2H_2O$?

15. The cobalt in 0.50-g samples believed to contain about 2.00% Co_3O_4 is to be precipitated with α-nitroso-β-naphthol $(C_{10}H_7NO_2)$. If a 10.0% excess of the

reagent is needed to minimize solubility losses, what volume of a 0.70% (w/v) solution of the reagent in glacial acetic acid should be employed? (The formal ratio of reagent to Co in the precipitate is 3:1.)

*16. The bismuth oxycarbonate $(BiO)_2CO_3$ in a commercial drug tablet was analyzed by dissolving a 0.200-g sample in dilute HNO_3 and precipitating the Bi^{3+} as $BiOCl$ by the addition of Cl^-. After drying at $100°C$, this precipitate was found to weigh 0.0512 g. Calculate the percent $(BiO)_2CO_3$ in the sample.

17. After appropriate preliminary steps, the uranium in a 0.434-g specimen containing the mineral carnotite $[K_2(UO_3)_2(VO_4)_2 \cdot 3H_2O]$ was precipitated as the oxalate. Ignition of the precipitate yielded 0.163 g U_3O_8.
 (a) Calculate the percent carnotite in the sample.
 (b) How many pounds of vanadium are contained in one ton of the rock?

*18. A 0.543-g sample containing CdS and inert impurities was ignited in air to convert the CdS quantitatively to $CdSO_4$. After ignition, the residue was found to weigh 0.674 g. Calculate the percent Cd in the sample.

19. A 0.424-g sample which contained epsomite $(MgSO_4 \cdot 7H_2O)$ and inert material was dehydrated quantitatively by heating in a stream of dry air. The air was passed through a tube containing a desiccant. The original weight of the tube was 21.472 g; after water absorption, its weight was 21.593 g. Calculate the percent $MgSO_4 \cdot 7H_2O$ in the sample, assuming that the mineral represented the only source of the water in the sample.

*20. A 1.778-g sample containing the mineral hydrocerussite $[2PbCO_3 \cdot Pb(OH)_2]$ was heated to a temperature sufficient to cause decomposition to PbO.

$$2PbCO_3 \cdot Pb(OH)_2(s) \rightarrow 3PbO(s) + 2CO_2 + H_2O$$

The ignited residue was found to weigh 1.549 g.
 (a) Calculate the percent hydrocerussite in the sample.
 (b) How many pounds of lead are contained in one ton of the material analyzed?

21. The zinc in a 5.14-g sample of a foot powder was freed of organic matter by wet-ashing with a perchloric–nitric acid mixture and precipitated as $ZnNH_4PO_4$. Ignition of the filtered precipitate yielded 0.317 g of $Zn_2P_2O_7$. Calculate the percentage of zinc undecylenate, $Zn(C_{11}H_{19}O_2)_2$, in the sample.

*22. Chloromycetin is an antibiotic with the formula $C_{11}H_{12}O_5N_2Cl_2$. A 1.03-g sample of an ophthalmic ointment was heated in a closed tube with metallic sodium to destroy the organic material and free the chloride; after dissolving the ignited mixture in water, the carbonaceous residue was removed by filtration. The chloride was then precipitated with $AgNO_3$ to give 0.0129 g of AgCl. Calculate the percentage of chloromycetin in the sample.

23. Twenty tablets of a therapeutic iron preparation, with a total weight of 21.3 g, were ground and mixed thoroughly until homogeneous. A 3.13-g sample of the powdered material was then dissolved in HNO_3; the solution was heated to oxidize the Fe to the $+3$ state, and ammonia was added to precipitate $Fe_2O_3 \cdot xH_2O$. After ignition, the Fe_2O_3 was found to weigh 0.334 g. On an average, how many milligrams of $FeSO_4 \cdot 7H_2O$ were contained per tablet?

*24. In order to determine the phosphorus content of a fertilizer, a 0.217-g sample was decomposed with hot concentrated nitric acid. The resulting solution was diluted, following which the PO_4^{3-} was precipitated as the quinoline salt of phosphomolybdic acid, $(C_9H_7N)_2H_3PO_4 \cdot 12MoO_3$. After filtration and drying, the precipitate was found to weigh 0.684 g. Calculate the percent P_2O_5 in the sample.

25. The nitrobenzene in a 0.677-g sample was determined by dilution with an HCl-methanol mixture followed by the introduction of 0.834 g of pure Sn. After being refluxed for one hour, the solution was cooled and filtered. The residual Sn was found to weigh 0.619 g. Calculate the percent $C_6H_5NO_2$ in the sample.

$$2C_6H_5NO_2 + 3Sn(s) + 12H^+ \rightarrow 2C_6H_5NH_2 + 4H_2O + 3Sn^{4+}$$

26. A random selection of 25 commercial aspirin tablets weighed 8.75 g. After the tablets had been ground to a fine powder and homogenized, a 0.384-g sample was weighed into a flask and refluxed 20 minutes with an aqueous Na_2CO_3 solution; this treatment converted the aspirin ($C_9H_8O_4$) to salicylic acid ($C_7H_6O_3$). Talc and other inert ingredients were then removed by filtration, iodine was added, and the solution was digested for one hour. The tetraiodophenylenequinone $(C_6H_2I_2O)_2$ produced was filtered, dried, and found to weigh 0.686 g. Calculate the average number of grains of aspirin per tablet (1.00 grain = 0.0648 g). The gravimetric factor should be based on the number of formula weights of aspirin needed to produce each formula weight of tetraiodophenylenequinone; see p. 142.

*27. A sulfuric acid solution was standardized by precipitating the SO_4^{2-} from a 100.0-ml aliquot as $BaSO_4$. The ignited $BaSO_4$ weighed 0.473 g. What was the formal H_2SO_4 concentration of the solution?

28. The KCNO content of a herbicide preparation was determined by dissolving a 2.14-g sample in water and adding an excess of semicarbazide hydrochloride. Reaction:

$$NH_2CONHNH_3^+ + CNO^- \rightarrow NH_2CONHNHCONH_2(s)$$

The resulting precipitate was filtered, washed, and dried; its weight was 0.217 g. Calculate the percent KCNO in the sample.

*29. A soil sample was found to contain 5.35% H_2O by oven drying. A 1.04-g sample of the dried soil yielded, after suitable treatment, 0.584 g SiO_2 and 1.97 g $CaCO_3$. Calculate the percent Si and percent Ca in the sample on a dried basis and on an as-received basis.

30. The total phosphorus content of a fertilizer was found to be 16.3% P_2O_5 on the basis of a dried sample. The sample, as received, was found to contain 3.47% water.
    (a) What was the percent P_2O_5 in the undried material?
    (b) What was the percent superphosphate, $Ca(H_2PO_4)_2 \cdot H_2O$, in the undried material?

*31. A 0.998-g sample that contained $Al_2(SO_4)_3$, $MgSO_4$, and inert materials was dissolved, and the Al^{3+} and Mg^{2+} were precipitated with 8-hydroxyquinoline. After filtration and drying at 300°C, the mixture of $Al(C_9H_6NO)_3$ and $Mg(C_9H_6NO)_2$ was found to weigh 0.8746 g. The residue was then ignited at a temperature that converted the two hydroxyquinolates to Al_2O_3 and MgO, which weighed 0.1067 g. Calculate the percent $Al_2(SO_4)_3$ and $MgSO_4$ in the sample.

32. A 1.34-g sample containing KCl, KI, and inert materials was dissolved and treated with an excess of $AgNO_3$. After filtration, washing, and drying, the combined AgI and AgCl weighed 1.047 g. The precipitate was then heated in a stream of Cl_2 to convert the AgI quantitatively to AgCl. The residue then weighed 0.843 g. Calculate the percent KCl and KI in the sample.

*33. A 0.894-g sample known to contain only $Al_2(SO_4)_3$ and $BeSO_4$ was dissolved and made ammoniacal. The hydrous oxides of the two cations were filtered, washed, and ignited to form BeO and Al_2O_3. The residue weighed 0.2347 g. Calculate the percent $Al_2(SO_4)_3$ and $BeSO_4$ in the sample.

34. An excess of $(NH_4)_2CrO_4$ was added to a 50.0-ml sample of an aqueous solution containing $Ba(NO_3)_2$ and $Pb(NO_3)_2$. The precipitate of $PbCrO_4$ and $BaCrO_4$ was filtered, washed, and dried; its weight was found to be 0.3696 g. When a second 50.0-ml sample was treated with dilute sulfuric acid, the combined weight of $BaSO_4$ and $PbSO_4$ produced was 0.3445 g. Calculate the mg/ml of each nitrate in the sample solution.

*35. A 2.621-g sample containing $NaBr$, $NaBrO_3$, and inert materials was dissolved in water and diluted to exactly 250 ml. Addition of $NaHSO_3$ and acid to a 50.0-ml portion of the solution resulted in a conversion of the BrO_3^- to Br^-.

$$BrO_3^- + 3HSO_3^- \rightarrow Br^- + 3SO_4^{2-} + 3H^+$$

An excess of $AgNO_3$ was added, which yielded 0.331 g of AgBr. Acidification of a second 50.0-ml portion of the solution with H_2SO_4 caused formation of bromine.

$$BrO_3^- + 5Br^- + 6H^+ \rightarrow 3Br_2 + 3H_2O$$

The ratio of Br^- to BrO_3^- in the original sample was such that an excess of Br^- remained. After removal of Br_2 by boiling, the residual Br^- was precipitated as AgBr, 0.102 g being produced. Calculate the percent NaBr and $NaBrO_3$ in the sample.

36. A 0.400-g sample containing only In_2O_3 and TeO_2 was analyzed by solution and precipitation with 8-hydroxyquinoline. The mixture of $In(C_9H_6ON)_3$ and $TeO(C_9H_6ON)_2$ was found to weigh 1.530 g. Calculate the percent of each oxide present in the sample.

*37. How many grams of ethyl oxalate $(C_2H_5)_2C_2O_4$ should be taken to precipitate calcium homogeneously as the oxalate from a 0.200-g sample that contains 15.0% CaO? (See Table 6-2 for the reaction.) Assume a 10.0% excess of the reagent.

38. How many grams of thioacetamide should be taken for the homogeneous precipitation of As_2S_3 from 100.0 ml of a solution that is 0.100 F in H_3AsO_3 if, after precipitation, the solution is to be 0.100 F in H_2S? (See Table 6-2 for the reaction.)

*39. A possible gravimetric method for determining Se is being considered in which the element is oxidized to SeO_4^{2-} and precipitated as Ag_2SeO_4 ($K_{sp} = 5.6 \times 10^{-8}$). Assume that samples containing about 0.100 g of Se are to be analyzed and that after conversion to SeO_4^{2-}, enough $AgNO_3$ is added to complete the precipitation and leave the solution 0.200 F in $AgNO_3$; the final volume is to be 80.0 ml.

    (a) Calculate the solubility loss, in terms of grams Se. (Assume that SeO_4^{2-} in the mother liquor does not react with H_3O^+.)

    (b) If the precipitate is to be washed with four 10.0-ml portions of water, what weight of Se will be lost in the wash water as a consequence of solubility of Ag_2SeO_4? (Assume that the wash water contains no $AgNO_3$ and becomes saturated with Ag_2SeO_4.)

    (c) What will be the relative error in the analysis resulting from solubility losses?

40. The solubility product for PbOHBr has a value of 2.0×10^{-15}.

$$PbOHBr(s) \rightleftarrows Pb^{2+} + OH^- + Br^-$$

The lead in 100.0 ml of a solution containing 0.400 g $Pb(NO_3)_2$ was determined by the addition of Br^- to a solution maintained at $[H_3O^+] = 1.00 \times 10^{-6}$; after precipitation was complete, the solution was 0.0200 F in KBr and had a volume of 200 ml.

(a) Calculate the weight of unprecipitated $Pb(NO_3)_2$ in the solvent and the resultant relative error in the analysis.

(b) If the precipitate were washed with four 10.0-ml portions of the solution buffered to $[H_3O^+] = 1.00 \times 10^{-6}$, what would be the weight of $Pb(NO_3)_2$ lost in the washing? (Assume that the wash water is free of Br^- and is saturated with PbOHBr.)

(c) What is the combined relative error in the analysis due to solubility in solvent and wash water?

AN INTRODUCTION TO VOLUMETRIC METHODS OF ANALYSIS

A quantitative analysis based upon the measurement of volume is called a volumetric or titrimetric method. Volumetric methods are much more widely used than gravimetric methods because they are usually more rapid and convenient; in addition, they are often as accurate.

Definition of Some Terms

Titration is the process by which the quantity of analyte in a solution is determined from the amount of a standard reagent it consumes. Ordinarily, a titration is performed by carefully adding the reagent of known concentration until reaction with the analyte is judged to be complete; the volume of standard reagent is then measured. Occasionally, it is convenient or necessary to add an excess of the reagent and then determine the excess by *back-titration* with a second reagent of known concentration.

The reagent of exactly known concentration that is used in a titration is called a *standard solution*. The accuracy with which its concentration is known sets a definite limit upon the accuracy of the method; for this reason, much care is taken in the preparation of standard solutions. The concentration of a standard solution is established either directly or indirectly:

1. by dissolving a carefully weighed quantity of the pure reagent and diluting to an exactly known volume,
2. by titrating a solution containing a weighed quantity of a pure compound with the reagent solution.

In either method, a highly purified chemical compound—called a *primary standard*—is required as the reference material. The process whereby the concentration of a standard solution is determined by titration of a primary standard is called a *standardization*.

The goal of every titration is the addition of standard solution in an amount that is chemically equivalent to the substance with which it reacts. This condition is achieved at the *equivalence point*. For example, the equivalence point in the titration of sodium chloride with silver nitrate is attained when exactly one formula weight of silver ion has been introduced for each formula weight of chloride ion present in the sample. In the titration of sulfuric acid with sodium hydroxide, the equivalence point occurs when two formula weights of the latter have been introduced for each formula weight of the former.

The equivalence point in a titration is a theoretical concept; in actual fact, its position can be estimated only by observing physical changes associated with equivalence. These changes manifest themselves at the *end point* of the titration. It is to be hoped that any volume difference between the end point and equivalence point will be small. Differences do exist, however, owing to inadequacies in the physical changes and our ability to observe them; a *titration error* is the result.

A common method of end-point detection in volumetric analysis involves the use of a supplementary chemical compound that exhibits a change in color as a result of concentration changes occurring near the equivalence point. Such a substance is called an *indicator*.

Reactions and Reagents Used in Volumetric Analysis

It is convenient to classify volumetric methods according to four reaction types, specifically, precipitation, complex formation, neutralization (acid-base), and oxidation-reduction. Each reaction type is unique in such matters as nature of equilibria involved; the indicators, reagents, and primary standards available; and the definition of equivalent weight.

PRIMARY STANDARDS

The accuracy of a volumetric analysis is critically dependent upon the primary standard used to establish, directly or indirectly, the concentration of the standard solution. Important requirements for a substance to serve as a good primary standard include the following:

1. Highest purity. Moreover, established methods should be available for confirming its purity.
2. Stability. It should not be attacked by constituents of the atmosphere.

3.   Absence of hydrate water. If the substance were hygroscopic or efflorescent, drying and weighing would be difficult.
4.   Ready availability at reasonable cost.
5.   Reasonably high equivalent weight. The weight of a compound required to standardize or prepare a solution of a given concentration increases directly with its equivalent weight. Since the relative error in weighing decreases with increasing weight, a high equivalent weight will tend to minimize weighing errors.

Few substances meet or even approach these requirements. As a result, the number of primary-standard substances available to the chemist is limited.

In some instances, it is necessary to use less pure substances in lieu of a primary standard. The assay (that is, the percent purity) of such a *secondary* standard must be established by careful analysis.[1]

STANDARD SOLUTIONS

An ideal standard solution for titrimetric analysis would have the following properties:

1.   Its concentration should remain constant for months or years after preparation to eliminate the need for restandardization.
2.   Its reaction with the analyte should be rapid in order that the waiting period after each addition of reagent does not become excessive.
3.   The reaction between the reagent and the analyte should be reasonably complete. As we shall presently show, satisfactory end points generally require this condition.
4.   The reaction of the reagent with the analyte must be such that it can be described by a balanced chemical equation; otherwise, the weight of the analyte cannot be calculated directly from the volumetric data. This requirement implies the absence of side reactions between the reagent and the unknown or with other constituents of the solution.
5.   A method must exist for detecting the equivalence point between the reagent and the analyte; that is, a satisfactory end point is required.

Few volumetric reagents currently in use meet all of these requirements perfectly.

End Points in Volumetric Methods

End points are based upon a physical property which changes in a characteristic way at or near the equivalence point in the titration. The most common end point involves a color change due to the reagent, the analyte, or an indicator substance. Other physical properties, such as electrical potential, conductivity, temperature, and refractive index, have also been employed to locate the equivalence point in titrations.

[1] Usage of the terms primary and secondary standard is a source of disagreement among chemists. Some reserve primary standard for the relatively few compounds that possess the properties listed at the start of this section (see, for example, H. A. Laitinen and W. B. Harris, *Chemical Analysis*, 2d ed., p. 101. New York: McGraw-Hill Book Company, Inc., 1975) and use secondary to denote species that have reliably known composition. Others prefer to consider all substances of known assay as primary standards and refer to solutions of known concentration as secondary standards. We prefer the former set of definitions.

CONCENTRATION CHANGES DURING TITRATION

End points are the result of marked alterations in at least one reactant concentration during the titration. For most (but not all) end points, changes that occur in the region surrounding the equivalence point are particularly important. To illustrate, the second and third columns of Table 7-1 contain data that show the

TABLE 7-1   Concentration Changes during a Titration[a]

Vol 0.100-F NaOH, ml	Concentration H_3O^+, mole/liter	Concentration OH^-, mole/liter	pH	pOH
0.0	1.0×10^{-1}	1.0×10^{-13}	1.00	13.00
40.91	1.0×10^{-2}	1.0×10^{-12}	2.00	12.00
49.01	1.0×10^{-3}	1.0×10^{-11}	3.00	11.00
49.90	1.0×10^{-4}	1.0×10^{-10}	4.00	10.00
49.990	1.0×10^{-5}	1.0×10^{-9}	5.00	9.00
49.9990	1.0×10^{-6}	1.0×10^{-8}	6.00	8.00
50.0000	1.0×10^{-7}	1.0×10^{-7}	7.00	7.00
50.0010	1.0×10^{-8}	1.0×10^{-6}	8.00	6.00
50.010	1.0×10^{-9}	1.0×10^{-5}	9.00	5.00
50.10	1.0×10^{-10}	1.0×10^{-4}	10.00	4.00
51.01	1.0×10^{-11}	1.0×10^{-3}	11.00	3.00
61.1	1.0×10^{-12}	1.0×10^{-2}	12.00	2.00

[a] While these data are useful for illustration purposes, it should be understood that their realization in the laboratory would not be possible because measurements of volumes and normalities to six significant figures are ordinarily impossible.

changes in hydronium and hydroxide ion concentrations that occur when 50.0 ml of 0.100-F hydrochloric acid are titrated with standard 0.100-F NaOH. To emphasize the *relative* changes in concentration that occur, the volumes selected for tabulation correspond to the increments needed to cause a tenfold decrease or increase in concentration of the reacting species. Thus, we see that 40.91 ml of base are needed to decrease the hydronium ion concentration from 0.100 M to 0.0100 M. An additional 8.1 ml are required to lower the concentration to 0.00100 M; 0.89 ml will cause another tenfold decrease and so on. Analogous increases in the hydroxide ion concentration occur simultaneously.

It is apparent from the data in Table 7-1 that the equivalence-point region is characterized by the largest changes in the relative concentrations of the reacting species. Most end points depend upon these large changes.

Figure 7-1 consists of plots of the concentration data in Table 7-1. The ordinate scale for these plots must perforce be large to encompass the enormous concentration changes experienced by the reactants. As a result, changes occurring in the region of most interest—that is, the equivalence point—are obscured.

To show more clearly the changes taking place in the region of interest, it is useful to express concentrations in logarithmic, rather than linear, terms. Further, since the concentrations are ordinarily less than unity, it is advantageous to employ a negative logarithm since this function will express the concentration

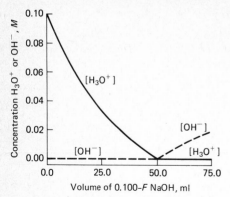

Figure 7-1 Changes in Reactant Concentrations during Titration of 50.00 ml of 0.100-F HCl with 0.100-F NaOH.

as a small positive number. The negative logarithm (to the base 10) of a molar concentration has come to be known as its *p-function*. Values for pH and pOH are given in the last two columns of Table 7-1.

Example. Calculate the p-functions for each ion in a solution that is 0.020 F in NaCl and 0.0054 F in HCl.

1. $$[H_3O^+] = 0.0054 = 5.4 \times 10^{-3}$$
$$pH = -\log [H_3O^+] = -\log (5.4 \times 10^{-3})$$
$$= -\log 5.4 - \log 10^{-3}$$
$$= -0.73 - (-3.00) = 2.27$$

Note that three figures are retained in the result even though $[H_3O^+]$ was given to only two (see p. 77).

2. $$[Na^+] = 0.020 = 2.0 \times 10^{-2}$$
$$pNa = -\log (2.0 \times 10^{-2}) = -\log 2.0 - \log 10^{-2}$$
$$= -0.30 + 2.00 = 1.70$$

3. $$[Cl^-] = 0.020 + 0.0054 = 0.0254$$
$$pCl = -\log (2.54 \times 10^{-2}) = -\log 2.54 - \log 2.00$$
$$= -0.405 + 2.00 = 1.60$$

Example. Calculate $[Ag^+]$ for a solution in which pAg is 6.372.

$$pAg = -\log [Ag^+] = 6.372$$
$$\log [Ag^+] = -6.372 = -7.000 + 0.628$$
$$[Ag^+] = \text{antilog} (-7.000) \times \text{antilog} (0.628)$$
$$= 10^{-7} \times 4.246 = 4.25 \times 10^{-7}$$

(For method of rounding, see page 78.)

If the molar concentration of an ion is greater than unity, its p-function will be a negative number. For example, in a 2.0-F solution of HCl

$$[H_3O^+] = 2.0$$
$$pH = -\log [H_3O^+] = -\log 2.0 = -0.30$$

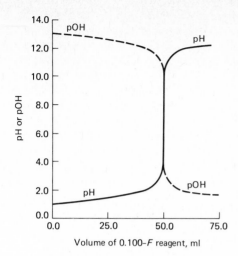

Figure 7-2   Titration Curves for 50.0 ml of 0.100-*F* HCl with 0.100-*F* NaOH.

If the values of pH and pOH given in Table 7-1 are plotted against the volume of reagent added, we obtain curves that give a much clearer picture of the changes that occur in the solution near the equivalence point; it is evident that a marked increase in pH and a corresponding decrease in pOH occurs here.

Curves such as either of those in Figure 7-2 are called *titration curves*. Plots of analogous data for titrations involving oxidation-reduction, precipitation, or complex-formation reactions have the same general characteristics. Typical examples are derived and discussed in later chapters.

The shape of the titration curve and especially the magnitude of the change in the equivalence-point region are of great interest to the analytical chemist, because these determine how easily and how accurately the equivalence point can be located. The ease with which the physical change used to signal an end point can be detected is usually directly related to the magnitude and the rate of change in p-values. In general, the most satisfactory end points occur where changes in p-functions are large and take place over a small range of reagent volumes.

Calculations Associated with Titrimetric Methods

In preparation for the discussion that follows, the student may find it helpful to review the material in Chapter 2 (pp. 10–17) on chemical units of weight and concentration.

Chemists frequently make use of the *equivalent weight* (or the *milli-equivalent weight*) as the basis for volumetric calculations; *normality* is the corresponding unit of concentration. The manner in which these units are defined depends upon the type of reaction which serves as the basis for the analysis—that is, whether neutralization, oxidation-reduction, precipitation, or complex formation is involved. Furthermore, the chemical behavior of the substance must be carefully specified if its equivalent weight is to be defined unambiguously. If the substance can react in more than one way, it will likewise have more than one equivalent or milliequivalent weight. *Thus, the definition of the equivalent weight or the milliequivalent weight for a substance is always based on its behavior in a specific chemical reaction. Evaluation of the equivalent weight is impossible if the reaction involved is not specifically stated. Likewise, the concentration of a solution cannot be expressed in terms of normality in the absence of this information.*

EQUIVALENT OR MILLIEQUIVALENT WEIGHTS IN NEUTRALIZATION REACTIONS

The *equivalent weight* (eq wt) of a substance participating in a neutralization reaction is that weight which either contributes or reacts with one gram formula weight of hydrogen ion *in that reaction*. The *milliequivalent weight* (meq wt) is $\frac{1}{1000}$ of the equivalent weight.

For strong acids or bases and for acids or bases containing only a single reactive hydrogen or hydroxide ion, the relationship between the equivalent weight and the formula weight can be readily determined. For example, the equivalent weight for potassium hydroxide and hydrochloric acid must be equal to their formula weights, respectively, since each has only a single reactive hydrogen ion or hydroxide ion. Similarly, we know that only one hydrogen in acetic acid, $HC_2H_3O_2$, is acidic; therefore, the formula weight and equivalent weight for this acid must also be identical. Barium hydroxide, $Ba(OH)_2$, is a strong base containing two hydroxide ions that are indistinguishable. In any acid-base reaction, this base will necessarily react with two hydrogen ions; thus, its equivalent weight will be one-half its formula weight. For sulfuric acid, dissociation of the second hydrogen ion in water is not complete; the hydrogen sulfate ion, however, is a sufficiently strong acid so that both hydrogens participate in all aqueous neutralization reactions. The equivalent weight of H_2SO_4 as an acid is, therefore, always one-half its formula weight in aqueous solution.

The situation becomes more complex for a polyfunctional acid, which contains two or more hydrogen ions with differing tendencies to dissociate. For example, certain indicators undergo a color change when only the first of the three protons in phosphoric acid has been neutralized; that is,

$$H_3PO_4 + OH^- \rightarrow H_2PO_4^- + H_2O$$

Other indicators change color after two hydrogen ions have reacted:

$$H_3PO_4 + 2OH^- \rightarrow HPO_4^{2-} + 2H_2O$$

In the first titration, the equivalent weight of phosphoric acid is equal to its formula weight; in the second, it is one-half its formula weight. (It is not practical to titrate the third proton. Thus, an equivalent weight for H_3PO_4 that is one-third the formula weight is not encountered.) *Without knowing which of these reactions is involved, an unambiguous definition of the equivalent weight for phosphoric acid is impossible.*

EQUIVALENT WEIGHT IN OXIDATION-REDUCTION REACTIONS

The equivalent weight of a participant in an oxidation-reduction reaction is that weight which, directly or indirectly, consumes or produces one mole of electrons. The numerical value for the equivalent weight is conveniently established by dividing the formula weight of the substance of interest by the change in oxidation number associated with its reaction. As an example, consider the oxidation of oxalate ion by permanganate:

$$5C_2O_4^{2-} + 2MnO_4^- + 16H^+ \rightarrow 10CO_2 + 2Mn^{2+} + 8H_2O$$

In this reaction, the change in oxidation number for manganese is 5, since the element passes from the $+7$ to the $+2$ state; the equivalent weight for MnO_4^- or Mn^{2+} is therefore one-fifth of the corresponding formula weights. Each carbon atom in the oxalate ion, on the other hand, is oxidized from the $+3$ to the $+4$ state. The equivalent weight of sodium oxalate is one-half its formula weight because it contains two carbon atoms. The equivalent weight of carbon dioxide, on the other hand, is equal to its formula weight because it contains but a single atom of carbon. These and other examples of equivalent weights are given in Table 7-2. Note that the oxidation state of the element in the compound

TABLE 7-2 Equivalent Weights of Some Manganese and Carbon-containing Species

Substance	Equivalent Weight[a]
Mn	$\dfrac{\text{gfw Mn}}{5}$
$KMnO_4$	$\dfrac{\text{gfw } KMnO_4}{5}$
$Ca(MnO_4)_2 \cdot 4H_2O$	$\dfrac{\text{gfw } Ca(MnO_4)_2 \cdot 4H_2O}{2 \times 5}$
Mn_2O_3	$\dfrac{\text{gfw } Mn_2O_3}{2 \times 5}$
CO_2	$\dfrac{\text{gfw } CO_2}{1}$
$C_2O_4^{2-}$	$\dfrac{\text{gfw } C_2O_4^{2-}}{2}$

[a] Based on: $5C_2O_4^{2-} + 2MnO_4^- + 16H^+ \rightarrow 10CO_2 + 2Mn^{2+} + 8H_2O$.

whose equivalent weight is being calculated is not important in determining the equivalent weight. Thus, for an example, the oxidation state of manganese in Mn_2O_3 is $+3$. Before the reaction just considered could be used to determine the Mn_2O_3 content of a sample, the manganese would have to be converted to permanganate ion by some subsidiary reagent prior to titration. Once in the $+7$ state, each manganese would be reduced to the $+2$ state in the titration step. The equivalent weight is thus gfw $Mn_2O_3/10$.

As in neutralization reactions, the equivalent weight for a given oxidizing or reducing agent is not invariant. Potassium permanganate, for example, reacts with reducing agents in at least four different ways, depending upon the conditions existing in the solution. The half-reactions are:

$$MnO_4^- + e \rightarrow MnO_4^{2-}$$
$$MnO_4^- + 3e + 2H_2O \rightarrow MnO_2(s) + 4OH^-$$
$$MnO_4^- + 4e + 3H_2P_2O_7^{2-} + 8H^+ \rightarrow Mn(H_2P_2O_7)_3^{3-} + 4H_2O$$
$$MnO_4^- + 5e + 8H^+ \rightarrow Mn^{2+} + 4H_2O$$

The changes in oxidation number for manganese are 1, 3, 4, and 5. The equivalent weight of potassium permanganate would be equal to the gram formula weight for the first reaction, and one-third, one-fourth, and one-fifth of the gram formula weight, respectively, for the others.

EQUIVALENT WEIGHTS IN PRECIPITATION AND COMPLEX-FORMATION REACTIONS

It is awkward to devise a definition for equivalent weight that is entirely free of ambiguity for compounds involved in precipitation or complex-formation reactions. As a result, many chemists prefer to avoid the use of the concept for reactions of this type and use formula weights exclusively instead. We are in sympathy with this practice. However, it is likely that the student will encounter situations where equivalent or milliequivalent weights are specified for substances involved in precipitation or complex formation; knowledge of their definition is thus important.

The equivalent weight of a participant in a precipitation or a complex-formation reaction is that weight which reacts with or provides one gram formula weight of the reacting cation if it is univalent, one-half of the gram formula weight if it is divalent, one-third of the gram formula weight if it is trivalent, and so on. Our earlier definitions of equivalent weight were based either on one mole of hydrogen ions or on one mole of electrons. Here one mole of a univalent cation or the equivalent thereof is used. *The cation referred to in this definition is always the cation directly involved in the reaction* and not necessarily the cation contained in the compound whose equivalent weight is being defined.

To illustrate the application of this definition, consider the reaction

$$Ag^+ + Cl^- \rightarrow AgCl(s)$$

Here the reacting cation is the univalent silver ion. The equivalent weights of some compounds that might be associated with this reaction are shown in Table 7-3.

The equivalent weight of the barium chloride dihydrate is one-half its formula weight, *not* because this weight contains one-half formula weight of divalent barium ions but rather because this is the weight that reacts with one formula weight of silver ions. A consideration of the last example will show why it is important to make such a fine distinction. Here we might be tempted to say

**TABLE 7-3   Equivalent Weights of Some Species Which React to
Form Precipitates**

Substance	Equivalent Weight[a]
Ag^+	gfw Ag^+
$AgNO_3$	gfw $AgNO_3$
Ag_2SO_4	$\dfrac{\text{gfw } Ag_2SO_4}{2}$
NaCl	gfw NaCl
$BaCl_2 \cdot 2H_2O$	$\dfrac{\text{gfw } BaCl_2 \cdot 2H_2O}{2}$
$AlCl_3$	$\dfrac{\text{gfw } AlCl_3}{3}$
AlOCl	gfw AlOCl

[a] Based on: $Ag^+ + Cl^- \rightarrow AgCl(s)$.

that the equivalent weight of the AlOCl is one-third its formula weight based upon the presence of one mole of trivalent aluminum. This assignment would be correct if Al^{3+} were the cation involved in the reaction. In the example, however, silver ion is the reacting cation. Therefore, the equivalent weight of AlOCl must be that weight which reacts with one formula weight of silver ions; that is, the formula weight and equivalent weight are identical.

As previously noted, a given compound may have more than one equivalent weight. For example, when silver is titrated with a solution of potassium cyanide, an end point can be detected for either of two reactions:

$$2CN^- + 2Ag^+ \rightarrow Ag[Ag(CN)_2](s)$$

or

$$2CN^- + Ag^+ \rightarrow Ag(CN)_2^-$$

In the first reaction, the equivalent weight of potassium cyanide would be identical to its formula weight; in the second, it would be *twice* the formula weight. The latter is an example of an equivalent weight that is *greater* than a formula weight.

EQUIVALENT WEIGHTS OF COMPOUNDS NOT PARTICIPATING DIRECTLY IN A VOLUMETRIC REACTION

There is frequent need to define the equivalent weight of an analyte that is only indirectly related to the actual reactants of the titration. To illustrate, lead can be determined by an indirect volumetric method in which the cation is first precipitated as the chromate from an acetic acid solution. The precipitate is filtered, washed free of excess precipitant, and redissolved in dilute hydrochloric acid to give a solution of lead and dichromate ions. The latter may then be determined by an oxidation-reduction titration with a standard iron(II) solution. The reactions are:

$$Pb^{2+} + CrO_4^{2-} \xrightarrow[HOAc]{dil} PbCrO_4(s) \quad \text{(precipitate filtered and washed)}$$

$$2PbCrO_4(s) + 2H^+ \xrightarrow[HC1]{dil} 2Pb^{2+} + Cr_2O_7^{2-} + H_2O \quad \text{(precipitate redissolved)}$$

$$Cr_2O_7^{2-} + 6Fe^{2+} + 14H^+ \rightarrow 2Cr^{3+} + 6Fe^{3+} + 7H_2O \quad \text{(titration)}$$

For the purpose of calculation, we must ascribe an equivalent weight to lead. *Since the titration is an oxidation-reduction process, the equivalent weight of lead will have to be based on a change of oxidation number.* Clearly, the lead exhibits no such change. It is, however, associated in a 1:1 ratio with chromium, and this element changes from the $+6$ to the $+3$ state in the titration. Therefore, we can say that a change in oxidation state of 3 is *associated* with each lead, and its equivalent weight in this sequence of reactions is one-third its atomic weight.

It is often helpful to make an inventory of the chemical relationships existing between the substance whose equivalent weight is sought and one of the participants in the titration. In this example, the equations reveal that

$$2Pb^{2+} \equiv 2CrO_4^{2-} \equiv Cr_2O_7^{2-} \equiv 6Fe^{2+} \equiv 6e$$

Thus, the quantity of each substance associated with the transfer of one mole of electrons is

$$\frac{2 \text{ gfw } Pb^{2+}}{6} \equiv \frac{2 \text{ gfw } CrO_4^{2-}}{6} \equiv \frac{1 \text{ gfw } Cr_2O_7^{2-}}{6} \equiv \frac{6 \text{ gfw } Fe^{2+}}{6} \equiv \frac{6 \text{ moles } e}{6}$$

Let us consider one more example. The nitrogen in the organic compound $C_9H_9N_3$ can be determined by quantitative conversion to ammonia, followed by titration with a standard solution of acid. The reactions are:

$$C_9H_9N_3 + \text{reagent} \rightarrow 3NH_3 + \text{products}$$
$$NH_3 + H^+ \rightarrow NH_4^+ \quad \text{(titration)}$$

Here titration involves a neutralization process; therefore, the equivalent weight for the analyte must be based on the consumption or production of hydrogen ions. Since each molecule of the analyte yields three ammonia molecules, it can be considered responsible for the consumption of three hydrogen ions. The equivalent weight, then, is the formula weight of $C_9H_9N_3$ divided by 3. Using the same reasoning, each nitrogen is converted to one ammonia molecule which reacts with one hydrogen ion; the equivalent weight of nitrogen, N, is thus equal to its formula weight.

CONCENTRATION UNITS USED IN VOLUMETRIC CALCULATIONS

Some of the terms by which chemists express the concentrations of solutions were discussed in Chapter 2; these included formal concentration, molar concentration, and various types of percentage composition. We now need two additional terms, *titer* and *normality*, which are commonly used to define the concentration of solutions used in volumetric analysis.

Titer. The titer of a solution is the weight of a substance that is chemically equivalent to one milliliter of that solution. Thus, a silver nitrate solution having a titer of 1.00 mg of chloride would contain just enough silver nitrate in each milliliter to react completely with that weight of chloride ion. The titer might also be expressed in terms of milligrams or grams of potassium chloride, barium chloride, sodium iodide, or any other compound that reacts with silver nitrate. The concentration of a reagent to be used for the routine analysis of many samples is advantageously expressed in terms of its titer.

Normality. The normality, N, of a solution expresses the number of milliequivalents of solute contained in one milliliter of solution or the number of equivalents contained in one liter. Thus, a 0.20-N silver nitrate solution contains 0.20 milliequivalent (meq) of this solute in each milliliter of solution or 0.20 equivalent (eq) per liter.

SOME IMPORTANT WEIGHT-VOLUME RELATIONSHIPS

The raw data from a volumetric analysis are ordinarily expressed in units of milliliters, grams, and normality. Volumetric calculations involve conversion of such information into units of milliequivalents followed by reconversion into the metric weight of some other chemical species. Two relationships, based on the foregoing definitions, are used for these transformations. The first involves converting the weight of a compound from units of grams to those of milliequivalents. This transformation is accomplished by dividing the weight of the substance by its milliequivalent weight. That is,

$$\text{no. meq A} = \frac{\text{wt A, g}}{\text{meq wt A, g/meq}}$$

Example. Calculate the number of milliequivalents of chlordane, $C_{10}H_6Cl_8$ (gfw = 410), in 0.500 g of the pure insecticide, assuming that all of the chlorine present is ultimately titrated with Ag^+.

$$\text{no. meq } C_{10}H_6Cl_8 = \frac{\text{wt } C_{10}H_6Cl_8 \text{ (g)}}{\text{meq wt } C_{10}H_6Cl_8 \text{ (g/meq)}}$$

Since each formula weight of chlordane provides eight Cl^- which react with eight Ag^+, we may write

$$\text{no. meq } C_{10}H_6Cl_8 = \frac{0.500 \text{ g } C_{10}H_6Cl_8}{0.410 \text{ g/8 meq}} = 9.76$$

The second relationship permits calculation of the number of milliequivalents of solute contained in a given volume, V, provided the normality of the solution is known. Since, by definition, the normality is equal to the number of milliequivalents in each milliliter, it follows that

$$\text{no. meq A} = V_A \text{ (ml)} \times N_A \text{ (meq/ml)}$$

Example. The number of milliequivalents involved in a titration that required 27.3 ml of 0.200-N $KMnO_4$ is given by

$$\text{no. meq } KMnO_4 = 27.3 \text{ ml} \times 0.200 \text{ meq/ml}$$
$$= 5.46$$

Further applications of these relationships are illustrated in the following examples.

Example. What weight of primary standard $K_2Cr_2O_7$ (gfw = 294.2) is needed to prepare exactly 2 liters of 0.1200-N reagent by the direct method? Titrations with dichromate involve the half-reaction

$$Cr_2O_7{}^{2-} + 14H^+ + 6e \rightleftarrows 2Cr^{3+} + 7H_2O$$

The number of milliequivalents of $K_2Cr_2O_7$ required is first calculated:

$$\text{no. meq } K_2Cr_2O_7 = \text{ml}_{K_2Cr_2O_7} \times N_{K_2Cr_2O_7}$$
$$= 2000 \text{ ml} \times 0.1200 \text{ meq/ml}$$
$$= 240.0$$

Conversion of a weight in milliequivalents to a weight in grams involves multiplication by the milliequivalent weight.

$$\text{wt } K_2Cr_2O_7 = 240.0 \text{ meq} \times \frac{294.2 \text{ g/fw}}{6 \text{ eq/fw}} \times \frac{1}{1000 \text{ meq/eq}}$$

$$= 11.768 = 11.77 \text{ g}$$

Example. What volume of 0.100-N HCl can be produced by diluting 150 ml of 1.24-N acid? The number of milliequivalents of HCl must be the same in the two solutions. Therefore,

$$\text{no. meq HCl in diluted solution} = \text{no. meq HCl in concentrated solution}$$
$$\text{ml of diluted solution} \times 0.100 \text{ meq/ml} = 150 \text{ ml} \times 1.24 \text{ meq/ml}$$

$$\text{ml of dilute solution} = \frac{150 \times 1.24}{0.100} = 1860$$

$$= 1.86 \times 10^3$$

Thus, a 0.100-N HCl solution would be obtained by diluting 150 ml of 1.24-N acid to 1.86 × 10³ ml.

A Fundamental Relationship between Quantities of Reacting Substances. By definition, one equivalent weight of an acid contributes one mole of hydrogen ions to a reaction. Also, one equivalent weight of a base consumes one mole of these ions. It then follows that at the equivalence point in a neutralization titration, the number of equivalents (or milliequivalents) of acid and of base will always be numerically equal. Similarly, at the equivalence point in an oxidation-reduction titration, the number of milliequivalents of oxidizing and reducing agent must also be equal. An identical relationship holds for precipitation and complex-formation titrations. To generalize, we may state that *at the equivalence point in any titration, the number of milliequivalents of standard is exactly equal to the number of milliequivalents of the substance with which it has reacted.* Nearly all volumetric calculations are based on this relationship.

CALCULATION OF CONCENTRATION OF STANDARD SOLUTIONS

The normality of a standard solution is computed either from a standardization titration or from the data related to its actual preparation.

Example. A $Ba(OH)_2$ solution was standardized by titration against 0.1280-*N* HCl, 31.76 ml of the base being required to neutralize 46.25 ml of the acid. Calculate the normality of the $Ba(OH)_2$ solution.

Provided the end point in the titration corresponds to the equivalence point,

$$\text{no. meq } Ba(OH)_2 = \text{no. meq HCl}$$

$$ml_{Ba(OH)_2} \times N_{Ba(OH)_2} = ml_{HCl} \times N_{HCl}$$

$$31.76 \text{ ml} \times N_{Ba(OH)_2} = 46.25 \text{ ml} \times 0.1280 \text{ meq/ml}$$

$$N_{Ba(OH)_2} = \frac{46.25 \times 0.1280}{31.76} = 0.1864 \text{ meq/ml}$$

Example. Calculate the normality of an iodine solution if 37.34 ml were required to titrate a 0.2040-g sample of primary standard As_2O_3 (gfw = 197.8). The reaction is

$$I_2 + H_2AsO_3^- + H_2O \rightarrow 2I^- + H_2AsO_4^- + 2H^+$$

At the equivalence point

$$\text{no. meq } I_2 = \text{no. meq } As_2O_3$$

The number of milliequivalents of I_2 can be computed from the volume and normality; the number of milliequivalents of As_2O_3 can be calculated from the weight taken for the titration. Thus, we may write

$$ml_{I_2} \times N_{I_2} = \frac{\text{wt } As_2O_3}{\text{meq wt } As_2O_3}$$

In its reaction with iodine, arsenic loses two electrons and is thus oxidized from the +3 to the +5 state. Therefore, a total change of 4 is associated with each As_2O_3 molecule, making its equivalent weight one-fourth its formula weight. Thus,

$$37.34 \text{ ml} \times N_{I_2} = \frac{0.2040 \text{ g}}{0.1978 \text{ g/4 meq}}$$

$$N_{I_2} = \frac{0.2040}{37.34 \times 0.04945} = 0.1105$$

Example. A standard solution of $AgNO_3$ (gfw = 169.9) is prepared by dissolving exactly 24.15 g of the pure solid and diluting to exactly 2.000 liters with water. What is the normality of the solution?

Since normality is the number of milliequivalents of solute per milliliter of solution, we may write

$$\text{no. meq AgNO}_3 = \frac{24.15\,g}{0.1699\,g/meq}$$

Therefore,

$$N = \frac{\text{no. meq}}{ml} = \frac{24.15/0.1699}{2000} = 0.07107$$

CALCULATION OF RESULTS FROM TITRATION DATA

Example. The organic matter in a 3.77-g sample of mercuric ointment was decomposed with HNO_3. After dilution, the Hg^{2+} was titrated with a 0.114-N solution of NH_4SCN; exactly 21.3 ml of the reagent were required. Calculate the percent Hg (gfw = 201) and the percent $Hg(NO_3)_2$ (gfw = 325) in the ointment.

This titration involves the formation of a stable neutral complex, $Hg(SCN)_2$; that is,

$$Hg^{2+} + 2SCN^- \rightarrow Hg(SCN)_2$$

At the equivalence point

$$\text{no. meq Hg}^{2+} = \text{no. meq NH}_4SCN$$
$$= 21.3\ ml \times 0.114\ meq/ml$$

Because this is a complex-formation reaction the milliequivalent weight of the reacting cation, Hg(II), is one-half its milliformula weight. Therefore,

$$\text{wt Hg} = 21.3\ ml \times 0.114\ \frac{meq}{ml} \times \frac{0.201\ g}{2\ meq}$$

and

$$\%\ Hg = \frac{21.3 \times 0.114 \times 0.201/2}{3.77} \times 100 = 6.47$$

The percentage of $Hg(NO_3)_2$ is calculated identically:

$$\text{no. meq Hg(NO}_3)_2 = \text{no. meq NH}_4SCN$$

and

$$\%\ Hg(NO_3)_2 = \frac{21.3 \times 0.114 \times 0.325/2}{3.77} \times 100 = 10.5$$

Example. A 0.804-g sample of an iron ore was dissolved in acid. The iron was then reduced to the +2 state and titrated with 47.2 ml of a 0.112-N $KMnO_4$ solution. Calculate the results of the analysis in terms of percent Fe (gfw = 55.8) as well as percent Fe_2O_3 (gfw = 160).

The titration involves oxidation of Fe^{2+} to Fe^{3+};

$$5Fe^{2+} + MnO_4^- + 8H^+ \rightarrow 5Fe^{3+} + Mn^{2+} + 4H_2O$$

At the equivalence point,

$$\text{no. meq Fe} = \text{no. meq KMnO}_4$$
$$= 47.2\ ml \times 0.112\ meq/ml$$

Since Fe^{2+} loses one electron in this reaction, its milliequivalent weight is identical to its milliformula weight, and

$$\text{g Fe} = 47.2\ ml \times 0.112\ \frac{meq}{ml} \times \frac{55.8\ g}{1000\ meq}$$

Therefore,

$$\% \text{ Fe} = \frac{47.2 \times 0.112 \times 55.8/1000}{0.804} \times 100$$

$$= 36.7$$

The percentage of Fe_2O_3 can be obtained in essentially the same way; thus, at the equivalence point,

$$\text{no. meq } Fe_2O_3 = \text{no. meq } KMnO_4$$

and by the same arguments,

$$\% \text{ Fe}_2\text{O}_3 = \frac{47.2 \times 0.112 \times 160/2000}{0.804} \times 100$$

$$= 52.6$$

Example. A 0.475-g sample containing $(NH_4)_2SO_4$ was dissolved in water and made alkaline with KOH. The liberated NH_3 was distilled into exactly 50.0 ml of 0.100-N HCl. The excess HCl was back-titrated with 11.1 ml of 0.121-N NaOH. The percent NH_3 (gfw = 17.0) as well as the percent $(NH_4)_2SO_4$ (gfw = 132) in the sample are required.

At the equivalence point, the number of milliequivalents of acid and base are equal. In this titration, however, two bases are involved, NaOH and NH_3. Thus,

$$\text{no. meq HCl} = \text{no. meq } NH_3 + \text{no. meq NaOH}$$

After rearranging,

$$\text{no. meq } NH_3 = \text{no. meq HCl} - \text{no. meq NaOH}$$

$$= (50.0 \times 0.100 - 11.1 \times 0.121)$$

Thus,

$$\% \text{ NH}_3 = \frac{(50.0 \times 0.100 - 11.1 \times 0.121) \times 17.0/1000}{0.475} \times 100$$

$$= 13.1$$

The number of milliequivalents of $(NH_4)_2SO_4$ is the same as the number of milliequivalents of NH_3 by definition. Therefore,

$$\% \text{ (NH}_4)_2\text{SO}_4 = \frac{(50.0 \times 0.100 - 11.1 \times 0.121) \times 132/2000}{0.475} \times 100$$

$$= 50.8$$

Here, the milliequivalent weight of $(NH_4)_2SO_4$ is one-half the milliformula weight because

$$(NH_4)_2SO_4 \equiv 2NH_3 \equiv 2H^+$$

Example. A standard 0.120-N AgNO$_3$ solution is to be used for the routine determination of salt in brines.

1. Express the titer of this solution as mg NaCl/ml.

$$\text{NaCl titer} = \frac{0.120 \text{ meq AgNO}_3}{\text{ml AgNO}_3} \times \frac{1.00 \text{ meq NaCl}}{1.00 \text{ meq AgNO}_3} \times \frac{58.4 \text{ mg NaCl}}{\text{meq NaCl}}$$

$$= \frac{7.01 \text{ mg NaCl}}{\text{ml AgNO}_3}$$

2. Calculate the milligrams of NaCl in each milliliter of brine if titration of a 50.0-ml sample required 21.4 ml of the $AgNO_3$ solution.

$$\frac{\text{mg NaCl}}{\text{ml brine}} = \frac{21.4 \text{ ml AgNO}_3}{50.0 \text{ ml brine}} \times \frac{7.01 \text{ mg NaCl}}{\text{ml AgNO}_3} = 3.00$$

Example. What is the formal concentration of a potassium hydroxide that has a titer of 3.50 mg H_2SO_4/ml?

$$F_{\text{KOH}} = \frac{3.50 \text{ mg H}_2\text{SO}_4}{\text{ml KOH}} \times \frac{1}{98.1 \text{ mg H}_2\text{SO}_4/\text{mfw H}_2\text{SO}_4} \times \frac{2 \text{ mfw KOH}}{\text{mfw H}_2\text{SO}_4}$$

$$= 0.0714 \text{ mfw KOH/ml}$$

PROBLEMS

1. Calculate the p-functions for each ion in
   *(a) a solution that is 0.0100 F in NaBr.
   (b) a solution that is 0.0100 F in $BaBr_2$.
   *(c) a solution that is 3.5×10^{-3} F in $Ba(OH)_2$.
   (d) a solution that is 0.040 F in HCl and 0.020 F in NaCl.
   *(e) a solution that is 5.2×10^{-3} F in $CaCl_2$ and 3.6×10^{-3} F in $BaCl_2$.
   (f) a solution that is 4.8×10^{-8} F in $Zn(NO_3)_2$ and 5.6×10^{-7} F in $Cd(NO_3)_2$.

2. Calculate the pPb and/or pI for a solution that is
   *(a) 0.0644 F in BaI_2.
   (b) saturated with PbI_2.
   *(c) formed by mixing 25.0 ml of 0.100-F Pb^{2+} with 25.0 ml of 0.150-F NaI.
   (d) formed by mixing 25.0 ml of 0.100-F Pb^{2+} with 25.0 ml of 0.150-F BaI_2.
   *(e) 1.8 F in $Pb(NO_3)_2$.
   (f) 0.500 F in CeI_3.

3. Convert the following p-functions into molar concentrations:
   *(a) pH $= 8.67$ *(e) pLi $= -0.321$
   (b) pOH $= 0.125$ (f) $pNO_3 = 7.77$
   *(c) pBr $= 0.034$ *(g) pMn $= 0.0025$
   (d) pCa $= 12.35$ (h) pCl $= 1.020$

*4. Classify the following reactions as to type, and indicate the equivalent weight for each of the substances listed on the right as a fraction or multiple of its gram formula weight:

   (a) $Ca(OH)_2(aq) + 2H_3O^+ \rightarrow Ca^{2+}$
                              $+ 4H_2O$ H_2SO_4, $Ca(OH)_2$, $Ca_2P_2O_7$
   (b) $I_2(aq) + Sn^{2+} \rightarrow 2I^- + Sn^{4+}$ I_2, KI, $SnCl_4$, $Sn_2O(NO_3)_2$
   (c) $2Ce^{3+} + 3C_2O_4^{2-} \rightarrow Ce_2(C_2O_4)_3(s)$ Ce^{3+}, $Na_2C_2O_4$, C, $NaHC_2O_4$
   (d) $Al^{3+} + 6F^- \rightarrow AlF_6^{3-}$ $Al(OH)_2Cl$, NaF, BaF_2, $C_6H_3F_3$

5. Classify the following reactions as to type, and indicate the equivalent weight for each of the substances listed on the right as a fraction or multiple of its gram formula weight:

   (a) $Fe^{3+} + 3C_2O_4^{2-} \rightarrow Fe(C_2O_4)_3^{3-}$ Fe, Fe_3O_4, $Na_2C_2O_4$
   (b) $(CH_3)_3N + H_3O^+ \rightarrow (CH_3)_3NH^+ + H_2O$ $(CH_3)_3NH^+$, N_2, H_2SO_4
   (c) $Hg^{2+} + 2CN^- \rightarrow Hg(CN)_2(aq)$ $Hg(NO_3)_2$, Hg, $Ba(CN)_2$
   (d) $4BiO^+ + 3N_2H_5^+ \rightarrow 3N_2(g) + 4Bi(s)$
                              $+ 4H_2O + 7H^+$ $BiCl_3$, N_2H_4, Bi_2O_3

*6. Classify the following reactions as to type, and indicate the milliequivalent weight for each of the substances listed on the right as a fraction or multiple of its gram formula weight:

(a) $Ag^+ + 2S_2O_3^{2-} \rightarrow Ag(S_2O_3)_2^{3-}$ Ag_2SO_4, Ag, $Na_2S_2O_3$

(b) $I_2 + 2S_2O_3^{2-} \rightarrow 2I^- + S_4O_6^{2-}$ I_2, $Na_2S_2O_3$, $Na_2S_4O_6$

(c) $La^{3+} + 3IO_3^- \rightarrow La(IO_3)_3(s)$ $LaCl_3$, $NaIO_3$, $AlOH(IO_3)_2$, I_2O_5

(d) $H_3PO_4 + 2OH^- \rightarrow HPO_4^{2-} + 2H_2O$ H_3PO_4, P_2O_5, $Ba(OH)_2$

7. Classify the following reactions as to type, and indicate the milliequivalent weight for each of the substances listed on the right as a fraction or multiple of its gram formula weight:

(a) $2MnO_4^- + 5U^{4+} + 2H_2O$
$\rightarrow 2Mn^{2+} + 5UO_2^{2+} + 4H^+$ $KMnO_4$, Mn_3O_4, UO_2Cl_2, U_3O_8

(b) $NH_4^+ + OH^- \rightarrow NH_3 + H_2O$ $(NH_4)_2SO_4$, NH_3, $NaOH$

(c) $Al^{3+} + 2C_2O_4^{2-} \rightarrow Al(C_2O_4)_2^-$ $AlOHCl_2$, $Na_2C_2O_4$, Al_2O_3, CO_2

(d) $Hg_2^{2+} + 2Cl^- \rightarrow Hg_2Cl_2(s)$ $Hg_2(NO_3)_2$, $HgSO_4$, $BaCl_2 \cdot 2H_2O$, HCl

*8. A solution contains 0.377 g of $H_2C_2O_4 \cdot 2H_2O$ in 500 ml. Calculate the concentration of this solution in terms of its

(a) formality.

(b) normality in an acid-base titration in which both protons react.

(c) normality as a reducing agent for Ce^{4+}.

$$2Ce^{4+} + H_2C_2O_4 + 6H^+ \rightarrow 2CO_2 + 2Ce^{3+} + 4H_2O$$

(d) normality when employed for titration of Ca^{2+}.

$$Ca^{2+} + H_2C_2O_4 \rightarrow CaC_2O_4(s) + 2H^+$$

(e) CaO titer [for reaction see part (d)].

9. A solution contains 2.13 g of NaCN in 600 ml. Calculate

(a) its formal concentration.

(b) its normality as a base.

$$CN^- + H_3O^+ \rightarrow HCN + H_2O$$

(c) its normality for the determination of Ni^{2+}.

$$Ni^{2+} + 4CN^- \rightarrow Ni(CN)_4^{2-}$$

(d) its normality for the determination of Ag^+.

$$Ag^+ + CN^- \rightleftarrows AgCN(s)$$

(e) its normality in a titration with basic $KMnO_4$.

$$CN^- + 2MnO_4^- + 2OH^- \rightarrow 2MnO_4^{2-} + CNO^- + H_2O$$

*10. A solution is 0.200 F with respect to permanganate. What is its normality as an oxidizing agent

(a) in a strongly alkaline environment, where MnO_4^{2-} is the reduction product?

(b) in a strongly acidic solution, where Mn^{2+} is produced?

(c) in a neutral solution, where $MnO_2(s)$ is produced?

(d) in the presence of pyrophosphate ion ($H_2P_2O_7^{2-}$), where the product is $Mn(H_2P_2O_7)_3^{3-}$?

*11. A solution contains 3.42 g of $K_4Fe(CN)_6 \cdot 3H_2O$ in 750 ml of solution. Calculate

(a) its formal concentration.

(b) its normality in the standardization of Ce^{4+}.

$$Ce^{4+} + Fe(CN)_6{}^{4-} \rightleftarrows Ce^{3+} + Fe(CN)_6{}^{3-}$$

(c) its normality for the determination of Zn^{2+}.

$$2Fe(CN)_6{}^{4-} + 3Zn^{2+} + 2K^+ \rightarrow K_2Zn_3[Fe(CN)_6]_2(s)$$

(d) its $Zn_2P_2O_7$ titer [for reaction see part (c)].

12. Potassium hydrogen iodate, $KH(IO_3)_2$, is available in high purity and can be employed for the standardization of several reagents. A solution containing 6.37 g per 100 ml was prepared. Calculate
    (a) its formal concentration.
    (b) its normality as an acid.
    (c) its normality as an oxidizing reagent, where I^- is the reaction product.
    (d) its normality as a reducing agent, where its reaction product is $IO_4{}^-$.
    (e) its normality as an oxidant in strong HCl solution where its reaction product is ICl (here I is in the $+1$ state).
    (f) its KSCN titer.
    $$2SCN^- + 3IO_3{}^- + 3Cl^- + 4H^+ \rightarrow 2HCN + 2SO_4{}^{2-} + 3ICl + H_2O$$

*13. How many milliequivalents of solute are contained in 20.0 ml of
    (a) 0.100-N $Ba(OH)_2$?
    (b) 0.100-F $Ba(OH)_2$?
    (c) 0.100-F NaOH?
    (d) 0.100-N KCN? $\qquad$ $Ni^{2+} + 4CN^- \rightleftarrows Ni(CN)_4{}^{2-}$
    (e) 0.100-F KCN? $\qquad$ $Ni^{2+} + 4CN^- \rightleftarrows Ni(CN)_4{}^{2-}$
    (f) 0.100-F $K_2Cr_2O_7$? $\qquad$ $Cr_2O_7{}^{2-} + 14H^+ + 6e \rightleftarrows 2Cr^{3+} + 7H_2O$

14. How many milliformula weights of solute are contained in 10.0 ml of
    (a) 0.200-F H_2SO_4?
    (b) 0.200-N H_2SO_4?
    (c) 0.200-N NaOH?
    (d) 0.200-N $Na_2S_2O_3$? $\qquad$ $Ag^+ + 2S_2O_3{}^{2-} \rightleftarrows Ag(S_2O_3)_2{}^{3-}$
    (e) 0.200-N $Na_2S_2O_3$? $\qquad$ $2S_2O_3{}^{2-} + I_2 \rightleftarrows S_4O_6{}^{2-} + 2I^-$
    (f) a solution of $AgNO_3$ having a titer of 8.21 mg $BaCl_2$?

*15. How many grams of solute are contained in
    (a) 13.2 ml of 0.200-N $KMnO_4$? $\qquad$ (reaction product Mn^{2+})
    (b) 1.50 liters of 0.150-N H_2SO_4?
    (c) 43.5 ml of 0.175-N $Hg(NO_3)_2$? $\qquad$ $Hg^{2+} + 2Br^- \rightleftarrows HgBr_2(aq)$
    (d) 4.00 liters of 0.0820-N $KBrO_3$? $\qquad$ $BrO_3{}^- + 3H_3AsO_3 \rightleftarrows Br^- + 3H_3AsO_4$
    (e) 250 ml of 0.100-N $Ba(OH)_2$?
    (f) 250 ml of 0.100-F $Ba(OH)_2$?

16. How many milligrams of solute are contained in
    (a) 5.00 liters of 0.250-N $K_2Cr_2O_7$? $\qquad$ (reaction product Cr^{3+})
    (b) 150 ml of 0.174-N H_3PO_4? $\qquad$ (reaction product $HPO_4{}^{2-}$)
    (c) 10.0 ml of 0.0310-N KI? $\qquad$ (reaction product I_2)
    (d) 10.0 ml of 0.0310-N KI? $\qquad$ (reaction product $HgI_4{}^{2-}$)
    (e) 10.0 ml of 0.0310-N KI? $\qquad$ (reaction product AgI)

*17. Describe the preparation of 3.00 liters of 0.0800-N H_2SO_4 from
    (a) 3.00-F H_2SO_4.
    (b) 13% (w/w) H_2SO_4 solution.
    (c) 0.185-N H_2SO_4.
    (d) the concentrated reagent ($d = 1.84$ g/ml; % $H_2SO_4 = 95$).

18. Describe the preparation of 800 ml of 0.0500-N KOH from
    (a) a 6.00-F solution.
    (b) a 3.61% (w/w) KOH solution.
    (c) a concentrated reagent ($d = 1.505$ g/ml; % KOH $= 50.0$).
    (d) a 0.186-N solution.

*19. Describe how 500.0 ml of the following solutions should be prepared for the analytical application given in parentheses:
    (a) 0.100-N KCl from the primary standard reagent.

$$2Cl^- + Hg^{2+} \rightarrow HgCl_2)$$

    (b) 0.200-N H_3AsO_3 from primary standard grade As_2O_3.

$$(H_3AsO_3 + I_2 + H_2O \rightarrow H_3AsO_4 + 2I^- + 2H^+)$$

    (c) 0.150-N HCl from constant-boiling HCl containing 20.2 g HCl per 100 g solution.

$$(H^+ + OH^- \rightarrow H_2O)$$

    (d) a $K_2Cr_2O_7$ solution having an iron titer of 8.00 mg/ml.

$$(Cr_2O_7^{2-} + 6Fe^{2+} + 14H^+ \rightarrow 2Cr^{3+} + 7H_2O)$$

    (e) 0.135-N $K_2Cr_2O_7$ from a 3.00-F solution. [See part (d) for reaction.]

20. Describe how 2.00 liters of the following solutions should be prepared for the analytical applications shown in parentheses.
    (a) 0.0500-N KCN from the solid.

$$(4CN^- + Ni^{2+} \rightarrow Ni(CN)_4^{2-})$$

    (b) 0.0400-N H_3PO_4 from the concentrated reagent that has a density of 1.69 g/ml and is 85.0% H_3PO_4 by weight.

$$(H_3PO_4 + 2OH^- \rightarrow HPO_4^{2-} + 2H_2O)$$

    (c) 0.125-N $Na_2B_4O_7$ from pure $Na_2B_4O_7 \cdot 10H_2O$.

$$(B_4O_7^{2-} + 2H^+ + 5H_2O \rightarrow 4H_3BO_3)$$

    (d) a $KBrO_3$ solution with a hydrazine titer of 0.60 mg/ml, from 0.200-N $KBrO_3$.

$$(3N_2H_4 + 2BrO_3^- \rightarrow 3N_2 + 2Br^- + 6H_2O)$$

*21. A solution of perchloric acid was standardized by dissolving 0.374 g of primary standard grade HgO in a solution of KBr.

$$HgO(s) + 4Br^- + H_2O \rightarrow HgBr_4^{2-} + 2OH^-$$

The liberated OH^- required 37.7 ml of the acid. Calculate the normality.

22. A 0.336-g sample of primary standard grade sodium carbonate required 28.6 ml of a sulfuric acid solution to reach the end point for the reaction

$$CO_3^{2-} + 2H^+ \rightarrow H_2O + CO_2(g)$$

What is the normality of the H_2SO_4? What is its formality?

*23. What is the normality of an $AgNO_3$ solution that has a titer of 5.63 mg $BaCl_2 \cdot 2H_2O$/ml?

24. What is the normality of a solution of $KMnO_4$ that has a titer of 11.0 mg Fe_2O_3/ml? Reaction:

$$MnO_4^- + 5Fe^{2+} + 8H^+ \rightarrow Mn^{2+} + 5Fe^{3+} + 4H_2O$$

*25. Exactly 40.0 ml of a solution of $HClO_4$ were added to a solution containing 0.479 g of primary standard Na_2CO_3. The solution was boiled to remove CO_2, and the excess $HClO_4$ was back-titrated with 8.70 ml of a NaOH solution. In a separate experiment, 25.0 ml of the NaOH were found to neutralize 27.4 ml of $HClO_4$. Calculate the normality of the $HClO_4$ and the NaOH.

26. A 0.339-g sample of sodium sulfate, which had an assay of 96.4% Na_2SO_4, was titrated with a solution of $BaCl_2$. [$Ba^{2+} + SO_4^{2-} \rightarrow BaSO_4(s)$] What was the normality of the $BaCl_2$ solution if the end point was observed when 35.7 ml of the reagent were added? What was the SO_4^{2-} titer of the solution?

*27. A 0.121-g sample of a sulfur-containing organic compound was burned in a stream of O_2, and the resulting SO_2 was absorbed in a solution of H_2O_2. ($H_2O_2 + SO_2 \rightarrow H_2SO_4$) The acid produced was titrated with 20.5 ml of 0.107-N KOH. Calculate the percent S in the compound.

28. A 0.241-g specimen containing the mineral hausmannite, Mn_3O_4, was analyzed by treatment to convert the Mn quantitatively to MnO_4^-. The resulting solution required 40.3 ml of a 0.0744-N Fe^{2+} solution to reduce the MnO_4^{2-} to Mn^{2+}. Calculate the percent Mn_3O_4 in the sample.

*29. The CO concentration of a gas was determined by passing a 2.00-liter sample through a heated tube containing I_2O_5.

$$5CO + I_2O_5 \rightarrow 5CO_2 + I_2$$

The I_2 formed was sublimed into 20.0 ml of 0.0106-N $Na_2S_2O_3$.

$$I_2 + 2S_2O_3^{2-} \rightarrow 2I^- + S_4O_6^{2-}$$

The excess $Na_2S_2O_3$ required 4.74 ml of 0.0113-N I_2 solution. Calculate the milligrams of CO per liter of sample.

30. The arsenic in a 1.22-g sample of a pesticide was converted to H_3AsO_4 by suitable treatment. The acid was then neutralized, and exactly 40.0 ml of 0.0789-N $AgNO_3$ were added to precipitate the As quantitatively as Ag_3AsO_4. The excess Ag^+ in the filtrate and washings from the precipitate was titrated with 11.2 ml of 0.100-N KSCN. Calculate the percent As_2O_3 in the sample.

*31. The ethyl acetate concentration in an alcoholic solution of the ester was determined by diluting a 10.0-ml sample to exactly 100 ml. A 20.0-ml portion of the diluted solution was refluxed with 40.0 ml of 0.0467-N KOH.

$$CH_3COOC_2H_5 + OH^- \rightarrow CH_3OO^- + C_2H_5OH$$

After cooling, the excess OH^- was back-titrated with 3.41 ml of 0.0504-N HCl. Calculate the grams of ethyl acetate per 100 ml of the original sample.

32. The thiourea in a 1.45-g organic sample was extracted into a dilute H_2SO_4 solution and titrated with 37.3 ml of 0.00937-N Hg^{2+}.

$$4(NH_2)_2CS + Hg^{2+} \rightarrow [(NH_2)_2CS]_4Hg^{2+}$$

Calculate the percent $(NH_2)_2CS$ in the sample.

PRECIPITATION TITRATIONS

Volumetric methods based upon the formation of sparingly soluble silver salts are among the oldest known; these procedures were and still are routinely employed for the analysis for silver as well as for the determination of such ions as chloride, bromide, iodide, and thiocyanate. Volumetric precipitation methods that do not involve silver as one of the reactants are relatively limited.

Titration Curves for Precipitation Reactions

In preparation for the discussion that follows, the student may wish to review the material in Chapter 3 on solubility of precipitates.

Titration curves are useful for deducing the properties required of an indicator for the titration as well as the titration error that is likely to be encountered. The following example demonstrates how precipitation titration curves are derived from solubility-product data.

Example. Derive a curve for the titration of 50.00 ml of 0.00500-F NaBr with 0.01000-F AgNO$_3$.

We will derive data for both pBr and pAg; ordinarily only one of the two (the one that affects the indicator behavior) is needed.

Initial Point. At the outset the solution is 0.00500 F in Br^- and 0.000 F in Ag^+. Thus, $pBr = -\log(5.00 \times 10^{-3}) = 2.30$. The pAg is indeterminate.

After Addition of 5.00 ml Reagent. Here the bromide ion concentration will have been decreased by precipitation as well as by dilution. Thus,

$$F_{NaBr} = \frac{50.00 \times 0.00500 - 5.00 \times 0.01000}{50.00 + 5.00} = 3.64 \times 10^{-3}$$

Here the first term in the numerator gives the number of milliformula weights of NaBr present originally and the second term the number of milliformula weights of $AgNO_3$ added.

Bromide ions in the solution arise from both the unreacted NaBr and the slight solubility of AgBr. Thus, the *molar* concentration of Br^- is larger than the residual concentration of NaBr by an amount equal to the formal solubility of the precipitate. That is,

$$[Br^-] = 3.64 \times 10^{-3} + [Ag^+]$$

Here the second term on the right accounts for the contribution of AgBr to the molar bromide concentration since one Ag^+ ion is formed for each Br^- ion from this source. Unless the concentration of NaBr is very small, this second term can be neglected. That is,

$$[Ag^+] \ll 3.64 \times 10^{-3}$$

and

$$[Br^-] \cong 3.64 \times 10^{-3}$$
$$pBr = -\log(3.64 \times 10^{-3}) = 2.439 = 2.44$$

A convenient way to find pAg is to take the negative logarithm of the solubility-product expression for AgBr. That is,

$$-\log([Ag^+][Br^-]) = -\log K_{sp} = -\log(5.2 \times 10^{-13})$$
$$-\log[Ag^+] - \log[Br^-] = -\log K_{sp} = 12.28$$

or

$$pAg + pBr = pK_{sp} = 12.28$$
$$pAg = 12.28 - 2.44 = 9.84$$

Other points up to chemical equivalence can be derived in this same way.

Equivalence Point. Here neither NaBr nor $AgNO_3$ is in excess; necessarily, then,

$$[Ag^+] = [Br^-]$$

Substitution into the solubility-product expression yields

$$[Ag^+] = [Br^-] = \sqrt{5.2 \times 10^{-13}} = 7.21 \times 10^{-7}$$
$$pAg = pBr = -\log(7.21 \times 10^{-7}) = 6.14$$

After Addition of 25.10 ml Reagent. An excess of $AgNO_3$ is now present; thus,

$$F_{AgNO_3} = \frac{25.10 \times 0.01000 - 50.00 \times 0.00500}{75.10} = 1.33 \times 10^{-5}$$

and

$$[Ag^+] = 1.33 \times 10^{-5} + [Br^-] \cong 1.33 \times 10^{-5}$$

Here the second term in the right-hand side of the equation accounts for Ag^+ ions resulting from the slight solubility of AgBr; it can ordinarily be neglected.

$$pAg = -\log (1.33 \times 10^{-5}) = 4.876 = 4.88$$
$$pBr = 12.28 - 4.88 = 7.40$$

Additional points defining the titration curve beyond the equivalence point can be obtained in an analogous way.

SIGNIFICANT FIGURES IN TITRATION-CURVE CALCULATIONS

In deriving titration curves, concentration data associated with the equivalence-point region are often of low precision because they are based upon small differences between large numbers. For example, in the calculation for F_{AgNO_3}, following introduction of 25.10 ml of 0.0100-F AgNO$_3$, the numerator (0.2510 − 0.2500) contains only two significant figures; at best, then, F_{AgNO_3} is known to two significant figures as well. To minimize the rounding error, however, we retained three digits in this calculation and rounded after calculating pAg.

In rounding p-functions it is important to recall (p. 77) that a logarithm consists of the characteristic (the numbers to the left of the decimal) and the mantissa and that the former simply locates the decimal point in the original number. *Thus, it is the mantissa only that should be rounded to the appropriate number of significant figures.*

For points away from the equivalence-point region in this example, we would be justified in carrying another significant figure. Thus, if we chose, we could report the initial pAg as 2.301. There is little point in reporting the extra digit, however, because it is the equivalence-point region of a titration curve that is of prime interest. Thus, here, and in derivations of other titration curves, we will round p-functions to two places to the right of the decimal. This limited precision suffices because it is the change in p-function that is of importance, and these changes are great enough so that they are not obscured by the uncertainty in the data.

FACTORS INFLUENCING THE SHARPNESS OF END POINTS

For end points that are sharp and thus easy to locate, it is usually necessary that in the equivalence-point region small additions of reagent produce large changes in p-function. It is therefore of interest to examine those variables that influence the magnitude of change in p-function during a titration.

Reagent Concentration. Table 8-1 is a compilation of data computed by the methods illustrated in the example just considered. Three concentrations of bromide and silver ion, differing by factors of 10, are assumed. The pAg data for the three sets are plotted in Figure 8-1 and clearly show the effect of concentration on titration curves. It is apparent that an increase in analyte and reagent concentration enhanced the change in pAg in the equivalence-point region; an analogous effect is observed when pBr is plotted rather than pAg.

TABLE 8-1   Changes in pAg and pBr during Titration with Solutions of Different Concentration

Volume AgNO₃, ml	50.0 ml 0.0500-F Br⁻ with 0.100-F AgNO₃		50.0 ml of 0.00500-F Br⁻ with 0.0100-F AgNO₃		50.0 ml of 0.000500-F Br⁻ with 0.00100-F AgNO₃	
	pAg	pBr	pAg	pBr	pAg	pBr
0.00	—	1.30	—	2.30	—	3.30
10.00	10.68	1.60	9.68	2.60	8.68	3.60
20.00	10.13	2.15	9.13	3.15	8.13	4.15
23.00	9.72	2.56	8.72	3.56	7.72	4.56
24.90	8.41	3.87	7.41	4.87	6.50	5.78[a]
24.95	8.10	4.18	7.10	5.18	6.33	5.95[a]
25.00	6.14	6.14	6.14	6.14	6.14	6.14
25.05	4.18	8.10	5.18	7.10	5.95	6.33[b]
25.10	3.88	8.40	4.88	7.40	5.78	6.50[b]
27.00	2.58	9.70	3.58	8.70	4.58	7.70
30.00	2.20	10.08	3.20	9.08	4.20	8.08

[a] Here the approximation that $[Ag^+] \ll F_{NaBr}$ could not be used.
[b] Here the approximation that $[Br^-] \ll F_{AgNO_3}$ could not be used.

These effects have practical significance for the titration of bromide ion. If the analyte concentration is sufficient to permit the use of a silver nitrate solution that is 0.1 F or stronger, easily detected end points are observed, and the titration error is minimal. On the other hand, with solutions that are 0.001 F or less, the change in pAg or pBr is so small that end-point detection becomes difficult; a large titration error is thus to be expected.

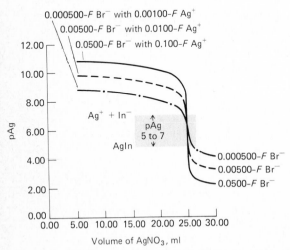

Figure 8-1   Effect of Reagent Concentration on Titration Curves. 50.00 ml of NaBr are titrated for each curve.

Although the influence of reagent concentration upon end-point sharpness has been demonstrated with a precipitation titration, the relationship applies to other reaction types as well.

Completeness of Reaction. Figure 8-2 shows how the solubility of the product influences titration curves for reactions in which 0.1-F silver nitrate serves as the reagent. Clearly, the greatest change in pAg occurs in the titration of iodide ion which, of all the anions considered, forms the least soluble silver salt and hence represents the most nearly complete reaction. The poorest break is observed for the reaction that is least complete—that is, in the titration of bromate ion. Reactions that produce silver salts with solubilities intermediate between these extremes yield titration curves with end-point breaks that are also intermediate. Again, we shall see that this effect is common to all reaction types.

TITRATION CURVES FOR MIXTURES

The methods developed in the previous section can be extended to mixtures that form precipitates of differing solubilities with the titrant. To illustrate, consider the titration of a 50.00-ml solution that is 0.0800 F in iodide ion and 0.1000 F in chloride with 0.2000-F silver nitrate.

Because silver iodide has a much smaller solubility than silver chloride, the initial additions of the reagent will result in formation of the iodide exclusively. Here, the titration curve should be similar to that for iodide shown in Figure 8-2. It is of interest, then, to determine the extent to which silver iodide precipitation occurs before appreciable formation of silver chloride takes place.

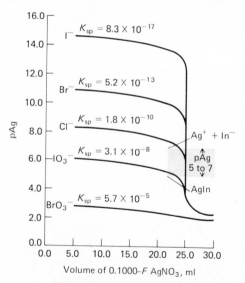

Figure 8-2 Effect of Completeness of Reaction on Titration Curves. For each curve, 50.00 ml of 0.0500-F solution of the anion are titrated with 0.1000-F AgNO$_3$.

With the first appearance of the chloride precipitate, solubility-product expressions for both precipitates are satisfied; division of one by the other provides a useful relationship.

$$\frac{[Ag^+][I^-]}{[Ag^+][Cl^-]} = \frac{8.3 \times 10^{-17}}{1.82 \times 10^{-10}} = 4.56 \times 10^{-7}$$

$$[I^-] = 4.56 \times 10^{-7}[Cl^-]$$

It is apparent that the iodide concentration will have been reduced to a minuscule fraction of the chloride ion concentration prior to the onset of precipitation by silver chloride. That is, formation of silver chloride will not occur until very near the equivalence point for iodide, or after addition of approximately 20 ml of reagent in this titration. Here, the chloride ion concentration, because of dilution, will be approximately

$$F_{Cl^-} \cong [Cl^-] = \frac{50.00 \times 0.1000}{70.00} = 0.0714$$

and

$$[I^-] = 4.56 \times 10^{-7} \times 0.0714 = 3.26 \times 10^{-8}$$

The percentage of iodide unprecipitated at this point can be calculated as follows:

original no. mfw I^- = 50.00 × 0.0800 = 4.00

$$\% \, I^- \text{ unprecipitated} = \frac{3.26 \times 10^{-8} \times 70.00 \times 100}{4.00} = 5.7 \times 10^{-5}$$

Thus, to within about 6×10^{-5} percent of the equivalence point for iodide, no silver chloride should form and the titration curve should be similar to that of the iodide alone. The first half of the titration curve shown by the solid line in Figure 8-3 was derived on this basis.

When chloride ion begins to precipitate, the rapid decrease in pAg is terminated abruptly. The pAg is most conveniently calculated from the solubility-product constant for silver chloride:

$$[Cl^-] \cong 0.0714$$

$$[Ag^+] = \frac{1.82 \times 10^{-10}}{0.0714} = 2.55 \times 10^{-9}$$

$$pAg = -\log (2.55 \times 10^{-9}) = 8.59$$

Further additions of silver nitrate decrease the chloride ion concentration and the curve then becomes that for chloride by itself. For example, after 25.00 ml of reagent have been added,

$$F_{Cl^-} \cong [Cl^-] = \frac{50.00 \times 0.1000 + 50.00 \times 0.0800 - 25.00 \times 0.2000}{75.00}$$

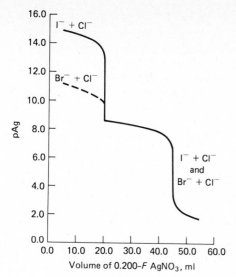

Figure 8-3 Titration Curves for 50.00 ml of Solutions That Were 0.100 F in Cl^- and 0.0800 F in Br^- or I^-.

Here the first two terms in the numerator give the number of milliformula weights of chloride and iodide respectively, and the third is the number of milliformula weights of titrant added. Thus,

$$[Cl^-] = 0.0533$$

$$[Ag^+] = \frac{1.82 \times 10^{-10}}{0.0533} = 3.41 \times 10^{-9}$$

$$pAg = 8.47$$

The remainder of the curve can be derived in the same way as a curve for chloride by itself.

Figure 8-3 illustrates that a titration curve for a mixture is a synthesis of the two individual titration curves. The two equivalence-point regions are marked by rapid decreases in pAg. It is to be expected that the change associated with the first equivalence point will become less distinct as the solubilities of the two precipitates approach one another. This effect is demonstrated by the curve for a mixture of bromide and chloride ions shown in the figure. In this titration the initial pAg values are lower, because solubility of silver bromide is greater than that of silver iodide. Beyond the equivalence point, however, the two titration curves become identical.

It is possible to obtain experimental curves similar to those shown in Figure 8-3 by measuring the potential of a silver electrode immersed in the solution; this technique, which is discussed in Chapter 17, permits the analysis of the individual components in mixtures.

CHEMICAL INDICATORS FOR PRECIPITATION TITRATIONS

A chemical indicator produces a visually detectable change—usually of color or turbidity—in the solution. The indicator functions by reacting competitively with one of the reactants or products of the titration. Thus, in the titration of species A with a reagent B and in the presence of an indicator (In) that can react with B, a chemical description of the solution throughout the titration will be

$$A + B \rightleftarrows AB$$

$$In + B \rightleftarrows InB$$

For indicator action, it is necessary, of course, that InB impart a significantly different appearance to the solution than In. In addition, the amount of InB required for an observable change should be so small that no appreciable consumption of B occurs when InB is formed. Finally, the equilibrium constant for the indicator reaction must be such that the ratio [InB]/[In] is shifted from a small to a large value as a consequence of the change in [B] (or pB) that occurs in the equivalence-point region. This last condition is most likely to be realized where the change in pB is large.

To illustrate, consider the application of a typical indicator, which exhibits a full color change as pAg varies from 7 to 5, to the three titrations described in Table 8-1 and Figure 8-1. It is clear that each titration requires a different volume of titrant to encompass this range. Thus, the data in the second column of Table 8-1 indicate that less than 0.10 ml of 0.1-F AgNO$_3$ is needed; that is, the color change will begin beyond 24.95 ml and be complete before 25.05 ml of silver have been added. An abrupt color change and a minimal titration error can be expected. In contrast, with the 0.001-F AgNO$_3$, the color change will be initiated at about 24.5 ml and will be complete at 25.8 ml. Location of the end point under these circumstances is impossible. For the titration with 0.01-F reagent, the end-point change requires somewhat less than 0.2 ml; here the indicator would be usable, but the uncertainty with respect to the equivalence point would be substantial.

Consider now the applicability and the effectiveness of this same indicator with regard to the titrations represented by the curves in Figure 8-2. With bromate and iodate ions, the indicator will exist largely as AgIn throughout the titration; thus, no color change is observed after the first addition of silver nitrate.

The solubilities of silver bromide and silver iodide are small enough to prevent formation of significant amounts of AgIn until the equivalence-point region is reached; thus, from Figure 8-2 it is seen that pAg remains above 7 until just before the equivalence point for the bromide titration and just beyond the equivalence point for the iodide. In both cases, the excess silver ion required to complete the color change corresponds to less than 0.01 ml of reagent; the titration error is thus negligible.

An indicator with a pAg range of 5 to 7 would not be satisfactory for a chloride titration because formation of appreciable amounts of AgIn would begin approximately one milliliter short of the equivalence point and extend over a range of about one milliliter; exact location of the end point would be impossible. On the other hand, an indicator with a pAg range of 4 to 6 would be perfectly satisfactory. No satisfactory chemical indicator exists for the iodate and bromate titrations because the pAg changes in the equivalence-point region are too small.

Examples of indicators employed for precipitation titrations with silver ions are described in the paragraphs that follow.

The Formation of a Second Precipitate; the Mohr Method. The formation of a second precipitate of distinctive color is the basis for end-point detection with the *Mohr method.* The procedure has been widely applied to the titration of chloride ion and bromide ion with standard silver nitrate. Chromate ion is the indicator, the end point being signaled by the appearance of brick-red silver chromate, Ag_2CrO_4.

The formal solubility of silver chromate is substantially greater than that for the silver halides. In a Mohr titration, then, no silver chromate will be produced until essentially all of the halide has been precipitated. Through control of the chromate ion concentration, it is possible to retard the formation of silver chromate until the silver ion concentration acquires a value that corresponds to the theoretical equivalence-point region for the halide titration.

Example. An uncertainty of ± 1 ppt corresponds to ± 0.025 ml in a 25-ml titration. In the titration of 0.05-F Cl^- with 0.100-F Ag^+, calculations such as those on page 172 reveal that the silver ion concentration is $4.78 \times 10^{-6} M$ after introduction of 24.975 ml and $3.81 \times 10^{-5} M$ when 25.025 ml have been added. What range of chromate ion concentrations is needed to cause formation of Ag_2CrO_4 within ± 0.025 ml of the equivalence point?

Precipitation of Ag_2CrO_4 can occur at 24.975 ml when

$$[CrO_4{}^{2-}][Ag^+]^2 \geq K_{sp}$$

or

$$[CrO_4{}^{2-}] = \frac{K_{sp}}{[Ag^+]^2} > \frac{1.1 \times 10^{-12}}{(4.78 \times 10^{-6})^2} = 0.048$$

and at 25.025 ml when

$$[CrO_4{}^{2-}] > \frac{1.1 \times 10^{-12}}{(3.81 \times 10^{-5})^2} = 7.6 \times 10^{-4}$$

These calculations suggest that the indicator concentration needs only to be kept within the rather large range between 0.0008 and 0.05 M for an accurate Mohr titration. In practice, however, at concentrations greater than about 0.005 M, the intense yellow color of the chromate ion masks detection of the red silver chromate. A concentration of chromate somewhat lower than 0.005 M is ordinarily employed.

The minimum amount of silver chromate that must be formed in order to be detectable by the eye cannot be obtained by calculation; indeed, such information can be determined only by experiment. On the average, an over-titration corresponding to approximately 0.05 ml of 0.1-N silver nitrate is necessary before the red precipitate is detected. In order to correct for the resulting titration error it is common practice to determine an *indicator blank* at the time of the analysis. Here the silver ion consumption of a chloride-free suspension of calcium carbonate is measured in about the same volume of solution and with the same amount of indicator. The blank titration mixture serves as a convenient color standard for subsequent titrations. An alternative that largely eliminates the indicator error is to use the Mohr method to standardize the silver nitrate solution against pure sodium chloride. The "working normality" obtained for the solution will compensate not only for the overconsumption of reagent but also for the acuity of the analyst in detecting the color change.

Attention must be paid to the acidity of the medium because the equilibrium

$$2CrO_4{}^{2-} + 2H^+ \rightleftarrows Cr_2O_7{}^{2-} + H_2O$$

is displaced to the right as the hydrogen ion concentration is increased. Since silver dichromate is considerably more soluble than the chromate, the indicator reaction in acid solution requires substantially larger silver ion concentrations, if indeed it occurs at all. If the medium is made strongly alkaline, there is danger that silver oxide will precipitate.

$$2Ag^+ + 2OH^- \rightleftarrows 2AgOH(s) \rightleftarrows Ag_2O(s) + H_2O$$

Thus, the determination of chloride by the Mohr method must be carried out in a medium that is neutral or nearly so (pH 7 to 10). The addition of sodium hydrogen carbonate, calcium carbonate, or borax to the solution is a convenient way of maintaining the hydrogen ion concentration within suitable limits.

Formation of a Colored Complex; the Volhard Method. A standard solution of thiocyanate may be used to titrate silver ion by the *Volhard method:*

$$Ag^+ + SCN^- \rightleftarrows AgSCN(s)$$

Iron(III) ion serves as the indicator, imparting a red coloration to the solution with the first slight excess of thiocyanate:

$$Fe^{3+} + SCN^- \rightleftarrows \underset{red}{Fe(SCN)^{2+}}$$

The titration must be carried out in acid solution to prevent precipitation of iron(III) as the hydrated oxide. As shown by the following example, the indicator concentration to reduce the titration error to zero is readily obtained.

Example. From experiment it has been found that the average observer can just detect the red color of $Fe(SCN)^{2+}$ when its concentration is 6.4×10^{-6} M. In the titration of 50.0 ml of 0.050-F Ag^+ with 0.100-F KSCN, what concentration of Fe^{3+} should be used to reduce the titration error to zero?

For a zero titration error, the $FeSCN^{2+}$ color should appear when the concentration of Ag^+ remaining in the solution is identical to the sum of the two thiocyanate species. That is, at the equivalence point

$$[Ag^+] = [SCN^-] + [Fe(SCN)^{2+}]$$
$$= [SCN^-] + 6.4 \times 10^{-6}$$

or

$$\frac{K_{sp}}{[SCN^-]} = \frac{1.1 \times 10^{-12}}{[SCN^-]} = [SCN^-] + 6.4 \times 10^{-6}$$

which rearranges to

$$[SCN^-]^2 + 6.4 \times 10^{-6}[SCN^-] - 1.1 \times 10^{-12} = 0$$
$$[SCN^-] = 1.7 \times 10^{-7}$$

The formation constant for $FeSCN^{2+}$ is

$$K_f = 1.4 \times 10^2 = \frac{[Fe(SCN)^{2+}]}{[Fe^{3+}][SCN^-]}$$

If we now substitute the $[SCN^-]$ necessary to give a detectable concentration of $FeSCN^{2+}$ at the equivalence point, we obtain

$$1.4 \times 10^2 = \frac{6.4 \times 10^{-6}}{[Fe^{3+}]1.7 \times 10^{-7}}$$
$$[Fe^{3+}] = 0.27$$

The indicator concentration is not critical in the Volhard titration. In fact, calculations similar to those just shown demonstrate that a titration error of one part in a thousand or less is, in theory, possible if the iron(III) concentration is held between 0.002 and 1.6 F. In practice, it is found that an indicator concentration greater than 0.2 M imparts sufficient color to the solution to make detection of the thiocyanate complex difficult. Therefore, lower concentrations (usually about 0.01 M) of iron(III) ion are employed.

Application of the Volhard Method to the Determination of Chloride Ions. The most important application of the Volhard method is for the indirect determination of chloride. A measured excess of standard silver nitrate solution is added to the chloride sample, and the excess silver ion is determined by back-titration with a standard thiocyanate solution. The requirement of a strongly acid environment represents a distinct advantage for the Volhard titration over other methods for chloride because such ions as carbonate, oxalate, and arsenate (which form slightly soluble silver salts in neutral media) do not interfere.

Silver chloride, in contrast to the other silver halides, is more soluble than silver thiocyanate. As a consequence, the reaction

$$AgCl(s) + SCN^- \rightleftarrows AgSCN(s) + Cl^- \tag{8-1}$$

causes the end point in the Volhard determination of chloride to fade; an over-consumption of thiocyanate ion and a negative error for the analysis may result. The magnitude of this error is dependent upon the indicator concentration.

Two general methods are employed to avoid the error resulting from the reaction between thiocyanate and silver chloride. The first involves the use of the maximum allowable indicator concentration [about 0.2-M iron(III) ion].[1] The more popular way involves isolation of the precipitated silver chloride before back-titration with the thiocyanate. Filtration, followed by titration of an aliquot of the filtrate, yields excellent results provided the precipitated silver chloride is first briefly digested. The time required for filtration is, of course, a disadvantage. Probably the most widely employed modification is that of Caldwell and Moyer,[2] which consists of coating the silver chloride with nitrobenzene, thereby substantially removing it from contact with the solution. The coating is accomplished by shaking the titration mixture with a few milliliters of the organic liquid prior to back-titration.

Adsorption Indicators. Adsorption indicators are organic compounds which are adsorbed on or desorbed from the surface of the precipitate formed during a titration. Ideally the adsorption or desorption occurs near the equivalence point and results not only in a color change but also a transfer of color from the solution to the solid or the reverse. Analytical procedures based upon adsorption indicators are sometimes called *Fajans methods* in honor of the scientist who was active in their development.

A typical adsorption indicator is the organic dye *fluorescein*, which is employed as an indicator for the titration of chloride ion with silver nitrate. In aqueous solution, fluorescein partially dissociates into hydrogen ions and negatively charged fluoresceinate ions that impart a yellowish-green color to the medium. The fluoresceinate ion forms a highly colored silver salt of limited solubility. In its application as an indicator, however, *the concentration of the dye is never large enough to exceed the solubility product for silver fluoresceinate*.

In the early stages of a Fajans titration for chloride with silver ions, the dye anion is not appreciably adsorbed by the precipitate; it is, in fact, repelled from the surface by the negative charge resulting from adsorbed chloride ions. When the equivalence point is passed, however, the precipitate particles become positively charged by virtue of the strong adsorption by excess silver ions; under these conditions, retention of fluoresceinate ions *in the counter-ion layer* occurs. The net result is the appearance of the red color of silver fluoresceinate *on the surface of the precipitate*. It is important to note that the color change is an *adsorption* (not a precipitation) process inasmuch as the solubility product of the silver fluoresceinate is never exceeded. The adsorption is reversible, the dye being desorbed upon back-titration with chloride ion.

The successful application of an adsorption indicator requires that the precipitate and the indicator have the following properties:

1. The particles of the precipitate must be of colloidal dimensions so that the quantity of indicator adsorbed is enhanced by the high specific surface area of the solid.

[1] E. H. Swift, G. M. Arcand, R. Lutwack, and D. J. Meier, *Anal. Chem.*, **22**, 306 (1950).
[2] J. R. Caldwell and H. V. Moyer, *Ind. Eng. Chem., Anal. Ed.*, **7**, 38 (1935).

2. The precipitate must strongly adsorb its own ions. We have seen (Chapter 6) that this property is characteristic of colloidal precipitates.

3. The indicator dye must be strongly held in the counter-ion layer by the primarily adsorbed ion. In general, this type of adsorption correlates with a low solubility of the salt formed between the dye and the lattice ion. At the same time, the solubility of this species must be sufficiently great to prevent its precipitation.

4. The pH of the solution must be maintained at a suitable level. The active form of most adsorption indicators is an ion that is the conjugate acid or base of the dye molecule and thus capable of combining with hydrogen or hydroxide ions to form the inactive parent molecule. Thus, the pH of the solution must be such as to assure that the ionic form of the indicator predominates.

Titrations involving adsorption indicators are rapid, accurate, and reliable. Their application, however, is limited to a relatively few precipitation reactions in which a colloidal precipitate is rapidly formed. In the presence of high electrolyte concentrations, end points with adsorption indicators tend to be less satisfactory, owing to coagulation of the precipitate and the consequent decrease in the surface on which adsorption can occur.

Most adsorption indicators are weak acids. Their use is thus confined to neutral or slightly acidic solutions where the indicator is present predominantly as the anion. A few cationic adsorption indicators are known; these are suitable for titrations in strongly acid solutions. For such indicators, adsorption of the dye and coloration of the precipitate occur in the presence of an excess of the anion of the precipitate (that is, when the precipitate particles possess a negative charge).

Finally, some adsorption indicators sensitize silver-containing precipitates toward photodecomposition, which may cause difficulties.

OTHER METHODS FOR END-POINT DETECTION

In Chapters 17 and 21 electroanalytical methods, which can be applied for detection of end points for some precipitation reactions, are described.

Applications of Precipitation Titrations

Most applications of precipitation titrations are based upon the use of a standard silver nitrate solution and are sometimes called *argentometric* methods as a consequence. Table 8-2 lists some typical applications of argentometry. Note that many of these analyses are based upon precipitation of the analyte with a measured excess of silver nitrate followed by a Volhard titration with standard potassium thiocyanate. Both of these reagents are obtainable in primary standard quality; however, potassium thiocyanate is somewhat hygroscopic, which makes it difficult to weigh accurately on humid days. Both silver nitrate and potassium thiocyanate solutions are stable indefinitely.

Table 8-3 lists some miscellaneous volumetric methods based on reagents other than silver nitrate.

Specific directions for argentometric titrations are found in Experiments 7 to 10 in Chapter 31.

TABLE 8-2 Typical Argentometric Precipitation Methods

Substance Determined	End Point	Remarks
AsO_4^{3-}, Br^-, I^-, CNO^-, SCN^-	Volhard	Removal of silver salt not required
CO_3^{2-}, CrO_4^{2-}, CN^-, Cl^-, $C_2O_4^{2-}$, PO_4^{3-}, S^{2-}	Volhard	Removal of silver salt required before back-titration of excess Ag^+
BH_4^-	Modified Volhard	Titration of excess Ag^+ following: $BH_4^- + 8Ag^+ + 8OH^- \rightleftarrows 8Ag(s) + H_2BO_3^- + 5H_2O$
Epoxide	Volhard	Titration of excess Cl^- following hydrohalogenation
K^+	Modified Volhard	Precipitation of K^+ with known excess of $B(C_6H_5)_4^-$, addition of excess Ag^+ which precipitates $AgB(C_6H_5)_4$ and back-titration of the excess
Br^-, Cl^-	Mohr	
Br^-, Cl^-, I^-, SeO_3^{2-}	Adsorption indicator	
$V(OH)_4^+$, fatty acids, mercaptans	Electroanalytical	Direct titration with Ag^+
Zn^{2+}	Modified Volhard	Precipitate as $ZnHg(SCN)_4$. Filter, dissolve in acid, add excess Ag^+; back-titrate excess Ag^+
F^-	Modified Volhard	Precipitate as $PbClF$. Filter, dissolve in acid, add excess Ag^+; back-titrate excess Ag^+

TABLE 8-3 Miscellaneous Volumetric Precipitation Methods

Reagent	Ion Determined	Reaction Product	End Point
$K_4Fe(CN)_6$	Zn^{2+}	$K_2Zn_3[Fe(CN)_6]_2$	Diphenylamine
$Pb(NO_3)_2$	SO_4^{2-} MoO_4^{2-}	$PbSO_4$ $PbMoO_4$	Erythrosin B Eosin A
$Pb(OAc)_2$	PO_4^{3-} $C_2O_4^{2-}$	$Pb_3(PO_4)_2$ PbC_2O_4	Dibromofluorescein Fluorescein
$Th(NO_3)_4$	F^-	ThF_4	Alizarin red
$Hg_2(NO_3)_2$	Cl^-, Br^-	Hg_2X_2	Bromophenol blue
$NaCl$	Hg_2^{2+}	Hg_2X_2	Bromophenol blue

PROBLEMS

1. Describe a method of preparation for each of the following standard solutions which are to be employed for the analyses shown in Table 8-3:
   *(a) 2.00 liters of 0.100-N $K_4Fe(CN)_6$ from the pure compound.
   (b) 500 ml of approximately 0.020-N $Th(NO_3)_4$ from $Th(NO_3)_4 \cdot 4H_2O$.
   *(c) 750 ml of approximately 0.060-N $Hg_2(NO_3)_2$ from the compound.
   (d) 1.5 liters of approximately 0.075-N Pb^{2+} from $Pb(CH_3COO)_2 \cdot 3H_2O$.

*2. A solution of NH_4SCN was standardized by dissolving a 0.103-g sample of primary standard grade $AgNO_3$ and titrating with iron(III) ion as indicator. Exactly 27.5 ml of the NH_4SCN were used.
   (a) Calculate the normality of the NH_4SCN.
   (b) What is the Hg titer of the solution?

$$Hg^{2+} + 2SCN^- \rightarrow Hg(SCN)_2$$

3. A solution of Na_2S was standardized by dissolving 0.213 g of pure zinc in acid, neutralizing, and titrating the Zn^{2+} with 36.3 ml of the reagent (product, ZnS).
   (a) Calculate the normality of the Na_2S solution.
   (b) What is its $Zn_2P_2O_7$ titer?

*4. A 0.510-g sample of a pesticide was decomposed by fusion with sodium carbonate and leaching the residue with hot water. The fluoride present in the sample was then precipitated as PbClF by addition of HCl and $Pb(NO_3)_2$. The precipitate was filtered, washed, and dissolved in 5% HNO_3. The Cl^- was precipitated by addition of 50.0 ml of 0.200-N $AgNO_3$. After coating the AgCl with nitrobenzene, the excess Ag^+ was back-titrated with 7.42 ml of 0.176-N NH_4SCN. Calculate the percent F and the percent Na_2SeF_6 in the sample.

5. A 0.986-g sample of a fertilizer was decomposed and treated in such a way as to yield an aqueous solution of HPO_4^{2-}. Addition of 40.0 ml of 0.204-N $AgNO_3$ resulted in quantitative precipitation of Ag_3PO_4. The excess Ag^+ in the filtrate and washings from this precipitate required 8.72 ml of 0.117-N KSCN. Calculate the percent P_2O_5 in the sample.

*6. The formaldehyde in a 5.00-g sample of a seed disinfectant was steam distilled and the aqueous distillate was collected in a 500-ml volumetric flask. After dilution to volume, a 25.0-ml aliquot was treated with 30.0 ml of 0.121-F KCN solution to convert the formaldehyde to potassium cyanohydrin.

$$K^+ + CH_2O + CN^- \rightarrow KOCH_2CN$$

The excess KCN was then removed by addition of 40.0 ml of 0.100-N $AgNO_3$.

$$2CN^- + 2Ag^+ \rightarrow Ag_2(CN)_2(s)$$

The excess Ag^+ in the filtrate and washings required a 16.1-ml titration with 0.134-N NH_4SCN. Calculate the percent CH_2O in the sample.

7. Monochloroacetic acid, which is used as a preservative in fruit juices, reacts quantitatively with aqueous $AgNO_3$:

$$ClCH_2COOH + Ag^+ + H_2O \rightarrow AgCl + HOCH_2COOH + H^+$$

After acidification with H_2SO_4, the $ClCH_2COOH$ in a 150.0-ml sample of fruit juice was extracted into diethyl ether. The acid was then transferred to an aqueous solution by extraction with 1-F NaOH. After acidification, 40.0 ml of a

AgNO$_3$ solution were added, and the filtrate and washings from the AgCl were titrated with 18.7 ml of 0.0515-N NH$_4$SCN. A blank carried through the identical procedure required 38.0 ml of the NH$_4$SCN solution. Calculate the milligrams ClCH$_2$COOH per 100 ml of sample.

*8. The residual fluoride on a 50.0-g sample of lettuce was determined by drying and igniting the sample in the presence of CaO. The resulting calcium fluoride was decomposed with acid in the presence of SiO$_2$, and the resulting SiF$_4$ was distilled. The F$^-$ in the distillate was converted to ThF$_4$(s) by titration with 7.62 ml of 0.00893-N Th(NO$_3$)$_4$; alizarin red served as an indicator. Calculate the parts per million F$^-$ in the sample.

9. A 0.213-g sample of an alloy was analyzed by dissolving the sample in acid and adding 40.0 ml of 0.0932-N K$_4$Fe(CN)$_6$. After allowing 2 minutes for the formation of the solid K$_2$Zn$_3$[Fe(CN)$_6$]$_3$, the excess K$_4$Fe(CN)$_6$ was back-titrated with 4.44 ml of 0.106-N Zn^{2+} employing diphenylamine and a small amount of K$_3$Fe(CN)$_6$ as indicator. Calculate the percent Zn in the alloy.

*10. The elemental Se, dispersed in a 5.00-ml sample of detergent for dandruff control, was determined by suspending the sample in a warm, ammoniacal solution that contained 45.0 ml of 0.0200-N AgNO$_3$:

$$6Ag^+ + 3Se(s) + 6NH_3 + 3H_2O \rightarrow 2Ag_2Se(s) + Ag_2SeO_3(s) + 6NH_4^+$$

The mixture was next treated with excess nitric acid which dissolved the Ag$_2$SeO$_3$ but not the Ag$_2$Se. The Ag$^+$ from the Ag$_2$SeO$_3$ and the excess AgNO$_3$ consumed 16.74 ml of 0.0137-N KSCN in a Volhard titration. How many milligrams of Se were contained per milliliter of sample?

11. A 7.50-g sample containing BaCl$_2$ and BaI$_2$ as well as inert materials was dissolved and diluted to 250 ml in a volumetric flask. A 25.0-ml aliquot was titrated with 0.0847-N AgNO$_3$. Bromophenol blue served as an adsorption indicator which gave a color change when both the I$^-$ and Cl$^-$ were precipitated quantitatively. A 50.0-ml aliquot was then titrated with the AgNO$_3$ employing eosin which is adsorbed after only the I$^-$ is titrated. The first titration required 41.2 ml of the AgNO$_3$ and the second 38.3 ml. Calculate the percent BaCl$_2$ and BaI$_2$ in the original sample.

*12. A 0.224-g sample that contained only BaCl$_2$ and KBr required 19.7 ml of 0.100-N AgNO$_3$ to reach the end point for a Mohr titration. Calculate the percent of each compound present in the sample.

13. For each of the following precipitation titrations, calculate the concentration of the cation and the anion at reagent volumes corresponding to the equivalence point as well as ± 10.0ml, ± 1.00 ml, and ± 0.10 ml of equivalence. Construct a titration curve from the data plotting the p-function of the cation versus reagent volume.

   *(a) 20.0 ml of 0.0400-F AgNO$_3$ with 0.0200-F NH$_4$SCN.
   (b) 30.0 ml of 0.0400-F AgNO$_3$ with 0.0200-F KI.
   *(c) 30.0 ml of 0.00100-F AgNO$_3$ with 0.00100-F NaCl.
   (d) 25.0 ml of 0.0400-F Na$_2$SO$_4$ with 0.0200-F Pb(NO$_3$)$_2$.
   *(e) 60.0 ml of 0.0300-F BaCl$_2$ with 0.0600-F Na$_2$SO$_4$.
   (f) 50.0 ml of 0.100-F NaI with 0.200-F TlNO$_3$ (K_{sp} for TlI $= 6.5 \times 10^{-8}$).

9

THEORY OF NEUTRALIZATION TITRATIONS FOR SIMPLE SYSTEMS

End-point detection in a neutralization titration is ordinarily based upon the abrupt change in pH that occurs in the vicinity of the equivalence point. The pH range within which such a change occurs varies from titration to titration and is determined both by the nature and the concentration of the analyte as well as the titrant. The selection of an appropriate indicator and the estimation of the titration error require knowledge of the pH changes which occur throughout the titration. Thus, we need to know how neutralization titration curves are derived.

This chapter is concerned with simple acids or bases which produce a single hydronium or hydroxide ion per molecule. In Chapter 10 titration curves for polyfunctional acids or bases and mixtures of acids will be considered.

In preparation for the discussion that follows, the student may find it helpful to review the material on acids and bases in Chapter 2 as well as the introductory treatment of acid-base equilibria in Chapter 3.

Standard Reagents for Neutralization Titrations

The standard reagents employed for neutralization titrations are always strong acids or strong bases because these react more completely than their weaker counterparts; in common with precipitation titrations, the more complete the reaction, the greater will be the change in p-function and thus the more satisfactory will be the end point.

Indicators for Acid-Base Titrations

A variety of compounds, both synthetic and naturally occurring, differ in color depending upon the pH of the solutions in which they are dissolved. Some of these substances have been used for thousands of years to indicate the alkalinity or acidity of water. They are also of importance to the modern chemist who employs them to estimate the pH of solutions and signal the end point in acid-base titrations.

THEORY OF INDICATOR BEHAVIOR

Acid-base indicators are generally organic compounds which behave as weak acids or bases. The dissociation or association reactions of indicators are accompanied by internal structural rearrangements that are responsible for the changes in color.

We can symbolize the typical reaction of an acid-base indicator as follows:

$$H_2O + \underset{\text{(acid color)}}{HIn} \rightleftarrows H_3O^+ + \underset{\text{(base color)}}{In^-}$$

or

$$\underset{\text{(base color)}}{In} + H_2O \rightleftarrows \underset{\text{(acid color)}}{InH^+} + OH^-$$

For the first indicator, HIn will be the major constituent in strongly acid solutions and will be responsible for the "acid color" of the indicator, whereas In^- will represent its "basic color." For the second indicator, the species In will predominate in basic solutions and thus be responsible for the "basic color" of this indicator, while InH^+ will constitute the "acid color."

Equilibrium expressions for these processes are:

$$\frac{[H_3O^+][In^-]}{[HIn]} = K_a$$

and

$$\frac{[InH^+][OH^-]}{[In]} = K_b$$

These expressions can be rearranged to give

$$\frac{[In^-]}{[HIn]} = \frac{K_a}{[H_3O^+]}$$

and

$$\frac{[InH^+]}{[In]} = \frac{K_b}{[OH^-]} = \frac{K_b[H_3O^+]}{K_w}$$

A solution containing an indicator will show a continuous change in the color with variations in pH. The human eye is not very sensitive to these changes, however. Typically, a five- to tenfold excess of one form is required before the color of that species appears predominant to the observer; further increases in the ratio have no detectable effect. It is only in the region where the ratio varies from a five- to a tenfold excess of one form to a similar excess of the other that the color of the solution appears to change. Thus, the subjective "color change" involves a major alteration in the position of the indicator equilibrium. Using HIn as an example, we may write that the indicator exhibits its pure acid color to the average observer when

$$\frac{[In^-]}{[HIn]} \leq \frac{1}{10}$$

and its basic color when

$$\frac{[In^-]}{[HIn]} \geq \frac{10}{1}$$

The color appears to be intermediate for ratios between these two values. These numerical estimates, of course, represent average behavior only. Some indicators require smaller ratio changes and others larger. Furthermore, considerable variation in the ability to judge colors exists among observers; indeed, a color-blind person may be unable to discern any change.

   If the two concentration ratios are substituted into the dissociation-constant expression for the indicator, the range of hydronium ion concentrations needed to effect the indicator color change can be evaluated. Thus, for the full acid color,

$$\frac{[H_3O^+][In^-]}{[HIn]} = \frac{[H_3O^+]1}{10} = K_a$$

$$[H_3O^+] = 10K_a$$

and similarly, for the full basic color,

$$\frac{[H_3O^+]\,10}{1} = K_a$$

$$[H_3O^+] = \tfrac{1}{10}\,K_a$$

To obtain the indicator range, we take the negative logarithms of the two expressions. That is,

$$\text{indicator pH range} = -\log 10K_a \text{ to } -\log \frac{K_a}{10}$$

$$= -1 + pK_a \text{ to } -(-1) + pK_a$$

$$= pK_a \pm 1$$

Thus, an indicator with an acid dissociation constant of 1×10^{-5} will show a complete color change when the pH of the solution in which it is dissolved changes from 4 to 6. A similar relationship is easily derived for an indicator of the basic type.

TYPES OF ACID-BASE INDICATORS

A list of compounds possessing acid-base indicator properties is large and includes a variety of organic structures. An indicator covering almost any desired pH range can ordinarily be found. A few common indicators are given in Table 9-1.

TABLE 9-1   Some Important Acid-Base Indicators[a]

Common Name	Transition Range (pH)	Color Change Acid	Color Change Base	Indicator Type[b]
Thymol blue	1.2–2.8	red	yellow	1
	8.0–9.6	yellow	blue	
Methyl yellow	2.9–4.0	red	yellow	2
Methyl orange	3.1–4.4	red	yellow	2
Bromocresol green	3.8–5.4	yellow	blue	1
Methyl red	4.2–6.3	red	yellow	2
Chlorophenol red	4.8–6.4	yellow	red	1
Bromothymol blue	6.0–7.6	yellow	blue	1
Phenol red	6.4–8.0	yellow	red	1
Neutral red	6.8–8.0	red	yellow-orange	2
Cresol purple	7.4–9.0	yellow	purple	1
	1.2–2.8	red	yellow	
Phenolphthalein	8.0–9.6	colorless	red	1
Thymolphthalein	9.3–10.5	colorless	blue	1
Alizarin yellow	10.1–12.0	colorless	violet	2

[a] Taken from I. M. Kolthoff and H. A. Laitinen, *pH and Electro Titrations*, p. 29. New York: John Wiley & Sons, Inc., 1929. With permission.

[b] (1) Acid type: $HIn + H_2O \rightleftharpoons H_3O^+ + In^-$.

(2) Base type: $In + H_2O \rightleftharpoons InH^+ + OH^-$.

The majority of acid-base indicators possess structural properties that permit classification into perhaps half a dozen categories.[1] Three of these classes are described in the following paragraphs.

Phthalein Indicators. Most phthalein indicators are colorless in moderately acidic solutions and exhibit a variety of colors in alkaline media. In strongly alkaline solutions their colors tend to fade slowly, which is an inconvenience in some applications. As a group, the phthaleins are sparingly soluble in water; ethanol is the ordinary solvent for indicator solutions.

[1] See I. M. Kolthoff and C. Rosenblum, *Acid-Base Indicators*, chapter 5. New York: The Macmillan Company, 1937.

The best-known phthalein indicator is *phenolphthalein*, whose structures may be represented as

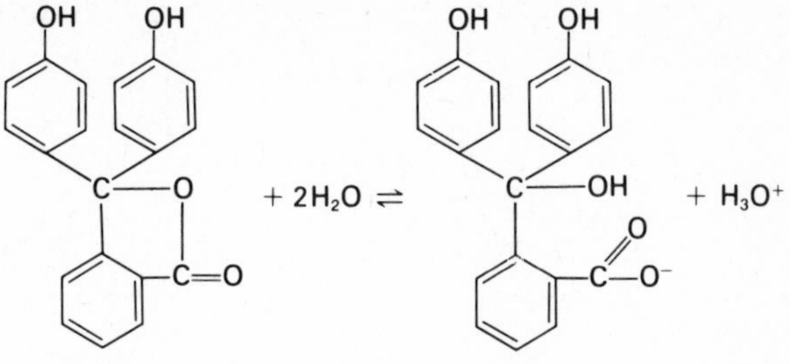

colorless   colorless

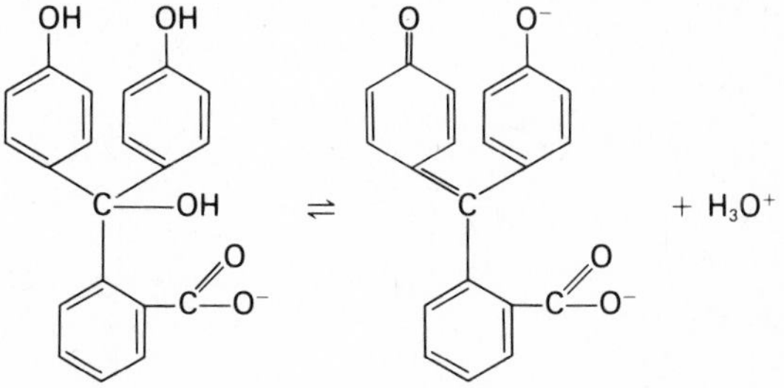

colorless   red

Note that the second equilibrium results in the formation of a quinoid ring, a structure that is often associated with color in molecules incorporating it. Significant concentrations of the colored ion appear in the pH range between 8.0 and 9.6; the pH at which the color is first detectable depends upon the concentration of the indicator and the visual acuity of the observer.

The other phthalein indicators differ in that the phenolic rings contain additional functional groups; *thymolphthalein*, for example, has two alkyl groups on each ring. The basic structural changes associated with the color change of this indicator are similar to those of phenolphthalein.

Sulfonphthalein Indicators. Many of the sulfonphthaleins exhibit two useful color-change ranges; one occurs in rather acidic solutions and the other in neutral or moderately basic media. In contrast to the phthaleins, the basic color shows good stability toward strong alkali.

The sodium salts of the sulfonphthaleins are ordinarily used for the preparation of indicator solutions owing to the appreciable acidity of the parent molecule. Solutions can be prepared directly from the sodium salt or indirectly by dissolving the sulfonphthalein itself in an appropriate volume of dilute aqueous sodium hydroxide.

The simplest sulfonphthalein indicator is *phenolsulfonphthalein*, known also as *phenol red*. The principal equilibria for this compound are

Only the second of the two color changes, occurring in the pH range between 6.4 and 8.0, is useful.

Substitution of halogens or alkyl groups for the hydrogens in the phenolic rings of the parent compound yields sulfonphthaleins that differ in color and pH range.

Azo Indicators. Most azo indicators exhibit a color change from red to yellow with increasing basicity; their transition ranges are generally on the acid side of neutrality. The most commonly encountered examples are *methyl orange* and *methyl red*; the behavior of the former is as follows:

Methyl red is similar to methyl orange except that the sulfonic acid group is replaced by a carboxylic acid group. Variations in the substituents on the amino nitrogen and in the rings give rise to a series of indicators with slightly different properties.

TITRATION ERRORS WITH ACID-BASE INDICATORS

Two types of titration errors can be distinguished. The first is a determinate error which occurs when the pH at which the indicator changes color differs from the pH at chemical equivalence. This type of error can usually be minimized through judicious indicator selection; often a blank will provide an appropriate correction if such is necessary.

The second type is an indeterminate error arising from the limited ability of the eye to distinguish reproducibly the point at which a color change occurs. The magnitude of this error will depend upon the change in pH per milliliter of reagent at the equivalence point, the concentration of the indicator, and the sensitivity of the eye to the two indicator colors. On the average, the visual uncertainty with an acid-base indicator corresponds to about ± 0.5 pH unit. By matching the color of the solution being titrated with that of a reference standard containing a similar amount of indicator at the appropriate pH, the uncertainty can often be reduced to ± 0.1 pH unit or less. It must be understood that these uncertainties are approximations that will vary considerably from indicator to indicator as well as from person to person.

VARIABLES THAT INFLUENCE THE BEHAVIOR OF INDICATORS

The pH interval over which a given indicator exhibits a color change is influenced by the temperature, the ionic strength of the medium, the presence of organic solvents, and the presence of colloidal particles. Some of these effects, particularly the last two, can cause the transition range to shift by one or more pH units.[2]

Titration Curves for Strong Acids or Strong Bases

When both reagent and analyte are strong, the net neutralization reaction can be expressed as

$$H_3O^+ + OH^- = 2H_2O$$

and derivation of a curve for such a titration is analogous to that for a precipitation titration.

TITRATION OF A STRONG ACID WITH A STRONG BASE

The hydronium ions in an aqueous solution of a strong acid come from two sources, namely, (1) the reaction of the solute with water and (2) the dissociation

[2] For a discussion of these effects, see H. A. Laitinen, *Chemical Analysis,* pp. 50–55. New York: McGraw-Hill Book Company, Inc., 1960.

of water itself. In any but the most dilute solutions, however, the contribution from the solute far exceeds that from the solvent. Thus, for a solution of HCl having a concentration greater than about $1 \times 10^{-6} F$, we may write

$$[H_3O^+] = F_{HCl} \tag{9-1}$$

If the concentration is less than $1 \times 10^{-6} F$, the more exact statement, obtained from charge-balance considerations, can be used to calculate $[H_3O^+]$.

$$[H_3O^+] = [Cl^-] + [OH^-]$$

$$= F_{HCl} + \frac{K_w}{[H_3O^+]} \tag{9-2}$$

Here the first term on the right side of this equation represents the hydronium ion contribution from the HCl, while the second is that from the water.

For a solution of a strong base such as sodium hydroxide, an analogous situation exists. That is,

$$[OH^-] = F_{NaOH}$$

Clearly, for the completely dissociated base $Ba(OH)_2$

$$[OH^-] = 2 \times F_{Ba(OH)_2}$$

A useful relationship for calculating the pH of basic solutions can be obtained by taking the negative logarithm of each side of the ion product-constant expression for water. That is,

$$-\log K_w = -\log ([H_3O^+][OH^-]) = -\log [H_3O^+] - \log [OH^-]$$
$$pK_w = pH + pOH$$

At 25°C, pK_w has a value of 14.00.

As will be seen from the following example, the derivation of a curve for the titration of a strong acid with a strong base is simple because the hydronium ion or the hydroxide ion concentration is obtained directly from the formal concentration of the acid or base that is present in excess.

Example.   Derive a curve for the titration of 50.00 ml of 0.0500-F HCl with 0.1000-F NaOH. Round pH data to two places to the right of the decimal point.

Initial pH.   The solution is $5.00 \times 10^{-2} F$ in HCl. Since HCl is completely dissociated,

$$[H_3O^+] = 5.00 \times 10^{-2}$$
$$pH = -\log (5.00 \times 10^{-2}) = -\log 5.00 - \log 10^{-2}$$
$$= -0.699 + 2 = 1.301 = 1.30$$

pH after Addition of 10.00 ml NaOH.   The volume of the solution is now 60.00 ml and part of the HCl has been neutralized. Thus,

$$[H_3O^+] = \frac{50.00 \times 0.0500 - 10.00 \times 0.1000}{60.00} = 2.50 \times 10^{-2}$$

$$pH = 2 - \log 2.50 = 1.60$$

Additional data, which define the curve short of the equivalence point, are derived in the same way. The results of such calculations are given in column 2 of Table 9-2.

pH after Addition of 25.00 ml NaOH.   Here the solution contains neither an excess of HCl nor of NaOH; thus, the pH is governed by the dissociation of water

$$[H_3O^+] = [OH^-] = \sqrt{K_w} = 1.00 \times 10^{-7}$$
$$pH = 7.00$$

pH after Addition of 25.10 ml NaOH.   Here

$$F_{NaOH} = \frac{25.10 \times 0.1000 - 50.00 \times 0.0500}{75.10} = 1.33 \times 10^{-4}$$

Provided the concentration of OH^- from the dissociation of water is negligible with respect to F_{NaOH},

$$[OH^-] = 1.33 \times 10^{-4}$$
$$pOH = 3.88$$
$$pH = 14.00 - 3.88 = 10.12$$

Additional data for this titration, calculated in the same way, are given in column 2 of Table 9-2.

TABLE 9-2 Changes in pH during the Titration of a Strong Acid with a Strong Base

Volume NaOH, ml	50.0 ml of 0.0500-*F* HCl with 0.1000-*F* NaOH pH	50.0 ml of 0.000500-*F* HCl with 0.001000-*F* NaOH pH
0.00	1.30	3.30
10.00	1.60	3.60
20.00	2.15	4.15
24.00	2.87	4.87
24.90	3.87	5.87
25.00	7.00	7.00
25.10	10.12	8.12
26.00	11.12	9.12
30.00	11.80	9.80

Effect of Concentration. The effects of reagent and analyte concentrations on neutralization titration curves are shown by the two sets of data in Table 9-2; these data are plotted in Figure 9-1.

For the titration with 0.1-*N* NaOH (curve *A*), the change in pH in the equivalence-point region is large. With 0.001-*F* NaOH, the change is markedly less but still pronounced.

Figure 9-1 shows that the selection of an indicator is not critical when the reagent concentration is approximately 0.1 *F*. Here, the volume differences among titrations with the three indicators are of the same magnitude as the uncertainties associated with the reading of the buret and are thus negligible. On the other hand, bromocresol green would be clearly unsuited for a titration

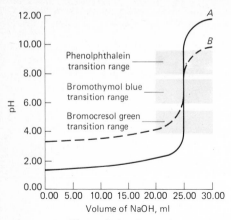

Figure 9-1 Curve for the Titration of HCl with Standard NaOH. Curve B: 50.00 ml 0.000500-F HCl with 0.001000-F NaOH. Curve A: 50.00 ml 0.0500-F HCl with 0.1000-F NaOH.

involving the 0.001-F reagent; not only would the color change occur continuously over a substantial range of titrant volumes, but the transition to the alkaline form would be essentially complete before the equivalence point was reached; a significant determinate error would result. The use of phenolphthalein would be subject to similar objections. Of the three indicators, only bromothymol blue would provide a satisfactory end point with a minimal determinate titration error.

TITRATION OF A STRONG BASE WITH A STRONG ACID

An analogous situation exists for the titration of strong bases with strong acids. In the region short of the equivalence point, the solution is highly alkaline, the hydroxide ion concentration being numerically equal to the normality of the base. The solution is neutral at the equivalence point for precisely the same reason noted previously. Finally, the solution becomes acidic in the region beyond the equivalence point; here the pH is computed from the excess of strong acid that has been introduced. A curve for the titration of a strong base with 0.1-F hydrochloric acid is shown in Figure 9-5 (p. 211). The choice of indicator is subject to the same considerations as noted for the titration of a strong acid with a strong base.

Titration Curves for Weak Acids or Weak Bases

In generating a titration curve for a weak acid, it is necessary to evaluate the pH for a solution of the acid itself, its conjugate base, and mixtures of these two solutes; derivation of a titration curve for a weak base involves similar calculations. The first two were considered on pages 30 to 36. Calculation of the pH for solutions containing appreciable quantities of conjugate acid-base pairs is treated in the section that follows.

THE pH OF SOLUTIONS CONTAINING CONJUGATE
ACID-BASE PAIRS

A solution containing a conjugate acid-base pair may be acidic, basic, or neutral, depending upon the position of two competitive equilibria that exist in the solution. For a weak acid HA and its sodium salt NaA, these equilibria are

$$HA + H_2O \rightleftarrows H_3O^+ + A^- \qquad K_a = \frac{[H_3O^+][A^-]}{[HA]} \qquad (9-3)$$

and

$$A^- + H_2O \rightleftarrows OH^- + HA \qquad K_b = \frac{K_w}{K_a} = \frac{[OH^-][HA]}{[A^-]} \qquad (9-4)$$

If the first equilibrium lies farther to the right than the second, the solution will be acidic. Conversely, if the second equilibrium is more favorable, the solution will be basic. From the two equilibrium-constant expressions, it is evident that the relative concentrations of the hydronium and hydroxide ions will depend upon not only the relative magnitudes of K_a and K_b but also upon the ratio between the concentrations of the acid and its conjugate base.

For a weak base and its conjugate acid, a similar situation prevails. Thus, for a solution of ammonia and ammonium chloride, the equilibria are

$$NH_3 + H_2O \rightleftarrows NH_4{}^+ + OH^-$$

$$K_b = \frac{[NH_4{}^+][OH^-]}{[NH_3]} = 1.76 \times 10^{-5}$$

and

$$NH_4{}^+ + H_2O \rightleftarrows NH_3 + H_3O^+$$

$$K_a = \frac{[H_3O^+][NH_3]}{[NH_4{}^+]} = \frac{K_w}{K_b}$$

$$= \frac{1.00 \times 10^{-14}}{1.76 \times 10^{-5}} = 5.68 \times 10^{-10}$$

We see that K_b is much larger than K_a; thus, the first equilibrium ordinarily predominates and the solutions will be basic. When the ratio of $[NH_4{}^+]$ to $[NH_3]$ becomes greater than about 200, however, the difference in equilibrium constants is offset, and such solutions are then acidic.

Calculation of the pH of a Solution Containing a Weak Acid and Its Conjugate Base. Consider a solution that has a formal acid concentration F_{HA} and a formal concentration of conjugate base F_{NaA}. The important equilibria and equilibrium-constant expressions are given in Equations 9-3 and 9-4.

The reaction shown in Equation 9-3 tends to decrease the concentration of HA by an amount equal to $[H_3O^+]$. Reaction 9-4, conversely, increases the concentration of HA by an amount equal to $[OH^-]$. Thus, the *molar* concentration of HA is given by

$$[HA] = F_{HA} - [H_3O^+] + [OH^-] \qquad (9-5)$$

Similarly, the reaction in Equation 9-3 increases $[A^-]$ by an amount equal to $[H_3O^+]$, while reaction 9-4 decreases $[A^-]$ by an amount corresponding to $[OH^-]$.

$$[A^-] = F_{NaA} + [H_3O^+] - [OH^-] \qquad (9\text{-}6)$$

Equations 9-5 and 9-6 are also readily derived from mass- and charge-balance considerations as follows. The $[HA]$ and $[A^-]$ in the solution arise from two sources, the HA and the NaA that have been introduced in known formal amounts. Thus,

$$F_{HA} + F_{NaA} = [HA] + [A^-] \qquad (9\text{-}7)$$

Electrical neutrality considerations require that

$$[Na^+] + [H_3O^+] = [A^-] + [OH^-]$$

but

$$[Na^+] = F_{NaA}$$

Therefore,

$$F_{NaA} + [H_3O^+] = [A^-] + [OH^-] \qquad (9\text{-}8)$$

which rearranges to Equation 9-6. That is,

$$[A^-] = F_{NaA} + [H_3O^+] - [OH^-]$$

Subtraction of Equation 9-8 from 9-7 and rearrangement gives

$$[HA] = F_{HA} - [H_3O^+] + [OH^-]$$

which is identical to Equation 9-5.

In most of the problems we shall encounter, the difference

$$[H_3O^+] - [OH^-]$$

will be much smaller than the two formal concentrations; under these circumstances, Equations 9-5 and 9-6 simplify to

$$[HA] \cong F_{HA} \qquad (9\text{-}9)$$
$$[A^-] \cong F_{NaA} \qquad (9\text{-}10)$$

Substitution of these quantities into the dissociation-constant expression and rearrangement yields

$$[H_3O^+] = K_a \frac{F_{HA}}{F_{NaA}} \qquad (9\text{-}11)[3]$$

The assumption leading to Equations 9-9 and 9-10 sometimes breaks down with acids or bases that have dissociation constants greater than about 10^{-3} or

---

[3] An alternative form of Equation 9-11 is frequently encountered in biological and biochemical texts. It is obtained by expressing each term in the form of its negative logarithm and inverting the concentration ratio to keep all signs positive. Thus,

$$-\log [H_3O^+] = -\log K_a - \log \frac{F_{HA}}{F_{NaA}}$$

Therefore,

$$pH = pK_a + \log \frac{F_{NaA}}{F_{HA}} \qquad (9\text{-}12)$$

This expression is known as the *Henderson-Hasselbach* equation.

where the formal concentration of either the acid or its conjugate base (or both) is very small. As in earlier calculations, the assumptions should be made. The provisional value for $[H_3O^+]$ and $[OH^-]$ should then be used to test the assumptions.[4]

Within the limits imposed by the assumptions made in its derivation, Equation 9-11 states that the hydronium ion concentration of a solution containing a weak acid and its conjugate base is dependent only upon the *ratio* between the formal concentrations of these two solutes. Furthermore, this ratio remains *independent of the dilution*, since the concentration of each component changes in a proportionate manner upon a change in volume. Thus, *the hydronium ion concentration of a solution containing appreciable quantities of a weak acid and its conjugate base is independent of dilution and depends only upon the formal ratio between the two solutes.* This independence of pH from dilution is one manifestation of the *buffering* properties of such solutions.

Example. What is the pH of a solution that is $0.400\,F$ in formic acid and $1.00\,F$ in sodium formate?

The equilibrium governing the hydronium ion concentration in this solution is

$$H_2O + HCOOH \rightleftarrows H_3O^+ + HCOO^-$$

for which

$$K_a = \frac{[H_3O^+][HCOO^-]}{[HCOOH]} = 1.77 \times 10^{-4}$$

$$[HCOO^-] \cong F_{HCOO^-} = 1.00$$
$$[HCOOH] \cong F_{HCOOH} = 0.400$$

and

$$\frac{[H_3O^+](1.00)}{0.400} = 1.77 \times 10^{-4}$$

Thus,

$$[H_3O^+] = 7.08 \times 10^{-5}$$
$$pH = -\log(7.08 \times 10^{-5}) = 4.150 = 4.15$$

If we compare the pH of this solution with that of a solution that is $0.400\,F$ in formic acid only, we find that the addition of sodium formate has had the effect of raising the pH from 2.08 to 4.15 as a result of common ion effect.

Calculation of the pH of a Solution of a Weak Base and Its Conjugate Acid. The calculation of pH for a solution of a weak base and its conjugate acid is completely analogous to that discussed in the preceding section.

Example. Calculate the pH of a solution that is $0.28\,F$ in NH_4Cl and $0.070\,F$ in NH_3. The equilibrium of interest is

$$NH_3 + H_2O \rightleftarrows NH_4^+ + OH^-$$

for which $K_b = 1.76 \times 10^{-5}$.

---

[4] Ordinarily, when $[H_3O^+]$ is significant, the solution will be acidic, and $[OH^-]$ can be neglected. Alternatively, if the solution is basic, only $[OH^-]$ will be important, and $[H_3O^+]$ is of no consequence.

Normally we may assume that

$$[NH_3] \cong F_{NH_3} = 0.070$$

$$[NH_4^+] \cong F_{NH_4^+} = 0.28$$

Substituting these values into the equilibrium-constant expression,

$$\frac{0.28[OH^-]}{0.070} = 1.76 \times 10^{-5}$$

$$[OH^-] = 4.40 \times 10^{-6}$$

$$pOH = -\log (4.40 \times 10^{-6}) = 5.36$$

$$pH = 14.00 - 5.36 = 8.64$$

EFFECT OF IONIC STRENGTH ON ACID-BASE EQUILIBRIA

As shown on page 109, the degree of dissociation of a weak acid (such as acetic acid) increases with increasing ionic strength. The cause of this phenomenon is analogous to that observed with solubility equilibria. Since the undissociated form of the acid bears no charge, its behavior is only slightly affected by changes in ionic strength; on the other hand, because they are influenced by the ions of electrolyte, the hydronium and acetate ions have less tendency to recombine.

The accompanying example illustrates how a correction for the effect of ionic strength can be made, provided activity coefficients for the ions of the acid are available.

Example. Use activities to calculate the pH of a buffer solution that is $0.0400\,F$ in formic acid and $0.100\,F$ in sodium formate; compare the result with that obtained using concentrations only.

It is reasonable to assume that the ionic strength of this solution is determined by the concentration of sodium formate; that is, that the concentration of H_3O^+ and formate ions from dissociation of the acid is small. Then

$$\mu = \tfrac{1}{2}(0.100 \times 1^2 + 0.100 \times 1^2) = 0.100$$

From Table 5-4 (p. 113), for $\mu = 0.1$

$$f_{H_3O^+} = 0.83$$

$$f_{HCOO^-} = 0.76$$

Thus,

$$K_a = \frac{[H_3O^+][HCOO^-]}{[HCOOH]} \times \frac{f_{H_3O^+} f_{HCOO^-}}{f_{HCOOH}} = 1.77 \times 10^{-4}$$

The *concentration equilibrium constant* K_a' is computed by combining the activity coefficients with K_a.

$$K_a' = \frac{1.74 \times 10^{-4} \times f_{HCOOH}}{f_{H_3O^+} f_{HCOO^-}} = \frac{1.77 \times 10^{-4} \times 1.0}{0.83 \times 0.76}$$

$$= 2.81 \times 10^{-4} = \frac{[H_3O^+][HCOO^-]}{[HCOOH]}$$

Note that the activity coefficient of the uncharged HCOOH is assumed to be 1.0. This assumption is generally valid for neutral species.

The concentration equilibrium constant K_a' is now used for the calculation.

$$[HCOO^-] = 0.100$$
$$[HCOOH] = 0.0400$$

$$\frac{[H_3O^+] \times 0.100}{0.0400} = 2.81 \times 10^{-4}$$

$$[H_3O^+] = 1.12 \times 10^{-4}$$
$$pH = -\log (1.12 \times 10^{-4}) = 3.95$$

Note that a pH of 4.15 would be obtained if this calculation was made without the use of activity coefficients.

BUFFER SOLUTIONS

A *buffer solution* is defined as a solution that resists changes in pH as a result of (1) dilution or (2) small additions of acids or bases. The most effective buffer solution contains large and approximately equal concentrations of a conjugate acid-base pair.

Effect of Dilution. The pH of a buffer solution remains essentially independent of solution until its concentrations are reduced to the point where the approximations used to develop Equations 9-9 and 9-10 become invalid. The following example illustrates the behavior of a typical buffer during dilution.

Example. Calculate the pH of the buffer of pH 4.15 described in the example on page 199 upon dilution by a factor of (1) 50, and (2) 10,000.

1. Upon dilution by a factor of 50

$$F_{HCOOH} = \frac{0.400}{50} = 8.00 \times 10^{-3}$$

$$F_{HCOONa} = \frac{1.00}{50} = 2.00 \times 10^{-2}$$

If we assume, as before, that $([H_3O^+] - [OH^-])$ is small with respect to the two formalities (p. 198), we obtain

$$\frac{[H_3O^+] \times 2.00 \times 10^{-2}}{8.00 \times 10^{-3}} = 1.77 \times 10^{-4}$$

$$[H_3O^+] = 7.08 \times 10^{-5}$$
$$pH = 4.15$$

Here

$$[OH^-] = \frac{1.00 \times 10^{-14}}{7.08 \times 10^{-5}} = 1.41 \times 10^{-10}$$

Thus, the assumption $(7.08 \times 10^{-5} - 1.41 \times 10^{-10}) \ll 8.00 \times 10^{-3}$ and 2.00×10^{-2} is reasonably good.

2. Upon dilution by a factor of 10,000

$$F_{HCOOH} = 4.00 \times 10^{-5}$$
$$F_{HCOONa} = 1.00 \times 10^{-4}$$

Here the more exact statements given by Equations 9-5 and 9-6 must be used. That is,

$$[HCOOH] = 4.00 \times 10^{-5} - [H_3O^+] + [OH^-]$$
$$[HCOO^-] = 1.00 \times 10^{-4} + [H_3O^+] - [OH^-]$$

The solution is acidic; thus, it seems probable that $[OH^-] \ll [H_3O^+]$ (see footnote 4, p. 199). Therefore,

$$[HCOOH] \cong 4.00 \times 10^{-5} - [H_3O^+]$$
$$[HCOO^-] \cong 1.00 \times 10^{-4} + [H_3O^+]$$

Substitution of these quantities into the dissociation-constant expression gives

$$\frac{[H_3O^+](1.00 \times 10^{-4} + [H_3O^+])}{4.00 \times 10^{-5} - [H_3O^+]} = 1.77 \times 10^{-4}$$

This equation rearranges to the quadratic form

$$[H_3O^+]^2 + 2.77 \times 10^{-4}[H_3O^+] - 7.08 \times 10^{-9} = 0$$

The solution is

$$[H_3O^+] = 2.36 \times 10^{-5}$$
$$pH = 4.63$$

Thus, a 10,000-fold dilution caused the pH to increase from 4.15 to 4.63, whereas a 50-fold dilution had essentially no effect.

Figure 9-2 contrasts the behavior of buffered and unbuffered solutions with dilution. For each the initial solute concentration is 1.00 F. The resistance of the buffered solution to pH changes is clear.

Addition of Acids and Bases to Buffers. The resistance of buffer mixtures to pH changes from added acids or bases is conveniently illustrated with examples.

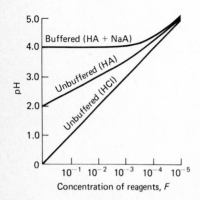

Figure 9-2   Effect of Dilution on the pH of Buffered and Unbuffered Solutions. The dissociation constant for HA is assumed to be 1.00 $\times$ 10^{-4}. Initial solute concentrations are 1.00 F.

Example. Calculate the pH change which takes place when 100 ml of (1) 0.0500-F NaOH and (2) 0.0500-F HCl are added to 400 ml of a buffer solution that is 0.200 F in NH_3 and 0.300 F in NH_4Cl.

The initial pH of the solution is obtained by assuming

$$[NH_3] \cong F_{NH_3} = 0.200$$

$$[NH_4^+] \cong F_{NH_4Cl} = 0.300$$

and substituting into the dissociation-constant expression for NH_3

$$\frac{0.300[OH^-]}{0.200} = 1.76 \times 10^{-5}$$

$$[OH^-] = 1.17 \times 10^{-5}$$

$$pH = 14.00 - (-\log 1.17 \times 10^{-5}) = 9.07$$

1. Addition of NaOH converts part of the NH_4^+ of the buffer to NH_3:

$$NH_4^+ + OH^- \rightarrow NH_3 + H_2O$$

The formal concentrations of NH_3 and NH_4Cl then become

$$F_{NH_3} = \frac{400 \times 0.200 + 100 \times 0.0500}{500} = 0.170$$

$$F_{NH_4Cl} = \frac{400 \times 0.300 - 100 \times 0.0500}{500} = 0.230$$

When substituted into the dissociation-constant expression these values yield

$$[OH^-] = \frac{1.76 \times 10^{-5} \times 0.170}{0.230} = 1.30 \times 10^{-5}$$

$$pH = 14.00 - (-\log 1.30 \times 10^{-5}) = 9.11$$

and

$$\Delta pH = 9.11 - 9.07 = 0.04$$

2. Addition of HCl converts part of the NH_3 to NH_4^+; thus,

$$F_{NH_3} = \frac{400 \times 0.200 - 100 \times 0.0500}{500} = 0.150$$

$$F_{NH_4^+} = \frac{400 \times 0.300 + 100 \times 0.0500}{500} = 0.250$$

$$[OH^-] = 1.76 \times 10^{-5} \times \frac{0.150}{0.250} = 1.06 \times 10^{-5}$$

$$pH = 14.00 - (-\log 1.06 \times 10^{-5}) = 9.02$$

$$\Delta pH = 9.02 - 9.07 = -0.05$$

Example. Contrast the behavior of an unbuffered solution with a pH 9.07 ($[OH^-] = 1.17 \times 10^{-5}$) when treated in the same way as the buffered solution described in the previous example.

1. After addition of NaOH,

$$[OH^-] = \frac{400 \times 1.17 \times 10^{-5} + 100 \times 0.0500}{500} = 1.00 \times 10^{-2}$$

$$pH = 14.00 - (-\log 1.00 \times 10^{-2}) = 12.00$$

$$\Delta pH = 12.00 - 9.07 = 2.93$$

2. After addition of HCl,

$$[H_3O^+] = \frac{100 \times 0.0500 - 400 \times 1.17 \times 10^{-5}}{500} = 1.00 \times 10^{-2}$$

$$pH = -\log 1.00 \times 10^{-2} = 2.00$$

$$\Delta pH = 2.00 - 9.07 = -7.07$$

Buffer Capacity. The foregoing examples demonstrate that a solution containing a conjugate acid-base pair shows remarkable resistance to changes in pH. The ability of a buffer to prevent a significant change in pH is directly related to the total concentration of the buffering species as well as their concentration ratios. For example, 400-ml portions of a solution formed by diluting the buffer in the foregoing example by a factor of 10 would exhibit changes of about ± 0.5 pH unit when treated with the same amounts of sodium hydroxide or hydrochloric acid; the change for the more concentrated buffer was only about 0.04.

The *buffer capacity* of a solution is defined as the number of equivalents of strong acid or base needed to cause 1.00 liter of the buffer to undergo a 1.00-unit change in pH. As we have seen, the buffer capacity is dependent upon the concentration of the conjugate acid-base pair. It is also dependent upon their concentration ratio, and reaches maximum when this ratio is unity.[5]

Preparation of Buffers. In principle, a buffer solution of any desired pH can be prepared by combining calculated quantities of a suitable conjugate acid-base pair.

[5] Consider a buffer solution which contains *a* moles of the acid HA and *b* moles of the conjugate base NaA. To this solution is added *n* moles of a strong acid. The formal concentrations of acid and conjugate base then become

$$F_{HA} = \frac{a + n}{v}$$

$$F_{NaA} = \frac{b - n}{v}$$

where *v* is the volume of solution. The pH of this solution (Note 1, p. 198) is given by

$$pH = pK_a + \log\frac{F_{NaA}}{F_{HA}} = pK_a + \log\frac{b - n}{a + n}$$

$$= pK_a + \log(b - n) - \log(a + n)$$

Differentiation of this expression with respect to *n* gives the rate of change in pH with added acid.

$$\frac{dpH}{dn} = 0.434 \times \frac{(-1)}{b - n} - 0.434 \times \frac{1}{a + n}$$

To find the minimum change in pH with *n*, we set the second derivative of this expression equal to zero. Thus,

$$\frac{d^2pH}{dn^2} = 0.434 \times \frac{(-1)}{(b - n)^2} + 0.434 \times \frac{1}{(a + n)^2} = 0$$

which rearranges to

$$\frac{1}{(b - n)^2} = \frac{1}{(a + n)^2}$$

Taking the square root of both sides of this equation and rearranging gives

$$b = a$$

Thus, the minimum rate of pH change with added acid occurs when the number of moles of conjugate base *b* is equal to the number of moles of conjugate acid *a*. The same relationship is obtained when a base is added to the buffer or when the buffer is a weak base and its conjugate acid.

Example. Describe the preparation of a pH 5.00 buffer from a 0.500-F solution of acetic acid ($K_a = 1.75 \times 10^{-5}$) and a 0.426-$F$ solution of NaOH.

The acetate–acetic acid ratio that yields a solution with a pH of 5.00 is obtained by rearrangement of the dissociation-constant expression

$$\frac{[OAc^-]}{[HOAc]} = \frac{K_{HOAc}}{[H_3O^+]} = \frac{1.75 \times 10^{-5}}{1.00 \times 10^{-5}}$$

$$= 1.75$$

In order to produce this ratio it will be necessary to mix V_{HA} ml of acetic acid with V_{OH^-} ml of 0.426-F base. The concentrations of acid and conjugate base following mixing will be

$$F_{OAc^-} = \frac{0.426 V_{OH^-}}{V_{HA} + V_{OH^-}} \cong [OAc^-]$$

$$F_{HOAc} = \frac{0.500 V_{HA} - 0.426 V_{OH^-}}{V_{HA} + V_{OH^-}} \cong [HOAc]$$

When these quantities are substituted for $[OAc^-]$ and $[HOAc]$ in the original ratio, the denominators cancel, and

$$\frac{0.426 V_{OH^-}}{0.500 V_{HA} - 0.426 V_{OH^-}} = 1.75$$

which rearranges to

$$\frac{V_{OH^-}}{V_{HA}} = 0.747$$

If 100 ml of buffer were needed,

$$V_{OH^-} + V_{HA} = 100$$
$$0.747 V_{HA} + V_{HA} = 100$$
$$V_{HA} = 57.2 \text{ ml}$$
$$V_{OH^-} = 100 - 57.2 = 42.8 \text{ ml}$$

Thus, by adding 42.8 ml of the sodium hydroxide to 57.2 ml of acetic acid, a buffer of the desired pH should, in principle, result.

Calculations such as the foregoing provide guidance for preparation of buffer solutions with *approximately* the desired pH; however, they generally cannot be expected to yield precise data because of the uncertainties that exist in the numerical values for many dissociation constants. More importantly, the ionic strength of a buffer is usually so high that good values for the activity coefficients of the ions in the solution cannot be obtained from the Debye-Hückel relationship. Thus, in the example just considered, the ionic strength is about 0.18; the concentration equilibrium constant K_a' (see p. 200) would, therefore, be significantly larger than 1.75×10^{-5} and quite uncertain (about 3×10^{-5}).

Empirically derived recipes for preparation of buffer solutions of known pH are available in chemical handbooks and reference works.[6] Because of their widespread employment, two buffer systems deserve specific mention. McIlvaine buffers cover a pH range from about 2 to 8 and are prepared by mixing solutions

[6] See, for example, L. Meites, Ed., *Handbook of Analytical Chemistry*, pp. **5**-12 and **11**-5 to **11**-7. New York: McGraw-Hill Book Company, Inc., 1963.

of citric acid with disodium hydrogen phosphate. Clark and Lubs buffers, which include a pH range from 2 to 10, make use of three systems: phthalic acid, potassium hydrogen phthalate; potassium dihydrogen phosphate, dipotassium hydrogen phosphate; and boric acid, sodium borate.

TITRATION CURVES FOR WEAK ACIDS

To derive a titration curve for a weak acid (or weak base) four types of calculations must be employed, corresponding to four distinct parts of the curve. (1) At the outset, the solution contains only a weak acid or a weak base, and the pH is calculated from the formal concentration of that solute. (2) After various increments of reagent (in quantities up to, but not including, an equivalent amount) have been added, the solution consists of a series of buffers; the pH of each can be calculated from the formal concentrations of the product and the residual weak acid or base. (3) At the equivalence point, the solution contains the conjugate of the weak acid or base being titrated, and the pH is calculated from the formal concentration of this product. (4) Finally, beyond the equivalence point, the excess of strong acid or base that is used as titrant represses the basic or acidic character of the product to such an extent that the pH is governed largely by the concentration of the excess reagent.

Derivation. To illustrate these calculations, consider derivation of a curve for the titration of 50.00 ml of 0.1000-F acetic acid ($K_a = 1.75 \times 10^{-5}$) with 0.1000-$F$ sodium hydroxide.

Example.

Initial pH. Here we must calculate the pH of a 0.1000-F solution of HOAc; using the method shown on page 34, a value of 2.88 is obtained.

pH after Addition of 10.00 ml of Reagent. A buffer solution consisting of NaOAc and HOAc has now been produced. The formal concentrations of the two constituents are given by

$$F_{HOAc} = \frac{50.00 \times 0.1000 - 10.00 \times 0.1000}{60.00} = \frac{4.000}{60.00}$$

$$F_{NaOAc} = \frac{10.00 \times 0.1000}{60.00} = \frac{1.000}{60.00}$$

Upon substituting these concentrations into the dissociation-constant expression for acetic acid, we obtain

$$\frac{[H_3O^+] \cdot 1.000/60.00}{4.000/60.00} = K_a = 1.75 \times 10^{-5}$$

$$[H_3O^+] = 7.00 \times 10^{-5}$$

$$pH = 4.16$$

Calculations similar to those above will delineate the curve throughout the buffer region. Data from such calculations are given in column 2 of Table 9-3. When the acid has been 50% neutralized (in this particular titration, after an

TABLE 9-3   Changes in pH during the Titration of a Weak Acid with a Strong Base

Volume NaOH, ml	50.00 ml of 0.1000-F HOAc with 0.1000-F NaOH pH	50.00 ml of 0.001000-F HOAc with 0.001000-F NaOH pH
0.00	2.88	3.91
10.00	4.16	4.30
25.00	4.76	4.80
40.00	5.36	5.38
49.00	6.45	6.45
49.90	7.46	7.46
50.00	8.73	7.73
50.10	10.00	8.00
51.00	11.00	9.00
60.00	11.96	9.96
75.00	12.30	10.30

addition of exactly 25.0 ml of base), note that the formal concentrations of acid and conjugate base are identical; within the limits of the usual approximations, so also are their molar concentrations. Thus, these terms cancel one another in the equilibrium-constant expression, and the hydronium ion concentration is numerically equal to the dissociation constant; that is, the pH is equal to the pK_a. Likewise, in the titration of a weak base, the hydroxide ion concentration is numerically equal to the dissociation constant of the base at the midpoint.

Example.

Equivalence-Point pH.   At the equivalence point in the titration, the acetic acid has been converted to sodium acetate. The solution is therefore similar to one formed by dissolving that base in water. After evaluating the formal concentration of the salt, the pH calculation is identical to that described on page 36 for a weak base. Here, the base, NaOAc, has a formal concentration of 0.0500. Thus,

$$[OH^-] \cong \sqrt{\frac{K_w}{K_a} \times 0.0500} = \sqrt{\frac{1.00 \times 10^{-14} \times 0.0500}{1.75 \times 10^{-5}}}$$

$$[OH^-] = 5.34 \times 10^{-6}$$

$$pH = 14.00 - (-\log 5.34 \times 10^{-6})$$

$$= 8.73$$

pH after Addition of 50.10 ml of Base.   After 50.10 ml of base have been added, hydroxide ions arise from both the excess of sodium hydroxide and the equilibrium between acetate ion and water. The contribution of the latter is small enough to be neglected, however, since the excess of strong base will tend to repress the equilibrium. This fact becomes evident when we consider that the hydroxide ion concentration was only 5.34×10^{-6} M at the equivalence point of the titration; once an excess of strong base has been added, the contribution from the reaction of the acetate will be even smaller. Thus,

$$[OH^-] \cong F_{NaOH} = \frac{50.10 \times 0.1000 - 50.00 \times 0.1000}{100.1} = 1.0 \times 10^{-4}$$

$$pOH = 14.00 - (-\log 1.0 \times 10^{-4}) = 10.00$$

Note that titration curves for a weak acid and a strong base become identical with those for a strong acid with a strong base in the region slightly beyond the equivalence point.

Effect of Concentration. The second and third columns of Table 9-3 contain pH data for the titration of 0.1000-F and of 0.001000-F acetic acid, respectively, with sodium hydroxide solutions of the same strengths. As illustrated by the following example, derivation of part of the curve for the dilute solution requires the use of more exact expressions for the concentration relationships.

Example. Calculate the pH of the solution that results when 50.00 ml of 0.001000-F acetic acid are mixed with 10.00 ml of 0.001000-F NaOH.

The formal concentrations are

$$F_{HOAc} = \frac{50.00 \times 0.001000 - 10.00 \times 0.001000}{60.00} = 6.67 \times 10^{-4}$$

$$F_{NaOAc} = \frac{10.00 \times 0.001000}{60.00} = 1.67 \times 10^{-4}$$

If we attempt to apply the same approximations as in the titration with the 0.1000-F reactants (p. 206), we find that

$$[H_3O^+] = 7.00 \times 10^{-5}$$

Note, however, that $[H_3O^+]$ is no longer small with respect to the two formal concentrations; therefore, Equations 9-5 and 9-6 must be used. That is,

$$[HOAc] = F_{HA} - [H_3O^+] + [\cancel{OH^-}]$$
$$[OAc^-] = F_{NaA} + [H_3O^+] - [\cancel{OH^-}]$$

Because the solution is acidic, $[OH^-]$ will be negligible. Substitution into the dissociation-constant expression yields

$$\frac{[H_3O^+](F_{NaA} + [H_3O^+])}{(F_{HA} - [H_3O^+])} = K_a$$

which rearranges to the quadratic form

$$[H_3O^+]^2 + (F_{NaA} + K_a)[H_3O^+] - K_a F_{HA} = 0$$

Thus, for this example

$$[H_3O^+]^2 + 1.84 \times 10^{-4}[H_3O^+] - 1.17 \times 10^{-8} = 0$$

and

$$[H_3O^+] = 5.00 \times 10^{-5}$$
$$pH = 4.30$$

The data for additions of 10.00 ml to 49.00 ml of base in column 3 of Table 9-3 were obtained by the technique just described.

Figure 9-3 is a plot of the data in Table 9-3. Note that the initial pH values are higher and the equivalence-point pH is lower for the more dilute solution than for the 0.1000-F solution. At intermediate titrant volumes, however, the pH

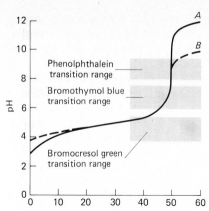

Figure 9-3 Curve for the Titration of Acetic Acid with NaOH. Curve *A*: 0.1000-*F* acid with 0.1000-*F* base. Curve *B*: 0.001000-*F* acid with 0.001000-*F* base.

values differ only slightly. Acetic acid–sodium acetate buffers exist within this region; Figure 9-3 is graphical confirmation of the fact that the pH of buffers is largely independent of dilution.

    Titration curves for 0.1000-*F* solutions of acids of differing strengths are shown in Figure 9-4.

    Clearly, the change in pH in the equivalence-point region becomes smaller as the acid becomes weaker—that is, as the reaction between the acid and the base becomes less complete. The relation between completeness of reaction and reagent concentrations illustrated by Figures 9-3 and 9-4 is analogous to these effects on precipitation titration curves (pp. 173 to 175).

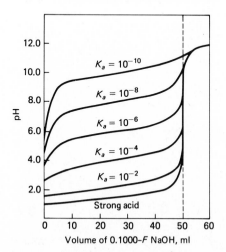

Figure 9-4 Influence of Acid Strength upon Titration Curves. Each curve represents titration of 50.0 ml of 0.1000-*F* acid with 0.1000-*F* NaOH.

Indicator Choice; Feasibility of Titration. Figures 9-3 and 9-4 clearly indicate that the choice of indicator for the titration of a weak acid is more limited than for a strong acid. For example, bromocresol green is totally unsuited for titration of 0.1000-F acetic acid; nor would bromothymol blue be satisfactory, since its full color change would occur over a range between about 47 and 50 ml of 0.100-F base. It might still be possible to employ bromothymol blue by titrating to the full basic color of the indicator rather than to the change in color; this technique would require a basic comparison standard containing the same concentration of indicator as the solution being titrated. An indicator such as phenolphthalein, which changes color in the basic region, would be more satisfactory for the titration.

The end-point pH change associated with the titration of 0.00100-F acetic acid (curve *B*, Figure 9-3) is so small that a significant titration error is likely to be introduced. By employing an indicator with a transition range between that of phenolphthalein and bromothymol blue, and by using a color comparison standard, it should be possible to establish the end point with a reproducibility of a few percent.

Figure 9-4 illustrates that similar problems exist as the strength of the acid being titrated decreases. Precision on the order of ± 2 ppt can be achieved in the titration of a 0.100-F acid solution with a dissociation constant of 10^{-8} provided a suitable color comparison standard is available. With more concentrated solutions, somewhat weaker acids can be titrated with reasonable precision.

TITRATION CURVES FOR WEAK BASES

The derivation of a curve for the titration of a weak base is analogous to that for a weak acid.

Example. A 50.00-ml aliquot of 0.0500-F NaCN is titrated with 0.1000-F HCl. Calculate the pH after the additions of (1) 0.00, (2) 10.00, (3) 25.00, and (4) 26.00 ml of acid.

1. 0.00 ml Reagent. Cyanide ion is a simple, weak base; calculation of the pH is performed as on page 35. That is,

$$[OH^-] \cong \sqrt{K_b F_{NaCN}}$$

where

$$K_b = \frac{K_w}{K_{HCN}} = \frac{1.00 \times 10^{-14}}{2.1 \times 10^{-9}} = 4.76 \times 10^{-6}$$

The pH is found to be 10.69.

2. 10.00 ml Reagent. Addition of acid produces a buffer with a composition given by

$$F_{NaCN} = \frac{50.00 \times 0.0500 - 10.00 \times 0.1000}{60.00} = \frac{1.500}{60.00}$$

$$F_{HCN} = \frac{10.00 \times 0.1000}{60.00} = \frac{1.000}{60.00}$$

These values are then substituted into the acid-dissociation constant of HCN to give

$$[H_3O^+] = \frac{2.1 \times 10^{-9} \times 1.000/60.00}{1.500/60.00} = 1.4 \times 10^{-9}$$

$$pH = 8.85$$

3. 25.00 ml Reagent. This volume corresponds to the equivalence point; the principal solute species is the weak acid HCN. Thus,

$$F_{HCN} = \frac{25.00 \times 0.1000}{75.00} = 0.03333$$

and, as on page 32,

$$[H_3O^+] \cong \sqrt{K_a F_{HCN}} = \sqrt{2.1 \times 10^{-9} \times 0.03333}$$

$$= 8.37 \times 10^{-6}$$

$$pH = 5.08$$

4. 26.00 ml Reagent. The excess of strong acid now present will repress the dissociation of the HCN to the point where its contribution to the pH is negligible. Thus,

$$[H_3O^+] = F_{HCl} = \frac{26.00 \times 0.100 - 50.00 \times 0.0500}{76.00}$$

$$= 1.32 \times 10^{-3}$$

$$pH = 2.88$$

Figure 9-5 shows titration curves derived for a series of weak bases of differing strengths. Clearly, indicators with *acidic* transition ranges must be employed for weaker bases.

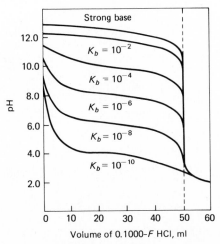

Figure 9-5 Influence of Base Strength upon Titration Curves. Each curve represents titration of 50.0 ml of 0.1000-*F* base with 0.1000-*F* HCl.

PROBLEMS

In all pH or pOH calculations round the results to two places to the right of the decimal.

*1. Calculate the pH of a solution produced by mixing 40.00 ml of 0.01500-F HCl with
  (a) 40.00 ml of water.
  (b) 20.00 ml of 0.02000-F NaOH.
  (c) 40.00 ml of 0.02000-F NaOH.
  (d) 20.00 ml of 0.02000-F Ba (OH)$_2$.
  (e) 40.00 ml of 0.01000-F hypochlorous acid, HClO.
  (f) 30.00 ml of 0.02000-F NaOH.

2. Calculate the pH of a solution prepared by mixing 20.00 ml of 0.02000-F KOH with
  (a) 20.00 ml of 0.02000-F Ba(OH)$_2$.   (d) 20.00 ml of 0.01000-F NH$_3$.
  (b) 20.00 ml of 0.01000-F HCl.   (e) 20.00 ml of water.
  (c) 5.00 ml of 0.02000-F H$_2$SO$_4$.   (f) 40.00 ml of 0.02000-F HCl.

*3. Calculate the pH of an aqueous solution that contains 2.50% HCl by weight and has a density of 1.012 g/ml.

4. Calculate the pH of an aqueous solution that contains 1.00% NaOH by weight and has a density of 1.011 g/ml.

*5. Calculate the pH of a solution prepared by mixing 25.00 ml of 0.04000-F sodium hypochlorite with
  (a) 25.00 ml of water.   (e) 24.90 ml of 0.04000-F HCl.
  (b) 25.00 ml of 0.05000-F HCl.   (f) 25.10 ml of 0.04000-F HCl.
  (c) 25.00 ml of 0.02000-F NaOH.   (g) 25.00 ml of 0.02000-F HOCl.
  (d) 50.00 ml of 0.02000-F HCl.

6. Calculate the pH of a solution prepared by mixing 30.00 ml of 0.0500-F mandelic acid with
  (a) 50.00 ml of water.   (e) 24.00 ml of 0.0600-F NaOH.
  (b) 10.00 ml of 0.0600-F NaOH.   (f) 10.00 ml of 0.0600-F Ba(OH)$_2$.
  (c) 10.00 ml of 0.0600-F HCl.   (g) 30.00 ml of 0.0500-F sodium
  (d) 25.00 ml of 0.0600-F NaOH.   mandelate.

*7. Calculate the pH of a solution prepared by mixing 20.00 ml of 0.1200-F salicylic acid with
  (a) 20.00 ml of water.   (e) 19.00 ml of 0.1200-F NaOH.
  (b) 20.00 ml of 0.0500-F NaOH.   (f) 14.00 ml of 0.1000-F sodium
  (c) 30.00 ml of 0.0800-F NaOH.   salicylate.
  (d) 20.00 ml of 0.2000-F NaOH.   (g) 20.00 ml of 0.5000-F HCl.

8. Calculate the pH of a solution prepared by mixing 400 ml of 0.2000-F ethanolamine with
  (a) 600 ml of water.   (e) 600 ml of 0.0500-F NaOH.
  (b) 600 ml of 0.1200-F HCl.   (f) 600 ml of 0.0500-F HCl.
  (c) 600 ml of 0.1334-F HCl.   (g) 600 ml of 0.0500-F ethanolamine
  (d) 600 ml of 0.1600-F HCl.   hydrochloride.

*9. What is the pH of a solution
  (a) containing 0.0140 g of benzoic acid and 0.0260 g of sodium benzoate per 100 ml?
  (b) prepared by adding 10.0 ml of 1.00-F HCl to 1.00 liter of 0.0500-F sodium benzoate?
  (c) prepared by dissolving 3.60 g of benzoic acid in 200 ml of 0.200-F NaOH?

*10. What is the pH of a solution prepared by
   (a) adding 0.0470 g of hydrazine to 75 ml of 0.0300-F HCl?
   (b) mixing 400 ml of 0.150-F hydrazine with 300 ml of 0.200-F HCl?
   (c) adding 2.75 g of HCl and 5.02 g of hydrazine to water and diluting to 750 ml?

*11. Calculate the pH of a buffer that is
   (a) 0.0405 F in NH_3 and 0.0216 F in $(NH_4)_2SO_4$.
   (b) 0.0176 F with respect to phenol and 0.0254 F with respect to sodium phenolate.
   (c) 1.00 F in trichloroacetic acid and 0.500 F in sodium trichloroacetate.
   (d) 0.164 F with respect to ethylamine and 0.272 F with respect to ethylamine hydrochloride.

12. Calculate the pH of a buffer that is
   (a) 0.0157 F with respect to aniline and 0.0123 F with respect to aniline hydrochloride.
   (b) 0.347 F with respect to sulfamic acid and 0.610 F with respect to sodium sulfamate.
   (c) 0.177 F in pyridine and 0.264 F in pyridine hydrochloride.
   (d) 0.700 F in formic acid and 0.660 F in sodium formate.

*13. Calculate the molar ratio of base to conjugate acid or acid to conjugate base in a buffer that has a pH of 9.80 and contains
   (a) NH_3 and NH_4Cl.
   (b) CH_3NH_2 and CH_3NH_3Cl.
   (c) HCN and NaCN.
   (d) HOCl and NaOCl.

14. Calculate the molar ratio of base to conjugate acid or acid to conjugate base in a buffer that has a pH of 4.75 and contains
   (a) HNO_2 and $NaNO_2$.
   (b) CH_3CH_2COOH and CH_3CH_2COONa.
   (c) C_5H_5N and C_5H_5NHCl.
   (d) $C_6H_5NH_2$ and $C_6H_5NH_3Cl$.

*15. Calculate the pH of a buffer solution that is 0.200 F in NH_3 and 0.100 F in NH_4Cl
   (a) employing concentrations.
   (b) employing activity coefficients.

16. Calculate the pH of a buffer solution that is 0.0500 F in sodium benzoate and 0.0800 F in benzoic acid
   (a) employing concentrations.
   (b) employing activity coefficients.

*17. Calculate the pH change that occurs when the following solutions are diluted by a factor of 100:
   (a) 0.0400-F HNO_2.
   (b) 0.0400-F $NaNO_2$ and 0.0400-F HNO_2.
   (c) 0.400-F $NaNO_2$ and 0.400-F HNO_2.

18. Calculate the pH change that occurs when the following solutions undergo a 50-fold dilution:
   (a) 0.0500-F NH_3.
   (b) 0.0500-F NH_3 and 0.0500-F NH_4Cl.
   (c) 5.00×10^{-4} F NH_3 and 5.00×10^{-4} F NH_4Cl.

*19. Calculate the change in pH that occurs when 50.0 ml of 0.0200-F HCl are added to 50.0 ml of each of the undiluted solutions in Problem 17.

20. Calculate the change in pH that occurs when 50.0 ml of 0.0200-F HCl are added to 50.0 ml of each of the undiluted solutions in Problem 18.

*21. Calculate the change in pH which occurs when 50.0 ml of 0.0200-F NaOH are added to 50.0 ml of each of the undiluted solutions in Problem 17.

22. Calculate the change in pH which occurs when 50.0 ml of 0.0200-F NaOH are added to 50.0 ml of each of the undiluted solutions in Problem 18.

*23. How many grams of NH_4Cl must be added to 500 ml of 0.137-F NH_3 to give a buffer of pH 10.34 (assume no volume change)?

24. How many grams of sodium acetate must be added to 2.00 liters of 0.316-F acetic acid to produce a buffer of pH 4.87 (assume no volume change)?

*25. Describe the preparation of 250 ml of a buffer of pH 4.50 by mixing 0.200-F sodium formate with 0.300-F formic acid.

26. Describe the preparation of 50.00 ml of a buffer of pH 9.75 by mixing 0.331-F ethanolamine with 0.222-F HCl.

*27. Describe the preparation of 300 ml of a buffer of pH 5.60 from 0.229-F pyridine and 0.100-F $HClO_4$.

28. Describe the preparation of 1.00 liter of a buffer of pH 5.55 from 0.375-F sodium acetate and 0.300-F HCl.

*29. Calculate the pH of the solution that results from the addition of 0.00, 5.00, 20.00, 35.00, 39.00, 40.00, 41.00, 45.00, and 50.00 ml of 0.1000-F HCl in the titration of
(a) 50.00 ml of 0.0800-F NH_3.
(b) 50.00 ml of 0.0800-F piperidine.
(c) 50.00 ml of 0.0800-F hydroxylamine.
(d) 50.00 ml of 0.0800-F sodium phenolate.

30. Calculate the pH of the solution that results from the addition of 0.00, 5.00, 20.00, 35.00, 39.00, 40.00, 41.00, 45.00, and 50.00 ml of 0.100-F NaOH in the titration of
(a) 50.00 ml of 0.0800-F propanoic acid.
(b) 50.00 ml of 0.0800-F hypochlorous acid.
(c) 50.00 ml of 0.0800-F trichloroacetic acid.
(d) 50.00 ml of 0.0800-F hydroxylamine hydrochloride.

TITRATION CURVES FOR COMPLEX ACID-BASE SYSTEMS

This chapter is concerned with the equilibria that exist in solutions containing two acids or two bases that differ in strength. Two types of solutions must be considered, specifically those in which the two solutes are chemically unrelated and those containing a single polyfunctional species which yields more than one mole of hydronium or hydroxide ions per formula weight. As in the previous two chapters, the titration curve forms the basis for discussion.

Titration Curves for a Mixture of a Weak and Strong Acid or a Weak and Strong Base

To illustrate the derivation of curves for mixtures of strong and weak acids or bases, consider the titration of a solution containing hydrochloric acid and a weak acid HA.

Throughout the early stages of this titration, the hydronium ion concentration can be expressed as

$$[H_3O^+] = F_{HCl} + [A^-] + [\cancel{OH^-}]$$

The first term on the right expresses the contribution of the strong acid to the acidity and the second, that from the weak acid. The third term represents the contribution of hydronium ions from dissociation of the solvent; this term is vanishingly small in most (if not all) acidic solutions and can thus be neglected in all calculations.

As shown by the following example, the hydrochloric acid represses the dissociation of the weak acid in the early stages of the titration. Thus, $[A^-] \ll F_{HCl}$, and the hydronium ion concentration is simply equal to the formal concentration of the strong acid.

Example. Calculate the pH of a mixture which is 0.1200 F in hydrochloric acid and 0.0800 F in a weak acid HA ($K_a = 1.00 \times 10^{-4}$).

$$[H_3O^+] = 0.1200 + [A^-]$$

Let us assume that $[A^-] \ll 0.1200$. Then $[H_3O^+] = 0.1200$ and the pH is 0.92. To check the assumption, the calculated $[H_3O^+]$ can be substituted into the dissociation constant for HA to give

$$\frac{[A^-]}{[HA]} = \frac{K_a}{[H_3O^+]} = \frac{1.00 \times 10^{-4}}{0.1200} = 8.33 \times 10^{-4}$$

Since

$$0.0800 = F_{HA} = [HA] + [A^-]$$

then

$$0.0800 = \frac{[A^-]}{8.33 \times 10^{-4}} + [A^-]$$

or

$$[A^-] = 6.66 \times 10^{-5}$$

Thus, the assumption is valid; $[A^-] \ll 0.1200$.

The approximation employed in this example can be shown to apply until most of the hydrochloric acid has been neutralized by the titrant. Thus, the curve in this region is *identical to the titration of a strong acid by itself*.

The presence of HA must be taken into account, however, when the concentration of hydrochloric acid becomes small.

Example. Calculate the pH of the solution formed by the addition of 29.00 ml of 0.1000-N NaOH to 25.00 ml of the solution described in the earlier example.

Here

$$F_{HCl} = \frac{25.00 \times 0.1200 - 29.00 \times 0.1000}{54.00} = 1.85 \times 10^{-3}$$

$$F_{HA} = \frac{25.00 \times 0.0800}{54.00} = 3.70 \times 10^{-2}$$

Thus,

$$[H_3O^+] = F_{HCl} + [A^-] = 1.85 \times 10^{-3} + [A^-] \tag{10-1}$$

In addition, we may write that

$$[HA] + [A^-] = F_{HA} = 3.70 \times 10^{-2} \qquad (10\text{-}2)$$

The acid dissociation-constant expression for HA can be rearranged to read

$$[HA] = \frac{[H_3O^+][A^-]}{1.00 \times 10^{-4}} \qquad (10\text{-}3)$$

Combining the last two equations gives

$$\frac{[H_3O^+][A^-]}{1.00 \times 10^{-4}} + [A^-] = 3.70 \times 10^{-2}$$

which can be rearranged to

$$[A^-] = \frac{3.70 \times 10^{-6}}{[H_3O^+] + 1.00 \times 10^{-4}}$$

Substitution for $[A^-]$ and F_{HCl} in Equation 10-1 yields

$$[H_3O^+] = 1.85 \times 10^{-3} + \frac{3.70 \times 10^{-6}}{[H_3O^+] + 1.00 \times 10^{-4}}$$

or

$$[H_3O^+]^2 - 1.75 \times 10^{-3}[H_3O^+] - 3.88 \times 10^{-6} = 0$$

$$[H_3O^+] = 3.03 \times 10^{-3}$$

$$pH = 2.52$$

Note that the contributions to the hydronium ion concentration from the HCl (1.85×10^{-3}) and from HA $(3.03 \times 10^{-3} - 1.85 \times 10^{-3})$ are of comparable magnitude.

When the amount of base added is equivalent to the amount of hydrochloric acid originally present, the solution is identical in all respects to one prepared by dissolving appropriate quantities of the weak acid and sodium chloride in a suitable amount of water. The salt, however, has no effect on the pH (neglecting the influence of increased ionic strength); thus, the remainder of the titration curve is that for a dilute solution of HA.

The shape of the curve for a mixture of weak and strong acids, and hence the information obtainable from it, will depend in large measure upon the strength of the weak acid. Figure 10-1 depicts the titration of mixtures containing hydrochloric acid and several weak acids. Note that a break in the titration curve at the first equivalence point is essentially nonexistent if the weak acid has a relatively large dissociation constant (curve A); for such mixtures, only the total acidity (that is, HCl + HA) can be obtained from the titration. Conversely, where the weak acid has a very small dissociation constant, only the strong acid content can be determined. For weak acids of intermediate strength, two useful end points may exist.

The determination of each component in a mixture that contains a strong base and a weak base is also possible provided the dissociation constant of the latter is somewhat smaller than 10^{-4}. The derivation of a curve for such a titration is analogous to that for a mixture of acids.

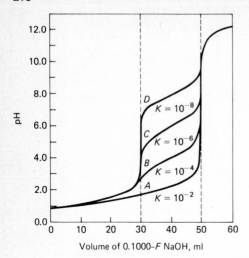

Volume of 0.1000-F NaOH, ml

Figure 10-1 Curves for the Titration of Strong Acid–Weak Acid Mixtures with 0.1000-F NaOH. Each titration involves 25.00 ml of a solution that is 0.1200 F with respect to HCl and 0.0800 F with respect to HA.

Equilibrium Calculations for Polyfunctional Acid-Base Systems

An examination of Appendix 4 reveals that many weak acids which are of interest to chemists contain two or more ionizable protons. For example, three equilibria are established in an aqueous solution of phosphoric acid:

$$H_3PO_4 + H_2O \rightleftarrows H_3O^+ + H_2PO_4^- \qquad K_1 = \frac{[H_3O^+][H_2PO_4^-]}{[H_3PO_4]}$$
$$= 7.11 \times 10^{-3}$$

$$H_2PO_4^- + H_2O \rightleftarrows H_3O^+ + HPO_4^{2-} \qquad K_2 = \frac{[H_3O^+][HPO_4^{2-}]}{[H_2PO_4^-]}$$
$$= 6.34 \times 10^{-8}$$

$$HPO_4^{2-} + H_2O \rightleftarrows H_3O^+ + PO_4^{3-} \qquad K_3 = \frac{[H_3O^+][PO_4^{3-}]}{[HPO_4^{2-}]}$$
$$= 4.2 \times 10^{-13}$$

With this acid, as with most others, each successive dissociation becomes less favorable; thus, $K_3 < K_2 < K_1$.

Polyfunctional bases are also common, an example being Na_2CO_3. Carbonate ion, the conjugate base of the hydrogen carbonate ion, is involved in the stepwise equilibria:

$$CO_3^{2-} + H_2O \rightleftarrows HCO_3^- + OH^- \qquad (K_b)_1 = \frac{[HCO_3^-][OH^-]}{[CO_3^{2-}]}$$
$$= 2.13 \times 10^{-4}$$

$$HCO_3^- + H_2O \rightleftarrows H_2CO_3 + OH^- \qquad (K_b)_2 = \frac{[H_2CO_3][OH^-]}{[HCO_3^-]}$$

$$= 2.25 \times 10^{-8}$$

Numerical values for the basic dissociation constants for Na_2CO_3 are related to the stepwise acid dissociation constants of the corresponding carbonic acids (p. 31). That is,

$$(K_b)_1 = \frac{K_w}{(K_a)_2}$$

and

$$(K_b)_2 = \frac{K_w}{(K_a)_1}$$

where $(K_a)_1$ and $(K_a)_2$ are, respectively, the first and second dissociation constants of H_2CO_3.

The pH of solutions containing a polyfunctional acid or a buffer mixture involving such an acid can be evaluated rigorously through use of the systematic approach to the multiequilibrium problem introduced earlier (p. 91). Solution of the several simultaneous equations that result from this treatment, however, is usually difficult and time-consuming without the aid of a computer. Fortunately, for acids or bases with successive dissociation constants that differ by a factor of about 10^3 (or more), a simplifying assumption can be applied which decreases the algebraic labor of these calculations. In most of the discussion that follows, we shall consider examples to which this approximation is applicable. As a consequence, only one new type of pH calculation, that for a solution of a salt of the type NaHA, will be required.

SOLUTIONS OF SALTS OF THE TYPE NaHA

A solute of the type NaHA is both an acid and a base. Thus, its acid dissociation reaction can be formulated as

$$HA^- + H_2O \rightleftarrows A^{2-} + H_3O^+$$

It is also, however, the conjugate base of H_2A. That is,

$$HA^- + H_2O \rightleftarrows H_2A + OH^-$$

One of these reactions produces hydronium ions and the other hydroxide ions. Whether an HA^- solution is acidic or basic will be determined by the relative magnitude of the equilibrium constants for these processes:

$$K_2 = \frac{[H_3O^+][A^{2-}]}{[HA^-]} \tag{10-4}$$

$$K_b = \frac{K_w}{K_1} = \frac{[H_2A][OH^-]}{[HA^-]} \tag{10-5}$$

If K_b is greater than K_2, the solution will be basic; otherwise, it will be acidic.

A substance, such as NaHA, that can behave both as an acid and as a base is said to be *amphiprotic*.

A solution of NaHA can be described in terms of material balance and charge balance:

$$F_{NaHA} = [HA^-] + [H_2A] + [A^{2-}] \tag{10-6}$$

and

$$[Na^+] + [H_3O^+] = [HA^-] + 2[A^{2-}] + [OH^-]$$

Since the sodium ion concentration is equal to the formal concentration of the salt, the last equation can be rewritten as

$$F_{NaHA} = [HA^-] + 2[A^{2-}] + [OH^-] - [H_3O^+] \tag{10-7}$$

Frequently neither the hydroxide nor the hydronium ion concentration can be neglected in solutions of this type; therefore, a fifth equation is needed to take care of the five unknowns. The ion-product constant for water will serve:

$$K_w = [H_3O^+][OH^-]$$

The derivation of a rigorous expression for the hydronium ion concentration from these five equations is difficult. However, a reasonable approximation, applicable to solutions of most acid salts, can be obtained. Equating Equation 10-7 with Equation 10-6 yields

$$[H_2A] = [A^{2-}] + [OH^-] - [H_3O^+]$$

With the aid of Equations 10-4 and 10-5, we can express $[H_2A]$ and $[A^{2-}]$ in terms of $[HA^-]$; that is,

$$\frac{K_w[HA^-]}{K_1[OH^-]} = \frac{K_2[HA^-]}{[H_3O^+]} + [OH^-] - [H_3O^+]$$

Replacement of $[OH^-]$ by the equivalent expression $K_w/[H_3O^+]$ gives

$$\frac{[H_3O^+][HA^-]}{K_1} = \frac{K_2[HA^-]}{[H_3O^+]} + \frac{K_w}{[H_3O^+]} - [H_3O^+]$$

Multiplication of both sides by $[H_3O^+]$ and rearrangement yields

$$[H_3O^+]^2 \left(\frac{[HA^-]}{K_1} + 1 \right) = K_2[HA^-] + K_w$$

Finally,

$$[H_3O^+] = \sqrt{\frac{K_1K_2[HA^-] + K_1K_w}{[HA^-] + K_1}} \tag{10-8}$$

Under most circumstances, it can be *assumed* that

$$[HA^-] \cong F_{NaHA} \tag{10-9}$$

and Equation 10-8 takes the form

$$[H_3O^+] = \sqrt{\frac{K_1K_2F_{NaHA} + K_1K_w}{F_{NaHA} + K_1}} \tag{10-10}$$

Equation 10-9 is a reasonable approximation provided the two equilibrium constants that contain $[HA^-]$ are small and the formal concentration of NaHA is not too low.

Frequently K_1 will be so much smaller than F_{NaHA} that it can be neglected in the denominator, thus permitting further simplification of Equation 10-10. Moreover, K_1K_w will often be much smaller than $K_1K_2F_{NaHA}$. Thus, provided $K_1 \ll F_{NaHA}$ and $K_1K_w \ll K_1K_2F_{NaHA}$, Equation 10-10 simplifies to

$$[H_3O^+] \cong \sqrt{K_1K_2} \qquad (10\text{-}11)$$

Note that solutions of this type tend to have a constant pH over a substantial range of concentrations.

Example.　Calculate the hydronium ion concentration of a 0.100-F NaHCO$_3$ solution.

We must first examine the assumptions required for Equation 10-11. The dissociation constants, K_1 and K_2, for H_2CO_3 are 4.45×10^{-7} and 4.7×10^{-11}, respectively. Obviously, K_1 is much smaller than F_{NaHA}; K_1K_w is equal to 4.45×10^{-21}, while $K_1K_2F_{NaHA}$ is about 2×10^{-18}. We can therefore use the simplified equation. Thus,

$$[H_3O^+] = \sqrt{4.45 \times 10^{-7} \times 4.7 \times 10^{-11}}$$

$$[H_3O^+] = 4.6 \times 10^{-9}$$

Example.　Calculate the hydronium ion concentration of a 1.0×10^{-3} F solution of Na$_2$HPO$_4$.

Here the pertinent dissociation constants (those containing $[HPO_4{}^{2-}]$) are K_2 and K_3 for H_3PO_4, 6.34×10^{-8} and 4.2×10^{-13}, respectively. Considering again the assumptions implicit in Equation 10-11, we find that 6.34×10^{-8} is indeed much smaller than the formal concentration of Na$_2$HPO$_4$; the denominator can again be simplified. On the other hand, K_2K_w is by no means much smaller than $K_2K_3F_{Na_2HPO_4}$, the former being about 6×10^{-22} and the latter about 3×10^{-23}. We should therefore use the partially simplified version of Equation 10-10:

$$[H_3O^+] = \sqrt{\frac{6.34 \times 10^{-8} \times 4.2 \times 10^{-13} \times 1.0 \times 10^{-3} + 6.34 \times 10^{-22}}{1.0 \times 10^{-3}}}$$

$$= 8.1 \times 10^{-10}$$

Use of Equation 10-11 would have yielded a value of 1.6×10^{-10} M.

Example.　Find the hydronium ion concentration of a 0.0100-F NaH$_2$PO$_4$ solution.

Here the two dissociation constants of importance (those containing $[H_2PO_4{}^-]$) are K_1 and K_2 for H_3PO_4; these have values of 7.11×10^{-3} and 6.34×10^{-8}. Because K_1 is not much smaller than $F_{NaH_2PO_4}$, the denominator of Equation 10-10 cannot be simplified. The numerator can be simplified to $K_1K_2F_{NaH_2PO_4}$. Thus, Equation 10-10 becomes

$$[H_3O^+] = \sqrt{\frac{7.11 \times 10^{-3} \times 6.34 \times 10^{-8} \times 1.0 \times 10^{-2}}{0.010 + 7.11 \times 10^{-3}}}$$

$$= 1.6 \times 10^{-5}$$

SOLUTIONS OF POLYFUNCTIONAL ACIDS

The dibasic weak acid H_2A dissociates in two steps that can be written as:

$$H_2A + H_2O \rightleftarrows H_3O^+ + HA^-$$
$$HA^- + H_2O \rightleftarrows H_3O^+ + A^{2-}$$

Application of the mass law yields the corresponding equilibrium constants

$$K_1 = \frac{[H_3O^+][HA^-]}{[H_2A]} \tag{10-12}$$

$$K_2 = \frac{[H_3O^+][A^{2-}]}{[HA^-]} \tag{10-13}$$

If the formal concentration of an H_2A solution is known, mass balance requires that

$$F_{H_2A} = [H_2A] + [HA^-] + [A^{2-}] \tag{10-14}$$

and charge balance requires that

$$[H_3O^+] \cong [HA^-] + 2[A^{2-}] + [OH^-] \tag{10-15}$$

Note that the concentration of hydroxide ions in the acidic solution is assumed to be negligible in Equation 10-15.

These four independent algebraic expressions would permit calculation of the four remaining concentrations $[H_2A]$, $[HA^-]$, $[A^{2-}]$, and $[H_3O^+]$. However, further simplification is possible if the ratio, K_1/K_2, is large; under these circumstances we can ordinarily assume that the hydronium ions produced by the first dissociation step repress the inherently less favorable second step to such an extent that the protons from the latter can be neglected. This assumption is tantamount to saying that $[A^{2-}]$ is much smaller than $[HA^-]$ and $[H_2A]$ in Equations 10-14 and 10-15. If this assumption can be made, Equation 10-14 becomes

$$F_{H_2A} \cong [H_2A] + [HA^-] \tag{10-16}$$

and Equation 10-15 simplifies to

$$[H_3O^+] \cong [HA^-] \tag{10-17}$$

Substitution of Equations 10-16 and 10-17 into Equation 10-12 yields

$$\frac{[H_3O^+]^2}{F_{H_2A} - [H_3O^+]} = K_1 \tag{10-18}$$

Equation 10-18 will be recognized as being identical to the one used previously to calculate the hydronium ion concentration for a simple weak acid (p. 32); as before, it may be solved either by the quadratic formula or by making the further assumption that $[H_3O^+]$ is small relative to F_{H_2A}.

The propriety of neglecting the second dissociation to the hydronium ion concentration can be judged by substituting Equation 10-17 into Equation 10-13; this gives

$$[A^{2-}] \cong K_2$$

and provides an approximate concentration for $[A^{2-}]$ against which the trial values for $[H_3O^+]$ or $[HA^-]$ can be compared. In general, it is only where K_1/K_2 is small or where the solution is very dilute that neglect of the second dissociation step will lead to serious error. If the assumption is not justified, a rigorous and tedious solution of the first four equations is required.

Example. Calculate the pH of a 0.100-F maleic acid solution. If we symbolize the acid by H_2M, we may write

$$H_2M + H_2O \rightleftarrows H_3O^+ + HM^- \qquad \frac{[H_3O^+][HM^-]}{[H_2M]} = K_1 = 1.20 \times 10^{-2}$$

$$HM^- + H_2O \rightleftarrows H_3O^+ + M^{2-} \qquad \frac{[H_3O^+][M^{2-}]}{[HM^-]} = K_2 = 5.96 \times 10^{-7}$$

If we neglect the effects of the second dissociation and employ Equation 10-18, we obtain

$$\frac{[H_3O^+]^2}{0.100 - [H_3O^+]} = K_1 = 1.20 \times 10^{-2}$$

The first dissociation constant for maleic acid is so large that the quadratic equation must be solved for $[H_3O^+]$. That is,

$$[H_3O^+]^2 + 1.20 \times 10^{-2}[H_3O^+] - 1.20 \times 10^{-3} = 0$$
$$[H_3O^+] = 2.92 \times 10^{-2} = [HM^-]$$
$$pH = -\log(2.92 \times 10^{-2}) = 1.54$$

A check of the assumption that $[M^{2-}]$ is much smaller than $[HM^-]$ is done by substituting the relationship $[H_3O^+] = [HM^-]$ into the expression for K_2. That is,

$$\frac{[\cancel{H_3O^+}][M^{2-}]}{[\cancel{HM^-}]} = 5.96 \times 10^{-7}$$

$$[M^{2-}] = 5.96 \times 10^{-7}$$

Clearly the approximation will not lead to a significant error.

Example. Calculate the hydronium ion concentration of a 0.0400-F H_2SO_4 solution.

One proton in H_2SO_4 is completely dissociated but the second is not. If the dissociation of HSO_4^- is assumed to be negligible, it follows that

$$[H_3O^+] = [HSO_4^-] = 0.0400$$

However, an estimate of $[SO_4^{2-}]$ based upon this approximation and the expression for K_2 reveals that

$$\frac{[\cancel{H_3O^+}][SO_4^{2-}]}{[\cancel{HSO_4^-}]} \cong 1.20 \times 10^{-2}$$

Clearly $[SO_4^{2-}]$ is *not* small with respect to $[HSO_4^-]$; a more rigorous solution is thus required.

From stoichiometric considerations it is necessary that

$$[H_3O^+] = 0.0400 + [SO_4^{2-}]$$

The two terms represent the contributions to $[H_3O^+]$ from the first and second dissociations, respectively. Rearrangement yields

$$[SO_4^{2-}] = [H_3O^+] - 0.0400$$

Mass-balance considerations require that

$$F_{H_2SO_4} = 0.0400 = [HSO_4^-] + [SO_4^{2-}]$$

Combining the last two equations and rearranging yields

$$[HSO_4^-] = 0.0800 - [H_3O^+]$$

Introduction of these equations for $[SO_4^{2-}]$ and $[HSO_4^-]$ into the expression for K_2 yields

$$\frac{[H_3O^+]([H_3O^+] - 0.0400)}{0.0800 - [H_3O^+]} = 1.20 \times 10^{-2}$$

which rearranges to

$$[H_3O^+]^2 - 0.0280[H_3O^+] - 9.60 \times 10^{-4} = 0$$
$$[H_3O^+] = 0.0480$$

SOLUTIONS OF POLYFUNCTIONAL BASES

Calculation of the pH for a solution of a polyfunctional base is analogous to that of a weak acid.

Example. Calculate the pH of a 0.100-F Na_2CO_3 solution.
The equilibria that must be considered are

$$CO_3^{2-} + H_2O \rightleftarrows HCO_3^- + OH^- \qquad \frac{[HCO_3^-][OH^-]}{[CO_3^{2-}]} = 2.13 \times 10^{-4}$$

$$HCO_3^- + H_2O \rightleftarrows H_2CO_3 + OH^- \qquad \frac{[H_2CO_3][OH^-]}{[HCO_3^-]} = 2.25 \times 10^{-8}$$

If the first reaction produces enough OH^- to repress the second equilibrium, then $[H_2CO_3] \ll [HCO_3^-]$ or $[CO_3^{2-}]$. Thus,

$$[HCO_3^-] = [OH^-]$$
$$[CO_3^{2-}] = F_{Na_2CO_3} - [OH^-] \cong F_{Na_2CO_3}$$

Substitution of these quantities into the first equilibrium-constant expression yields

$$[OH^-]^2 = 2.13 \times 10^{-4} \times 0.100 = 2.13 \times 10^{-5}$$
$$[OH^-] = 4.62 \times 10^{-3}$$
$$pH = 14.00 - (-\log 4.62 \times 10^{-3}) = 11.66$$

To test the assumption that $[H_2CO_3]$ is negligible, we substitute $[HCO_3^-] = [OH^-]$ into the second equilibrium-constant expression

$$\frac{[H_2CO_3][\cancel{OH^-}]}{[\cancel{HCO_3^-}]} = 2.25 \times 10^{-8}$$

Clearly $[H_2CO_3]$ is much smaller than the concentrations of the other two carbonate species. The assumption that $[CO_3{}^{2-}] \cong F_{Na_2CO_3}$ introduces an error of less than 5%.

BUFFER SOLUTIONS INVOLVING POLYFUNCTIONAL ACIDS

Two buffer systems can be prepared from a weak dibasic acid and its salts. The first consists of the free acid H_2A and the salt NaHA; the second makes use of NaHA and its conjugate base Na_2A. The pH of the latter system will be the higher, since the acid dissociation constant for HA^- is always less than that for H_2A.

It is a simple matter to write enough equations to permit a rigorous evaluation of the hydronium ion concentration for either of these systems. Ordinarily, however, it is permissible to introduce the simplifying assumption that only the equilibrium that contains terms for the principal solute species is important in determining the hydronium ion concentration of the solution. Thus, for a buffer consisting of H_2A and NaHA, the dissociation equilibrium for H_2A is considered; the further reaction of HA^- to yield A^{2-} is neglected. With this simplification, the hydronium ion concentration is calculated by the method described on page 197 for a simple buffer solution. As before, it is an easy matter to check the validity of the assumption by calculating an approximate concentration for A^{2-} and comparing this value with the concentrations of H_2A and HA^-.

Example. Calculate the hydronium ion concentration of a buffer solution that is 0.100 F in phthalic acid (H_2P) and 0.200 F in potassium hydrogen phthalate (KHP).

The principal equilibrium in this solution is the dissociation of H_2P:

$$H_2P + H_2O \rightleftarrows H_3O^+ + HP^- \qquad \frac{[H_3O^+][HP^-]}{[H_2P]} = K_1 = 1.12 \times 10^{-3}$$

The further dissociation of HP^- is assumed to be negligible; that is, $[P^{2-}] \ll [HP^-]$ or $[H_2P]$. Then, as on page 198,

$$[H_2P] \cong F_{H_2P} = 0.100$$

$$[HP^-] \cong F_{KHP} = 0.200$$

$$[H_3O^+] = \frac{1.12 \times 10^{-3} \times 0.100}{0.200} = 5.60 \times 10^{-4}$$

An approximate value for $[P^{2-}]$, obtained by substituting values for $[H_3O^+]$ and $[HP^-]$ into K_2, verifies that neglect of the second dissociation step was justified.

For a buffer prepared from NaHA and Na_2A, the second dissociation reaction is assumed to predominate, and the reaction

$$HA^- + H_2O \rightleftarrows H_2A + OH^-$$

is disregarded. If it can be assumed that the concentration of H_2A is negligible compared with that of HA^- and A^{2-}, the hydronium ion concentration can be calculated from the second dissociation constant, again employing the techniques for a simple buffer solution. To test the assumption, an estimate of the concentration of H_2A is compared with the concentrations of HA^- and A^{2-}.

Example. Calculate the hydronium ion concentration of a buffer that is 0.0500 F in potassium hydrogen phthalate and 0.150 F in potassium phthalate.

Here, the principal equilibrium is the dissociation of HP^-.

$$HP^- + H_2O \rightleftarrows H_3O^+ + P^{2-} \qquad \frac{[H_3O^+][P^{2-}]}{[HP^-]} = K_2 = 3.91 \times 10^{-6}$$

Provided the concentration of H_2P in this solution is negligible, then

$$[HP^-] \cong F_{KHP} = 0.0500$$
$$[P^{2-}] \cong F_{K_2P} = 0.150$$

$$[H_3O^+] = \frac{3.91 \times 10^{-6} \times 0.0500}{0.150} = 1.30 \times 10^{-6}$$

To check the first assumption, an approximate value for $[H_2P]$ is calculated by substituting numerical values for $[H_3O^+]$ and for $[HP^-]$ into K_1.

$$\frac{(1.30 \times 10^{-6})(0.0500)}{[H_2P]} = 1.12 \times 10^{-3}$$
$$[H_2P] \cong 6 \times 10^{-5}$$

This result justifies the assumption that $[H_2P]$ is much smaller than $[HP^-]$ and $[P^{2-}]$—that is, that the basic dissociation reaction of HP^- can be neglected.

In all but a few situations, the assumption of a single principal equilibrium, as invoked in these examples, will provide a satisfactory estimate for the pH of buffer mixtures derived from polybasic acids. Appreciable errors will occur, however, where the concentration of the acid or the salt is very low or where the two dissociation constants are numerically close to one another. A more laborious and rigorous calculation is then required.

Derivation of Titration Curves for Polyfunctional Acids

Titration curves for solutions of polyfunctional acids are ordinarily more complex than those that we have considered so far. Multiple inflection points are observed when the component acids differ appreciably in strength; more than one useful equivalence point may exist if the change in pH is sufficiently abrupt in the regions of these inflections.

The approximate techniques just described permit the derivation, with reasonable accuracy, of theoretical titration curves for polybasic acids, provided the ratio K_1/K_2 is somewhat greater than 10^3. If the ratio is smaller than this figure, the error, particularly in the region of the first equivalence point, becomes prohibitive, and a more rigorous treatment of the equilibrium relationships is required.

In the example that follows, we shall derive points on the titration curve of maleic acid, a bifunctional weak organic acid with the empirical formula $H_2C_4H_2O_4$. Employing H_2M to symbolize the free acid, the two dissociation equilibria are

$$H_2M + H_2O \rightleftarrows H_3O^+ + HM^- \qquad K_1 = 1.20 \times 10^{-2}$$
$$HM^- + H_2O \rightleftarrows H_3O^+ + M^{2-} \qquad K_2 = 5.96 \times 10^{-7}$$

Because the ratio K_1/K_2 is large (2×10^4), it is feasible to neglect the second dissociation step when deriving points in the early part of the curve; that is, we assume that $[M^{2-}] \ll [HM^-]$ and $[H_2M]$ in this region. It can be shown that, to within a few tenths of a milliliter of the first equivalence point, this assumption does not lead to serious error. Shortly beyond the first equivalence point, the second equilibrium is sufficiently dominant so that the basic reaction of HM^-,

$$HM^- + H_2O \rightleftarrows OH^- + H_2M$$

does not significantly influence the pH. Here we are assuming that $[H_2M] \ll [HM^-]$ and $[M^{2-}]$.

Example. Derive a curve for the titration of 25.00 ml of 0.1000-F maleic acid with 0.1000-F NaOH.

   Initial pH. This calculation is shown in the example on page 223. The pH is 1.54.

   pH after Addition of 5.00 ml Base. With the addition of base, a buffer is formed consisting of the weak acid H_2M and its conjugate base HM^-. To the extent that dissociation of HM^- to give M^{2-} is negligible, the solution can be treated as a simple buffer system. Then

$$F_{H_2M} \cong \frac{25.00 \times 0.1000 - 5.00 \times 0.1000}{30.00} = 6.67 \times 10^{-2}$$

$$F_{NaHM} \cong \frac{5.00 \times 0.1000}{30.00} = 1.67 \times 10^{-2}$$

Using Equations 9-9 and 9-10 (p. 198),

$$[H_2M] \cong 6.67 \times 10^{-2}$$

$$[HM^-] \cong 1.67 \times 10^{-2}$$

When these values are substituted into the equilibrium-constant expression, a tentative value of 4.8×10^{-2} mole/liter is obtained for $[H_3O^+]$. It is clear, however, that the approximation $[H_3O^+] \ll F_{H_2M}$ or F_{HM^-} is not valid; therefore, Equations 9-7 and 9-8 (p. 198) must be used, and

$$[HM^-] = 1.67 \times 10^{-2} + [H_3O^+] - [\cancel{OH^-}]$$

$$[H_2M] = 6.67 \times 10^{-2} - [H_3O^+] + [\cancel{OH^-}]$$

Because the solution is quite acidic, the approximation that $[OH^-]$ is very small is surely valid. Substitution of these expressions into the dissociation-constant relationship gives

$$\frac{[H_3O^+](1.67 \times 10^{-2} + [H_3O^+])}{6.67 \times 10^{-2} - [H_3O^+]} = 1.20 \times 10^{-2} = K_1$$

or

$$[H_3O^+]^2 + 2.87 \times 10^{-2}[H_3O^+] - 8.00 \times 10^{-4} = 0$$

and

$$[H_3O^+] = 1.74 \times 10^{-2}$$

and

$$pH = 1.76$$

Additional points in the first buffer region can be computed in a similar way.

First Equivalence Point. At the first equivalence point

$$[HM^-] \cong \frac{2.500}{50.00} = 5.00 \times 10^{-2}$$

Simplification of the numerator in Equation 10-10 is clearly justified. On the other hand, the concentration of HM^- is relatively close to the value for K_1. Hence,

$$[H_3O^+] \cong \sqrt{\frac{K_1 K_2 F_{HM^-}}{K_1 + F_{HM^-}}} = \sqrt{\frac{(1.20 \times 10^{-2})(5.96 \times 10^{-7})(5.00 \times 10^{-2})}{(5.00 \times 10^{-2} + 1.20 \times 10^{-2})}}$$

$$= 7.60 \times 10^{-5}$$

$$pH = 4.12$$

Second Buffer Region. Further additions of base to the solution create a new buffer system consisting of HM^- and M^{2-}. When enough base has been added so that the reaction of HM^- with water to give OH^- may be neglected (a few tenths of a milliliter), the pH of the mixture is readily obtained from K_2. With the introduction of 25.50 ml of NaOH, for example,

$$F_{Na_2M} \cong \frac{(25.50 - 25.00)0.1000}{50.50} = \frac{0.050}{50.50}$$

and the formal concentration of NaHM will be equal to

$$F_{NaHM} \cong \frac{(25.00 \times 0.1000) - (25.50 - 25.00)0.1000}{50.50}$$

$$= \frac{2.45}{50.50}$$

Thus,

$$\frac{[H_3O^+](0.050/50.50)}{(2.45/50.50)} = 5.96 \times 10^{-7}$$

$$[H_3O^+] = 2.92 \times 10^{-5}$$

The assumption that $[H_3O^+]$ is small with respect to the two formal concentrations is valid and

$$pH = 4.54$$

Second Equivalence Point. After 50.0 ml of 0.1000-*F* sodium hydroxide have been added, the solution is 0.0333 *F* in Na_2M. Reaction of the base M^{2-} with water is the predominant equilibrium in the system and the only one that must be taken into account. Thus,

$$M^{2-} + H_2O \rightleftarrows OH^- + HM^-$$

$$K_b = \frac{K_w}{K_2} = \frac{1.00 \times 10^{-14}}{5.96 \times 10^{-7}} = \frac{[OH^-][HM^-]}{[M^{2-}]}$$

$$[OH^-] \cong [HM^-]$$

$$[M^{2-}] = 0.0333 - [OH^-] \cong 0.0333$$

$$\frac{[OH^-]^2}{0.0333} = \frac{1.00 \times 10^{-14}}{5.96 \times 10^{-7}}$$

$$[OH^-] = 2.36 \times 10^{-5}$$

$$pH = 14.00 - (-\log 2.36 \times 10^{-5}) = 9.37$$

pH beyond the Second Equivalence Point. Further additions of sodium hydroxide repress the basic dissociation of M^{2-}. The pH is calculated from the concentration of NaOH added in excess of that required for the complete neutralization of H_2M.

Figure 10-2 depicts the titration of 0.1000-*F* maleic acid with 0.1000-*F* NaOH. This curve was derived by use of the calculations shown in the preceding example. Two end points are apparent; in principle, the judicious choice of indicator would permit either the first or both acidic hydrogens of maleic acid to be titrated. The second end point is clearly the better one to use, however, inasmuch as the pH change is more pronounced.

Figure 10-3 shows titration curves for three other dibasic acids. These curves illustrate that a well-defined end point corresponding to the first equivalence point is observed only where the degree of dissociation of the two acids is sufficiently different.

The ratio of K_1 to K_2 for oxalic acid (curve *B*) is approximately 1000. The curve for this titration shows an inflection corresponding to the first equivalence point. However, the magnitude of the pH change is too small to permit the location of this equivalence point by means of an indicator; thus, only the second equivalence point can be used for accurate analyses.

Curve *A* illustrates the theoretical titration curve for the tribasic phosphoric acid. Here the ratio K_1 to K_2 is approximately 10^5, a figure that is about 100 times greater than that for oxalic acid; two well-defined end points are observed, either of which is satisfactory for analytical purposes. If an indicator with an acid transition range is used, one equivalent of base will be consumed per mole of acid. With an indicator exhibiting a color change in the basic region, two equivalents of base will be used. The third hydrogen of phosphoric acid is so slightly dissociated ($K_3 = 4.2 \times 10^{-13}$) that it does not yield an end point of any practical value. The buffering effect of the third dissociation is noticeable,

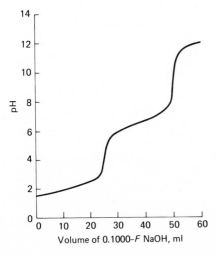

Figure 10-2 Titration Curve for 25.00 ml of 0.1000-*F* Maleic Acid, H_2M.

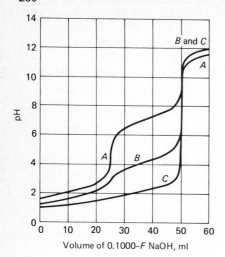

Volume of 0.1000-F NaOH, ml

Figure 10-3 Curves for the Titration of Polybasic Acids. A 0.1000-F NaOH solution is used to titrate 25.00 ml of 0.1000-F H_3PO_4 (curve A), 0.1000-F oxalic acid (curve B), and 0.1000-F H_2SO_4 (curve C).

however, causing the pH for curve A to be lower than that for the other two in the region beyond their second equivalence points.

   In general, the titration of polyfunctional acids or bases yields individual end points that are of practical value only where the ratio of the two dissociation constants is at least 10^4. If the ratio is much less than this, the pH change at the first equivalence point will be too small for accurate detection—only the second end point will prove satisfactory for analysis.

Titration Curves for Polyfunctional Bases

   The derivation of a titration curve for a polyfunctional base involves no new principles. To illustrate, consider the titration of a sodium carbonate solution with standard hydrochloric acid. The important equilibrium constants are

$$CO_3^{2-} + H_2O \rightleftarrows OH^- + HCO_3^- \qquad (K_b)_1 = \frac{K_w}{(K_a)_2} = 2.13 \times 10^{-4}$$

$$HCO_3^- + H_2O \rightleftarrows OH^- + H_2CO_3 \qquad (K_b)_2 = \frac{K_w}{(K_a)_1} = 2.25 \times 10^{-8}$$

The reaction of carbonate ion with water governs the initial pH of the solution; the method shown on page 224 is used in the pH calculation. With the first additions of acid, a carbonate–hydrogen carbonate buffer is established; the hydroxide ion concentration is calculated from $(K_b)_1$ [or the hydronium ion concentration from $(K_a)_2$]. Sodium hydrogen carbonate is the principal solute species at the first equivalence point; Equation 10-11 will give the hydronium ion concentration of this solution. With the further introductions of acid, a hydrogen carbonate–carbonic acid buffer governs the pH; here the hydroxide ion concentration is obtained from $(K_b)_2$ [or the hydronium ion concentration from $(K_a)_1$]. At the second equivalence point, the solution consists of carbonic acid and sodium chloride; the hydronium ion concentration is estimated in the usual way for a simple weak acid, using $(K_a)_1$. Finally, when excess hydrochloric acid has been introduced, the dissociation of the weak acid is repressed to a point where the hydronium ion concentration is essentially that of the formal concentration of the strong acid.

Figure 10-4 illustrates a titration curve for a solution of sodium carbonate. Two end points are observed, the second being appreciably sharper than the first. It is apparent that mixtures of sodium carbonate and sodium hydrogen carbonate can be analyzed by neutralization methods. Thus, titration to a phenolphthalein end point would yield the number of milliformula weights of carbonate present, while titration to a bromocresol green color change would require an amount of acid equal to twice the number of milliformula weights of carbonate plus the number of milliformula weights of bicarbonate in the sample.

Titration Curves for Amphiprotic Species

As noted on page 220, an amphiprotic substance behaves both as a weak acid and as a weak base when dissolved in a suitable solvent. If either its acidic or its basic character predominates sufficiently, titration of the species with a strong base or a strong acid may be feasible. For example, in sodium dihydrogen phosphate solution, the following equilibria exist:

$$H_2PO_4^- + H_2O \rightleftarrows H_3O^+ + HPO_4^- \qquad K_2 = 6.34 \times 10^{-8}$$

and

$$H_2PO_4^- + H_2O \rightleftarrows OH^- + H_3PO_4 \qquad K_b = \frac{K_w}{K_1} = 1.41 \times 10^{-12}$$

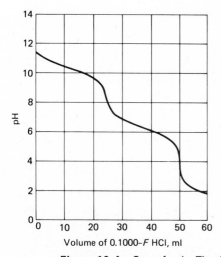

Figure 10-4   Curve for the Titration of 25.00 ml of 0.1000-F Na_2CO_3 with 0.1000-F HCl.

We see that K_b is too small to permit titration of $H_2PO_4^-$ with an acid; on the other hand, K_2 is large enough for a successful titration of the ion with a standard base solution. A different situation prevails in solutions containing disodium hydrogen phosphate. Here,

$$HPO_4^{2-} + H_2O \rightleftharpoons H_3O^+ + PO_4^{3-} \qquad K_3 = 4.2 \times 10^{-13}$$

and titration with base would be unsatisfactory. On the other hand,

$$HPO_4^{2-} + H_2O \rightleftharpoons OH^- + H_2PO_4^- \qquad K_b = \frac{K_w}{K_2} = 1.58 \times 10^{-7}$$

and HPO_4^{2-} could be titrated with standard hydrochloric acid.

The simple amino acids represent an important class of amphiprotic compounds which owe their acid-base properties to the presence of both a weakly acid and a weakly basic functional group. In an aqueous solution of a typical amino acid such as glycine, three important equilibria exist:

$$NH_2CH_2COOH \rightleftharpoons NH_3^+CH_2COO^- \tag{10-19}$$

$$NH_3^+CH_2COO^- + H_2O \rightleftharpoons NH_2COO^- + H_3O^+ \qquad K_a = 2 \times 10^{-10} \tag{10-20}$$

$$NH_3^+CH_2COO^- + H_2O \rightleftharpoons NH_3^+COOH + OH^- \qquad K_b = 2 \times 10^{-12} \tag{10-21}$$

The first reaction constitutes a kind of internal acid-base reaction and is analogous to the reaction one would observe between carboxylic acid and an amine. That is,

$$R_1NH_2 + R_2COOH \rightleftharpoons R_1NH_3^+ + R_2COO^- \tag{10-22}$$

The typical aliphatic amine has a basic ionization constant of 10^{-4} to 10^{-5} (see Appendix 5) while many carboxylic acids have acidic dissociation constants of about the same magnitude. The consequence is that both reactions 10-19 and 10-22 proceed far to the right with the product or products being the predominant species in the solution.

The amino acid species in Equation 10-19, bearing both a positive and a negative charge, is called a *zwitterion*. As shown by Equations 10-20 and 10-21, the zwitterion of glycine is a slightly stronger acid than a base. Thus, an aqueous solution of glycine is slightly acidic.

The zwitterion of an amino acid, containing as it does a positive and a negative charge, has no tendency to migrate in an electric field; on the other hand, the singly charged anionic or the cationic species is attracted to the positive and the negative electrodes, respectively. No *net* migration of the amino acid occurs in an electric field when the pH of the solvent is such that the concentrations of the anionic and cationic forms are identical. The pH at which no net migration occurs is called the *isoelectric point* and is an important physical constant for characterizing amino acids. The isoelectric point is readily related to the ionization constants for the species. Thus, for glycine

$$\frac{[H_3O^+][NH_2COO^-]}{[NH_3^+CH_2COO^-]} = K_a$$

$$\frac{[OH^-][NH_3^+COOH]}{[NH_3^+CH_2COO^-]} = K_b$$

At the isoelectric point,

$$[NH_2CH_2COO^-] = [NH_3{}^+CH_2COOH]$$

Thus, division of K_a by K_b gives

$$\frac{[H_3O^+][\cancel{NH_2CH_2COO^-}]}{[OH^-][\cancel{NH_3{}^+CH_2COOH}]} = \frac{[H_3O^+]}{[OH^-]} = \frac{K_a}{K_b}$$

Substitution of $K_w/[H_3O^+]$ for $[OH^-]$ and rearrangement yields

$$[H_3O^+] = \sqrt{\frac{K_a K_w}{K_b}}$$

The isoelectric point for glycine occurs at a pH of 6.0.

$$[H_3O^+] = \left[\frac{2 \times 10^{-10}}{2 \times 10^{-12}} \times 1 \times 10^{-14}\right]^{1/2} = 1 \times 10^{-6}$$

The magnitude of K_a and K_b for simple amino acids is generally so small that determination by direct neutralization titration is impossible. Addition of formaldehyde, however, removes the base functional group and leaves the carboxylic acid available for titration with a standard base. For example, with glycine

$$NH_3{}^+CH_2COO^- + CH_2O \rightarrow CH_2{=}NCH_2COOH + H_2O$$

The titration curve for the product is that of a typical carboxylic acid.

Composition of Solutions of a Polybasic Acid as a Function of pH

In order to understand clearly the compositional changes that occur in the course of a titration involving a polybasic acid, it is instructive to plot the *relative* amount of the free acid as well as each of its anions as a function of the pH of the solution. For this purpose, we shall define so-called *a*-values for each of the anion-containing species. An *a*-value is the fraction of the total weak acid concentration represented by a particular species. Turning again to the maleic acid system, if we define C_T as the sum of the concentrations of the maleate-containing species, then mass balance requires that

$$C_T = [H_2M] + [HM^-] + [M^{2-}]$$

and by definition,

$$a_0 = \frac{[H_2M]}{C_T}$$

$$a_1 = \frac{[HM^-]}{C_T}$$

$$a_2 = \frac{[M^{2-}]}{C_T}$$

The sum of the a-values for a system must equal unity; that is,

$$a_0 + a_1 + a_2 = 1$$

We can readily express a_0, a_1, and a_2 in terms of $[H_3O^+]$, K_1, and K_2. Rearrangement of the dissociation-constant expressions gives

$$[HM^-] = \frac{K_1[H_2M]}{[H_3O^+]}$$

$$[M^{2-}] = \frac{K_1K_2[H_2M]}{[H_3O^+]^2}$$

After substituting these quantities, the mass-balance equation becomes

$$C_T = [H_2M] + \frac{K_1[H_2M]}{[H_3O^+]} + \frac{K_1K_2[H_2M]}{[H_3O^+]^2}$$

which can be converted to

$$[H_2M] = \frac{C_T[H_3O^+]^2}{[H_3O^+]^2 + K_1[H_3O^+] + K_1K_2}$$

Substituting this value for $[H_2M]$ into the equation defining a_0 gives

$$a_0 = \frac{[H_3O^+]^2}{[H_3O^+]^2 + K_1[H_3O^+] + K_1K_2} \tag{10-23}$$

By similar manipulation, it is easily shown that

$$a_1 = \frac{K_1[H_3O^+]}{[H_3O^+]^2 + K_1[H_3O^+] + K_1K_2} \tag{10-24}$$

$$a_2 = \frac{K_1K_2}{[H_3O^+]^2 + K_1[H_3O^+] + K_1K_2} \tag{10-25}$$

Note that the denominator is the same for each expression; calculation of a-values at any desired pH is thus relatively simple. Furthermore, the equations illustrate that the fractional amount of each species at any fixed pH is independent of the total concentration, C_T.

The three curves plotted in Figure 10-5 show the a-values for each maleate-containing species as a function of pH. A consideration of these curves in conjunction with the titration curve for maleic acid (Figure 10-2) gives a clear picture of all concentration changes that occur during the course of the titration. For example, Figure 10-2 reveals that before the addition of any base, the pH of the solution is 1.5. Referring to Figure 10-5, we see that at this pH, a_0 for H_2M is roughly 0.7 while a_1 for HM^- is approximately 0.3. For all practical purposes, a_2 is zero. Thus, approximately 70% of the maleic acid exists in the undissociated form and 30% as HM^-. With addition of base, the pH rises, as does the fraction of HM^-. At the first equivalence point (pH = 4.12), essentially all of the maleate is present as HM^- ($a_1 \rightarrow 1$). Beyond the first equivalence point, HM^- decreases and M^{2-} increases. At the second equivalence point (pH = 9.37), it is evident that essentially all of the maleate exists as M^{2-}.

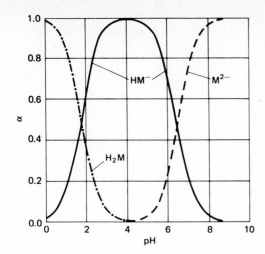

Figure 10-5 Composition of H_2M Solutions as a Function of pH.

PROBLEMS

*1. Calculate the pH of a solution that is 0.0600 F with respect to
   (a) hydrogen sulfide.
   (b) sulfuric acid.
   (c) sodium sulfide.
   (d) ethylenediamine.

2. Calculate the pH of a solution that is 0.0600 F with respect to
   (a) phosphoric acid.
   (b) oxalic acid.
   (c) sodium sulfite.
   (d) trisodium phosphate.

*3. Calculate the pH of a solution that is 0.0400 F with respect to
   (a) sodium hydrogen sulfide.
   (b) sodium hydrogen oxalate.
   (c) sodium hydrogen sulfite.
   (d) ethylenediamine hydrochloride ($NH_2C_2H_4NH_3Cl$).

4. Calculate the pH of a solution that is 0.0400 F with respect to
   (a) sodium hydrogen fumarate.
   (b) sodium hydrogen sulfate.
   (c) disodium hydrogen arsenate.
   (d) sodium dihydrogen arsenate.

*5. Calculate the pH of a solution that is
   (a) 0.0100 F in HCl and 0.0200 F in picric acid.
   (b) 0.0100 F in HCl and 0.0200 F in benzoic acid.
   (c) 0.0100 F in NaOH and 0.100 F in Na_2CO_3.
   (d) 0.0100 F in NaOH and 0.100 F in NH_3.

6. Calculate the pH of a solution that is
   (a) 0.0100 F in $HClO_4$ and 0.0300 F in monochloroacetic acid.
   (b) 0.0100 F in HCl and 0.0150 F in H_2SO_4.
   (c) 0.0100 F in NaOH and 0.0300 F in Na_2S.
   (d) 0.0100 F in NaOH and 0.0300 F in sodium acetate.

*7. Identify the principal conjugate acid-base pair, and calculate the ratio between them in a solution that is buffered to pH 6.00 with species derived from
(a) H_3AsO_4.
(b) citric acid.
(c) malonic acid.
(d) tartaric acid.

8. Identify the principal conjugate acid-base pair, and calculate the ratio between them in a solution that is buffered to pH 9.00 with species derived from
(a) H_2S.
(b) ethylenediamine dihydrochloride.
(c) H_3AsO_4.
(d) H_2CO_3.

*9. Calculate the pH of a solution that is
(a) 0.0500 F in H_3AsO_4 and 0.0200 F in NaH_2AsO_4.
(b) 0.0300 F in NaH_2AsO_4 and 0.0500 F in Na_2HAsO_4.
(c) 0.0600 F in Na_2CO_3 and 0.0300 F in $NaHCO_3$.
(d) 0.0400 F in H_3PO_4 and 0.0200 F in Na_2HPO_4.
(e) 0.0500 F in $NaHSO_4$ and 0.0400 F in Na_2SO_4.

10. Calculate the pH of a solution that is
(a) 0.240 F in H_3PO_3 and 0.480 F in NaH_2PO_3.
(b) 0.0670 F in Na_2SO_3 and 0.0315 F in $NaHSO_3$.
(c) 0.640 F in $HOC_2H_4NH_2$ and 0.750 F in $HOC_2H_4NH_3Cl$.
(d) 0.240 F in $H_2C_2O_4$ (oxalic acid) and 0.360 F in $Na_2C_2O_4$.
(e) 0.0100 F in $Na_2C_2O_4$ and 0.0400 F in $NaHC_2O_4$.

*11. What is the pH of the buffer formed when 50.0 ml of 0.200-F NaH_2PO_4 are mixed with
(a) 50.0 ml of 0.120-F HCl?
(b) 50.0 ml of 0.120-F NaOH?

12. What is the pH of the buffer formed by adding 100 ml of 0.150-F potassium hydrogen phthalate to
(a) 100 ml of 0.0800-F NaOH?
(b) 100 ml of 0.0800-F HCl?

*13. Describe the preparation of 1.00 liter of a buffer with a pH of 9.60 from 0.300-F Na_2CO_3 and 0.200-F HCl.

14. Describe how to prepare 1.00 liter of a buffer of pH 7.00 from 0.200-F H_3PO_4 and 0.160-F NaOH.

15. Describe the preparation of 1.00 liter of a buffer of pH 6.00 from 0.500-F Na_3AsO_4 and 0.400-F HCl.

*16. How many grams of $Na_2HPO_4 \cdot 2H_2O$ must be added to 400 ml of 0.200-F H_3PO_4 to give a buffer of pH 7.30?

17. How many grams of dipotassium phthalate must be added to 750 ml of 0.0500-F phthalic acid to give a buffer of pH 5.75?

18. Derive a curve for the titration of 50.00 ml of 0.1000-F solution of compound A with 0.2000-F solution of compound B. For each titration calculate the pH after the following additions of compound B: 0.00, 12.50, 20.00, 24.00, 25.00, 26.00, 37.50, 45.00, 49.00, 50.00, 51.00, and 60.0.

A	B
*(a) Na_2CO_3	HCl
(b) ethylenediamine	HCl
*(c) H_2SO_4	NaOH
(d) $H_2C_2O_4$	NaOH

19. For pH values of 2.00, 4.00, 6.00, 8.00, 10.00, and 12.00, calculate a for each species in an aqueous solution of
*(a) phthalic acid.
(b) phosphoric acid.
*(c) citric acid.
(d) arsenic acid.

APPLICATIONS OF NEUTRALIZATION TITRATIONS

Volumetric methods based upon neutralization involve the titration of hydronium or hydroxide ions produced directly or indirectly from the sample. Neutralization methods are extensively employed in chemical analysis. For most applications, water serves conveniently as the solvent; it should be recognized, however, that the acidic or basic character of a solute is determined in part by the nature of the solvent in which it is dissolved and that the substitution of some other solvent for water may be sufficient to permit a titration that cannot be successfully performed in an aqueous environment. Titrations in nonaqueous media are discussed in Chapter 12.

Reagents for Neutralization Reactions

In Chapter 9 we noted that the most pronounced pH changes in the equivalence-point region occur when strong acids and strong bases are involved in the titration. It is for this reason that standard solutions for neutralization titrations are prepared from such acids and bases.

PREPARATION OF STANDARD ACID SOLUTIONS

Hydrochloric acid is the most commonly used standard acid for volumetric analysis. Dilute solutions of the reagent are indefinitely stable and can be used in the presence of most cations without complicating precipitation reactions. It is reported that 0.1-N solutions can be boiled for as long as 1 hr without loss of acid provided that water lost by evaporation is periodically replaced; 0.5-N solutions can be boiled for at least 10 min without significant loss.

Solutions of perchloric acid and sulfuric acid are also stable and can serve as standard reagents in titrations where the presence of chloride ion would cause precipitation difficulties. Standard solutions of nitric acid are seldom used because of their oxidizing properties.

A standard acid solution is ordinarily prepared by diluting an appropriate volume of the concentrated reagent and standardizing against a primary standard base. Less frequently, the composition of the concentrated acid is established through careful density measurement, following which a weighed quantity is diluted to an exact volume (tables relating density of reagents to composition are found in most chemistry or chemical engineering handbooks). A stock solution with an exactly known hydrochloric acid concentration can also be prepared by distillation of the concentrated reagent; under controlled conditions, the final quarter of the distillate has a fixed and known composition, its acid content being dependent only upon the atmospheric pressure. For a pressure, P, lying between 670 and 780 mm of mercury, the weight in air of the distillate that contains exactly one equivalent of acid is given by

$$\text{g constant boiling HCl} = 164.673 + 0.02039P \qquad (11\text{-}1)$$

Standard solutions can be prepared by diluting a calculated weight of the acid to an exactly known volume.

STANDARDIZATION OF ACIDS

Dilute solutions of hydrochloric or sulfuric acid can be standardized gravimetrically by weighing the silver chloride or barium sulfate produced from a known volume of the reagent. These methods, of course, assume a stoichiometric relationship between the anion and the hydronium ion. More commonly, acids are standardized against a primary standard base.

Sodium Carbonate. Sodium carbonate is a frequently used standard for acid solutions. Primary-standard grade sodium carbonate is available commercially; it can also be prepared by heating purified sodium hydrogen carbonate at 270 to 300°C for 1 hr:

$$2NaHCO_3 \rightarrow Na_2CO_3 + H_2O + CO_2(g)$$

As shown in Figure 10-4, two end points are observed in the titration of sodium carbonate. The first, corresponding to conversion of carbonate to hydrogen carbonate, occurs at a pH of about 8.4. The second, involving the formation of carbonic acid, is observed at about pH 4.0; the second end point is always used for standardization because of the larger change in pH associated with it. An even sharper end-point change can be achieved by boiling the solution briefly to decompose the reaction product, carbonic acid. The sample is titrated to the first appearance of the acid color of the indicator (such as bromocresol green or methyl orange). At this point the solution contains a large amount of carbonic acid and a small amount of unreacted hydrogen carbonate. Boiling effectively destroys this buffer by eliminating the carbonic acid:

$$H_2CO_3 \rightarrow CO_2(g) + H_2O$$

As a result, the solution again acquires an alkaline pH owing to the residual hydrogen carbonate ion. The titration is completed after the solution has cooled. Now, however, considerably larger changes in pH attend the final additions of acid; a sharper indicator transition is thus observed.

As an alternative, the acid can be introduced in an amount sufficient to provide a slight excess over that needed to convert the sodium carbonate to carbonic acid. The solution is boiled, as before, to remove carbon dioxide; after cooling, the excess acid is back-titrated with a dilute solution of base. Any indicator suitable for a strong acid–strong base titration can be employed. The volume ratio between the acid and the base must be established by an independent titration.

Other Primary Standards for Acids. *Tris*(hydroxymethyl)aminomethane $(HOCH_3)_3CNH_2$, known also as TRIS or THAM, is available in primary-standard purity from commercial sources. It possesses the advantage of a substantially larger equivalent weight than sodium carbonate.

Other standards include sodium tetraborate, mercury(II) oxide, and calcium oxalate; details concerning their use can be found in standard reference works.[1]

PREPARATION OF STANDARD SOLUTIONS OF BASE

Sodium hydroxide is the most common basic reagent, although potassium hydroxide and barium hydroxide are also used. None of these is obtainable in primary-standard purity; after preparation to approximate strength, their solutions must be standardized.

Standard base solutions are reasonably stable so long as they are protected from prolonged exposure to glass and from contact with the atmosphere. Sodium hydroxide reacts slowly with glass to form silicates. Thus, a standard solution that is to be employed for longer than a week or two should be stored in a polyethylene bottle or a glass bottle that has been coated with paraffin.

Effect of Carbon Dioxide upon Standard Base Solutions. In solution as well as the solid state, the hydroxides of sodium, potassium, and barium avidly react with atmospheric carbon dioxide to produce the corresponding carbonates:

$$CO_2 + 2OH^- \rightarrow CO_3^{2-} + H_2O$$

---

[1] See, for example, I. M. Kolthoff and V. A. Stenger, *Volumetric Analysis*, vol. 2, pp. 74–94. New York: Interscience Publishers, Inc., 1947; L. Meites, *Handbook of Analytical Chemistry*, p. 3-34. New York: McGraw-Hill Book Company, Inc., 1963.

The absorption of the gas by a standardized solution of a base does not necessarily alter its acid titer. For example, if potassium or sodium hydroxide solutions are employed where circumstances permit the use of an indicator with an acid transition range (for example, bromocresol green), each carbonate ion in the reagent will have reacted with two hydronium ions of the analyte (see Figure 10-4). That is,

$$CO_3{}^{2-} + 2H_3O^+ \rightarrow H_2CO_3 + 2H_2O$$

This consumption is chemically equivalent to the amount of base used to form the carbonate, and no error will be incurred.

Unfortunately, most applications of standard base require the use of an indicator with a basic transition range (phenolphthalein, for example). Here each carbonate ion will consume but one hydronium ion at the color change of the indicator:

$$CO_3{}^{2-} + H_3O^+ \rightarrow HCO_3{}^- + H_2O$$

The effective normality of the base is thus diminished and a determinate error (called a *carbonate error*) will result.

When carbon dioxide is absorbed by standard barium hydroxide, precipitation of barium carbonate occurs:

$$CO_2 + Ba^{2+} + 2OH^- \rightarrow BaCO_3(s) + H_2O$$

The acid titer is thus decreased regardless of the indicator employed in the application of the base; a carbonate error is the consequence.

The solid reagents used to prepare standard solutions of base represent a further source of carbonate ion. The extent of contamination is frequently great, owing to the absorption of atmospheric carbon dioxide by the solid. As a result, even freshly prepared solutions of base may contain significant quantities of carbonate. Its presence will not cause a carbonate error provided the analysis is performed with the same indicator that was used for standardization; the versatility of the reagent is lost, however, by the presence of carbonate ion.

Several methods exist for the preparation of carbonate-free hydroxide solutions. Barium hydroxide may be employed as the reagent; the sparingly soluble carbonate can be diminished further by the addition of a neutral barium salt, such as the chloride or the nitrate. A barium salt can also be used to eliminate the carbonate from a potassium or sodium hydroxide solution. The presence of barium ion is frequently undesirable, however, owing to its tendency to form slightly soluble salts with anions that may be present in the analyte.

Carbonate-free solutions of the alkali-metal hydroxides may be prepared by direct solution of the freshly cleaned metals. Most chemists think that the possibilities for fire and explosion during the solution process represent an unacceptable risk.

The preferred method for preparing sodium hydroxide solutions takes advantage of the very low solubility of sodium carbonate in concentrated solutions of the alkali. An approximately 50% aqueous solution of sodium hydroxide is prepared (or purchased from commercial sources); after the sodium carbonate has settled, a portion of the supernatant liquid is decanted and diluted to the desired concentration. Details for this procedure are given in Chapter 31, Experiment 11. Alternatively, the concentrated sodium hydroxide solution can be filtered to eliminate the sodium carbonate.

A carbonate-free base solution must be prepared from water that contains no carbon dioxide. Distilled water, which is frequently supersaturated with respect to carbon dioxide, should be boiled briefly to eliminate the gas; the water is allowed to cool to room temperature before the introduction of base since hot alkali solutions rapidly absorb carbon dioxide.

Figure 11-1 shows an arrangement for preventing the uptake of atmospheric carbon dioxide by basic solutions during storage. Air entering the vessel is passed over a solid absorbent for CO_2, such as soda lime or Ascarite.[2] The contamination that occurs as the solution is transferred from this storage bottle to the buret is negligible. Absorption during a titration can be minimized by covering the open end of the buret with a small test tube or beaker.

If a standard solution of base is to be used for no longer than a week or so, storage in a tightly stoppered polyethylene bottle will usually provide sufficient protection against the uptake of atmospheric carbon dioxide. Care should be taken to keep the bottle stoppered except during the brief periods when the contents are being transferred to a buret.

The parts of any ground-glass fitting will freeze upon prolonged exposure to solutions of the alkalis. For this reason, glass-stoppered containers should not be used for the storage of strong bases; similarly, burets equipped with glass stopcocks should be promptly drained and thoroughly cleaned after being used

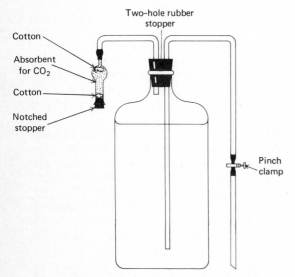

Figure 11-1 Arrangement for the Storage of Standard Base Solutions.

[2] Arthur H. Thomas Co., Philadelphia, Pa. Ascarite consists of sodium hydroxide deposited on asbestos.

to dispense these reagents. A better alternative is the employment of burets equipped with Teflon stopcocks.

STANDARDIZATION OF BASES

Several excellent primary standards are available for the standardization of bases. Most are weak organic acids that require the use of an indicator with a basic transition range.

Potassium Hydrogen Phthalate, $KHC_8H_4O_4$. Potassium hydrogen phthalate possesses many qualities that are desirable in a primary standard. It is a non-hygroscopic crystalline solid with a high equivalent weight. For most purposes, the commercial analytical-grade salt can be used without the need for further purification. Potassium hydrogen phthalate of certificated purity is available from the National Bureau of Standards for the most exacting work.

Other Primary Standards for Bases. Benzoic acid is obtainable in primary-standard purity and can be used for the standardization of bases. Because its solubility in water is limited, the reagent is ordinarily dissolved in ethanol prior to dilution and titration. The dihydrate of oxalic acid has also been recommended as a primary standard for bases.

Potassium hydrogen iodate, $KH(IO_3)_2$, is an excellent primary standard with a high equivalent weight. It is also a strong acid; as a result, virtually any indicator with a pH transition range between 4 and 10 can be used.

Typical Applications of Neutralization Titrations

The most obvious application of neutralization methods is for the determination of the innumerable inorganic, organic, and biological species that possess inherent acidic or basic properties. Equally important, however, are the many applications which involve conversion of the analyte to an acid or a base by suitable chemical treatment followed by titration with a standard strong base or acid.

Two types of end points are commonly used for neutralization titrations. The first, which we have treated in the previous chapters, is based upon the color change of an indicator. The second involves the direct measurement of pH throughout the titration by means of a glass-calomel electrode system; here the potential of the glass electrode is directly proportional to pH. Appropriate plots of the data permit establishment of the end point. The potentiometric procedure is considered in Chapter 17.

ELEMENTAL ANALYSIS

Several important elements that occur in organic and biological systems are most conveniently determined by methods that involve an acid-base titration as the final step. Generally, the elements susceptible to this type of analysis are nonmetallic. Principal among these are carbon, nitrogen, chlorine, bromine, sulfur, phosphorus, and fluorine; in addition, similar methods exist for several less commonly encountered species. In each instance, the element is converted to an inorganic acid or base that can then be titrated. A few examples follow.

Nitrogen. Nitrogen occurs in many materials important to mankind including proteins, synthetic drugs, fertilizers, explosives, and potable water supplies. The analysis for nitrogen is thus of signal importance to research and to industry.

The *Dumas method* and the *Kjeldahl method* are of particular importance for the determination of organically bound nitrogen. The former is discussed in Chapter 28. The latter, which is based upon a neutralization titration, is appropriately considered here.

The Kjeldahl method, which was first described in 1883, is one of the most widely used of all chemical analyses. It requires no special equipment and is readily adapted to the routine analysis of large numbers of samples. The Kjeldahl method (or one of its modifications) is the standard means for determining protein nitrogen in grains, meats, and other biological materials.

In essence, the sample is oxidized in hot, concentrated sulfuric acid, during which time the bound nitrogen is converted to ammonium ion. The solution is then treated with an excess of strong base; following distillation, the liberated ammonia is titrated.

The critical step in the Kjeldahl method is the oxidation with sulfuric acid. The carbon and hydrogen in the sample are converted to carbon dioxide and water, respectively. The fate of nitrogen, however, depends upon its state of combination in the original compound. If it existed as an amine or an amide, as in proteinaceous matter, conversion to ammonium ion is nearly always quantitative. On the other hand, nitrogen present in higher oxidation states, such as nitro, azo, and azoxy groups, will be converted to the elemental state or to nitrogen oxides during the oxidation step and will not be retained in the sulfuric acid. Low results from this source can be prevented by treating the sample with a reducing agent prior to the digestion step; conversion of the element to an oxidation state which will yield ammonium ion upon digestion with the sulfuric acid is thus assured. One prereduction calls for the addition of salicylic acid and sodium thiosulfate to the concentrated sulfuric acid solution containing the sample; the digestion is then performed in the usual way.

Certain aromatic heterocyclic compounds, such as pyridine and its derivatives, are particularly resistant to complete oxidation by sulfuric acid. Thus, unless special precautions are followed, low results attend the analysis for nitrogen in samples containing these species (see Figure 4-1, p. 46).

The oxidation process is the most time-consuming step in the Kjeldahl method; an hour or more may be needed for refractory samples. Of the many modifications aimed at improving the kinetics of the process, the one proposed by Gunning is now almost universally employed. Here a neutral salt, such as potassium sulfate, is added to increase the boiling point of the sulfuric acid solution and thus the temperature at which the oxidation occurs. Care is needed, however, because oxidation of the ammonium ion can occur if the salt concentration is too great. This problem is enhanced if evaporation of the sulfuric acid during digestion is excessive.

Attempts to hasten the Kjeldahl oxidation by introducing such stronger oxidizing agents as perchloric acid, potassium permanganate, and hydrogen peroxide fail because the ammonium ions are partially oxidized to volatile nitrogen oxides.

Many substances catalyze the oxidation step. Mercury, copper, and selenium, either combined or in the elemental state, are effective. Mercury ion, if present, must be precipitated with hydrogen sulfide prior to the distillation step; otherwise, some ammonia will be retained as an ammine complex.

Figure 11-2 illustrates a typical distillation arrangement for the Kjeldahl method. The long-necked flask, which is used for both oxidation and distillation, is called a *Kjeldahl flask*. After the oxidation is judged complete, the contents of the flask are cooled, diluted with water, and then made basic to liberate the ammonia:

$$NH_4^+ + OH^- \rightarrow NH_3(g) + H_2O$$

To avoid losses of ammonia during neutralization, a concentrated sodium hydroxide solution, more dense than the diluted oxidation mixture, is carefully poured down the side of the flask to form a layer on its bottom. The flask is quickly joined to the distillation apparatus; only then are the two layers mixed by gently swirling the flask.

In addition to the Kjeldahl flask, the apparatus shown in Figure 11-2 includes a spray trap that prevents droplets of the strongly alkaline solution

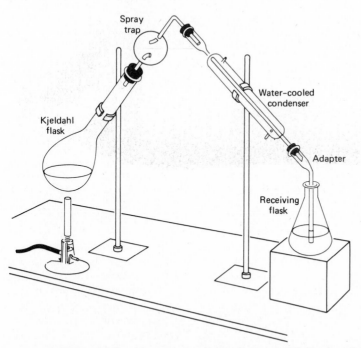

Figure 11-2   Apparatus for the Distillation and Collection of Ammonia.

from being carried over in the vapor stream. A water-cooled condenser is provided. During distillation the end of the adapter tube extends below the surface of an acidic solution in the receiving flask.

Two methods exist for titration of the collected ammonia. In one, the receiver contains a known quantity of standard acid. After distillation is complete, the excess acid is back-titrated with standard base. An indicator with an acidic transition interval is required, owing to the presence of ammonium ions at equivalence. A convenient alternative that requires only one standard solution involves use of an unmeasured excess of boric acid in the receiving flask; its reaction with ammonia is

$$HBO_2 + NH_3 \rightleftarrows NH_4^+ + BO_2^-$$

The borate ion produced is a reasonably strong base and can be titrated with a standard solution of hydrochloric acid:

$$BO_2^- + H_3O^+ \rightleftarrows HBO_2 + H_2O$$

At the equivalence point the solution contains boric acid and ammonium ions; an indicator with an acidic transition interval is thus required.

Details of the Kjeldahl method are found in Chapter 31, Experiment 13.

Sulfur. Sulfur in organic and biological materials is conveniently determined by burning the sample in a stream of oxygen; apparatus for this step in the analysis is described in Chapter 28. The sulfur dioxide (and sulfur trioxide) formed during the oxidation is collected in a dilute solution of hydrogen peroxide:

$$SO_2(g) + H_2O_2 \rightarrow H_2SO_4$$

The sulfuric acid is then titrated with standard base.

Other Elements. Table 11-1 lists other elements which can be determined by neutralization methods.

TABLE 11-1   Elemental Analysis Based on Neutralization Titrations

Element	Converted to	Absorption or Precipitation Products	Titration
N	NH_3	$NH_3(g) + H_3O^+ \rightarrow NH_4^+ + H_2O$	Excess HCl with NaOH
S	SO_2	$SO_2(g) + H_2O_2 \rightarrow H_2SO_4$	NaOH
C	CO_2	$CO_2(g) + Ba(OH)_2 \rightarrow BaCO_3(s) + H_2O$	Excess $Ba(OH)_2$ with HCl
Cl(Br)	HCl	$HCl(g) + H_2O \rightarrow Cl^- + H_3O^+$	NaOH
F	SiF_4	$SiF_4(g) + H_2O \rightarrow H_2SiF_6$	NaOH
P	H_3PO_4	$12H_2MoO_4 + 3NH_4^+ + H_3PO_4 \rightarrow$ $(NH_4)_3PO_4 \cdot 12MoO_3(s) + 12H_2O + 3H^+$ $(NH_4)_3PO_4 \cdot 12MoO_3(s) + 26OH^- \rightarrow$ $HPO_4^{2-} + 12MoO_4^{2-} + 14H_2O + 3NH_3(g)$	Excess NaOH with HCl

DETERMINATION OF INORGANIC SUBSTANCES

Numerous inorganic species can be determined by titration with strong acids or bases. A few examples follow.

Ammonium Salts. Ammonium salts can be conveniently determined by liberating ammonia with strong base and distillation in the Kjeldahl apparatus shown in Figure 11-2. The ammonia is then collected and titrated as in the Kjeldahl method.

Nitrates and Nitrites. The method just considered can also be applied to the determination of inorganic nitrate or nitrite by reducing these species to ammonium ion. Devarda's alloy (50% Cu, 45% Al, 5% Zn) is commonly used as the reducing agent. Granules of the alloy are introduced to a strongly alkaline solution of the sample in a Kjeldahl flask. The ammonia is distilled after reaction is complete. Arnd's alloy (60% Cu, 40% Mg) can also be used as the reducing agent.

Carbonate and Carbonate Mixtures. The qualitative and quantitative determination of constituents in a solution containing sodium carbonate, sodium hydrogen carbonate, and sodium hydroxide, alone or admixed, provides interesting examples of neutralization titrations. No more than two of these three constituents can exist in appreciable amount in any solution because reaction will eliminate the third. Thus, the mixing of sodium hydroxide with sodium hydrogen carbonate results in the formation of sodium carbonate until one or the other of the original reactants is exhausted. If the sodium hydroxide is used up, the solution will contain sodium carbonate and sodium hydrogen carbonate; if sodium hydrogen carbonate is used up, sodium carbonate and sodium hydroxide will remain. Finally, if equiformal amounts are mixed, the principal solute species will be sodium carbonate.

The analysis of such mixtures requires two titrations with standard acid. An indicator with a transition in the vicinity of pH 8 to 9 is used for one; an acid-range indicator is used for the other. The composition of the solution can be deduced from the relative volumes of acid needed to titrate equal volumes of the sample (see Table 11-2 and Figure 10-4). Once the composition has been established, the volume data can be used to establish the concentration of each component in the sample.

TABLE 11-2   Volume Relationship in the Analysis of Mixtures Containing Carbonate, Hydrogen Carbonate, and Hydroxide Ions

Constituents Present	Relationship between Volume of Acid Needed to Reach a Phenolphthalein End Point, V_{ph}, and a Bromocresol Green End Point, V_{bg}
NaOH	$V_{ph} = V_{bg}$
Na_2CO_3	$V_{ph} = \frac{1}{2}V_{bg}$
$NaHCO_3$	$V_{ph} = 0$
NaOH, Na_2CO_3	$V_{ph} > \frac{1}{2}V_{bg}$
Na_2CO_3, $NaHCO_3$	$V_{ph} < \frac{1}{2}V_{bg}$

Example. A solution contained one or more of the following species: Na_2CO_3, $NaHCO_3$, and NaOH. A 50.0-ml portion required 22.1 ml of 0.100-N HCl when titrated to a phenolphthalein end point. A second 50.0-ml sample was titrated with bromocresol green as the indicator, the solution being boiled near the second equivalence point in order to remove CO_2. Here exactly 48.4 ml of the HCl were used. What was the formal composition of the original solution?

Had the solution contained only NaOH, the volume of acid required would have been the same, regardless of indicator (that is, $V_{ph} = V_{bg}$). In fact, however, the second titration required a total of 48.4 ml. Since less than half of this amount was involved in the first titration, the solution must have contained some $NaHCO_3$ in addition to Na_2CO_3. We can now calculate the concentration of the two constituents. When the phenolphthalein end point was reached, the CO_3^{2-} originally present was converted to HCO_3^-. Thus,

$$\text{no. mfw } Na_2CO_3 = 22.1 \times 0.100 = 2.21$$

The titration from the phenolphthalein to the bromocresol green end point (48.4 − 22.1 = 26.3 ml) involved both the hydrogen carbonate originally present as well as that formed by titration of the carbonate. Thus,

$$\text{no. mfw } NaHCO_3 + \text{no. mfw } Na_2CO_3 = 26.3 \times 0.100$$

Hence,

$$\text{no. mfw } NaHCO_3 = 2.63 - 2.21 = 0.42$$

The formal concentrations are readily calculated from these data:

$$F_{Na_2CO_3} = \frac{2.21}{50.0} = 0.0442 \text{ mfw/ml}$$

$$F_{NaHCO_3} = \frac{0.42}{50.0} = 0.0084 \text{ mfw/ml}$$

In practice, the titration used in this example is not entirely satisfactory because the pH change corresponding to the hydrogen carbonate equivalence point is not sufficient to give a sharp color change with a chemical indicator (see Figure 10-4). Titration to a color match with a solution containing an approximately equivalent amount of sodium hydrogen carbonate is helpful; nevertheless, errors of 1% or more must be expected.

The limited solubility of barium carbonate can be used to improve the titration of carbonate-hydroxide or carbonate–hydrogen carbonate mixtures. The *Winkler* method for the determination of carbonate-hydroxide mixtures involves titration of both components in an aliquot with an acid-range indicator. An excess of neutral barium chloride is then added to a second aliquot to precipitate the carbonate ion, following which the hydroxide ion is titrated to a phenolphthalein end point. Provided the concentration of the excess barium ion is about 0.1 F, the solubility of barium carbonate is too low to interfere with the titration.

An accurate analysis of a carbonate–hydrogen carbonate mixture can be achieved by establishing the total equivalents through titration of an aliquot to the acidic end point with an indicator such as bromocresol green. The hydrogen carbonate in a second aliquot is converted to carbonate by the addition of a known excess of standard base. After a large excess of barium chloride has been introduced, the excess base is titrated with standard acid to a phenolphthalein end point.

The presence of solid barium carbonate does not hamper end-point detection in either of these methods.

DETERMINATION OF ORGANIC FUNCTIONAL GROUPS

Neutralization titrations are convenient for the direct or indirect determination of several organic functional groups. Brief descriptions of methods for the more common groups follow.

Carboxylic and Sulfonic Acid Groups. The carboxylic and the sulfonic acid groups are the two most common structures that impart acidity to organic compounds. Most carboxylic acids have dissociation constants that range between 10^{-4} and 10^{-6} and are thus readily titrated. An indicator that changes color in the basic range is required; phenolphthalein is widely used for this purpose.

Many carboxylic acids are not sufficiently soluble in water to permit a direct titration in this medium. Where this problem exists, the acid is often dissolved in ethanol and titrated with aqueous base. Alternatively, the acid may be dissolved in an excess of standard base; the unreacted base is then back-titrated with standard acid.

Sulfonic acids are generally strong acids and readily soluble in water. Their titration with base is, therefore, straightforward.

Neutralization titrations are often employed to determine equivalent weights of purified organic acids; the data serve as an aid in qualitative identification.

Amine Groups. Aliphatic amines generally have basic dissociation constants on the order of 10^{-5} and can thus be titrated directly with a solution of a strong acid. Aromatic amines such as aniline and its derivatives, on the other hand, are usually too weak for titration in aqueous medium ($K_b \sim 10^{-10}$); the same is true for cyclic amines with aromatic character, such as pyridine and its derivatives. Saturated cyclic amines, such as piperdine, on the other hand, are often similar to aliphatic amines.

Many amines that are not susceptible to neutralization titration in aqueous media are readily determined in a nonaqueous solvent that enhances their basicity (see Chapter 12).

Ester Groups. Esters are commonly determined by saponification with a measured quantity of standard base:

$$R_1COOR_2 + OH^- \rightarrow R_1COO^- + HOR_2$$

The excess base is then titrated with standard acid.

Esters vary widely in their reactivity toward alkali. Some require several hours of heating with a base to complete the saponification. A few react rapidly enough to permit direct titration with standard base. Typically, the ester is refluxed with standard 0.5-N base for 1 to 2 hr. After cooling, the excess base is determined with standard acid.

Hydroxyl Groups. Hydroxyl groups in organic compounds can be determined by esterification with various carboxylic acid anhydrides or chlorides; the two most common reagents are acetic anhydride and phthalic anhydride. With acetic anhydride, the reaction is

$$(CH_3CO)_2O + ROH \rightarrow CH_3COOR + CH_3COOH$$

The acetylation is ordinarily carried out by mixing the sample with a carefully measured volume of acetic anhydride in pyridine. After heating, water is added to hydrolyze the unreacted anhydride:

$$(CH_3CO)_2O + H_2O \rightarrow 2CH_3COOH$$

The acetic acid is then titrated with a standard solution of alcoholic sodium or potassium hydroxide. A blank is carried through the analysis to establish the original amount of anhydride.

Most amines are converted quantitatively to amides by acetic anhydride; often, however, a correction for this source of interference is possible by direct titration of another portion of the sample with standard acid.

Carbonyl Groups. Many aldehydes and ketones can be determined with a standard solution of hydroxylamine hydrochloride. The reaction, which produces an oxime, is

$$\underset{R_2}{\overset{R_1}{>}}C{=}O + NH_2OH \cdot HCl \rightarrow \underset{R_2}{\overset{R_1}{>}}C{=}NOH + HCl$$

where R_2 may be an atom of hydrogen. The liberated hydrochloric acid is titrated with base. Here again the conditions necessary for quantitative reaction vary. Typically, 30 min suffice for aldehydes. Many ketones require refluxing with the reagent for an hour or more.

PROBLEMS

*1. Describe the preparation of 2.00 liters of
   (a) 0.15-N KOH from the solid.
   (b) 0.015-N Ba(OH)$_2$ · 8H$_2$O from the solid.
   (c) 0.200-N HCl from a reagent having a density of 1.0579 g/ml and containing 11.50% HCl.
   (d) 0.150-N reagent from constant-boiling HCl, which was distilled at 750 mm Hg.

2. Describe the preparation of 500 ml of
   (a) 0.250-N H$_2$SO$_4$ from a reagent having a density of 1.1539 g/ml and containing 21.8% H$_2$SO$_4$.
   (b) 0.30-N NaOH from the solid.
   (c) 0.500-N HCl from constant-boiling HCl, which was distilled at 770 mm Hg.
   (d) 0.0800-N Na$_2$CO$_3$ from the pure solid. (To be used with an indicator having an acidic range.)

*3. The following data were obtained for the standardization of HCl against samples of sodium tetraborate $Na_2B_4O_7 \cdot 10H_2O$:

$$B_4O_7^{2-} + 2H_3O^+ + 3H_2O \rightarrow 4H_3BO_3$$

g Taken	ml HCl
0.6442	33.74
0.7102	37.56
0.5934	31.26

(a) Calculate the mean normality for the set.
(b) Calculate the standard deviation for the normality.

4. (a) Calculate the mean normality of a $Ba(OH)_2$ solution from the following standardization data:

g $KH(IO_3)_2$ Taken	ml $Ba(OH)_2$
0.2574	26.77
0.2733	28.45
0.2885	30.11

(b) Calculate the standard deviation for the normality.

5. Suggest a range of sample weights for the indicated primary standard if it is desired to use between 35 and 45 ml of titrant.
  *(a) 0.030-N $HClO_4$ vs. Na_2CO_3 (CO_2 product)
  (b) 0.075-N HCl vs. $Na_2C_2O_4$

$$(Na_2C_2O_4 \rightarrow Na_2CO_3 + CO; \ CO_3^{2-} + 2H^+ \rightarrow H_2O + CO_2)$$

  *(c) 0.20-N NaOH vs. benzoic acid
  (d) 0.030-F $Ba(OH)_2$ vs. $KH(IO_3)_2$
  *(e) 0.010-N $HClO_4$ vs. THAM
  (f) 0.080-N H_2SO_4 vs. $Na_2B_4O_7 \cdot 10H_2O$ (see Problem 3)

*6. A 25.0-ml aliquot of dilute H_2SO_4 yielded 0.347 g of $BaSO_4$. Calculate the normality of the acid.

7. A 50.0-ml aliquot of dilute HCl yielded 0.477 g of AgCl. What was the normality of the acid?

*8. A 50.0-ml sample of a white dinner wine required 21.4 ml of 0.0377-N NaOH to achieve a phenolphthalein end point. Express the acidity of the wine in terms of grams tartaric acid ($H_2C_4H_4O_6$, gfw = 150) per 100 ml (assume that two hydrogens of the acid are titrated).

9. A 25.0-ml sample of a household cleaning solution was diluted to exactly 250 ml in a volumetric flask. A 50.0-ml aliquot of this solution required 40.3 ml of 0.250-N HCl to reach a bromocresol green end point. Calculate the weight-volume percent of NH_3 in the sample (assuming that all of its alkalinity results from that constituent).

*10. A 0.229-g sample of a recrystallized organic acid required a 29.8-ml titration with 0.100-N NaOH to reach a phenolphthalein end point. What is the equivalent weight of the acid?

11. In order to establish the identity of the cation in a pure carbonate, a 0.140-g sample was dissolved in 50.0 ml of 0.114-N HCl and boiled to remove CO_2. The excess HCl was back-titrated with 24.2 ml of 0.0980-N NaOH. Identify the carbonate.

*12. The active ingredient in Antabuse, a drug used for treatment of chronic alcoholism, is tetraethylthiuram disulfide

$$\underset{\substack{\parallel \\ }}{\overset{S}{}} \quad \underset{}{\overset{S}{}}$$

$$(C_2H_5)_2NC\overset{S}{\overset{\parallel}{S}}SC\overset{S}{\overset{\parallel}{N}}(C_2H_5)_2$$

(gfw = 296). The sulfur in a 0.432-g sample of an Antabuse preparation was oxidized to SO_2, which was absorbed in H_2O_2 to give H_2SO_4. The acid was titrated with 22.1 ml of 0.0373-N base. Calculate the percent active ingredient in the preparation.

13. Neohetramine, $C_{16}H_{21}ON_4$ (gfw = 285), is a common antihistamine. A 0.124-g sample containing this compound was analyzed by the Kjeldahl method. The ammonia produced was collected in HBO_2; the resulting BO_2^- was titrated with 26.1 ml of 0.0147-N HCl. Calculate the percent neohetramine in the sample.

*14. To obtain the percent protein in a wheat product, the percent nitrogen present is generally multiplied by 5.70. A 0.909-g sample of a wheat flour was analyzed by the Kjeldahl procedure. The ammonia formed was distilled into 50.0 ml of 0.0506-N HCl; a 7.46-ml back-titration with 0.0491-N base was required. Calculate the percent protein in the flour.

15. The formaldehyde content of a pesticide preparation was determined by weighing 2.87 g of the liquid sample into a flask containing 50.0 ml of 0.996-N NaOH and 50 ml of 3% H_2O_2. Upon heating, the following reaction took place:

$$OH^- + HCHO + H_2O_2 \rightarrow HCOO^- + 2H_2O$$

After cooling, the excess base was titrated with 23.3 ml of 1.01-N H_2SO_4. Calculate the percent HCHO in the sample.

*16. A 3.00-liter sample of urban air was bubbled through a solution containing 50.00 ml of 0.0116-N Ba(OH)$_2$; $BaCO_3$ precipitated. The excess base was back-titrated to a phenolphthalein end point with 23.6 ml of 0.0108-N HCl. Calculate the parts per million CO_2 in the air (that is, the ml $CO_2/10^6$ ml air) if the density of CO_2 is 1.98 g/liter.

17. Air was bubbled at a rate of 30 liters/min through a trap containing 75 ml of 1% H_2O_2 ($H_2O_2 + SO_2 \rightarrow H_2SO_4$). After 10.0 min, the H_2SO_4 was titrated with 11.1 ml of 0.00204-N NaOH. Calculate the parts per million SO_2 (that is, ml $SO_2/10^6$ ml air) if the density of SO_2 is 0.00285 g/ml.

*18. What should be the normality of a Ba(OH)$_2$ solution if its titer is to be 1.00 mg HCl/ml?

19. What is the H_3AsO_4 titer of a 0.0676-N NaOH solution if phenolphthalein is to serve as the indicator?

*20. A 0.816-g sample containing dimethylphthalate, $C_6H_4(COOCH_3)_2$ (gfw = 194), and unreactive species was saponified by refluxing with 50.0 ml of 0.103-N NaOH. After reaction was complete, the excess NaOH was back-titrated with 24.2 ml of 0.164-N HCl. Calculate the percent dimethylphthalate in the sample.

21. A 50.0-ml sample containing methylethylketone $CH_3COC_2H_5$ and unreactive components was treated with an excess of hydroxylamine hydrochloride ($NH_2OH \cdot HCl$). After oxime formation was complete, the liberated HCl was titrated with 19.1 ml of 0.0112-N NaOH. Calculate the milligrams of the ketone per liter of sample.

*22. A 50.0-ml sample containing acetaldehyde, ethyl acetate, and unreactive substances was diluted to 250 ml with H_2O. A 50.0-ml aliquot of the diluted sample was mixed with 40.0 ml of 0.0545-N NaOH and 20 ml of 3% H_2O_2. Upon refluxing for a half hour, quantitative saponification of the ester occurred; at the same time the acetaldehyde was oxidized to sodium acetate.

$$H_2O_2 + CH_3CHO + OH^- \rightarrow CH_3COO^- + 2H_2O$$

The unreacted NaOH consumed 10.1 ml of 0.0251-N HCl.

   A 25.0-ml aliquot of the sample solution was treated with $NH_2OH \cdot HCl$ to convert the acetaldehyde to the corresponding oxime. The liberated HCl consumed 10.9 ml of the standard base.

   Calculate the weight-volume percent of acetaldehyde and ethyl acetate in the sample.

23. A 1.21-g sample containing $(NH_4)_2SO_4$, NH_4NO_3, and nonreactive substances was dissolved and diluted to 200 ml in a volumetric flask. A 50.00-ml aliquot was made basic with strong alkali and the liberated NH_3 distilled into 30.0 ml of 0.0842-N HCl. The excess HCl required 10.1 ml of 0.0880-N NaOH.

   A 25.0-ml aliquot of the sample was made alkaline after the addition of Devarda's alloy; reduction of the NO_3^- to NH_3 occurred. The NH_3 from both NH_4^+ and NO_3^- was then distilled into 30.00 ml of the standard acid and back-titrated with 14.1 ml of the base.

   Calculate the percent $(NH_4)_2SO_4$ and NH_4NO_3 in the sample.

*24. A 0.141-g sample of a phosphorus-containing compound was digested in a mixture of HNO_3 and H_2SO_4, which resulted in formation of CO_2, H_2O, and H_3PO_4. Addition of ammonium molybdate yielded a solid having the composition $(NH_4)_3PO_4 \cdot 12MoO_3$. The precipitate was filtered, washed, and dissolved in 50.0 ml of 0.200-F NaOH.

$$(NH_4)_3PO_4 \cdot 12MoO_3(s) + 26OH^- \rightarrow$$
$$HPO_4^{2-} + 12MoO_4^{2-} + 14H_2O + 3NH_3(g)$$

After boiling the solution to remove the NH_3, the excess NaOH was titrated with 14.1 ml of 0.174-N HCl. Calculate the percent P in the sample.

25. A 0.841-g sample containing $NaHC_2O_4$, $H_2C_2O_4 \cdot 2H_2O$, and unreactive species was titrated to a phenolphthalein end point with 30.70 ml of 0.114-N NaOH. The resulting solution was then evaporated to dryness and ignited to decompose the sodium oxalate.

$$Na_2C_2O_4 \rightarrow Na_2CO_3 + CO$$

The residue was boiled with 50.0 ml of 0.130-N HCl, and, after cooling, the excess acid was back-titrated with 3.33 ml of the base. Calculate the percent $Na_2C_2O_4$ and $H_2C_2O_4 \cdot 2H_2O$ in the original sample.

*26. A 1.21-g sample of commercial KOH, which was contaminated by K_2CO_3, was dissolved in water, and the resulting solution was diluted to 500 ml. A 50.0-ml aliquot of this solution was treated with 40.0 ml of 0.0530-N HCl and boiled to remove CO_2. The excess acid consumed 4.74 ml of 0.0498-N NaOH (phenolphthalein indicator). An excess of neutral $BaCl_2$ was added to another 50.0-ml aliquot to precipitate the carbonate as $BaCO_3$. The solution was then titrated with 28.3 ml of the acid to a phenolphthalein end point. Calculate the percent KOH, K_2CO_3, and water in the sample, assuming that these are the only compounds present.

27. A 0.500-g sample containing $NaHCO_3$, Na_2CO_3, and H_2O was dissolved and diluted to exactly 250.0 ml. A 25.0-ml aliquot of the sample was then boiled with 50.00 ml of 0.0125-N HCl. After cooling, the excess acid in the solution required 2.34 ml of 0.0106-N NaOH when titrated to a phenolphthalein end point. A second 25.0-ml aliquot was then treated with an excess of $BaCl_2$ and 25.0 ml of the base; precipitation of all of the carbonate resulted, and 7.63 ml of the HCl were required to titrate the excess base. Calculate the composition of the mixture.

*28. Calculate the volume of 0.0612-N HCl needed to titrate
  (a) 20.0 ml of 0.0555-F Na_3PO_4 to a thymolphthalein end point.
  (b) 25.0 ml of 0.0555-F Na_3PO_4 to a bromocresol green end point.
  (c) 40.0 ml of a solution that is 0.0210 F in Na_3PO_4 and 0.0165 F in Na_2HPO_4 to a bromocresol green end point.
  (d) 20.0 ml of a solution that is 0.0210 F in Na_3PO_4 and 0.0165 F in NaOH to a thymolphthalein end point.

29. Calculate the volume of 0.0773-N NaOH needed to titrate
  (a) 25.0 ml of a solution that is 0.0300 F in HCl and 0.0100 F in H_3PO_4 to a bromocresol green end point.
  (b) the solution in (a) to a thymolphthalein end point.
  (c) 30.0 ml of a solution that is 0.0640 F in NaH_2PO_4 to a thymolphthalein end point.
  (d) 25.0 ml of a solution that is 0.0200 F in H_3PO_4 and 0.0300 F in NaH_2PO_4 to a thymolphthalein end point.

*30. A series of solutions containing NaOH, Na_2CO_3, and $NaHCO_3$, alone or in compatible combination, was titrated with standard 0.120-N HCl. Tabulated below are the acid volumes needed to titrate 25.00-ml portions of each solution to (1) a phenolphthalein end point and (2) a bromocresol green end point. Use this information to deduce the composition of the solutions. In addition, calculate the number of milligrams of each solute per milliliter of solution.

	(1)	(2)
(a)	22.42	22.44
(b)	15.67	42.13
(c)	29.64	36.42
(d)	16.12	32.23
(e)	0.00	33.33

31. A series of solutions containing NaOH, Na_3AsO_4, and Na_2HAsO_4, alone or in compatible combination, was titrated with standard 0.0860-N HCl. Tabulated below are the acid volumes needed to titrate 25.00-ml portions of each solution to (1) a phenolphthalein end point and (2) a bromocresol green end point. Use this information to deduce the composition of the solutions. In addition, calculate the number of milligrams of each solute per milliliter of solution.

	(1)	(2)
(a)	0.00	18.15
(b)	21.00	28.15
(c)	19.80	39.61
(d)	18.04	18.03
(e)	16.00	37.37

*32. A series of solutions can contain HCl, H_3PO_4, or NaH_2PO_4, alone or in any compatible combination of these solutes. Tabulated below are the volumes of 0.1200-N NaOH needed to titrate 25.00-ml portions of each solution to (1) a bromocresol green end point and (2) a thymolphthalein end point. Use this information to deduce the composition of the solutions. In addition, calculate the number of milligrams of each solute per milliliter of solution.

	(1)	(2)
(a)	18.72	23.60
(b)	7.93	7.95
(c)	0.00	16.77
(d)	13.12	35.19
(e)	13.33	26.65

33. A series of solutions can contain HCl, maleic acid, and sodium hydrogen maleate, alone or in any compatible combination of these solutes. Tabulated below are the volumes of 0.0994-N NaOH needed to titrate 25.00-ml portions of each solution to (1) a bromocresol green end point and (2) a thymolphthalein end point. Use this information to deduce the composition of the solutions. In addition, calculate the number of milligrams of each solute per milliliter of solution.

	(1)	(2)
(a)	0.00	27.67
(b)	7.34	23.34
(c)	7.34	14.68
(d)	9.99	29.00
(e)	12.70	12.70

12

ACID-BASE TITRATIONS IN NONAQUEOUS MEDIA

In Chapter 9 we noted that acids or bases with dissociation constants smaller than about 1×10^{-8} cannot be titrated because their reactions are not sufficiently complete to yield a satisfactory end point. Many of these same species become titratable, however, in nonaqueous solvent systems that emphasize the acidic or basic character that they do possess. Nonaqueous solvent systems have the further advantage of permitting the titration of numerous substances that are sparingly soluble in water.[1]

Unfortunately, the quantitative information and data required for the derivation of titration curves are usually lacking for nonaqueous solvents; as a consequence, conclusions regarding the feasibility of titrations must be based on qualitative concepts only. Here the Brønsted theory (p. 8) is often of considerable help.

[1] For more extensive discussions of nonaqueous neutralization titrations, see J. Kucharsky and L. Safarik, *Titrations in Non-Aqueous Solvents*. New York: American Elsevier Publishing Company, Inc., 1963; J. S. Fritz and G. S. Hammond, *Quantitative Organic Analysis*, chapter 3. New York: John Wiley & Sons, Inc., 1957; H. A. Laitinen and W. E. Harris, *Chemical Analysis*, 2d ed., pp. 56–92 and 112–121. New York: McGraw-Hill Book Company, 1975; W. Huber, *Titrations in Nonaqueous Solvents*. New York: Academic Press, Inc., 1967.

Solvents for Nonaqueous Titrations

It is convenient to classify solvents into three categories, depending upon their character.

Amphiprotic solvents possess both acidic and basic properties and undergo self-dissociation or *autoprotolysis*. Although water is the most common amphiprotic solvent, many other substances exhibit analogous behavior. Thus,

$$2H_2O \rightleftarrows H_3O^+ + OH^-$$

$$2C_2H_5OH \rightleftarrows C_2H_5OH_2^+ + C_2H_5O^-$$

$$2HOAc \rightleftarrows H_2OAc^+ + OAc^-$$

$$2NH_3 \rightleftarrows NH_4^+ + NH_2^-$$

or, in general,

$$2SH \rightleftarrows SH_2^+ + S^-$$

where SH represents the amphiprotic solvent molecule and SH_2^+ the solvated proton; the base thus corresponds to the anion S^-. Table 12-1 lists autoprotolysis constants for several common solvents.

TABLE 12-1 Autoprotolysis Constants for Some Common Solvents at 25°C

Solvent	K_s	Dielectric Constant
Water	1.01×10^{-14}	78.5
Methanol	2×10^{-17}	32.6
Ethanol	8×10^{-20}	24.3
Formic acid	6×10^{-7}	58.5
Acetic acid	3.6×10^{-15}	6.2
Sulfuric acid	1.4×10^{-4}	>84
Ammonia[a]	1×10^{-33}	22
Ethylenediamine	5×10^{-16}	14.2

[a] At $-50°C$.

In contrast to water and the alcohols, some amphiprotic solvents, such as acetic acid, sulfuric acid, or formic acid, have considerably stronger acidic than basic properties; others, such as ammonia or ethylenediamine, are stronger bases than acids.

Aprotic, or inert, solvents have no appreciable acidic or basic character and do not undergo autoprotolysis to any detectable extent. Benzene, carbon tetrachloride, and pentane fall into this category.

Finally, there exist a number of solvents, such as ketones, ethers, esters, and pyridine derivatives, with basic properties but essentially no acidic tendencies. Solvents of this type do not undergo autoprotolysis.

NEUTRALIZATION REACTIONS IN AMPHIPROTIC SOLVENTS

Alteration of the solvent often exerts a profound effect on the completeness of a neutralization reaction. The effects of amphiprotic solvents on the completeness of acid-base reactions are considered in this section.

The Completeness of Neutralization Reactions. In water, the titration of a weak base B with a standard strong acid can be formulated as

$$B + H_3O^+ \rightleftarrows BH^+ + H_2O \tag{12-1}$$

and the magnitude of the equilibrium constant for this reaction can be used as a measure of its completeness; that is,

$$K_{equil} = \frac{[BH^+]}{[B][H_3O^+]} = \frac{K_b}{K_w} \tag{12-2}$$

Note that the equilibrium constant for the reaction is equal to the quotient of the basic dissociation constant of B and the ion-product constant for water.

In an analogous fashion, completeness of the reaction between a weak acid HA and a strong base can be expressed by the equilibrium constant

$$K_{equil} = \frac{[A^-]}{[HA][OH^-]} = \frac{K_a}{K_w} \tag{12-3}$$

Similar relations can be derived for reactions in nonaqueous solvents. For example, when the weak base B is titrated with a strong acid in anhydrous formic acid, we may write

$$B + HCOOH_2^+ \rightleftarrows BH^+ + HCOOH \tag{12-4}$$

where $HCOOH_2^+$ represents the solvated proton analogous to H_3O^+ in an aqueous solution. Here,

$$K_{equil} = \frac{[BH^+]}{[B][HCOOH_2^+]} = \frac{K_b'}{K_s} \tag{12-5}$$

where K_b' is the dissociation constant for the base *in formic acid*; that is,

$$B + HCOOH \rightleftarrows BH^+ + HCOO^- \qquad K_b' = \frac{[BH^+][HCOO^-]}{[B]} \tag{12-6}$$

The constant K_s is the autoprotolysis constant for formic acid:

$$2HCOOH \rightleftarrows HCOOH_2^+ + HCOO^- \qquad K_s = [HCOOH_2^+][HCOO^-] \tag{12-7}$$

As with the ion-product constant for water, the concentration of the solvent HCOOH is essentially invariant and is thus included in K_s.

The titration of a weak acid HA with sodium ethoxide in ethanol is readily formulated as

$$HA + C_2H_5O^- \rightleftarrows A^- + C_2H_5OH \tag{12-8}$$

Here the standard strong base is a solution of sodium ethoxide (C_2H_5ONa) in ethanol. In common with the previous examples, the completeness of the reaction can be measured with the equilibrium constant

$$K_{equil} = \frac{[A^-]}{[HA][C_2H_5O^-]} = \frac{K'_a}{K_s} \qquad (12\text{-}9)$$

where K'_a is the dissociation constant for the acid in ethanol,

$$HA + C_2H_5OH \rightleftarrows C_2H_5OH_2^+ + A^- \qquad K'_a = \frac{[A^-][C_2H_5OH_2^+]}{[HA]} \qquad (12\text{-}10)$$

and K_s is the autoprotolysis constant for ethanol,

$$2C_2H_5OH \rightleftarrows C_2H_5OH_2^+ + C_2H_5O^- \qquad K_s = [C_2H_5OH_2^+][C_2H_5O^-] \qquad (12\text{-}11)$$

These examples demonstrate that completeness of the reaction is a function of both the dissociation constant of the substance being titrated and the autoprotolysis constant of the solvent. The appearance of both constants in the equation can be best understood, perhaps, by considering each neutralization reaction as representing a competition for protons. Thus, for example, the extent of reaction 12-4 is governed by the success with which solvent molecules HCOOH compete with base molecules B for a stoichiometrically limited number of hydrogen ions H^+. The effectiveness of each participant in this competition is measured by its dissociation constant K_s and K'_b, respectively. Similarly, reaction 12-8 can be thought of as a competition between the ions A^- and $C_2H_5O^-$ for H^+, with the effectiveness of each being measured by K'_a and K_s.

From this discussion it is clearly advantageous to perform acid-base titrations in solvents that have low autoprotolysis constants. Furthermore, acid-base reactions are more complete in those solvents in which K'_a or K'_b is large. These two considerations, which are not entirely independent of one another, govern the choice of amphiprotic solvents for nonaqueous titrations.

Effect of Acidity or Basicity of Solvents on Solute Behavior. A number of amphiprotic solvents, including formic acid, acetic acid, and sulfuric acid, are considerably better proton donors than proton acceptors and are therefore classed as acidic solvents. In such solvents, the basic properties of a solute are magnified, while its acidic properties are diminished. Thus, for example, aniline, $C_6H_5NH_2$, cannot be titrated in aqueous solution because its basic dissociation constant is only about 10^{-10}. In glacial acetic acid, however, aniline is an appreciably stronger base because of its enhanced tendency to react with the solvent, Thus, the equilibrium constant, K'_b, for the reaction

$$C_6H_5NH_2 + HOAc \rightleftarrows C_6H_5NH_3^+ + OAc^-$$

is significantly larger than K_b for the analogous reaction in water:

$$C_6H_5NH_2 + H_2O \rightleftarrows C_6H_5NH_3^+ + OH^-$$

While an acidic solvent tends to increase the basicity of bases, it has the opposite effect on acids, making them weaker. Thus, hydrochloric acid, a strong acid in water, is only partially dissociated in glacial acetic acid; an acid that is weak in water becomes even weaker in an acidic solvent.

Solvents such as ethylenediamine and liquid ammonia have a strong affinity for protons and are therefore classified as basic solvents. In these media, the acid properties of a solute are enhanced. Thus, phenol, with an acid dissociation constant of about 10^{-10} in water, becomes sufficiently strong in ethylenediamine to permit its titration with a standard base. The strengths of bases are, of course, diminished in solvents of this type.

Water and the aliphatic alcohols, such as methanol and ethanol, are examples of neutral amphiprotic solvents. These solvents possess less pronounced capacities as proton donors or acceptors than those just considered. It is important to realize, however, that this name does not necessarily imply an exact equality between acid and base character.

Effect of Dielectric Constant on Behavior of Solutes. The dielectric constant of a solvent measures its capacity for separating oppositely charged particles. In a solvent with a high dielectric constant, such as water ($D_{H_2O} = 78.5$), a minimum of work is required to separate a positively charged ion from one with a negative charge; for a solvent with a low dielectric constant, such as acetic acid ($D_{HOAc} = 6.2$), a greater amount of energy is required to accomplish the process. Methanol and ethanol have dielectric constants of 33 and 24, respectively, and are intermediate in their behavior. The dielectric constants for several solvents are shown in Table 12-1.

The dielectric constant of the solvent plays an important role in determining the strengths of solute acids or bases insofar as the ionization process produces oppositely charged species. For example, when an uncharged weak acid HA is dissolved in an amphiprotic solvent SH, the dissociation process requires the separation of the charged particles SH_2^+ and A^-:

$$HA + SH \rightleftarrows SH_2^+ + A^-$$

The same would be true upon solution of an uncharged base B:

$$B + SH \rightleftarrows BH^+ + S^-$$

Reactions of this sort would be expected to proceed further to the right in a solvent such as water than in methanol or ethanol because less work is required for the dissociation process. The magnitude of this effect can be large; for example, the dissociation constant for acetic acid in water is approximately 10^{-5}, whereas in ethanol it is somewhat smaller than 10^{-10}. Other acids of this type show similar decreases.

The strength of an acid or base is not affected significantly by the dielectric constant of the medium when the dissociation reaction does not involve a charge separation. For example, the following equilibria would not be altered by changes in dielectric constant of the solvent SH:

$$BH^+ + SH \rightleftarrows SH_2^+ + B$$

$$A^- + SH \rightleftarrows HA + S^-$$

Choice of Amphiprotic Solvents for Neutralization Titrations. We have shown that the completeness of a neutralization reaction is directly dependent upon the ionization constant of the solute acid or base and inversely related to the autoprotolysis constant of the solvent. Furthermore, the first of these factors, the ionization constant, is dependent upon the acidic or basic properties and the dielectric constant of the solvent. Thus, the most advantageous choice of solvent for a given titration hinges upon three interrelated properties:

1. Its autoprotolysis constant, a numerically small value being desirable.
2. Its properties as a proton donor or acceptor. For the titration of a weak base, a solvent with strong donor tendencies is helpful (that is, an acidic solvent); for the analysis of a weak acid, a solvent that is a good proton acceptor is desirable.
3. Its dielectric constant, a high value being most useful.

In addition, of course, the solvent must be one in which the sample is reasonably soluble.

Glacial acetic acid is often chosen as solvent for the titration of very weak bases because it tends to donate protons and thus enhances the strengths of dissolved bases. Its autoprotolysis constant (3.6×10^{-15}) is also somewhat more favorable than that for water. On the other hand, its low dielectric constant partially offsets these advantages. The two favorable properties outweigh the single disadvantage, however, and acetic acid is a generally superior solvent for the titration of weak bases; it will be clearly inferior to water for titration of weak acids because of its weakness as a proton acceptor.

It is profitable to consider formic acid in the role of an acidic solvent. Like acetic acid, it is a much better proton donor than water; furthermore, in contrast to acetic acid, formic acid has a dielectric constant that is comparable with water. On these two counts, then, it would appear to be an ideal solvent for the titration of weak bases. Unfortunately, however, its autoprotolysis constant is much larger than that of water or acetic acid. As a consequence, despite its two very desirable properties, formic acid appears to offer little advantage compared with water.

Methanol and ethanol have been widely applied as solvents for acid-base titrations. Both are classified as neutral solvents because their proton donor and acceptor properties do not differ markedly from water. Both have advantageous autoprotolysis constants. On the other hand, their low dielectric constants frequently offset the advantage of their small autoprotolysis constants. For example, in ethanol, the dissociation constants of most uncharged acids, such as benzoic acid, are about 10^{-6} as great as in water; at the same time the ratio of autoprotolysis constants is smaller by nearly the same factor (8×10^{-6}). Thus, the ratio K_a'/K_s is only slightly more favorable in ethanol than in water, and the improvement in end points gained by use of this solvent is modest. On the other hand, significant gain is realized by the use of ethanol for the titration of a charged weak acid such as the ammonium ion. Here, no charge separation is involved in the dissociation process:

$$NH_4^+ + C_2H_5OH \rightleftharpoons NH_3 + C_2H_5OH_2^+$$

In contrast to benzoic acid, dissociation of the acid NH_4^+ is not significantly decreased in ethanol. The reaction of NH_4^+ with a strong base is, however, much more complete in ethanol because of the low autoprotolysis constant of the solvent. As a consequence, NH_4^+ can be titrated satisfactorily in ethanol but not in water.

Example. Calculate the percentage of unreacted NH_4^+ at the equivalence point in the titration of 0.20-F NH_4^+ with (1) 0.20-F NaOH in an aqueous solution and (2) 0.20-F C_2H_5ONa in anhydrous ethanol.

The autoprotolysis constants for the solvents are 1×10^{-14} and 8×10^{-20}, respectively; the acid dissociation constant for NH_4^+ is approximately 6×10^{-10} in water and 1×10^{-10} in ethanol.

For the titration in water,

$$NH_4^+ + OH^- \rightleftarrows NH_3 + H_2O$$

the equilibrium constant is given by

$$K_{equil} = \frac{[NH_3]}{[NH_4^+][OH^-]} = \frac{K_a}{K_w} = \frac{6 \times 10^{-10}}{1 \times 10^{-14}} = 6 \times 10^4$$

Similarly, for the titration in ethanol,

$$NH_4^+ + C_2H_5O^- \rightleftarrows NH_3 + C_2H_5OH$$

$$K_{equil} = \frac{[NH_3]}{[NH_4^+][C_2H_5O^-]} = \frac{K_a'}{K_s} = \frac{1 \times 10^{-10}}{8 \times 10^{-20}} = 1.2 \times 10^9$$

At the equivalence point in each titration, the formal NH_3 concentration will be 0.10. In the aqueous titration,

$$[NH_4^+] = [OH^-]$$
$$[NH_3] = 0.10 - [NH_4^+]$$

while in the ethanol solution,

$$[NH_4^+] = [C_2H_5O^-]$$
$$[NH_3] = 0.10 - [NH_4^+]$$

If we further assume that in both media $[NH_4^+] \ll [NH_3]$, then

(1)
$$\frac{0.10}{[OH^-]^2} \cong 6 \times 10^4$$

$$[OH^-] = [NH_4^+] = 1 \times 10^{-3}$$

(2)
$$\frac{0.10}{[OH^-]^2} \cong 1.2 \times 10^9$$

$$[OH^-] = [NH_4^+] = 9 \times 10^{-6}$$

Thus, at the equivalence point in an aqueous titration, approximately 1% of the NH_4^+ remains unreacted; the comparable quantity in anhydrous ethanol is only about 0.01%.

Several amphiprotic basic solvents have been employed for the titration of very weak acids. Among the most basic of these, and thus best from the standpoint of enhancing the acidity of the solute, is ethylenediamine, $NH_2CH_2CH_2NH_2$;

its autoprotolysis constant is about 5×10^{-16}. The advantageous acceptor property and autoprotolysis constant of this solvent is partially offset by its low dielectric constant. Dimethylformamide, $HCON(CH_3)_2$, a weaker base than ethylenediamine, has also proved useful; its dielectric constant is 27.

NEUTRALIZATION REACTIONS IN APROTIC SOLVENTS AND MIXED SOLVENTS

Solvents without amphiprotic properties offer the advantage that they do not compete for protons with the reactants in a titration; that is these solvents have autoprotolysis constants that approach zero. Thus, neutralization reactions should be more nearly complete when carried out in solvents of this type.

Inorganic solutes tend to be sparingly soluble in aprotic solvents. As a consequence, many mixtures of aprotic solvents with more polar solvents have been investigated. Examples include benzene–methanol and ethylene glycol–hydrocarbon. Unfortunately, fundamental knowledge regarding the properties and behavior of acids and bases in such systems is meager; the application of a mixed solvent to a specific problem can thus be determined only empirically.

END-POINT DETECTION IN NONAQUEOUS TITRATIONS

Undoubtedly the most popular method of end-point detection for nonaqueous titration involves measuring the potential of a glass electrode that responds to the concentration of the solvated proton. The glass electrode is discussed in detail in Chapter 17.

Many acid-base indicators used for aqueous titrations are also applicable in nonaqueous solvents. To be sure, their behavior in an aqueous environment cannot be extrapolated to predict their properties in nonaqueous solutions; the limited information available with respect to these properties in solvents other than water makes the choice of indicator largely a matter of experience and empirical observation.

Applications of Nonaqueous Titrations

The problems to which nonaqueous acid-base titrations have been applied are numerous[2]; the following will suggest the potentialities of the technique.

TITRATIONS IN GLACIAL ACETIC ACID

Many bases that are too weak for titration in water are readily determined in glacial acetic acid. The titrant is a standard solution of perchloric acid, which is stronger than either hydrochloric or sulfuric acid in this solvent. In glacial acetic

---

[2] C. A. Streuli, *Anal. Chem.*, **34**, 302R (1962); **36**, 363R (1964). G. A. Harlow and D. H. Morman, *Anal. Chem.*, **38**, 485R (1966); **40**, 418R (1968). J. J. Logowski, *Anal. Chem.*, **42**, 305R (1970); **44**, 524R (1972); **46**, 460R (1974).

acid, sodium acetate behaves as a base just as sodium hydroxide does in an aqueous environment; standard solutions of sodium acetate can thus be used for back-titration when needed.

Coefficient of Expansion of Glacial Acetic Acid Solutions. In common with most organic solvents, acetic acid has a significantly greater coefficient of expansion than water (0.11% per degree Celsius compared with 0.025%). Therefore, greater care must be exercised to eliminate errors arising from temperature fluctuations during volumetric measurements. A common practice is to note the temperature of the perchloric acid reagent at the time of its standardization; the temperature is then recorded at the time the reagent is used for an analysis, and the volume is corrected to the temperature of standardization with the equation

$$V_{std} = V[1 + 0.0011(T_{std} - T)] \qquad (12\text{-}12)$$

where T_{std} is the temperature of the reagent at the time of standardization, T is its temperature when used, V is the measured volume of the reagent, and V_{std} is the corrected volume.

Effect of Water. Water acts as a weak base in acetic acid and tends to compete with the solvent for protons. As a consequence, its presence leads to smaller pH changes in the equivalence-point region of a neutralization titration and causes less satisfactory end points. The amount of water that can be tolerated during a titration varies; for very weak bases, nearly anhydrous conditions are required. On the other hand, for the titration of bases that are relatively strong in the solvent, as much as 3.0% of water by volume is not harmful.

Fortunately, nearly anhydrous acetic acid can be obtained by the introduction of acetic anhydride. Any water present reacts with the anhydride to form acetic acid; ordinarily, an excess of acetic anhydride is avoided because it may interfere in the titration.

End Points in Acetic Acid. Two acid-base indicators, crystal violet and methyl violet, have been found useful for titrations in acetic acid; both exhibit complex color changes. For example, methyl violet changes from violet through green to yellow as the solution becomes more acidic; the disappearance of the violet color signals the end point. The color change is not as obvious as might be desired; with practice, however, significant titration errors can be avoided when stronger bases are titrated. With very weak bases it is impractical to use the visual end point; here the potentiometric end point with a glass electrode must be employed (Chapter 17).

Primary Standards for Solutions of Perchloric Acid. Potassium hydrogen phthalate is the most commonly used primary standard for the standardization of acetic acid solutions of perchloric acid. The properties of this substance were discussed in Chapter 11 in connection with its use for the standardization of aqueous base solutions. It will be noted that potassium hydrogen phthalate is a sufficiently strong base in glacial acetic acid, so that it can now be employed to standardize an acid solution. The reaction, of course, involves conversion of the acid salt to the undissociated acid.

Sodium carbonate has also been employed for standardization of perchloric acid solutions in acetic acid.

Applications of Titrations in Acetic Acid. The use of anhydrous acetic acid permits the determination of many inorganic and organic salts which are not amenable to titrations in aqueous solution. For example, the sodium salts of inorganic anions such as chloride, bromide, iodide, nitrate, chlorate, and sulfate have all been determined by titration in glacial acetic acid; standard solutions of perchloric acid in acetic acid or anhydrous dioxane were used as reagents. The ammonium and alkali metal salts of most carboxylic acids can also be determined in this medium; typical examples include ammonium benzoate, sodium salicylate, sodium acetate, potassium tartrate, and sodium citrate. Curve *A* in Figure 12-1 is a typical titration curve; here a solution of sodium acetate in anhydrous acetic acid was titrated with standard perchloric acid solution in the same solvent. A glass electrode (see Chapter 17) was employed to measure the quantity $(K + pH)$ where K is constant throughout the titration. For comparison, the theoretical curve (B) for the same concentration of sodium acetate in water is also shown.

Anhydrous acetic acid has proved to be a particularly useful medium for the titration of organic compounds containing the amine or amide functional groups. Moreover, this solvent, in contrast to water, permits the direct titration of most amino acids with a standard acid. It was noted earlier (p. 231) that in aqueous media, these compounds exist largely as zwitterions which are not strong enough acids or bases for titration. In glacial acetic acid, however, the

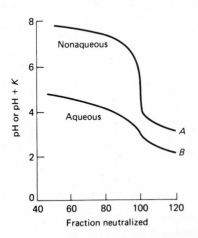

Figure 12-1 Titration Curves for 0.075-*F* Sodium Acetate with 0.100-*N* HClO$_4$. Curve *A*: experimental curve with sample and reagent dissolved in anhydrous acetic acid; ordinate scale is (pH + *K*) where *K* is an empirical constant. Curve *B*: theoretical titration curve for aqueous solution; ordinate is pH.

dissociation of the carboxylic acid group is essentially completely repressed, leaving the amine group available for titration with perchloric acid.

In many of the foregoing applications, a standard solution of sodium acetate in acetic acid is employed as a base to allow back-titration. Laboratory procedures in Chapter 31 (Experiments 14–16) illustrate the application of these reagents.

TITRATION IN BASIC SOLVENTS

A number of basic solvents have been employed to determine acids that are too weak to be titrated in water. These solvents include ethylenediamine, pyridine, dimethylformamide, acetone, methyl isobutyl ketone, acetonitrile, and a 1:1 mixture of ethylene glycol and *i*-propyl alcohol. Several strongly basic solutes have been used in conjunction with these solvents. A standard solution of sodium methoxide in a benzene-methanol mixture has proved useful for the titration of weak acids dissolved in several of these solvents.[3] Sodium amino-ethoxide in ethylenediamine makes possible the determination of a number of phenols as well as most weak carboxylic acids.[4] Solutions of the strong base tetrabutylammonium hydroxide, $(C_4H_9)_4NOH$, may be prepared in benzene-methanol, isopropyl alcohol, or ethanol to give standard base solutions suitable for nonaqueous titrations.

Most carboxylic acids are easily determined in basic solvents; many phenols are also susceptible to analysis in such media. A number of salts of weak bases, including ammonium salts as well as those of aliphatic and aromatic amines, give satisfactory end points when titrated in ethylenediamine or dimethyl-formamide. Certain enols and imides can also be titrated in these more basic solvents.

Figure 12-2 shows an experimental curve for the titration of mixtures of acids dissolved in methyl isobutyl ketone; the titrant was a 0.2-N solution of tetrabutyl ammonium hydroxide in isopropyl alcohol. The abscissa is the potential of a glass electrode system (Chapter 17), which varies linearly with pH. Note that perchloric acid is enough stronger than hydrochloric acid in this solvent to give separate end points for each. Note also that phenol, which is far too weak to be titrated in water, gives a sharp, well-defined end point in the nonaqueous medium.

TITRATIONS IN APROTIC OR NEUTRAL SOLVENTS

Aprotic solvents by themselves are seldom useful because of the low solubility of reagents and samples in these substances. Often, however, they find use when mixed with other neutral solvents. For example, 1:1 mixtures of ethylene glycol or propylene glycol with a hydrocarbon or a chlorinated hydrocarbon are excellent solvents for alkali salts of organic acids. Standard perchloric acid solutions in the same solvents are used for titration of these salts.[5] Titration of weak acids in methanol and ethanol is also common.

---

[3] J. S. Fritz and N. M. Lisicki, *Anal. Chem.*, **23**, 589 (1951).
[4] M. L. Moss, J. H. Elliott, and R. T. Hall, *Anal. Chem.*, **20**, 784 (1948).
[5] S. R. Palit, *Anal. Chem.*, **18**, 246 (1949).

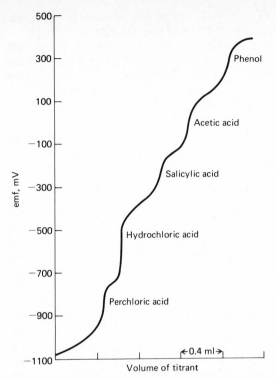

Figure 12-2 Titration of a Mixture of Acids in Methyl Isobutyl Ketone. [D. B. Bruss and G. E. A. Wyld, *Anal. Chem.*, **29**, 234 (1957). With permission of the American Chemical Society.]

PROBLEMS

1. Write autoprotolysis expressions for each of the following solvents:
   *(a) H_2O.   (d) CH_3OH.
   (b) C_2H_5OH.   *(e) $HCOOH$.
   *(c) H_2SO_4.   (f) $NH_2C_2H_4NH_2$.

2. Calculate the pH of each of the solutions in Problem 1 where pH $= -\log[H_3O^+]$, $-\log[C_2H_5OH_2^+]$, and so forth.

*3. (a) Derive a curve for the titration of 50.0 ml of 0.0500-N $HClO_4$ with 0.100-N C_2H_5ONa when both reagents are dissolved in anhydrous ethanol. Assume that both the acid and the base are completely dissociated in this solvent. Calculate the pH ($-\log[C_2H_5OH_2^+]$) after the addition of 0.00, 12.5, 24.0, 24.9, 25.0, 25.1, 26.0, and 30.0 ml of the base.
   (b) Compare the change in pH from 49.9 to 50.1 ml in (a) with the pH change for the same range if water was the solvent and the base was NaOH.

4. Perform the calculations in Problem 3 for a titration in anhydrous methanol with $NaOCH_3$ as the base.

*5. Calculate the pH of each of the following solutions in water and in ethanol (pH $= -\log[C_2H_5OH_2^+]$). The dissociation constant for acetic acid in anhydrous ethanol is 5.6×10^{-11}.

    (a) a 0.0500-F solution of acetic acid

    (b) a solution that is 0.0500 F in acetic acid and 0.0500 F in sodium acetate

    (c) a 0.0500-F solution of sodium acetate

6. Calculate the pH of each of the following solutions in water and in methanol (pH $= -\log [CH_3OH_2^+]$). The dissociation constant for acetic acid is 3.0×10^{-10} in anhydrous CH_3OH.

    (a) 0.200-F acetic acid

    (b) 0.200-F acetic acid and 0.100-F sodium acetate

    (c) 0.200-F sodium acetate

*7. Calculate the pH of each of the following solutions in water and in anhydrous ethanol (pH $= -\log [C_2H_5OH_2^+]$); use 4.0×10^{-14} for the basic dissociation constant of aniline, $C_6H_5NH_2$, in ethanol.

    (a) 0.0100-F $C_6H_5NH_2$

    (b) 0.0100-F $C_6H_5NH_3^+$ and 0.0200-F $C_6H_5NH_2$

    (c) 0.0100-F $C_6H_5NH_3^+$

8. A 0.100-N solution of sodium ethoxide (C_2H_5ONa) in anhydrous ethanol was employed to titrate 50.0 ml of 0.100-F ethanolic solution of acetic acid.

    (a) Derive a curve for this titration, calculating the pH ($-\log [C_2H_5OH_2^+]$) after the addition of: 0.0, 10.0, 25.0, 40.0, 49.0, 49.9, 50.0, 50.1, 51.0, and 60.0 ml of base. (See Problem 5 for the necessary equilibrium constants.)

    (b) Compare the data in (a) with the analogous data for the titration of 0.100-N aqueous acetic acid with 0.100-N aqueous NaOH.

*9. A 0.100-N solution of sodium ethoxide (C_2H_5ONa) in anhydrous ethanol was employed to titrate 50.0 ml of 0.100-F ethanolic solution of anilinium chloride ($C_6H_5NH_3^+Cl^-$).

    (a) Derive a curve for this titration, calculating the pH ($-\log [C_2H_5OH_2^+]$) after the addition of: 0.0, 10.0, 25.0, 40.0, 49.0, 49.9, 50.0, 50.1, 51.0, and 60.0 ml of base. (See Problem 7 for the necessary equilibrium constants.)

    (b) Compare the data in (a) with the analogous data for the titration of 0.100-N aqueous anilinium chloride with 0.100-N aqueous NaOH.

COMPLEX-FORMATION TITRATIONS

Many metal ions react with electron-pair donors to form coordination compounds or complex ions. The donor species, or *ligand*, must have at least one pair of unshared electrons available for bond formation. The water molecule, ammonia, and halide ions are common ligands.

While exceptions exist, a specific cation ordinarily forms coordination bonds to a maximum of two, four, or six, its *coordination number* being a statement of this maximum. The species formed as a result of coordination can be electrically positive, neutral, or negative. Thus, for example, copper(II), with a coordination number of four, forms a cationic amine complex $Cu(NH_3)_4{}^{2+}$, a neutral complex with glycine $Cu(NH_2CH_2COO)_2$, and an anionic complex with chloride ion, $CuCl_4{}^{2-}$.

Titrimetric methods based upon complex formation (sometimes called *complexometric methods*) have been used for at least a century. The truly remarkable growth in their analytical applications is of more recent origin, however, and is based upon a particular class of coordination compounds called *chelates*. A chelate is produced when a metal ion coordinates with two (or more) donor groups of a single ligand. The copper complex of glycine, mentioned earlier, is an example. Here copper is bonded by both the oxygen of the carboxylate group and the nitrogen of the amine group. A chelating agent that has two donor groups available for coordination bonding is called *bidentate*, while one with three groups is called *terdentate*. Quadri- quinque-, and sexadentate chelating agents are known.

From the standpoint of volumetric analysis, the great virtue of a reagent that yields a chelate over one that simply forms a complex with the metal ion lies in the fact that chelation is essentially a single-step process, whereas complex formation may involve the production of one or more intermediate species. Consider, for example, the equilibrium that exists between the metal ion M, with a coordination number of four, and the quadridentate ligand D.[1]

$$M + D \rightleftarrows MD$$

The equilibrium expression for this process is

$$K_f = \frac{[MD]}{[M][D]}$$

where K_f is the *formation constant*.

Similarly, the equilibrium between M and the bidentate ligand B can be represented by

$$M + 2B \rightleftarrows MB_2$$

This equation, however, as well as the formation constant referring to it, is the summation of a two-step process that involves formation of the intermediate MB

$$M + B \rightleftarrows MB$$
$$MB + B \rightleftarrows MB_2$$

for which

$$K_1 = \frac{[MB]}{[M][B]} \quad \text{and} \quad K_2 = \frac{[MB_2]}{[MB][B]}$$

The product of K_1 and K_2 yields the equilibrium constant for the overall process[2]

$$\beta_2 = K_1 K_2 = \frac{[MB]}{[M][B]} \times \frac{[MB_2]}{[MB][B]} = \frac{[MB_2]}{[M][B]^2}$$

In a like manner, the reaction between M and the unidentate ligand A involves the overall equilibrium

$$M + 4A \rightleftarrows MA_4$$

and the equilibrium constant β_4 for the formation of MA_4 from M and A is numerically equal to the product of the equilibrium constants for the four constituent processes.

Each of the titration curves depicted in Figure 13-1 is based upon a reaction that has an overall equilibrium constant of 10^{20}. Curve *A* is derived for the formation of MD in a single step. Curve *B* involves the formation of MB_2 in a two-step process for which K_1 is 10^{12} and K_2 is 10^8. Curve *C* represents forma-

---

[1] The electrostatic charges associated with M and D determine the charge of the product but are not important in terms of the present argument.

[2] It is customary to use the symbol β_i to indicate an overall formation constant. Thus, for example, $\beta_2 = K_1 K_2$, $\beta_3 = K_1 K_2 K_3$, $\beta_4 = K_1 K_2 K_3 K_4$, and so forth.

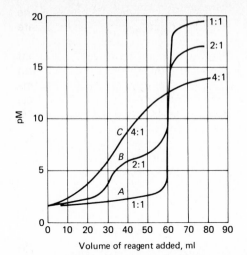

Figure 13-1 Curves for Complex-Formation Titrations. Titration of 60.0 ml of 0.020-F M with (curve A) 0.020-F solution of the quadridentate ligand D to give MD as product, (curve B) 0.040-F solution of the bidentate ligand B to give MB_2, and (curve C) 0.080-F solution of the unidentate ligand A to give MA_4. The overall formation constant for each product is 1.0×10^{20}.

tion of MA_4 for which the equilibrium constants for the four individual steps are 10^8, 10^6, 10^4, and 10^2, respectively. These curves demonstrate the clear superiority of a ligand that combines with a metal ion in a 1:1 ratio because the change in pM in the equivalence-point region is largest with such a system. Thus, polydentate ligands, which generally combine in lower ratios with metal ions, are ordinarily superior reagents for complex-formation titrations.

Regardless of reaction type, the error associated with a titration decreases with increasing completeness of the reaction upon which the titration is based. In this respect, polydentate ligands usually offer a distinct advantage over unidentate titrants; formation constants associated with reactions between metal ions and the former tend to be larger than those involving the latter.

Titrations with Inorganic Complexing Reagents

Complexometric titrations are among the oldest of volumetric methods.[3] For example, the titration of iodide ion with mercury(II)

$$Hg^{2+} + 4I^- \rightleftarrows HgI_4^{2-}$$

was first reported in 1834; the determination of cyanide based upon formation of the dicyanoargentate(I) ion, $Ag(CN)_2^-$, was described by Liebig in 1851. Table 13-1 lists typical nonchelating complexing reagents as well as some of the uses to which they have been put.

[3] For further information, see I. M. Kolthoff and V. A. Stenger, *Volumetric Analysis*, vol. 2, pp. 282, 331. New York: Interscience Publishers, Inc., 1947.

TABLE 13-1 Typical Inorganic Complex-Formation Titrations[a]

Titrant	Analyte	Remarks
$Hg(NO_3)_2$	Br^-, Cl^-, SCN^-, CN^-, thiourea	Products are neutral mercury(II) complexes; various indicators used
$AgNO_3$	CN^-	Product is $Ag(CN)_2^-$; indicator is I^-; titrate to first turbidity of AgI
$NiSO_4$	CN^-	Product is $Ni(CN)_4^{2-}$; indicator is AgI; titrate to first turbidity of AgI
KCN	Cu^{2+}, Hg^{2+}, Ni^{2+}	Products $Cu(CN)_4^{2-}$, $Hg(CN)_2$, $Ni(CN)_4^{2-}$; various indicators used

[a] For further applications and selected references, see L. Meites, *Handbook of Analytical Chemistry*, p. 3-226. New York: McGraw-Hill Book Company, Inc., 1963.

Titrations with Aminopolycarboxylic Acids

Several tertiary amines that also contain carboxylic acid groups form remarkably stable complexes with many metal ions. Their potential as analytical reagents was first recognized by Schwarzenbach in 1945; since that time, they have been extensively investigated. The revival of interest in volumetric complex-formation methods can be traced to these reagents.[4]

REAGENTS

Ethylenediaminetetraacetic acid (also called ethylenedinitrilotetraacetate and often abbreviated EDTA) has the structure

$$
\begin{array}{ccc}
\text{HOOC—CH}_2 & & \text{CH}_2\text{—COOH} \\
& \diagdown \qquad \diagup & \\
& \text{N—CH}_2\text{—CH}_2\text{—N} & \\
& \diagup \qquad \diagdown & \\
\text{HOOC—CH}_2 & & \text{CH}_2\text{—COOH}
\end{array}
$$

EDTA is the most widely used of the polyaminocarboxylic acids. It is a weak acid for which $pK_1 = 2.0$, $pK_2 = 2.67$, $pK_3 = 6.16$, and $pK_4 = 10.26$. These values indicate that the first two protons are lost much more readily than the remaining two. In addition to the four acidic hydrogens, each nitrogen atom has an unshared pair of electrons; the molecule thus has six potential sites for bonding with a metal ion and may be considered to be a sexadentate ligand.

The abbreviations H_4Y, H_3Y^-, H_2Y^{2-}, HY^{3-}, and Y^{4-} are often employed in referring to EDTA and its ions.

Both the free acid H_4Y and the dihydrate of the disodium salt $Na_2H_2Y \cdot 2H_2O$ are available in reagent quality. The former can serve as a primary standard after drying for several hours at 130 to 145°C. It is then dissolved in the minimum amount of base required for complete solution.

[4] These reagents are the subject of several excellent monographs. See, for example, G. Schwarzenbach and H. Flashka (trans. by H. M. N. H. Irving), *Complexometric Titrations*, 2d ed. London: Methuen & Co., Ltd., 1969; F. J. Welcher, *The Analytical Uses of Ethylenediaminetetraacetic Acid*. Princeton, N.J.: D. Van Nostrand Co., Inc., 1958; A. Ringbom, *Complexation in Analytical Chemistry*. New York: Interscience Publishers, Inc., 1963.

Under normal atmospheric conditions, the dihydrate $Na_2H_2Y \cdot 2H_2O$ contains 0.3% moisture in excess of the stoichiometric amount. For all but the most exacting work, this excess is sufficiently reproducible to permit use of a corrected weight for the salt in the direct preparation of a standard solution. If necessary, the pure dihydrate can be prepared by drying at 80°C in an atmosphere of 50% relative humidity for several days.

Another common reagent is nitrilotriacetic acid (abbreviated NTA), which has the structure

$$HOOC-CH_2 \qquad CH_2-COOH$$
$$N$$
$$CH_2-COOH$$

Aqueous solutions of this quadridentate ligand are prepared from the free acid.

Other related substances have also been investigated but have not been as widely employed for volumetric analysis. We shall confine our discussion to the applications of EDTA.

Composition of EDTA Solutions as a Function of pH. Five EDTA-containing species can exist in an aqueous solution; the relative amount of each species in a given solution is dependent upon the pH. A convenient way of showing this relationship involves plotting a-values (see p. 233) for the various species as a function of pH, where

$$a_0 = \frac{[H_4Y]}{C_T}$$

$$a_1 = \frac{[H_3Y^-]}{C_T}$$

and so forth. It will be recalled that C_T is the sum of the equilibrium concentrations of all species. Thus, a-values give the mole fractions of any particular form. Figure 13-2 depicts the manner in which the several a-values for EDTA depend upon pH. It is apparent that H_2Y^{2-} is the predominant species in moderately acid media (pH 3 to 6). In the range between pH 6 and 10, HY^{3-} is the major constituent. Only at pH values greater than 10 does Y^{4-} become a major component of the solution.

These relationships strongly influence the equilibria that exist in solutions containing EDTA and various cations.

COMPLEXES OF EDTA AND METAL IONS

A particularly valuable property of EDTA as a titrant is that it combines with metal ions in a 1:1 ratio regardless of the charge on the cation. In moderately acid solution, these reactions can be formulated as[5]

$$M^{2+} + H_2Y^{2-} \rightleftarrows MY^{2-} + 2H^+$$
$$M^{3+} + H_2Y^{2-} \rightleftarrows MY^- + 2H^+$$
$$M^{4+} + H_2Y^{2-} \rightleftarrows MY + 2H^+$$

[5] In this chapter and those that follow, we shall revert to the use of H^+ as a convenient shorthand notation for H_3O^+; from time to time we shall refer to the species represented by H^+ as the hydrogen ion.

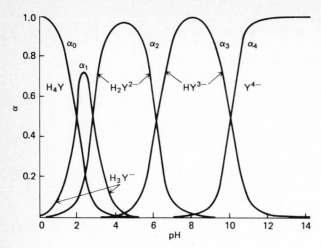

Figure 13-2 Composition of EDTA Solutions as a Function of pH.

Figure 13-2 indicates that reactions performed in neutral to moderately basic solutions are better written as

$$M^{n+} + HY^{3-} \rightleftarrows MY^{(n-4)+} + H^+$$

EDTA is remarkable from the standpoint of the high stability of its metal adducts as well as the ubiquity of its complexes. All cations react at least to some extent; with the exception of the alkali metals, the complexes formed are sufficiently stable to form the basis for volumetric analysis. This great stability undoubtedly results from the several complexing groups within the molecule that give rise to structures that effectively surround and isolate the cation. One form of the complex is depicted in Figure 13-3. Note that all six ligand groups in the EDTA are involved in bonding the divalent metal ion.

Table 13-2 lists formation constants K_{MY} for common EDTA complexes.

TABLE 13-2   Formation Constants for EDTA Complexes[a]

Cation	K_{MY}	$\log K_{MY}$	Cation	K_{MY}	$\log K_{MY}$
Ag^+	2.1×10^7	7.32	Cu^{2+}	6.3×10^{18}	18.80
Mg^{2+}	4.9×10^8	8.69	Zn^{2+}	3.2×10^{16}	16.50
Ca^{2+}	5.0×10^{10}	10.70	Cd^{2+}	2.9×10^{16}	16.46
Sr^{2+}	4.3×10^8	8.63	Hg^{2+}	6.3×10^{21}	21.80
Ba^{2+}	5.8×10^7	7.76	Pb^{2+}	1.1×10^{18}	18.04
Mn^{2+}	6.2×10^{13}	13.79	Al^{3+}	1.3×10^{16}	16.13
Fe^{2+}	2.1×10^{14}	14.33	Fe^{3+}	1.3×10^{25}	25.1
Co^{2+}	2.0×10^{16}	16.31	V^{3+}	7.9×10^{25}	25.9
Ni^{2+}	4.2×10^{18}	18.62	Th^{4+}	1.6×10^{23}	23.2

[a] Data from G. Schwarzenbach, *Complexometric Titrations*, p. 8. New York: Interscience Publishers, Inc., 1957. With permission. (Constants valid at 20°C and ionic strength of 0.1.)

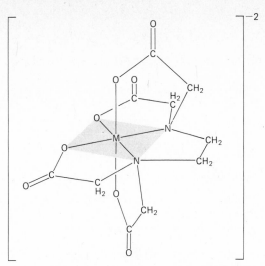

Figure 13-3 Structure of a Metal-EDTA Chelate.

Note that the constant refers to the equilibrium involving the species Y^{4-} with the metal ion. That is,

$$M^{n+} + Y^{4-} \rightleftarrows MY^{(n-4)+} \qquad K_{MY} = \frac{[MY^{(n-4)+}]}{[M^{n+}][Y^{4-}]} \qquad (13\text{-}1)$$

DERIVATION OF EDTA TITRATION CURVES

The general approach to derivation of a titration curve for a metal-ion EDTA reaction does not differ fundamentally from that used for precipitation or neutralization titrations. Here, however, it is ordinarily necessary to account for more than one equilibrium; as a result, the computational detail is somewhat greater than in the earlier examples.

Effect of pH. Inspection of metal-ion EDTA equilibria reveals that the extent of complex formation will depend upon the pH of the environment. Thus, an alkaline medium is needed for titrations involving cations, such as calcium and magnesium, which form weak complexes. On the other hand, titrations involving cations that form more stable complexes, such as zinc and nickel, can be performed successfully in moderately acidic solutions.

Because of this pH dependence, EDTA titrations are generally carried out in solutions that are buffered to a constant and predetermined pH. The restriction of a constant pH permits a considerable simplification in computation.

In the derivation of EDTA titration curves for buffered solutions, it is convenient to employ the term a_4, defined as

$$a_4 = \frac{[Y^{4-}]}{C_T} \qquad (13\text{-}2)$$

In this expression, C_T is the total concentration of *uncomplexed* EDTA. That is,

$$C_T = [Y^{4-}] + [HY^{3-}] + [H_2Y^{2-}] + [H_3Y^-] + [H_4Y]$$

Thus, a_4 represents the fraction of the total uncomplexed reagent that exists as Y^{4-}.

A derivation similar to that shown on page 234 reveals that a_4 depends only upon the pH and the four acid dissociation constants of EDTA, K_1, K_2, K_3, and K_4. Thus,

$$a_4 = \frac{K_1K_2K_3K_4}{[H^+]^4 + K_1[H^+]^3 + K_1K_2[H^+]^2 + K_1K_2K_3[H^+] + K_1K_2K_3K_4} \quad (13\text{-}3)$$

Table 13-3 lists values of a_4 for selected pH levels; these were calculated

TABLE 13-3   Values for a_4 for EDTA in Solutions of Various pH

pH	a_4	pH	a_4
2.0	3.7×10^{-14}	7.0	4.8×10^{-4}
3.0	2.5×10^{-11}	8.0	5.4×10^{-3}
4.0	3.6×10^{-9}	9.0	5.2×10^{-2}
5.0	3.5×10^{-7}	10.0	3.5×10^{-1}
6.0	2.2×10^{-5}	11.0	8.5×10^{-1}
		12.0	9.8×10^{-1}

with Equation 13-3. Substitution of a_4C_T for $[Y^{4-}]$ (Equation 13-2) into the formation-constant expression (Equation 13-1) and rearrangement yields

$$K'_{MY} = a_4K_{MY} = \frac{[MY^{(n-4)+}]}{[M^{n+}]C_T} \quad (13\text{-}4)$$

where K'_{MY} is a *conditional* or *effective* formation constant that describes equilibrium conditions *only at the pH for which a_4 is applicable.*

Conditional constants provide a simple means by which the equilibrium concentrations of metal ion and complex can be calculated at any point in a titration curve. Note that the expression for the conditional constant differs from that of the formation constant only in that the term C_T replaces the equilibrium concentration of the completely dissociated anion, $[Y^{4-}]$. This difference is significant, however, because C_T is more readily determined from the stoichiometry of the reaction than is $[Y^{4-}]$.

Example.   Derive a curve for the titration of 50.0 ml of 0.0100-*F* Ca^{2+} with 0.0100-*F* EDTA in a solution buffered to a constant pH of 10.0.

Calculation of Conditional Constant.   The conditional formation constant for the calcium-EDTA complex at pH 10 can be obtained from the formation constant of the complex (Table 13-2) and the a_4-value for EDTA at pH 10 (Table 13-3). Thus,

$$K'_{CaY} = a_4K_{CaY} = 0.35 \times 5.0 \times 10^{10}$$
$$= 1.75 \times 10^{10}$$

Preequivalence-Point Values for pCa. Before the equivalence point has been reached, the molar concentration of Ca^{2+} will be equal to the sum of the contributions from the untitrated excess of the ion and that which results from dissociation of the complex; the latter quantity will be equal to C_T. It is ordinarily reasonable to assume that C_T is small with respect to the formal concentration of uncomplexed calcium ion. Thus, for example, after the addition of 25.0 ml of reagent,

$$[Ca^{2+}] = \frac{50.0 \times 0.0100 - 25.0 \times 0.0100}{75.0} + C_T \cong 3.33 \times 10^{-3}$$

and

$$pCa = 2.48$$

Equivalence-Point pCa. Here the solution is $0.00500\ F$ in CaY^{2-}; the only source of Ca^{2+} ions will be from dissociation of this complex. It is evident that the calcium ion concentration must be identical to the sum of the concentrations of the uncomplexed EDTA ions, C_T. Thus,

$$[Ca^{2+}] = C_T$$

$$[CaY^{2-}] = 0.00500 - [Ca^{2+}] \cong 0.00500$$

The conditional formation constant for CaY^{2-} at pH 10.0 is

$$\frac{[CaY^{2-}]}{[Ca^{2+}]C_T} = 1.75 \times 10^{10}$$

Thus, upon substitution,

$$\frac{0.00500}{[Ca^{2+}]^2} = 1.75 \times 10^{10}$$

$$[Ca^{2+}] = 5.35 \times 10^{-7}$$

$$pCa = 6.27$$

Postequivalence-Point pCa. Beyond the equivalence point, formal concentrations of CaY^{2-} and EDTA are directly obtained. For example, after 60.0 ml of reagent have been added

$$F_{CaY^{2-}} = \frac{50.0 \times 0.0100}{110} = 4.55 \times 10^{-3}$$

$$F_{EDTA} = \frac{10.0 \times 0.0100}{110} = 9.1 \times 10^{-4}$$

As an approximation, we may write

$$[CaY^{2-}] = 4.55 \times 10^{-3} - [Ca^{2+}] \cong 4.55 \times 10^{-3}$$

$$C_T = 9.1 \times 10^{-4} + [Ca^{2+}] \cong 9.1 \times 10^{-4}$$

and

$$\frac{4.55 \times 10^{-3}}{[Ca^{2+}] \times 9.1 \times 10^{-4}} = 1.75 \times 10^{10} = K'_{CaY}$$

$$[Ca^{2+}] = 2.86 \times 10^{-10}$$

$$pCa = 9.54$$

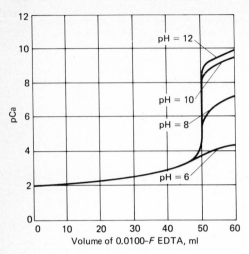

Figure 13-4 Influence of pH upon the Titration of 0.0100-F Ca²⁺ with 0.0100-F EDTA.

Figure 13-4 shows titration curves for calcium ion in solutions buffered to various pH levels. It is apparent that appreciable changes in pCa can be achieved only if the pH of the solution is maintained at about 8 or greater. As shown in Figure 13-5, however, cations with larger formation constants provide good end points even in acidic solutions. Figure 13-6 shows the minimum permissible pH for satisfactory end points in the titration of various metal ions in the absence of competing complexing agents. Note that many divalent heavy-metal cations can be titrated in a moderately acidic environment and iron(III) in strongly acidic medium.

Effect of Other Complexing Agents on EDTA Titrations. Many EDTA titrations are complicated by the tendency on the part of the ion being titrated to precipitate as a basic oxide or hydroxide at the pH required for a satisfactory

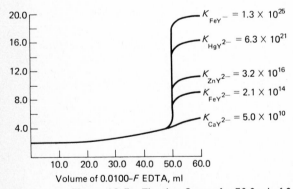

Figure 13-5 Titration Curves for 50.0 ml of 0.0100-F Cation Solutions at pH 6.0.

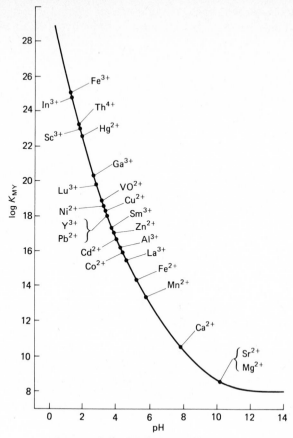

Figure 13-6 Minimum pH Needed for Satisfactory Titration of Various Cations with EDTA. [From C. N. Reilley and R. W. Schmid, *Anal. Chem.*, **30**, 947 (1958). With permission of the American Chemical Society.]

titration. In order to keep the metal ion in solution, particularly in the early stages of the titration, it is necessary to include an auxiliary complexing agent. Thus, for example, the titration of zinc(II) is ordinarily performed in the presence of high concentrations of ammonia and ammonium chloride. These species buffer the solution to an acceptable pH. In addition, the ammonia prevents precipitation of zinc hydroxide by forming ammine complexes. The titration reaction with EDTA is thus best represented as

$$Zn(NH_3)_4{}^{2+} + HY^{3-} \rightleftarrows ZnY^{2-} + 3NH_3 + NH_4{}^+$$

In a system such as this, the completeness of reaction, and thus the quality of the end point, depends not only upon pH but also upon the concentration of ammonia present.

The influence of an auxiliary complexing reagent can be treated in much the same way that account was taken of pH effects on the concentration of EDTA species. A quantity β is defined such that

$$\beta = \frac{[M^{n+}]}{C_M} \qquad (13\text{-}5)$$

The quantity C_M is the sum of the concentrations of species containing the metal ion *exclusive* of that combined with EDTA. For solutions containing zinc(II) and ammonia, then,

$$C_M = [Zn^{2+}] + [Zn(NH_3)^{2+}] + [Zn(NH_3)_2{}^{2+}]$$
$$+ [Zn(NH_3)_3{}^{2+}] + [Zn(NH_3)_4{}^{2+}] \qquad (13\text{-}6)$$

The value of β can be readily expressed in terms of the ammonia concentration and the formation constants for the various complexes. Thus, for example,

$$K_1 = \frac{[Zn(NH_3)^{2+}]}{[Zn^{2+}][NH_3]}$$

or

$$[Zn(NH_3)^{2+}] = K_1[Zn^{2+}][NH_3]$$

Similarly, it is readily shown that

$$[Zn(NH_3)_2{}^{2+}] = K_1 K_2 [Zn^{2+}][NH_3]^2$$

and so forth.

Substitution of these values into Equation 13-6 gives

$$C_M = [Zn^{2+}](1 + K_1[NH_3] + K_1 K_2[NH_3]^2$$
$$+ K_1 K_2 K_3[NH_3]^3 + K_1 K_2 K_3 K_4[NH_3]^4)$$

Combining this expression with Equation 13-5 (here $[M^{n+}] = [Zn^{2+}]$) yields

$$\beta = \frac{1}{1 + K_1[NH_3] + K_1 K_2[NH_3]^2 + K_1 K_2 K_3[NH_3]^3 + K_1 K_2 K_3 K_4[NH_3]^4}$$
$$(13\text{-}7)$$

Finally, a conditional constant for the equilibrium between EDTA and zinc(II) in an ammonia–ammonium chloride buffer is obtained by substituting Equation 13-5 into Equation 13-4 and rearranging

$$K''_{ZnY} = a_4 \beta K_{ZnY} = \frac{[ZnY^{2-}]}{C_M C_T} \qquad (13\text{-}8)$$

where K''_{ZnY} is a new conditional constant that applies at a single pH and a single concentration of ammonia. The following example will show how this conditional constant is employed for derivation of titration curves.

Example. Calculate the pZn of solutions prepared by mixing 40.0, 50.0, and 60.0 ml of 0.00100-F EDTA with 50.0 ml of 0.00100-F Zn^{2+}. Assume that both Zn^{2+} and the EDTA solutions are 0.100 F in NH_3 and 0.176 F in NH_4Cl to provide a constant pH of 9.0. Formation constants for the Zn-NH_3 complexes are: $K_1 = 1.9 \times 10^2$, $K_1 K_2 = 4.2 \times 10^4$, $K_1 K_2 K_3 = 1.04 \times 10^7$, and $K_1 K_2 K_3 K_4 = 1.14 \times 10^9$.

1. *Calculation of a Conditional Constant.* A value for β can be obtained by assuming that $[NH_3] = F_{NH_3}$; the stepwise formation constants K_1, K_2, K_3, and K_4 as well as F_{NH_3} are then substituted into Equation 13-7 to give

$$\beta = \frac{1}{1 + 19 + 420 + 1.04 \times 10^4 + 1.14 \times 10^5} = 8.0 \times 10^{-6}$$

A value for K_{ZnY} is found in Table 13-2, and α_4 for pH 9.0 is given in Table 13-3. Substituting into Equation 13-8, we find

$$K''_{ZnY} = 5.2 \times 10^{-2} \times 8.0 \times 10^{-6} \times 3.2 \times 10^{16} = 1.33 \times 10^{10}$$

2. *Calculation of pZn after Addition of 40.0 ml of EDTA.* The concentration of unreacted Zn^{2+} at this point is given to an excellent approximation by

$$C_M \cong \frac{50.0 \times 0.00100 - 40.0 \times 0.00100}{90.0} = 1.11 \times 10^{-4}$$

Here we assume that there is negligible dissociation of the several complexes. The quantity C_M, which is the sum of the equilibrium concentrations of all the Zn species not complexed by EDTA, can be employed in Equation 13-5 to give $[Zn^{2+}]$. Thus,

$$[Zn^{2+}] = C_M\beta = (1.11 \times 10^{-4})(8.0 \times 10^{-6})$$
$$= 8.9 \times 10^{-10}$$
$$pZn = 9.05$$

3. *Calculation of pZn after Addition of 50.0 ml of EDTA.* At the equivalence point for the titration, the ZnY^{2-} concentration is 5.00×10^{-4} F. The sum of the concentrations of the various zinc species not combined with EDTA will equal the sum of the concentrations of the uncomplexed EDTA species. That is,

$$C_M = C_T$$

and

$$[ZnY^{2-}] = 5.00 \times 10^{-4} - C_M \cong 5.00 \times 10^{-4}$$

Substituting into Equation 13-8,

$$K''_{ZnY} = 1.33 \times 10^{10} = \frac{5.00 \times 10^{-4}}{C_M{}^2}$$
$$C_M = 1.94 \times 10^{-7}$$

Employing Equation 13-5, we obtain

$$[Zn^{2+}] = C_M\beta = (1.94 \times 10^{-7})(8.0 \times 10^{-6})$$
$$= 1.55 \times 10^{-12}$$
$$pZn = 11.81$$

4. *Calculation of pZn after Addition of 60.0 ml of EDTA.* The solution now contains an excess of EDTA; thus,

$$F_{EDTA} = C_T = \frac{60.0 \times 0.00100 - 50.0 \times 0.00100}{110} = 9.1 \times 10^{-5}$$

and

$$[ZnY^{2-}] = \frac{50.0 \times 0.00100}{110} = 4.55 \times 10^{-4}$$

Substituting into Equation 13-8,

$$C_M = \frac{4.55 \times 10^{-4}}{(9.1 \times 10^{-5})(1.33 \times 10^{10})} = 3.76 \times 10^{-10}$$

and

$$[Zn^{2+}] = C_M\beta = (3.76 \times 10^{-10})(8.0 \times 10^{-6}) = 3.01 \times 10^{-15}$$
$$pZn = 14.52$$

Figure 13-7 shows two theoretical curves for the titration of zinc(II) with EDTA at a pH of 9.00. The equilibrium concentration of ammonia was 0.100 F for the one titration and 0.0100 F for the other. It is seen that the presence of ammonia has the effect of decreasing the change in pZn in the equivalence-point region. Thus, it is desirable to keep the concentration of any auxiliary reagent to the minimum required to prevent hydroxide formation. Note that the magnitude of β does not affect pZn beyond the equivalence point. On the other hand, it will be recalled (Figure 13-4) that α_4, and thus pH, plays an important role in defining this region of the titration curve.

END POINTS FOR EDTA TITRATIONS

Metal-Ion Indicators. A number of metal-ion indicators have been developed for use in complexometric titrations with EDTA and other chelating reagents. In general, these indicators are organic dyes that form colored chelates with metal ions in a pM range that is characteristic of the particular cation and dye. The complexes are often intensely colored, being discernible to the eye in the range of 10^{-6} to 10^{-7} M.

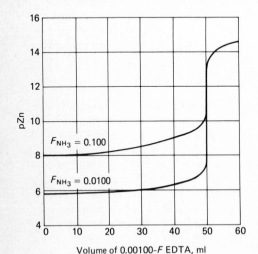

Volume of 0.00100-F EDTA, ml

Figure 13-7 Influence of Ammonia Concentration upon the Titration of 0.0010-F Zn^{2+} with 0.00100-F EDTA. Solutions are buffered to a pH of 9.0.

Most metal-ion indicators will also bond with protons to give species that impart to solutions colors that resemble those of the metal complexes. Thus, these dyes function as acid-base indicators as well and are useful as indicators for metal ions only in pH ranges in which competition with the proton is negligible.

These properties are exhibited by *Eriochrome black T*, a widely used metal-ion indicator. In acidic and moderately basic solutions, the predominant acid-base equilibrium exhibited by the indicator is as follows:

or

$$H_2In^- \rightleftharpoons HIn^{2-} + H^+$$
red   blue

Each of the two species, H_2In^- and HIn^{2-}, exists in several resonance forms. At very high pH levels, HIn^{2-} further dissociates to In^{3-}, which is orange.

The metal complexes of Eriochrome black T are generally red. To observe a color change with this indicator, then, it is necessary to adjust the pH to 7 or above so that the blue form of the indicator HIn^{2-} predominates. The end-point reaction is then

$$MIn^- + HY^{3-} \rightleftharpoons HIn^{2-} + MY^{2-}$$
red   blue

Eriochrome black T forms red complexes with more than two dozen different metal ions, but with only certain of these cations are the stabilities of the products appropriate for end-point detection. To be suitable for a titration with EDTA, the formation constant of the metal-indicator complex should be less than one-tenth that of the metal-EDTA complex; otherwise, a premature end point will be observed. On the other hand, if this ratio becomes too small, as is the situation with calcium ion, late end points are observed. The applicability of a given indicator to an EDTA titration can be determined from the change in pM in the equivalence-point region provided the formation constant for the metal-indicator complex is known.[6]

[6] For a discussion of the principles of indicator choice in complex-formation titrations, see C. N. Reilley and R. W. Schmid, *Anal. Chem.*, **31**, 887 (1959).

A limitation of Eriochrome black T is that its solutions decompose slowly with standing. It is claimed that solutions of calmagite, an indicator that for all practical purposes is identical in behavior, do not suffer this disadvantage. The structure of Calmagite is similar to Eriochrome black T.

Many other metal indicators have been developed for EDTA titrations.[7] Most behave in an analogous fashion. In contrast to Eriochrome black T, some can be employed in strongly acidic media.

TITRATION METHODS EMPLOYING EDTA

Several procedures are employed in the application of EDTA to volumetric analysis. The most common of these are considered in the following paragraphs.

Direct Titration Procedure. Welcher[7] lists 25 metal ions that can be determined by direct titration with EDTA using metal-ion indicators for end-point detection. Direct titration procedures are limited to those reactions for which a method for end-point detection exists and to those metal ions that react rapidly with EDTA. When direct methods fail, it is often possible to achieve an analysis by a back-titration or a displacement titration.

Back-Titration. Back-titration procedures are useful for the analysis of cations that form very stable EDTA complexes and for which a satisfactory indicator is not available. In such an analysis, the excess EDTA is determined by back-titration with a standard magnesium solution, using Eriochrome black T or Calmagite as the indicator. The metal-EDTA complex must be more stable than the magnesium-EDTA complex; otherwise, the back-titrant would displace the metal ion.

This technique is also useful for the titration of metals in the presence of anions that would otherwise form slightly soluble precipitates with the cation

[7] F. J. Welcher, *The Analytical Use of Ethylenediaminetetraacetic Acid*, chapter 3. New York: D. Van Nostrand Co., Inc.. 1958.

8. A 50.0-ml aliquot of a solution containing iron(II) and iron(III) required 13.7 ml of 0.0120-F EDTA when titrated at a pH of 2.0 and 29.6 ml when titrated at a pH of 6.0. Express the concentration of the solution in terms of the parts per million of each solute.

*9. A 24-hr urine specimen was diluted to exactly 2.00 liters. After being buffered to pH 10, a 10.0-ml aliquot was titrated with 26.8 ml of 0.00347-F EDTA. The calcium in a second 10.0-ml aliquot was isolated as $CaC_2O_4(s)$, redissolved in acid, and titrated with 11.6 ml of the EDTA solution.

  Assuming that 15–300 mg of magnesium and 50–400 mg of calcium are normal, did this specimen fall within these ranges?

10. An EDTA solution was prepared by dissolving approximately 4 g of the disodium salt in approximately 1 liter of water. An average of 42.3 ml of this solution was required to titrate 50.0-ml aliquots of a standard that contained 0.768 g of $MgCO_3$ per liter.

  Titration of a 25.0-ml mineral water sample at pH 10 required 18.8 ml of the EDTA solution.

  A 50.0-ml aliquot of the mineral water was rendered strongly alkaline to precipitate the magnesium as $Mg(OH)_2$. Titration with a calcium specific indicator required 31.5 ml of the EDTA solution. Calculate

  (a) the titer of the EDTA solution in terms of mg $MgCO_3$/ml as well as mg $CaCO_3$/ml.

  (b) the ppm of $CaCO_3$ in the mineral water sample.

  (c) the ppm of $MgCO_3$ in the water.

*11. The sulfate in a 1.515-g sample was homogeneously precipitated by adding 50.0 ml of 0.0264-F Ba-EDTA solution and slowly increasing the acid concentration to liberate Ba^{2+}. When precipitation was complete, the solution was buffered to pH 10 and diluted to exactly 250 ml. A 25.0-ml aliquot of the supernatant required a 28.7-ml titration with standard 0.0154-F Mg^{2+} solution. Express the results of this analysis in terms of percent $Na_2SO_4 \cdot 10H_2O$.

12. Calamine, which is used for relief of skin irritations, is a mixture of zinc and iron oxides. A 1.022-g specimen of dried calamine was dissolved in acid and diluted to exactly 250 ml. Potassium fluoride was added to a 10.0-ml aliquot of the diluted solution to mask the iron; after suitable adjustment of the pH, Zn^{2+} consumed 38.7 ml of 0.0129-F EDTA. A second 50.0-ml aliquot was suitably buffered and titrated with 2.40 ml of 0.00272-F ZnY^{2-} solution. Reaction:

$$Fe^{3+} + ZnY^{2-} = FeY^- + Zn^{2+}$$

Calculate the respective percentages of ZnO and Fe_2O_3 in the sample.

*13. A 3.65-g sample containing bromate and bromide was dissolved in sufficient water to give exactly 250 ml. After acidification, silver nitrate was introduced to a 25.0-ml aliquot; the AgBr produced was filtered, washed, and then redissolved in an ammoniacal solution of potassium tetracyanonickelate(II). Reaction:

$$Ni(CN)_4{}^{2-} + 2AgBr(s) = 2Ag(CN)_2{}^- + Ni^{2+} + 2Br^-$$

The liberated nickel ion required 26.7 ml of 0.0208-F EDTA.

  The bromate in a 10.00-ml aliquot was reduced to bromide with arsenic(III) prior to the addition of silver nitrate; 21.9 ml of the EDTA solution were needed to titrate the nickel ion that was subsequently released. Calculate the respective percentages of NaBr and $NaBrO_3$ in the sample.

14. The chromium ($d = 7.10$ g/cm^3) plated on a 9.75 cm^2 surface was dissolved with hydrochloric acid; the solution was then diluted to exactly 100 ml. A 25.0-ml aliquot was buffered to pH 5 and 50.00 ml of 0.00862-F EDTA were added. Titration of the excess chelating reagent required 7.36 ml of 0.01044-F Zn^{2+}. Calculate the average thickness of the chromium plating.

*15. The silver ion in a 25.0-ml sample was converted to dicyanoargentate(I) ion by the addition of 30.0 ml 0.0800-F Ni(CN)$_4^{2-}$. Reaction:

$$Ni(CN)_4^{2-} + 2Ag^+ = 2Ag(CN)_2^- + Ni^{2+}$$

The liberated nickel ion was subsequently titrated with 43.7 ml of 0.0240-F EDTA. Calculate the concentration of the silver solution.

16. The potassium ion in a 250-ml mineral water sample was precipitated with sodium tetraphenylboron

$$K^+ + (C_6H_5)_4B^- \rightarrow KB(C_6H_5)_4(s)$$

The precipitate was filtered, washed, and then redissolved in an organic solvent. An excess of the mercury(II)-EDTA chelate was added

$$4HgY^{2-} + (C_6H_5)_4B^- + 4H_2O = H_3BO_3 + 4C_6H_5Hg^+ + 4HY^{3-} + OH^-$$

The liberated EDTA was titrated with 29.64 ml of 0.0558-F Mg^{2+}. Calculate the potassium ion concentration in parts per million.

*17. A 0.408-g sample containing lead, magnesium, and zinc was dissolved and treated with cyanide to complex the zinc

$$Zn^{2+} + 4CN^- = Zn(CN)_4^{2-}$$

Titration of the lead and magnesium required 42.2 ml of 0.0206-F EDTA. The lead was next masked with BAL (2,3-dimercaptopropanol) and the released EDTA was titrated with 19.3 ml of 0.00765-F magnesium solution. Finally, formaldehyde was introduced to demask the zinc

$$Zn(CN)_4^{2-} + 4HCHO + 4H_2O = Zn^{2+} + 4HOCH_2CN + 4OH^-$$

which was subsequently titrated with 28.6 ml of 0.0206-F EDTA. Calculate the percentages of the three metals in the sample.

18. Chromel is an alloy composed of nickel, iron, and chromium. A 0.647-g sample was dissolved and diluted to exactly 250 ml. When 50.0 ml of 0.0518-F EDTA were mixed with an equal volume of the diluted sample, all three ions were chelated, and a 5.11-ml back-titration with 0.0624-F copper(II) was required. The chromium in a second 50.0-ml aliquot was masked through the addition of hexamethylenetetramine; titration of the iron and nickel required 36.2 ml of 0.0518-F EDTA. Iron and chromium were masked with pyrophosphate in a third 50.0-ml aliquot; 25.9 ml of the EDTA solution were needed to react with the nickel. Calculate the respective percentages of nickel, chromium, and iron in the alloy.

*19. A 0.328-g sample of brass was dissolved in nitric acid. The sparingly soluble metastannic acid was removed by filtration, and the combined filtrate and washings were then diluted to exactly 500 ml.

A 10.00-ml aliquot was suitably buffered; titration of the lead, zinc, and copper in this aliquot required 37.5 ml of 0.00250-F EDTA.

The copper in a 25.00-ml aliquot was masked with thiosulfate; the lead and zinc were then titrated with 27.6 ml of the EDTA solution.

Cyanide ion was used to mask the copper and zinc in a 100-ml aliquot; 10.80 ml of the EDTA solution were needed to titrate the lead ion.

Determine the composition of the brass sample; evaluate the percentage of tin by difference.

*20. Calculate conditional constants for the formation of the EDTA complex of Mn^{2+} at (a) pH 6.0, (b) pH 8.0, and (c) pH 10.0.

21. Calculate conditional constants for the formation of the EDTA complex of Sr^{2+} at (a) pH 7.0, (b) pH 9.0, and (c) pH 11.0.

*22. Formation constants for the successive ammine complexes of cadmium are 320, 91, 20, and 6.2, respectively. Calculate conditional constants for the reaction of Cd^{2+} with Y^{4-} when
(a) the formal NH_3 concentration is 0.050 and the pH is 9.0.
(b) the formal NH_3 concentration is 0.050 and the pH is 11.0.
(c) the formal NH_3 concentration is 0.50 and the pH is 9.0.
(d) the formal NH_3 concentration is 0.50 and the pH is 11.0.

23. Formation constants for the successive ethylenediamine (en) complexes of nickel are 4.6×10^7, 2.5×10^6, and 3.4×10^4. Calculate the conditional constant for the reaction of Ni^{2+} with Y^{4-} when
(a) the concentration of en is 0.025 F and the pH is 9.0.
(b) the concentration of en is 0.025 F and the pH is 11.0.
(c) the concentration of en is 0.25 F and the pH is 9.0.
(d) the concentration of en is 0.25 F and the pH is 11.0.

*24. Derive a titration curve for 50.00 ml of 0.01000-F Sr^{2+} with 0.02000-F EDTA in a solution buffered to pH 11.0. Calculate pSr values after addition of 0.00, 10.00, 24.00, 24.90, 25.00, 25.10, 26.00, and 30.00 ml of titrant.

25. Derive a titration curve for 50.00 ml of 0.0150-F Fe^{2+} with 0.0300-F EDTA in a solution buffered to pH 7.0. Calculate pFe values after addition of 0.00, 10.00, 24.00, 24.90, 25.00, 25.10, 26.00, and 30.00 ml of titrant.

*26. Derive a curve for the titration of 25.0 ml of 0.0400-F Co^{2+} with 0.0500-F Na_2H_2Y in a solution maintained at a pH of 9.00 with an NH_3/NH_4^+ buffer. Assume that the NH_3 concentration is constant at 0.0400 M throughout. Calculate values for pCo after the addition of 0.00, 5.00, 10.00, 18.0, 20.0, 22.0, and 30.0 ml of reagent.

27. Derive a curve for the titration of 25.0 ml of 0.0200-F Ni^{2+} with 0.0100-F Na_2H_2Y in a solution maintained at a pH of 10.0 with an NH_3/NH_4^+ buffer. Consider that the concentration of NH_3 remains 0.100 M throughout. Calculate values for pNi after the addition of 0, 10.0, 25.0, 40.0, 45.0, 50.0, 55.0, and 60.0 ml of reagent.

chapter

14

EQUILIBRIA IN OXIDATION-REDUCTION SYSTEMS

Oxidation-reduction, or *redox*, processes involve the transfer of electrons from one reactant to another. Volumetric methods based upon electron transfer are more numerous and more varied than those for any other reaction type.

Oxidation involves the loss of electrons by a substance and *reduction* the gain of electrons. In any oxidation-reduction reaction, the molar ratio between the substance oxidized and the substance reduced is such that the number of electrons lost by one species is equal to the number gained by the other. This fact must always be taken into account when balancing equations for oxidation-reduction reactions.

Oxidizing agents or *oxidants* possess a strong affinity for electrons and cause other substances to be oxidized by abstracting electrons from them. In the process, the oxidizing agent accepts electrons and is thereby reduced. *Reducing agents* or *reductants* have little affinity for electrons and, in fact, readily give up electrons thereby causing some other species to be reduced. A consequence of this electron transfer is the oxidation of the reducing agent.

Separation of an oxidation-reduction reaction into its component parts (that is, into *half-reactions*) is a convenient way of indicating clearly the species that gains electrons and the one that loses them. Thus, the overall reaction

$$5Fe^{2+} + MnO_4^- + 8H^+ \rightleftarrows 5Fe^{3+} + Mn^{2+} + 4H_2O$$

is obtained by combining the half-reaction for the oxidation of iron(II)

$$5Fe^{2+} \rightleftarrows 5Fe^{3+} + 5e$$

with that for the reduction of permanganate

$$MnO_4^- + 5e + 8H^+ \rightleftarrows Mn^{2+} + 4H_2O$$

Note that it was necessary to multiply the first half-reaction through by 5 to eliminate the electrons from the overall equation.

The rules for balancing half-reactions are the same as those for ordinary reactions; that is, the number of atoms of each element as well as the net charge on both sides of the equation must be equal.

Oxidation-reduction reactions may result from the direct transfer of electrons from the donor to the receiver. Thus, upon immersing a strip of zinc in a copper(II) sulfate solution, copper(II) ions migrate to, and are reduced at, the surface of the zinc

$$Cu^{2+} + 2e \rightleftarrows Cu(s)$$

and a chemically equivalent quantity of zinc is oxidized

$$Zn(s) \rightleftarrows Zn^{2+} + 2e$$

The overall process is given by the sum of these half-reactions. A unique aspect of oxidation-reduction reactions is that the transfer of electrons —and hence the same net reaction —can be accomplished when the donor and acceptor are remote from each other. With the arrangement illustrated in Figure 14-1, direct

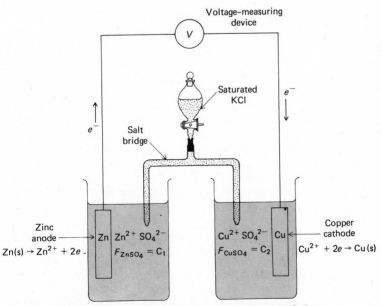

Figure 14-1 Schematic Diagram of a Galvanic Cell.

contact between zinc and copper(II) is prevented by a *salt bridge*, which consists of a saturated potassium chloride solution in a U-tube. Despite the separation of reactants, electrons are transferred from the zinc metal to the copper ions. The external metallic conductor provides the means for this transfer. The oxidation-reduction reaction can be expected to continue until the copper(II) and zinc ion concentrations achieve levels corresponding to equilibrium for the reaction

$$Zn(s) + Cu^{2+} \rightleftarrows Zn^{2+} + Cu(s)$$

When this condition is reached, no further net flow of electrons will occur. *It is important to note that the overall reaction and its position of equilibrium are the same regardless of which of the two ways the process is carried out.*

Fundamentals of Electrochemistry

The device illustrated in Figure 14-1 is, of course, an *electrochemical cell*; it develops an electrical potential from the tendency of the reacting species to transfer electrons and thereby approach the condition of equilibrium. The potential existing between the zinc and the copper electrodes is a measure of driving force of the reaction and is easily evaluated with a suitable voltage-measuring device, *V*, placed in the circuit as shown. We shall see that the voltage produced by an electrochemical cell is directly related to the equilibrium constant for the particular oxidation-reduction process involved as well as the extent to which the existing concentrations of the participants differ from their equilibrium values. Measurement of such potentials, in fact, constitutes an important source of numerical values for equilibrium constants for oxidation-reduction reactions. It is desirable, therefore, to examine more closely the construction and behavior of electrochemical cells as well as the measurement of the potentials they develop.

CELLS

A cell consists of a pair of conductors or electrodes, usually metallic, each of which is immersed in an electrolyte. When the electrodes are arranged as shown in Figure 14-1 and a current passes through the conductor, a chemical oxidation occurs at the surface of one electrode and a reduction at the surface of the other.

A *galvanic*, or *voltaic*, cell is one that provides electrical energy. In contrast, an *electrolytic* cell requires an external source of electrical energy for operation. The cell shown in Figure 14-1 provides a spontaneous flow of electrons from the zinc electrode to the copper electrode through the external circuit; it is thus a galvanic cell.

The same cell could also be operated as an electrolytic cell. Here it would be necessary to introduce a battery or some other electrical source in the external circuit to force electrons to flow in the opposite direction through the cell. Under these circumstances, zinc would deposit, and copper would dissolve. These processes would consume energy from the battery.

Reversible and Irreversible Cells. Frequently, as with the cell in Figure 14-1, altering the direction of current simply causes a reversal of the chemical reactions that occur at the electrodes. Such a cell is said to be electrochemically *reversible*. With other cells, reversing the current flow results in entirely different reactions at one or both electrodes. Such cells are called *irreversible*.

Passage of Current through a Cell. Electrical current is transported by three distinctly different processes in various parts of the cell shown in Figure 14-1. In the electrodes and the external conductor, electrons serve as carriers, moving from the zinc to the copper. Within the solutions, current flow involves migration of both positive and negative ions. Thus, zinc ions, hydronium ions, and other positive species migrate away from the zinc electrode as it is oxidized; likewise, negative ions are attracted toward this electrode by the excess positive ions produced by the electrochemical process. In the salt bridge, the current is carried largely by potassium ions moving in the direction of the copper electrode and chloride ions migrating toward the zinc electrode.

The third type of current-transport process occurs at the two electrode surfaces. Here an oxidation or a reduction reaction provides the mechanism whereby the ionic conduction of the solution is coupled with the electron conduction of the electrodes to give a completed circuit through which current may flow.

ELECTRODE PROCESSES

We have thus far considered a cell as being composed of two *half-cells*, each of which is associated with the process occurring at one of the electrodes. It must be stressed, however, that it is impossible to operate one half-cell independently of a second, nor can the potential of an individual half-cell be measured without reference to another.

Anode and Cathode. In *any* electrochemical cell, the *electrode at which oxidation occurs* is called the *anode*, whereas the *electrode at which reduction takes place* is the *cathode*.

Common cathode reactions are illustrated by the following equations:

$$Ag^+ + e \rightarrow Ag(s)$$
$$2H^+ + 2e \rightarrow H_2(g)$$
$$Fe^{3+} + e \rightarrow Fe^{2+}$$
$$NO_3^- + 10H^+ + 8e \rightarrow NH_4^+ + 3H_2O$$

All of these reactions can occur at the surface of an inert cathode such as platinum. Hydrogen is frequently evolved from solutions that contain no other easily reducible species.

Typical anodic reactions include:

$$Cd(s) \rightarrow Cd^{2+} + 2e$$
$$2Cl^- \rightarrow Cl_2(g) + 2e$$
$$Fe^{2+} \rightarrow Fe^{3+} + e$$
$$2H_2O \rightarrow O_2(g) + 4H^+ + 4e$$

The last is often observed when a solution contains no easily oxidized species.

Electrode Signs. Electrodes are frequently provided with positive and negative signs to indicate the direction of electron flow during the operation of a cell. This practice leads to ambiguity, particularly when attempts are made to relate these signs to anodic or cathodic processes. In a galvanic cell, the negative electrode is the one from which electrons flow to the external circuit (see Figure 14-1). In an electrolytic cell, the negative electrode is the one into which electrons are forced from an external source. Thus, in a cell used for removal of zinc by electrolysis, the negative electrode is the one upon which metallic zinc is deposited, that is, the *cathode*. Clearly, the sign assumed by, say, the anode depends upon whether we are considering an electrolytic or a galvanic cell. Irrespective of sign, however, the anode is the electrode at which oxidation occurs, and the cathode is the site for reduction. Use of these terms, rather than positive and negative signs, in describing electrode function is unambiguous and much preferred.

ELECTRODE POTENTIAL

Turning again to the galvanic cell shown in Figure 14-1, the driving force for the reaction

$$Zn(s) + Cu^{2+} \rightleftarrows Cu(s) + Zn^{2+}$$

manifests itself as a measurable electrical force or voltage between the two electrodes. This force is the sum of two potentials called *half-cell potentials* or single electrode potentials; one of these is associated with the half-reaction occurring at the anode and the other with the half-reaction taking place at the cathode.

It is not difficult to see how we might obtain information regarding the *relative* magnitudes of half-cell potentials. For example, if we replaced the left-hand side of the cell in Figure 14-1 with a cadmium electrode immersed in a solution of cadmium sulfate, the voltmeter reading would be approximately 0.4 V less than that of the original cell. Since the copper cathode remains unchanged, we logically ascribe this voltage decrease to the different anode reaction and thus conclude that the half-cell potential for the oxidation of cadmium is about 0.4 V less than that for the oxidation of zinc. Substitutions in the left-hand side of the cell would enable us to compare the chemical driving forces of other half-reactions to that of copper by means of similar electrical measurements.

Such a technique does not provide absolute values for half-cell potentials. Indeed, there is no way to determine such quantities, since all voltage-measuring devices determine only *differences* in potential. In order to measure a potential difference, one lead of such a device is connected to the electrode in question; the second lead must then be brought in contact with the solution via another conductor. This latter contact, however, inevitably involves a solid-solution interface and hence acts as a second half-cell at which a chemical reaction *must also take place if the current flow required for the potential measurement is to occur*. Thus, we do not obtain an absolute measurement of the desired half-cell potential but rather a sum consisting of the potential of interest and the half-cell potential for the contact between the voltage-measuring device and the solution.

The inability to measure absolute potentials for half-cell processes is not a serious handicap because relative half-cell potentials are just as useful. Relative potentials can be combined to give cell potentials; in addition, they are useful for the calculation of equilibrium constants.

The Standard Hydrogen Electrode. To obtain consistent relative half-cell potential data, it is necessary to compare all electrode potentials to a common reference. This reference electrode should be relatively easy to construct, exhibit reversible behavior, and give constant and reproducible potentials for a given set of experimental conditions.

The standard hydrogen electrode (SHE), illustrated in Figure 14-2, meets these requirements and is universally recognized as the ultimate reference. It is a typical *gas electrode.*

A hydrogen electrode consists of a piece of platinum foil (coated or *platinized* with finely divided platinum to increase its surface area) immersed in a solution of constant hydrogen ion activity and continuously swept by a stream of hydrogen at constant pressure. The platinum takes no part in the electrochemical reaction and serves only as the site for the transfer of electrons. The half-cell reaction responsible for the transmission of current across the interface is

$$H_2(g) \rightleftarrows 2H^+ + 2e$$

This process undoubtedly involves the solution of molecular hydrogen as a preliminary step; the overall equilibrium thus consists of two stepwise equilibria

$$H_2(g) \rightleftarrows H_2(sat'd)$$
$$H_2(sat'd) \rightleftarrows 2H^+ + 2e$$

The continuous stream of gas at constant pressure provides the solution with a constant molecular hydrogen concentration.

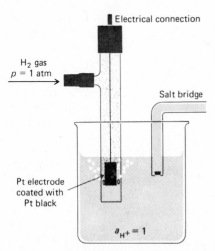

Figure 14-2   The Standard Hydrogen Electrode.

The hydrogen electrode may act as anode or cathode, depending upon the half-cell with which it is coupled. In the former, hydrogen is oxidized to hydrogen ions; in the latter, the reverse reaction takes place. Under proper conditions, then, the hydrogen electrode is electrochemically reversible.

The potential of a hydrogen electrode depends upon temperature, the hydrogen ion concentration (more correctly, the activity) in the solution, and the pressure of the hydrogen at the surface of the electrode. Values for these parameters must be carefully defined in order for the half-cell process to serve as a reference. Specifications for the *standard* (known also as a *normal*) *hydrogen electrode* call for a hydrogen ion activity of unity and a partial pressure of one atmosphere for hydrogen. *By convention, the potential of the standard hydrogen electrode is assigned the value of exactly zero volts at all temperatures.*

Several secondary reference electrodes, which are more convenient to use on a routine basis, find extensive employment. Some of these are described in Chapter 17.

Measurement of Electrode Potentials. A cell such as that shown in Figure 14-3 can be used to measure relative half-cell potentials. Here one-half of the cell consists of the standard hydrogen electrode, and the other half is the electrode whose potential is to be determined. Connecting the two half-cells is a salt bridge to provide a conducting medium that also prevents direct interaction between the contents of the two half-cell compartments. Ordinarily, the effect of a salt bridge upon the cell voltage is vanishingly small.

The left-hand cell in Figure 14-3 consists of a pure metal in contact with a solution of its ions. The electrode reaction is given by

$$M(s) \rightleftarrows M^{2+} + 2e$$

or the reverse. If the metal is cadmium and the solution is approximately one molar in cadmium ions, the voltage indicated by the measuring device, V, will

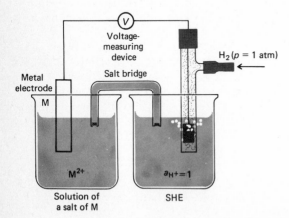

Figure 14-3 Schematic Diagram of an Arrangement for Measurement of Electrode Potentials against the Standard Hydrogen Electrode.

be about 0.4 V. Moreover, the cadmium will behave as the anode so that electrons pass from this electrode to the hydrogen electrode via the external circuit. Thus, the half-cell reactions for the galvanic cell can be written as

$$Cd(s) \rightleftarrows Cd^{2+} + 2e \qquad \text{anode}$$
$$2H^+ + 2e \rightleftarrows H_2(g) \qquad \text{cathode}$$

The overall cell reaction is the sum of these, or

$$Cd(s) + 2H^+ \rightleftarrows Cd^{2+} + H_2(g)$$

If the cadmium electrode was replaced by a zinc electrode immersed in a solution that is one molar in zinc ions, a potential of slightly less than 0.8 V would be observed. Again the metal electrode would behave as the anode. The larger voltage developed reflects the greater tendency of zinc to be oxidized. The difference between this potential and the one for cadmium is a quantitative measure of the relative strengths of these two metals as reducing agents.

Since a value of zero is assigned as the potential for the standard hydrogen electrode, the potentials for the two half-cells thus become 0.4 and 0.8 V with respect to this reference. It must be understood, however, that these are actually the potentials of electrochemical cells in which the standard hydrogen electrode serves as a common reference.

If the half-cell in Figure 14-3 consisted of a copper electrode in a one-molar solution of copper(II) ions, a potential of about 0.3 V would develop. However, in distinct contrast to the previous two examples, copper would tend to deposit, and external electron flow, if allowed, would be from the hydrogen electrode to the copper electrode. The spontaneous cell reaction, then, is the reverse of the two cells considered earlier:

$$Cu^{2+} + H_2(g) \rightleftarrows Cu(s) + 2H^+$$

Thus, metallic copper is a much less effective reducing agent than either zinc or cadmium, *or hydrogen ion*. As before, the observed potential is a quantitative measure of this strength.

When the potential for the copper electrode is compared with those for zinc and cadmium, there emerges the further necessity of specifying the difference in direction of the half-reaction with respect to the reference electrode. Differentiation is conveniently done by assigning a positive or negative sign to the potentials, thereby making the sign of the potential for the copper half-reaction opposite to that of the other two. The choice as to which potential will be positive and which will be negative is *purely arbitrary*; whatever sign convention is chosen, however, must be used consistently.

Sign Conventions for Electrode Potentials. It is not surprising that the arbitrariness in the specifying of signs has led to much controversy and confusion in the course of the development of electrochemistry. In 1953 the International Union of Pure and Applied Chemistry (IUPAC), meeting in Stockholm, attempted to resolve this controversy. The sign convention adopted at this meeting is sometimes called the IUPAC or Stockholm convention; there appears to be hope for its general adoption in years to come. We shall always use the IUPAC sign convention.

Any sign convention must be based upon half-cell processes written in a single way—that is, as oxidations or as reductions. According to the IUPAC convention, the term *electrode potential* (or more exactly, *relative electrode potential*) is reserved exclusively for half-reactions written as reductions. There is no objection to the use of the term *oxidation potential* to connote an electrode process in the opposite sense, but an oxidation potential should not be called an electrode potential.

The sign of the electrode potential is determined by the actual sign of the electrode of interest when it is coupled with a standard hydrogen electrode in a galvanic cell. Thus, a zinc or a cadmium electrode will behave as an anode from which electrons flow through the external circuit to the standard hydrogen electrode. These metal electrodes are thus the negative terminals of galvanic cells, and their electrode potentials are *assigned* negative values. That is,

$$Zn^{2+} + 2e \rightleftarrows Zn(s) \qquad E = -0.8 \text{ V}$$
$$Cd^{2+} + 2e \rightleftarrows Cd(s) \qquad E = -0.4 \text{ V}$$

The potential for the copper electrode, on the other hand, is given a positive sign because the copper behaves as a cathode in a galvanic cell constructed from this electrode and the hydrogen electrode; electrons flow toward the copper electrode through the exterior circuit. It is thus the positive terminal of the galvanic cell and

$$Cu^{2+} + 2e \rightarrow Cu(s) \qquad E = +0.3 \text{ V}$$

It is important to note that each half-reaction describing the electrode potential *is written as a reduction*. For the first two, however, the spontaneous reactions are oxidations. It is evident, then, that the *sign of the electrode potential will indicate whether or not the reduction is spontaneous with respect to the standard hydrogen electrode*. That is, the positive sign for the copper electrode potential means that the reaction

$$Cu^{2+} + H_2 \rightleftarrows 2H^+ + Cu(s)$$

proceeds toward the right under ordinary conditions. The negative electrode potential for zinc, on the other hand, means that the analogous reaction

$$Zn^{2+} + H_2 \rightleftarrows 2H^+ + Zn(s)$$

does not ordinarily occur; indeed, the equilibrium favors the species on the left.

The IUPAC convention was adopted in 1953, but electrode-potential data given in many texts and reference works are not always in accord with it. For example, in the prime source of oxidation-potential data compiled by Latimer,[1] one finds

$$Zn(s) \rightleftarrows Zn^{2+} + 2e \qquad E = +0.76 \text{ V}$$
$$Cu(s) \rightleftarrows Cu^{2+} + 2e \qquad E = -0.34 \text{ V}$$

---

[1] W. M. Latimer, *The Oxidation States of the Elements and Their Potentials in Aqueous Solutions.* 2d ed. Englewood Cliffs, N.J.: Prentice-Hall, Inc., 1952.

To convert these oxidation potentials to electrode potentials as defined by the IUPAC convention, one must mentally (1) express the half-reactions as reductions and (2) change the signs of the potentials.

The sign convention employed in a table of standard potentials may not be explicitly stated. This information is readily determined, however, by referring to a half-reaction with which one is familiar and noting the direction of the reaction and the sign of the potential. Whatever changes, if any, are required to convert to the IUPAC convention are then applied to the remainder of the data in the table. For example, all one needs to remember is that, under the IUPAC convention, strong oxidizing agents such as chlorine have large positive electrode potentials. That is, the reaction

$$Cl_2(g) + 2e \rightleftarrows 2Cl^- \qquad E = +1.36 \text{ V}$$

tends to occur spontaneously with respect to the standard hydrogen electrode. The sign and direction of this reaction in a given table can then serve as a key to any changes that may be needed to convert all data to the IUPAC convention.

Effect of Concentration on Electrode Potentials. The electrode potential, because it is a measure of the chemical driving force of a half-reaction, is affected by concentration. Thus, the tendency of copper(II) ions to be reduced to the elemental state is substantially greater from a concentrated than from a dilute solution. Therefore, the electrode potential for this process must also be greater in the more concentrated solution. In general, the concentrations of reactants and products in a half-reaction have a marked effect on the electrode potential; the quantitative aspects of this effect must now be examined.

Consider the generalized reversible half-cell reaction

$$aA + bB + \cdots + ne \rightleftarrows cC + dD + \cdots$$

where the capital roman letters represent formulas of reacting species (whether charged or uncharged), e represents electrons, and the lower-case italic letters indicate the number of moles of each species participating in the reaction. It can be shown theoretically as well as experimentally that the potential E for this electrode process is governed by the relation

$$E = E^\circ - \frac{RT}{nF} \ln \frac{[C]^c [D]^{d \cdots}}{[A]^a [B]^{b \cdots}} \qquad (14\text{-}1)$$

where

E° = a constant, called the *standard electrode potential*, characteristic of the particular half-reaction

R = the gas constant = 8.314 volt coulomb $^\circ K^{-1}$ mole^{-1}

T = the absolute temperature

n = number of electrons participating in the reaction as defined by the equation describing the half-cell reaction

F = the faraday = 96,493 coulombs

$\ln$ = the natural logarithm = 2.303 $\log_{10}$

Upon substituting numerical values for the various constants and converting to base 10 logarithms, Equation 14-1 becomes, at 25°C,

$$E = E^0 - \frac{0.0591}{n} \log \frac{[C]^c[D]^{d \cdots}}{[A]^a[B]^{b \cdots}} \qquad (14\text{-}2)$$

The letters in brackets represent the activities of the various species. For many calculations, substitution of concentrations for activities does not introduce an excessive error. Thus, where the species is a solute,

$$[\ ] \cong \text{concentration in moles per liter}$$

If the reactant is a gas,

$$[\ ] \cong \text{partial pressure of the gas, in atmospheres}$$

If the reactant exists in a second phase as a pure solid or liquid, then, by definition,

$$[\ ] = 1$$

The basis of the last statement is the same as that described earlier, namely, that the concentration of a compound in a pure solid (or a pure liquid) is constant; therefore, as long as some solid is present, its effect on an equilibrium is invariant and can be included in the constant E^0. Similarly, although it may be involved in the half-cell process, the concentration of water as a solvent is ordinarily so large with respect to the other reactants that, for all practical purposes, it can be considered to remain unchanged. Its contribution is customarily included in the constant E^0. Thus, if water appears in Equation 14-2,

$$[H_2O] = 1$$

Equation 14-1 is called the *Nernst equation* in honor of a nineteenth-century electrochemist. Application of the Nernst equation is illustrated in the following examples.

1. $$Zn^{2+} + 2e \rightleftarrows Zn(s)$$

$$E = E^0 - \frac{0.0591}{2} \log \frac{1}{[Zn^{2+}]}$$

The activity of the elemental zinc is unity, by definition; the electrode potential then varies with the logarithm of the reciprocal of the molar zinc ion concentration in the solution as shown.

2. $$Fe^{3+} + e \rightleftarrows Fe^{2+}$$

$$E = E^0 - \frac{0.0591}{1} \log \frac{[Fe^{2+}]}{[Fe^{3+}]}$$

This electrode potential can be measured with an inert metal electrode immersed in a solution containing iron(II) and iron(III). The potential is dependent upon the ratio between the molar concentrations of these ions.

3. $$2H^+ + 2e \rightleftarrows H_2(g)$$

$$E = E^0 - \frac{0.0591}{2} \log \frac{p_{H_2}}{[H^+]^2}$$

In this example, p_{H_2} represents the partial pressure of hydrogen, expressed in atmospheres, at the surface of the electrode. Ordinarily, p_{H_2} will be very close to atmospheric pressure.

4. $$Cr_2O_7{}^{2-} + 14H^+ + 6e \rightleftarrows 2Cr^{3+} + 7H_2O$$

$$E = E^\circ - \frac{0.0591}{6} \log \frac{[Cr^{3+}]^2}{[Cr_2O_7{}^{2-}][H^+]^{14}}$$

Here the potential depends not only on the concentration of the chromium(III) and dichromate ions but also on the pH of the solution.

5. $$AgCl(s) + e \rightleftarrows Ag(s) + Cl^-$$

$$E = E^\circ - \frac{0.0591}{1} \log \frac{[Cl^-]}{1}$$

This half-reaction, which describes the behavior of a silver electrode immersed in a solution of chloride ions that is *saturated* in silver chloride, is the sum of two reactions, namely,

$$AgCl(s) \rightleftarrows Ag^+ + Cl^-$$
$$Ag^+ + e \rightleftarrows Ag(s)$$

Here the activities of both metallic silver and silver chloride are equal to unity, by definition. The potential of the silver electrode will depend only upon chloride ion concentration.

We need to make one additional observation with respect to the application of the Nernst equation. Although it is not obvious from this presentation, the quotient in the logarithm term is a unitless quantity. Each of the individual terms in the quotient is, in fact, a ratio of the present activity of the species compared to its activity in the standard state, which has been arbitrarily assigned a value of unity (see footnote 1, p. 25). Thus, for the half-reaction in the third example,

$$E = E^\circ - \frac{0.0591}{2} \log \frac{p_{H_2}/(p_{H_2})_0}{[H^+]^2/[H^+]_0{}^2} = E^\circ - \frac{0.0591}{2} \log \frac{p_{H_2}/1.00}{[H^+]^2/(1.00)^2}$$

Here $(p_{H_2})_0$ is the partial pressure of hydrogen in its standard state, which is 1.00 atm. Similarly, $[H^+]_0$ represents the activity of hydrogen ion in its standard state, 1.00 mole/liter. Hence, the units cancel.

The Standard Electrode Potential, E°. An examination of Equation 14-1 or Equation 14-2 reveals that the constant E° is equal to the half-cell potential when the logarithmic term is zero. This condition occurs whenever the activity quotient is equal to unity, one such instance being when the activities of all reactants and products are one. Thus, the *standard potential may be defined as the electrode potential of a half-cell reaction (versus SHE) when all reactants and products are at unit activity.*

The standard electrode potential is an important physical constant that gives a quantitative description of the relative driving force for a half-cell reaction. Several facts regarding this constant should be kept in mind. First, an electrode potential is temperature-dependent; if it is to have significance, the temperature at which it is determined must be specified. Second, the standard electrode potential is a relative quantity in the sense that it is really the potential of an electrochemical cell in which the anode is a carefully specified reference electrode —that is, the standard hydrogen electrode, whose potential is assigned a value of zero volts. Third, a standard electrode potential is given the sign of the conductor that is in contact with the half-cell of interest under the specified conditions of activity and when the cell is functioning as a galvanic cell. Finally, the standard potential is a measure of the intensity of the driving force for a half-reaction. As such, it is independent of the notation employed to express the half-cell process. Thus, the potential for the process

$$Ag^+ + e \rightleftarrows Ag(s) \qquad E^\circ = +0.799 \text{ V}$$

while dependent upon the *activity* of silver ions, is the same regardless of whether we write the half-reaction as above or as

$$100Ag^+ + 100e \rightleftarrows 100Ag(s) \qquad E^\circ = +0.799 \text{ V}$$

To be sure, the Nernst equation must be consistent with the half-reaction as it has been written. For the first of these,

$$E = 0.799 - \frac{0.0591}{1} \log \frac{1}{[Ag^+]}$$

and for the other,

$$E = 0.799 - \frac{0.0591}{100} \log \frac{1}{[Ag^+]^{100}}$$

Standard electrode potentials are available for numerous half-reactions. Many have been determined directly from voltage measurements of cells in which the standard hydrogen electrode actually constitutes the other electrode. It is possible, however, to calculate E° values from equilibrium studies of oxidation-reduction systems and from thermochemical data relating to such reactions. Many of the values found in the literature were so obtained.[2]

A few standard electrode potentials are given in Table 14-1; a more comprehensive list is found in the table of Appendix 2. In these tabulations, the species in the upper left part of the table are most easily reduced, as indicated by the large positive E° values; they are therefore the most effective oxidizing agents. Proceeding down the table, each succeeding species is a less effective acceptor of electrons than the one above it. The half-cell reactions at the bottom of the table have little tendency to take place as written. On the other hand, they do tend to occur in the opposite sense, as oxidations. The most effective reducing agents, then, are those species that appear in the lower right-hand portion of a table of standard electrode potentials.

[2] Two authoritative sources for standard potential data are L. Meites, *Handbook of Analytical Chemistry*, pp. 5-6 to 5-14. New York: McGraw-Hill Book Company, 1963; W. M. Latimer, *The Oxidation States of the Elements and Their Potentials in Aqueous Solutions*, 2d ed. Englewood Cliffs, N.J.: Prentice-Hall, Inc., 1952.

TABLE 14-1   Standard Electrode Potentials[a]

Reaction	E^0 at 25°C, V
$MnO_4^- + 8H^+ + 5e \rightleftarrows Mn^{2+} + 4H_2O$	$+1.51$
$Cl_2(g) + 2e \rightleftarrows 2Cl^-$	$+1.359$
$Ag^+ + e \rightleftarrows Ag(s)$	$+0.799$
$Fe^{3+} + e \rightleftarrows Fe^{2+}$	$+0.771$
$I_3^- + 2e \rightleftarrows 3I^-$	$+0.536$
$AgCl(s) + e \rightleftarrows Ag(s) + Cl^-$	$+0.222$
$Ag(S_2O_3)_2^{3-} + e \rightleftarrows Ag(s) + 2S_2O_3^{2-}$	$+0.010$
$2H^+ + 2e \rightleftarrows H_2(g)$	0.000
$Cd^{2+} + 2e \rightleftarrows Cd(s)$	-0.403
$Cr^{3+} + e \rightleftarrows Cr^{2+}$	-0.41
$Zn^{2+} + 2e \rightleftarrows Zn(s)$	-0.763

[a] See Appendix 2 for a more extensive list.

A compilation of standard potentials provides the chemist with a qualitative picture regarding the extent and direction of electron-transfer reactions between the tabulated species. On the basis of Table 14-1, for example, we see that zinc is more easily oxidized than cadmium, and we conclude that a piece of zinc, immersed in a solution of cadmium ions, will cause the deposition of metallic cadmium. On the other hand, cadmium has little tendency to reduce zinc ions. Also from Table 14-1, we see that iron(III) is a better oxidizing agent than the triiodide ion and may therefore predict that in a solution containing an equilibrium mixture of iron(III), iodide, iron(II), and triiodide ions, the latter pair will predominate.

Calculation of Half-Cell Potentials from E^0 Values. Some applications of the Nernst equation to the calculation of half-cell potentials are illustrated in the following examples.

Example.   What is the potential for a half-cell consisting of a cadmium electrode immersed in a solution that is 0.0100 F in Cd^{2+}?
From Table 14-1, we find

$$Cd^{2+} + 2e \rightleftarrows Cd(s) \qquad E^0 = -0.403 \text{ V}$$

Thus,

$$E = E^0 - \frac{0.0591}{2} \log \frac{1}{[Cd^{2+}]}$$

Substituting the Cd^{2+} concentration into this equation gives

$$E = -0.403 - \frac{0.0591}{2} \log \frac{1}{0.0100}$$

$$= -0.403 - \frac{0.0591}{2} (+2.0)$$

$$= -0.462 \text{ V}$$

The sign for the potential indicates the direction of the half-reaction when this half-cell is coupled with the standard hydrogen electrode. The fact that it is negative shows that the reverse reaction,

$$Cd(s) + 2H^+ \rightleftarrows H_2(g) + Cd^{2+}$$

occurs spontaneously. Note that the calculated potential is a larger negative number than the standard electrode potential itself. This follows from mass-law considerations, since the half-reaction, as written, has less tendency to occur with the lower cadmium ion concentration.

Example.   Calculate the potential for a platinum electrode immersed in a solution prepared by saturating a 0.0100-F solution of KBr with Br_2.

Here the half-reaction is

$$Br_2(l) + 2e \rightleftarrows 2Br^- \qquad E^0 = 1.065 \text{ V}$$

Note that the term (l) in the equation is employed to indicate that the aqueous solution is kept saturated by the presence of an excess of *liquid* Br_2. Thus, the overall process is the sum of the two equilibria

$$Br_2(l) \rightleftarrows Br_2(\text{sat'd aq})$$

$$Br_2(\text{sat'd aq}) + 2e \rightleftarrows 2Br^-$$

The Nernst equation for the overall process is

$$E = 1.065 - \frac{0.0591}{2} \log \frac{[Br^-]^2}{1.00}$$

Here the activity of Br_2 in the pure liquid is constant and given the value of 1.00 by definition. Thus,

$$E = 1.065 - \frac{0.0591}{2} \log (0.0100)^2$$

$$= 1.065 - \frac{0.0591}{2} (-4.00)$$

$$= 1.183 \text{ V}$$

Example.   Calculate the potential for a platinum electrode immersed in a solution that is 0.0100 F in KBr and 1.00×10^{-3} F in Br_2.

Here the half-reaction used in the preceding example does not apply *because the solution is no longer saturated in Br_2*. In Appendix 2, however, we find the half-reaction

$$Br_2(aq) + 2e \rightleftarrows 2Br^- \qquad E^0 = 1.087 \text{ V}$$

The term (aq) implies that all of the Br_2 present is in solution. That is, 1.087 is the electrode potential for the half-reaction when the Br^- and the Br_2 *solution* activities are 1.00 mole/liter. It turns out, however, that the solubility of Br_2 in water at 25°C is only about 0.18 mole/liter. Therefore, the recorded potential of 1.087 is based on a *hypothetical system that cannot be realized experimentally*. Nevertheless, this potential is useful because it provides a means by which potentials for undersaturated systems can be obtained. Thus,

$$E = 1.087 - \frac{0.0591}{2} \log \frac{(1.00 \times 10^{-2})^2}{1.00 \times 10^{-3}}$$

$$= 1.087 - \frac{0.0591}{2} \log 0.100$$

$$= 1.117 \text{ V}$$

Here the Br_2 activity is 1.00×10^{-3} rather than 1.00, as is the case when the solution is saturated.

Electrode Potentials in Presence of Precipitation and Complex-Forming Reagents. Reagents that react with the participants of an electrode process have a marked effect on the potential for that process. For example, the standard electrode potential for the reaction $Ag^+ + e \rightleftarrows Ag(s)$ is $+0.799$ V. This potential is the output of a cell consisting of a silver electrode immersed in a solution containing silver ions at unit activity and a standard hydrogen electrode. Addition of chloride ions to the solution of the former will materially alter the silver ion concentration and hence the electrode potential. The effect is illustrated in the following example.

Example. Calculate the potential of a silver electrode in a solution that is saturated with silver chloride and has a chloride ion activity of exactly 1.00.

$$Ag^+ + e \rightleftarrows Ag(s) \qquad E^0 = +0.799 \text{ V}$$

$$E = +0.799 - 0.0591 \log \frac{1}{[Ag^+]}$$

We may calculate $[Ag^+]$ from the solubility-product constant

$$[Ag^+] = \frac{K_{sp}}{[Cl^-]}$$

Substituting into the Nernst equation,

$$E = +0.799 - \frac{0.0591}{1} \log \frac{[Cl^-]}{K_{sp}}$$

This equation may be rewritten as

$$E = +0.799 + 0.0591 \log K_{sp} - 0.0591 \log [Cl^-] \qquad (14\text{-}3)$$

If we substitute 1.00 for $[Cl^-]$ and use 1.75×10^{-10} for K_{sp}, the solubility product for AgCl at $25.0°C$, we obtain

$$E = +0.222 \text{ V}$$

The example just considered shows that the half-cell potential for the reduction of silver ion becomes smaller in the presence of chloride ions. Qualitatively, this is the expected effect, since decreases in the concentration of silver ions diminish the tendency for their reduction.

Equation 14-3 relates the potential of a silver electrode to the chloride ion concentration of a solution that is also saturated with silver chloride. *When the chloride ion activity is unity*, the potential is the sum of two constants. This sum can be considered the standard electrode potential for the half-reaction

$$AgCl(s) + e \rightleftarrows Ag + Cl^- \qquad E^\circ = +0.222 \text{ V}$$

where

$$E^\circ = +0.799 + 0.0591 \log K_{sp}$$

The Nernst relationship for the silver electrode in a solution saturated with silver chloride can then be written as

$$E = E^\circ - 0.0591 \log [Cl^-] = 0.222 - 0.0591 \log [Cl^-]$$

Thus, when in contact with a solution *saturated with silver chloride*, the potential of a silver electrode can be described *either* in terms of the silver ion concentration (using the standard electrode potential for the simple silver half-reaction) or in terms of the chloride ion concentration (using the standard electrode potential for the silver–silver chloride half-reaction).

The potential of a silver electrode in a solution that contains an ion that forms a soluble complex with silver ion can be treated in a fashion analogous to the foregoing. For example, in a solution containing thiosulfate ion, the half-reaction can be written as

$$Ag(S_2O_3)_2{}^{3-} + e \rightleftarrows Ag(s) + 2S_2O_3{}^{2-}$$

The standard electrode potential for this half-reaction will be the electrode potential when both the complex and the complexing anion are at unit activity. Using the same approach as in the previous example, we find

$$E^\circ = +0.799 + 0.0591 \log \frac{1}{K_f}$$

where K_f is the formation constant for the complex ion.

Data for the potential of the silver electrode in the presence of selected ions are given in the table of standard electrode potentials in Appendix 2. Similar information is also provided for other electrode systems. Such data often simplify the calculation of half-cell potentials.

CELLS AND CELL POTENTIALS

Schematic Representation of Cells. Chemists frequently use a shorthand notation to describe electrochemical cells. For example, the cell shown in Figure 14-1 can be represented as

$$Zn|ZnSO_4(C_1)|\,|CuSO_4(C_2)|Cu$$

where C_1 and C_2 are formal concentrations of the two salts. Customarily, the anode is listed on the left. Single vertical lines represent phase boundaries in the cell. Potential differences arising across these are included in the measured cell potential. A double vertical line represents the two phase boundaries that exist at either end of the salt bridge. One of these boundaries is between the $ZnSO_4$

solution and saturated KCl; the other involves the KCl solution and the $CuSO_4$ solution. A potential, called a *liquid junction potential*, develops at each interface, owing to the different rates at which ions diffuse across the junction. This potential may be as large as several hundredths of a volt. The potentials at the two interfaces of a salt bridge, however, tend to cancel each other so that the net contribution to the total junction potential is a few millivolts or less. For the purposes of this chapter, the junction potential can be neglected without serious error.

Calculation of Cell Potentials. One of the important uses of standard electrode potentials is the calculation of the potential obtainable from a galvanic cell or the potential required to operate an electrolytic cell. These calculated potentials (sometimes called *thermodynamic potentials*) are theoretical in the sense that they refer to cells in which there is essentially no passage of current; additional factors must be taken into account where a current flow is involved.

One obtains the electromotive force of a cell by combining half-cell potentials as follows:

$$E_{cell} = E_{cathode} - E_{anode}$$

where E_{anode} and $E_{cathode}$ are the *electrode potentials* for the two half-reactions constituting the cell.

Consider the cell

$$Zn|ZnSO_4(a_{Zn^{2+}} = 1.00)||CuSO_4(a_{Cu^{2+}} = 1.00)|Cu$$

The overall cell process consists of the oxidation of elemental zinc to zinc ions and the reduction of copper(II) ions to the metallic state. Since the activities of the two ions are specified as unity, the standard potentials are also the electrode potentials. The cell diagram also specifies that the zinc electrode is the anode. Thus, using E^0 data from Table 14-1,

$$E_{cell} = +0.0337 - (-0.763) = +1.100 \text{ V}$$

The positive sign for the cell potential indicates that the reaction

$$Zn(s) + Cu^{2+} \rightarrow Zn^{2+} + Cu(s)$$

occurs spontaneously and that this is a galvanic cell.

The foregoing cell, diagrammed as

$$Cu|Cu^{2+}(a_{Cu^{2+}} = 1.00)||Zn^{2+}(a_{Zn^{2+}} = 1.00)|Zn$$

implies that the copper electrode is now the anode. Thus,

$$E_{cell} = -0.763 - (+0.337) = -1.100 \text{ V}$$

The negative sign indicates the nonspontaneity of the reaction

$$Cu(s) + Zn^{2+} \rightarrow Cu^{2+} + Zn(s)$$

To cause this reaction to occur would require the application of an external potential greater than 1.100 V.

Example. Calculate the theoretical potential of the cell

$$\text{Pt, } H_2(0.80 \text{ atm}) | HCl(0.20 \, F), \text{ AgCl(sat'd)} | Ag$$

The two half-cell reactions and standard electrode potentials are

$$2H^+ + 2e \rightleftarrows H_2(g) \qquad E^0 = 0.000 \text{ V}$$

$$AgCl(s) + e \rightleftarrows Ag(s) + Cl^- \qquad E^0 = +0.222 \text{ V}$$

If we make the approximation that the activities of the various species are identical to their molar concentrations, then for the hydrogen electrode,

$$E = 0.000 - \frac{0.0591}{2} \log \frac{0.80}{(0.20)^2} = -0.038 \text{ V}$$

and for the silver–silver chloride electrode,

$$E = +0.222 - 0.0591 \log 0.20 = +0.263 \text{ V}$$

The cell diagram specifies the hydrogen electrode as the anode and the silver electrode as the cathode. Therefore,

$$E_{cell} = +0.263 - (-0.038) = +0.301 \text{ V}$$

The positive sign indicates that the reaction

$$H_2(g) + 2AgCl(s) \rightarrow 2H^+ + 2Ag(s) + 2Cl^-$$

will occur if current is drawn from this cell.

Example. Calculate the potential required to initiate deposition of metallic copper from a solution that is 0.010 F in $CuSO_4$ and contains sufficient sulfuric acid to give a hydrogen ion concentration of $1.0 \times 10^{-4} \, M$.

The deposition of copper necessarily occurs at the cathode. Since no easily oxidizable species are present, the anode reaction will involve oxidation of H_2O to give O_2. From the table of standard potentials, we find

$$Cu^{2+} + 2e \rightleftarrows Cu(s) \qquad E^0 = +0.337 \text{ V}$$

$$\tfrac{1}{2}O_2(g) + 2H^+ + 2e \rightleftarrows H_2O \qquad E^0 = +1.23 \text{ V}$$

Thus, for the copper electrode,

$$E = +0.337 - \frac{0.0591}{2} \log \frac{1}{0.010} = +0.278 \text{ V}$$

Assuming that O_2 is evolved at 1.00 atm, the potential for the oxygen electrode is

$$E = +1.23 - \frac{0.0591}{2} \log \frac{1}{(1.0 \times 10^{-4})^2(1.00)^{1/2}} = +0.99 \text{ V}$$

The cell potential is then

$$E_{cell} = +0.278 - 0.99 = -0.71 \text{ V}$$

Thus, to initiate the reaction

$$Cu^{2+} + H_2O \rightarrow \tfrac{1}{2}O_2(g) + 2H^+ + Cu(s)$$

would require the application of a potential greater than 0.71 V.

EQUILIBRIUM CONSTANTS FROM STANDARD ELECTRODE POTENTIALS

Changes in Potential during Discharge of a Cell. Consider again the galvanic cell based upon the reaction

$$Zn(s) + Cu^{2+} \rightleftarrows Zn^{2+} + Cu(s)$$

At all times, the cell potential is given by

$$E_{cell} = E_{cathode} - E_{anode}$$

As current is drawn from this cell, however, the zinc ion concentration increases, and the copper(II) concentration decreases. These changes have the effect of rendering the electrode potential less negative for the zinc half-cell and less positive for the copper half-cell. The net effect, then, is a decrease in the potential of the cell. Ultimately, the concentrations attain values such that there remains no further tendency for the transfer of electrons. The potential of the cell thus becomes zero, and the system is in equilibrium. That is, at equilibrium

$$E_{cell} = E_{cathode} - E_{anode} = 0$$

or

$$E_{cathode} = E_{anode}$$

This equation is an important and general relationship: *For an oxidation-reduction system at chemical equilibrium, the electrode potentials (that is, the reduction potentials) of all half-reactions of the system will be equal.*

Calculation of Equilibrium Constants. Consider the oxidation-reduction equilibrium

$$aA_{red} + bB_{ox} \rightleftarrows aA_{ox} + bB_{red}$$

for which the half-reactions may be written

$$aA_{ox} + ne \rightleftarrows aA_{red}$$
$$bB_{ox} + ne \rightleftarrows bB_{red}$$

When the components of this system are at chemical equilibrium

$$E_A = E_B$$

where E_A and E_B are the electrode potentials for the two half-cells. When this equality is expressed in terms of the Nernst equation, we find that *at equilibrium,*

$$E^0_A - \frac{0.0591}{n} \log \frac{[A_{red}]^a}{[A_{ox}]^a} = E^0_B - \frac{0.0591}{n} \log \frac{[B_{red}]^b}{[B_{ox}]^b}$$

which yields, upon rearrangement and combination of the log terms,

$$E^0_B - E^0_A = \frac{0.0591}{n} \log \frac{[A_{ox}]^a [B_{red}]^b}{[A_{red}]^a [B_{ox}]^b}$$

This relationship was derived for *equilibrium conditions*. Thus, the concentration terms are equilibrium concentrations; *hence the quotient is the equilibrium constant for the reaction*. That is,

$$\log K_{eq} = \frac{n(E^0_B - E^0_A)}{0.0591} \tag{14-4}$$

Example. Calculate the equilibrium constant for the reaction

$$MnO_4^- + 5Fe^{2+} + 8H^+ \rightleftarrows Mn^{2+} + 5Fe^{3+} + 4H_2O$$

From the table of standard electrode potentials (Appendix 2), we find

$$MnO_4^- + 8H^+ + 5e \rightleftarrows Mn^{2+} + 4H_2O \qquad E^0_{MnO_4^-} = +1.51 \text{ V}$$

$$5Fe^{3+} + 5e \rightleftarrows 5Fe^{2+} \qquad E^0_{Fe^{3+}} = +0.771 \text{ V}$$

Note that the half-reactions must be written to involve the same number of moles of species as in the balanced chemical equation for which K_{eq} is sought. Here the components of the second half-reaction are multiplied by 5.

Since at equilibrium

$$E_{Fe^{3+}} = E_{MnO_4^-}$$

then

$$E^0_{Fe^{3+}} - \frac{0.0591}{5} \log \frac{[Fe^{2+}]^5}{[Fe^{3+}]^5} = E^0_{MnO_4^-} - \frac{0.0591}{5} \log \frac{[Mn^{2+}]}{[MnO_4^-][H^+]^8}$$

Rearranging,

$$\frac{0.0591}{5} \log \frac{[Mn^{2+}][Fe^{3+}]^5}{[MnO_4^-][Fe^{2+}]^5[H^+]^8} = E^0_{MnO_4^-} - E^0_{Fe^{3+}}$$

or

$$\log K_{eq} = \frac{5(1.51 - 0.77)}{0.0591}$$

$$= 62.7$$

$$K_{eq} = 10^{62.7} = 10^{0.7} \times 10^{62} = 5 \times 10^{62}$$

Example. A piece of copper is placed in a 0.050-*F* solution of $AgNO_3$. What is the equilibrium composition of the solution? The reaction is

$$Cu(s) + 2Ag^+ \rightleftarrows Cu^{2+} + 2Ag(s)$$

We first calculate the equilibrium constant for the reaction and then use this to determine the solution composition. From the table of standard potentials, we find

$$2Ag^+ + 2e \rightleftarrows 2Ag(s) \qquad E^0_{Ag^+} = +0.799 \text{ V}$$

$$Cu^{2+} + 2e \rightleftarrows Cu(s) \qquad E^0_{Cu^{2+}} = +0.337 \text{ V}$$

Since at equilibrium

$$E_{Cu^{2+}} = E_{Ag^+}$$

then

$$E^0_{Cu^{2+}} - \frac{0.0591}{2} \log \frac{1}{[Cu^{2+}]} = E^0_{Ag^+} - \frac{0.0591}{2} \log \frac{1}{[Ag^+]^2}$$

$$\log \frac{[Cu^{2+}]}{[Ag^+]^2} = \frac{2(E^0_{Ag^+} - E^0_{Cu^{2+}})}{0.0591} = 15.63$$

and

$$\frac{[Cu^{2+}]}{[Ag^+]^2} = K_{eq} = 4.3 \times 10^{15}$$

The magnitude of the equilibrium constant suggests that nearly all Ag^+ is used up. The molar concentration of Cu^{2+} is thus given by

$$[Cu^{2+}] = \tfrac{1}{2}(0.050 - [Ag^+])$$

Since the reaction is nearly complete, it is probably safe to assume that $[Ag^+]$ is small. Then

$$[Cu^{2+}] \cong \tfrac{1}{2}(0.050) = 0.025$$

Substituting,

$$\frac{0.025}{[Ag^+]^2} = 4.3 \times 10^{15}$$

$$[Ag^+] = 2.4 \times 10^{-9}$$

$$[Cu^{2+}] = \tfrac{1}{2}(0.050 - 2.4 \times 10^{-9}) \cong 0.025$$

DETERMINATION OF DISSOCIATION, SOLUBILITY PRODUCT, AND FORMATION CONSTANTS BY POTENTIAL MEASUREMENTS

We have seen that the potential of an electrode is determined by the concentrations of those species that participate in the electrode reaction; as a consequence, measurement of a half-cell potential often provides a convenient way to determine the concentration of solute species. One important virtue of this approach is that the measurement can be made without affecting appreciably any equilibria that may exist in the solution. For example, the potential of a silver electrode in a solution containing the cyanide complex of silver depends only upon the silver ion activity. With suitable equipment, this potential can be measured with a negligible passage of current. Thus, the concentration of silver ions in the solution is not sensibly altered during the measurement; the position of the equilibrium

$$Ag^+ + 2CN^- \rightleftarrows Ag(CN)_2^-$$

is likewise undisturbed.

Example. The solubility-product constant for the sparingly soluble CuX_2 can be determined by saturating a 0.0100-F solution of NaX with solid CuX_2. After equilibrium is achieved, this solution is made part of the cell

$$Cu|CuX_2(sat'd), NaX(0.0100\ F)|\ |\ SHE$$

where SHE is the standard hydrogen electrode.

Assume that the potential of this cell is found to be 0.0103 V, with the copper electrode behaving as the anode, as indicated. We may then write

$$E_{cell} = E_{cathode} - E_{anode}$$

$$0.0103 = 0.000 - \left(E^0_{Cu^{2+}} - \frac{0.0591}{2} \log \frac{1}{[Cu^{2+}]}\right)$$

$$0.0103 = -0.337 - \frac{0.0591}{2} \log [Cu^{2+}]$$

The solution to this equation is

$$[Cu^{2+}] = 1.7 \times 10^{-12}$$

Since the X^- concentration is 0.0100 mole/liter,

$$K_{sp} = (1.7 \times 10^{-12})(0.0100)^2$$
$$= 1.7 \times 10^{-16}$$

Any electrode system that is sensitive to the hydrogen ion concentration of a solution can, in theory at least, be used to determine the dissociation constants of acids and bases. All half-cells in which hydrogen ion is a participant fall in this category. However, relatively few of these have been applied to the problem.

Example. Calculate the dissociation constant of the acid HP if the cell

$$\text{Pt, } H_2(1.00 \text{ atm})|HP(0.010 \text{ } F), \text{ NaP}(0.030 \text{ } F)| \text{ } | \text{ SHE}$$

develops a potential of 0.295 V.

As indicated by the diagram, the standard hydrogen electrode is the cathode Thus, $E_{cathode} = 0.000$ V. For the other half of the cell, we may write

$$2H^+ + 2e \rightleftarrows H_2(g)$$

$$E_{anode} = 0.000 - \frac{0.0591}{2} \log \frac{p_{H_2}}{[H^+]^2}$$

and

$$E_{cell} = E_{cathode} - E_{anode}$$

$$0.295 = 0.000 - \left(0.000 - \frac{0.0591}{2} \log \frac{p_{H_2}}{[H^+]^2}\right)$$

$$= \frac{0.0591}{2} \log \frac{1.00}{[H^+]^2} = -\frac{2 \times 0.0591}{2} \log [H^+]$$

After rearrangement,

$$[H^+] = 1.0 \times 10^{-5}$$

$$K_a = \frac{[H^+][P^-]}{[HP]}$$

Since the HP present is largely undissociated,

$$K_a \cong \frac{(1.0 \times 10^{-5})(0.030)}{0.010} = 3.0 \times 10^{-5}$$

Formation constants for complex ions can be determined in an analogous fashion. Thus, to evaluate the constant for the $Ag(CN)_2^-$ complex, we might measure the potential of the cell

$$Ag|Ag(CN)_2^-(C_1), \text{ } CN^-(C_2)| \text{ } | \text{ SHE}$$

Values for C_1 and C_2 would be known from the quantities used to prepare the cell. Together with the measured cell potential, these quantities would then be substituted into the Nernst equation for the silver half-cell; rearrangement would permit evaluation of the desired constant.

SOME LIMITATIONS TO THE USE OF STANDARD ELECTRODE POTENTIALS

The examples in the previous section illustrate typical applications of the Nernst equation to problems that are important to the analytical chemist. It frequently happens, however, that marked differences exist between calculated and experimentally measured potentials; the intelligent application of electrode-potential calculations requires an appreciation of the sources for uncertainty and their effects upon a calculated result.

Use of Concentrations instead of Activities. It is usually expedient to substitute molar concentrations for activities in the Nernst equation even though differences between these quantities tend to increase with increasing ionic strength (Chapter 5). The typical electrochemical cell possesses a high ionic strength; as a consequence, potentials based on molar concentrations are likely to differ somewhat from those obtained by direct measurement.

Effect of Other Equilibria. The application of standard electrode potential data to many systems of interest is further complicated by solvolysis, dissociation, association, and complex-formation equilibria involving species that appear in the Nernst equation. These phenomena can be accounted for only if their existence is known and equilibrium-constant data are available. This information is frequently lacking; the chemist is then forced to hope that neglect of such effects will not seriously influence his calculations.

Formal Potential. In order to compensate partially for activity effects and errors resulting from side reactions, Swift[3] has proposed the use of a quantity called the *formal potential* in place of the standard electrode potential. The formal potential of a system is the potential of the half-cell with respect to the standard hydrogen electrode when the concentration of each reactant and product is one formal and the concentrations of any other constituents of the solution are carefully specified. Formal potentials are available for many of the systems tabulated in Appendix 2. Thus, for example, the formal potential for the reduction of iron(III) to iron(II) is $+0.731$ V in one formal perchloric acid and $+0.700$ V in one formal hydrochloric acid, as compared with the standard potential of 0.771 V. The lower formal potential in perchloric acid results from the fact that the activity coefficient of iron(III) is smaller than that for iron(II) in the highly ionic medium. The even larger effect of HCl results from the greater stability of chloride complexes of iron(III) than those for iron(II).

If formal potentials are employed in place of standard electrode potentials in the Nernst equation, better agreement is obtained between calculated and experimental potentials, provided, of course, that the electrolyte concentration of the solution approximates that for which the formal potential was measured. Needless to say, application of formal potentials to systems differing greatly as

[3] E. H. Swift, *A System of Chemical Analysis*, p. 50. San Francisco: W. H. Freeman and Company, 1939.

to kind and concentration of electrolyte can cause errors greater than those encountered when standard potentials are used. Henceforth, we shall use whichever is the more appropriate.

Reaction Rate. Calculations from standard potentials will indicate whether or not a given oxidation-reduction reaction is complete enough for application to an analytical problem. Unfortunately, however, such calculations provide no information with respect to the rate at which the equilibrium state will be approached; frequently a reaction that appears extremely favorable from equilibrium considerations may be totally unacceptable from a kinetic standpoint. The oxidation of arsenic(III) with a solution of quadrivalent cerium in dilute sulfuric acid is an example:

$$H_2AsO_3^- + 2Ce^{4+} + H_2O \rightleftarrows H_2AsO_4^- + 2Ce^{3+} + 2H^+$$

The respective formal potentials for the two systems are

$$2Ce^{4+} + 2e \rightleftarrows 2Ce^{3+} \qquad\qquad E^f = +1.4\ V$$

$$H_2AsO_4^- + 2H^+ + 2e \rightleftarrows H_2O + H_2AsO_3^- \qquad E^f = +0.56\ V$$

and an equilibrium constant of about 10^{28} can be deduced from these data. Despite this very favorable equilibrium constant, solutions of arsenic(III) cannot be titrated with cerium(IV) unless a catalyst is introduced because several hours are required for the attainment of equilibrium.

PROBLEMS

*1. Complete and balance the following equations, adding H^+, OH^-, or H_2O as required.
   (a) $Tl^{3+} + Ag(s) + Br^- \rightleftarrows TlBr(s) + AgBr(s)$
   (b) $Fe^{2+} + UO_2^{2+} \rightleftarrows Fe^{3+} + U^{4+}$
   (c) $N_2(g) + H_2(g) \rightleftarrows N_2H_5^+$
   (d) $Cr_2O_7^{2-} + I^- \rightleftarrows Cr^{3+} + I_3^-$
   (e) $Ce^{3+} + O_2(g) \rightleftarrows Ce^{4+} + H_2O_2$
   (f) $IO_3^- + I^- \rightleftarrows I_2(s)$

2. Complete and balance the following equations, adding H^+, OH^-, or H_2O as required.
   (a) $CuI(s) + I_3^- \rightleftarrows Cu^{2+} + I^-$
   (b) $I_3^- + S_2O_3^{2-} \rightleftarrows I^- + S_4O_6^{2-}$
   (c) $MnO_4^- + H_2SO_3 \rightleftarrows Mn^{2+} + SO_4^{2-}$
   (d) $MnO_2(s) \rightleftarrows MnO_4^- + Mn^{2+}$
   (e) $IO_3^- + H_3AsO_3 + Cl^- \rightleftarrows ICl_2^- + H_3AsO_4$
   (f) $SO_4^{2-} + MnO_4^- \rightleftarrows S_2O_8^{2-} + Mn^{2+}$

*3. Identify the oxidizing agent and the reducing agent on the left side of each equation in Problem 1; write a balanced equation for each half-reaction.

4. Identify the oxidizing agent and the reducing agent on the left side of each equation in Problem 2; write a balanced equation for each half-reaction.

*5. Calculate equilibrium constants for each of the reactions in Problem 1. (For $Tl^{3+} + Br^- + 2e \rightleftarrows TlBr(s)$, $E^0 = 1.44V$.)

6. Calculate equilibrium constants for each of the reactions in Problem 2.

*7. Calculate the electrode potential of a mercury electrode immersed in
  (a) 0.0400-F $Hg(NO_3)_2$.
  (b) 0.0400-F $Hg_2(NO_3)_2$.
  (c) 0.0400-F KCl that is saturated with Hg_2Cl_2.
  (d) 0.0400-F $Hg(SCN)_2$. ($Hg^{2+} + 2SCN^- \rightarrow Hg(SCN)_2$; $K_f = 1.8 \times 10^{17}$).

8. Calculate the electrode potential for a copper electrode immersed in
  (a) 0.0200-F Cu^{2+}.
  (b) 0.0200-F Cu^+.
  (c) 0.0300-F KI that is saturated with CuI.
  (d) 0.001-F NaOH that is saturated with $Cu(OH)_2$.

*9. Calculate the electrode potential for a platinum electrode immersed in a solution that is
  (a) 0.075 F in $Fe_2(SO_4)_3$ and 0.060 F in $FeSO_4$.
  (b) 0.244 M in V^{3+}, 0.414 M in VO^{2+}, and 1.00×10^{-5} F in NaOH.
  (c) 0.111 F in KI and 0.200 F in KI_3.
  (d) 0.117 F in $K_4Fe(CN)_6$ and 0.333 F in $K_3Fe(CN)_6$.
  (e) 0.0731 F in $SbONO_3$, 0.0100 F in HNO_3 and saturated with Sb_2O_5.
  (f) 0.0731 F in $SbONO_3$, 1.00×10^{-5} F in HNO_3 and saturated with Sb_2O_5.

10. Calculate the electrode potential for a platinum electrode immersed in a solution that is
  (a) 0.313 F in $Tl_2(SO_4)_3$ and 0.209 F in Tl_2SO_4.
  (b) saturated with hydrogen at 1.00 atm and has a pH of 3.50.
  (c) 0.0774 M in UO_2^{2+}, 0.0507 M in U^{4+}, and 1.00×10^{-4} F in $HClO_4$.
  (d) 0.0627 M in $S_2O_3^{2-}$ and 0.0714 M in $S_4O_6^{2-}$.
  (e) 0.0540 M in $Cr_2O_7^{2-}$, 0.149 M in Cr^{3+}, and 0.100 F in $HClO_4$.

*11. Indicate whether each of the following half-cells would behave as anode or as cathode when coupled with a standard hydrogen electrode in a galvanic cell. Calculate the potential of the cell.
  (a) Pb | $Pb^{2+}(0.200 \times 10^{-3}$ $M)$
  (b) Pt | $Sn^{4+}(0.200$ $M)$, $Sn^{2+}(0.100$ $M)$
  (c) Pt | $Sn^{4+}(1.0 \times 10^{-6}$ $M)$, $Sn^{2+}(0.50$ $M)$
  (d) Pt | $Ti^{3+}(0.300$ $M)$, $TiO^{2+}(0.100$ $M)$, $H^+(0.200$ $M)$
  (e) Ag | AgBr(sat'd), KBr(1.00×10^{-4} $F)$
  (f) Ag | $AgNO_3(0.0100$ $F)$, KBr(0.400 $F)$

12. Indicate whether each of the following half-cells would behave as anode or as cathode when coupled with a standard hydrogen electrode in a galvanic cell. Calculate the potential of the cell.
  (a) Pt | $V^{3+}(0.50$ $M)$, $V^{2+}(1.00 \times 10^{-6}$ $M)$
  (b) Ag | $AgNO_3(0.0100$ $F)$, $Na_2S_2O_3(0.0200$ $F)$
  (c) Ag | $AgNO_3(0.100$ $F)$, $Na_2S_2O_3(0.200$ $F)$
  (d) Bi | $BiCl_4^-(0.010$ $M)$, $Cl^-(0.500$ $M)$
  (e) Ag | $Ag(CN)_2^-(0.200$ $M)$, $CN^-(1.00 \times 10^{-6}$ $M)$

*13. Indicate which direction the reactions in Problem 1 will proceed if all of the species are initially at unit activity. (For $Tl^{3+} + Br^- + 2e \rightleftarrows TlBr(s)$, $E^0 = 1.44V$.)

14. Indicate which direction the reactions in Problem 2 will proceed if all of the species are initially at unit activity.

*15. Calculate the theoretical cell potential for each of the following. Is the cell, as written, galvanic or electrolytic?
  (a) Pb|$PbSO_4$(sat'd), $SO_4^{2-}(0.200$ $M)$| |$Sn^{2+}(0.150$ $M)$, $Sn^{4+}(0.250$ $M)$|Pt
  (b) Pt|$Fe^{3+}(0.0100$ $M)$, $Fe^{2+}(0.00100$ $M)$| |$Ag^+(0.0350$ $M)$|Ag

    (c) $Cu|CuI(sat'd), KI(0.0100\ F)|\ |KI(0.200\ F), CuI(sat'd)|Cu$

    (d) $Pt|UO_2^{2+}(0.100\ M), U^{4+}(0.0100\ M), H^+(1.00\times10^{-6}\ M)|\ |$
$$AgCl(sat'd), KCl(1.00\times10^{-4}\ M)|Ag$$

    (e) $Hg|Hg_2Cl_2(sat'd), Cl^-(0.0500\ M)|\ |V^{2+}(0.200\ M), V^{3+}(0.300\ M)|Pt$

    (f) $Pt|VO^{2+}(0.250\ M), V^{3+}(0.100\ M), H^+(1.00\times10^{-3}\ M)|\ |$
$$Tl^{3+}(0.100\ M), Tl^+(0.0500\ M)|Pt$$

16.   Calculate the theoretical cell potential for each of the following. Is the cell, as written, galvanic or electrolytic?

    (a) $Ag|AgBr(sat'd), Br^-(0.0400\ M)|\ |H^+(1.00\times10^{-4}\ M)|H_2(0.90\ atm), Pt$

    (b) $Pt|Cr^{3+}(0.0500\ M), Cr^{2+}(0.0250\ M)|\ |Ni^{2+}(0.0100\ M)|Ni$

    (c) $Ag|Ag(CN)_2^-(0.240\ M), CN^-(0.100\ M)|\ |$
$$Br_2(1.00\times10^{-3}\ M), KBr(0.200\ F)|Pt$$

    (d) $Ag|AgCl(sat'd), HCl(5.00\times10^{-3}\ M)|H_2(0.300\ atm), Pt$

    (e) $Hg|Hg_2Cl_2(sat'd), HCl(0.0050\ F)|\ |HCl(1.50\ F), Hg_2Cl_2(sat'd)|Hg$

    (f) $Pt|TiO^{2+}(0.200\ F), Ti^{3+}(0.100\ F), H^+(2.00\times10^{-3}\ M)|\ |$
$$SO_4^{2-}(0.200\ M), PbSO_4(sat'd)|Pb$$

*17.   The solubility-product constant for Ag_2SO_3 is 1.5×10^{-14}. Calculate E^0 for the process

$$Ag_2SO_3(s) + 2e \rightarrow 2Ag(s) + SO_3^{2-}$$

18.   The solubility-product constant for $Ni_2P_2O_7$ is 1.7×10^{-13}. Calculate E^0 for the process

$$Ni_2P_2O_7(s) + 4e \rightarrow 2Ni(s) + P_2O_7^{4-}$$

*19.   Calculate the solubility product of Hg_2SO_4, given the standard potentials

$$Hg_2SO_4(s) + 2e \rightleftarrows 2Hg(l) + SO_4^{2-} \qquad E^0 = 0.615\ V$$

$$Hg_2^{2+} + 2e \rightleftarrows Hg(l) \qquad E^0 = 0.789\ V$$

20.   Calculate the solubility product of Ag_2MoO_4, given the standard potentials

$$Ag_2MoO_4(s) + 2e \rightleftarrows 2Ag(s) + MoO_4^{2-} \qquad E^0 = 0.486\ V$$

$$Ag^+ + e \rightleftarrows Ag(s) \qquad E^0 = 0.799\ V$$

*21.   Compute E^0 for the process

$$ZnY^{2-} + 2e \rightleftarrows Zn(s) + Y^{4-}$$

where Y^{4-} is the completely deprotonated anion of EDTA. The formation constant for ZnY^{2-} has a value of 3.2×10^{16}.

22.   Calculate E^0 for the process

$$VY^- + e \rightarrow VY^{2-}$$

if the formation constant for the EDTA complex of V^{2+} is 5.0×10^{12} and for the V^{3+} complex is 7.9×10^{25}.

*23.   A silver electrode immersed in $1.00\times10^{-2}\ F\ Na_2SeO_3$ that is saturated with Ag_2SeO_3 acts as a cathode when coupled with a standard hydrogen electrode. Calculate K_{sp} for Ag_2SeO_3 if this cell develops a potential of 0.450 V.

24.   A lead electrode, immersed in a solution that has a pH of 8.00, is $2.00\times10^{-3}\ F$ in KBr, is saturated with PbOHBr, and acts as an anode when coupled with a standard hydrogen electrode. Calculate K_{sp} for PbOHBr if the cell develops a potential of 0.303 V.

*25. The potential of the following cell was found to be 0.605 V:

$$SHE \mid \mid Hg(OAc)_2 (2.50 \times 10^{-3} \ M), \ OAc^- (0.0500 \ M) \mid Hg$$

where $Hg(OAc)_2$ is the neutral acetate complex of Hg^{2+}. Calculate its formation constant.

26. The potential for the following cell was 1.072 V:

$$Zn \mid X^- (0.150 \ M), \ ZnX_4{}^{2-} (6.00 \times 10^{-2} \ M) \mid \mid SHE$$

Calculate the formation constant for $ZnX_4{}^-$.

*27. The following cell was employed to determine the formation constant of the citrate (Cit^{3-}) complex of Cu(II):

$$Cu \mid CuCit^- (0.0400 \ F), \ Na_3Cit (0.100 \ F), \ H^+ (1.00 \times 10^{-6} \ M) \mid \mid SHE$$

(Note that Cit^{3-} is the conjugate base of $HCit^{2-}$, H_2Cit^-, and H_3Cit.) Its potential was found to be 0.091 V. Calculate the formation constant for $CuCit^-$.

28. To determine the formation constant for the EDTA complex MY^{2-}, 25.0 ml of 0.100-F MCl_2 were mixed with 25.0 ml of 0.200-F Na_2H_2Y. The solution was then diluted to 100.0 ml with a pH 9.00 buffer. An electrode of M was found to behave as a cathode in this solution and develop a potential of 0.373 V when coupled with a standard hydrogen electrode. Calculate the formation constant for MY^{2-} ($E_M^0 = 0.889$ V).

*29. The following cell was employed to determine the dissociation constant of the weak acid HA:

$$Pt, \ H_2 (1.00 \ atm) \mid NaA (0.250 \ F), \ HA (0.150 \ F) \mid \mid SHE$$

The potential was 0.310 V. Calculate K_a.

30. The following cell was employed to determine the dissociation constant of the amine RNH_2:

$$Pt, \ H_2 (1.00 \ atm) \mid RNH_2 (0.0540 \ F), \ RNH_3Cl (0.0750 \ F) \mid \mid SHE$$

where RNH_3Cl is the chloride salt of the amine. The potential of the cell was 0.481 V. Calculate K_b.

THEORY OF OXIDATION-REDUCTION TITRATIONS

The successful application of an oxidation-reduction reaction to volumetric analysis requires, among other things, the means for detecting the equivalence point. We must therefore examine the changes that occur in the course of such a titration and pay particular attention to those changes that are most pronounced in the region of the equivalence point.

Titration Curves

In the titration curves considered thus far, the negative logarithm of the concentration (the p-function) of one of the reacting species has been plotted against the volume of reagent added. The species chosen for each curve has been one to which the indicator for the reaction is sensitive. Most of the indicators used for oxidation-reduction titrations are themselves oxidizing or reducing agents that respond to changes in the potential of the system rather than to changes in concentration of any particular reactant or product. For this reason, the usual practice is to plot the electrode potential for the system as the ordinate of the curve for an oxidation-reduction titration rather than a p-function for a reactant.

To develop a clear understanding of what is meant by the term "electrode potential for the system," consider the titration of iron(II) with cerium(IV):

$$Ce^{4+} + Fe^{2+} \rightleftarrows Ce^{3+} + Fe^{3+}$$

Equilibrium is attained after each addition of titrant. After the first addition of cerium(IV), then, all four species will be present in amounts dictated by the equilibrium constant for the reaction. Recall, now (p. 309), that at equilibrium, the electrode potentials for the two half-reactions are identical; that is, at any point in the titration,

$$E_{Ce^{4+}} = E_{Fe^{3+}} = E_{system}$$

and it is this potential that we call the potential of the system. If a reversible oxidation-reduction indicator is present in the solution as well, its potential must also be the same as that for the system. That is, the ratio between the two forms of the indicator changes until

$$E_{In} = E_{Ce^{4+}} = E_{Fe^{3+}} = E_{system}$$

The potential of a system can be measured experimentally by determining the emf of a suitable cell. Thus, for the titration of iron(II) with cerium(IV), the analyte solution could be made part of the cell:

$$SHE \,||\, Ce^{4+}, Ce^{3+}, Fe^{3+}, Fe^{2+} | Pt$$

Here the potential of the platinum electrode (versus the SHE) is determined both by the affinity of Fe^{3+} for electrons

$$Fe^{3+} + e \rightleftarrows Fe^{2+}$$

as well as that of Ce^{4+}

$$Ce^{4+} + e \rightleftarrows Ce^{3+}$$

At equilibrium, the concentration ratios of oxidized and reduced forms of each species are such that these two affinities (and thus their electrode potentials) are identical. Note that the concentration ratios for both species vary continuously as the titration proceeds; so also must E_{system}. It is the characteristic variation in this parameter that provides a means of end-point detection.

In deriving E_{system} data for a titration curve, either $E_{Ce^{4+}}$ or $E_{Fe^{3+}}$ can be employed. We choose the more convenient one for any particular calculation. Short of the equivalence point, the concentrations of iron(II), iron(III), and cerium(III) are readily deduced from the amount of titrant that has been added; the concentration of cerium(IV), however, is vanishingly small. Thus, application of the Nernst equation for the iron(III)/(II) couple directly provides a value for the potential of the system. The corresponding expression involving the cerium(IV)/(III) couple would give the same answer; however, it would first be necessary to calculate a value for the concentration of cerium(IV), which in turn would require evaluation of the equilibrium constant for the reaction. The situation is reversed after excess titrant has been introduced. Here, the concentrations of cerium(IV), cerium(III), and iron(III) are immediately available, while the concentration for iron(II) would require a preliminary calculation involving the equilibrium constant. Thus, the potential of the system is most directly evaluated from the cerium(IV)/(III) couple.

EQUIVALENCE-POINT POTENTIAL

The potential of an oxidation-reduction system at the equivalence point is of particular importance from the standpoint of indicator selection. Calculation of the equivalence potential is also unique in that there is insufficient stoichiometric information to permit direct use of the Nernst equation for either half-cell process. Using the iron(II)/cerium(IV) titration as an example, values for the formal concentrations of cerium(III) and iron(III) at the equivalence point are easily calculated. On the other hand, we know only that the concentrations of cerium(IV) and iron(II) are small and numerically equal. As before, the concentrations of the minor constituents can be calculated from the equilibrium-constant expression. Alternatively, we may write that, in common with any other point in the titration, the system potential at equivalence, E_{eq}, is given by

$$E_{eq} = E^0_{Ce^{4+}} - 0.0591 \log \frac{[Ce^{3+}]}{[Ce^{4+}]}$$

and also by

$$E_{eq} = E^0_{Fe^{3+}} - 0.0591 \log \frac{[Fe^{2+}]}{[Fe^{3+}]}$$

Upon adding these expressions, we obtain

$$2E_{eq} = E^0_{Ce^{4+}} + E^0_{Fe^{3+}} - 0.0591 \log \frac{[Ce^{3+}][Fe^{2+}]}{[Ce^{4+}][Fe^{3+}]}$$

Note that the ratio in this expression is *not* the equilibrium constant. We know from stoichiometric considerations that *at the equivalence point,*

$$[Fe^{3+}] = [Ce^{3+}]$$

$$[Fe^{2+}] = [Ce^{4+}]$$

Thus,

$$2E_{eq} = E^0_{Ce^{4+}} + E^0_{Fe^{3+}} - 0.0591 \log \frac{[\cancel{Ce^{3+}}][\cancel{Fe^{2+}}]}{[\cancel{Ce^{4+}}][\cancel{Fe^{3+}}]}$$

and

$$E_{eq} = \frac{E^0_{Ce^{4+}} + E^0_{Fe^{3+}}}{2}$$

The equilibrium concentrations of the reacting species can be readily calculated from the equivalence-point potential.

Example. Calculate the concentration of the various reactants and products at the equivalence point in the titration of a 0.100-N solution of Fe^{2+} with 0.100-N Ce^{4+} at 25°C if both solutions are 1.0 F in H_2SO_4.

Here it is convenient to substitute formal potentials (see p. 313) into the derived expression for the equivalence-point potential. That is,

$$E_{eq} = \frac{+1.44 + 0.68}{2} = 1.06 \text{ V}$$

The Nernst equation allows us to evaluate the molar ratio of Fe^{2+} to Fe^{3+} at the equivalence point

$$+1.06 = +0.68 - 0.0591 \log \frac{[Fe^{2+}]}{[Fe^{3+}]}$$

$$\log \frac{[Fe^{2+}]}{[Fe^{3+}]} = -\frac{0.38}{0.0591} = -6.4$$

$$\frac{[Fe^{2+}]}{[Fe^{3+}]} = 4 \times 10^{-7}$$

It is clear that most of the Fe^{2+} has been converted to Fe^{3+} at the equivalence point. As a consequence of dilution, then, the Fe^{3+} concentration will be essentially equal to one-half the original Fe^{2+} concentration. That is,

$$[Fe^{3+}] = \frac{0.100}{2} - [Fe^{2+}] \cong 0.050$$

Thus,

$$[Fe^{2+}] = 4 \times 10^{-7} \times 0.050$$
$$= 2 \times 10^{-8}$$

Finally, the stoichiometry of the reaction requires that

$$[Ce^{4+}] = [Fe^{2+}] \cong 2 \times 10^{-8}$$
$$[Ce^{3+}] = [Fe^{3+}] \cong 0.050$$

Note that identical results would be obtained if the Nernst equation had been applied to the cerium(IV)-cerium(III) system.

Example. Derive the equivalence-point potential for the somewhat more complicated reaction

$$5Fe^{2+} + MnO_4^- + 8H^+ \rightleftarrows 5Fe^{3+} + Mn^{2+} + 4H_2O$$

The respective half-reactions may be written as

$$Fe^{3+} + e \rightleftarrows Fe^{2+}$$
$$MnO_4^- + 8H^+ + 5e \rightleftarrows Mn^{2+} + 4H_2O$$

The potential of this system is given by either

$$E = E^0_{Fe^{3+}} - \frac{0.0591}{1} \log \frac{[Fe^{2+}]}{[Fe^{3+}]}$$

or

$$E = E^0_{MnO_4^-} - \frac{0.0591}{5} \log \frac{[Mn^{2+}]}{[MnO_4^-][H^+]^8}$$

In order to combine the logarithmic terms so that the concentrations of various species cancel, it is necessary to multiply the permanganate half-reaction equation by 5.

$$5E_{eq} = 5E^0_{MnO_4^-} - 0.0591 \log \frac{[Mn^{2+}]}{[MnO_4^-][H^+]^8}$$

After addition we find that

$$6E_{eq} = E^0_{Fe^{3+}} + 5E^0_{MnO_4^-} - 0.0591 \log \frac{[Fe^{2+}][Mn^{2+}]}{[Fe^{3+}][MnO_4^-][H^+]^8}$$

The stoichiometry at the equivalence point requires that

$$[Fe^{3+}] = 5[Mn^{2+}]$$
$$[Fe^{2+}] = 5[MnO_4^-]$$

Substituting and rearranging,

$$E_{eq} = \frac{E^0_{Fe^{3+}} + 5E^0_{MnO_4^-}}{6} - \frac{0.0591}{6} \log \frac{5[MnO_4^-][Mn^{2+}]}{5[Mn^{2+}][MnO_4^-][H^+]^8}$$

$$= \frac{E^0_{Fe^{3+}} + 5E^0_{MnO_4^-}}{6} - \frac{0.0591}{6} \log \frac{1}{[H^+]^8}$$

Note that the equivalence-point potential for this titration is dependent upon the pH.

Example. Derive an expression for the equivalence-point potential for the reaction

$$6Fe^{2+} + Cr_2O_7^{2-} + 14H^+ \rightleftarrows 6Fe^{3+} + 2Cr^{3+} + 7H_2O$$

Proceeding as before, we obtain the expression

$$7E_{eq} = E^0_{Fe^{3+}} + 6E^0_{Cr_2O_7^{2-}} - 0.0591 \log \frac{[Fe^{2+}][Cr^{3+}]^2}{[Fe^{3+}][Cr_2O_7^{2-}][H^+]^{14}}$$

At the equivalence point,

$$[Fe^{2+}] = 6[Cr_2O_7^{2-}]$$
$$[Fe^{3+}] = 3[Cr^{3+}]$$

Substitution of these quantities into the previous equation reveals that

$$E_{eq} = \frac{E^0_{Fe^{3+}} + 6E^0_{Cr_2O_7^{2-}}}{7} - \frac{0.0591}{7} \log \frac{2[Cr^{3+}]}{[H^+]^{14}}$$

Note the equivalence-point potential in the last example is dependent not only upon the concentration of hydrogen ion but also upon that of a product ion. In general, the equivalence-point potential will depend upon the concentration of one of the participants in the reaction whenever there exists a molar ratio other than unity between the species containing that participant as a reactant and as a product.

VARIATION IN POTENTIAL AS A FUNCTION OF REAGENT VOLUME

The shape of the titration curve for an oxidation-reduction depends upon the nature of the system under consideration. The derivation of typical curves will illustrate the effects of several important variables.

Example. Derive a curve for the titration of 50.00 ml of 0.0500-N iron(II) with 0.1000-N cerium(IV). Assume that both solutions are 1.0 F in H_2SO_4.

Here we will use formal rather than standard potentials.

1. *Initial Potential.*   The solution contains no cerium ions. It will have a small but unknown concentration of iron(III) due to air oxidation of iron(II). Thus, we cannot calculate an initial potential that has any significance.

2. *Addition of 5.00 ml of Cerium(IV).* With the introduction of a volume of oxidant, the solution acquires appreciable concentrations of three of the participating ions; that for the fourth, cerium(IV), will be small.

The concentration of cerium(III) is given by its formal concentration less that for the unreacted cerium(IV) or

$$[Ce^{3+}] = \frac{5.00 \times 0.1000}{55.00} - [Ce^{4+}] \cong \frac{0.500}{55.00}$$

The approximation appears reasonable since the equilibrium constant for the reaction is large. Similarly,

$$[Fe^{3+}] = \frac{5.00 \times 0.1000}{55.00} - [Ce^{4+}] \cong \frac{0.500}{55.00}$$

$$[Fe^{2+}] = \frac{50.00 \times 0.0500 - 5.00 \times 0.100}{55.00} + [Ce^{4+}]$$

$$\cong \frac{2.00}{55.00}$$

As we have shown, the potential for the system can be calculated with the aid of *either of the two equations*

$$E = E^0_{Ce^{4+}} - 0.0591 \log \frac{[Ce^{3+}]}{[Ce^{4+}]}$$

$$= E^0_{Fe^{3+}} - 0.0591 \log \frac{[Fe^{2+}]}{[Fe^{3+}]}$$

The second equation is the more convenient for this calculation since the two concentrations that appear in it are known within acceptable limits. Therefore, substituting for the iron(III) and iron(II) concentrations,

$$E = +0.68 - 0.0591 \log \frac{2.00/55.00}{0.500/55.00}$$

$$= +0.64 \text{ V}$$

Had we used the formal potential for the cerium(IV)-cerium(III) system and the equilibrium concentrations of these ions, an identical potential would have been obtained.

Further values for the potentials needed to define a curve short of the end point can be calculated in a fashion strictly analogous to that in step 2. Table 15-1 contains a number of these data. The student should check one or two to be sure he understands how they were obtained.

3. *Equivalence-Point Potential.* We have seen that the potential at the equivalence point in this titration has a value of 1.06 V.

4. *Addition of 25.10 ml of Reagent.* The solution now contains an excess of quadrivalent cerium in addition to equivalent quantities of iron(III) and cerium(III) ions. The concentration of iron(II) will be very small. Therefore,

$$[Fe^{3+}] = \frac{25.00 \times 0.1000}{75.10} - [Fe^{2+}] \cong \frac{2.500}{75.10}$$

$$[Ce^{3+}] = \frac{25.00 \times 0.1000}{75.10} - [Fe^{2+}] \cong \frac{2.500}{75.10}$$

$$[Ce^{4+}] = \frac{25.10 \times 0.1000 - 50.00 \times 0.0500}{75.10} + [Fe^{2+}] \cong \frac{0.010}{75.10}$$

TABLE 15-1   Electrode Potentials during Titrations of 50.0 ml of 0.0500-N Iron(II) Solutions[a]

Volume of 0.100-N Reagent, ml	Potential vs. Standard Hydrogen Electrode, V	
	Titration with Ce^{4+}	Titration with MnO_4^-
5.00	0.64	0.64
15.00	0.69	0.69
20.00	0.72	0.72
24.00	0.76	0.76
24.90	0.82	0.82
25.00	1.06 ← equivalence point → 1.37	
25.10	1.30	1.48
26.00	1.36	1.49
30.00	1.40	1.50

[a] H_2SO_4 concentration $= 1.0\ F$ throughout.

These approximations should be reasonable in view of the favorable equilibrium constant. As before, the desired potential could be calculated from the iron(III)-iron(II) system. At this stage in the titration, however, it is more convenient to use the cerium(IV)-cerium(III) potential, since the concentrations of these species are immediately available. Thus,

$$E = +1.44 - 0.0591 \log \frac{[Ce^{3+}]}{[Ce^{4+}]}$$

$$= +1.44 - 0.0591 \log \frac{2.500/\cancel{75.10}}{0.010/\cancel{75.10}}$$

$$= +1.30\ V$$

The additional postequivalence-point potentials shown in Table 15-1 were calculated in a similar fashion.

The titration for iron(II) with cerium(IV) appears as curve A in Figure 15-1. Its shape is similar to the curves encountered in neutralization, precipitation, and complex-formation titrations, the equivalence point being signaled by a large change in the ordinate function. A titration involving 0.01-F reactants will yield a curve that is, for all practical purposes, identical with the one that was derived, since the electrode potentials are independent of dilution.

The titration curve just considered is symmetric about the equivalence point because the molar ratio of oxidant to reductant is equal to unity. As demonstrated by the following example, an asymmetric curve results if the ratio differs from this value.

Example.   Derive a curve for the titration of 50.00 ml of 0.0500-N iron(II) with 0.1000-N $KMnO_4$. For convenience, assume that the solution is 1.0 F in H_2SO_4 throughout.

No formal electrode potential is available for MnO_4^-; we must, therefore, employ its standard potential of 1.51 V.

The reaction is

$$5Fe^{2+} + MnO_4^- + 8H^+ \rightleftarrows 5Fe^{3+} + Mn^{2+} + 4H_2O$$

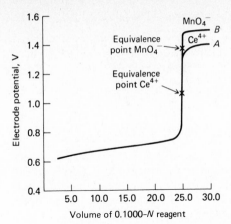

Figure 15-1 Titration Curves for 50.00 ml of 0.0500-N Fe(III) with (curve A) 0.1000-N Ce(IV) and (curve B) 0.1000-N KMnO$_4$.

It should be noted that molar rather than normal concentrations are employed in the Nernst equations. For manganese species in the foregoing reaction the molarity is equal to one-fifth of the normality.

 1. *Preequivalence-Point Potentials.* The preequivalence-point potentials are most easily calculated from the concentrations of iron(II) and iron(III) in the solution. Their values will be identical to those computed for the titration with cerium(IV) ion.

 2. *Equivalence-Point Potential.* The equivalence-point potential for this reaction is given by the equation

$$E = \frac{E^0_{Fe^{3+}} + 5E^0_{MnO_4^-}}{6} - \frac{0.0591}{6} \log \frac{1}{[H^+]^8}$$

A calculation similar to that on page 223 reveals that in a 1.0-F solution of H_2SO_4

$$[H^+] \cong 1.0$$

Therefore,

$$E = \frac{0.68 + 5(+1.51)}{6} - \frac{0.0591}{6} \log \frac{1}{(1.0)^8}$$

$$= 1.37 \text{ V}$$

 3. *Postequivalence-Point Potentials.* When 25.10 ml of the 0.1000-N KMnO$_4$ have been added, stoichiometry requires that

$$[Fe^{3+}] = \frac{50.00 \times 0.0500}{75.10} - [Fe^{2+}] \cong \frac{2.500}{75.10}$$

$$[Mn^{2+}] = \frac{1}{5}\left(\frac{50.00 \times 0.0500}{75.10} - [Fe^{2+}]\right) \cong \frac{0.500}{75.10}$$

$$[MnO_4^-] = \frac{1}{5}\left(\frac{25.10 \times 0.1000 - 50.00 \times 0.0500}{75.10} + [Fe^{+2}]\right) \cong \frac{2.0 \times 10^{-3}}{75.10}$$

It is now advantageous to calculate the electrode potential from the standard potential of the manganese system. That is,

$$E = 1.51 - \frac{0.0591}{5} \log \frac{[Mn^{2+}]}{[H^+]^8[MnO_4^-]}$$

$$= 1.51 - \frac{0.0591}{5} \log \frac{0.500/\cancel{75.10}}{(1.0)^8(2.0 \times 10^{-3})/\cancel{75.10}}$$

$$= 1.48$$

Table 15-1 contains additional data obtained in this way.

Figure 15-1 depicts curves for the titration of iron(II) with both permanganate and cerium(IV). Note that the plots for both are alike to within 99.9% of the equivalence point; however, the equivalence-point potentials are quite different. Further, the permanganate curve is markedly asymmetric, the potential increasing only slightly beyond the equivalence point. Finally, the total change in potential associated with equivalence is somewhat greater with the permanganate titration, owing to the more favorable equilibrium constant for this reaction.

Effect of Concentration on Redox Titration Curves. It is of importance to note that the ordinate function, E_{system}, in the preceding calculations is ordinarily determined by the logarithm of a *ratio* of concentrations; it is, therefore, *independent* of dilution over a considerable range.[1] As a consequence, curves for oxidation-reduction titrations, in distinct contrast to those for other reaction types, tend to be independent of analyte and reagent concentrations.

Effect of the Completeness of Reaction on Redox Titration Curves. The change in the ordinate function in the equivalence-point region of an oxidation-reduction titration becomes larger as the reaction becomes more complete; this effect is identical with that encountered for other reaction types. Thus, in Figure 15-2, data are plotted for titrations involving a hypothetical analyte that has a standard potential of 0.2 V with several reagents that have standard potentials ranging from 0.4 to 1.2 V; the corresponding equilibrium constants for the reaction (Chapter 14) range from about 2×10^3 to 9×10^{16}. Clearly, an increase in completeness is accompanied by an increase in the electrode potential for the system.

The curves in Figure 15-2 were derived for reactions in which the oxidant and reductant each exhibit a one-electron change; where both reactants undergo a two-electron transfer, the change in potential in the region of 24.9 to 25.1 ml is larger by about 0.14 V.

TITRATION OF MIXTURES

Solutions containing two oxidizing agents or two reducing agents will yield titration curves that contain two inflection points, provided the standard potentials for the two species are sufficiently different. If this difference is greater than

---

[1] When the solution becomes sufficiently dilute so that the reaction is incomplete, E_{system} will in fact vary with further dilution. The magnitude of this effect can be determined by dispensing with the usual approximation that the formal and molar concentrations are identical.

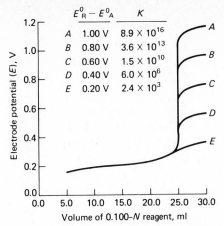

Figure 15-2   Titration of 50.0 ml of 0.0500-N A. E^0_A is assumed to be 0.200 V. From the top, E^0_R for the reagent is 1.20, 1.00, 0.80, 0.60, 0.40 V, respectively. Both the reagent and analyte are assumed to undergo a one-electron change.

about 0.2 V, the end points are usually distinct enough to permit analysis for each component. This situation is quite comparable to the titration of two acids having different dissociation constants or of two ions forming precipitates of different solubilities with the same reagent.

In addition, the behavior of a few redox systems is analogous to that of polyfunctional acids or bases. For example, consider the two half-reactions

$$VO^{2+} + 2H^+ + e \rightleftarrows V^{3+} + H_2O \qquad E^0 = +0.361 \text{ V}$$

$$V(OH)_4^+ + 2H^+ + e \rightleftarrows VO^{2+} + 3H_2O \qquad E^0 = +1.00 \text{ V}$$

The curve for the titration of V^{3+} with a strong oxidizing agent such as permanganate will have two inflection points; the first will correspond to oxidation of the V^{3+} to VO^{2+} and the second to oxidation of VO^{2+} to $V(OH)_4^+$. The stepwise oxidation of molybdenum(III), first to the $+5$ oxidation state and subsequently to the $+6$ state, is another common example. Here, again, satisfactory inflections occur in the curves because the potential difference between the pertinent half-reactions is about 0.4 V.

Derivation of titration curves for either type of mixture is not difficult if the difference in standard potential is sufficiently great. An example is the titration of a solution containing iron(II) and titanium(III) ions with potassium permanganate. The standard potentials for these systems are

$$TiO^{2+} + 2H^+ + e \rightleftarrows Ti^{3+} + H_2O \qquad E^0 = +0.1 \text{ V}$$

$$Fe^{3+} + e \rightleftarrows Fe^{2+} \qquad E^0 = +0.77 \text{ V}$$

The first additions of permanganate are used up by the more readily oxidized titanium(III) ion; as long as an appreciable concentration of this species remains in solution, the potential of the system cannot become high enough to alter greatly the concentration of iron(II) ions. Thus, the first part of the titration curve can be defined from the stoichiometric proportions of titanium(III) and titanium(IV) ion, with the relationship

$$E = +0.1 - 0.0591 \log \frac{[Ti^{3+}]}{[TiO^{2+}][H^+]^2}$$

For all practical purposes, then, the first part of this curve is identical to the titration curve for titanium(III) ion by itself. Beyond the first equivalence point, the solution will contain both iron(II) and iron(III) ions in appreciable concentrations, and the points on the curve can be most conveniently obtained from the relationship

$$E = +0.77 - 0.0591 \log \frac{[Fe^{2+}]}{[Fe^{3+}]}$$

Throughout this region and beyond the second equivalence point, the curve will be essentially identical to that for the titration of iron(II) ion alone (see Figure 15-1). This, then, leaves undefined only the potential at the first equivalence point. A convenient way of estimating its value is to add the Nernst equations for the iron(II) and titanium(III) potentials. Since the potentials for all oxidation-reduction systems in solution will be identical at equilibrium, we can write that

$$2E = +0.1 + 0.77 - 0.0591 \log \frac{[Ti^{3+}][Fe^{2+}]}{[TiO^{2+}][Fe^{3+}][H^+]^2}$$

The principal source of iron(III) ions in the solution at this point is the reaction

$$2H^+ + TiO^{2+} + Fe^{2+} \rightleftarrows Fe^{3+} + Ti^{3+} + H_2O$$

Thus, we may write

$$[Fe^{3+}] \cong [Ti^{3+}]$$

Substitution of this equality into the previous equation for the potential yields

$$E = \frac{+0.87}{2} - \frac{0.0591}{2} \log \frac{[Fe^{2+}]}{[TiO^{2+}][H^+]^2}$$

Finally, if $[TiO^{2+}]$ and $[Fe^{2+}]$ are assumed to be essentially identical to their formal concentrations, we can compute the equivalence-point potential.

A titration curve for a mixture of iron(II) and titanium(III) ions is shown in Figure 15-3.

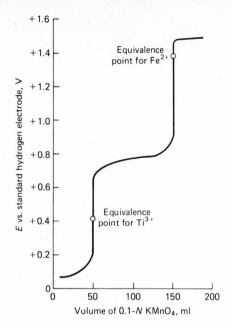

Figure 15-3 Curve for the Titration of a 50.0-ml Sample That Is 0.100 F with Respect to Ti^{3+} and 0.200 F with Respect to Fe^{2+} with 0.100-N MnO_4^-. The concentration of H^+ ions is assumed to be 1.00 M throughout.

Oxidation-Reduction Indicators

We have seen that the equivalence point in an oxidation-reduction titration is characterized by a marked change in the electrode potential of the system. Several methods exist for detecting such a change, and these can serve to signal the end point in the titration.

CHEMICAL INDICATORS

Indicators for oxidation-reduction titrations are of two types. *Specific indicators* owe their behavior to a reaction with one of the participants in the titration. *True oxidation-reduction indicators*, on the other hand, respond to the potential of the system rather than to the appearance or disappearance of a particular species during the titration.

Specific Indicators. Perhaps the best-known specific indicator is starch, which forms a dark-blue complex with triiodide ion. This complex serves to signal the end point in titrations in which iodine is either produced or consumed. Another specific indicator is potassium thiocyanate, which may be employed, for example, in the titration of iron(III) with solutions of titanium(III) sulfate; the end point involves disappearance of the iron(III)-thiocyanate complex, owing to a marked decrease in the iron(III) concentration at the equivalence point.

True Oxidation-Reduction Indicators. True oxidation-reduction indicators enjoy wider application, since their behavior is dependent only upon the change in the potential of the system.

The half-reaction responsible for color change in a typical true oxidation-reduction indicator can be written as

$$In_{ox} + ne \rightleftarrows In_{red}$$

If the indicator reaction is reversible, we may write

$$E = E^0 - \frac{0.0591}{n} \log \frac{[In_{red}]}{[In_{ox}]}$$

Typically, a change from the color of the oxidized form of the indicator to the reduced involves a change in the ratio of reactant concentration of about 100. That is, when

$$\frac{[In_{red}]}{[In_{ox}]} \geqq \frac{1}{10}$$

changes to

$$\frac{[In_{red}]}{[In_{ox}]} \leqq 10$$

The conditions for the full color change for a typical indicator can be found by substituting these values into the Nernst equation[2]

$$E = E^0 \pm \frac{0.0591}{n} \tag{15-1}$$

Equation 15-1 suggests that a typical indicator will exhibit a detectable color change when the titrant causes a shift of about $(0.118/n)$ V in the potential of the system. With many indicators, $n = 2$; a change of 0.059 V is thus sufficient.

The potential at which a color transition will occur depends upon the standard potential for the particular indicator system. Table 15-2 shows that indicators which function in any desired potential range up to about +1.25 V are available. Structures for and reactions of a few of these indicators are considered in the paragraphs that follow.

Iron(II) Complexes of the Orthophenanthrolines. A class of organic compounds known as 1,10-phenanthrolines (or orthophenanthrolines) forms stable complexes with iron(II) and certain other ions. The parent compound has a pair of nitrogen atoms located in such positions that each can form a covalent

---

[2] It should be noted that protons are involved in the reduction of some indicators; for these, the transition potentials will be pH-dependent.

TABLE 15-2   A Selected List of Oxidation-Reduction Indicators[a]

Indicators	Color		Transition Potential, V	Conditions
	Oxidized	Reduced		
5-Nitro-1, 10-phenanthroline iron(II) complex	Pale blue	Red-violet	+1.25	1-F H_2SO_4
2,3′-Diphenylamine dicarboxylic acid	Blue-violet	Colorless	+1.12	7–10-F H_2SO_4
1,10-Phenanthroline iron(II) complex	Pale blue	Red	+1.11	1-F H_2SO_4
Erioglaucin A	Bluish red	Yellow-green	+0.98	0.5-F H_2SO_4
Diphenylamine sulfonic acid	Red-violet	Colorless	+0.85	Dilute acid
Diphenylamine	Violet	Colorless	+0.76	Dilute acid
p-Ethoxychrysoidine	Yellow	Red	+0.76	Dilute acid
Methylene blue	Blue	Colorless	+0.53	1-F acid
Indigo tetrasulfonate	Blue	Colorless	+0.36	1-F acid
Phenosafranine	Red	Colorless	+0.28	1-F acid

[a] Data taken in part from I. M. Kolthoff and V. A. Stenger, *Volumetric Analysis*, 2d ed., vol. I, p. 140. New York: Interscience Publishers, Inc., 1942.

bond with the iron(II) ion. Three orthophenanthroline molecules combine with each iron ion to yield a complex with the structure

This complex is sometimes called "ferroin"; for convenience its formula will be written as $(Ph)_3Fe^{2+}$.

The complexed iron in the ferroin undergoes a reversible oxidation-reduction reaction that may be written

$$(Ph)_3Fe^{3+} + e \rightleftarrows (Ph)_3Fe^{2+} \qquad E^0 = +1.06 \text{ V}$$
$$\text{pale blue} \qquad\qquad \text{red}$$

The iron(III) complex is a pale blue; in practice, the color change associated with the reduction is actually from nearly colorless to red. Because of the difference in color intensity, the end point is usually taken when only about 10% of the indicator is in the iron(II) form. The transition potential is thus approximately +1.11 V in one formal sulfuric acid.

Of all the oxidation-reduction indicators, ferroin approaches most closely the ideal substance. Its color change is very sharp, and its solutions are readily prepared and stable. In contrast to many indicators, the oxidized form is remarkably inert toward strong oxidizing agents. The indicator reaction is rapid and reversible. At temperatures above 60°C, ferroin is decomposed.

A number of substituted phenanthrolines have been investigated for their indicator properties, and some have proved to be as useful as the parent compound. Among these, the 5-nitro and the 5-methyl derivatives are noteworthy, with transition potentials of +1.25 V and +1.02 V, respectively.

Diphenylamine and Its Derivatives. One of the first true redox indicators to be discovered was diphenylamine, $C_{12}H_{11}N$. This compound was recommended by Knop in 1924 for the titration of iron(II) with potassium dichromate.

In the presence of a strong oxidizing agent, diphenylamine is believed to undergo the following reactions:

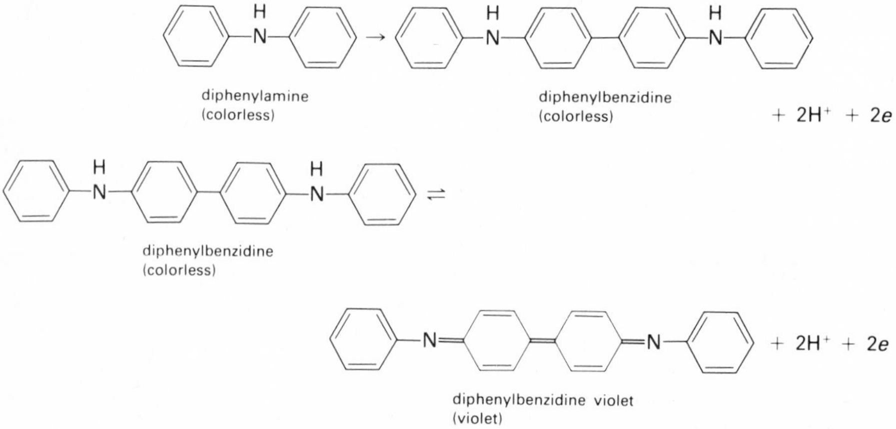

diphenylamine   diphenylbenzidine
(colorless)   (colorless) $+ 2H^+ + 2e$

diphenylbenzidine
(colorless)

$+ 2H^+ + 2e$

diphenylbenzidine violet
(violet)

The first reaction, involving the formation of the colorless diphenylbenzidine, is nonreversible; the second, however, giving a violet product, can be reversed and constitutes the actual indicator reaction.

The reduction potential for the second reaction is about +0.76 V. Despite the fact that hydrogen ions appear to be involved, variations in acidity have little effect upon the magnitude of this potential. This may be due to association of hydrogen ions with the colored product.

There are drawbacks in the application of diphenylamine as an indicator. Because of its low solubility in water, for example, the reagent must be prepared in rather concentrated sulfuric acid solutions. Further, the oxidation product forms a sparingly soluble precipitate with tungstate ion, which precludes its use in the presence of this element. Finally, the indicator reaction is slowed by mercury(II) ions.

The sulfonic acid derivative of diphenylamine does not suffer from these disadvantages:

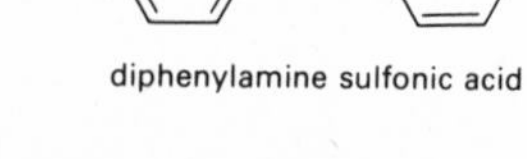

diphenylamine sulfonic acid

The barium or sodium salt of this acid may be used to prepare aqueous indicator solutions; these salts behave in essentially the same manner as the parent substance. The color change is somewhat sharper, passing from colorless through green to a deep violet. The transition potential is about $+0.8$ V and again is independent of acid concentration. The sulfonic acid derivative is now widely used in redox titrations.

It might be surmised from the two equations for the indicator reaction of diphenylamine that diphenylbenzidine should behave in an identical fashion and consume less oxidizing agent in its reaction. Unfortunately, the low solubility of diphenylbenzidine in water and sulfuric acid has precluded its widespread use. As might be expected, the sulfonic acid derivative of diphenylbenzidine has proved to be a satisfactory indicator.

Starch-Iodine Solution. Most commonly, starch is used as a specific indicator in oxidation-reduction titrations in which iodine is a reactant. As pointed out by Kolthoff and Stenger,[3] however, a solution of starch containing a little iodine or iodide ion can function as a true redox indicator as well. In the presence of a strong oxidizing agent, the iodide-iodide ratio is high, and the blue color of the iodine-starch complex is seen. With a strong reducing agent, on the other hand, iodide ion predominates, and the blue color disappears. Thus, the indicator system changes from colorless to blue in the titration of many strong reducing agents with various oxidizing agents. This color change is quite independent of the chemical composition of the reactants, depending only upon the potential of the system at the equivalence point.

CHOICE OF CHEMICAL INDICATORS

Referring again to Figure 15-2, it is apparent that all of the indicators in Table 15-2 except for the first and the last could be employed with reagent A. On the other hand, only indigo tetrasulfonate could be employed with reagent D. The change in potential with reagent E is too small to be satisfactorily detected by an indicator. For a reaction involving a one-electron transfer, the difference in standard potentials for the reagent and analyte must exceed about 0.4 V if the titration error is to be kept minimal; for a two-electron transfer, a difference of about 0.25 V is necessary.

POTENTIOMETRIC END POINTS

End points in many oxidation-reduction titrations are readily observed by making the solution of the analyte a part of the cell

reference electrode | |analyte solution|Pt

The potential of this cell, which is measured, will then vary in a way analogous to that shown in Figures 15-1 and 15-2. The end point can be determined from a plot of the measured potential as a function of titrant volume.

[3] I. M. Kolthoff and V. A. Stenger, *Volumetric Analysis*, 2d ed., vol. I, p. 105. New York: Interscience Publishers, Inc., 1942.

The reference electrode could be a standard hydrogen electrode. Usually, however, it is more convenient to use a secondary reference electrode. The experimental titration curves will then take the same form as those in Figures 15-1 and 15-2 but will be displaced on the vertical axis by an amount corresponding to the difference between the potential of the reference electrode and the standard hydrogen electrode. The potentiometric end point is considered in detail in Chapter 17.

SUMMARY

Conclusions based upon the calculations in this chapter are helpful to the chemist as guides to the choice of reaction conditions and indicators for oxidation-reduction titrations. Thus, for example, the curves shown in Figures 15-1 and 15-2 clearly define the range of potentials within which an indicator must exhibit a color change for a successful titration. Nevertheless, it is important to emphasize that these calculations are theoretical and that they may not necessarily take into account all factors that determine the applicability and feasibility of a volumetric method. Also to be considered should be the rates at which both the principal and the indicator reactions occur, the effects of electrolyte concentration, pH, and complexing agents, the presence of colored components other than the indicator in the solution, and the variation in color perception among individuals. The state of chemistry has not advanced to the point where the effects of these variables can be completely determined by computation. Theoretical calculations can and will eliminate useless experiments and act as guides to the ones most likely to be profitable. The final test must always come in the laboratory.

PROBLEMS

*1. Calculate the electrode potential of the system at the equivalence point for each of the following reactions. Where necessary, assume that at equivalence $[H^+] = 0.100$.
  (a) $2Ti^{2+} + Sn^{4+} \rightleftarrows 2Ti^{3+} + Sn^{2+}$
  (b) $2Ce^{4+} + H_2SeO_3 + H_2O \rightleftarrows SeO_4^{2-} + 2Ce^{3+} + 4H^+$ (in 1-F H_2SO_4)
  (c) $2MnO_4^- + 10HNO_2 + 2H_2O \rightleftarrows 2Mn^{2+} + 10NO_3^- + 4H^+$

2. Calculate the electrode potential of the system at the equivalence point for each of the following reactions. Where necessary, assume that at equivalence $[H^+] = 0.100$.
  (a) $Cr^{2+} + Fe(CN)_6^{3-} \rightleftarrows Cr^{3+} + Fe(CN)_6^{4-}$
  (b) $Tl^{3+} + 2V^{3+} + 2H_2O \rightleftarrows Tl^+ + 2VO^{2+} + 4H^+$
  (c) $5U^{4+} + 2MnO_4^- + 2H_2O \rightleftarrows 5UO_2^{2+} + 2Mn^{2+} + 4H^+$

3. Calculate the electrode potential of the system at the equivalence point for each of the following reactions. Assume that the reactant solutions were 0.100 N and that at equivalence $[H_3O^+] = 1.00 \times 10^{-8}$ and $[I^-] = 0.300$.
  *(a) $I_3^- + 2Ti^{3+} + 2H_2O \rightleftarrows 3I^- + 2TiO^{2+} + 2H^+$
  (b) $I_3^- + Sn^{2+} \rightleftarrows 3I^- + Sn^{4+}$

*4. Calculate equilibrium constants for each of the reactions in Problem 1.

5. Calculate equilibrium constants for each of the reactions in Problem 2.

6. Calculate the equivalence point concentration of each of the first-mentioned species in Problem 1 if 0.100-N solutions were employed.

7. Calculate the equivalence point concentration of each of the first-mentioned species in Problem 2 if 0.100-N solutions were employed.

8. Construct curves for the following titrations. Calculate potentials after addition of 10.00, 25.00, 49.00, 49.90, 50.00, 50.10, 51.00, and 60.00 ml of the reagent. Where necessary, assume that $[H^+] = 1.00$ throughout.

  *(a) 50.00 ml of 0.1000-N V^{2+} with 0.1000-N Sn^{4+}

  (b) 50.00 ml of 0.1000-N $Fe(CN)_6^{3-}$ with 0.1000-N Cr^{2+}

  *(c) 50.00 ml of 0.1000-N $Fe(CN)_6^{4-}$ with 0.1000-N Tl^{3+}

  (d) 50.00 ml of 0.1000-N Fe^{3+} with 0.1000-N Sn^{2+}

  *(e) 50.00 ml of 0.01000-N U^{4+} with 0.01000-N MnO_4^-

  (f) 50.00 ml of 0.01000-N Sn^{2+} with 0.01000-N I_3^-
       (Assume that $[I^-] = 0.500$ M throughout the titration.)

APPLICATIONS OF OXIDATION-REDUCTION TITRATIONS

Oxidation-reduction titrations are based upon the reaction of the analyte with a standard solution of an oxidizing agent or a reducing agent. For the titration to be meaningful, the analyte must exist entirely in a single oxidation state at the outset of the titration. An auxiliary oxidizing or reducing agent is frequently required to assure this condition.

In this chapter auxiliary reagents for the adjustment of oxidation state are considered first; following this, the properties and applications of some common standard oxidizing and reducing agents are described.

Auxiliary Reagents

The steps preliminary to an oxidation-reduction titration often leave the analyte in more than one oxidation state. For example, when an iron alloy is dissolved in acid, a mixture of iron(II) and iron(III) results. Before the iron can be titrated, therefore, a reagent must be added that will convert the element quantitatively either to the divalent state for titration with a standard oxidizing agent or to the trivalent state for titration with a reducing agent. Such an auxiliary reagent must possess unique properties. It must be a sufficiently strong oxidizing or reducing agent to convert the analyte quantitatively to the desired oxidation state. Nevertheless, it should not be so strong as to convert other components of the solution into states that will also react with the titrant. Another requirement is that the unused portion of the reagent be easily removed from the solution, for it will almost inevitably interfere with the titration by reacting with the standard solution. Thus, a reagent that would convert iron quantitatively to the divalent state for titration with a standard solution of permanganate would, of necessity, be a good reducing agent. Any excess remaining after the reduction would surely consume permanganate if it was not removed from the solution.

Reagents that find general application to the pretreatment of samples are described in the following paragraphs.

OXIDIZING REAGENTS

Sodium Bismuthate. Sodium bismuthate is an extremely powerful oxidizing agent capable, for example, of converting manganese(II) quantitatively to permanganate. It exists as a sparingly soluble solid of somewhat uncertain composition; its formula is usually written as $NaBiO_3$. Upon reaction, the bismuth(V) is converted to the more common trivalent state. Ordinarily, the solution to be oxidized is boiled in contact with an excess of the solid. The unused reagent is then removed by filtration.

Ammonium Peroxodisulfate. In acid solutions, ammonium peroxodisulfate, $(NH_4)_2S_2O_8$, is a potent oxidizing agent that will convert chromium to dichromate, cerium(III) to the quadrivalent state, and manganese(II) ion to permanganate. The half-reaction is

$$S_2O_8^{2-} + 2e \rightleftarrows 2SO_4^{2-} \qquad E^0 = 2.01 \text{ V}$$

The oxidations are catalyzed by traces of silver ion. The excess reagent is readily decomposed by boiling the solution for a few minutes.

$$2S_2O_8^{2-} + 2H_2O \rightleftarrows 4SO_4^{2-} + O_2(g) + 4H^+$$

Sodium and Hydrogen Peroxide. Peroxide is a convenient oxidizing agent. Both the solid sodium salt and dilute solutions of the acid are used. The half-reaction for peroxide in acidic solution is

$$H_2O_2 + 2H^+ + 2e \rightleftarrows 2H_2O \qquad E^0 = 1.77 \text{ V}$$

Any excess reagent is readily decomposed by brief boiling.

$$2H_2O_2 \rightleftarrows 2H_2O + O_2(g)$$

REDUCING REAGENTS

Metals. An examination of standard electrode potential data reveals a number of good reducing agents among the pure metals.[1] Such elements as

---

[1] For a discussion of metal reductants, the reader should see I. M. Kolthoff and R. Belcher, *Volumetric Analysis*, vol. 3, pp. 11–23. New York: Interscience Publishers, Inc., 1957; W. I. Stephen, *Ind. Chemist*, **28**, 13, 55, 107 (1952).

zinc, cadmium, aluminum, lead, nickel, copper, mercury, and silver have proved useful for prereduction purposes. Where sticks or coils of the metal are used, the excess reductant is simply lifted from the solution and washed thoroughly. If granular or powdered forms of the metal are employed, filtration may be required. An alternative method involves the use of a *reductor*, an example of which is illustrated in Figure 16-1. Here the solution is passed through a column packed with granules of the metallic reducing agent.

Of the several simple metallic reductors, that containing silver (called a *Walden reductor*) is perhaps the most widely used because of its somewhat selective reducing properties. Some applications of this reductor are given in Table 16-1. The silver reductor is nearly always used with hydrochloric acid

TABLE 16-1   Uses of the Walden Reductor and the Jones Reductor[a]

Walden Reductor $Ag(s) + Cl^- \rightleftarrows AgCl(s) + e$	Jones Reductor $Zn(s) \rightleftarrows Zn^{2+} + 2e$
$e + Fe^{3+} \rightarrow Fe^{2+}$	$e + Fe^{3+} \rightarrow Fe^{2+}$
$e + Cu^{2+} \rightarrow Cu^+$	Cu^{2+} reduced to metallic Cu
$e + H_2MoO_4 + 2H^+ \rightarrow MoO_2^+ + 2H_2O$	$3e + H_2MoO_4 + 6H^+ \rightarrow Mo^{3+} + 4H_2O$
$2e + UO_2^{2+} + 4H^+ \rightarrow U^{4+} + 2H_2O$	$2e + UO_2^{2+} + 4H^+ \rightarrow U^{4+} + 2H_2O$ $3e + UO_2^{2+} + 4H^+ \rightarrow U^{3+} + 2H_2O$[b]
$e + V(OH)_4^+ + 2H^+ \rightarrow VO^{2+} + 3H_2O$	$3e + V(OH)_4^+ + 4H^+ \rightarrow V^{2+} + 4H_2O$
TiO^{2+} not reduced	$e + TiO^{2+} + 2H^+ \rightarrow Ti^{3+} + H_2O$
Cr^{3+} not reduced	$e + Cr^{3+} \rightarrow Cr^{2+}$

[a] Taken from I. M. Kolthoff and R. Belcher, *Volumetric Analysis*, vol. 3, p. 12. New York: Interscience Publishers, Inc., 1957. With permission.

[b] A mixture of oxidation states is obtained. This does not, however, preclude the use of the Jones reductor for the analysis of uranium, since any U^{3+} formed can be converted to U^{4+} by shaking the solution with air for a few minutes.

solutions because the reducing properties of the metal are enhanced by formation of silver chloride.

One disadvantage of very reactive reductants such as cadmium and zinc is their tendency to evolve hydrogen when in contact with acids. This parasitic reaction not only consumes reductant but also introduces large quantities of the metallic ion into the solution. The problem of hydrogen evolution can be largely overcome by amalgamation of the reducing metal with mercury. Thus, for example, zinc amalgam is produced by treating granules of the metal briefly with mercury(II) chloride solution.

$$2Zn + Hg^{2+} \rightarrow Zn^{2+} + Zn(Hg)$$

The presence of the mercury so inhibits the formation of hydrogen that zinc and cadmium can be used as reductants even in quite acidic solutions.

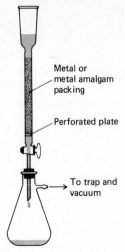

Metal or
metal amalgam
packing

Perforated plate

To trap and
vacuum

Figure 16-1   A Metal or Metal Amalgam Reductor.

Figure 16-1 illustrates a typical amalgam reductor (called a *Jones reductor*), in which amalgamated zinc acts as the reducing agent. The column is 40 to 50 cm long and has a diameter of about 2 cm. The packing is covered with liquid at all times to protect against air oxidation and the resultant formation of basic salts which tend to cause clogging.

The Jones reductor is the most widely used amalgam reductor. Table 16-1 lists its principal applications. The preparation and use of the Jones reductor is described in Chapter 31, Experiment 18.

Gaseous Reductants. Both hydrogen sulfide and sulfur dioxide are reasonably effective reducing reagents and have been used for prereduction. Excesses of these reagents can be readily eliminated by boiling the acidified solution.

The reactions of both reagents are often slow, half an hour or more being required to complete the reduction and rid the solution of excess reagent. In addition to this time disadvantage, both gases are noxious and toxic. The employment of other reductants is much preferred.

Potassium Permanganate

Potassium permanganate, a powerful oxidant, is perhaps the most widely used of all standard oxidizing agents. The color of a permanganate solution is so intense that an indicator is not ordinarily required. The reagent is readily available at modest cost. On the other hand, the tendency of permanganate to oxidize chloride ion is a disadvantage because hydrochloric acid is such a useful solvent. The multiplicity of possible reaction products can, at times, cause uncertainty regarding the stoichiometry of a permanganate oxidation. Finally, permanganate solutions have limited stability.

REACTIONS OF PERMANGANATE ION

Potassium permanganate is most commonly employed with solutions that are 0.1 N or greater in mineral acid. Under these conditions, the product is manganese(II) ion:

$$MnO_4^- + 8H^+ + 5e \rightleftarrows Mn^{2+} + 4H_2O \qquad E^0 = 1.51 \text{ V}$$

Although the mechanisms involved in the formation of manganese(II) are frequently complicated, the oxidation of most substances proceeds rapidly in acidic solution. Notable exceptions include the reaction with oxalic acid which requires elevated temperatures and with arsenic(III) oxide for which a catalyst is needed.

In solutions that are weakly acidic (above pH 4), neutral, or weakly alkaline, permanganate usually undergoes a three-electron reduction, a brown precipitate of manganese dioxide, MnO_2, being formed. Titrations of certain species with permanganate can be carried out to advantage under these conditions. For example, cyanide is oxidized to cyanate; sulfide, sulfite, and thiosulfate are converted to sulfate; manganese(II) is oxidized to manganese dioxide; and hydrazine is oxidized to nitrogen.

Solutions of manganese(III) are not stable owing to the disproportionation reaction

$$2Mn^{3+} + 2H_2O \rightleftarrows MnO_2(s) + Mn^{2+} + 4H^+$$

However, manganese(III) ion forms several complexes that are sufficiently stable to permit existence of the +3 state in aqueous solution. Lingane has made use of this property to titrate manganese(II) with permanganate in highly concentrated solutions of pyrophosphate ion[2]; the reaction may be expressed as

$$MnO_4^- + 4Mn^{2+} + 15H_2P_2O_7^{2-} + 8H^+ \rightleftarrows 5Mn(H_2P_2O_7)_3^{3-} + 4H_2O$$

The titration is carried out in a pH range between 4 and 7.

In solutions that are greater than 1 N in sodium hydroxide, permanganate ion undergoes a one-electron reduction to manganate ion, MnO_4^{2-}. Alkaline oxidations with permanganate have proved useful for the determination of certain organic compounds.

END POINT

One of the most obvious properties of potassium permanganate is its intense purple color, which commonly serves as the indicator for titrations. As little as 0.01 to 0.02 ml of a 0.02-F (0.1-N) solution is sufficient to impart a perceptible color to 100 ml of water. For very dilute permanganate solutions, diphenylamine sulfonic acid or the orthophenanthroline-iron(II) complex (Chapter 15) will give a sharper end point.

The permanganate end point is not permanent and gradually fades. Decolorization results from the reaction of the excess permanganate with the relatively large concentration of manganese(II) ion that is present at the end point.

$$2MnO_4^- + 3Mn^{2+} + 2H_2O \rightleftarrows 5MnO_2(s) + 4H^+$$

[2] J. J. Lingane and R. Karplus, *Ind. Eng. Chem., Anal. Ed.*, **18**, 191 (1946).

The equilibrium constant for this reaction, which is readily calculated from the standard potentials for the two half-reactions, has a numerical value of about 10^{47}. Thus, even in a highly acidic solution, the concentration of permanganate in equilibrium with manganese(II) ion is small. Fortunately, the rate at which this equilibrium is attained is slow, with the result that the end point fades only gradually.

The intense color of a permanganate solution complicates the measurement of reagent volumes in a buret. It is frequently more practical to use the surface of the liquid rather than the bottom of the meniscus as the point of reference.

STABILITY OF PERMANGANATE SOLUTIONS

Aqueous solutions of permanganate are not completely stable because of the tendency of that ion to oxidize water. The process may be depicted by the equation

$$4MnO_4^- + 2H_2O \rightleftarrows 4MnO_2(s) + 3O_2(g) + 4OH^-$$

Although the constant for this equilibrium indicates that the products are favored in neutral solution, the reaction is so slow that a permanganate solution that has been properly prepared is moderately stable. The decomposition has been shown to be catalyzed by light, heat, acids, bases, manganese(II) ion, and manganese dioxide. In order to obtain a stable reagent for analysis, it is necessary to minimize the influence of these effects.

The decomposition of permanganate solutions is greatly accelerated in the presence of manganese dioxide. Since it is also a product of the decomposition, this solid has an *autocatalytic* effect upon the process.

Photochemical catalysis of the decomposition is often observed when a permanganate solution is allowed to stand in a buret for an extended period. Manganese dioxide forms as a brown stain and serves to show that the concentration of the reagent has undergone change.

In general, the heating of acidic solutions containing an excess of permanganate should be avoided because of the decomposition error that cannot adequately be compensated for with a blank. At the same time, it is perfectly acceptable to titrate hot, acidic solutions of reductants directly with the reagent, since at no time during the titration is the oxidant concentration large enough to cause a measurable uncertainty.

PREPARATION, STANDARDIZATION, AND STORAGE OF PERMANGANATE SOLUTIONS

Solid potassium permanganate, which is ordinarily used for the preparation of permanganate solutions, is inevitably contaminated with manganese dioxide to some extent; as a result, preparation of standard permanganate solutions directly by weight is not possible.

A permanganate solution possessing reasonable stability can be obtained provided a number of precautions are observed. Perhaps the most important variable affecting stability is the catalytic influence of manganese dioxide. This contaminant occurs in the starting material and is also produced when permanganate oxidizes organic matter in the water used to prepare the solution. Removal of manganese dioxide by filtration markedly enhances the stability of standard permanganate solutions. Sufficient time must be allowed for complete oxidation of contaminants in the water before filtration; the solution may be boiled to hasten the oxidation. Paper cannot be used for the filtration since it reacts with the permanganate to form the undesirable dioxide.

Standardized solutions should be stored in the dark. If any solid is detected in the solution, filtration and restandardization are necessary. In any event, restandardization every one to two weeks is a good precautionary measure.

Standardization against Sodium Oxalate. In acidic solution, permanganate oxidizes oxalic acid to carbon dioxide and water.

$$2MnO_4^- + 5H_2C_2O_4 + 6H^+ \rightleftarrows 2Mn^{2+} + 10CO_2(g) + 8H_2O$$

This reaction is complex and proceeds slowly at room temperature; even at elevated temperatures it is not rapid unless catalyzed by manganese(II) ion. Thus, several seconds are required to decolorize a hot oxalic acid solution at the outset of a permanganate titration. Later, when the concentration of manganese(II) ion has become appreciable, decolorization becomes rapid as a consequence of the autocatalysis.

The stoichiometry of the reaction between permanganate ion and oxalic acid has been investigated in great detail by McBride[3] and more recently by Fowler and Bright.[4] The former devised a procedure in which the oxalic acid is titrated slowly at a temperature between 60 and 90°C until the faint pink color of the permanganate persists. Fowler and Bright have demonstrated, however, that this titration consumes 0.1 to 0.4% too little permanganate, due perhaps to air oxidation of a small part of the oxalic acid

$$H_2C_2O_4 + O_2(g) \rightleftarrows H_2O_2 + 2CO_2$$

In the hot solution, the peroxide is postulated to decompose spontaneously to oxygen and water.

Fowler and Bright devised a scheme for standardization in which 90 to 95% of the required permanganate is added rapidly to the cool oxalic acid solution. After all of this reagent has reacted, the solution is heated to 55 to 60°C and titrated as before. While it minimizes the air oxidation of oxalic acid and gives data that appear to be in exact accord with the theoretical stoichiometry, this method suffers from the disadvantage of requiring a knowledge of the approximate normality of the permanganate solution to make the proper initial addition of the reagent. In this respect, the Fowler-Bright procedure is not as convenient as the McBride method.

For many purposes, the method of McBride will give perfectly adequate data (usually 0.2 to 0.3% too high). If a more accurate standardization is required, it is convenient to perform one titration by this procedure to obtain the

[3] R. S. McBride, *J. Amer. Chem. Soc.*, **34**, 393 (1912).
[4] R. M. Fowler and H. A. Bright, *J. Res. Nat. Bur. Stand.*, **15**, 493 (1935).

approximate normality of the solution. Then a pair of titrations employing the Fowler and Bright method can be made. Directions for both procedures are found in Chapter 31, Experiment 19.

Other Primary Standards. Several other primary standards can be employed to establish the normality of permanganate solutions. These include arsenic(III) oxide, potassium iodide, and metallic iron. Detailed procedures for the use of these standards are described by Kolthoff and Belcher.[5]

APPLICATIONS OF PERMANGANATE TITRATIONS TO ACIDIC SOLUTIONS

Table 16-2 indicates the multiplicity of analyses that make use of standard permanganate solutions in acidic media. Most of these reactions are rapid enough for direct titrations. Two typical applications, the determination of iron and the determination of calcium, are discussed in the paragraphs that follow.

Determination of Iron in an Ore. The common iron ores are hematite (Fe_2O_3), magnetite (Fe_3O_4), and limonite ($3Fe_2O_3 \cdot 3H_2O$). Volumetric methods for analysis of iron samples containing these substances consist of three steps: (1) solution of the sample, (2) reduction of the iron to the divalent state, and (3) titration with a standard oxidant.

Iron ores are often completely decomposed in concentrated hydrochloric acid. The rate of attack by this reagent is increased by the presence of a small amount of tin(II) chloride, which probably acts by reducing slightly soluble iron(III) oxides on the surface of the particles to more soluble iron(II) species. Because iron(III) tends to form stable chloride complexes, hydrochloric acid is a much more efficient solvent than either sulfuric or nitric acid.

Most iron ores contain silicates that, in some instances, are not decomposed by treatment with hydrochloric acid. Where decomposition is complete, the white residue of hydrated silica that remains behind in no way interferes with the analysis. Incomplete decomposition is indicated by a dark residue remaining after prolonged treatment with the acid. Since this solid residue may contain iron, it must be broken down by more drastic treatment. The usual procedure entails filtration and ignition of the solid, followed by fusion with sodium carbonate. This process converts the cationic components of the residue into carbonates, which can then be dissolved in acid and combined with the solution containing the bulk of the sample.

Part or all of the iron will exist in the trivalent state after solution is complete. Prereduction of the sample must, therefore, precede titration with the oxidant. Any of the methods described earlier may be used—the Jones reductor, for example. Zinc amalgam, however, will also reduce other elements commonly associated with iron, including titanium, niobium, vanadium, chromium, uranium, tungsten, molybdenum, and arsenic. In their lower oxidation states,

[5] I. M. Kolthoff and R. Belcher. *Volumetric Analysis*, vol. 3, pp. 41–59. New York: Interscience Publishers, Inc., 1957.

TABLE 16-2   Applications of Potassium Permanganate in Acidic Solution

Substance Sought	Half-Reaction	Condition
I	$I^- + HCN \rightleftarrows ICN + H^+ + 2e$	In 0.1-F HCN with ferroin indicator
Br	$2Br^- \rightleftarrows Br_2 + 2e$	Boiling H_2SO_4 solution
As	$H_3AsO_3 + H_2O \rightleftarrows H_3AsO_4 + 2H^+ + 2e$	KI, or ICl catalyst in HCl solution
Sb	$H_3SbO_3 + H_2O \rightleftarrows H_3SbO_4 + 2H^+ + 2e$	HCl solution
Sn	$Sn^{2+} \rightleftarrows Sn^{4+} + 2e$	Prereduction with Zn
H_2O_2	$H_2O_2 \rightleftarrows O_2(g) + 2H^+ + 2e$	
Fe	$Fe^{2+} \rightleftarrows Fe^{3+} + e$	Prereduction with $SnCl_2$ or with Jones or Walden reductor
$Fe(CN)_6^{4-}$	$Fe(CN)_6^{4-} \rightleftarrows Fe(CN)_6^{3-} + e$	
V	$VO^{2+} + 3H_2O \rightleftarrows V(OH)_4^+ + 2H^+ + e$	Prereduction with Bi amalgam or SO_2
Mo	$Mo^{3+} + 4H_2O \rightleftarrows MoO_4^{2-} + 8H^+ + 3e$	Prereduction with Jones reductor
W	$W^{3+} + 4H_2O \rightleftarrows WO_4^{2-} + 8H^+ + 3e$	Prereduction with Zn or Cd
U	$U^{4+} + 2H_2O \rightleftarrows UO_2^{2+} + 4H^+ + 2e$	Prereduction with Jones reductor
Ti	$Ti^{3+} + H_2O \rightleftarrows TiO^{2+} + 2H^+ + e$	Prereduction with Jones reductor
Nb	$Nb^{3+} + H_2O \rightleftarrows NbO^{3+} + 2H^+ + 2e$	Prereduction with Jones reductor
$H_2C_2O_4$	$H_2C_2O_4 \rightleftarrows 2CO_2 + 2H^+ + 2e$	
Mg, Ca, Zn, Co, La, Th, Ba, Sr, Ce, Ag, Pb	$H_2C_2O_4 \rightleftarrows 2CO_2 + 2H^+ + 2e$	Sparingly soluble metal oxalates filtered, washed, and dissolved in acid; liberated oxalic acid titrated
HNO_2	$HNO_2 + H_2O \rightleftarrows NO_3^- + 3H^+ + 2e$	15-min reaction time; excess $KMnO_4$ back-titrated
K	$K_2NaCo(NO_2)_6 + 6H_2O \rightleftarrows$ $Co^{2+} + 6NO_3^- + 12H^+ + 2K^+ + Na^+ + 11e$	Precipitated as $K_2NaCo(NO_2)_6$; filtered and dissolved in $KMnO_4$; excess $KMnO_4$ back-titrated
Na	$U^{4+} + 2H_2O \rightleftarrows UO_2^{2+} + 4H^+ + 2e$	Precipitated as $NaZn(UO_2)_3(OAc)_9$; filtered, washed, dissolved; U determined as above

these too react with permanganate; their presence, if undetected, will cause high results for iron.

The Walden reductor has the advantage of being inert with respect to titanium and trivalent chromium, both of which are likely to occur in significant amounts. Vanadium continues to interfere.

Perhaps the most satisfactory of all prereductants for iron is tin(II) chloride. The only other common elements reduced by this reagent are vanadium, copper, molybdenum, tungsten, and arsenic. The excess reducing agent is removed from solution by the addition of mercury(II) chloride:

$$Sn^{2+} + 2HgCl_2 \rightleftarrows Hg_2Cl_2(s) + Sn^{4+} + 2Cl^-$$

The slightly soluble mercury(I) chloride produced will not consume permanganate, nor will the excess mercury(II) chloride reoxidize divalent iron. Care must be exerted, however, to prevent occurrence of the alternative reaction,

$$Sn^{2+} + HgCl_2 \rightleftarrows Hg(s) + Sn^{4+} + 2Cl^-$$

Metallic mercury reacts with permanganate and causes a high result for iron. Formation of mercury is favored by an appreciable excess of tin(II); it is prevented by careful control of this excess and the rapid addition of excess mercury(II) chloride. A proper reduction is indicated by the appearance of a slight, white precipitate after addition of this reagent. A gray precipitate indicates the presence of mercury; the total absence of precipitate indicates that an insufficient amount of tin(II) chloride was added. In either event, the sample must be discarded.

The reaction of iron(II) with permanganate proceeds smoothly and rapidly to completion. In the presence of hydrochloric acid, however, high results are obtained owing to the oxidation of chloride ion by permanganate. This reaction, which normally does not proceed rapidly enough to cause serious errors, is induced by the presence of divalent iron. Its effects are avoided by preliminary removal of chloride by evaporation with sulfuric acid or by use of the *Zimmermann-Reinhardt reagent*. The latter is a solution of manganese(II) in fairly concentrated sulfuric and phosphoric acids. Manganese(II) inhibits the oxidation of chloride ion, while phosphoric acid forms complexes with the iron(III) produced in the titration and prevents the intense yellow color of iron(III) chloride complexes from interfering with the end point.

Determination of Calcium. Calcium is conveniently precipitated as calcium oxalate. After being filtered and washed, the precipitate is redissolved in dilute acid; the liberated oxalic acid is then titrated with a standard solution of permanganate or some other oxidizing agent. This method is applicable to samples that contain magnesium and the alkali metals. Most other cations must be absent, however, as they either precipitate or coprecipitate as oxalates and cause positive errors in the analysis.

To obtain satisfactory results from this procedure, the mole ratio of calcium to oxalate must be exactly 1 in the precipitate and thus in the solution at the time of titration. Several precautions are necessary to assure this condition.

For example, calcium oxalate formed in a neutral or ammoniacal solution is likely to be contaminated with calcium hydroxide or a basic calcium oxalate; the presence of either causes low results. This problem can be eliminated by adding the oxalate to an acidic solution of the sample and slowly forming the precipitate by the dropwise addition of ammonia. Losses due to the solubility of calcium oxalate are negligible at a pH of about 4, provided washing is restricted to freeing the precipitate of excess oxalic acid. A precipitate formed in this way is coarsely crystalline and readily filtered.

A potential source of positive error in the analysis arises from coprecipitation of sodium oxalate, which occurs when the sodium ion concentration exceeds that of calcium. The error from this source can be eliminated by double precipitation.

Magnesium, if present in high concentrations, may also contaminate the calcium oxalate precipitate. This interference is minimized if the excess of oxalate is sufficient to allow formation of a soluble magnesium complex and if filtration is performed promptly after the completion of precipitation. When the magnesium content exceeds that of calcium in the sample, a double precipitation may be required.

Quadrivalent Cerium

A solution of cerium(IV) in sulfuric acid, which is very nearly as potent an oxidizing reagent as permanganate, can be substituted for the latter in most of the applications described in the previous section. The reagent is indefinitely stable and does not oxidize chloride ion at a detectable rate. Further, only a single reduction product, trivalent cerium, is possible; thus, the stoichiometry of the reaction is less subject to uncertainty. In these respects, cerium(IV) possesses considerable advantages over permanganate. On the other hand, the color of quadrivalent cerium solutions is not sufficiently intense to serve as an indicator. In addition, the reagent cannot be used in neutral or basic solutions. A final disadvantage is the relatively high cost of cerium compounds.

PROPERTIES OF QUADRIVALENT CERIUM SOLUTIONS

The electrode potential for a cerium(IV) solution depends upon the acid that is used for preparation. Solutions of the reagent prepared with sulfuric acid are roughly comparable in oxidizing power to those of permanganate; solutions containing nitric or perchloric acid are appreciably more potent. To a lesser degree, the electrode potential is also influenced by the concentration of the acid.

Composition of Cerium(IV) Solutions. Acid solutions of quadrivalent cerium have highly complex compositions. The exact nature of the cerium-containing species present has not yet been established. Much of that which is known has come from studies concerned with the effects of various acids and their concentrations upon the potential of the Ce(IV)/Ce(III) couple and upon the color of solutions of the reagent. Table 16-3 provides some typical potential

TABLE 16-3   Formal Electrode Potentials for Cerium(IV)

Acid Concentration, N	Formal Potential vs. Standard Hydrogen Electrode, V		
	$HClO_4$ Solution	HNO_3 Solution	H_2SO_4 Solution
1	+1.70	+1.61	+1.44
2	1.71	1.62	1.44
4	1.75	1.61	1.43
8	1.87	1.56	1.42

data. Note that stronger oxidizing properties are exhibited by the cerium(IV) species in perchloric acid than by those in nitric or sulfuric acids. In all three media the formal reduction potential varies with acid concentration. These data suggest that cerium(IV) ions form stable complexes with nitrate and sulfate ions. In addition, such species as $Ce(OH)^{3+}$ and $Ce(OH)_2^{2+}$ exist in perchloric acid solution. Finally, the presence of a dimeric cerium(IV) ion, particularly in highly concentrated solutions, has been reported. All evidence indicates that the concentration of the simple hydrated ion, $Ce(H_2O)_x^{4+}$, is small in any of these solutions.

Stability of Cerium(IV) Solutions. Sulfuric acid solutions of quadrivalent cerium are remarkably stable, remaining constant in titer for years. Solutions heated to 100°C for considerable periods do not change appreciably. Perchloric and nitric acid solutions of the reagent are by no means as stable; these decompose water and decrease in normality by 0.3 to 1% during storage for one month. The decomposition reaction is catalyzed by light.

The oxidation of chloride is so slow that other reducing agents can be titrated without error in the presence of high concentrations of this ion. Hydrochloric acid solutions of cerium(IV), however, are not stable enough for use as standard solutions.

Indicators for Cerium(IV) Titrations. Several of the redox indicators discussed in Chapter 15 are suitable for use in titrations with cerium(IV) solutions. The various phenanthrolines deserve particular mention because their transition potentials frequently correspond to the equivalence-point potentials for these reactions.

PREPARATION AND STANDARDIZATION OF CERIUM(IV) SOLUTIONS

Preparation of Solutions. Several cerium(IV) salts are commercially available[6]; the most common are listed in Table 16-4. Cerium(IV) ammonium nitrate of primary-standard quality can be purchased; thus, standard solutions

---

[6] For further information regarding the preparation, standardization, and use of cerium(IV) solutions, see G. Frederick Smith, *Cerate Oxidimetry*. Columbus, Ohio: The G. Frederick Smith Chemical Co., 1942; I. M. Kolthoff and R. Belcher, *Volumetric Analysis*, vol. 3, pp. 121–167. New York: Interscience Publishers, Inc., 1957.

TABLE 16-4 Analytically Useful Cerium(IV) Compounds

Name	Formula	Equivalent Weight
Cerium(IV) ammonium nitrate	$Ce(NO_3)_4 \cdot 2NH_4NO_3$	548.2
Cerium(IV) ammonium sulfate	$Ce(SO_4)_2 \cdot 2(NH_4)_2SO_4 \cdot 2H_2O$	632.6
Cerium(IV) hydroxide	$Ce(OH)_4$	208.2
Cerium(IV) hydrogen sulfate	$Ce(HSO_4)_4$	528.4

can be prepared directly by weight. More frequently, solutions of approximately the desired normality are prepared from one of the less expensive reagent-grade salts and then standardized. A stable and entirely satisfactory sulfuric acid solution of quadrivalent cerium is obtained from cerium(IV) ammonium nitrate without removal of the ammonium or nitrate ions.

Solutions of cerium(IV) tend to react with water, even in acid solutions, to produce slightly soluble basic salts. The acidity of solutions containing cerium(IV) must be 0.1 N or greater to prevent this precipitation. Neutral or basic solutions cannot be titrated with the reagent.

Standardization against Arsenic(III) Oxide. Arsenic(III) oxide is perhaps the most satisfactory primary standard for solutions of quadrivalent cerium. In the absence of a catalyst, the reaction is so slow that iron(II) can be titrated without interference in the presence of trivalent arsenic. Fortunately, good catalysts are available that permit standardization with this useful reagent. The best of these is osmium tetroxide, which is effective even at very low concentrations (10^{-5} F). Iodine monochloride also catalyzes the reaction. Directions for standardization with arsenic(III) oxide are found in Chapter 31, Experiment 22.

Standardization against Sodium Oxalate. Several methods exist for the standardization of sulfuric acid solutions of cerium(IV) against sodium oxalate.[7] The directions found in Chapter 31 call for titration at 50°C in hydrochloric acid solution. Iodine monochloride is used as a catalyst; orthophenanthroline is the indicator.

APPLICATIONS OF QUADRIVALENT CERIUM SOLUTIONS

Many applications of cerium(IV) solutions are found in the literature. In general, these parallel the uses of permanganate given in Table 16-2.[8] Directions for the determination of iron are given in Chapter 31, Experiment 23.

[7] I. M. Kolthoff and R. Belcher, *Volumetric Analysis*, vol. 3, pp. 132–134. New York: Interscience Publishers, Inc., 1957.
[8] Information about these methods can be found in I. M. Kolthoff and R. Belcher, *Volumetric Analysis*, vol. 3, pp. 136–158. New York: Interscience Publishers, Inc., 1957.

Potassium Dichromate

In its analytical applications, dichromate ion is reduced to the trivalent state:

$$Cr_2O_7^{2-} + 14H^+ + 6e \rightleftarrows 2Cr^{3+} + 7H_2O \qquad E^0 = 1.33 \text{ V}$$

Potassium dichromate is more limited in application than either potassium permanganate or quadrivalent cerium owing to its lesser oxidizing properties and the slowness of some of its reactions. Despite these handicaps, the reagent is nonetheless useful because its solutions are indefinitely stable and are also inert toward hydrochloric acid. Further, the solid reagent can be obtained in high purity and at modest cost; standard solutions may be prepared directly by weight.

Standard solutions of potassium dichromate may be boiled for long periods without decomposition.

PREPARATION AND PROPERTIES OF DICHROMATE SOLUTIONS

For most purposes, commercial reagent–grade or primary standard–grade potassium dichromate can be used to prepare standard solutions with no prior treatment other than drying at 150 to 200°C. If desired, two or three recrystallizations of the solid from water will assure a high-quality, primary-standard substance.

Although solutions of dichromate are orange, the color is not sufficiently intense for end-point determination. Diphenylamine sulfonic acid (Chapter 15) is an excellent indicator for titrations with the reagent; the color change is from the green of the chromium(III) ion to the violet color of the oxidized form of the indicator. An indicator blank is not readily obtained because dichromate oxidizes the indicator only slowly in the absence of other oxidation-reduction systems. Ordinarily, however, the error resulting from neglect of the blank is vanishingly small. The reaction of diphenylamine sulfonic acid is reversible, and back-titration of small excesses of dichromate with iron(II) is possible. In the presence of large oxidant concentrations and at low acidities (above pH 2), the indicator is irreversibly oxidized to yellow or red compounds.

APPLICATIONS OF DICHROMATE

Determination of Iron. The principal use of dichromate involves titration of iron(II),

$$6Fe^{2+} + Cr_2O_7^{2-} + 14H^+ \rightleftarrows 6Fe^{3+} + 2Cr^{3+} + 7H_2O$$

Moderate amounts of hydrochloric acid do not affect the accuracy of the titration. Detailed procedures for the analysis of iron are found in Chapter 31, Experiment 25.

Other Applications. A common method for the determination of oxidizing agents calls for treatment of the sample with a known excess of iron(II), followed by titration of the excess with standard dichromate. This technique has been successfully applied to the determination of nitrate, chlorate, permanganate, dichromate, and organic peroxides, among others.

Potassium Bromate

Potassium bromate is a strong oxidizing agent which is readily reduced to bromide ion, the half-reaction for the process being

$$BrO_3^- + 6H^+ + 6e \rightleftarrows Br^- + 3H_2O \qquad E^0 = 1.44 \text{ V}$$

In acidic solution, however, bromate ion is capable of reacting with bromide ion to give bromine. That is,

$$BrO_3^- + 5Br^- + 6H^+ \rightleftarrows 3Br_2(aq) + 3H_2O$$

The numerical value of the equilibrium constant for this reaction is

$$K = \frac{[Br_2]^3}{[BrO_3^-][Br^-]^5[H^+]^6} = 3 \times 10^{36}$$

Note that the position of equilibrium bears a sixth-power relationship to the hydrogen ion concentration. The magnitude of the constant is such that at pH 1, bromate is quantitatively reduced to bromine by an excess of bromide ion; in neutral solution, on the other hand, the amount of bromine produced under the same circumstances is inconsequential.

When used to titrate acidic solutions of most reducing reagents, bromate is initially reduced quantitatively to bromide ion. After reaction with the analyte is complete, however, oxidation of bromide to bromine begins. The first appearance of bromine, then, can serve as an indicator for the titration.

Potassium bromate of primary-standard quality is available commercially. Solutions of the reagent are stable indefinitely.

INDICATORS FOR BROMATE TITRATIONS

Several organic indicators, such as methyl orange and methyl red (Chapter 9), are readily brominated to yield products differing in color from the original compounds. Unfortunately, these reactions are totally nonreversible. This behavior precludes any sort of back-titration and, more important, makes direct titration more difficult because of the great need to avoid local excesses of the reagent.

Three indicators, a-naphthoflavone, p-ethoxychrysoidine, and quinoline yellow, are reversible with respect to bromine and make employment of the reagent more attractive; they are commercially available.

DIRECT TITRATION

A variety of substances such as arsenic(III), antimony(III), iron(II), hydrogen peroxide, alkyl sulfides and disulfides, and oxalic acid can be determined by direct titration with standard bromate solution. In most applications the medium is at least 1 F in hydrochloric acid.

APPLICATIONS OF BROMATE AS A BROMINATING REAGENT

A standard solution of potassium bromate can serve as a convenient and stable source of bromine for titrimetric purposes. In this application, an excess of bromide ion is added to an acidic solution of the analyte. The addition of standard bromate releases an equivalent quantity of bromine for reaction with the analyte. This indirect approach eliminates the principal disadvantage associated with use of standard bromine solutions, namely, lack of stability. Note that each bromate forms three atoms of bromine which in turn require six electrons for reduction to bromide. That is,

$$BrO_3^- \equiv 3Br_2 \equiv 6e$$

The equivalent weight of potassium bromate is again one-sixth of its formula weight, as in the direct application of the reagent.

Organic compounds react with bromine either by substitution or by addition. The former involves replacement of hydrogen in an aromatic ring by atoms of the halogen. For example, three hydrogen atoms are replaced when phenol is brominated:

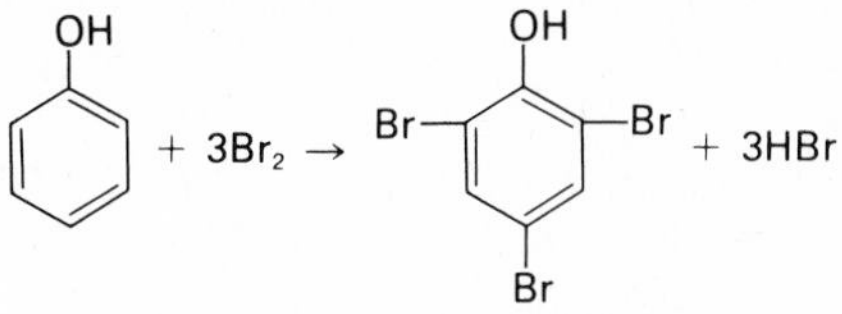

Since this process requires six equivalents of bromine, the equivalent weight of phenol is one-sixth of its formula weight. Addition reactions involve the opening of an olefinic double bond. For example, two equivalents of bromine react with ethylene:

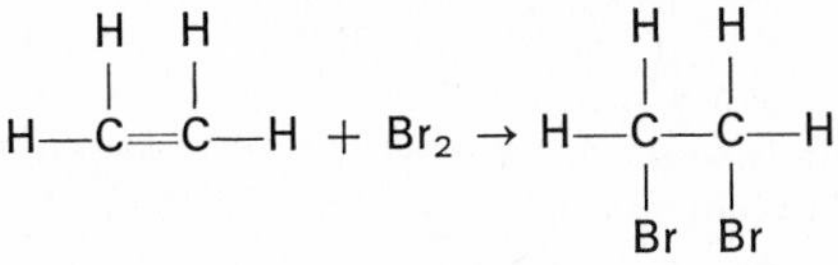

The equivalent weight of ethylene will clearly be equal to one-half its formula weight in this reaction.

Most organic compounds do not react sufficiently rapidly for a direct titration. Instead, a measured excess of a bromate-bromide solution is added to the sample, and the mixture is acidified; after bromination is judged complete, the excess bromine is determined by back-titration with standard arsenious acid. If desired, the analysis may be completed by adding an excess of potassium iodide and titrating the iodine liberated with a standard solution of sodium thiosulfate (see pp. 362–366). Although the iodometric procedure appears complicated, it is, in practice, quite simple.

Reaction vessels must be tightly stoppered during the bromination step; furthermore, care must be taken to minimize volatilization losses during the back-titration.

Substitution Reactions. Bromination methods have been applied successfully to the analysis of aromatic compounds that contain strong ortho-para directing substituents in the ring, particularly amines and phenols. Table 16-5

TABLE 16-5 Some Organic Compounds That Can Be Analyzed by Bromine Substitution[a]

Compound	Reaction Time, min	Equivalents Br$_2$/mole Compound	Accuracy (percent theoretical)
Phenol	5–30	6	99.89
p-Chlorophenol	30	4	99.87
Salicylic acid	30	6	99.86
Acetylsalicylic acid	30	6	99.84
m-Cresol	1	6	99.75
β-Naphthol	15–20	2	99.89
Aniline	5–10	6	99.92
o-Nitroaniline	30	4	99.86
Sulfanilic acid	30	6	99.98
m-Toluidine	5–10	6	99.87

[a] Data from A. R. Day and W. T. Taggart, *Ind. Eng. Chem.*, **20,** 545 (1928). With permission of the American Chemical Society.

lists typical examples. Many organic substances interfere with the method. These include unsaturated compounds that can add bromine; readily oxidized groups, such as mercaptans, sulfides, and quinones; and aliphatic compounds containing carbonyl, carboxylate, and other groups that lead to unwanted substitution reactions. A procedure for the analysis of phenol by bromination is found in Chapter 31, Experiment 32.

An important application of the bromination technique is the titration of 8-hydroxyquinoline.

The reaction, which is sufficiently rapid to allow direct titration in hydrochloric acid solution with methyl red as an indicator, is of particular interest because

8-hydroxyquinoline is an excellent precipitant for cations (Chapter 6). Using the analysis of aluminum as an example, the reactions for its determination are

$$Al^{3+} + 3HOC_9H_6N \xrightarrow{\text{pH 4-9}} Al(OC_9H_6N)_3(s) + 3H^+$$

8-hydroxyquinoline (filter and wash)

$$Al(OC_9H_6N)_3(s) + 3H^+ \xrightarrow[\text{HCl}]{\text{hot 4 } F} 3HOC_9H_6N + Al^{3+}$$

$$3HOC_9H_6N + 6Br_2 \rightarrow 3HOC_9H_4NBr_2 + 6HBr$$

Each mole of aluminum is indirectly responsible for the reaction of 6 moles (or 12 equivalents) of bromine; the equivalent weight of the metal is therefore one-twelfth of its formula weight.

Addition Reactions. Olefinic unsaturation can be determined by the addition of bromine to the double bond. A variety of methods, many involving the use of bromate-bromide mixtures, are found in the literature.[9] Most of these are applied to the estimation of unsaturation in fats, oils, and petroleum products.

Potassium Iodate

Potassium iodate is commercially available in a high state of purity and can be used without further treatment, other than drying, for the direct preparation of standard solutions. Iodate solutions, which are stable indefinitely, have a number of interesting and important uses in analytical chemistry.

REACTIONS OF IODATE

The reaction of iodate with iodide is analogous to the bromate-bromide reaction. Thus, the position of equilibrium for the reaction

$$IO_3^- + 5I^- + 6H^+ \rightleftarrows 3I_2 + 3H_2O$$

lies far to the right in acidic solutions and far to the left in basic media.

This reaction is a convenient source for known amounts of iodine. A measured quantity of iodate is mixed with an excess of iodide in a solution that is 0.1 to 1 N in acid. Exactly six equivalents of iodine are liberated for each mole of iodate; the resulting solution can then be used for the standardization of thiosulfate solutions (p. 364) or for other analytical purposes.

In strongly acidic solutions, iodate will oxidize iodide or iodine to the +1 state, provided some anion such as chloride, bromide, or cyanide is present to stabilize this oxidation state. For example, in solutions that are greater than 3 N in hydrochloric acid, the following reaction proceeds essentially to completion:

$$IO_3^- + 2I_2 + 10Cl^- + 6H^+ \rightleftarrows 5ICl_2^- + 3H_2O$$

[9] For example, see A. Polgar and J. L. Jungnickel, *Organic Analysis*, vol. 3. New York: Interscience Publishers, Inc., 1956.

In solutions of hydrogen cyanide, lower acidities (1 to 2 N) are sufficient for the quantitative formation of +1 iodine, since iodine cyanide, ICN, is more stable than ICl_2^-.

A number of important iodate titrations are performed in strong hydrochloric acid or hydrogen cyanide solutions. In these reactions the iodate is initially reduced to iodine. As the reducing agent is consumed, however, the reaction described in the previous paragraph takes place. Generally, the end point for the process is signaled by the complete disappearance of the iodine. The equivalent weight for the iodate is one-fourth of its formula weight since the final reaction product is iodine in the +1 state. When used in this manner, iodate is a less powerful oxidant than either permanganate or cerium(IV) ion.

END POINTS IN IODATE TITRATIONS

The disappearance of iodine from the solution is often sufficient to indicate the end point in an iodate titration. Iodine is nearly always formed in the initial stages of the reaction; only at the equivalence point is it completely oxidized to the +1 state. For titrations carried out in the presence of hydrogen cyanide, starch can serve as the indicator, the equivalence point being signaled by the disappearance of the familiar blue complex. The indicator fails to function, however, in the highly acidic media required for the production of ICl_2^-. Here, a small quantity of an immiscible organic solvent is used as the indicator. A few milliliters of carbon tetrachloride, chloroform, or benzene are added at the start of the titration. After each addition of iodate, the mixture is shaken thoroughly; the organic layer is examined after the phases have separated. The bulk of any unreacted iodine remains in the organic layer and imparts a violet-red color to it. The titration is judged complete when the minimum amount of iodate needed to discharge this color has been added. This excellent method for detecting iodine is quite comparable in sensitivity to the starch-iodine color; it suffers, however, from the disadvantage of being more time-consuming.

Directions for the iodate titration of iodide and iodine in solutions are given in Chapter 31, Experiment 30; further applications can be found in reference sources.[10]

Periodic Acid

Aqueous solutions of +7 iodine are highly complex.[11] In strongly acid solutions, paraperiodic acid, H_5IO_6, and its conjugate base predominate, although the metaperiodates HIO_4 and IO_4^- are undoubtedly present as well. Solutions of periodic acid are strong oxidizing agents. Thus,

$$H_5IO_6 + H^+ + 2e \rightleftarrows IO_3^- + 3H_2O \qquad E^0 = 1.6 \text{ V}$$

[10] For details on the application of iodate titrations, see I. M. Kolthoff and R. Belcher, *Volumetric Analysis*, vol. 3, pp. 449–473. New York: Interscience Publishers, Inc., 1957; R. Lang in W. Böttger, Ed., *Newer Methods of Volumetric Analysis*, pp. 69–98. New York: D. Van Nostrand, Inc., 1938.

[11] For discussion of the composition of acidic solutions of periodate, see C. E. Crouthamel, A. M. Hayes, and D. S. Martin, *J. Amer. Chem. Soc.*, **73**, 82 (1951).

PREPARATION AND PROPERTIES OF PERIODIC ACID SOLUTIONS

Several periodates are available for the preparation of standard solutions. Among these is paraperiodic acid itself, a crystalline, readily soluble, hygroscopic solid. An even more useful compound is sodium metaperiodate, $NaIO_4$, which is soluble in water to the extent of 0.06 F at 25°C. Sodium paraperiodate, Na_5IO_6, is not sufficiently soluble for the preparation of standard solutions; however, it is readily converted to the more soluble metaperiodate by recrystallization from hot concentrated nitric acid. Potassium metaperiodate can be used as a primary standard for the preparation of periodate solutions.[12] At room temperature its solubility is only about 5 g/liter; at elevated temperatures, however, it is readily dissolved and converted to a more soluble form by the addition of base.

Periodate solutions vary considerably in stability, depending on their mode of preparation and storage. A solution prepared by dissolving sodium metaperiodate in water decomposes at the rate of several percent per week. On the other hand, a solution of potassium metaperiodate in excess alkali was found to change no more than 0.3 to 0.4% in 100 days. The most stable periodate solutions appear to be those containing an excess of sulfuric acid; these decrease in normality by less than 0.1% in four months.[12]

STANDARDIZATION OF PERIODATE SOLUTIONS

Periodate solutions are most conveniently standardized by adding excess iodide ion and a slightly alkaline buffer, such as borax or hydrogen carbonate, to an aliquot of the reagent; iodine is liberated by the following reaction:

$$H_4IO_6^- + 2I^- \rightarrow IO_3^- + I_2 + 2OH^- + H_2O$$

As long as the solution is kept neutral, further reduction of iodate does not occur, and the liberated iodine can be titrated directly with a standard arsenite solution.

$$I_2 + HAsO_3^{2-} + H_2O \rightleftarrows HAsO_4^{2-} + 2I^- + 2H^+$$

A standard sodium thiosulfate solution (p. 363) can also be used for this titration.

APPLICATIONS OF PERIODIC ACID

The most important applications of periodic acid involve the selective oxidation of organic compounds containing certain combinations of functional groups.[13]

[12] H. H. Willard and L. H. Greathouse, *J. Amer. Chem. Soc.*, **60**, 2869 (1938).
[13] The use of periodic acid for this purpose was first studied by L. Malaprade, *Compt. rend.*, **186**, 382 (1928).

Ordinarily, these oxidations are performed at room temperature in the presence of a measured excess of the periodate; 30 min to 1 hr is required for most oxidations. After the oxidation is complete, the excess periodate is determined by the methods described for standardization. Alternatively, a reaction product such as ammonia, formaldehyde, or a carboxylic acid may be determined; here the exact quantity of periodate used need not be known.

Periodate oxidations are usually carried out in aqueous solution although solvents such as methanol, ethanol, or dioxane may be added to enhance the solubility of the sample.

Compounds Attacked by Periodate. At room temperature organic compounds containing aldehyde, ketone, or alcohol functional groups *on adjacent carbon atoms* are rapidly oxidized by periodic acid. In addition, primary and secondary *a*-hydroxyl amines are readily attacked; *a*-diamines are not. With a few exceptions, other organic compounds do not react at a significant rate. Thus, compounds containing isolated hydroxyl, carboxyl, or amine groups are not affected by periodic acid; nor are compounds with a carboxylic acid group either isolated from or adjacent to any of the reactive groups. At elevated temperatures, the extraordinary selectivity of periodic acid tends to disappear.

Periodate oxidations of organic compounds follow a regular and predictable pattern; the following rules apply:

1. Attack of adjacent functional groups always results in rupture of the carbon-to-carbon bond between these groups.
2. A carbon atom containing a hydroxyl group is oxidized to an aldehyde or ketone.
3. A carbonyl group is converted to a carboxylic acid group.
4. A carbon atom containing an amine group loses ammonia (or a substituted amine) and is itself converted to an aldehyde.

The following half-reactions will illustrate these rules:

$$CH_3\text{---}\underset{\underset{H}{|}}{\overset{\overset{OH}{|}}{C}}\text{---}\underset{\underset{H}{|}}{\overset{\overset{OH}{|}}{C}}\text{---}H \rightarrow CH_3\overset{O}{\overset{||}{C}}\text{---}H + H\text{---}\overset{O}{\overset{||}{C}}\text{---}H + 2H^+ + 2e$$

propylene glycol

$$H\text{---}\underset{\underset{H}{|}}{\overset{\overset{OH}{|}}{C}}\text{---}\underset{\underset{H}{|}}{\overset{\overset{OH}{|}}{C}}\text{---}\underset{\underset{H}{|}}{\overset{\overset{OH}{|}}{C}}\text{---}H + H_2O \rightarrow 2H\text{---}\overset{O}{\overset{||}{C}}\text{---}H + H\text{---}\overset{O}{\overset{||}{C}}\text{---}OH + 4H^+ + 4e$$

glycerol

For purposes of predicting the reaction products, the first step in the glycerol oxidation can be thought of as producing 1 mole of formaldehyde and 1 mole of an *a*-hydroxyaldehyde (glycolic aldehyde). The latter is then further oxidized and by the second rule produces a second mole of formaldehyde plus 1 mole of formic acid.

$$CH_3-\underset{biacetyl}{\overset{O\quad O}{\overset{\|\quad\|}{C-C}}}-CH_3 + 2H_2O \rightarrow 2CH_3\overset{O}{\overset{\|}{C}}-OH + 2H^+ + 2e$$

$$CH_3-\underset{acetoin}{\overset{O\quad OH}{\overset{\|\quad|}{C-C}}}-CH_3 + H_2O \rightarrow CH_3\overset{O}{\overset{\|}{C}}-OH + CH_3\overset{O}{\overset{\|}{C}}-H + H^+ + e$$

$$\underset{\underset{ethanolamine}{\overset{|\quad|}{H\quad H}}}{\overset{OH\ NH_2}{\overset{|\quad|}{H-C-C-H}}} + H_2O \rightarrow 2H-\overset{O}{\overset{\|}{C}}-H + NH_3 + 2H^+ + 2e$$

Analysis of Glycerol. As noted in the preceding paragraph, glycerol is cleanly oxidized by periodic acid, 2 moles of formaldehyde and 1 mole of formic acid being produced. About one-half hour is required to complete the reaction; the glycerol is readily estimated from the amount of periodate consumed. A procedure for this analysis is found in Chapter 31, Experiment 31.

Determination of α-Hydroxylamines. As mentioned earlier, periodate oxidation of compounds having hydroxyl and amino groups on adjacent carbon atoms results in the formation of aldehydes and the liberation of ammonia. The latter is readily distilled from the alkaline oxidation mixture and determined by neutralization titration (Chapter 11). This procedure is particularly useful in the analysis of mixtures of the various amino acids that occur in proteins. Only serine, threonine, β-hydroxyglutamic acid, and hydroxylysine have the requisite structure for liberation of ammonia; the method is thus selective for these compounds.

Iodimetric Methods

Many volumetric analyses are based on the half-reaction

$$I_3^- + 2e \rightleftarrows 3I^- \qquad E^\circ = 0.536 \text{ V}$$

These analyses fall into two categories. The first comprises procedures that use a standard solution of iodine to titrate easily oxidized substances. These *direct* or *iodimetric methods* have limited applicability since iodine is a relatively weak oxidizing agent. *Indirect* or *iodometric methods* employ a standard solution of sodium thiosulfate or arsenious acid to titrate the iodine liberated when an oxidizing substance is allowed to react with an unmeasured excess of potassium iodide. The quantity of iodine formed is chemically equivalent to the amount of the oxidizing agent and thus serves as the basis for the analysis.

Iodine, which is a relatively weak oxidant, finds use for the selective determination of strong reducing agents. The availability of a sensitive and reversible indicator for iodine is a great advantage. Disadvantages include the low stability of iodine solutions and the incompleteness of reactions between iodine and many reductants.

PREPARATION AND PROPERTIES OF IODINE SOLUTIONS

A saturated aqueous iodine solution is only about 0.001 F at room temperature. Much higher concentrations can be achieved, however, in the presence of iodide ion, owing to formation of the soluble triiodide complex.

$$I_2(s) + I^- \rightleftarrows I_3^- \qquad K = 7.1 \times 10^2$$

Because the concentration of the species I_2 is low in such solutions, it would be more proper to refer to them as *triiodide solutions*. As a practical matter, however, they are usually called *iodine solutions* because of the convenience this affords in writing equations and describing stoichiometric behavior.

Preparation of Solutions. Standard iodine solutions can be directly prepared from commercial reagent grades of the element. Where necessary, the solid is readily purified by sublimation. It is usually more convenient, however, to prepare an iodine solution in approximately the desired concentration and standardize it against a suitable primary standard.

The rate at which iodine dissolves in potassium iodide solution is slow, particularly where the iodide concentration is low. As a consequence, it is necessary to dissolve the solid completely in a small amount of a concentrated iodide solution before diluting to the desired volume. All of the element must be dissolved before dilution; otherwise, the normality of the resulting reagent will increase continuously as the remaining iodine slowly passes into solution.

Stability. Iodine solutions require restandardization every few days. This lack of stability has several sources, one being the volatility of the solute. Even though the excess of iodide is large, a measurable amount of iodine is lost in a relatively short period from an open container.

Iodine will slowly attack rubber or cork stoppers as well as other organic substances. Thus, reasonable precautions must be taken to protect standard solutions of the reagent from contact with these materials. Contact with organic dust and fumes must also be avoided.

Finally, changes in iodine normality result from air oxidation of iodide ions in the solution:

$$4I^- + O_2 + 4H^+ \rightleftarrows 2I_2 + 2H_2O$$

This reaction is catalyzed by light, heat, and acids; consequently, it is good practice to store the reagent in a dark, cool place. In contrast to the other effects, air oxidation of iodide causes an increase in normality.

Completeness of Iodine Oxidations. Because iodine is such a weak oxidizing agent, the chemist frequently must take full advantage of those experimental variables that enhance its reduction to iodide by the analyte. Two effects, pH and the presence of complexing agents, are of particular importance.

In acidic solutions the pH has little influence upon the electrode potential of the iodine-iodide couple since hydrogen ions do not participate in the half-reaction. Many of the substances that react with iodine, however, produce hydrogen ions as they are oxidized. The position of equilibrium may therefore be markedly influenced by pH. The arsenic(III)-arsenic(V) system provides an important example. Its electrode potential differs by only 0.02 V from that for the iodide-iodine half-reaction:

$$H_3AsO_4 + 2H^+ + 2e \rightleftarrows H_3AsO_3 + H_2O \qquad E^0 = 0.559 \text{ V}$$

In a strongly acidic medium, arsenic(V) will quantitatively oxidize iodide to iodine. On the other hand, in a nearly neutral solution, trivalent arsenic can be titrated successfully with iodine. Although iodine oxidations often become more nearly complete with lowered acidity, care must be taken to prevent the formation of hypoiodite, which tends to occur in alkaline solutions:

$$I_2 + OH^- \rightleftarrows HOI + I^-$$

The hypoiodite may subsequently disproportionate to iodate and iodide:

$$3HOI + 3OH^- \rightleftarrows IO_3^- + 2I^- + 3H_2O$$

The occurrence of these reactions will cause serious errors in an iodimetric analysis. In some titrations, the reaction of iodate and hypoiodite with the reducing reagent is so slow that overconsumption of iodine is observed. In others, the presence of these two species can alter the reaction products; the oxidation of thiosulfate is an important example of such behavior. Thus, solutions to be titrated with iodine cannot have pH values much higher than 9. Occasionally, a pH greater than 7 is detrimental.

Complexing reagents are also used to force certain iodine oxidations toward completion. For example, it is readily seen that the reduction potential of iodine is too low to permit the quantitative oxidation of iron(II) to the trivalent state. In the presence of reagents that strongly complex iron(III), however, complete conversion is achieved (see p. 305 for effect of complexing agents on electrode potentials); pyrophosphate ion and ethylenediaminetetraacetate are useful for this purpose.

End Points for Iodine Titrations. Several sensitive methods exist for determining the end point in an iodine titration. The color of the triiodide ion itself is often sufficiently intense for the titration of colorless solutions. Thus, a concentration of about 5×10^{-6} F triiodide can just be detected by the eye; in a typical titration, this concentration would be produced by an overtitration of less than one drop of 0.1-N iodine solution.

A greater sensitivity can be obtained, at the sacrifice of convenience, by adding a few milliliters of an immiscible organic solvent such as chloroform or carbon tetrachloride to the solution. The bulk of any iodine present is transferred to the organic layer by shaking and imparts an intense violet color to it. When this end point is used, the titration is carried out in a glass-stoppered flask; after each addition of reagent, the flask is shaken vigorously and then upended so that the organic layer collects in the narrow neck for examination.

The most widely used indicator for iodimetry is an aqueous suspension of starch which imparts an intense blue color to a solution containing a trace of triiodide ion. The nature of the colored species has been the subject of much speculation and controversy.[14] It is now believed that the iodine is held as an adsorption complex within the helical chain of the macromolecule, β-amylose, a component of most starches. Another component, a-amylose, is undesirable because it produces a red coloration with iodine that is not readily reversible. Interference from a-amylose is seldom serious, however, because the substance tends to settle rapidly from aqueous suspension. Other starch fractions do not appear to form colored complexes with iodine. Potato, arrowroot, and rice starches contain large proportions of a- and β-amylose and can be employed as indicators. Corn starch is not suitable because of its high content of the former. The so-called *soluble starch* that is commercially available consists principally of β-amylose, the a-fraction having been removed. Indicator solutions are readily prepared from this product.

Aqueous starch suspensions decompose within a few days, primarily because of bacterial action. The decomposition products may consume iodine as well as interfere with the indicator properties of the preparation. The rate of decomposition can be greatly inhibited by preparing and storing the indicator under sterile conditions and by the introduction of mercury(II) iodide or chloroform to act as a bacteriostat. Alternatively, a fresh indicator suspension can be prepared each day an iodine titration is to be performed.

Starch added to a solution containing a high concentration of iodine is decomposed to products whose indicator properties are not entirely reversible. Thus, addition of the indicator to a solution containing an excess of iodine should be postponed until most of the iodine has been titrated, as indicated by a light yellow color of the solution.

STANDARDIZATION OF IODINE SOLUTIONS

Iodine solutions are most commonly standardized against arsenious oxide, sodium thiosulfate, or potassium antimony(III) tartrate.

Arsenious oxide, As_2O_3, is available commercially in primary-standard quality. The oxide dissolves only slowly in water or in the common acids; solution occurs rapidly in 1-N NaOH, however.

$$As_2O_3(s) + 4OH^- \rightarrow 2HAsO_3^{2-} + H_2O$$

In strongly alkaline solution, arsenic(III) is readily air-oxidized to arsenic(V), whereas neutral or slightly acidic solutions are indefinitely stable toward oxygen. Thus, after the arsenious oxide has been dissolved in base, hydrochloric acid

[14] See R. E. Rundle, J. F. Foster. and R. R. Baldwin, *J. Amer. Chem. Soc.*, **66,** 2116 (1944).

should be added immediately until the solution is slightly acidic. Standard solutions of arsenious acid are useful for periodic standardization of iodine solutions.

The equilibrium constant for the reaction

$$H_3AsO_3 + I_2 + H_2O \rightleftarrows H_3AsO_4 + 2I^- + 2H^+$$

has been found experimentally[15] to be 1.6×10^{-1}. From the magnitude of this constant, it is clear that a quantitative oxidation of arsenious acid can be expected only if the reaction is forced to the right by keeping the concentration of the products small. From a practical standpoint, the hydrogen ion concentration is most readily controlled. Kolthoff[16] has shown that a quantitative reaction occurs even at a pH of 3.5, although the approach to equilibrium is prohibitively slow. In practice, it is advisable to maintain the pH at values in excess of 5. The pH of the medium may be as high as 11 without adverse effect on the results if the arsenic(III) is titrated with iodine. The reverse titration, however, requires a pH less than 9 in order to avoid hydrolysis of the iodine (p. 359); the iodate and hypoiodite produced react only slowly with arsenic(III).

The iodimetric titration of arsenious acid must be carried out in a buffered system to use up the hydrogen ions formed in the reaction; otherwise, the pH may decrease below the tolerable limit. Buffering is conveniently accomplished by acidifying the sample slightly and then saturating with sodium hydrogen carbonate. The carbonic acid–hydrogen carbonate buffer so established will hold the pH in a range between 7 and 8.

APPLICATIONS OF STANDARD IODINE

Common analyses that make use of iodine as an oxidizing reagent are summarized in Table 16-6.

TABLE 16-6   Analysis with Standard Iodine Solutions

Substance Analyzed	Half-Reaction
As	$H_3AsO_3 + H_2O \rightleftarrows H_3AsO_4 + 2H^+ + 2e$
Sb	$H_3SbO_3 + H_2O \rightleftarrows H_3SbO_4 + 2H^+ + 2e$
Sn	$Sn^{2+} \rightleftarrows Sn^{4+} + 2e$
H_2S	$H_2S \rightleftarrows S(s) + 2H^+ + 2e$
SO_2	$SO_3^{2-} + H_2O \rightleftarrows SO_4^{2-} + 2H^+ + 2e$
$S_2O_3^{2-}$	$2S_2O_3^{2-} \rightleftarrows S_4O_6^{2-} + 2e$
N_2H_4	$N_2H_4 \rightleftarrows N_2 + 4H^+ + 4e$
Te	$Te(s) + 3H_2O \rightleftarrows H_2TeO_3 + 4H^+ + 4e$
Cd^{2+}, Zn^{2+}, Hg^{2+}, Pb^{2+}, etc.	$M^{2+} + H_2S \rightleftarrows 2H^+ + MS(s)$ (filter and wash)
	$MS(s) \rightleftarrows M^{2+} + S + 2e$

[15] H. A. Liebhafsky, *J. Phys. Chem.*, **35**, 1648 (1931).
[16] I. M. Kolthoff and R. Belcher, *Volumetric Analysis*, vol. 3, p. 217. New York: Interscience Publishers, Inc., 1957.

Determination of Arsenic, Antimony, and Tin. Arsenic, antimony, and tin are conveniently determined by titration with iodine. The reaction of the reagent with arsenious acid was discussed in a previous section.

In acidic solutions, divalent tin is rapidly oxidized to the quadrivalent state by iodine. The ease with which tin(II) is oxidized by air represents the principal difficulty in the use of this reaction. The problem is eliminated by covering the solution with an inert gas during the titration and by using oxygen-free solutions throughout the analysis. Prereduction of quadrivalent tin is conveniently accomplished with metallic lead or nickel.

The reaction of trivalent antimony with iodine is quite analogous to that of trivalent arsenic. Here, however, an additional step is required to prevent precipitation of such basic salts as antimony oxychloride, SbOCl, as the solution is neutralized. These species react incompletely with iodine and cause erroneously low results. The problem is readily overcome by the addition of tartaric acid prior to dilution. The tartrate complex ($SbOC_4H_4O_6^-$) that forms is rapidly and completely oxidized by iodine.

Determination of Antimony in Stibnite. The analysis of stibnite, a common antimony ore, illustrates the application of a direct iodimetric method. Stibnite is primarily antimony sulfide that contains silica and other contaminants. Provided the material is free of iron and arsenic, determination of its antimony content is a straightforward process. The sample is decomposed in hot, concentrated hydrochloric acid to eliminate the sulfide as H_2S. Some care is required in this step to prevent losses of the volatile antimony trichloride. The addition of potassium chloride increases the tendency for formation of non-volatile chloride complexes. These probably have the formulas $SbCl_4^-$ and $SbCl_6^{3-}$.

A procedure for determining antimony in a stibnite is found in Chapter 31, Experiment 27.

Iodometric Methods

Iodide ion is a moderately effective reducing agent that has been widely employed for the analysis of oxidants. Iodide solutions are not ordinarily used for direct titrations, however, owing to their susceptibility to air oxidation. Instead, a standard solution of sodium thiosulfate (or occasionally arsenious acid) is used to titrate the iodine liberated by reaction of the analyte with an unmeasured excess of potassium iodide.

THE REACTION OF IODINE WITH THIOSULFATE ION

The reaction between iodine and thiosulfate ion is described by the equation

$$2S_2O_3^{2-} + I_2 \rightleftarrows S_4O_6^{2-} + 2I^-$$

The production of the tetrathionate ion requires the loss of two electrons from two thiosulfate ions; the equivalent weight of thiosulfate in this reaction must therefore be equal to its gram formula weight.

The quantitative conversion of thiosulfate to tetrathionate ion is almost unique with iodine. Other oxidizing reagents tend to carry the oxidation, wholly or in part, to sulfate ion. The reaction with hypoiodous acid provides an important example of this stoichiometry:

$$4HOI + S_2O_3{}^{2-} + H_2O \rightleftarrows 2SO_4{}^{2-} + 4I^- + 6H^+$$

The presence of hypoiodite in slightly alkaline solutions of iodine (p. 359), then, will seriously upset the stoichiometry of the iodine-thiosulfate reaction, causing too little thiosulfate or too much iodine to be used in the titration.

The equilibrium constant for the reaction

$$I_2 + H_2O \rightleftarrows HOI + I^- + H^+$$

is small (about 3×10^{-13}).[17] It can be shown, however, that hypoiodite formation should become significant in media where the pH is greater than 7. Thus, in the titration of 25 ml of 0.1-N iodine with 0.1-N thiosulfate, Kolthoff has demonstrated[18] an error of about 4% in the presence of 0.5 g of sodium carbonate; this error becomes 10% or more in solutions containing about 2 g of this salt. Kolthoff recommends that the pH should always be less than 7.6 for titration of 0.1-N solutions, 6.5 or less for 0.01-N solutions, and less than 5 for 0.001-N solutions.

The titration of highly acidic iodine solutions with thiosulfate yields quantitative results, provided care is taken to prevent air oxidation of iodide ion.

The end point in the titration is readily established by means of a starch solution. The properties of this indicator have been described earlier. It should be emphasized that starch is partially decomposed in the presence of a large excess of iodine. For this reason, the indicator is never added to an iodine solution until the bulk of that substance has been reduced. The change in color of the iodine from a red to a faint yellow signals the proper time for the addition of the indicator.

PREPARATION AND PROPERTIES OF STANDARD THIOSULFATE SOLUTIONS

Stability of Thiosulfate Solutions. Principal among the variables affecting the stability of thiosulfate solutions are the pH, the presence of microorganisms and impurities, the concentration of the solution, the presence of atmospheric oxygen, and exposure to sunlight. Generally, the decrease in iodine titer may amount to as much as several percent in a few weeks. Occasionally, however, increases in normality are observed. Proper attention to detail will yield standard thiosulfate solutions that need only occasional restandardization.

The following reaction occurs at an appreciable rate when the pH of a thiosulfate solution is 5 or less:

$$S_2O_3{}^{2-} + H^+ \rightleftarrows HS_2O_3{}^- \rightarrow HSO_3{}^- + S(s)$$

---

[17] W. C. Bray and E. L. Connolly, *J. Amer. Chem. Soc.*, **33**, 1485 (1911).
[18] I. M. Kolthoff and R. Belcher, *Volumetric Analysis*, vol. 3, pp. 214–215. New York: Interscience Publishers, Inc., 1957.

The velocity of this reaction increases with the hydrogen ion concentration; in strongly acidic solution, elemental sulfur forms within a few seconds. The hydrogen sulfite ion produced is also oxidized by iodine, reacting with twice the quantity of that reagent as the thiosulfate from which it was derived. Clearly, thiosulfate solutions cannot be allowed to stand in contact with acid. On the other hand, iodine solutions that are 3 to 4 F in acid may be titrated without error as long as care is taken to introduce the thiosulfate slowly and with good mixing. Under these conditions, the thiosulfate is so rapidly oxidized by the iodine that the slower acid decomposition cannot occur to any measurable extent.

Experiments indicate that the stability of thiosulfate solutions is at a maximum in the pH range between 9 and 10, although, for most purposes, a pH of 7 is adequate. Addition of small amounts of bases such as sodium carbonate, borax, or disodium hydrogen phosphate is frequently recommended to preserve standard solutions of the reagent. If this procedure is followed, the iodine solutions to be titrated must be made sufficiently acidic to neutralize the added base. Otherwise, hypoiodite formation may occur before the equivalence point is attained and cause the partial oxidation of the thiosulfate to sulfate.

The most important single cause of instability can be traced to certain bacteria that metabolize the thiosulfate ion, converting it to sulfite, sulfate, and elemental sulfur.[19] Solutions that are free of bacteria are remarkably stable; it is common practice, therefore, to impose reasonably sterile conditions in preparation of standard solutions. Substances such as chloroform, sodium benzoate, or mercury(II) iodide can be added to inhibit bacterial growth. Bacterial activity appears to be at a minimum at a pH between 9 and 10, which accounts, at least in part, for the maximum stability of thiosulfate solutions in this range.

Many other variables affect the stability of thiosulfate solutions. Decomposition is reported to be catalyzed by copper(II) ions as well as by the decomposition products themselves. Solutions that have become turbid from the formation of sulfur should be discarded. Exposure to sunlight increases the rate of decomposition as does atmospheric oxygen. Finally, the decomposition rate is greater in more dilute solutions.

STANDARDIZATION OF THIOSULFATE SOLUTIONS

Primary standards for thiosulfate solutions are oxidizing agents that liberate a known amount of iodine when treated with excess of iodide ion. The iodine is then titrated with the solution to be standardized.

Potassium Iodate. Iodate ion reacts rapidly with iodide in slightly acidic solution to give iodine.

$$IO_3^- + 5I^- + 6H^+ \rightleftarrows 3I_2 + 3H_2O$$

$$3I_2 + 6e \rightarrow 6I^-$$

---

[19] M. Kilpatrick, Jr., and M. L. Kilpatrick, *J. Amer. Chem. Soc.*, **45**, 2132 (1923); F. O. Rice, M. Kilpatrick, Jr., and W. Lemkin, *J. Amer. Chem. Soc.*, **45**, 1361 (1923).

Three moles of iodine are furnished by each formula weight of potassium iodate for the standardization; thus, its equivalent weight is one-sixth its formula weight, since a six-electron change is associated with the reduction of three iodine molecules, the species actually titrated. That is,

$$IO_3^- \equiv 3I_2 \equiv 6I^-$$

The sole disadvantage of potassium iodate as a primary standard is its low equivalent weight (35.67). Only slightly more than 0.1 g can be taken for standardization of a 0.1-N thiosulfate solution; the relative error normally incurred in weighing this quantity may, under some circumstances, be somewhat greater than desirable for a standardization. This problem can be circumvented by dissolving a larger quantity of the solid in a known volume and taking aliquots of the resulting solution.

Other Primary Standards. Other primary standards for sodium thiosulfate include potassium dichromate, potassium bromate, potassium hydrogen iodate [$KH(IO_3)_2$], potassium ferricyanide, and metallic copper. Details for the use of these may be found in various reference works.[20]

SOURCES OF ERROR IN IODOMETRIC METHODS

1. Air Oxidation of Iodide Ion. Considered solely from the standpoint of equilibrium, the reaction

$$4I^- + O_2(g) + 4H^+ \rightleftarrows 2I_2 + 2H_2O$$

should necessitate the exclusion of atmospheric oxygen from all iodometric titration mixtures. Fortunately, however, the rate at which the oxidation occurs is so slow that this precaution is ordinarily unnecessary.

Kinetic studies have shown that the velocity of air oxidation increases greatly with the hydrogen ion concentration and is appreciable in media that are greater than 0.4 to 0.5 F in hydrogen ion. Light also catalyzes the process. Thus, storage in a dark place is recommended if time is required for completion of the reaction between iodide ion and an oxidizing agent.

Traces of copper(II) ion and nitrogen oxides are known to catalyze the air oxidation of iodide ion. The latter are potential sources of interference to any iodometric analysis that makes use of nitric acid in the preliminary steps.

Where air oxidation is believed to be a problem, recourse must be made to an inert atmosphere over the titration mixture. This is conveniently accomplished by periodically adding small portions (300 mg) of sodium hydrogen carbonate to the acidic solution as the titration progresses; alternatively, carbon dioxide can be introduced from a tank or in the form of small pieces of dry ice.

2. Volatilization of Liberated Iodine. Errors due to the volatilization of liberated iodine are avoided by using stoppered containers when solutions must

[20] For example, see I. M. Kolthoff and R. Belcher, *Volumetric Analysis*, vol. 3, pp. 234–243. New York: Interscience Publishers, Inc., 1957.

stand, by maintaining a goodly excess of iodide ion, and by avoiding elevated temperatures.

3. Decomposition of Thiosulfate Solutions. The decomposition of thiosulfate solutions was discussed in a preceding section.

4. Alteration in Stoichiometry of the Iodine-Thiosulfate Reaction. As noted earlier, alteration in stoichiometry is encountered when basic solutions are titrated.

5. Premature Addition of Starch Indicator. See page 360 for control of the addition of starch indicator.

APPLICATIONS OF THE INDIRECT IODOMETRIC METHOD

Numerous substances can be determined iodometrically; some of the more common applications are summarized in Table 16-7. Detailed instructions for the iodometric determination of copper, a typical example of the indirect procedure, are found in Chapter 31.

Determination of Copper. A cursory examination of electrode potentials suggests that an analysis based upon reduction of copper(II) by iodide would not be feasible. That is,

$$Cu^{2+} + e \rightleftarrows Cu^+ \qquad E^0 = 0.15 \text{ V}$$

$$I_2 + 2e \rightleftarrows 2I^- \qquad E^0 = 0.54 \text{ V}$$

In fact, however, the reduction is quantitative in the presence of a reasonable excess of iodide by virtue of the low solubility of copper(I) iodide. Thus, when the more appropriate half-reaction

$$Cu^{2+} + I^- + e \rightleftarrows CuI(s) \qquad E^0 = 0.86 \text{ V}$$

is employed, it becomes apparent that the equilibrium

$$2Cu^{2+} + 4I^- \rightleftarrows 2CuI(s) + I_2$$

is reasonably favorable. Here the iodide ion serves not only as a reducing agent for copper(II) ion but also as a precipitant for copper(I).

Much systematic experimentation has been devoted to establishing ideal conditions for a copper analysis.[21] These studies have revealed that the solution should be at least 4% with respect to potassium iodide. Further, a pH less than 4 is expedient, because at higher pH, formation of basic copper(II) species causes a slower and less complete oxidation of iodide ion. In the presence of copper(I) ion, hydrogen ion concentrations greater than about 0.3 M must be avoided to prevent air oxidation of iodide ion.

[21] See E. W. Hammock and E. H. Swift, *Anal. Chem.*, **21**, 975 (1949).

TABLE 16-7   **Some Applications of the Indirect Iodometric Method**
$$2I^- \rightleftarrows I_2 + 2e$$

Substance	Half-Reaction	Special Conditions
IO_4^-	$IO_4^- + 8H^+ + 7e \rightleftarrows \frac{1}{2}I_2 + 4H_2O$	Acidic solution
	$IO_4^- + 2H^+ + 2e \rightleftarrows IO_3^- + H_2O$	Neutral solution; titration with arsenic(III)
IO_3^-	$IO_3^- + 6H^+ + 5e \rightleftarrows \frac{1}{2}I_2 + 3H_2O$	
BrO_3^-	$BrO_3^- + 6H^+ + 6e \rightleftarrows Br^- + 3H_2O$	
ClO_3^-	$ClO_3^- + 6H^+ + 6e \rightleftarrows Cl^- + 3H_2O$	Strong acid; slow reaction
$HClO$	$HClO + H^+ + 2e \rightleftarrows Cl^- + H_2O$	
Cl_2	$Cl_2 + 2e \rightleftarrows 2Cl^-$	
Br_2	$Br_2 + 2e \rightleftarrows 2Br^-$	
I^-	$I^- + 3Cl_2 + 3H_2O \rightleftarrows IO_3^- + 6Cl^- + 6H^+$	Excess Cl_2 removed by boiling
	$IO_3^- + 6H^+ + 5e \rightleftarrows \frac{1}{2}I_2 + 3H_2O$	
NO_2^-	$HNO_2 + H^+ + e \rightleftarrows NO + H_2O$	
H_3AsO_4	$H_3AsO_4 + 2H^+ + 2e \rightleftarrows H_3AsO_3 + H_2O$	Strong HCl
H_3SbO_4	$H_3SbO_4 + 2H^+ + 2e \rightleftarrows H_3SbO_3 + H_2O$	Strong HCl
$Fe(CN)_6^{3-}$	$Fe(CN)_6^{3-} + e \rightleftarrows Fe(CN)_6^{4-}$	
MnO_4^-	$MnO_4^- + 8H^+ + 5e \rightleftarrows Mn^{2+} + 4H_2O$	
Ce^{4+}	$Ce^{4+} + e \rightleftarrows Ce^{3+}$	
$Cr_2O_7^{2-}$	$Cr_2O_7^{2-} + 14H^+ + 6e \rightleftarrows 2Cr^{3+} + 7H_2O$	
Pb^{2+}, Ba^{2+}, Sr^{2+}	$Pb^{2+} + CrO_4^{2-} \rightleftarrows PbCrO_4(s)$	Filter and wash; dissolve in $HClO_4$
	$2PbCrO_4(s) + 2H^+ \rightleftarrows 2Pb^{2+} + Cr_2O_7^{2-} + H_2O$	
	$Cr_2O_7^{2-} + 14H^+ + 6e \rightleftarrows 2Cr^{3+} + 7H_2O$	
Fe^{3+}	$Fe^{3+} + e \rightleftarrows Fe^{2+}$	Strong HCl solution; slow reaction
Cu^{2+}	$Cu^{2+} + I^- + e \rightleftarrows CuI(s)$	Cu^+ precipitates as CuI
O_2	$O_2 + 4Mn(OH)_2(s) + 2H_2O \rightleftarrows 4Mn(OH)_3(s)$	Basic solution + iodide; solution finally acidified
	$Mn(OH)_3(s) + 3H^+ + e \rightleftarrows Mn^{2+} + 3H_2O$	
O_3	$O_3 + 2H^+ + 2e \rightleftarrows O_2 + H_2O$	Neutral solution
H_2O_2	$H_2O_2 + 2H^+ + 2e \rightleftarrows 2H_2O$	Molybdate catalyst
Organic peroxides	$ROOH + 2H^+ + 2e \rightleftarrows ROH + H_2O$	
MnO_2	$MnO_2(s) + 4H^+ + 2e \rightleftarrows Mn^{2+} + 2H_2O$	

It has been found experimentally that the titration of iodine by thiosulfate in the presence of copper(I) iodide tends to yield slightly low results because small but appreciable quantities of iodine are physically adsorbed upon the solid. The adsorbed iodine is released only slowly, even in the presence of thiosulfate ion; transient and premature end points result. This difficulty is largely overcome by the addition of thiocyanate ion, which also forms a sparingly soluble copper(I) salt. Part of the copper(I) iodide is converted to the corresponding thiocyanate at the surface of the solid:

$$CuI(s) + SCN^- \rightleftarrows CuSCN(s) + I^-$$

Accompanying this reaction is the release of the adsorbed iodine, thus making it available for titration. Early addition of thiocyanate must be avoided, however, because of the tendency for that ion to reduce iodine slowly.

The iodometric method is convenient for the assay of copper in an ore. Ordinarily, samples dissolve readily in hot, concentrated nitric acid. Care must be taken to volatilize any nitrogen oxides formed in the process since these catalyze the air oxidation of iodide. Some samples require the addition of hydrochloric acid to complete the solution step. The chloride ion must, however, be removed by evaporation with sulfuric acid because iodide ion will not reduce copper(II) quantitatively from its chloride complexes.

Of the elements ordinarily associated with copper in nature, only iron, arsenic, and antimony interfere with the iodometric procedure. Fortunately, difficulties caused by these elements are readily eliminated. Iron is rendered nonreactive by the addition of such complexing agents as fluoride or pyrophosphate; because these form more stable complexes with iron(III) than with iron(II), the potential for this system is altered to the point where oxidation of iodide cannot occur appreciably. Interference by arsenic and antimony is prevented by converting these elements to the +5 state during solution of the sample. Ordinarily, the hot nitric acid employed will convert the elements to the desired oxidation state although a small amount of bromine water can be added in case of doubt; the excess bromine is then expelled by boiling. As has been pointed out, arsenic in the +5 state does not oxidize iodide ion, provided the solution is not too acidic. Antimony exhibits a similar behavior. Thus, by maintaining the pH of the solution at 3 or greater, interference by these elements is avoided. We have seen, however, that oxidation of iodide by copper is incomplete at pH values greater than 4. Thus, when copper is to be determined in the presence of arsenic or antimony, control of the pH between 3 and 4 is essential. Ammonium hydrogen fluoride, NH_4HF_2, is a convenient buffer for this purpose. The anion of the salt dissociates as follows:

$$HF_2^- \rightleftarrows HF + F^- \qquad K = 0.26$$

$$HF \rightleftarrows H^+ + F^- \qquad K = 6.7 \times 10^{-4}$$

The first dissociation provides equal quantities of hydrogen fluoride and fluoride ion which then buffer the solution to a pH somewhat greater than 3. In addition to acting as a buffer, the salt also serves as a source of fluoride ions to complex any iron(III) that may be present. A procedure for the iodometric determination of copper in an ore is given in Chapter 31.

The iodometric method is also applicable to the determination of copper in brass, an alloy consisting principally of copper, zinc, lead, and tin. Several other elements, iron and nickel, for example, may be tolerated in minor amounts.

The method, which is described in detail in Chapter 31, is relatively simple and applicable to the analysis of brasses containing less than 2% of iron. It involves solution in nitric acid, removal of nitrate by fuming with sulfuric acid, adjustment of the pH by neutralization with ammonia, acidification with a measured quantity of phosphoric acid, and finally the iodometric determination of the copper.

It is instructive to consider the fate of each major constituent during the course of this treatment. Tin is oxidized to the +4 state by the nitric acid and precipitates slowly as the slightly soluble hydrous tin(IV) oxide, $SnO_2 \cdot 4H_2O$. This precipitate, which tends to form as a colloid, is sometimes called *metastannic acid*. It has a tendency to adsorb copper(II) and other cations from the solution. Lead, zinc, and copper are oxidized to soluble divalent salts by the nitric acid, and iron is converted to the trivalent state. Evaporation and fuming with sulfuric acid redissolves the metastannic acid but may cause part of the lead to precipitate as the sulfate; the copper, zinc, and iron are unaffected. Upon dilution with water, lead is nearly completely precipitated as lead sulfate while the other elements remain in solution. None, except copper and iron, is reduced by iodide. Interference from the iron is eliminated by complexing with phosphate ion.

Determination of Oxygen. The Winkler method for the determination of dissolved oxygen in natural water is an interesting example of iodometry. In this procedure, the sample is first treated with an excess of manganese(II), sodium iodide, and sodium hydroxide. The white manganese(II) hydroxide that forms reacts rapidly with oxygen to form brown manganese(III) hydroxide. That is,

$$4Mn(OH)_2(s) + O_2 + 2H_2O \rightarrow 4Mn(OH)_3(s)$$

When acidified, the manganese(III) oxidizes iodide to iodine. Thus,

$$2Mn(OH)_3(s) + 2I^- + 6H^+ \rightarrow I_2 + 3H_2O + 2Mn^{2+}$$

The liberated iodine is titrated in the usual way.

PROBLEMS

*1. Write balanced equations to describe the following processes:
   (a) the oxidation of manganese(II) to permanganate by ammonium peroxo-disulfate.
   (b) the oxidation of cerium(III) to cerium(IV) by sodium bismuthate.
   (c) the oxidation of uranium(IV) to uranium(VI) by H_2O_2.
   (d) the reaction of $V(OH)_4^+$ in a silver reductor.
   (e) the oxidation of SCN^- to SO_4^{2-} by IO_3^- in 1-F HCl.
   (f) the titration of $H_2C_2O_4$ with a standard solution of $KBrO_3$, the equivalence point being signaled by the first appearance of Br_2.
   (g) the titration of H_2O_2 with $KMnO_4$.
   (h) the reaction between KI and ClO_3^- in acidic solution.

2. Write balanced equations to describe the following processes:
   (a) the oxidation of manganese(II) to permanganate by periodic acid.
   (b) the reduction of iron(III) to iron(II) by sulfur dioxide.
   (c) the reaction of molybdic acid in a Jones reductor.
   (d) the oxidation of nitrous acid by a solution of potassium permanganate.
   (e) the reaction of aniline with a mixture of $KBrO_3$ and KBr in acidic solution.
   (f) the oxidation of dimethyl sulfide, $(CH_3)_2S$, to the sulfoxide, $(CH_3)_2SO$, by a standard solution of $KBrO_3$. Assume that the equivalence point is signaled by the first appearance of Br_2.
   (g) the air oxidation of $HAsO_3^{2-}$ to $HAsO_4^{2-}$.
   (h) the reaction of KI with HNO_2 in acidic solution.

*3. Indicate the number of moles of each product that is formed from one mole of the following compounds when oxidized by periodic acid at room temperature:

(a) $CH_2OH(CHOH)_4CH_2OH$

(b)
$$\underset{\displaystyle HC-C-H}{\overset{\displaystyle O \quad O}{\overset{\displaystyle \| \quad \|}{}}}$$

(c)
$$\underset{\displaystyle \underset{H \quad H}{\overset{|}{} \quad \overset{|}{}}}{\overset{H \; NH_2 \; O}{\overset{| \quad | \quad \|}{HC-C-CH}}}$$

4. Indicate the number of moles of each product that is formed from one mole of the following compounds when oxidized by periodic acid at room temperature:

(a)
$$\underset{\displaystyle \underset{H \quad H}{| \quad |}}{\overset{H \; NH_2 \; O}{\overset{| \quad | \quad \|}{H-C-C-C-CH_3}}}$$

(b)
$$\underset{\displaystyle CH_3C-CH}{\overset{O \quad O}{\overset{\| \quad \|}{}}}$$

(c)
$$\underset{\displaystyle \underset{H}{|}}{\overset{OH \; O}{\overset{| \quad \|}{CH_3C-CH}}}$$

*5. Describe the preparation of 2.50 liters of 0.100-N $K_2Cr_2O_7$ from the pure salt.

6. Describe the preparation of 500 ml of 0.200-N KIO_3 to be used for the titration of a reducing agent in a 1-F HCl solution.

*7. Describe the preparation of 750 ml of 0.150-N $KBrO_3$ to be used as a source of bromine.

8. Describe the preparation of 2.00 liters of 0.170-N I_3^- from pure I_2 and KI.

*9. What is the Fe_2O_3 titer of the solution described in Problem 5?

10. What is the KCNS titer of the solution described in Problem 6 (Reaction: $CNS^- + 4H_2O \rightleftarrows SO_4^{2-} + HCN + 7H^+ + 6e$)?

*11. What is the phenol titer of the solution described in Problem 7?

12. What is the Sb_2O_3 titer of the solution described in Problem 8?

*13. To standardize a solution of $Na_2S_2O_3$, 0.151 g of $K_2Cr_2O_7$ was dissolved in dilute HCl. An excess of KI was added, following which the liberated I_2 was titrated with 46.1 ml of the reagent. Calculate the normality of the $Na_2S_2O_3$.

14. To standardize a solution of $Na_2S_2O_3$, 0.101 g of $KBrO_3$ was dissolved in dilute HCl. An excess of KI was added, following which the liberated I_2 was titrated with 39.7 ml of the reagent. Calculate the normality of the $Na_2S_2O_3$.

*15. Arsenic(III) oxide occurs in nature as the mineral claudetite. A 0.210-g sample of impure claudetite was found to require 29.3 ml of 0.0520-N iodine solution. Calculate the percentage of As_2O_3 in the sample.

16. The potassium chlorate in a 0.134-g sample of high explosive was determined by reaction with 50.0 ml of 0.0960-N Fe^{2+}:

$$ClO_3^- + 6Fe^{2+} + 6H^+ \rightarrow Cl^- + 3H_2O + 6Fe^{3+}$$

When reaction was complete, the excess iron(II) was back-titrated with 13.3 ml of 0.0836-N Ce^{4+}. Calculate the percentage of $KClO_3$ in the sample.

*17. The antimony(III) in a 1.080-g stibnite sample required a 41.6-ml titration with 0.0653-N I_2. Express the results of this analysis in terms of
   (a) percent antimony.
   (b) percent Sb_2S_3.

18. A 0.223-g sample of limestone was dissolved in dilute HCl. Ammonium oxalate was then introduced, and the pH of the resulting solution was adjusted to permit the quantitative precipitation of calcium oxalate. The solid was isolated by filtration, washed free of excess oxalate, and then redissolved in dilute sulfuric acid. Titration of the liberated oxalic acid required 26.7 ml of 0.120-N $KMnO_4$. Calculate the percentage of calcium oxide in the sample.

*19. An 8.13-g sample of an ant-control preparation was decomposed by wet-ashing with H_2SO_4 and HNO_3. The arsenic in the residue was reduced to the trivalent state with hydrazine. After removal of the excess reducing agent, the As(III) required a 23.7-ml titration with 0.0485-N I_2 in a faintly alkaline medium. Express the results of this analysis in terms of the percent As_2O_3 in the original sample.

20. A 4.97-g sample containing the mineral tellurite was brought into solution and then treated with a 50.0-ml aliquot of 0.0943-N dichromate. Reaction:

$$3TeO_2 + Cr_2O_7^{2-} + 8H^+ \rightarrow 3H_2TeO_4 + 2Cr^{3+} + H_2O$$

Upon completion of the reaction, the excess dichromate required a 10.9-ml back-titration with 0.113-N Fe^{2+}. Calculate the percentage of TeO_2 in the sample.

*21. A 25.0-ml aliquot of a solution containing thallium(I) ion was treated with potassium chromate. The Tl_2CrO_4 was filtered, washed free of excess precipitating agent, and then redissolved in dilute sulfuric acid. The dichromate ion produced was then titrated with 40.6 ml of 0.100-N iron(II) ammonium sulfate solution. What was the weight of thallium in the sample?

$$2Tl^+ + CrO_4^{2-} \rightarrow Tl_2CrO_4(s)$$

$$2Tl_2CrO_4(s) + 2H^+ \rightarrow 4Tl^+ + Cr_2O_7^{2-} + H_2O$$

$$Cr_2O_7^{2-} + 6Fe^{2+} + 14H^+ \rightarrow 6Fe^{3+} + 2Cr^{3+} + 7H_2O$$

22. A 0.306-g sample of an impure aluminum salt was dissolved in dilute acid, treated with an excess of ammonium oxalate, and slowly made alkaline through the addition of aqueous ammonia. The precipitated aluminum oxalate was filtered, washed, redissolved in dilute acid, and titrated with 36.0 ml of 0.121-N $KMnO_4$. Calculate the percentage of aluminum in the sample.

*23. The chromium in a 1.87-g sample of chromite ($FeO \cdot Cr_2O_3$) was oxidized to the +6 state by fusion with sodium peroxide. The fused mass was treated with water and boiled to destroy the excess peroxide. After acidification, the sample was treated with 50.0 ml of 0.160-N Fe^{2+}. A back-titration of 2.97 ml of 0.0500-N $K_2Cr_2O_7$ was required to oxidize the excess iron(II).
   (a) What was the percentage of chromite in the sample?
   (b) What was the percentage of chromium in the sample?

24. A 0.240-g sample of pyrolusite (MnO_2) was reduced to Mn^{2+} with 25.0 ml of 0.100-N sodium arsenite. The excess arsenite was then titrated with 3.12 ml of 0.107-N permanganate solution in the presence of a catalytic quantity of iodate. What was the percentage of pyrolusite in the original sample?

*25. In the presence of fluoride ion, Mn^{2+} can be titrated with MnO_4^-, both reactants being converted to a complex of Mn(III). A 0.545-g sample containing Mn_3O_4 was dissolved, and all manganese was converted to Mn^{2+}. Titration in the presence of fluoride ion consumed 31.1 ml of $KMnO_4$ that was 0.117 N against oxalate.

   (a) Write a balanced equation for the reaction, assuming that the complex is MnF_4^-.

   (b) What was the normality of the $KMnO_4$ in this titration?

   (c) What was the percent Mn_3O_4 in the sample?

26. A 50.0-ml aliquot containing uranium(VI) was passed through a Jones reductor to reduce the U to the $+3$ state. Aeration converted the U(III) to U(IV), following which the latter was titrated with 29.3 ml of 0.0773-N $K_2Cr_2O_7$. What weight of uranium was contained in each liter of the sample solution?

*27. A 2.50-g sample of stainless steel was dissolved in HCl (this treatment converts the Cr present to Cr^{3+}) and diluted to 500 ml in a volumetric flask. One 50.0-ml aliquot was passed through a silver reductor (Table 16-1) and then titrated with 29.6 ml of 0.0960-N potassium permanganate solution. A second 50.0-ml aliquot was passed through a Jones reductor (Table 16-1) into 50 ml of 0.100-F Fe^{3+}. Titration of the resulting solution required 14.7 ml of the standard permanganate solution. Calculate the percentages of iron and chromium in the alloy.

28. A 2.55-g sample containing both iron and vanadium was dissolved under conditions that converted the elements to Fe(III) and V(V). The solution was diluted to 500 ml; a 50.0-ml aliquot was passed through a Walden reductor and titrated with 17.7 ml of 0.100-N Ce^{4+}. A second 50.0-ml aliquot was passed through a Jones reductor and required 44.6 ml of the Ce^{4+} to reach an end point. Calculate the percent Fe_2O_3 and V_2O_5 in the sample. See Table 16-1 for the reductor reactions.

*29. A 0.649-g sample of alkali metal sulfates was dissolved and diluted to exactly 100 ml. The sodium in a 25.0-ml aliquot was precipitated as $NaZn(UO_2)_3(OAc)_9 \cdot 6H_2O$. After filtration and washing, the precipitate was dissolved in acid and the uranium reduced to U^{3+} in a Jones reductor. After aeration to convert the U^{3+} to U^{4+}, the solution was titrated with 33.3 ml of 0.108-N Ce^{4+}. Calculate the percent Na_2SO_4 in the sample.

30. A 10.9-g sample of an insecticide preparation was wet-ashed with a sulfuric–nitric acid mixture to destroy its organic components. The copper in the sample was then precipitated as the basic chromate, $CuCrO_4 \cdot 2CuO \cdot 2H_2O$, with an excess of potassium chromate. The solid was filtered, washed free of excess reagent, and then redissolved in acid:

$$2(CuCrO_4 \cdot 2CuO \cdot 2H_2O)(s) + 10H^+ \rightarrow 6Cu^{2+} + Cr_2O_7^{2-} + 9H_2O$$

Titration of the liberated dichromate required 31.4 ml of 0.116-N Fe^{2+}. Express the results of this analysis in terms of the percentage of copper(II) oleate, $Cu(C_{18}H_{33}O_2)_2$ (gfw = 626), present in the sample.

*31. The barium ion in a 1.50-g sample was precipitated with 50.0 ml of 0.0500-N potassium iodate solution. The $Ba(IO_3)_2$ was filtered, after which the filtrate and washings were treated with an excess of potassium iodide. The iodine liberated required 10.7 ml of 0.0396-N $Na_2S_2O_3$ solution. Express the results of this analysis in terms of percentage barium oxide.

32. A potassium permanganate solution was found to be 0.0913 N in terms of standardization against sodium oxalate in acidic solution. The oxidizing agent was then employed in a Volhard determination for manganese:

$$2MnO_4^- + 3Mn^{2+} + 4OH^- \rightarrow 5MnO_2(s) + 2H_2O$$

Calculate the percentage of manganese in a mineral specimen if a 0.467-g sample required a 36.2-ml titration with the permanganate solution.

*33. The H_2S and SO_2 concentrations of a gas were determined by passage through three absorber solutions connected in series. The first contained an ammoniacal solution of Cd^{2+} to trap the sulfide as CdS. The second contained 10.0 ml of 0.0396-N I_2 to oxidize the SO_2 to SO_4^{2-}. The third contained 2.00 ml of 0.0345-N thiosulfate solution to retain any I_2 carried over from the second absorber. A 25.0-liter gas sample was passed through the apparatus, followed by an additional amount of pure N_2 to sweep the last traces of SO_2 from the first to the second absorber.

The solution from the first absorber was made acidic, and 20.0 ml of the 0.0396-N I_2 were added. The excess I_2 was back-titrated with 7.45 ml of the thiosulfate solution.

The solutions in the second and third absorbers were combined, and the residual iodine was titrated with 2.44 ml of the thiosulfate solution.

Calculate the concentrations of SO_2 and H_2S in mg/liter of the sample.

34. A 25.0-ml sample of a household bleach was diluted to 250.0 ml in a volumetric flask. Calculate the weight-volume percentage of NaClO in the sample if 36.3 ml of 0.0961-N $Na_2S_2O_3$ were needed to titrate the iodine liberated when excess KI was introduced to a 50.0-ml aliquot of the diluted sample.

*35. A sensitive method for I^- in the presence of Cl^- and Br^- entails oxidation of the I^- to IO_3^- with Br_2. The Br_2 is then removed by boiling or by reduction with formate ion. The iodate is determined by addition of excess iodide and titration of the resulting iodine. A 1.20-g sample of mixed halides was dissolved and diluted to 500 ml. A 50.0-ml aliquot, analyzed by the foregoing procedure, required 20.6 ml of 0.0555-N thiosulfate. Calculate the percent KI in the sample.

36. A solution containing both $NaIO_3$ and $NaIO_4$ was analyzed by buffering a 25.0-ml sample with borax and adding an excess of iodide. In this slightly alkaline medium, IO_4^- was reduced to IO_3^-, liberating an equivalent quantity of iodine that required 15.5 ml of 0.0802-N thiosulfate. A 10.0-ml portion of the sample was made strongly acidic with HCl after addition of excess KI; 36.7 ml of the thiosulfate were needed to titrate the iodine liberated by both the IO_4^- and the IO_3^- (p. 353). Calculate the milligrams $NaIO_3$ and $NaIO_4$ per milliliter of the sample.

*37. The Winkler method for dissolved oxygen in water is based upon the rapid oxidation of solid $Mn(OH)_2$ to $Mn(OH)_3$ in alkaline medium. When acidified, the Mn(III) readily releases iodine from iodide.

A 250-ml water sample, in a stoppered vessel, was treated with 1.00 ml of a concentrated solution of NaI and NaOH and 1.00 ml of a manganese(II) solution. Oxidation of the $Mn(OH)_2$ was complete in about 1 min. The precipitates were then dissolved by addition of 2.00 ml of concentrated H_2SO_4, whereupon an amount of iodine equivalent to the $Mn(OH)_3$ (and hence to the dissolved O_2) was liberated. A 25.0-ml aliquot (of the 254 ml) was titrated with 12.7 ml of 0.00962-N thiosulfate. Calculate the milligrams O_2 per milliliter sample (assume that the concentrated reagents are O_2 free and take their dilutions of the sample into account).

38. The CO concentration in a 3.21-liter sample of air was obtained by passage over iodine pentoxide heated at 150°C:

$$I_2O_5 + 5CO \rightleftarrows 5CO_2 + I_2$$

The I_2 distilled at this temperature and was collected in a solution of iodide ion. The resulting triiodide was titrated with 7.76 ml of 0.00221-N thiosulfate. Calculate the parts per million CO in the gas, assuming an air density of 1.20×10^{-3} g/ml.

*39. Titration of a 3.06-g sample of a hair-setting preparation required 24.7 ml of 0.1028-N I_2. Calculate the percent thioglycolic acid (gfw = 92.1) in the sample. Reaction:

$$2HSCH_2COOH + I_2 \rightarrow HOOC-CH_2-S-S-CH_2-COOH + 2HI$$

40. A square of photographic film 2.0 cm on an edge was suspended in a 5% solution of $Na_2S_2O_3$ to dissolve the silver halides. After removal and washing of the film, the solution was treated with an excess of Br_2 to oxidize the iodide present to IO_3^- and destroy the excess thiosulfate ion. The solution was boiled to remove the bromine, and an excess of iodide was added. The liberated iodine was titrated with 13.7 ml of 0.0352-N thiosulfate solution.
(a) Write balanced equations for the reactions involved in the method.
(b) Calculate the milligrams of AgI per square centimeter of film.

*41. In slightly alkaline media, thiocyanate ion is quantitatively oxidized to sulfate and hydrogen cyanide by iodine (see Problem 10 for the half-reaction). A 0.517-g sample containing $Ba(SCN)_2$ was dissolved in a bicarbonate solution; 50.0 ml of 0.107-N iodine were added, and the mixture was allowed to stand for 5 min. The solution was then acidified, and the excess I_2 was titrated with 16.3 ml of 0.0965-N thiosulfate.
(a) Write a balanced equation for the oxidation of SCN^- by I_2.
(b) Calculate the percent $Ba(SCN)_2$ in the sample.

42. The ethyl mercaptan concentration of a mixture was determined by shaking a 1.65-g sample with 50.0 ml of 0.119-N iodine in a tightly stoppered flask:

$$2C_2H_5SH + I_2 \rightarrow C_2H_5SSC_2H_5 + 2I^- + 2H^+$$

The excess iodine was back-titrated with 16.7 ml of 0.132-N thiosulfate. Calculate the percent ethyl mercaptan.

*43. A 0.646-g sample containing $BaCl_2 \cdot 2H_2O$ was dissolved and an excess of a K_2CrO_4 solution added. After a suitable period, the $BaCrO_4$ was filtered, washed, and redissolved in HCl to convert the CrO_4^{2-} to $Cr_2O_7^{2-}$. An excess of KI was added, and the liberated iodine was titrated with 48.7 ml of 0.137-N thiosulfate. Calculate the percent $BaCl_2 \cdot 2H_2O$.

POTENTIOMETRIC METHODS

In Chapter 15 it was shown that the potential of an electrode can vary with the concentration (or, more correctly, the activity) of one or more of the species of the solution in which it is immersed. Thus, potential measurements can often be used to provide quantitative information about the composition of solutions.

The potentiometric method consists of measuring the potential between a pair of suitable electrodes immersed in the solution to be analyzed. The apparatus required consists of an indicator electrode, a reference electrode, and a device for measuring potential.

Potential Measurement

The electromotive force developed by a galvanic cell cannot be measured accurately by placing a simple dc voltmeter across the electrodes, because a significant current is required for operation of the meter. When this current is drawn from the cell, a voltage decrease occurs due, in part, to changes in the concentrations of the reacting species as the cell discharges. In addition, the cell has an internal resistance that results in an ohmic potential drop (equal to the current times the resistance), which opposes that of the two electrodes.

Thus, the measured potential is less than the actual cell potential. A truly significant value for the output of a cell can be attained only if the measurement is made with a negligible passage of current. A *potentiometer* is one type of instrument that meets this specification.

POTENTIAL MEASUREMENTS WITH A POTENTIOMETER

Figure 17-1 is a schematic diagram of a simple potentiometer. The working battery P is connected to the terminals of a linear voltage divider AB, whose electrical resistance, R_{AC}, from one end A to any point C, is directly proportional to the length AC of the resistor; that is, $R_{AC} = kAC$, where k is a proportionality constant. Included also is a sensitive current-detecting device, G, which may be either a sensitive galvanometer or a dc amplifier; a tapping key, K, by which the circuit can be momentarily closed; and a double-pole, double-throw switch to permit insertion of the unknown cell E_x or a standard cell of known potential E_s into the circuit. The potential of the working battery, E_P, is greater than that of E_x or of E_s.

Principles of the Potentiometer. A current I flows continuously from the battery through AB, causing a potential drop between A and B that, from Ohm's law, is given by $E_{AB} = IR_{AB}$. The current passing between A and C is also I. Thus, the potential drop between A and C is given by $E_{AC} = IR_{AC}$. We have noted, however, that the resistance along the resistor is linear. Thus, we may write

$$E_{AB} = IR_{AB} = kIAB$$
$$E_{AC} = IR_{AC} = kIAC$$

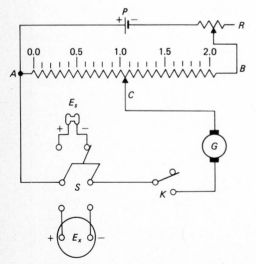

Figure 17-1   Diagram of a Potentiometer with a Linear Voltage Divider, *AB*.

Dividing the first equation by the second and rearranging yields

$$E_{AC} = E_{AB}\frac{AC}{AB} \tag{17-1}$$

If E_{AC} is greater than E_x (or E_s), electrons will be forced from right to left through the unknown cell when K is closed; the direction of flow will be reversed if E_{AC} is smaller. When E_{AC} is exactly equal to E_x no current will flow in the circuit containing G and K. Note, however, that current continues to flow from P through AB under these circumstances.

Applying Equation 17-1 when the unknown cell and the standard cell are placed in the circuit yields

$$E_x = E_{AC_x} = E_{AB}\frac{AC_x}{AB}$$

and

$$E_s = E_{AC_s} = E_{AB}\frac{AC_s}{AB}$$

where AC_x and AC_s represent the linear distances corresponding to balance for the two cases, respectively. Dividing the two equations and rearranging gives

$$E_x = E_s\frac{AC_x}{AC_s} \tag{17-2}$$

Thus, E_x may be obtained from the two measured distances and the known emf of the standard cell.

The Weston Standard Cell. A common standard cell for potentiometers is the *Weston cell*, which can be represented by

$$Cd(Hg) \mid CdSO_4 \cdot \tfrac{8}{3}H_2O(sat'd), Hg_2SO_4(sat'd) \mid Hg$$

where Cd(Hg) represents a solution of cadmium in mercury. The half-reactions, as they occur when current is drawn, are

$$Cd(Hg) \rightarrow Cd^{2+} + Hg(l) + 2e$$

$$Hg_2{}^{2+} + 2e \rightarrow 2Hg(l)$$

The potential of the Weston cell is 1.0183 V at 25°C.

For convenience, the scale reading of AC_x is ordinarily calibrated to give the potential directly in volts by first switching the Weston cell into the circuit and positioning the contact so that AC_s corresponds numerically to the output of this standard cell. The potential across AB is then adjusted by means of R until no current is indicated by the galvanometer. With the potentiometer adjusted thus, AC_s and E_s will be numerically equal; from Equation 17-2, then, AC_x and E_x will also be identical when balance is achieved with the unknown cell in the circuit.

The need for a working battery P may be questioned. In principle, there is nothing to prevent a direct measurement of E_x by replacing P with the standard cell E_s. It must be remembered, however, that current is being continuously drawn from P; a standard cell would not maintain a constant potential for long under such usage.

Accuracy of a Potentiometer. The accuracy of a potential measurement with a potentiometer depends upon the constancy of B, the linearity of AB, and the accuracy with which AC can be measured. Ordinarily, however, the ultimate precision of a good-quality instrument is determined by the sensitivity of the current-measuring device relative to the resistance of the circuit. Suppose, for example, that G is a galvanometer whose resistance plus that of the unknown cell is 1000 ohms, a typical figure. Further, if the galvanometer is just capable of detecting a current of 1 μA (10^{-6} A), application of Ohm's law shows that the minimum distinguishable voltage difference will be $10^{-6} \times 1000 = 10^{-3}$ V, or 1 mV. If the galvanometer is sensitive to 10^{-7} A, a difference of 0.1 mV will be detectable. A sensitivity of this order is found in an ordinary pointer-type galvanometer. More refined galvanometers with sensitivities of up to 10^{-10} A are available and permit the accurate measurement of potentials of cells having resistances as high as a megohm or more (1 megohm $= 10^6$ ohms).

Potentiometer with Electronic Amplification. The potentials of cells with resistances of several hundred megohms can be accurately measured by a potentiometer circuit like that in Figure 17-1 if the galvanometer G is replaced by an electronic circuit that amplifies the out-of-balance current by several orders of magnitude. The amplified current can then be detected by a rugged milliammeter. Commercial instruments that incorporate this feature are usually called *pH meters*, because they have been designed to measure the potentials of cells that contain a high-resistance, pH-sensitive glass electrode. The slide wires of such instruments are calibrated both in millivolts and in the linearly related pH units. Meters of this type can be conveniently employed for measuring the potential of both high- and low-resistance cells.

Direct-Reading Instruments for Potential Measurements. Electronic volt-meters can be designed to operate with such small currents (10^{-12} to 10^{-14} A) that the potential of a cell is unaffected when measured by such a device. A number of instrument manufacturers sell direct-reading voltmeters that employ electronic amplification to produce a signal that can be displayed in digital form or on a meter calibrated in both pH and millivolt units. These instruments are applicable to both high- and low-resistance cells and are also called pH meters.

Potentiometric-type pH meters offer somewhat greater accuracy and easier maintenance than most direct-reading instruments. On the other hand, the latter type is entirely satisfactory for many applications.

Reference Electrodes

In many electroanalytical methods, it is desirable that the half-cell potential of one electrode be known, constant, and completely insensitive to the composition of the solution under study. An electrode that fits this description is called a *reference electrode*. Employed in conjunction with the reference electrode will be an *indicator electrode*, whose response is dependent upon the analyte concentration.

A reference electrode should be easy to assemble and should maintain an essentially constant and reproducible potential with passage of small currents. Several electrode systems meet these requirements.

CALOMEL ELECTRODES

Calomel half-cells may be represented as follows:

$$| \, |Hg_2Cl_2(\text{sat'd}), \, KCl(xF)|Hg$$

where x represents the formal concentration of potassium chloride in the solution. The electrode reaction is given by the equation

$$Hg_2Cl_2(s) + 2e \rightleftarrows 2Hg + 2Cl^-$$

The potential of this cell will vary with the chloride concentration x, and this quantity must be specified in describing the electrode.

Table 17-1 lists the composition and the electrode potentials for the three

TABLE 17-1   Specifications of Calomel Electrodes

Name	Concentration of Hg_2Cl_2	KCl	Reduction Potential (V) vs. Standard Hydrogen Electrode ($Hg_2Cl_2(s) + 2e \rightleftarrows 2Hg + 2Cl^-$)
Saturated	Saturated	Saturated	$+0.242 - 7.6 \times 10^{-4}(t - 25)$
Normal	Saturated	1.0 F	$+0.280 - 2.4 \times 10^{-4}(t - 25)$
Decinormal	Saturated	0.1 F	$+0.334 - 7 \times 10^{-5}(t - 25)$

most commonly encountered calomel electrodes. Note that each solution is saturated with mercury(I) chloride and that the cells differ only with respect to the potassium chloride concentration. Note also that the potential of the normal calomel electrode is greater than the standard potential for the half-reaction because the chloride ion *activity* in a 1-F solution of potassium chloride is significantly smaller than 1. The last column in Table 17-1 gives expressions that permit calculation of the electrode potentials of calomel half-cells at temperatures other than 25°C.

The saturated calomel electrode (SCE) is most commonly used by the analytical chemist because of the ease with which it can be prepared. Compared with the other two, its temperature coefficient is somewhat larger; this is a disadvantage only where substantial changes in temperature occur during the measurement process.

A simple, easily constructed saturated calomel electrode is shown in Figure 17-2. The salt bridge, simply a tube filled with saturated potassium chloride, provides electric contact with the solution surrounding the indicator electrode. A fritted disk or a wad of cotton at one end of the salt bridge is often

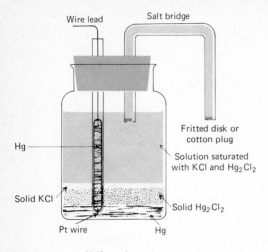

Half-reaction:
$Hg_2Cl_2(s) + 2e \rightleftharpoons 2Hg + 2Cl^-$

Figure 17-2   Diagram of a Saturated Calomel Electrode.

employed to prevent siphoning of cell liquid and contamination of the solutions by foreign ions.

Several convenient calomel electrodes are available commercially. Typical of these is the one illustrated in Figure 17-3 which consists of a tube 5 to 15 cm in length and 0.5 to 1.0 cm in diameter. Mercury–mercury(I) chloride paste is contained in an inner tube, which is connected to the saturated potassium chloride solution in the outer tube through a small opening. Contact with the

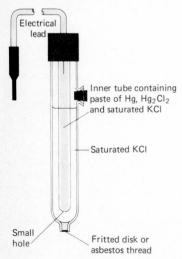

Figure 17-3   Diagram of a Typical Saturated Calomel Electrode.

second half-cell is made by means of a fritted disk or porous fiber sealed in the end of the outer tubing. An electrode such as this has a relatively high resistance (2000 to 3000 ohms) and a limited current-carrying capacity.

SILVER–SILVER CHLORIDE ELECTRODES

A system analogous to the calomel electrode consists of a silver electrode immersed in a solution of potassium chloride that is also saturated with silver chloride:

$$| \, |AgCl(sat'd), \, KCl(xF)|Ag$$

The half-reaction is

$$AgCl(s) + e \rightleftarrows Ag + Cl^-$$

Normally, this electrode is prepared with a saturated potassium chloride solution; its potential, at 25°C, is +0.197 V with respect to the standard hydrogen electrode.

A simple and easily constructed silver–silver chloride electrode is shown in Figure 17-4. The electrode is contained in a Pyrex tube fitted with a 10-mm fritted-glass disk (such tubing is obtainable from Corning Glass Works). A layer of agar gel saturated with potassium chloride is formed on top of the disk to prevent loss of solution from the half-cell. The plug can be prepared by suspending about 5 g of pure agar in 100 ml of boiling water and then adding about 35 g of potassium chloride. A portion of this suspension, while still warm, is poured into the tube; upon cooling, it solidifies to a gel with low electrical resistance. A layer of solid potassium chloride is placed on the gel, and the tube is filled with saturated solution of the salt. A drop or two of 1-F silver nitrate is then added, and a heavy-gauge (1- to 2-mm diameter) silver wire is inserted in the solution.

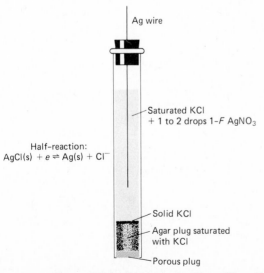

Ag wire

Saturated KCl
+ 1 to 2 drops 1-F AgNO₃

Half-reaction:
AgCl(s) + e ⇌ Ag(s) + Cl⁻

Solid KCl

Agar plug saturated
with KCl

Porous plug

Figure 17-4   Diagram of a Silver–Silver Chloride Electrode.

Indicator Electrodes

Indicator electrodes for potentiometric measurements are of two basic types, namely, metallic indicator electrodes and membrane electrodes.

METALLIC INDICATOR ELECTRODES

Several metals—such as silver, copper, mercury, lead, and cadmium—that show reversible half-reactions are satisfactory for construction of indicator electrodes. The potentials of these metals reproducibly and predictably reflect the activities of their ions in solution. In contrast, some metals are not very satisfactory indicator electrodes because they tend to develop nonreproducible potentials that are influenced by strains or crystal deformations in their structures and by oxide coatings on their surfaces. Metals in this category include iron, nickel, cobalt, tungsten, and chromium.

In addition to serving as indicator electrodes for their own cations, metal electrodes are also indirectly responsive to anions that form slightly soluble precipitates with cations. For this application, it is only necessary to saturate the solution under study with the sparingly soluble salt. For example, the potential of a silver electrode will accurately reflect the concentration of chloride ion in a solution that is saturated with silver chloride. Here, two equilibria govern the electrode behavior:

$$AgCl(s) \rightleftarrows Ag^+ + Cl^-$$

$$Ag^+ + e \rightleftarrows Ag(s) \qquad E^0_{Ag} = 0.779 \text{ V}$$

Combining these equations yields

$$AgCl(s) + e \rightleftarrows Ag(s) + Cl^- \qquad E^0_{AgCl} = 0.222 \text{ V}$$

As we have shown (Chapter 14),

$$E^0_{AgCl} = E^0_{Ag} + 0.0591 \log K_{sp}$$

The Nernst equation applied to the half-reaction involving silver chloride gives the relationship between electrode potential and anion concentration. Thus,

$$E = 0.222 - 0.0591 \log [Cl^-]$$

A silver electrode serving as an indicator for chloride ion is an example of an *electrode of the second order* because it measures the concentration of an ion that is not directly involved in the electron-transfer process. The same electrode used as an indicator for silver ion functions as an *electrode of the first order* because its potential is directly dependent upon a participant in the electrode process.

The use of a mercury electrode for the indirect determination of the concentration of the EDTA ion Y^{4-} can be cited as another example of a second-order electrode. The equilibria existing in solution containing Y^{4-}, a small and *constant* concentration of HgY^{2-}, and a mercury electrode are

$$Hg^{2+} + Y^{4-} \rightleftarrows HgY^{2-}$$

$$Hg^{2+} + 2e \rightleftarrows Hg(l) \qquad E^0_{Hg} = 0.854 \text{ V}$$

and

$$HgY^{2-} + 2e \rightleftarrows Hg(l) + Y^{4-}$$

To obtain the relationship between E^0_{Hg} and $E^0_{HgY^{2-}}$, we combine the formation constant expression for HgY^{2-} (p. 275) with the Nernst relationship for the mercury electrode. That is,

$$E = E^0_{Hg} - \frac{0.0591}{2} \log \frac{1}{[Hg^{2+}]} = E^0_{Hg} - \frac{0.0591}{2} \log \frac{K_f[Y^{4-}]}{[HgY^{2-}]}$$

$$E = E^0_{Hg} - \frac{0.0591}{2} \log K_f - \frac{0.0591}{2} \log \frac{[Y^{4-}]}{[HgY^{2-}]}$$

When $[Y^{4-}]$ and $[HgY^{2-}]$ are unity, $E = E^0_{HgY^{2-}}$ by definition. Thus,

$$E^0_{HgY^{2-}} = E^0_{Hg} - \frac{0.0591}{2} \log K_f$$

$$= 0.854 - \frac{0.0591}{2} \log 6.3 \times 10^{21} = 0.21 \text{ V}$$

and we may write

$$E_{Hg} = 0.21 - \frac{0.0591}{2} \log \frac{[Y^{4-}]}{[HgY^{2-}]} \tag{17-3}$$

Since HgY^{2-} is exceptionally stable, essentially all mercury(II) is complexed, and the concentration of HgY^{2-} remains substantially constant over wide concentration ranges of Y^{4-}. Thus, the potential varies only with $[Y^{4-}]$, and the electrode is second-order for this ion; under these circumstances, Equation 17-3 becomes

$$E_{Hg} = Q - \frac{0.0591}{2} \log [Y^{4-}] \tag{17-4}$$

where Q is a constant equal to

$$Q = 0.21 - \frac{0.0591}{2} \log \frac{1}{[HgY^{2-}]}$$

An electrode of this type is useful for establishing end points for EDTA titrations. In this application, a small, fixed amount of HgY^{2-} is introduced into the analyte solution before the titration is begun.

For oxidation-reduction reactions, the indicator electrode is most commonly an inert metal such as platinum or gold. The potential developed at such an electrode is a function of the concentration ratio between the oxidized and reduced forms of one or more of the species in the solution.

Metallic indicator electrodes are constructed from coils of wire, flat metal plates, or heavy cylindrical billets. Generally, exposure of a large surface area to the solution will assure rapid attainment of equilibrium. Thorough cleaning of the metal surface before use is often important; a brief dip in concentrated nitric acid followed by several rinsings with distilled water is satisfactory for many metals.

Membrane Electrodes[1]

For many years, the most convenient method for determining pH has involved measurement of the potential that develops across a thin glass membrane that separates two solutions with different hydrogen ion concentrations. The phenomenon, first reported by Cremer,[2] has been extensively studied by many investigators; as a result, the sensitivity and selectivity of glass membranes to pH is reasonably well understood. Furthermore, membrane electrodes have now been developed for the direct potentiometric determination of other ions, such as K^+, Na^+, Li^+, F^-, and Ca^{2+}.[3]

It is convenient to divide membrane electrodes into four categories, based upon the membrane composition. These include (1) glass electrodes, (2) liquid-membrane electrodes, (3) solid-state or precipitate electrodes, and (4) gas-sensing membrane electrodes. We shall consider the properties and behavior of the glass electrode in greater detail because of both its present and its historical importance.

THE GLASS ELECTRODE FOR pH MEASUREMENTS

Figure 17-5 shows a modern cell for the measurement of pH. It consists of commercially available calomel and glass electrodes immersed in the solution whose pH is to be measured. The calomel reference electrode is similar to the one described earlier in the chapter. The glass electrode is manufactured by

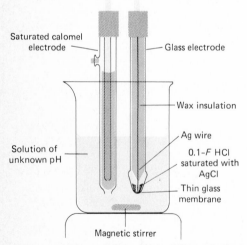

Figure 17-5   Typical Electrode System for Measuring pH.

[1] For further information on this topic, see R. A. Durst, Ed., *Ion-Selective Electrodes*, National Bureau of Standards Special Publication 314. Washington, D.C.: U.S. Government Printing Office, 1969.

[2] M. Cremer, *Z. Biol.*, **47**, 562 (1906).

[3] See G. A. Rechnitz, *Chem. Eng. News*, **45** (25), 146 (1967).

sealing a thin, pH-sensitive glass tip to the end of a heavy-walled glass tubing. The resulting bulb is filled with a solution of hydrochloric acid (often 0.1 F) that is saturated with silver chloride. A silver wire is immersed in the solution and is connected via an external lead to one terminal of a potential-measuring device; the calomel electrode is connected to the other terminal. Note that the cell contains two reference electrodes each of whose potentials is *independent of pH*; one is the external calomel electrode, and the other is the internal silver–silver chloride reference electrode (p. 381), which is a *part* of the glass electrode but is not the pH-sensitive component. In fact, it is the thin membrane at the tip of the electrode that responds to pH changes.

Figure 17-6 is a schematic representation of the cell shown in Figure 17-5. Note that the glass electrode consists of a silver–silver chloride reference electrode immersed in a solution of a fixed hydronium ion concentration and a glass membrane that separates these contents from the analyte solution.

It is found experimentally that the potential of a cell such as that in Figure 17-6 depends on the hydrogen ion activities a_1 and a_2 at 25°C as follows:

$$E = Q + 0.0591 \log \frac{a_1}{a_2}$$

where Q is a constant whose value is given by

$$Q = E_{Ag, AgCl} - E_{SCE} + j$$

where $E_{Ag, AgCl}$ and E_{SCE} are the potentials of the two reference electrodes and j is the *asymmetry potential*; the source and properties of j will be considered later.

In practice, the hydrogen ion activity of the internal solution a_2 is fixed by its method of preparation and is constant. We may write, therefore,

$$E = L + 0.0591 \log a_1 \qquad (17\text{-}5)$$
$$= L - 0.0591 \text{ pH} \qquad (17\text{-}6)$$

where L is a new constant containing the logarithmic function of a_2.

It is important to emphasize that the potentials E_{SCE}, $E_{Ag, AgCl}$, and j remain constant during a pH measurement. Thus, the source of the pH-dependent variation in E must lie in *the glass membrane*. That is, when a_1 and a_2 differ, the two surfaces of the membrane must differ in potential by some amount $V_1 - V_2$. The only function of the two reference electrodes is to make observation of this difference possible.

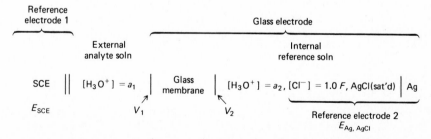

Figure 17-6 Schematic Diagram of a Glass-Calomel Cell for Measurement of pH.

Composition of pH-sensitive Glass Membranes. Not all glass membranes show the pH response given by Equation 17-6; indeed, quartz and Pyrex are virtually insensitive to pH variations. Much systematic investigation concerned with the effect of glass composition on sensitivity of membranes to protons and other cations has been carried out during the past quarter century, and several compositions are now used commercially.[4] For many years Corning 015 glass, composed approximately of 22% Na_2O, 6% CaO, and 72% SiO_2, has been widely used. This glass shows an excellent specificity toward hydrogen ions up to a pH of about 9. At higher pH values, the membrane becomes sensitive to sodium and other alkali ions as well; a pH value calculated from Equation 17-6 is likely to be low by several tenths of a unit in this region, particularly in solutions containing high concentrations of sodium ion.

Hygroscopicity of Glass Membranes. It has been shown that the surfaces of a glass membrane must be hydrated in order to function as a pH electrode. Nonhygroscopic glasses show no pH function. Even Corning 015 glass shows little pH response after dehydration by storage over a desiccant; however, its sensitivity is restored after standing for a few hours in water. This hydration involves absorption of approximately 50 mg of water per cubic centimeter of glass.

It has also been demonstrated experimentally that hydration of a pH-sensitive glass membrane is accompanied by an exchange reaction between univalent cations of the glass and protons of the solution. The process involves singly charged cations almost exclusively inasmuch as the di- and trivalent cations in the silicate structure are much more strongly bonded. Typically, then, the ion-exchange reaction can be written as

$$\underset{\text{soln}}{H^+} + \underset{\text{solid}}{Na^+Gl^-} \rightleftarrows \underset{\text{soln}}{Na^+} + \underset{\text{solid}}{H^+Gl^-} \qquad (17\text{-}7)$$

The equilibrium constant for this reaction is so large that the surface of a hydrated glass membrane will consist almost entirely of silicic acid (H^+Gl^-), except in very basic media where the hydrogen ion concentration is extremely small and the sodium ion concentration is large.

As a glass surface is exposed to water for longer and longer periods, the layer of silicic acid gel becomes as thick as 10^{-4} to 10^{-5} mm. At the outer surface of the gel, all fixed, singly charged sites are occupied by hydrogen ions. From the surface to the interior of the gel there is a continuous decrease in the number of protons and a corresponding increase in the number of sodium ions. A schematic representation of the two surfaces of a glass membrane is shown in Figure 17-7.

Resistance of Glass Membranes. A typical commercial glass electrode, which ranges in thickness between 0.03 and 0.1 mm, has an electrical resistance of 50 to 500 megohms. Current conduction through the dry glass region is ionic

[4] For a summary of this work, see J. O. Isard in G. Eisenman, *Glass Electrodes for Hydrogen and Other Cations*, chapter 3. New York: Marcel Dekker, Inc., 1967.

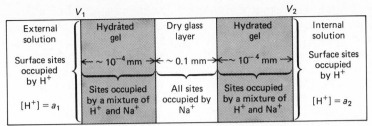

Boundary potential, $E_b = V_1 - V_2$

Figure 17-7 Schematic Representation of a Well-Soaked Glass Membrane.

and involves movement of the alkali ions from one site to another. Within the two gel layers the current is carried by both alkali and hydrogen ions. At each gel-solution interface, current passage involves the transfer of protons. The direction of transfer will be from gel to solution at the one,

$$H^+Gl^- \rightleftarrows Gl^- + H^+ \qquad (17\text{-}8)$$
$$\text{solid} \qquad \text{solid} \quad \text{soln}$$

and from the solution to gel at the other,

$$H^+ + Gl^- \rightleftarrows H^+Gl^- \qquad (17\text{-}9)$$
$$\text{soln} \quad \text{solid} \qquad \text{solid}$$

Theory of the Glass-Electrode Potential.[5] The potential across a glass membrane consists of a *boundary potential* and a *diffusion potential*. Under ideal conditions only the former is affected by pH.

The boundary potential for the membrane of a glass electrode contains two components, each associated with one of the gel-solution interfaces. If V_1 is the component arising at the interface between the external solution and the gel (see Figure 17-9), and if V_2 is the corresponding potential at the inner surface, then the boundary potential E_b for the membrane is given by

$$E_b = V_1 - V_2 \qquad (17\text{-}10)$$

The potential V_1 is determined by the hydrogen ion activities both in the external solution and upon the surface of the gel; it can be considered a measure of the driving force of the reaction shown by Equation 17-9. In the same way, the potential V_2 is related to the hydrogen ion activities in the internal solution and on the corresponding gel surface.

It can be demonstrated from thermodynamic considerations[6] that V_1 and V_2 are related to hydrogen ion activities at each interface as follows:

$$V_1 = j_1 + \frac{RT}{F} \ln \frac{a_1}{a_1'} \qquad (17\text{-}11)$$

$$V_2 = j_2 + \frac{RT}{F} \ln \frac{a_2}{a_2'} \qquad (17\text{-}12)$$

---

[5] For a comprehensive discussion of this topic, see G. Eisenman, *Glass Electrodes for Hydrogen and Other Cations*, chapters 4–6. New York: Marcel Dekker, Inc., 1967.

[6] G. Eisenman, *Biophys. J.*, **2** *Part 2*, 259 (1962).

where R, T, and F have their usual meanings; a_1 and a_2 are activities of the hydrogen ion in the *solutions* on either side of the membrane; and a_1' and a_2' are the hydrogen ion activities in each of the *gel layers* contacting the two solutions. If the two gel surfaces have the same number of sites from which protons can leave, then the two constants j_1 and j_2 will be identical; so also will be the two activities a_1' and a_2' in the gel layers, provided that all original sodium ions on the surface have been replaced by protons (that is, insofar as the equilibrium shown in Equation 17-7 lies far to the right). If these equalities are justified, substitution of Equations 17-11 and 17-12 into 17-10 yields

$$E_b = V_1 - V_2 = \frac{RT}{F} \ln \frac{a_1}{a_2} \qquad (17\text{-}13)$$

Thus, *provided the two gel surfaces are identical*, the boundary potential E_b depends only upon the activities of the hydrogen ion in the *solutions* on either side of the membrane. If one of these activities a_2 is kept constant, then Equation 17-13 further simplifies to

$$E_b = \text{constant} + \frac{RT}{F} \ln a_1 \qquad (17\text{-}14)$$

and the potential becomes a measure of the hydrogen ion activity in the external solution.

In addition to the boundary potential, a so-called *diffusion potential* also develops in each of the two gel layers. Its source lies in the difference between the mobilities of hydrogen ions and the alkali metal ions in the membrane. The two diffusion potentials are equal and opposite in sign insofar as the two solution-gel interfaces are the same. Under these conditions, then, the net diffusion potential is zero, and the emf across the membrane depends only on the boundary potential as given by Equation 17-13.

Asymmetry Potential. If identical solutions and identical reference electrodes are placed on either side of the membrane shown in Figure 17-7, $V_1 - V_2$ should be zero. However, it is found that a small potential, called the *asymmetry potential*, often does develop when this experiment is performed. Moreover, the asymmetry potential associated with a given glass electrode changes slowly with time.

The causes for the asymmetry potential are obscure; they undoubtedly include such factors as differences in strains established within the two surfaces during manufacture of the membrane, mechanical and chemical attack, or other contamination of the outer surface during use. The effect of the asymmetry potential on a pH measurement is eliminated by frequent calibration of the electrode against a standard buffer of known pH.

The Alkaline Error. In solutions of pH 9 or greater, some glass membranes respond not only to changes in hydrogen ion concentration but also to the concentration of alkali metal ions. The magnitude of the resulting error for four compositions of glass membranes is indicated on the right-hand side of Figure 17-8. In each of these curves, the sodium ion concentration was maintained at

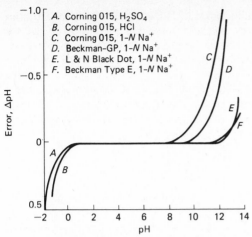

Figure 17-8   Acid and Alkaline Error of Selected Glass Electrodes at 25°C. (From R. G. Bates, *Determination of pH: Theory and Practice*, p. 316. New York: John Wiley & Sons, Inc., 1964. With permission.)

1 M, and the pH was varied. Note that the pH error is negative, which suggests that the electrode is responding to sodium ions as well as to protons. This observation is confirmed by data obtained for solutions with different sodium ion concentrations. Thus, at pH 12 the alkaline error for Corning 015 glass is about -0.7 pH when the sodium ion concentration is 1 M (Figure 17-8) and only about -0.3 pH in solutions that are 0.1 M in sodium ions.

All singly charged cations cause alkaline errors; their magnitudes vary according to the kind of metallic ion and the composition of the glass.

The alkaline error can be satisfactorily explained by assuming an exchange equilibrium between the hydrogen ions on the surface of the glass and the cations in the solution. This process can be formulated as

$$\underset{\text{gel}}{H^+Gl^-} + \underset{\text{soln}}{B^+} \rightleftarrows \underset{\text{gel}}{B^+Gl^-} + \underset{\text{soln}}{H^+}$$

where B^+ represents a singly charged cation, such as sodium ion. The equilibrium constant for this reaction is

$$K_{ex} = \frac{a_1 b_1'}{a_1' b_1} \tag{17-15}$$

where a_1 and b_1 are activities of H^+ and B^+ in solution, and a_1' and b_1' are the activities of the same ions in the surface of the gel. The constant K_{ex} depends upon the composition of the glass membrane and is ordinarily small. Thus, the fraction of the surface sites populated by cations other than hydrogen is small except when the hydrogen ion concentration is very low and the concentration of B^+ is high.

The effect of an alkali metal ion on the emf across a membrane can be described in quantitative terms by rewriting Equation 17-15 in the form

$$\frac{a_1}{a_1'} = \frac{(a_1 + K_{ex} b_1)}{(a_1' + b_1')} \tag{17-16}$$

Substitution of Equation 17-16 into Equation 17-11 and subtraction of Equation 17-12 gives

$$E_b = V_1 - V_2 = k_1 - k_2 + \frac{RT}{F} \ln \frac{(a_1 + K_{ex}b_1)a'_2}{(a'_1 + b'_1)a_2} \qquad (17\text{-}17)$$

If the number of sites on the two sides of the membrane is the same, the activity of hydrogen ions on the internal surface is approximately equal to the sum of the activities of the two cations on the external surface; that is, $a'_2 \cong (a'_1 + b'_1)$. Further, $k_1 = k_2$; thus, Equation 17-17 reduces to

$$E_b = \frac{RT}{F} \ln \frac{(a_1 + K_{ex}b_1)}{a_2}$$

Finally, where a_2 is constant,

$$E_b = \text{constant} + \frac{RT}{F} \ln (a_1 + K_{ex}b_1) \qquad (17\text{-}18)$$

It has been shown that not only the boundary potential but also the diffusion potential on one side of the membrane is changed when some of the surface sites are occupied by cations other than hydrogen. Under such circumstances the diffusion potentials do not subtract out as they did in the previous case (see p. 388), and E_b does not adequately describe the total emf that develops in the presence of another cation. It has also been shown that this effect can be accounted for by modifying Equation 17-18 as follows:[7]

$$E = \text{constant} + \frac{RT}{F} \ln \left[a_1 + K_{ex} \left(\frac{U_B}{U_H} \right) b_1 \right] \qquad (17\text{-}19)$$

where U_B and U_H are measures of the mobilities of B^+ and H^+ in the gel.

For many glasses the term $K_{ex}(U_B/U_H)b_1$ is small with respect to the hydrogen ion activity a_1 so long as the pH of the solution is less than about 9. Under these conditions, Equation 17-19 simplifies to Equation 17-14. With high concentrations of singly charged cations and at higher pH levels, however, this second term plays an increasingly important part in determining E. The composition of the glass membrane determines the magnitude of $K_{ex}(U_B/U_H)b_1$; this parameter is relatively small for glasses that have been designed for work in strongly basic solutions. The membrane of the Beckman Type E glass electrode, illustrated in Figure 17-8, is of this type.

The Acid Error. As shown in Figure 17-8, the typical glass electrode exhibits an error, opposite in sign to the alkaline error, in solutions of pH less than 1. As a consequence, pH readings tend to be too high in this region. The magnitude of the error depends upon a variety of factors and is generally not very reproducible. The causes of the acid error are not well understood.

[7] See G. Eisenman, *Glass Electrodes for Hydrogen and Other Cations*, chapters 4 and 5. New York: Marcel Dekker, Inc., 1967.

GLASS ELECTRODES FOR THE DETERMINATION OF OTHER CATIONS

The existence of the alkaline error in early glass electrodes led to studies concerning the effect of glass composition on the magnitude of this error. One consequence of this work has been the development of glasses for which the term $K_{ex}(U_B/U_H)b_1$ in Equation 17-19 is small enough so that the alkaline error is negligible below a pH of about 12. Other studies have been directed toward finding glass compositions in which this term is greatly enhanced in the interests of developing glass electrodes for the determination of cations other than hydrogen. This application requires that the hydrogen ion activity a_1 in Equation 17-19 be negligible with respect to the second term containing the activity b_1 of the other cation. Under these circumstances the potential of the electrode would be *independent* of pH but would vary with pB in the typical way.

A number of investigators have demonstrated that the presence of Al_2O_3 or B_2O_3 in glass causes the desired effect. Eisenman and co-workers carried out a systematic study of glasses containing Na_2O, Al_2O_3, and SiO_2 in various proportions; they have demonstrated clearly that it is practical to prepare membranes for the selective measurement of several cations in the presence of others.[8] Glass electrodes for potassium ion and sodium ion are now available commercially.

Figure 17-9 illustrates how two types of glass electrodes respond to pH

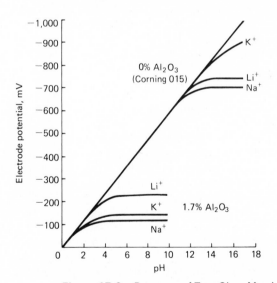

Figure 17-9   Response of Two Glass Membranes in the Presence of Alkali Metal Ions (0.1 *M* in each case). (Curves taken from G. Eisenman in C. N. Reilley, Ed., *Advances in Analytical Chemistry and Instrumentation*, vol. 4, p. 213. New York: John Wiley & Sons, Inc., 1965. With permission.)

---

[8] For a summary of this work, see G. Eisenman in C. E. Reilley, Ed., *Advances in Analytical Chemistry and Instrumentation*, vol. 4, p. 213. New York: John Wiley & Sons, Inc. 1965.

changes in solutions that are 0.1 F in various alkali metal ions. The one electrode was fabricated from Corning 015 glass, a composition that contains no Al_2O_3; at pH levels below 9 the term $K_{ex}(U_B/U_H)b_1$ in Equation 17-19 is small with respect to a_1. At higher pH values the second term becomes important; note that its magnitude depends upon the type of alkali ion present.

With the glass containing Al_2O_3, both potential-determining terms in Equation 17-19 are clearly important at low pH values, but at levels above about pH 5 the potential becomes *independent* of pH. Thus, $a_1 \ll K_{ex}(U_B/U_H)b_1$ in this region, with the result that the potential can be expected to vary linearly with pNa, pK, or pLi. Experimentally, this prediction is verified; such a glass can be employed to measure the concentration of these ions.

Sources for the difference in behavior between the two glasses shown in Figure 17-9 (toward Na^+, for example) include the relative affinity of anionic surface sites of the gel for hydrogen ions as compared with sodium ions and (to a lesser extent) the mobility ratios of the ions in the two glasses. Above pH 4, where glass containing Al_2O_3 is no longer sensitive to hydrogen ions, these anionic sites are occupied entirely by the metallic ion rather than by the hydrogen ion, and the electrode behaves as a perfect sodium electrode. With lithium ion, however, saturation occurs only at lower hydrogen ion concentrations (pH $\sim$ 6). Thus, the surface has less affinity for lithium than for sodium ion.

Studies such as these have led to the development of glass membranes that are suitable for the direct potentiometric measurement of the concentrations of Na^+, K^+, NH_4^+, Rb^+, Cs^+, Li^+, and Ag^+. Some of these glasses are reasonably selective, as may be seen in Table 17-2.

The development of membranes that are selective for cations other than hydrogen has been rapid during the past decade, and much work in the field is currently in progress. New compositions with useful analytical properties will doubtless appear in the near future.

LIQUID-MEMBRANE ELECTRODES

Liquid membranes owe their response to the potential that is established across the interface between the solution to be analyzed and an immiscible liquid that selectively bonds with the ion being determined. Liquid-membrane electrodes permit the direct potentiometric determination of the activities of several polyvalent cations and certain anions as well.

A liquid-membrane electrode differs from a glass electrode only in the respect that the solution of known and fixed activity is separated from the analyte by a thin layer of an immiscible organic liquid instead of a thin glass membrane. As shown schematically in Figure 17-10, a porous, hydrophobic (that is, water-repelling), plastic disk serves to hold the organic layer between the two aqueous solutions. By wick action, the pores of the disk or membrane are maintained full of the organic liquid from the reservoir in the outer of the two concentric tubes. The inner tube contains an aqueous standard solution of MCl_2, where M^{2+} is the cation whose activity is to be determined. This solution is also saturated with AgCl to form a Ag-AgCl reference electrode with the silver lead wire.

TABLE 17-2 Properties of Certain Cation-sensitive Glasses[a]

Principal Cation to Be Measured	Glass Composition	Selectivity Characteristics[b]	Remarks
Li+	15Li$_2$O 25Al$_2$O$_3$ 60SiO$_2$	$K_{Li^+,Na^+} \approx 3$, $K_{Li^+,K^+} > 1000$	Best for Li+ in presence of H+ and Na+
Na+	11Na$_2$O 18Al$_2$O$_3$ 71SiO$_2$	$K_{Na^+,K^+} \approx 2800$ at pH 11 $K_{Na^+,K^+} \approx 300$ at pH 7	Nernst type of response to $\sim 10^{-5}$-M Na+
	10.4Li$_2$O 22.6Al$_2$O$_3$ 67SiO$_2$	$K_{Na^+,K^+} \approx 10^5$	Highly Na+ selective, but very time-dependent
K+	27Na$_2$O 5Al$_2$O$_3$ 68SiO$_2$	$K_{K^+,Na^+} \approx 20$	Nernst type of response to $< 10^{-4}$-M K+
Ag+	28.8Na$_2$O 19.1Al$_2$O$_3$ 52.1SiO$_2$	$K_{Ag^+,H^+} \approx 10^5$	Highly sensitive and selective to Ag+, but poor stability
	11Na$_2$O 18Al$_2$O$_3$ 71SiO$_2$	$K_{Ag^+,Na^+} > 1000$	Less selective for Ag+ but more reliable

[a] Reprinted from *Chemical & Engineering News*, vol. 45, June 12, 1967, page 149. Copyright 1967 by the American Chemical Society and reprinted by permission of the copyright owner.

[b] The constant $K_{i,j}$, the selectivity ratio, is a measure of the relative response of the electrode to the two species i and j. For example, for the sodium electrode, K_{Na^+,K^+} indicates that the electrode is 2800 times as sensitive to sodium ion as to potassium. Thus, 2800 times the concentration of potassium ion is required to produce the same potential as a unit concentration of sodium ion. Another method for expressing selectivity is to define a constant which is the reciprocal of the one just described; that is, $K_{Na^+,K^+} = \frac{1}{2800} = 3.6 \times 10^{-4}$. Here one can say that a 3.6×10^{-4} M solution of sodium ion will produce the same effect as a 1-M solution of potassium ions. Unfortunately, both types of constants are found in the literature (often without definition). Thus, the data in Table 17-3 employ a constant which is the reciprocal of the one employed in this table.

The organic liquid is a nonvolatile, water-immiscible, organic ion exchanger that contains acidic, basic, or chelating functional groups. When in contact with an aqueous solution of a divalent cation, an exchange equilibrium is established that can be represented as

$$\underset{\substack{\text{organic} \\ \text{phase}}}{RH_{2x}} + \underset{\substack{\text{aqueous} \\ \text{phase}}}{xM^{2+}} \rightleftarrows \underset{\substack{\text{organic} \\ \text{phase}}}{RM_x} + \underset{\substack{\text{aqueous} \\ \text{phase}}}{2xH^+}$$

By repeated treatment, the exchange liquid can be converted essentially completely to the cationic form, RM_x; it is this form that is employed in electrodes for the determination of M^{2+}.

In order to determine the pM of a solution, the electrode shown in Figure 17-10 is immersed in a solution of the analyte which also contains a reference electrode—usually a saturated calomel electrode. The potential between the

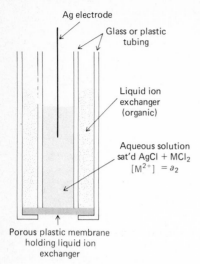

Ag electrode

Glass or plastic tubing

Liquid ion exchanger (organic)

Aqueous solution sat'd AgCl + MCl_2 $[M^{2+}] = a_2$

Porous plastic membrane holding liquid ion exchanger

Figure 17-10 Liquid-Membrane Electrode Sensitive to M^{2+}.

external and the internal reference electrodes is proportional to the pM of the analyte.

Figure 17-11 shows construction details of a commercial liquid-membrane electrode that is selective for calcium ion. The ion exchanger is an aliphatic diester of phosphoric acid dissolved in a polar solvent. The chain lengths of the aliphatic groups in the former range from 8 to 16 carbon atoms. The diester contains a single dissociable proton; thus, two molecules are required to bond a divalent cation, here, calcium. It is this affinity for calcium ions that imparts the selective properties to the electrode. The internal aqueous solution in contact with the exchanger contains a fixed concentration of calcium chloride; a silver–silver chloride reference electrode is immersed in this solution. When used for a calcium ion determination, the porous disk containing the ion-exchange liquid separates the solution to be analyzed from the reference calcium

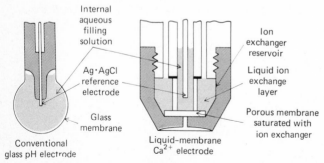

Internal aqueous filling solution

Ion exchanger reservoir

Ag·AgCl reference electrode

Liquid ion exchange layer

Glass membrane

Porous membrane saturated with ion exchanger

Conventional glass pH electrode

Liquid–membrane Ca^{2+} electrode

Figure 17-11 Comparison of a Liquid-Membrane Calcium Ion Electrode with a Glass Electrode. (Sketch courtesy of Orion Research, Inc., Cambridge, Mass.)

chloride solution. The equilibrium established at each interface can be represented as

$$[(RO)_2POO]_2Ca \rightleftarrows 2(RO)_2POO^- + Ca^{2+}$$
$$\quad\quad\text{organic}\quad\quad\quad\quad\text{organic}\quad\quad\text{aqueous}$$

By a set of arguments similar to those given for a glass membrane (p. 387), the potential for the electrode is

$$E = K + \frac{0.0591}{2} \log a_1$$

where a_1 here is the activity of Ca^{2+}.

The performance of the calcium electrode just described is reported to be independent of pH in the range between 5.5 and 11. At lower pH levels, hydrogen ions undoubtedly exchange with calcium ions on the exchanger to a significant extent; the electrode then becomes pH- as well as pCa-dependent. Its sensitivity to calcium ion exceeds that for magnesium ion by a factor of 50 and that for sodium or potassium by a factor of 1000. It can be employed to measure calcium ion activities as low as 10^{-5} M.

The calcium ion membrane electrode has proved to be a valuable tool for physiological studies because this ion plays important roles in nerve conduction, bone formation, muscle contraction, cardiac conduction and contraction, and renal tubular function. It is of interest that at least some of these processes are influenced more by calcium ion activity than calcium concentration; activity, of course, is the parameter measured by the electrode.

Another specific ion electrode of great consequence in physiological studies is that for potassium because the transport of nerve signals appears to involve movement of this ion across nerve membranes. Study of the process requires an electrode which can detect small concentrations of potassium ion in the presence of much larger concentrations of sodium. A number of liquid-membrane electrodes show promise of meeting these needs. One employs a solution of valinomycin in diphenyl ether. Valinomycin is an antibiotic with a cyclical ether structure that has a much stronger affinity for potassium ions than sodium. The consequence is a liquid membrane that has a selectivity ratio of potassium to sodium of over 10,000.[9]

Table 17-3 lists commercially available liquid-membrane electrodes. The electrodes that are sensitive to anions employ a solution of an anion exchanger in an organic solvent.

SOLID-STATE OR PRECIPITATE ELECTRODES

Considerable work has been devoted recently to the development of solid membranes that are selective toward anions in the way that some glasses behave toward specific cations. As we have seen, the selectivity of a glass membrane owes its origin to the presence of anionic sites on its surface that show particular affinity toward certain positively charged ions. By analogy, a membrane having similar cationic sites might be expected to respond selectively toward anions.

[9] M. S. Frant and J. W. Ross, Jr., *Science*, **167**, 987 (1970).

TABLE 17-3 Commercial Liquid-Membrane Electrodes[a]

Analyte Ion	Concentration Range, M	Preferred pH	Selectivity Constant for Interferences[b]
Ca^{2+}	10^0–10^{-5}	6–8	Zn^{2+}, 3.2; Fe^{2+}, 0.8; Pb^{2+}, 0.63; Mg^{2+}, 0.014
Cl^-	10^{-1}–10^{-5}	2–11	I^-, 17; NO_3^-, 4.2; Br^-, 1.6; HCO_3^-, 0.19; SO_4^{2-}, 0.14; F^-, 0.10
BF_4^-	10^{-1}–10^{-5}	3–10	NO_3^-, 0.1; Br^-, 0.04; OAc^-, 0.004; HCO_3^-, 0.004; Cl^-, 0.001
NO_3^-	10^{-1}–10^{-5}	3–10	I^-, 20; Br^-, 0.1; NO_2^-, 0.04; Cl^-, 0.004; SO_4^{2-}, 0.00003; CO_3^{2-}, 0.0002; ClO_4^-, 1000; F^-, 0.00006
ClO_4^-	10^{-1}–10^{-5}	3–10	I^-, 0.012; NO_3^-, 0.0015; Br^-, 0.00056; F^-, 0.00025; Cl^-, 0.00022
K^+	10^0–10^{-5}	3–10	Cs^+, 1.0; NH_4^+, 0.03; H^+, 0.01; Ag^+, 0.001; Na^+, 0.002; Li^+, 0.001
$Ca^{2+} + Mg^{2+}$	10^0–10^{-8}	5–8	Zn^{2+}, 3.5; Fe^{2+}, 3.5; Cu^{2+}, 3.1; Ni^{2+}, 1.35; Ba^{2+}, 0.94; Na^+, 0.015

[a] *Analytical Methods Guide*, 6th ed. Cambridge Mass.: Orion Research, Inc., August, 1973.
[b] Note that these selectivity constants are the reciprocals of the ones in Table 17-2 (see footnote [b] of Table 17-2).

To exploit this possibility, attempts have been made to prepare membranes of salts containing the anion of interest and a cation that selectively precipitates that anion from aqueous solutions; for example, barium sulfate has been proposed for sulfate ion and silver halides for the various halide ions. The problem encountered in this approach has been in finding methods for fabricating membranes from the desired salt with adequate physical strength, conductivity, and resistance to abrasion and corrosion.

A solid-state electrode selective for fluoride ion has been described.[10] The membrane consists of a single crystal of lanthanum fluoride that has been doped with a rare earth to increase its electrical conductivity. The membrane, supported between a reference solution and the solution to be measured, shows the theoretical response to changes in fluoride ion activity (that is, $E = K + 0.0591$ log a_{F^-}) to as low as 10^{-6} M. The electrode is purported to be selective for fluoride over other common anions by several orders of magnitude. Only hydroxide ion appears to offer serious interference.

Membranes prepared from cast pellets of silver halides have been successfully used in electrodes for the selective determination of chloride, bromide, and iodide ions. Other commercially available solid-state electrodes are listed in Table 17-4.

GAS-SENSING ELECTRODES

Figure 17-12 is a schematic diagram of a so-called gas-sensing electrode, which consists of a reference electrode, a membrane electrode, and an electrolyte

[10] M. S. Frant and J. W. Ross., Jr., *Science*, **154,** 1553 (1966).

TABLE 17-4   Commercial Solid-State Electrodes[a]

Analyte Ion	Concentration Range, M	Preferred pH	Interferences
Br^-	10^0–5×10^{-6}	2–12	Max: $[S^{2-}] = 10^{-7}\ M$; $[I^-] = 2 \times 10^{-4}[Br^-]$
Cd^{2+}	10^0–10^{-7}	3–7	Max: $[Ag^+]$, $[Hg^{2+}]$, $[Cu^{2+}] = 10^{-7}\ M$
Cl^-	10^0–5×10^{-5}	2–11	Max: $[S^{2-}] = 10^{-7}\ M$; traces of Br^-, I^-, CN^- do not interfere
Cu^{2+}	10^0–10^{-7}	3–7	Max: $[S^{2-}]$, $[Ag^+]$, $[Hg^{2+}] = 10^{-7}\ M$; high levels of Cl^-, Br^-, Fe^{3+}, Cd^{2+} interfere
CN^-	10^0–10^{-6}	11–13	Max: $[S^{2-}] = 10^{-7}\ M$; $[I^-] = 0.1\ [CN^-]$; $[Br^-] = 5 \times 10^3[CN^-]$; $[Cl^-] = 10^6[CN^-]$
F^-	10^0–10^{-6}	5–8	Max: $[OH^-] = 0.1[F^-]$
I^-	10^0–2×10^{-7}	3–12	Max: $[S^{2-}] = 10^{-7}\ M$
Pb^{2+}	10^0–10^{-7}	4–7	Max: $[Ag^+]$, $[Hg^{2+}]$, $[Cu^{2+}] = 10^{-7}\ M$; high levels of Cd^{2+} and Fe^{3+} interfere
Ag^+	10^0–10^{-7}	2–9	Max: $[Hg^{2+}] = 10^{-7}\ M$
S^{2-}	10^0–10^{-7}	13–14	Max: $[Hg^{2+}] = 10^{-7}\ M$
Na^+	10^0–10^{-6}	9–10	Selectivity constants: Li^+, 0.002; K^+, 0.001; Rb^+, 0.00003; Cs^+, 0.0015; NH_4^+, 0.00003; Tl^+, 0.0002; Ag^+, 350; H^+, 100
SCN^-	10^0–5×10^{-6}	2–12	Max: $[OH^-] = [SCN^-]$; $[Br^-] = 0.003[SCN^-]$; $[Cl^-] = 20[SCN^-]$; $[NH_3] = 0.13[SCN^-]$; $[S_2O_3^{2-}] = 0.01[SCN^-]$; $[CN^-] = 0.007[SCN^-]$; $[I^-]$, $[S^{2-}] = 10^{-7}\ M$

[a] From *Analytical Methods Guide*, 6th ed. Cambridge, Mass.: Orion Research, Inc., August, 1973.

solution housed in a cylindrical, plastic tube. A thin, replaceable, gas-permeable membrane that serves to separate the internal electrolyte solution from the analyte solution is attached to one end of the tube. The membrane is described by its manufacturers as being a thin, microporous film manufactured from a hydrophobic plastic. Water and electrolytes are prevented from entering and passing through the pores of the film owing to its water-repellent properties. Thus, the pores contain only air or other gases that the membrane is exposed to and not liquid water. When a solution containing a gaseous analyte such as sulfur dioxide is placed in contact with the membrane, distillation of the SO_2 into the pores occurs as shown by the reaction

$$\underset{\substack{\text{external} \\ \text{solution}}}{SO_2(aq)} \rightleftarrows \underset{\substack{\text{membrane} \\ \text{pores}}}{SO_2(g)}$$

Because the pores are numerous, a state of equilibrium is rapidly approached.

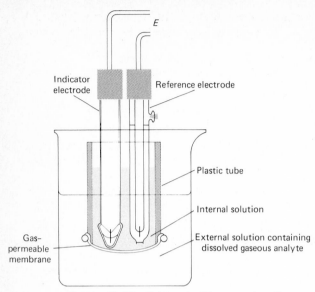

Figure 17-12 Schematic Diagram of a Gas-sensing Electrode. A glass-calomel electrode system is depicted (see Figure 17-5); other ion-specific and/or reference electrodes can be used as well.

The SO_2 in the pores, however, is also in contact with the internal solution and a second reaction can readily take place; that is,

$$\underset{\substack{\text{membrane} \\ \text{pores}}}{SO_2(g)} \rightleftarrows \underset{\substack{\text{internal} \\ \text{solution}}}{SO_2(aq)}$$

As a consequence of the two reactions, the film of internal solution adjacent to the membrane rapidly (in a few seconds to a few minutes) equilibrates with the external solution. Further, another equilibrium is established that causes the pH of the internal surface film to change, namely,

$$SO_2(aq) + 2H_2O \rightleftarrows HSO_3^- + H_3O^+$$

A glass electrode immersed in this film (see Figure 17-14) is then employed to detect the pH change.

The overall reaction for the process just described is obtained by adding the three chemical equations to give

$$\underset{\substack{\text{external} \\ \text{solution}}}{SO_2(aq)} + 2H_2O \rightleftarrows \underset{\substack{\text{internal} \\ \text{solution}}}{H_3O^+ + HSO_3^-}$$

The equilibrium constant for the reaction is given by

$$\frac{[H_3O^+][HSO_3^-]}{[SO_2(aq)]_{ext}} = K$$

If now the concentration of HSO_3^- in the internal solution is made relatively high so that its concentration is not altered significantly by the SO_2 which distills, then

$$\frac{[H_3O^+]}{[SO_2(aq)]_{ext}} = K[HSO_3^-] = K_g$$

which may be rewritten as

$$a_1 = [H_3O^+] = K_g[SO_2(aq)]_{ext}$$

where a_1 is the internal hydrogen ion activity. The potential of a cell consisting of a reference electrode and any indicator electrode is given by Equation 17-23 (p. 402); that is,

$$-\log a_1 = \frac{E_{obs} - K'}{0.0591/n}$$

where a_1 is the activity of the ion to which the indicator electrode is sensitive and K' is a constant dependent upon the characteristics of both the indicator and reference electrodes. In this cell the indicator electrode is a pH-sensitive glass electrode for which $n = 1.00$. Thus, combining the last two equations gives

$$-\log a_1 = -\log (K_g[SO_2(aq)]_{ext}) = \frac{E_{obs} - K'}{0.0591}$$

or

$$-\log [SO_2(aq)]_{ext} = \frac{E_{obs} - K'}{0.0591} + \log K_g$$

Finally, combining the two constants K' and K_g gives

$$-\log [SO_2(aq)]_{ext} = pSO_2 = \frac{E_{obs} - K''}{0.0591}$$

where

$$K'' = K' - 0.0591 \log K_g$$

The constant K'' can be evaluated by measuring E_{obs} for a solution having a known SO_2 concentration. Thus, the potential of the cell consisting of the internal reference and indicator electrode is determined by the SO_2 concentration of the external solution. Note that *no electrode comes directly in contact* with the analyte solutions; it would perhaps, therefore, be better to call the device a gas-sensing cell rather than a gas-sensing electrode. Note also that the only species that will interfere with the measurement are dissolved gases that can pass through the membrane and that will alter the pH of the internal solution.

The possibility exists for increasing the selectivity of the gas-sensing electrode by employing an internal electrode that is sensitive to a species other than hydrogen ion; for example, a nitrate-sensing electrode (see Table 17-3) could be used to provide a cell that would be sensitive to nitrogen dioxide. Here the equilibrium would be

$$\underset{\substack{\text{external} \\ \text{solution}}}{2NO_2(aq)} + \underset{\substack{\text{internal} \\ \text{solution}}}{H_2O} \rightleftarrows \underset{\text{internal solution}}{NO_3^- + NO_2^- + 2H^+}$$

This electrode should permit the determination of NO_2 in the presence of gases such as SO_2, CO_2, and NH_3 which also alter the pH of the internal solution.

Gas-sensing electrodes are commercially available for SO_2, NO_2, and NH_3. Others will undoubtedly be developed in the near future. Some possible gas-sensing electrodes are suggested in Table 17-5.

TABLE 17-5   Some Possible Gas-sensing Electrode Systems[a]

Analyte Species	Diffusing Species	Equilibria in the Internal Solution	Sensing Electrode
NH_3 or NH_4^+	NH_3	$NH_3 + H_2O \rightleftarrows NH_4^+ + OH^-$ $xNH_3 + M^{n+} \rightleftarrows M(NH_3)_x^{n+}$	H^+ $M = Ag^+$, Cd^{2+}, Cu^{2+}
SO_2, H_2SO_3 or SO_3^{2-}	SO_2	$SO_2 + H_2O \rightleftarrows H^+ + HSO_3^-$	H^+
NO_2^-, NO_2	$NO_2 + NO$	$2NO_2 + H_2O \rightleftarrows NO_3^- + NO_2^- + 2H^+$	H^+, NO_3^-
S^{2-}, HS^-, H_2S	H_2S	$H_2S + 2H_2O \rightleftarrows S^{2-} + 2H^+$	S^{2-}
CN^-, HCN	HCN	$2HCN + Ag^+ \rightarrow Ag(CN)_2^- + 2H^+$	Ag^+
F^-, HF	HF	$HF \rightleftarrows H^+ + F^-$	F^-
$HOAc$, OAc^-	$HOAc$	$HOAc \rightleftarrows H^+ + OAc^-$	H^+
CO_2, H_2CO_3, HCO_3^-, CO_3^{2-}	CO_2	$CO_2 + H_2O \rightleftarrows H^+ + HCO_3^-$	H^+
X_2, OX^-, X^- ($X = Cl^-$, Br^-, I^-)	X_2 or HX	$X_2 + H_2O \rightleftarrows 2H^+ + XO^- + X^-$	H^+, X^-

[a] From *Orion Research News Letter*, vol. V, no. 2, 1973, Orion Research, Inc., Cambridge, Mass.

Direct Potentiometric Measurements

Direct potentiometric measurements are frequently useful for completing an analysis. The technique is simple and involves only the comparison of the potential of the indicator electrode in a solution of the analyte with its potential when it is immersed in a standard solution of the ion of interest. Insofar as the indicator is selective for the analyte, preliminary separations are not required. Finally, the method is readily adapted to the continuous and automatic recording of analytical data.

LIQUID-JUNCTION POTENTIAL

In contrast to other electroanalytical methods, uncertainties in the liquid-junction potential set limits on the accuracy that can be attained from direct potentiometric measurements. Therefore, we need to consider the effect of this phenomenon.

The junction potential arises from an unequal distribution of cations and anions across the boundary between two electrolyte solutions; this difference results from differences in the rates with which various charged species migrate under the effects of diffusion.

Consider, for example, the events occurring at the interface between a 1-F and a 0.01-F hydrochloric acid solution. We may symbolize this interface as

$$HCl(1\ F)\ |\ HCl(0.01\ F)$$

Both hydrogen ions and chloride ions tend to diffuse across this boundary from the more concentrated to the more dilute solution, the driving force for this migration being proportional to the concentration difference. The rate at which ions move under the influence of a fixed force varies considerably (that is, their *mobilities* are different); in the present example, hydrogen ions are several times more mobile than chloride ions. As a consequence, there is a tendency for the hydrogen ions to outstrip the chloride ions as diffusion takes place; a separation of charge is the net result.

The more dilute side of the boundary becomes positively charged, owing to the more rapid migration of hydrogen ions; the concentrated side therefore acquires a negative charge from the slower-moving chloride ions. The charge that develops tends to counteract the differences in mobilities between the two ions; as a consequence, an equilibrium condition soon develops. The potential difference resulting from this charge separation may amount to several hundredths of a volt or occasionally more.

In a simple system such as this, where the behavior of only two ions need be considered, the magnitude of the junction potential can be calculated from a knowledge of the mobilities of the ions involved. However, it is seldom that a cell of analytical importance has a sufficiently simple composition to permit such a computation.

It is an experimental fact that the magnitude of the liquid-junction potential can be greatly decreased by interposition of a concentrated electrolyte solution (a *salt bridge*) between the two solutions. The effectiveness of this contrivance improves as the concentration of the salt in the bridge increases and as the mobilities of the ions of the salt approach one another in magnitude. A saturated potassium chloride solution is good from both standpoints, its concentration being somewhat greater than 4 F at room temperature and the mobility of its ions differing by only 4%. With such a bridge the junction potential typically amounts to a few millivolts or less, a negligible quantity for most electroanalytical methods except direct potentiometry.

EQUATIONS FOR DIRECT POTENTIOMETRY

The observed potential of a cell employed for a direct potentiometric measurement can be expressed in terms of the reference-electrode potential, the indicator-electrode potential, and a junction potential. That is,

$$E_{obs} = E_{ref} - E_{ind} + E_j \qquad (17\text{-}20)[11]$$

The liquid-junction potential occurs between the electrode solution of the external reference-electrode system and the analyte. As mentioned earlier, this junction potential is normally minimized by the use of a salt bridge, but its effect cannot be neglected in direct potentiometric measurements.

We have noted that the potential of the indicator electrode is generally related to the activity a_1 of the ion of interest, M^{n+}, by a Nernst-type relationship. Thus, at 25°C

$$E_{ind} = j + \frac{0.0591}{n} \log a_1 \qquad (17\text{-}21)$$

where j is a constant. Combination of Equation 17-20 with Equation 17-21 and rearrangement gives

$$pM = -\log a_1 = \frac{E_{obs} - (E_{ref} + E_j - j)}{0.0591/n} \qquad (17\text{-}22)$$

or

$$pM = -\log a_1 = \frac{E_{obs} - K'}{0.0591/n} \qquad (17\text{-}23)$$

where the constant K' contains the three constants, at least one of which (the junction potential) cannot be evaluated by theoretical calculations. For some electrodes, j is the standard potential for the indicator electrode. With membrane electrodes, however, j is an unknown asymmetry potential (p. 388) that changes with time.

Several methods for performing analyses by direct potentiometry have been developed. All are based, directly or indirectly, upon Equation 17-23.[12]

ELECTRODE CALIBRATION METHOD

In the electrode calibration procedure, K' in Equation 17-23 is determined by measuring E_{obs} for one or more standard solutions of known pM. The assumption is then made that K' does not change during measurement of the analyte solution. Generally, the calibration operation is performed at the time that pM for the unknown is determined; recalibration is ordinarily desirable if measurements extend over several hours.

The electrode calibration method offers the advantages of simplicity, speed, and applicability to the continuous monitoring of pM. Two important

[11] In some instances, the indicator electrode behaves as the cathode and the reference electrode as the anode; under these circumstances, the signs of the two electrodes will be reversed.

[12] For additional information, see R. A. Durst, *Ion-Selective Electrodes*, chapter 11. National Bureau of Standards Special Publication 314. Washington, D.C.: U.S. Government Printing Office, 1969.

disadvantages attend its use, however. One of these is that the results of an analysis are in terms of activities rather than concentrations; the other is that the accuracy of an analysis by this procedure is limited by the existence of an inherent uncertainty in the measurement that can never be totally eliminated.

Activity versus Concentration. Electrode response is related to activity rather than concentration of an analyte. Ordinarily, however, the scientist is interested in concentration, and the determination of this quantity from a potentiometric measurement requires activity coefficient data. More often than not, however, activity coefficients will not be available because the ionic strength of the solution is either unknown or so high that the Debye-Hückel equation (Chapter 5) is not applicable. Unfortunately, the assumption that activity and concentration are identical may lead to errors of several percent, particularly when the analyte is polyvalent.

The difference between activity and concentration is shown in Figure 17-13 where the lower curve gives the change in potential of a calcium electrode as a function of chloride concentration (note that the concentration scale is logarithmic). The nonlinearity of the curve is due to the increase in ionic strength and the consequent decrease in the activity coefficient of the calcium ion as the electrolyte concentration becomes larger. When these concentrations are converted to activities, the upper curve is obtained; note that this straight line has the theoretical slope of 0.0296 (0.0591/2).

We have seen (p. 111) that activity coefficients for singly charged ions are less affected by changes in ionic strength than those for species with multiple charges. Thus, the effect shown in Figure 17-13 will be less pronounced for electrodes that respond to H^+, Na^+, and other singly charged ions.

In potentiometric pH measurements, the pH of the standard buffer employed for calibration is generally based on the activity of hydronium ions. Thus, the resulting hydrogen ion analysis is also on an activity scale. If the unknown sample has a high ionic strength, the hydronium ion concentration will differ appreciably from the activity measured.

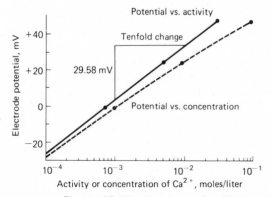

Figure 17-13   Response of a Calcium Ion Electrode to Variations of Concentration and Activity in Solutions Prepared from Pure Calcium Chloride. (Sketch courtesy of Orion Research, Inc., Cambridge, Mass.)

Inherent Error in the Electrode Calibration Procedure. A serious disadvantage of the electrode calibration method is the existence of an inherent uncertainty that results from the assumption that K' in Equation 17-23 remains constant after calibration.

This assumption can seldom, if ever, be exactly true, because the electrolyte composition of the unknown will almost inevitably differ from that of the solutions employed for calibration. The junction potential contained in K' will vary slightly as a consequence, even when a salt bridge is used. Often, this uncertainty will be of the order of one millivolt; unfortunately, because of the nature of the potential-activity relationship, such an uncertainty has an amplified effect on the inherent accuracy of the analysis. The magnitude of the uncertainty in analyte concentration can be estimated by differentiating Equation 17-23, holding E_{obs} constant:

$$\frac{\log_{10} e}{a_1} \, da_1 = \frac{0.434 \, da_1}{a_1} = \frac{dK'}{0.0591/n}$$

or

$$\frac{da_1}{a_1} = \frac{n \, dK'}{0.0257}$$

Upon replacing da_1 and dK' with the finite increments Δa_1 and $\Delta K'$, and multiplying both sides of the equation by 100, we obtain

$$\frac{\Delta a_1}{a_1} \times 100 = \text{percent relative error} = 3900 \, n \, \Delta K'$$

The quantity $\Delta a_1/a_1$ is the *relative* error in a_1 associated with an *absolute* uncertainty, $\Delta K'$ in K'. If, for example, $\Delta K'$ is ± 0.001 V, a relative error of about $\pm 4n\%$ can be expected. *It is important to appreciate that this uncertainty is characteristic of all measurements of cells that contain a salt bridge, and that this error cannot be eliminated by even the most careful measurements of cell potentials*; nor does it appear possible to devise a method for completely eliminating the uncertainty in K' that causes this error.

CALIBRATION CURVES FOR DIRECT POTENTIOMETRY

An obvious way of correcting potentiometric measurements to give results in terms of concentration is to make use of an empirical calibration such as the lower curve in Figure 17-13. For this approach to be successful, however, it is essential that the ionic composition of the standard approximate that of the analyte—a condition that is difficult to realize experimentally for complex samples.

For samples with electrolyte concentrations that are not too great, it is often helpful to swamp both the samples and the calibration standards with a measured excess of an inert electrolyte. Under these circumstances, the added effect of the electrolyte in the sample becomes negligible, and the empirical calibration curve yields results in terms of concentration. A procedure of this kind has been employed for the potentiometric determination of fluoride in public water supplies. Here both samples and standards are diluted on a 1:1

basis with a solution containing sodium chloride, a citrate buffer, and an acetate buffer; the diluent is sufficiently concentrated so that the samples and standards do not differ significantly in ionic strength. The procedure permits a rapid measurement of fluoride ion in the 1 ppm range with a precision of about 5% relative.

STANDARD ADDITION METHOD

In the standard addition method the potential of the electrode system is measured before and after addition of a small volume of a standard to a known volume of the unknown. The assumption is made that this addition does not alter the ionic strength and thus the activity coefficient f of the analyte.

If the potentials before and after the addition are E_1 and E_2, respectively, and the unknown is singly charged, we may write with the aid of Equation 17-23

$$-\log C_x f = \frac{E_1 - K'}{0.0591}$$

where $C_x f$ is equal to a_1, the activity of the analyte in the sample, and C_x is its molar concentration. If now V_s ml of a standard with a molar concentration of C_s is added to V_x ml of the sample, the potential will become

$$-\log \frac{C_x V_x + C_s V_s}{V_x + V_s} f = \frac{E_2 - K'}{0.0591}$$

These two equations can be combined and rearranged to give

$$C_x = \frac{C_s V_s}{V_x + V_s} \left(10^{-\Delta E/0.0591} - \frac{V_x}{V_x + V_s} \right)^{-1}$$

where $\Delta E = E_2 - E_1$.

In applying the standard addition method, it is always wise to establish experimentally that the electrode response is the theoretical $0.0591/n$ volt per decade change in concentration. This check can be accomplished by measuring the electrode potential after two successive additions of the standard to the unknown.

The standard addition method has been applied to the determination of chloride and fluoride in samples of the commercial phosphors.[13] Here two indicator electrodes and a reference electrode were used and the added standard contained known quantities of the two anions. The relative standard deviation for the measurement of replicate samples was found to be 0.7% for fluoride and 0.4% for chloride. Ordinarily, the relative errors for the method appeared to range between 1 and 2%.

POTENTIOMETRIC pH MEASUREMENTS WITH A GLASS ELECTRODE[14]

The glass electrode is unquestionably the most important indicator electrode for hydronium ion and has almost completely displaced all other electrodes for

[13] L. G. Bruton, *Anal. Chem.*, **43**, 579 (1971).
[14] For a detailed discussion of potentiometric pH measurements, see R. G. Bates, *Determination of pH, Theory and Practice*. New York: John Wiley & Sons, Inc., 1964.

pH measurements. It is convenient to use and subject to few of the interferences affecting other electrodes.

Glass electrodes are available at a relatively low cost and in many shapes and sizes. A common variety is illustrated in Figure 17-5; as shown here, the reference electrode is usually a commercial saturated calomel electrode.

The glass-calomel electrode system is a remarkably versatile tool for the measurement of pH under many conditions. The electrode can be used without interference in solutions containing strong oxidants, reductants, proteins, and gases; the pH of viscous or even semisolid fluids can be determined. Electrodes for special applications are available. Included among these are small electrodes for pH measurements in a drop (or less) of solution or in a cavity of a tooth, microelectrodes which permit the measurement of pH inside a living cell, systems for insertion in a flowing liquid stream to provide a continuous monitoring of pH, and a small glass electrode that can be swallowed to indicate the acidity of the stomach contents (the calomel electrode is kept in the mouth).

Summary of Errors Affecting pH Measurements with the Glass Electrode. The ubiquity of the pH meter and the general applicability of the glass electrode tend to lull the chemist into the attitude that any measurement obtained with such an instrument is surely correct. It is well to guard against this sense of security since there are distinct limitations to the electrode system. These have been discussed in earlier sections and include:

1. *The alkaline error.* The ordinary glass electrode becomes somewhat sensitive to alkali-metal ions at pH values greater than 9.
2. *The acid error.* At a pH less than zero, values obtained with a glass electrode tend to be somewhat high.
3. *Dehydration.* Dehydration of the electrode may cause unstable performance and errors.
4. *Errors in unbuffered neutral solutions.* Equilibrium between the electrode surface layer and the solution is achieved only slowly in poorly buffered, approximately neutral solutions. Since the measured potential is determined by the surface layer of liquid, errors will arise unless time is allowed for equilibrium to be established; this process may take several minutes. In determining the pH of poorly buffered solutions, the glass electrode should be thoroughly rinsed with water before use. Then, if unknown is plentiful, the electrodes should be placed in successive portions until a constant pH reading is obtained. Good stirring is also helpful; several minutes should be allowed for the attainment of steady readings.
5. *Variation in junction potential.* It should be reemphasized that this fundamental uncertainty in the measurement of pH is one for which a correction cannot be applied. Absolute values more reliable than 0.01 pH unit are generally unobtainable. Even reliability to 0.03 pH unit requires considerable care. On the other hand, it is often possible to detect pH *differences* between similar solutions or pH *changes* in a single solution that are as small as 0.001 unit. For this reason many pH meters are designed to permit readings to less than 0.01 pH unit.
6. *Error in the pH of the buffer solution.* Since the glass electrode must be frequently calibrated, any inaccuracies in the preparation or changes in composition of the buffer during storage will be propagated as errors in pH measurements.

Potentiometric Titrations

The potential of a suitable indicator electrode is conveniently employed to establish the equivalence point for a titration (a *potentiometric titration*). A potentiometric titration often provides different information from a direct potentiometric measurement. For example, direct measurement with a pH-sensitive electrode reveals the hydronium ion concentration of a solution of acetic acid. A potentiometric titration of the same solution, on the other hand, provides information as to the amount of acetic acid contained in the sample.

The potentiometric end point is widely applicable and provides inherently more accurate data than the corresponding indicator method. It is particularly useful for titration of colored or opaque solutions and for detecting the presence of unsuspected species in a solution. Unfortunately, it is more time-consuming than a titration performed with an indicator.

Figure 17-14 shows a typical apparatus for performing a potentiometric titration. Ordinarily, the titration involves measuring and recording a cell potential (or a pH reading) after each addition of reagent. The titrant is added in large increments at the outset; as the end point is approached (as indicated by larger potential changes per addition), the increments are made smaller.

Sufficient time must be allowed for the attainment of equilibrium after each addition of reagent. Precipitation titrations may require several minutes for equilibration, particularly in the vicinity of the equivalence point. A close approach to equilibrium is indicated when the measured potential ceases to drift by more

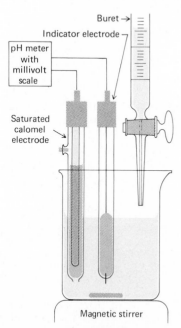

Figure 17-14 Apparatus for a Potentiometric Titration.

than a few millivolts. Good stirring is frequently effective in hastening the achievement of equilibrium.

The first two columns of Table 17-6 consist of a typical set of potentiometric titration data obtained with the apparatus illustrated in Figure 17-14. The data are plotted as curve (a) in Figure 17-15. Note that this experimental plot closely resembles titration curves derived from theoretical considerations (p. 175).

END-POINT DETERMINATION

Several methods can be used to determine the end point for a potentiometric titration. The most straightforward involves a direct plot of potential versus reagent volume as in curve (a), Figure 17-15. The midpoint in the steeply rising portion of the curve is then estimated visually and taken as the end point. Various mechanical methods to aid in the establishment of the midpoint have been

TABLE 17-6   Potentiometric Titration Data for 2.433 meq of Chloride with 0.1000-N Silver Nitrate

Vol AgNO$_3$, ml	E vs. SCE, V	$\Delta E/\Delta V$, V/ml	$\Delta^2 E/\Delta V^2$, V/ml^2
5.0	0.062		
		0.002	
15.0	0.085		
		0.004	
20.0	0.107		
		0.008	
22.0	0.123		
		0.015	
23.0	0.138		
		0.016	
23.50	0.146		
		0.050	
23.80	0.161		
		0.065	
24.00	0.174		
		0.09	
24.10	0.183		
		0.11	
24.20	0.194		0.28
		0.39	
24.30	0.233		0.44
		0.83	
24.40	0.316		−0.59
		0.24	
24.50	0.340		−0.13
		0.11	
24.60	0.351		−0.04
		0.07	
24.70	0.358		
		0.050	
25.00	0.373		
		0.024	
25.5	0.385		
		0.022	
26.0	0.396		
		0.015	
28.0	0.426		

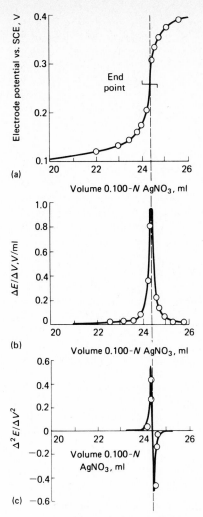

Figure 17-15 (a) Potentiometric Titration Curve of 2.433 meq of Cl⁻ with 0.100-*N* AgNO₃. (b) First Derivative Curve. (c) Second Derivative Curve.

proposed; it is doubtful, however, that these significantly improve the accuracy.

A second approach is to calculate the change in potential per unit change in volume of reagent (that is, $\Delta E/\Delta V$) as has been done in column 3 of Table 17-6. A plot of this parameter as a function of average volume leads to a sharp maximum at the end point [see curve (b), Figure 17-15]. Alternatively, the ratio can be evaluated during the titration and recorded directly in lieu of the potential itself. Thus, in column 3 of Table 17-6, it is seen that the maximum is located between 24.3 and 24.4 ml; selection of 24.35 ml would be adequate for most purposes.

Lingane[15] has shown that the volume can be fixed more exactly by estimating the point where the second derivative of the voltage with respect to volume (that is, $\Delta^2 E/\Delta V^2$) becomes zero. This calculation is easy if equal increments of solution are added in the vicinity of the equivalence point. Values of $\Delta^2 E/\Delta V^2$, calculated by subtracting the corresponding data for $\Delta E/\Delta V$, are given in the fourth column of Table 17-6. The function must become zero at some point between the two volumes where a change in sign occurs. The volume corresponding to this point can be obtained by interpolation. Curve (c) of Figure 17-15 is a plot of $\Delta^2 E/\Delta V^2$.

All of the methods we have considered are based on the assumption that the titration curve is symmetric about the equivalence point and that the inflection in the curve thus corresponds to that point. This assumption is perfectly valid, provided the participants in the chemical process react with one another in an equimolar ratio, and also provided the electrode process is perfectly reversible. Where these provisions are not met, an asymmetric curve results, as shown in Figure 15-1 (p. 325). Note that the curve for the oxidation of iron(II) by cerium(IV) is symmetrical about the equivalence point. On the other hand, 5 moles of iron(II) are oxidized by each mole of permanganate and a highly nonsymmetric titration curve results. Ordinarily, the change in potential within the equivalence-point region of these curves is large enough so that a negligible titration error is introduced if the midpoint of the steeply rising portion is chosen as the end point. Only when unusual accuracy is desired or where very dilute solutions are employed must account be taken of this source of uncertainty. If needed, a correction can be determined empirically by titration of a standard. Alternatively, when the error is due to a nonsymmetrical reaction, the correct location of the equivalence can be calculated from theoretical considerations.[16]

Another procedure for end-point detection involves titrating to the calculated theoretical potential or, better, to an empirical potential obtained by titration of standards. This procedure demands that the equivalence-point behavior of the system be reproducible.

PRECIPITATION TITRATIONS

Electrode Systems. The indicator electrode for a precipitation titration is frequently the metal from which the reacting cation was derived. Membrane electrodes that are sensitive to one of the ions involved in the titration may also be employed.

Titration Curves. Theoretical curves for a potentiometric titration are readily derived. For example, we may describe the potential of the silver electrode during the titration of chloride with silver ion by the expression

$$E_{Ag} = 0.222 - 0.0591 \log [Cl^-]$$

[15] J. J. Lingane, *Electroanalytical Chemistry*, 2d ed., p. 93. New York: Intersceince Publishers, Inc., 1958.
[16] See I. M. Kolthoff and N. H. Furman, *Potentiometric Titrations*, 2d ed., chapters 2 and 3. New York: John Wiley & Sons, Inc., 1931.

where 0.222 is the standard potential for the half-reaction

$$AgCl(s) + e \rightleftarrows Ag(s) + Cl^-$$

Alternatively, the potential is also given by

$$E_{Ag} = 0.799 - 0.0591 \log \frac{1}{[Ag^+]}$$

where 0.799 is the standard potential for the formation of silver from silver ion.

The first relationship is the more useful for deriving points prior to the equivalence point because the formal concentration of chloride ion will be known, and we may write

$$[Cl^-] \cong F_{Cl^-}$$

Beyond the equivalence point, the second equation is more easily used; here we may assume

$$[Ag^+] \cong F_{Ag^+}$$

If the calomel electrode serves as the anode, combination of its potential with that of the silver electrode gives

$$E_{cell} = E_{Ag} - 0.242$$

A theoretically derived curve will depart slightly from the experimental because it is not possible to calculate a value for the junction potential.

Titration Curves for Mixtures. Multiple end points are often revealed by a potentiometric titration when the sample contains more than one reacting species. The titration of halide mixtures with silver nitrate is an important example. Figure 8-3 (p. 177) is a theoretically derived titration curve for a mixture of chloride and iodide with silver nitrate. Because the measured potential is directly proportional to the negative logarithm of the silver ion concentration, the experimental potentiometric curve should have the same shape although the ordinate units would be different.

Experimental curves for the titration of halide mixtures do not show the sharp discontinuity at the first equivalence point that appears in Figure 8-3. More important, the volume of silver nitrate required to reach the first end point is generally somewhat greater than theoretical; the total volume, however, approaches the correct amount. These observations apply to the titration of chloride-bromide, chloride-iodide, and bromide-iodide mixtures and can be explained by assuming that the more soluble silver halide is coprecipitated during formation of the less soluble compound. An overconsumption of reagent in the first part of the titration thus occurs.

Despite the coprecipitation error, the potentiometric method is useful for the analysis of halide mixtures. When approximately equal quantities are present, relative errors can be kept to about 1 to 2%.[17]

---

[17] For further details, see I. M. Kolthoff and N. H. Furman, *Potentiometric Titrations*, 2d ed., pp. 154–158. New York: John Wiley & Sons, Inc., 1931.

COMPLEX-FORMATION TITRATIONS

Both metal electrodes and membrane electrodes have been applied to the detection of end points in reactions that involve formation of a soluble complex. The mercury electrode, illustrated in Figure 17-16, is particularly useful for EDTA titrations.[18] It will function as an indicator electrode for the titration of cations forming complexes that are less stable than HgY^{2-} (see p. 383). Reilley and Schmid have undertaken a systematic study of this end point and describe conditions that are necessary for various EDTA titrations.[19]

NEUTRALIZATION TITRATIONS

In Chapters 9 and 10 we considered theoretical curves for various neutralization titrations in some detail. The shapes of these curves can be closely approximated experimentally. Often, however, the experimental curves will be displaced from the theoretical curves because the latter are usually derived employing concentrations rather than activities. A study of the theoretical curves will show that the small error inherent in the potentiometric measurement of pH is of no consequence insofar as locating the end point is concerned.

Potentiometric acid-base titrations are particularly useful for the analysis of mixtures of acids or polyprotic acids (or bases) since discrimination between the end points can often be made. An approximate numerical value for the dissociation constant of the reacting species can also be estimated from potentiometric titration curves. In theory, this quantity can be obtained from any point along the curve; as a practical matter, it is most easily found from the

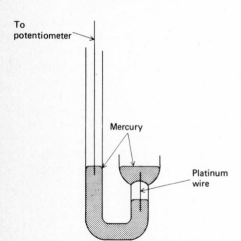

To
potentiometer

Mercury

Platinum
wire

Figure 17-16   A Typical Mercury Electrode.

[18] These electrodes can be obtained from Kontes Manufacturing Corp., Vineland, N.J.
[19] C. N. Reilley and R. W. Schmid, *Anal. Chem.* **30**, 947 (1958).

pH at the point of half-neutralization. For example, in the titration of the weak acid HA, we may ordinarily assume that at the midpoint,

$$[HA] \cong [A^-]$$

and, therefore,

$$K_a = \frac{[H^+][\cancel{A^-}]}{[\cancel{HA}]} = [H^+]$$

or

$$pK_a = pH$$

It is important to note that a dissociation constant determined in this way may differ from that shown in a table of dissociation constants by a factor of 2 or more because the latter is based upon activities while the former is not. Thus, if we write the dissociation-constant expression in its more exact form we obtain

$$K_a = \frac{a_{H_3O^+} \cdot a_{A^-}}{a_{HA}} = \frac{a_{H_3O^+} \cdot [A^-] \cdot f_{A^-}}{[HA] \cdot f_{HA}}$$

Again we may assume that $[A^-]$ and $[HA]$ are approximately equal; furthermore, the potential of a glass electrode gives a good approximation for $a_{H_3O^+}$. Therefore,

$$K_a = \frac{a_{H_3O^+} \cdot [\cancel{A^-}] \cdot f_{A^-}}{[\cancel{HA}] \cdot f_{HA}} = \frac{a_{H_3O^+} \cdot f_{A^-}}{f_{HA}}$$

Taking the negative logarithm of the two sides of the equation yields

$$-\log K_a = -\log a_{H_3O^+} - \log \frac{f_{A^-}}{f_{HA}}$$

Upon converting to p-functions and rearranging,

$$(pK_a)_{expt} = pH = pK_a + \log \frac{f_{A^-}}{f_{HA}}$$

Thus, the true pK_a will differ from the experimental pK_a by the logarithm of the ratio of activities. Ordinarily, the ionic strength during a titration will be 0.1 or greater. Thus, the ratio of f_{A^-} to f_{HA} would be at least 0.75 (see Table 5-4) if HA is uncharged. For solutes such as $H_2PO_4^-$ and HPO_4^{2-}, this ratio will be even larger.

A value of the equivalent weight and the approximate dissociation constant of a pure sample of an unknown acid can be obtained from a single potentiometric titration; this information is frequently sufficient to identify the acid.

OXIDATION-REDUCTION TITRATIONS

The derivation of theoretical titration curves for oxidation-reduction processes was considered in Chapter 15. In each example, an electrode potential related to the concentration ratio of the oxidized and reduced forms of either of the reactants was determined as a function of the titrant volume. These curves can be duplicated experimentally (here again the curves may be displaced on the ordinate scale because of the effect of ionic strength), provided an indicator electrode responsive to at least one of the couples involved in the reaction is available. Such electrodes exist for most, but not all, of the reagents described in Chapter 16.

Indicator electrodes for oxidation-reduction titrations are generally constructed from platinum, gold, mercury, or silver. The metal chosen must be unreactive with respect to the components of the reaction—it is merely a medium for electron transfer. Without question, the platinum electrode is most widely used for oxidation-reduction titrations. Curves similar to those shown in Figure 15-1 can be obtained experimentally with a platinum-calomel electrode system. The end point can be established by the methods already discussed.

DIFFERENTIAL TITRATIONS

We have seen that a derivative curve generated from the data of a conventional potentiometric titration curve (b) (in Figure 17-15) reaches a distinct maximum in the vicinity of the equivalence point. It is also possible to acquire titration data directly in derivative form by means of suitable apparatus.

A differential titration requires the use of two identical indicator electrodes, one of which is well shielded from the bulk of the solution. Figure 17-17 illustrates a typical arrangement. Here one of the electrodes is contained in a sidearm test tube. Contact with the bulk of the solution is made through a small (~1 mm) hole in the bottom of the tube. Because of this restricted access, the composition of the solution surrounding the shielded electrode will not be immediately affected by an addition of titrant to the bulk of the solution. The

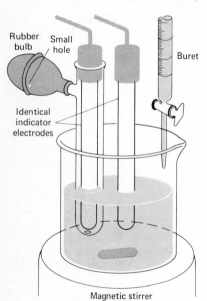

Figure 17-17 Apparatus for Differential Potentiometric Titrations.

resulting difference in solution composition gives rise to the difference in potential ΔE between the electrodes. After each potential measurement, the solution is homogenized by squeezing the rubber bulb several times, whereupon ΔE again becomes zero. If the volume of solution in the tube that shields the electrode is kept small (say, 1 to 5 ml), the error arising from failure of the final addition of reagent to react with this portion of the solution can be shown to be negligibly small.

For oxidation-reduction titrations, two platinum wires can be employed, with one enclosed in an ordinary medicine dropper.

The main advantage of a differential method is the elimination of the reference electrode and salt bridge. The end points are ordinarily sharply defined maxima similar to (b) in Figure 17-15.

AUTOMATIC TITRATIONS

In recent years several automatic titrators based on the potentiometric principle have come on the market. These are useful where many routine analyses are required. Such instruments cannot yield more accurate results than those obtained by manual potentiometric techniques; however, they do decrease the time needed to perform the titrations and thus may offer some economic advantages.

Basically, two types of automatic titrators are available. The first yields a titration curve of potential versus reagent volume, or, in some instances, $\Delta E/\Delta V$ or $\Delta^2 E/\Delta V^2$ against volume. The end point is then obtained from the curve by inspection. In the second type, the titration is stopped automatically when the potential of the electrode system reaches a predetermined value; the volume of reagent delivered is then read at the operator's convenience or printed on a tape.

Automatic titrators normally employ a buret system with a solenoid-operated valve to control the flow; alternatively, use is made of a syringe, the plunger of which is activated by a motor-driven micrometer screw. With both types, the rate of reagent addition must be either very slow throughout to prevent overrunning the end point, or some means must be provided to add reagent in smaller and smaller increments as the end point is approached. The latter method is to be preferred because of the shorter time required for titration. A number of instruments have been designed that anticipate the end point and add reagent in the same stepwise manner employed by a human operator.[20]

PROBLEMS

*1. (a) Calculate the standard potential for the reaction

$$CuSCN(s) + e \rightleftarrows Cu(s) + SCN^-$$

(b) Give a schematic representation of a cell with a copper indicator electrode as an anode and a SCE as a cathode that could be used for the determination of SCN^-.

[20] For a description of such instruments, see J. J. Lingane, *Electroanalytical Chemistry*, 2d ed., chapter 8. New York: Interscience Publishers, Inc., 1958.

(c) Derive an equation relating the measured potential of the cell in (b) to pSCN (assume that the junction potential is zero).

(d) Calculate the pSCN of a thiocyanate-containing solution that was saturated with CuSCN and employed in conjunction with a copper electrode in the type of cell in (b) if the resulting potential was 0.076 V.

2. (a) Calculate the standard potential for the reaction

$$Ag_2S(s) + 2e \rightleftarrows 2Ag(s) + S^{2-}$$

(b) Give a schematic representation of a cell with a silver indicator electrode as the cathode and a SCE as an anode that could be used for determining S^{2-}.

(c) Derive an equation relating the measured potential of the cell in (b) to pS (assume that the junction potential is zero).

(d) Calculate the pS of a solution that was saturated with Ag_2S and then employed with the type of cell in (b) if the resulting potential was 0.538 V.

*3. For each of the following cells, give a schematic representation that might be suitable and an equation for the relationship between the cell potential and the desired quantity. Assume that the junction potential is negligible, and specify any necessary concentrations as 1.00×10^{-4} M. In each instance, treat the indicator electrode as the cathode.

(a) A cell with a Hg indicator electrode for the determination of pCl.

(b) A cell with a Ag indicator electrode for the determination of pCO_3.

(c) A cell with a platinum electrode for the determination of pSn(IV).

4. For each of the following cells, give a schematic representation that might be suitable and an equation for the relationship between the cell potential and the desired quantity. Assume that the junction potential is negligible, and specify any necessary concentrations as 1.00×10^{-4} M. In each instance, treat the indicator electrode as the cathode.

(a) A cell with a Pb electrode for the determination of $pCrO_4$.

(b) A cell with a Ag indicator electrode for the determination of $pAsO_4$.

(c) A cell with a platinum electrode for the determination of pTl(III).

*5. The following cell was employed for the determination of $pCrO_4$:

$$SCE \,||\, Ag_2CrO_4(sat'd), CrO_4^{2-}(xM)\,|\, Ag$$

Calculate $pCrO_4$ when the cell potential was 0.402 V.

6. Calculate the potential of the cell (neglecting the junction potential)

$$indicator\ electrode \,||\, SCE$$

where the indicator electrode is mercury in a solution that is

(a) 3.12×10^{-5} M Hg^{2+}.

(b) 3.12×10^{-5} M Hg_2^{2+}.

(c) saturated with Hg_2Br_2 and is 5.00×10^{-4} M in Br^-.

(d) 3.00×10^{-6} F in $Hg(NO_3)_2$ and 0.020 F in KCl.

$$Hg^{2+} + 2Cl^- \rightleftarrows HgCl_2 \qquad K_{form} = 6.1 \times 10^{12}$$

(e) 3.00×10^{-6} F in $Hg(NO_3)_2$ and 0.0020 F in KCl.

*7. The formation constant for the soluble mercury(II) acetate is

$$Hg^{2+} + 2OAc^- \rightleftarrows Hg(OAc)_2 \qquad K_{form} = 2.5 \times 10^8$$

Calculate the standard potential for the reaction

$$Hg(OAc)_2 + 2e \rightleftarrows Hg(l) + 2OAc^-$$

8. Calculate the standard potential for the half-reaction

$$HgY^{2-} + 2e \rightleftarrows Hg(l) + Y^{4-}$$

where Y^{4-} is the anion of EDTA. The formation constant for the reaction $Hg^{2+} + Y^{4-} \rightleftarrows HgY^{2-}$ is $K_{form} = 6.3 \times 10^{21}$.

*9. Calculate the potential of the cell (neglecting the junction potential)

$$Hg|HgY^{2-}(7.24 \times 10^{-4} \, M), Y^{4-}(xM)| \, | \, SCE$$

where Y^{4-} is the EDTA anion, and the formation constant for HgY^{2-} has a value of 6.3×10^{21}. (For $HgY^{2-} + 2e \rightleftarrows Hg(l) + Y^{4-}$, $E^0 = 0.210$.) The concentration of Y^{4-} is
   (a) $2.00 \times 10^{-1} \, M$.
   (b) $2.00 \times 10^{-3} \, M$.
   (c) $2.00 \times 10^{-5} \, M$.

10. Calculate the potential of the cell

$$Hg|HgY^{2-}(7.24 \times 10^{-4} \, M), EDTA(C_T = 2.00 \times 10^{-2} \, F)| \, | \, SCE$$

   if the pH of the solution on the left is (a) 7.00, (b) 8.00, (c) 9.00. See Table 13-3 for values of α_4 and Problem 9 for E^0.

*11. Calculate the potential of the cell

$$Hg|HgY^{2-}(7.24 \times 10^{-4} \, M), EDTA(5.0 \times 10^{-4} \, F), Ca^{2+}(xF)| \, | \, SCE$$

   The solution is sufficiently alkaline so that essentially all of the EDTA is complexed by the calcium. The formal calcium ion concentration is (a) 0.0200; (b) 0.0400; (c) 0.100. See Problem 9 for E^0.

12. An alkaline solution containing Ca^{2+} was made $3.37 \times 10^{-4} \, M$ in HgY^{2-} and $2.26 \times 10^{-4} \, F$ in EDTA. This solution formed part of the cell

$$Hg|HgY^{2-}(3.37 \times 10^{-4} \, M), EDTA(2.26 \times 10^{-4} \, F), Ca^{2+}(xF)| \, | \, SCE$$

   Calculate the pCa of the solution if the potential was (a) -0.269 V; (b) -0.174 V; (c) -0.100 V. See Problem 9 for E^0. Assume that the solution is sufficiently alkaline that essentially all of the EDTA is present as CaY^{2-}.

13. The following cell was found to have a potential of 0.2094 V when the solution in the left compartment was a buffer of pH 4.006:

$$glass \; electrode|H^+(a = x)| \, | \, SCE$$

   The following potentials were obtained when the buffered solution was replaced with unknowns. Calculate the pH and the hydrogen ion activity of each unknown.
   *(a) 0.3011 V
   *(b) 0.0710 V
   (c) 0.4829 V
   (d) 0.6163 V

*14. Assume a difference in junction potential of -0.0005 V between calibration and pH measurements described in Problem 13. Calculate the percent relative error in pH and the percent relative error in the hydrogen ion activity which would result.

15. The following cell was found to have a potential of 0.367 V:

$$membrane \; electrode \; for \; Mg^{2+}|Mg^{2+}(a = 9.62 \times 10^{-3} \, M)| \, | \, SCE$$

(a) When the solution of known magnesium activity was replaced with an unknown solution, the potential was found to be 0.544 V. What was the pMg of this unknown solution?

(b) Assuming an uncertainty of ± 0.002 V in the junction potential, what is the range of Mg^{2+} activities within which the true value might be expected?

*16. The following cell was found to have a potential of 0.971 V:

$$Cd|CdX_2(sat'd), X^-(0.0100\ M)|\ |\ SCE$$

Calculate the solubility product of CdX_2, neglecting the junction potential.

17. The following cell was found to have a potential of 0.741 V:

$$Pt, H_2(1.00\ atm)|HA(0.300\ F), NaA(0.200\ F)|\ |\ SCE$$

Calculate the dissociation constant of HA, neglecting the junction potential.

18. A 40.0-ml aliquot of 0.100-N HNO_2 is diluted to 75.0 ml and titrated with 0.080-N Ce^{4+}. The hydrogen ion concentration is kept at 1.00 throughout the titration. (Use 1.44 V for the formal potential of the cerium system.)

*(a) Calculate the potential of the indicator electrode with respect to a saturated calomel reference electrode after the addition of 5.00, 10.0, 15.0, 25.0, 40.0, 49.0, 50.0, 51.0, 55.0, and 60.0 ml of cerium(IV).

(b) Draw a titration curve for these data.

19. Calculate the potential of a silver cathode (vs. SCE) after the addition of 5.00, 15.0, 25.0, 30.0, 35.0, 39.0, 40.0, 41.0, 45.0, and 50.0 ml of 0.100-F $AgNO_3$ to 50.0 ml of 0.0800-F KSeCN. Construct a titration curve from these data. (K_{sp} for AgSeCN = 4.0×10^{-16})

20. Calculate the equivalence-point cell potential for each of the following potentiometric titrations. Treat the indicator-electrode system as the cathode. The reference electrode is a saturated calomel electrode. In each instance, assume that the solutions of the reagent and substance titrated are 0.100 N at the outset.

*(a) The titration of SCN^- with a standard solution of $Hg_2(NO_3)_2$ [for $Hg_2(SCN)_2$, $K_{sp} = 3.0 \times 10^{-20}$], employing a mercury indicator electrode.

*(b) The titration of Sn^{2+} with I_3^-, employing a platinum indicator electrode (assume that $[I^-] = 0.400$ at the equivalence point).

(c) The titration of I^- with $Pb(NO_3)_2$ using a lead electrode.

(d) The titration of U^{4+} with MnO_4^-, employing a platinum electrode (assume that $[H^+] = 0.200$ at the equivalence point).

21. Quinhydrone is an equimolar mixture of quinone (Q) and hydroquinone (H_2Q). These two compounds react reversibly at a platinum electrode

$$Q + 2H^+ + 2e \rightleftarrows H_2Q \qquad E^0 = 0.699\ V$$

The pH of a solution can be determined by saturating it with quinhydrone and making it a part of the cell

$$Pt|quinhydrone(sat'd), H^+(xM)|\ |\ SCE$$

Such a cell was found to have a potential of -0.313 V. What was the pH of the solution, assuming that the junction potential was zero?

EFFECTS OF
CURRENT PASSAGE
IN ELECTROCHEMICAL CELLS

The electrochemical methods considered in the three chapters that follow differ from the potentiometric procedure in two regards. First, these methods depend upon the passage of current through the cell; in contrast, every effort is made to minimize the currents associated with potentiometric measurements. Second, junction potentials are unimportant and, as a practical matter, can be neglected in treating the theory of the methods.

When a current is passed through a cell, three phenomena are likely to be encountered that influence the overall cell potential, namely, ohmic potential, concentration polarization, and kinetic polarization. This chapter treats these phenomena.

Ohmic Potential; *IR* Drop

Current passage through either a galvanic or an electrolytic cell requires a driving force or a potential to overcome the resistance of the ions to movement toward the anode or the cathode. Just as in metallic conduction, this force follows Ohm's law and is equal to the product of the current in amperes and the resistance of the cell in ohms. The force is generally referred to as the *ohmic potential*, or the *IR drop*.

In general, the net effect of IR drop is to increase the potential required to operate an electrolytic cell and to decrease the measured potential of a galvanic cell. Therefore, the IR drop is always *subtracted* from the theoretical cell potential. That is,

$$E_{cell} = E_{cathode} - E_{anode} - IR \qquad (18\text{-}1)$$

Example. 1. Calculate the potential when 0.100 A is drawn from the galvanic cell

$$Cd|Cd^{2+}(1.00\ M)| \ |Cu^{2+}(1.00\ M)|Cu$$

Assume a cell resistance of 4.00 ohms.

Since both cation concentrations are 1.00 M, the respective half-cell potentials are equal to the standard potentials; thus,

$$E = E^0_{Cu} - E^0_{Cd} = 0.337 - (-0.403) = 0.740\ V$$

$$E_{cell} = 0.740 - IR = 0.740 - 0.100 \times 4.00$$

$$= 0.340\ V$$

Thus, the emf of this cell drops dramatically at the first instant the current is drawn.

2. Calculate the potential required to cause a current of 0.100 A to pass in the reverse direction in the foregoing cell.

$$E = E^0_{Cd} - E^0_{Cu} = -0.403 - 0.337 = -0.740\ V$$

$$E_{cell} = -0.740 - 0.100 \times 4.00 = -1.140\ V$$

Here an external potential greater than 1.140 V would be needed to cause Cd^{2+} to deposit and Cu to dissolve at a rate corresponding to 0.100 A.

Polarization Effects

The linear relationship between the potential and the instantaneous current being drawn from or forced through a cell (Equation 18-1) is frequently observed experimentally when I is small; at high currents, however, marked departures from linear behavior occur. Under these circumstances, the cell is said to be *polarized* (see Figure 18-1). Thus, a polarized electrolytic cell requires application of potentials larger than theoretical for a given current flow; similarly, a polarized galvanic cell develops potentials that are smaller than predicted. Under some conditions, the polarization of a cell may be so extreme that the current becomes essentially independent of the voltage; under these circumstances, polarization is said to be complete.

Polarization is an electrode phenomenon; either or both electrodes in a cell can be affected. Included among the factors influencing the extent of polarization are the size, shape, and composition of the electrodes; the composition of the electrolyte solution; the temperature and the rate of stirring; the magnitude of the current; and the physical states of the species involved in the cell reaction. Some of these factors are sufficiently understood to permit quantitative statements concerning their effects upon cell processes. Others, however, can be accounted for on an empirical basis only.

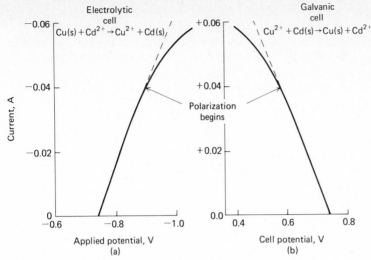

Figure 18-1 Current-Voltage Curves for Cell: (a) Cu|Cu²⁺(1.00 *M*)| |Cd²⁺(1.00 *M*)|Cd, (b) Cd|Cd²⁺(1.00 *M*)| |Cu²⁺(1.00 *M*)|Cu.

For purposes of discussion, polarization phenomena are conveniently classified into the two categories of *concentration polarization* and *kinetic polarization* (or *overvoltage*).

CONCENTRATION POLARIZATION

When the reaction at an electrode is rapid and reversible, the concentration of the reacting species in the layer of solution immediately adjacent to the electrode is always that which would be predicted from the Nernst equation. Thus, the cadmium ion concentration in the immediate vicinity of a cadmium electrode will always be given by

$$E = E_{Cd}^o - \frac{0.0591}{2} \log \frac{1}{[Cd^{2+}]}$$

irrespective of the concentration of this cation in the bulk of the solution. Since the reduction of cadmium ions is rapid and reversible, the concentration of this ion *in the film of liquid surrounding the electrode is determined at any instant by the potential of the cadmium electrode* at that instant. If the potential changes, there is an essentially instantaneous alteration of concentration in the film as required by the Nernst equation. This change will involve either deposition of cadmium or dissolution of the electrode.

In contrast to this substantially instantaneous surface process, the rate at which equilibrium between the electrode and the bulk of the solution is attained can be very slow, depending upon the magnitude of the current passing through the cell as well as the volume of the solution and its solute concentration.

When a sufficient potential is applied to an electrolytic cell such as the one just considered, cadmium ions are reduced, and an instantaneous current flows. For current to continue at a level predicted by Equation 18-1, however, an additional supply of the cation must be transported into the surface film surrounding the electrode at a suitable rate. If the demand for reactant cannot be met by mass transfer, concentration polarization will set in, and lowered currents must result. This type of polarization, then, occurs when the rate of transfer of reactive species between the bulk of the solution and the electrode surface is inadequate to maintain the current at the level required by Ohm's law. A departure from linearity such as that shown in curve (a) of Figure 18-1 is the consequence. Inadequate material transport will also cause concentration polarization in a galvanic cell such as that described by curve (b) in Figure 18-1; here, however, the current would be limited by the transport of copper ions.

Ions or molecules can be transported through a solution by (1) diffusion, (2) electrostatic attraction or repulsion, and (3) mechanical or convection forces. We must therefore consider briefly the variables that influence these forces as they relate to electrode processes.

Whenever a concentration gradient develops in a solution, molecules or ions diffuse from the more concentrated to the more dilute regions. The rate at which transfer occurs is proportional to the concentration difference. In an electrolysis a gradient is established as a result of ions being removed from the film of solution next to the cathode. Diffusion then occurs, the rate being expressed by the relationship

$$\text{rate of diffusion to cathode surface} = k(C - C_0) \qquad (18\text{-}2)$$

where C is the reactant concentration in the bulk of the solution, C_0 is its equilibrium concentration at the cathode surface, and k is a proportionality constant. *The value of C_0 is fixed by the potential of the electrode and can be calculated from the Nernst equation.* As larger potentials are applied to the electrode, C_0 becomes smaller and smaller and the diffusion rate greater and greater.

Electrostatic forces also influence the rate at which an ionic reactant migrates to or from an electrode surface. The electrostatic attraction (or repulsion) between a particular ionic species and the electrode becomes smaller as the total electrolyte concentration of the solution is increased. It may approach zero when the reactive species is but a small fraction of the total concentration of ions with a given charge.

Clearly, reactants can be transported to an electrode by mechanical means. Thus, stirring or agitation will aid in decreasing concentration polarization. Convection currents due to temperature or density differences are also effective.

To summarize, then, concentration polarization occurs when the forces of diffusion, electrostatic attraction, and mechanical mixing are insufficient to transport the reactant to or from an electrode surface at a rate demanded by the theoretical current flow. Concentration polarization causes the potential of a galvanic cell to be smaller than the value predicted on the basis of the thermodynamic potential and the IR drop. In an electrolytic cell a potential more negative than theoretical is required in order to maintain a given current.

Concentration polarization is important in several electroanalytical methods. In some applications steps are taken to eliminate it; in others, however, it is essential to the method, and every effort is made to promote its occurrence.

Experimentally, the degree of concentration polarization can be influenced by (1) the reactant concentration, (2) the total electrolyte concentration, (3) mechanical agitation, and (4) the size of the electrodes; as the area toward which a reactant is transported becomes greater, polarization effects become smaller.

KINETIC POLARIZATION

Kinetic polarization results when the rate at which the electrochemical reaction occurring at one or both electrodes is slow; here an additional potential (the *overvoltage*) is required to overcome the energy barrier to the half-reaction. In contrast to concentration polarization, the current is controlled by the *rate of the electron-transfer process* rather than by the rate of mass transfer.

While exceptions can be cited, some empirical generalizations can be made regarding the magnitude of overvoltage.

1. Overvoltage increases with current density (current density is defined as the amperes per square centimeter of electrode surface).
2. It usually decreases with increases in temperature.
3. Overvoltage varies with the chemical composition of the electrode, often being most pronounced with softer metals such as tin, lead, zinc, and particularly mercury.
4. Overvoltage is most marked for electrode processes that yield gaseous products; it is frequently negligible where a metal is being deposited or where an ion is undergoing a change of oxidation state.
5. The magnitude of overvoltage in any given situation cannot be specified exactly because it is determined by a number of uncontrollable variables.

The high overvoltage associated with the formation of hydrogen and oxygen is of particular interest to the chemist. Table 18-1 presents data that depict the extent of the phenomenon under specific conditions. The difference between the overvoltage of the gases on smooth and platinized platinum electrodes is especially noteworthy. This difference is due primarily to the much larger surface area associated with the platinum-black coating on the latter (see p. 295), which results in a smaller *real* current density than is apparent from the dimensions of the electrode. A platinized surface is always employed in construction of hydrogen reference electrodes to lower the current density and render the overvoltage negligible.

The high overvoltage associated with the formation of hydrogen permits the electrolytic deposition of several metals that require potentials at which hydrogen would otherwise be expected to interfere. For example, it is readily shown from their standard potentials that rapid formation of hydrogen should occur well before a potential sufficient for the deposition of zinc from a neutral solution is reached. In fact, however, a quantitative deposition of the metal can be achieved, provided a mercury or a copper electrode is used; because of the high overvoltage of hydrogen on these metals, little or no gas is evolved during the process.

The magnitude of overvoltage can, at best, be only crudely approximated from empirical information available in the literature. Calculation of cell potentials in which overvoltage plays a part cannot be very accurate. As with *IR* drop, the overvoltage is subtracted from the theoretical cell potential.

TABLE 18-1   **Kinetic Polarization Effects for Hydrogen and Oxygen Formation at Various Electrodes at 25°C[a]**

Electrode Com-position	Overvoltage, V (Current Density 0.001 A/cm^2)		Overvoltage, V (Current Density 0.01 A/cm^2)		Overvoltage, V (Current Density 1 A/cm^2)	
	H_2	O_2	H_2	O_2	H_2	O_2
Smooth Pt	0.024	0.721	0.068	0.85	0.676	1.49
Platinized Pt	0.015	0.348	0.030	0.521	0.048	0.76
Au	0.241	0.673	0.391	0.963	0.798	1.63
Cu	0.479	0.422	0.584	0.580	1.269	0.793
Ni	0.563	0.353	0.747	0.519	1.241	0.853
Hg	0.9[b]		1.1[c]		1.1[d]	
Zn	0.716		0.746		1.229	
Sn	0.856		1.077		1.231	
Pb	0.52		1.090		1.262	
Bi	0.78		1.05		1.23	

[a] National Academy of Sciences, *International Critical Tables of Numerical Data*, vol. 6, pp. 339–340. New York: McGraw-Hill Book Company, Inc., 1929. With permission.

[b] 0.556 V at 0.000077 A/cm^2; 0.929 V at 0.00154 A/cm^2.

[c] 1.063 V at 0.00769 A/cm^2.

[d] 1.126 V at 1.153 A/cm^2.

ELECTROGRAVIMETRIC METHODS

Electrolytic precipitation has been used for over a century for the gravimetric determination of metals. In most applications the metal is deposited on a weighed platinum cathode, and the increase in weight is determined. Important exceptions to this procedure include the anodic depositions of lead as lead dioxide on platinum and chloride as silver chloride on silver.

In most early applications, an electrodeposition was carried out by forcing a sufficiently large current through the cell to complete the reduction in a brief period. Current flow was ordinarily maintained by increasing the applied potential as the electrolysis proceeded. A disadvantage to the use of a constant current is that much of the selectivity inherent to electrodeposition is lost. On the other hand, if the working electrode is held at a fixed, predetermined emf, many useful separations can be performed. Both constant-current and constant cathode-potential methods are considered in the pages that follow, as well as a procedure which does not require an external source of power.

Current-Voltage Relationship during an Electrolysis

It is useful to consider the relationship between current, voltage, and time in an electrolytic cell when it is operated in three different modes: (1) the applied potential is held constant, (2) the cell current is kept constant, and (3) the potential of one of the electrodes (the working electrode) is held constant.

OPERATION OF A CELL AT A FIXED APPLIED POTENTIAL

To illustrate current-voltage relationships during an electrolysis at fixed potential, consider a cell consisting of two platinum electrodes, each with a surface area of 100 cm² , immersed in a solution that is 0.100 M with respect to copper(II) ion and 1.00 M with respect to hydrogen ion. The cell has a resistance of 0.50 ohm. When current is forced through the cell, copper is deposited upon the cathode, and oxygen is evolved at a partial pressure of 1.00 atm at the anode. The overall cell reaction can be expressed as

$$Cu^{2+} + H_2O \rightarrow Cu(s) + \tfrac{1}{2}O_2(g) + 2H^+$$

Decomposition Potential. From standard potential data for the half-reactions

$$Cu^{2+} + 2e \rightleftarrows Cu(s) \qquad E^0 = 0.34 \text{ V}$$
$$\tfrac{1}{2}O_2 + 2H^+ + 2e \rightleftarrows H_2O \qquad E^0 = 1.23 \text{ V}$$

a value of -0.92 V is obtained for the theoretical decomposition potential. No current flow would be expected at less negative potentials; at greater applied potentials, the current would theoretically be determined by the magnitude of the cell resistance.

The dotted lines in Figure 19-1 illustrate the theoretical current-voltage behavior of this cell; the decomposition potential is seen to be the intersection of two straight lines. In actuality, the current-voltage relationship for this system will more closely resemble the solid curve in Figure 19-1. Here the oxygen overvoltage at the anode has the effect of displacing the curve to more negative potentials. Moreover, a small current flows as soon as a potential is applied. A part of this current is due to the reduction of trace impurities such as oxygen and iron(III), which are inevitably present in the solution. In addition, however, small amounts of copper are deposited prior to impressment of the decomposition potential. This apparent departure from theoretical behavior arises because it was assumed that activity of copper was unity in calculating the decomposition potential. Experiments have shown, however, that the activity of a metal deposit only partially covering a platinum surface is less than 1;[1] under these circumstances, then, the behavior of the cathode is more correctly described by

$$E = E^0_{Cu^{2+}} - \frac{0.0591}{2} \log \frac{[Cu^{2+}]}{[Cu]}$$

where [Cu] is infinitely small at the outset and approaches 1 only after sufficient current has passed to coat the platinum surface completely.

Current-Changes with Time. In order for the cell reaction to occur at an appreciable rate, it is necessary to impress a potential considerably in excess of

[1] See L. B. Rogers et al., *J. Electrochem. Soc.*, **95**, 25, 33, 129 (1949); **98**, 447, 452, 457 (1951).

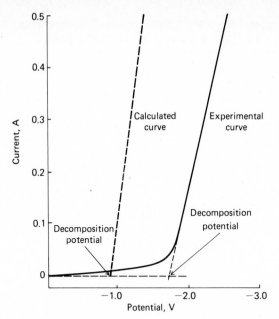

Figure 19-1 Current-Voltage Curve for the Electrolysis of a Copper(II) Solution.

the calculated theoretical value of -0.92 V. For example, operation at a current level of 1.0 A requires application of an additional 0.5 V to overcome the 0.50-ohm resistance of the cell. Moreover, account must be taken of the overvoltage of oxygen on the platinum anode. Since the anode area is 100 cm², a current density of 0.010 A/cm² will be involved under the proposed conditions of operation. From Table 18-1 we find that an overvoltage of about 0.85 V can be expected. Thus, a reasonable estimate of the potential required for the operation of this cell would be

$$E_{cell} = E_{cathode} - E_{anode} - IR - E_{overvoltage} \qquad (19\text{-}1)$$
$$= -0.92 - 0.5 - 0.85$$
$$= -2.3 \text{ V}$$

Let us now consider the changes that occur as the electrolysis proceeds at this fixed potential. As a consequence of the cell reaction, there will be a decrease in the copper ion concentration and a corresponding increase in the hydrogen ion concentration. These changes will cause both E_{anode} and $E_{cathode}$ to become less positive or more negative. Thus, when the copper concentration throughout the solution has been lowered to 10^{-6} M, the theoretical cathode potential will have decreased from $+0.31$ to $+0.16$ V; the anode potential will have changed only slightly to -1.22 V (the hydrogen ion concentration will have changed from 1.0 to 1.2 M as a result of the anode reaction). The sum of these potentials is -1.06 V as compared with the initial value of -0.92 V. Since the cell is to be operated at a fixed applied potential, it is apparent from Equation 19-1 that changes in $E_{cathode}$ and E_{anode} must be offset by corresponding decreases in IR and $E_{overvoltage}$. Long before the copper(II) concentration has been diminished to 10^{-6} M, however, the cathode will have been affected by concentration polarization; the onset of this phenomenon will have an even more profound effect on IR and $E_{overvoltage}$.

Concentration polarization occurs when copper(II) ions can no longer be brought to the electrode surface at a sufficient rate to carry the theoretical current. As a result, the current must fall, and thus decrease the magnitude of IR in Equation 19-1. The overvoltage term will also fall, since it is dependent upon current density. Figure 19-2 illustrates these changes in the cell being considered. The current drops very rapidly after a few minutes and eventually approaches zero as the electrolysis nears completion.

Cathode-Potential Changes. Even more important than the changes in current, IR drop, and overvoltage are the changes in the cathode potential induced by the onset of polarization. Recall that $E_{applied}$ in Equation 19-1 is fixed at -2.3 V; with the onset of concentration polarization, $E_{overvoltage}$ and IR drop decrease. Therefore, E_{anode}, $E_{cathode}$, or both, must become smaller. The potential of the anode, however, is stabilized at the equilibrium potential for the oxidation of water because this reactant is always present in plentiful supply at the electrode surface. Consequently, it is the cathode potential that must change to more negative values as the IR drop and the overvoltage decrease; that is, an electrode affected by polarization suffers a negative potential change. The effect is shown graphically in Figure 19-3.

The sharp change in cathode potential accompanying concentration polarization may have several consequences. First, from a prior calculation it

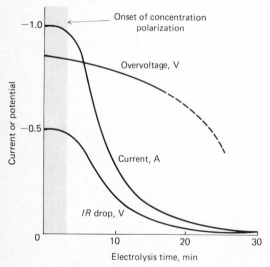

Figure 19-2 Hypothetical Cell Behavior during Electrolysis. The potential is constant at -2.3 V. The cell has a resistance of 0.5 ohm. The area of each electrode is 100 cm². Initial concentrations: $[Cu^{2+}] = 0.100$, $[H^+] = 1.0$.

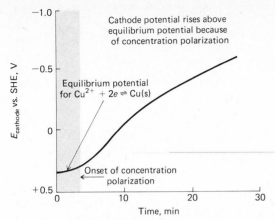

Figure 19-3 Variation in Cathode Potential during the Electrodeposition of Copper. Data are computed from Figure 19-2, assuming a fixed potential of -2.3 V and an anode potential of -1.2 V.

was established that a cathode potential of $+0.16$ V (versus the standard hydrogen electrode) was required for quantitative deposition of copper (that is, to lower the copper concentration to 10^{-6} M); from Figure 19-3 it is obvious that a value considerably more negative than this is assured. Given sufficient time, then, essentially complete removal of copper(II) ion may be expected. Second, the possibility exists for the occurrence of additional electrode reactions at the more negative cathode potentials resulting from electrode polarization. Certainly, if other ions that are reduced in the range between 0 and -0.5 V are also present, codeposition must be expected. Cobalt(II) ion, with a reduction potential of -0.25 V, and cadmium ion, which has a standard potential of -0.4 V, are examples. Another possible electrode process at the more negative potential is the formation of hydrogen. In this cell, the reduction of hydrogen ion would commence at a cathode potential of about zero volt were it not for the high overvoltage of hydrogen on the copper-plated cathode. According to Table 18-1, reduction of hydrogen ions may be expected at about -0.5 V; therefore, formation of elemental hydrogen might well be observed near the end of this electrolysis. Gas formation during an electrodeposition is often an undesirable phenomenon.

Loss of specificity, then, is a serious limitation of an electrolysis performed at constant cell potential. It is possible, of course, to reduce the negative shift in cathode potential by decreasing the applied potential chosen for the deposition. The consequence, however, is a diminution of the IR term in Equation 19-1 and a concomitant increase in the time required to complete the analysis.

At best, an electrolysis at constant cell potential can only be employed to separate an easily reduced cation from those which are more difficult to reduce than hydrogen ion. Evolution of hydrogen can also be expected near the end of the electrolysis unless certain precautions are taken to prevent it.

The formation of hydrogen during an electrodeposition frequently results in a deposit with unsatisfactory physical qualities. To avoid this problem, a substance that is more easily reduced than hydrogen ion but more difficultly reduced than the analyte is often introduced. For example, nitrate ion serves this function in the analysis of copper. Here the formation of ammonium ion occurs to the exclusion of gas evolution. The reaction is

$$NO_3^- + 10H^+ + 8e \rightleftarrows NH_4^+ + 3H_2O$$

CONSTANT-CURRENT ELECTROLYSIS

The analytical electrodeposition under consideration, as well as others, can be carried out by maintaining the current, rather than the applied potential, at a more or less constant level. Here periodic increases in the applied potential are required as the electrolysis proceeds.

In the preceding section, it was shown that concentration polarization at the cathode causes a decrease in current. Initially, this effect can be offset by increasing the applied potential; the enhanced electrostatic attraction will cause copper ions to migrate more rapidly, thus maintaining a constant current. With time, however, the solution becomes sufficiently depleted in copper ions so that the forces of diffusion and electrostatic attraction cannot keep the electrode surface supplied with sufficient copper ion to maintain the desired current. When this occurs, further increases in $E_{applied}$ cause a rapid change in the cathode potential (Equation 19-1); codeposition of hydrogen (or other reducible species) then takes place. The cathode potential ultimately becomes stabilized at a level fixed by the standard potential and the overvoltage for the new electrode reaction; further large increases in the cell potential are no longer necessary to maintain a constant current. Copper continues to deposit as copper(II) ions reach the electrode surface; the contribution of this process to the total current, however, becomes smaller and smaller as the deposition becomes more and more nearly complete. The alternative process, such as reduction of hydrogen ion, soon predominates. The changes in cathode potential under conditions of constant current are shown in Figure 19-4.

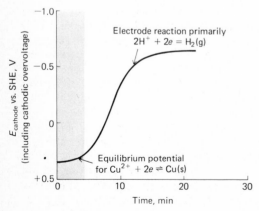

Figure 19-4   Changes in Cathode Potential during the Deposition of Copper with a Constant Current of 1.0 A.

CONSTANT CATHODE-POTENTIAL ELECTROLYSIS

From the Nernst equation it is seen that a tenfold decrease in the concentration of an ion being deposited requires a negative shift in potential of only $0.0591/n$ V. Electrogravimetric methods, therefore, are potentially highly selective. In the present example, the copper concentration is decreased from 0.1 M to 10^{-6} M as the cathode potential changes from an initial value of $+0.31$ to $+0.16$ V. In theory, then, it should be feasible to separate copper from any element that does not deposit within this 0.15-V potential range. Species that deposit quantitatively at potentials more positive than $+0.31$ V could be eliminated with a prereduction; ions that require potentials smaller than $+0.16$ V would not interfere with the copper deposition. Thus, if we are willing to accept a hundred-thousandfold lowering of concentration as a quantitative separation, it follows that univalent ions differing in standard potentials by 0.3 V or greater can, theoretically, be separated quantitatively by electrodeposition provided their initial concentrations are about the same. Correspondingly, 0.15- and 0.1-V differences are required for divalent and trivalent ions.

An approach to these theoretical separation values, within a reasonable electrolysis period, requires a more sophisticated technique than the ones thus far discussed because concentration polarization at the cathode, if unchecked, will prevent all but the crudest of separations. The change in cathode potential is governed by the decrease in IR drop. Thus, where relatively large currents are passed at the onset, the change in cathode potential can be expected to be large. On the other hand, if the cell is operated at low current levels so that the variation in cathode potential is lessened, the time required for completion of the deposition may become prohibitively long. An obvious answer to this dilemma is to initiate the electrolysis with an applied cell potential that is sufficiently high to ensure a reasonable current flow; as concentration polarization sets in, the applied potential is then continuously decreased to keep the cathode potential at the level necessary to accomplish the desired separation. Unfortunately, it is not feasible to predict the required changes in applied potential on a theoretical basis because of uncertainties in variables affecting the deposition, such as overvoltage effects and conductivity changes. Nor, indeed, does it help to measure the potential across the working electrodes, since this measures only the overall cell potential. The alternative is to measure the cathode potential against a third electrode whose potential in the solution is known and constant—that is, a reference electrode. The potential impressed across the working electrodes can then be adjusted to the level that will impart the desired potential to the cathode with respect to the reference. This technique is called *controlled cathode electrolysis*.

Experimental details for performing a controlled cathode-potential electrolysis are presented in a later section. For the present it is sufficient to note that the potential difference between the reference electrode and the cathode is measured with a potentiometer. The potential applied between the working electrodes is controlled with a voltage divider so that the cathode potential is maintained at a level suitable for the separation. Figure 19-5 shows a diagram of a typical apparatus. Calculation of the approximate cathode potential required for a separation with such an apparatus is illustrated in the following example.

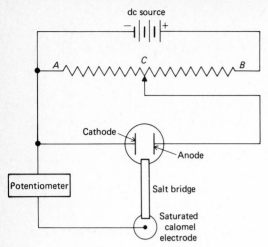

Figure 19-5 Apparatus for Electrolysis at a Controlled Cathode Potential. Contact C is continuously adjusted to maintain the cathode potential at the desired level.

Example. A solution is approximately 0.1 F in both zinc and cadmium ions. Calculate the cathode potential (with respect to a saturated calomel electrode) that should be used in order to separate these ions by electrodeposition.

From the table of standard electrode potentials, we find

$$Zn^{2+} + 2e \rightleftarrows Zn(s) \qquad\qquad E^0 = -0.76 \text{ V}$$

$$Cd^{2+} + 2e \rightleftarrows Cd(s) \qquad\qquad E^0 = -0.40 \text{ V}$$

$$Hg_2Cl_2(s) + 2e \rightleftarrows 2Hg(l) + 2Cl^-(\text{sat'd KCl}) \qquad E^0 = +0.24 \text{ V}$$

If we consider that quantitative removal has been accomplished when $[Cd^{2+}] = 10^{-6} M$, then the cathode potential needed to achieve this condition will be

$$E = -0.40 - \frac{0.0591}{2} \log \frac{1}{10^{-6}}$$
$$= -0.58 \text{ V}$$

Deposition of Zn requires a cathode potential of

$$E = -0.76 - \frac{0.0591}{2} \log \frac{1}{0.1}$$
$$= -0.79 \text{ V}$$

Thus, if the cathode is maintained between -0.58 and -0.79 V (with respect to the standard hydrogen electrode), a quantitative separation of cadmium should occur. If -0.70 V is selected, the potential of the cathode against the saturated calomel electrode should be

$$E_{\text{vs. SCE}} = -0.70 - (+0.24) = -0.94$$

To maintain this potential, a considerably larger emf is required across AC in Figure 19-5; its magnitude will depend upon the potential of the anode, the resistance of the solution, and overvoltage effects, if any.

An apparatus of the type shown in Figure 19-5 can be operated at relatively high initial applied potentials to give high currents. As the electrolysis progresses, however, a lowering of the applied potential across *AC* is required. This decrease, in turn, diminishes the current flow. Completion of the electrolysis will be indicated by the approach of the current to zero. The changes that occur in a typical constant cathode-potential electrolysis are depicted in Figure 19-6. In contrast to the electrolytic methods described earlier, this technique demands constant attention during operation. Usually some provision is made for automatic control; otherwise, the operator time required represents a major disadvantage to the controlled cathode-potential method.

Effects of Experimental Variables

In addition to potential and current, control over a number of other experimental variables is important in electrogravimetric analysis.

PHYSICAL VARIABLES THAT INFLUENCE THE PROPERTIES OF A DEPOSIT

For gravimetric purposes, an electrolytic deposit should be strongly adherent, dense, and smooth so that the processes of washing, drying, and weighing can be performed without mechanical loss or without reaction with the atmosphere. Good metallic deposits are fine-grained and have a metallic luster; spongy, powdery, or flaky precipitates are likely to be less pure and less adherent.

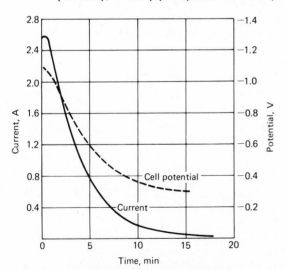

Figure 19-6 Changes in Applied Potential and Current during a Controlled Cathode-Potential Electrolysis. Deposition of copper upon a cathode maintained at −0.36 V vs. SCE. [Experimental data from J. J. Lingane, *Anal. Chim. Acta*, **2**, 589 (1949). With permission.]

The principal factors that influence the physical characteristics of deposits include competing electrode processes, current density, temperature, and stirring.

Gas Evolution. If a gas is formed during an electrodeposition, a spongy and irregular deposit is usually obtained. In cathodic reactions the usual offender is hydrogen; care must always be taken to prevent its formation by control of the cathode potential or by the addition of a so-called *depolarizer*. We have noted that the nitrate ion functions as a depolarizer in a copper analysis (p. 430).

Current Density. Electrolytic precipitates resemble chemical precipitates in that crystal size decreases as the rate of formation increases—that is, as the current density increases. Here, however, small crystal size is a desirable characteristic; metallic deposits that are smooth, strong, and adherent consist of very fine crystals.

While moderately high current densities generally give more satisfactory deposits, extremes should be avoided; very high current densities often lead to irregular precipitates with little physical strength, which develop as treelike structures from a few spots on the electrode. In addition, very high currents also lead to concentration polarization and concomitant gas formation. Ordinarily, a current density between 0.01 and 0.1 A/cm^2 is suitable for electroanalytical work.

Stirring. Stirring tends to reduce concentration polarization and is ordinarily desirable in an electrolysis.

Temperature. Although temperature may play an important part in determining the characteristics of a deposit, prediction of its effect is seldom possible. On the positive side, higher temperatures tend to inhibit concentration polarization by increasing the mobility of the ions and reducing the viscosity of the solvent. At the same time, elevated temperatures also tend to minimize over-voltage effects; increased gas formation may be observed under these circumstances. The best temperature for a given electrolysis can be determined only by experiment.

CHEMICAL VARIABLES

The success or failure of an electrolytic determination is often influenced by the chemical environment from which deposition occurs. The pH of the medium and the presence of complexing agents deserve particular mention.

Effect of pH. The pH of the solution may determine whether or not a given metal can be deposited completely. No problem is encountered with such easily reduced species as copper(II) ion or silver ion, both of which can be quantitatively removed without difficulty from quite acidic media. On the other hand, less readily reduced elements cannot be deposited from acidic solution, owing to the simultaneous evolution of hydrogen; thus, for example, neutral or alkaline media are required for the electrolytic deposition of nickel or cadmium.

Proper pH control sometimes permits the quantitative separation of cations. For example, copper is readily separated electrolytically from nickel, cadmium, or zinc in acidic solutions. Even if extreme concentration polarization occurs during the deposition of copper, the resulting change in cathode potential cannot become great enough to cause codeposition of the other metals. Hydrogen evolution or nitrate reduction will stabilize the cathode potential at a value less negative than that required to initiate deposition of these metals.

Effect of Complexing Agents. It is found empirically that many metals form smoother and more adherent films when deposited from solutions in which their ions exist primarily as complexes. The best metallic surfaces are frequently produced from solutions containing large amounts of cyanide or ammonia. The reasons for this effect are not obvious.

Deposition of a metal from a solution in which its ion exists as a complex requires a higher applied potential than in the absence of the ligand. The magnitude of this potential shift is readily calculated, provided the formation constant for the complex ion is known. The data in Table 19-1 show that these

TABLE 19-1 **Effect of Cyanide Concentration on the Cathode Potential Required for the Deposition of Certain Metals from 0.1-F Solutions**

Ion	Calculated Equilibrium Potential		
	No CN⁻ Present	In 0.1-M CN⁻	In 1-M CN⁻
Zn^{2+}	−0.79	−1.16	−1.28
Cd^{2+}	−0.43	−0.81	−0.93
Cu^{2+}	+0.31	−0.99	−1.15
Ag^+	+0.74	−0.38	−0.50

effects can be large and must be taken into account when considering the feasibility of an electrolytic determination or separation. Thus, while copper is readily separated from zinc or cadmium in acidic solution, simultaneous deposition of all three occurs in the presence of appreciable quantities of cyanide ion. The greater potential shifts for silver and copper can be directly attributed to the higher stability of their cyanide complexes.

Occasionally, electrolytic separation of ions that would ordinarily codeposit can be achieved by selective complex formation. For example, the copper in a steel sample can be deposited electrolytically from a solution containing phosphate or fluoride ions. Even though a large amount of iron is present, reduction of iron(III) ion does not occur because of the great stability of its complexes with these anions.

ANODIC DEPOSITS

Most electrogravimetric methods involve reduction of a metallic ion at the cathode. Occasionally, however, precipitates formed on an anode can also be used for analytical purposes. For example, lead is frequently oxidized to lead dioxide in nitric acid solution:

$$Pb^{2+} + 2H_2O \rightleftarrows PbO_2(s) + 4H^+ + 2e$$

The physical properties of the deposit make it a suitable weighing form for lead; a gravimetric determination is thus possible. Similarly, cobalt can be deposited and weighed as Co_2O_3.

Instrumentation

The apparatus for an analytical electrodeposition consists of a suitable cell and a direct-current power supply.

CELLS

Figure 19-7 shows a typical cell employed for the deposition of a metal on a solid electrode. Tall-form beakers are ordinarily employed, and mechanical stirring is provided to minimize concentration polarization; frequently the anode is rotated with an electric motor.

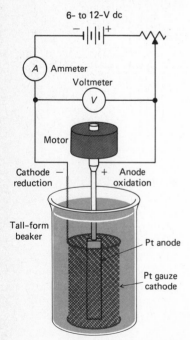

Figure 19-7 Apparatus for Electrodeposition of Metals.

make use of this relatively unrefined technique. In general, the analyte must be the only component in the solution that is more readily reduced than the hydrogen ion. Any potential interference should be eliminated by a chemical precipitation or prevented from depositing by complexation with a ligand that does not appreciably influence the electrochemical behavior of the analyte.

Constant-current deposition with a mercury cathode is also useful in removing easily reduced ions from solution prior to completion of an analysis by some other method. The deposition of interfering heavy metals prior to the quantitative determination of the alkali metals is an example of this application.

The equipment necessary for constant-current electrolysis is simple when contrasted to that required for constant cathode-potential methods.

Table 19-2 lists the common elements that can be determined by electro-

TABLE 19-2 Common Elements That Can Be Determined by Electrogravimetric Methods

Ion	Weighed As	Conditions
Cd^{2+}	Cd	Alkaline cyanide solution
Co^{2+}	Co	Ammoniacal sulfate solution
Cu^{2+}	Cu	HNO_3-H_2SO_4 solution
Fe^{3+}	Fe	$(NH_4)_2C_2O_4$ solution
Pb^{2+}	PbO_2	HNO_3 solution
Ni^{2+}	Ni	Ammoniacal sulfate solution
Ag^+	Ag	Cyanide solution
Sn^{2+}	Sn	$(NH_4)_2C_2O_4$-$H_2C_2O_4$ solution
Zn^{2+}	Zn	Ammoniacal or strong NaOH solution

gravimetric procedures for which control over the cathode potential is not required.

Controlled Electrode-Potential Methods[3]

The controlled cathode-potential method is a potent tool for the direct analysis of solutions containing a mixture of the metallic elements. Such control permits quantitative separation of elements with standard potentials that differ by only a few tenths of a volt. For example, Lingane and Jones[4] developed a method for the successive determination of copper, bismuth, lead, and tin. The first three can be deposited from a nearly neutral tartrate solution. Copper is first reduced quantitatively by maintaining the cathode potential at -0.2 V with respect to a saturated calomel electrode. After weighing, the copper-plated cathode is

---

[3] This method was first suggested by H. J. S. Sand, *Trans. Chem. Soc.*, **91**, 373 (1907). For many of its applications, see H. J. S. Sand, *Electrochemistry and Electrochemical Analysis*, vol. 2. Glasgow: Blackie & Son, Ltd., 1940. An excellent discussion of applications of automatic control to the method can be found in J. J. Lingane, *Electroanalytical Chemistry*, 2d ed., chapters 13–16. New York: Interscience Publishers, Inc., 1958. See also G. A. Rechnitz, *Controlled-Potential Analysis*. New York: The Macmillan Company, 1963.

[4] J. J. Lingane and S. L. Jones, *Anal. Chem.*, **23**, 1798 (1951).

returned to the solution, and bismuth is removed at a potential of -0.4 V. Lead is then deposited quantitatively by increasing the cathode potential to -0.6 V. Throughout these depositions the tin is retained in solution as a very stable tartrate complex. Acidification of the solution after deposition of the lead is sufficient to decompose the complex by converting tartrate ion to the undissociated acid; tin can then be readily deposited at a potential of -0.65 V. This method can be extended to include zinc and cadmium as well. Here the solution is made ammoniacal after removal of the copper, bismuth, and lead. Cadmium and zinc are then successively deposited and weighed. Finally, the tin is determined after acidification, as before.

A procedure such as this is particularly attractive for use with a potentiostat because of the small operator-time required for the complete analysis.

Table 19-3 lists some other separations that have been performed by the controlled-cathode method.

TABLE 19-3   Some Applications of Controlled Cathode-Potential Electrolysis

Element Determined	Other Elements That May Be Present
Ag	Cu and heavy metals
Cu	Bi, Sb, Pb, Sn, Ni, Cd, Zn
Bi	Cu, Pb, Zn, Sb, Cd, Sn
Sb	Pb, Sn
Sn	Cd, Zn, Mn, Fe
Pb	Cd, Sn, Ni, Zn, Mn, Al, Fe
Cd	Zn
Ni	Zn, Al, Fe

Spontaneous or Internal Electrolysis

An electrogravimetric analysis can occasionally be accomplished within a short-circuited galvanic cell. Under these circumstances, no external power is required; the deposition takes place as a consequence of the energetics of the cell reaction. For example, copper(II) ions in a solution will quantitatively deposit upon a platinum cathode when external contact is made with a zinc anode immersed in a solution of zinc ions. The cell reaction may be represented as

$$Zn(s) + Cu^{2+} \rightleftarrows Zn^{2+} + Cu(s)$$

If the reaction is allowed to proceed to equilibrium, substantially all copper(II) ions are removed from the solution. This technique is termed *internal electrolysis* or, perhaps more aptly, *spontaneous electrolysis*. Aside from simplicity insofar as equipment is concerned, it enjoys the advantage of being somewhat more selective than an ordinary electrolysis without cathode-potential control; through suitable choice of anode system the codeposition of many elements can be eliminated. Thus, for example, employment of a lead anode for the deposition of copper prevents interference from all species with more negative potentials than the lead ion–lead couple.

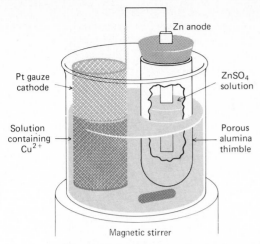

Figure 19-9 Apparatus for the Spontaneous Electrogravimetric Determination of Copper.

APPARATUS

Figure 19-9 illustrates a typical arrangement for the spontaneous electrolytic determination of copper. The element is deposited on a weighed platinum gauze cathode. A stirrer is employed to circulate the solution around the cathode.

The anode is a piece of zinc immersed in a zinc sulfate solution. This electrode must be isolated from the solution being analyzed to prevent the deposition of copper directly on the zinc. The anode may be conveniently isolated with a paper or a porous ceramic cup. A solution of zinc sulfate or some other electrolyte is placed in the cup.

Electrolysis is initiated by connecting the platinum and zinc electrodes with an external wire and continued until removal of the copper is judged complete.

A constant source of concern in a spontaneous electrolysis is the internal resistance of the cell, since this variable controls the rate at which deposition occurs. If the resistance becomes very high, the time required for completion of the reaction becomes prohibitively long. This problem does not arise in an ordinary electrolysis, where the effects of a high cell resistance can be readily overcome by increasing the applied potential. With spontaneous deposition, large currents can be obtained only if the resistance is kept small. Therefore, the apparatus for this method must always be designed with the view of minimizing the ohmic drop through the use of large electrodes, good stirring, and reasonably high concentrations of electrolyte. Depositions can be completed in less than an hour by this method under ideal conditions; often, however, several hours are required. Long deposition times are not necessarily a serious handicap, since the cell requires no attention during operation.

APPLICATIONS

Table 19-4 lists some of the applications of the internal-electrolysis method.

TABLE 19-4   Applications of the Internal-Electrolysis Method

Element Determined	Anode	Noninterfering Elements
Ag	Cu, $CuSO_4$	Cu, Fe, Ni, Zn
Cu	Zn, $ZnCl_2$	Ni, Zn
Bi	Mg, $MgCl_2$	—
Pb	Zn, $ZnCl_2$	Zn
Ni	Mg, $MgSO_4$	—
Co	Mg, NH_4Cl, HCl	—
Cd	Zn, $ZnCl_2$	Zn
Zn	Mg, NH_4Cl, HCl	—

PROBLEMS

*1. Bismuth is to be deposited at a cathode from a solution that is 0.150 M in BiO^+ and 0.600 F in $HClO_4$. Oxygen is evolved at a pressure of 0.800 atm at a 20-cm^2 platinum anode. The cell has a resistance of 1.30 ohms.
    (a) Calculate the thermodynamic (zero current) potential of the cell.
    (b) Calculate the IR drop if a current of 0.200 A is to be used.
    (c) Estimate the O_2 overvoltage.
    (d) Estimate the total applied potential required to begin operation at the specified conditions.
    (e) What potential will be required when the BiO^+ concentration is 0.0800 M?

2. Nickel is to be deposited at a cathode from a solution that is 0.200 M in Ni^{2+} and 0.400 F in $HClO_4$. Oxygen is evolved at a pressure of 0.800 atm at a 15-cm^2 platinum anode. The cell has a resistance of 2.10 ohms.
    (a) Calculate the thermodynamic (zero current) potential of the cell.
    (b) Calculate the IR drop if a current of 0.150 A is to be used.
    (c) Estimate the O_2 overvoltage.
    (d) Estimate the total applied potential required to begin operation at the specified conditions.
    (e) What potential will be required when the Ni^{2+} concentration is 0.100 M?

*3. It is desired to separate and determine bismuth and lead in a solution that is 0.0800 M in BiO^+, 0.0500 M in Pb^{2+}, and 1.00 F in $HClO_4$.
    (a) Using 1.00×10^{-6} M as the criterion for quantitative removal, determine whether or not a separation is feasible by controlled cathode-potential electrolysis.
    (b) If a separation is feasible, evaluate the range (vs. SCE) within which the cathode potential should be controlled.
    (c) What potential should be employed to deposit the second ion quantitatively after removal of the first?

4. It is desired to separate and determine bismuth, copper, and silver in a solution that is 0.0800 M in BiO^+, 0.242 M in Cu^{2+}, 0.106 M in Ag^+, and 1.00 F in $HClO_4$.
    (a) Using 1.00×10^{-6} M as the criterion for quantitative removal, determine whether or not separation of the three species is feasible by controlled cathode-potential electrolysis.
    (b) If separations are feasible, evaluate the range (vs. SCE) within which the cathode potential should be controlled for each.
    (c) What potential should be employed to deposit the third ion quantitatively after removal of the first two?

*5. Halide ions can be deposited at a silver anode, the reaction being

$$Ag(s) + X^- \rightarrow AgX(s) + e$$

(a) Determine whether or not it is theoretically feasible to separate I^- and Br^- ions from a solution that is 0.0400 M in each ion by controlling the silver anode potential. Take 1.00 × 10⁻⁶ M as the criterion of quantitative removal of one ion.

(b) Is a separation of Cl^- and I^- theoretically feasible?

(c) If a separation is feasible in either (a) or (b), what range of anode potentials (vs. SCE) should be employed?

6. What cathode potential (vs. SCE) would be required to lower the total nickel concentration of the following solutions to 1 × 10⁻⁵ F:

(a) a perchloric acid solution of Ni^{2+}?

(b) a solution having an equilibrium CN^- concentration of 0.0100 M?

$$Ni(CN)_4{}^{2-} + 2e \rightleftarrows Ni(s) + 4CN^- \qquad E^0 = -0.82$$

(c) a solution having an equilibrium Y^{4-} concentration of 1.00 × 10⁻² M where Y^{4-} is the anion of EDTA?

*7. What cathode potential (vs. SCE) would be required to lower the Hg^{2+} concentration of the following solutions to 1.00 × 10⁻⁶ F:

(a) an aqueous solution of Hg^{2+}?

(b) a solution with an equilibrium SCN^- concentration of 0.100 M?

$$Hg^{2+} + 2SCN^- \rightleftarrows Hg(SCN)_2(aq) \qquad K_f = 1.8 \times 10^7$$

(c) a solution with an equilibrium Br^- concentration of 0.250 M?

$$HgBr_4{}^- + 2e \rightleftarrows Hg(l) + 4Br^- \qquad E^0 = 0.223 \text{ V}$$

*8. The cathode compartment of an internal-electrolysis cell (see Figure 19-9) contained 50.0 ml of 0.200-F Cu^{2+} and a copper electrode. A zinc electrode immersed in 25.0 ml of 5.00 × 10⁻⁴ F Zn^{2+} served as the anode. The cell resistance was 7.5 ohms.

(a) What was the initial potential of this cell if no current was drawn?

(b) What was the initial current when the electrodes were short-circuited with a conductor?

(c) What was the cell potential when the Cu^{2+} concentration had been decreased to 1.00 × 10⁻⁵ M?

(d) What was the theoretical current when the Cu^{2+} concentration was 1.00 × 10⁻⁵ M if the cell resistance did not change? Is it likely that the observed current would be this large? Explain.

9. An internal-electrolysis cell (see Figure 19-9) consisted of a silver cathode immersed in 50.0 ml of 0.200-M $AgNO_3$ and a copper anode in 20 ml of 1.00 × 10⁻⁴ F Cu^{2+}. The resistance of the cell was 8.0 ohms.

(a) What was the initial potential of the cell if no current was drawn?

(b) What was the potential after the silver ion concentration was lowered to 1.00 × 10⁻³ M?

(c) What was the initial current when the electrodes were short-circuited?

(d) What was the theoretical current when the Ag^+ concentration had been decreased to 1.00 × 10⁻³ M?

10. The cathode compartment of an internal-electrolysis cell (see Figure 19-9) contained 100 ml of 0.100-M Co^{2+} and a cobalt electrode. A magnesium electrode immersed in 20.0 ml of 1.00×10^{-3} M Mg^{2+} served as the anode. The cell resistance was found to be 7.5 ohms.
   (a) What was the initial potential of this cell if no current was drawn?
   (b) What was the initial current when the electrodes were short-circuited with a conductor?
   (c) What was the cell potential when the Co^{2+} concentration had been lowered to 1.00×10^{-5} M?
   (d) What was the theoretical current when the Co^{2+} concentration was 1.00×10^{-5} M if the cell resistance did not change? Is it likely that the observed current would be this large? Explain.

COULOMETRIC METHODS OF ANALYSIS

Coulometry encompasses a group of analytical methods which involve measuring of the quantity of electricity (in coulombs) needed to convert the analyte to a different chemical state. In common with the gravimetric method, coulometry offers the advantage that the proportionality constant between the measured quantity in the analysis and concentration can be derived from known physical constants; thus, a calibration or standardization step is not ordinarily required. Coulometric methods are often as accurate as gravimetric or volumetric procedures; they are usually faster and more convenient than the former. Additionally, coulometric procedures are readily adapted to automation.[1]

The Measurement of Quantity of Electrons

UNITS

The quantity of electricity is measured in terms of the *coulomb* and the *faraday*. The coulomb is that amount of electricity that flows during the passage of a

---

[1] For summaries of coulometric methods, see H. L. Kies, *J. Electroanal. Chem.*, **4,** 257 (1962); J. J. Lingane, *Electroanalytical Chemistry*, 2d ed., chapters 19–21. New York: Interscience Publishers, Inc., 1958; G. W. C. Milner and G. Phillips, *Coulometry in Analytical Chemistry*. New York: Pergamon Press, 1967.

constant current of one ampere for one second. Thus, for a constant current of i amperes flowing for t seconds, the number of coulombs, q, is given by the expression

$$q = i \times t \qquad (20\text{-}1)$$

For a variable current, the number of coulombs is given by the integral

$$q = \int_0^t idt \qquad (20\text{-}2)$$

The faraday is the quantity of electricity that will produce one equivalent of chemical change at an electrode. Since the equivalent in an oxidation-reduction reaction corresponds to the change brought about by one mole of electrons, the faraday is equal to 6.02×10^{23} electrons. One faraday is also equal to 96,493 coulombs.

Example. A constant current of 0.800 A was passed through a solution for 15.2 min. Calculate the grams of copper deposited at the cathode and the grams of O_2 evolved at the anode, assuming that these are the only products formed.

From the equivalent weights are determined from consideration of the two half-reactions

$$Cu^{2+} + 2e \rightarrow Cu(s)$$
$$2H_2O \rightarrow 4e + O_2(g) + 4H^+$$

From Equation 20-1, we find

$$\text{quantity of electricity} = 0.800 \text{ A} \times 15.2 \text{ min} \times 60 \text{ sec/min}$$
$$= 729.6 \text{ A sec} = 729.6 \text{ coulombs}$$

or

$$\frac{729.6 \text{ coulombs}}{96,493 \text{ coulombs/faraday}} = 7.56 \times 10^{-3} \text{ faraday}$$

From the definition of the faraday, 7.56×10^{-3} eq of copper is deposited on the cathode; a similar quantity of oxygen is evolved at the anode. Therefore,

$$\text{wt Cu} = 7.56 \times 10^{-3} \text{ eq Cu} \times \frac{63.5 \text{ g Cu/mole}}{2 \text{ eq Cu/mole}}$$
$$= 0.240 \text{ g}$$

and

$$\text{wt O}_2 = 7.56 \times 10^{-3} \text{ eq O}_2 \times \frac{32.0 \text{ g O}_2\text{/mole}}{4 \text{ eq O}_2\text{/mole}}$$
$$= 0.0605 \text{ g}$$

INSTRUMENTS FOR MEASURING QUANTITY OF ELECTRICITY

Several devices can be used for the accurate measurement of coulombs. Such devices are ordinarily placed in series with the power supply and the coulometric cell (see Figure 20-1).

Coulomb Measurements with Constant Current. Some coulometric methods employ a power supply which provides a constant current flow through the cell for extended periods. Here, i and t in Equation 20-1 are measured

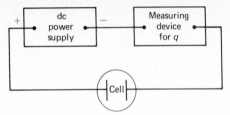

Figure 20-1 Components of Instruments for Coulometric Analyses.

independently. That is, the measuring device shown in Figure 20-1 consists of an electric stopclock to determine t, a standard resistor in series with the power supply, and a potentiometer to measure the potential drop across the resistor. Substitution for i (Ohm's law) in Equation 20-1 permits calculation of q. That is,

$$q = \frac{E}{R} t$$

where E is the measured potential drop across the resistor of resistance R ohms. An instrument employing this arrangement is shown in Figure 20-4.

Chemical Coulometer. In some applications, the current is not held constant; here an integrating device is required to determine q (Equation 20-2). Chemical coulometers function in this way.

Figure 20-2 shows a common type of chemical coulometer, which was designed by Lingane.[2] It consists of a thermostatted tube equipped with a stopcock and a pair of platinum electrodes. The tube is connected to a buret by a rubber tubing; both are filled with 0.5-F K_2SO_4. Passage of current through this device causes the liberation of hydrogen at the cathode and oxygen at the anode. Both gases are collected, and their total volume is measured by determining the volume of liquid displaced. The water jacket and thermometer provide a means of ascertaining the gas temperature.

Example. The quantity of Fe^{3+} in a solution was determined by quantitative reduction to Fe^{2+} at a platinum electrode. When current ceased, the volume of hydrogen and oxygen formed in a coulometer connected in series with the working cell was 39.3 ml at 23°C; the barometric pressure was 765 mm Hg. Calculate the milligrams $Fe_2(SO_4)_3$ in the solution.

Converting the gas volume to standard conditions gives

$$V = 39.3 \text{ ml} \times \frac{765 \text{ mm Hg}}{760 \text{ mm Hg}} \times \frac{273°\text{K}}{296°\text{K}} = 36.5 \text{ ml}$$

The reactions in the coulometer are

$$4H^+ + 4e \rightarrow 2H_2(g)$$
$$2H_2O \rightleftarrows O_2(g) + 4H^+ + 4e$$

[2] J. J. Lingane, *J. Amer. Chem. Soc.*, **67,** 1916 (1945).

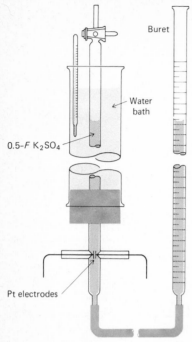

Figure 20-2 A Hydrogen-Oxygen Coulometer.

Thus 3 moles of gas are formed by passage of 4 moles of electrons, or 0.750 mole of gas is produced per faraday. Therefore,

$$\text{no. faraday} = \frac{36.5 \text{ ml gas}}{22{,}400 \text{ ml/mole}} \times \frac{1}{0.750 \text{ mole gas/faraday}}$$

$$= 2.17 \times 10^{-3}$$

$$\text{wt Fe}_2(\text{SO}_4)_3 = 2.17 \times 10^{-3} \text{ faraday} \times \frac{1.00 \text{ eq}}{\text{faraday}}$$

$$\times \frac{400 \text{ g Fe}_2(\text{SO}_4)_3}{\text{mole}} \times \frac{1}{2 \text{ eq/mole Fe}_2(\text{SO}_4)_3}$$

$$= 0.435 \text{ g}$$

Several other chemical coulometers have been developed. One, for example, is based upon the oxidation of iodide ion to triiodide ion by the current to be measured; the triiodide is then titrated with a standard solution of thiosulfate.

Electronic or Electromechanical Integrators. A strip-chart recorder can also be employed to measure coulombs when the current is variable. The area under the curve can be measured with a planimeter. Alternatively, the area can be cut out with scissors, and its weight can be compared with a piece of chart paper of known area.

Several electronic or electromechanical current integrators are available commercially, which provide a measure of the current-time integral and thus the quantity of electricity.

Types of Coulometric Methods

Two general techniques are used for coulometric analyses. The first involves maintaining the potential of the working electrode at a constant level such that quantitative oxidation or reduction of the analyte occurs without involvement of less reactive species in the sample or solvent. Here the current is initially high but decreases rapidly and approaches zero as the analyte is removed from the solution (see Figure 19-6). The quantity of electricity required is measured with a chemical coulometer or by integration of the current-time curve. A second coulometric technique makes use of a constant current that is passed until an indicator signals completion of the reaction. The quantity of electricity required to attain the end point is then calculated from the magnitude of the current and the time of its passage. The latter method has a wider variety of applications than the former; it is frequently called a *coulometric titration*.

A fundamental requirement of all coulometric methods is that the species determined interact with 100% current efficiency. That is, each faraday of electricity must bring about a chemical change corresponding to one equivalent in the analyte. This requirement does not, however, imply that the species must necessarily participate directly in the electron-transfer process at the electrode. Indeed, more often than not, the substance being determined is involved wholly or in part in a reaction that is secondary to the electrode reaction. For example, at the outset of the oxidation of iron(II) at a platinum anode, all current transfer results from the reaction

$$Fe^{2+} \rightleftarrows Fe^{3+} + e$$

As the concentration of iron(II) decreases, however, concentration polarization may cause the anode potential to rise until decomposition of water occurs as a competing process. That is,

$$2H_2O \rightleftarrows O_2(g) + 4H^+ + 4e$$

The current required to complete the oxidation of iron(II) would then exceed that demanded by theory. To avoid the consequent error, an excess of cerium(III) can be introduced at the start of the electrolysis. This ion is oxidized at a lower anode potential than is water:

$$Ce^{3+} \rightleftarrows Ce^{4+} + e$$

The cerium(IV) produced diffuses rapidly from the electrode surface, where it can then oxidize an equivalent amount of iron(II):

$$Ce^{4+} + Fe^{2+} \rightarrow Ce^{3+} + Fe^{3+}$$

The net effect is an electrochemical oxidation of iron(II) with 100% current efficiency even though only a fraction of the iron(II) ions are directly oxidized at the electrode surface.

The coulometric determination of chloride provides another example of an indirect process. Here a silver electrode serves as the anode and produces silver ions when current is passed. These cations diffuse into the solution and precipitate the chloride. A current efficiency of 100% with respect to the chloride ion is achieved even though this species is neither oxidized nor reduced in the cell.

Coulometric Methods at Constant Electrode Potential

Coulometric methods employing a controlled potential were first suggested by Hickling[3] in 1942 and have been further investigated by Lingane[4] and others. The techniques are similar to electrogravimetric methods using potential control (see Chapter 19); They differ only in that a quantity of electricity is measured rather than a weight of deposit. In contrast to a coulometric titration, a single reaction at the working electrode is required, although the species being determined need not react directly.

APPARATUS AND METHODS

A controlled-potential coulometric analysis frequently requires a potentiostat (p. 438) to maintain the potential at a predetermined level. Occasionally a potentiostat is not required; by limiting the overall potential applied to the cell, the working electrode potential is kept below some maximum level where interference could occur. In either case, the current varies in the manner shown in Figure 19-6. Thus, a device which permits the calculation of coulombs from the current-time integral must be employed.

APPLICATIONS OF CONTROLLED-POTENTIAL COULOMETRIC METHODS

A coulometric analysis with controlled potential possesses all advantages of a controlled cathode-potential electrogravimetric method (Chapter 19) and, in addition, is not subject to the limitation imposed by the need for a weighable product. The technique can therefore be applied to systems that yield deposits with poor physical properties as well as to reactions that yield no solid product. For example, arsenic may be determined coulometrically by electrolytic oxidation of arsenious acid (H_3AsO_3) to arsenic acid (H_3AsO_4) at a platinum anode. Similarly, the analytical conversion of iron(II) to iron(III) can be accomplished with suitable control of the anode potential. Other species having more than one stable oxidation state can also be analyzed in this way.

The coulometric method has been advantageously applied to the deposition of metals at a mercury cathode (see p. 438) for which a gravimetric completion of the analysis is inconvenient. Excellent methods have been described for the

[3] A. Hickling, *Trans. Faraday Soc.*, **38**, 27 (1942).
[4] For a good discussion of this method, see J. J. Lingane, *Electroanalytical Chemistry*, 2d ed., pp. 450–483. New York: Interscience Publishers, Inc., 1958.

analysis of lead in the presence of cadmium, copper in the presence of bismuth, and nickel in the presence of cobalt.

The controlled-potential coulometric procedure also offers possibilities for the electrolytic determination of organic compounds. For example, Meites and Meites[5] have demonstrated that trichloroacetic acid and picric acid are quantitatively reduced at a mercury cathode whose potential is suitably controlled:

$$Cl_3CCOO^- + H^+ + 2e \rightleftarrows Cl_2HCCOO^- + Cl^-$$

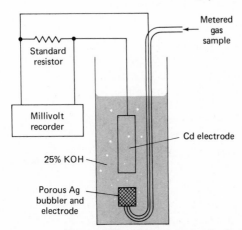

Coulometric measurements permit the analysis of these compounds with an accuracy of a few tenths of a percent.

Variable-current coulometric methods are frequently used to monitor continuously and automatically the concentration of constituents of gas or liquid streams. An important example is the determination of small concentrations of oxygen in gases or liquids.[6] A schematic diagram of the apparatus is shown in Figure 20-3. The cathode is a porous silver bubbler which also serves to break up the incoming gas into small bubbles; the reduction of oxygen takes place quantitatively within the pores. That is,

$$O_2(g) + 2H_2O + 4e \rightleftarrows 4OH^-$$

The anode is a heavy cadmium sheet; here the half-cell reaction is

$$Cd(s) + 2OH^- \rightleftarrows Cd(OH)_2(s) + 2e$$

Figure 20-3 An Instrument for Continuously Recording the O_2 Content of a Gas Stream.

[5] T. Meites and L. Meites, *Anal. Chem.*, **27**, 1531 (1955); **28**, 103 (1956).
[6] For further details, see F. A. Keidel, *Ind. Eng. Chem.*, **52**, 491 (1960).

Note that a galvanic cell is formed so that no external power supply is required. The current produced is passed through a standard resistor and the potential drop recorded on a millivolt recorder. The oxygen concentration can then be calculated by Equation 20-2. The chart paper, of course, can be made to read the instantaneous oxygen concentration directly. The instrument is reported to provide oxygen concentration data in the range from 1 ppm to 1%.

Coulometric Titrations

A coulometric titration involves the electrolytic generation of a reagent that in turn reacts with the analyte. The electrode reaction may involve only reagent preparation, as in the formation of silver ion for the precipitation of halides. In other titrations the substance being determined may also be directly involved at the generator electrode; the coulometric oxidation of iron(II)—in part by electrolytically generated cerium(IV) and in part by direct electrode reaction— is an example. Under any circumstances, the net process must approach 100% current efficiency with respect to a single chemical change in the analyte.

In contrast to the controlled-potential method, the current during a coulometric titration is carefully maintained at a constant and accurately known level; the product of this current in amperes and the time in seconds required to reach the equivalence point for the reaction yields the number of coulombs and thus the number of equivalents involved in the electrolysis. The constant-current aspect of this operation precludes the quantitative oxidation or reduction of the unknown species entirely at the generator electrode; as the solution is depleted of analyte, concentration polarization is inevitable. The electrode potential must then rise if a constant current is to be maintained. Unless this potential rise produces a reagent that can react with the analyte, the current efficiency will be less than 100%. Thus, in a coulometric titration, at least part (and frequently all) of the analytical reaction occurs away from the surface of the working electrode.

A coulometric titration, in common with the more conventional titration, requires some means of detecting the point of chemical equivalence. Most of the end points applicable to volumetric analysis are equally satisfactory here; the color change of indicators, potentiometric, amperometric (p. 484), and conductance measurements have all been successfully applied.

The analogy between a volumetric and a coulometric titration extends well beyond the common requirement of an observable end point. In both, the amount of unknown is determined through evaluation of its combining capacity—in the one case for a standard solution and in the other for a quantity of electricity. Similar demands are made of the reactions; that is, they must be rapid, essentially complete, and free of side reactions.

It is of interest to compare volumetric and coulometric analyses from the standpoint of equipment and methods. Figure 20-4 shows a block diagram for a coulometric apparatus; included is a constant-current source, an electric timer, a switch that simultaneously activates the stopclock and the generator circuit, and a device for measuring current. The source and the magnitude of the current are analogous to the titrant and its normality. The electric clock and switch

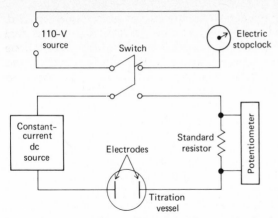

Figure 20-4   Schematic Diagram of a Coulometric Titration Apparatus.

correspond to the buret, the switch performing the same function as a stopcock. During the early phases of a coulometric titration, the switch is kept closed for extended periods; as the end point is approached, however, small additions of "reagent" are achieved by closing the switch for shorter and shorter intervals. The similarity to the operation of a buret is obvious.

Some real advantages can be claimed for a coulometric titration in comparison with the classical volumetric process. Principal among these is the elimination of problems associated with the preparation, standardization, and storage of standard solutions. This advantage is particularly important with labile reagents such as chlorine, bromine, or titanium(II) ion. Owing to their instability, these species are inconvenient as volumetric reagents; their utilization in coulometric analysis is straightforward, however, since they undergo reaction almost immediately after being generated.

Where small quantities of reagent are required, a coulometric titration offers a considerable advantage. By proper choice of current, micro quantities of a substance can be introduced with ease and accuracy, whereas the equivalent volumetric process requires small volumes of very dilute solutions, a recourse that is always difficult.

A single constant-current source can be employed to generate precipitation, oxidation-reduction, or neutralization reagents. Furthermore, the coulometric method is readily adapted to automatic titrations, since current control is easily accomplished.

Coulometric titrations are subject to five potential sources of error: (1) variation in the current during electrolysis, (2) departure of the process from 100% current efficiency, (3) error in the measurement of current, (4) error in the measurement of time, and (5) titration error due to the difference between the equivalence point and the end point. The last of these difficulties is common to volumetric methods as well; where the indicator error is the limiting factor, the two methods are likely to be comparable in reliability.

With simple instrumentation, currents constant to 0.2 to 0.5% relative are easily achieved; with somewhat more sophisticated apparatus, control to 0.01%

is obtainable. In general, then, errors due to fluctuations in current need not be serious.

Although generalizations concerning the magnitude of uncertainty associated with the electrode process are difficult, current efficiency does not appear to be the factor limiting the accuracy of many coulometric titrations.

Errors in measurement of current can be kept small. It is not difficult to determine the magnitude of even the smallest currents to 0.01% or better. Errors in the measurement of time, then, frequently represent the limiting factor in the accuracy of a coulometric titration. With a good-quality electric stopclock, however, relative errors of 0.1% or smaller can be achieved.

To summarize, then, the current-time measurements required for a coulometric titration are inherently as accurate or more accurate than the comparable volume-normality measurements of classical volumetric analysis, particularly where small quantities of reagent are involved. Often, however, the accuracy of a titration is not limited by these measurements but by the sensitivity of the end point; in this respect, the two procedures are equivalent.

APPARATUS AND METHODS

The apparatus for a coulometric titration can be relatively simple in comparison to the instrumentation required for the controlled-potential method. The basic components are shown in Figure 20-4 and discussed in the sections that follow.

Constant-Current Sources. Many constant-current sources for coulometric titrations have been described in the literature. These vary considerably in their complexity and performance characteristics. We shall consider only the simplest type; it is capable of delivering currents of about 20 mA (milliampere) that are constant to approximately 0.5%. Devices yielding currents of an ampere or greater and which vary by no more than 0.01% over extended periods of time are considerably more complex.[7]

A diagram for a constant-current source is shown in Figure 20-5. The power supply consists of two or more high-capacity, 45-V B-batteries, the current from which passes through a calibrated standard resistance R_1. A potentiometer is connected across R_1 to permit accurate measurement of the potential drop, from which the current is calculated by Ohm's law. The resistance of R_1 should be chosen so that IR_1 is about one volt; with this arrangement, a precise determination of I can be obtained with even a relatively simple potentiometer. The variable resistance R_2 has a maximum value of about 20,000 ohms.

When the circuit is completed by throwing the switch to the number 2 position, the current I passing through the cell is

$$I = \frac{E_B + E_{cell}}{R_1 + R_2 + R_B + R_{cell}}$$

[7] For a description of constant-current sources, see J. J. Lingane. *Electroanalytical Chemistry*, 2d ed., pp. 499–511. New York: Interscience Publishers, Inc., 1958.

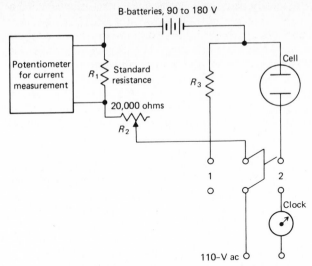

Figure 20-5 A Simple Apparatus for Coulometric Titrations.

where E_B is the potential of the B-batteries and E_{cell} comprises the cathode and anode potentials of the titration cell plus any overvoltage or junction potentials associated with its operation. The resistances of the batteries and cell are symbolized by R_B and R_{cell}, respectively.

The potential of dry cells remains reasonably constant for short periods of time, provided the current drawn is not too large; it is safe to assume, therefore, that E_B as well as R_B will remain unchanged during any given titration. Variations in I, then, arise only from changes in E_{cell} and R_{cell}. Ordinarily, however, R_{cell} will be on the order of 10 to 20 ohms while R_2 is perhaps 10,000 ohms. Thus even if R_{cell} were to change by as much as 10 ohms, which is highly unlikely, the effect on the current would be less than 1 ppt.

Changes in E_{cell} during a titration usually have a greater effect on the current, for the cell potential may be altered by as much as 0.5 V during the electrolysis. This change will cause a variation of 0.5 to 0.6% in I if E_B is 90 V; the same change would cause a variation of only about 0.3% if E_B was 180 V. Experience has shown these to be fairly realistic figures for this simple power source, *provided current is drawn more or less continuously from the battery*. To achieve this condition, the switching arrangement shown in Figure 20-5 causes the imposition of a resistance R_3 in the circuit whenever current is not passing through the cell; the magnitude of R_3 is selected to be comparable with R_{cell}.

Measurement of Time. The electrolysis time for a titration is best measured with an electric stopclock actuated by the same switch used to operate the cell. Completion of a typical titration requires between 100 and 500 sec; the clock should therefore be accurate to a few tenths of a second. An ordinary electric stopclock is not very satisfactory because the motor tends to coast when the current is shut off and also lag when started. While the error resulting from one start-stop sequence may be small, a coulometric titration may involve many such operations, and the accumulated error from this source can become appreciable. Stopclocks with solenoid-operated brakes eliminate this problem, but they are substantially more expensive than the simple laboratory timer.

Another error that may arise in connection with electric timing is caused by variations in the frequency of the 110-V power supply used to operate the clock. Ordinarily, these variations need be considered only when accuracies greater than 0.2% relative are sought.

Cells for Coulometric Titrations. A typical coulometric titration cell is shown in Figure 20-6. It consists of a generator electrode, at which the reagent is formed, and a second electrode to complete the circuit. The generator electrode, which should have a relatively large surface area, is often a rectangular strip or a wire coil of platinum; a gauze electrode such as that shown in Figure 19-7 can also be employed.

The products formed at the second electrode frequently represent potential sources of interference. For example, the anodic generation of oxidizing agents is frequently accompanied by the evolution of hydrogen from the cathode; unless this gas is allowed to escape from the solution, reaction with the oxidizing agent becomes a likelihood. To eliminate this type of difficulty, the second electrode is isolated by a sintered disk or some other porous medium.

External Generation of Reagent. Occasionally a coulometric titration cannot be used because of side reactions between the generator electrode and some

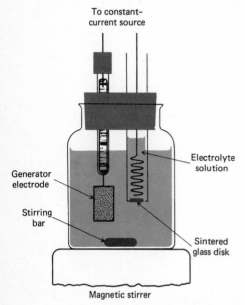

Figure 20-6 A Typical Coulometric Titration Cell.

other constituent in the solution. For example, the coulometric titration of acids involves the formation of base at the cathode:

$$2e + 2H_2O \rightleftarrows H_2(g) + 2OH^-$$

In the presence of easily reduced substances, a reaction other than the desired one can occur at the generator electrode; departures from the required 100% current efficiency thus result. Several ingenious devices for external reagent generation have been developed to overcome this problem; Figure 20-7 shows the essential features of a design by DeFord, Pitts, and Johns.[8] During electrolysis, an electrolyte solution such as sodium sulfate is fed through the tubing at a rate of about 0.2 ml/sec. The hydrogen ions formed at the anode are washed down one arm of the T-tube (along with an equivalent number of sulfate ions), while the hydroxide ions produced at the cathode are transported through the other. The apparatus is so arranged that flow of the electrolyte is discontinued whenever the electrolysis current is shut off, and a flush-out system is provided to rinse the residual reagent from the tube into the titration vessel.

Both electrode reactions shown in the illustration proceed with 100% current efficiency; thus, the solution emerging from the left arm of the apparatus can be used for the titration of acids and that from the right arm for the titration of bases. A current of about 250 mA is satisfactory for the titration of various acids or bases in the range between 0.2 and 2 meq; end points are detected with a glass-calomel electrode system. This apparatus has also been used for the electrolytic generation of iodine from iodide solutions.

APPLICATIONS OF COULOMETRIC TITRATIONS[9]

Coulometric titrations have been developed for all types of volumetric reactions. Typical applications are described in the following paragraphs.

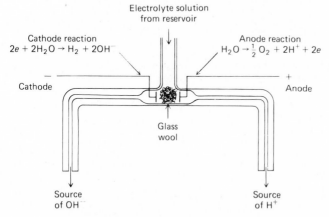

Figure 20-7   Cell for the External Generation of Acid and Base.

[8] D. D. DeFord, J. N. Pitts, and C. J. Johns, *Anal. Chem.*, **23,** 938 (1951).

[9] Applications of the coulometric procedure are summarized in J. J. Lingane, *Electroanalytical Chemistry*, 2d ed., pp. 536–613. New York: Interscience Publishers, Inc., 1958; and in H. L. Kies, *J. Electroanal. Chem.*, **4,** 257 (1962).

Neutralization Titrations. Both weak and strong acids can be titrated with a high degree of accuracy, using electrogenerated hydroxide ions. The most convenient method, where applicable, involves generation of hydroxide ion at a platinum cathode within the solution. In this application the platinum anode must be isolated by some sort of diaphragm (see Figure 20-6) to eliminate potential interference from the hydrogen ions produced. A convenient alternative involves the addition of chloride or bromide ions to the solution to be analyzed and the use of a silver wire as the anode; the reaction at this electrode then becomes

$$Ag(s) + Br^- \rightleftarrows AgBr(s) + e$$

Clearly, this anode product will not interfere with the neutralization reaction.

Both potentiometric and indicator end points can be employed for these titrations; the problems associated with the estimation of the equivalence point are identical to those encountered in the corresponding volumetric analysis. A real advantage to the coulometric method is that the carbonate problem is far less troublesome; it is necessary only to eliminate carbon dioxide from the solution to be analyzed, by aeration with a carbon dioxide–free gas, before beginning the analysis. Thus, the problems associated with the preparation and storage of a carbonate-free solution of standard base are avoided.

Coulometric titration of strong and weak bases can be performed with hydrogen ions generated at a platinum anode:

$$H_2O \rightleftarrows \tfrac{1}{2}O_2(g) + 2H^+ + 2e$$

The generation may be carried out either internally or externally; again, if the former method is used, the cathode must be isolated from the solution to prevent interference from the hydroxide ions produced at that electrode.

Precipitation and Complex-Formation Titrations. Numerous coulometric precipitation titrations are based upon anodically generated silver ions (see Table 20-1). A cell, such as that shown in Figure 20-6, can be employed with a

TABLE 20-1 **Typical Applications of Coulometric Titrations Involving Neutralization, Precipitation, and Complex-Formation Reactions**

Species Determined	Generator Electrode Reaction	Secondary Analytical Reaction
Acids	$2H_2O + 2e \rightleftarrows 2OH^- + H_2$	$OH^- + H^+ \rightleftarrows H_2O$
Bases	$H_2O \rightleftarrows 2H^+ + \tfrac{1}{2}O_2 + 2e$	$H^+ + OH^- \rightleftarrows H_2O$
Cl^-, Br^-, I^-	$Ag(s) \rightleftarrows Ag^+ + e$	$Ag^+ + Cl^- \rightleftarrows AgCl(s)$ etc.
Mercaptans	$Ag(s) \rightleftarrows Ag^+ + e$	$Ag^+ + RSH \rightleftarrows AgSR(s) + H^+$
Cl^-, Br^-, I^-	$2Hg(l) \rightleftarrows Hg_2^{2+} + 2e$	$Hg_2^{2+} + 2Cl^- \rightleftarrows Hg_2Cl_2(s)$ etc.
Zn^{2+}	$Fe(CN)_6^{3-} + e \rightleftarrows Fe(CN)_6^{4-}$	$3Zn^{2+} + 2K^+ + 2Fe(CN)_6^{4-} \rightleftarrows$ $K_2Zn_3[Fe(CN)_6]_2(s)$
$Ca^{2+}, Cu^{2+},$ Zn^{2+} and Pb^{2+}	$HgNH_3Y^{2-} + NH_4^+ + 2e \rightleftarrows$ $Hg(l) + 2NH_3 + HY^{3-}$ (where Y^{4-} is ethylene-diaminetetraacetate ion)	$HY^{3-} + Ca^{2+} \rightleftarrows CaY^{2-} + H^+$ etc.

generator electrode constructed from a piece of heavy silver wire. An adsorption indicator or a potentiometric method can be used for end-point detection. Similar applications employing mercury(I) ions formed at a mercury anode have been described.

The coulometric method has also been applied to the titration of several cations with ethylenediaminetetraacetate ion (HY^{3-}) generated at a mercury cathode.[10] In this application an excess of the mercury(II)-EDTA complex is mixed with an ammoniacal solution of the sample. The EDTA anion is then released by electrochemical reduction of the mercury(II):

$$HgNH_3Y^{2-} + NH_4^+ + 2e \rightarrow Hg(l) + 2NH_3 + HY^{3-}$$

The liberated HY^{3-} then reacts with the cation being determined. For example, with Ca^{2+} (and most other divalent cations) the reaction is

$$Ca^{2+} + HY^{3-} + NH_3 \rightarrow CaY^{2-} + NH_4^+$$

Because the mercury chelate is more stable than the corresponding complexes with calcium, zinc, lead, or copper, complexation of these ions cannot occur until the electrode process frees the complexing agent.

Oxidation-Reduction Titrations. Table 20-2 lists oxidizing and reducing

TABLE 20-2 Typical Applications of Coulometric Titrations Involving Oxidation-Reduction Reactions

Reagent	Generator Electrode Reaction	Substance Determined
Br_2	$2Br^- \rightleftarrows Br_2 + 2e$	As(III), Sb(III), U(IV), Tl(I), I^-, SCN^-, NH_3, N_2H_4, NH_2OH, phenol, aniline, mustard gas, 8-hydroxyquinoline
Cl_2	$2Cl^- \rightleftarrows Cl_2 + 2e$	As(III), I^-
I_2	$2I^- \rightleftarrows I_2 + 2e$	As(III), Sb(III), $S_2O_3^{2-}$, H_2S
Ce^{4+}	$Ce^{3+} \rightleftarrows Ce^{4+} + e$	Fe(II), Ti(III), U(IV), As(III), I^-, $Fe(CN)_6^{4-}$
Mn^{3+}	$Mn^{2+} \rightleftarrows Mn^{3+} + e$	$H_2C_2O_4$, Fe(II), As(III)
Ag^{2+}	$Ag^+ \rightleftarrows Ag^{2+} + e$	Ce(III), V(IV), $H_2C_2O_4$, As(III)
Fe^{2+}	$Fe^{3+} + e \rightleftarrows Fe^{2+}$	Cr(VI), Mn(VII), V(V), Ce(IV)
Ti^{3+}	$TiO^{2+} + 2H^+ + e \rightleftarrows Ti^{3+} + H_2O$	Fe(III), V(V), Ce(IV), U(VI)
$CuCl_3^{2-}$	$Cu^{2+} + 3Cl^- + e \rightleftarrows CuCl_3^{2-}$	V(V), Cr(VI), IO_3^-
U^{4+}	$UO_2^{2+} + 4H^+ + 2e \rightleftarrows U^{4+} + 2H_2O$	Cr(VI), Ce(IV)

agents that can be generated coulometrically and the analyses to which they have been applied. Electrogenerated bromine is particularly useful among the oxidizing agents, and many interesting methods based upon this substance have been developed. Of importance also are some of the unusual reagents not encountered in volumetric titrations because of the instability of their solutions; these include dipositive silver ion, tripositive manganese, and unipositive copper as the chloride complex.

[10] C. N. Reilley and W. W. Porterfield, *Anal. Chem.*, **28**, 443 (1956).

PROBLEMS

*1. The cadmium and zinc in a 1.06-g sample were dissolved and subsequently deposited from an ammoniacal solution with a mercury cathode. When the cathode potential was maintained at -0.95 V (vs. SCE), only the cadmium deposited. When the current ceased at this potential, a hydrogen-oxygen coulometer in series with the cell was found to have evolved 44.6 ml of gas (corrected for water vapor) at a temperature of $21.0°C$ and a barometric pressure of 773 mm of Hg. The potential was raised to about -1.3 V, whereupon zinc ion was reduced. Upon completion of this electrolysis, an additional 31.3 ml of gas were produced under the same conditions.

Calculate the percent Cd and Zn in the ore.

2. A 1.74-g sample of a solid containing $BaBr_2$, KI, and inert species was dissolved, made ammoniacal, and placed in a cell equipped with a silver anode. By maintaining the potential at -0.06 V (vs. SCE), I^- was quantitatively precipitated as AgI without interference from Br^-. The volume of H_2 and O_2 formed in a gas coulometer in series with the cell was 39.7 ml (corrected for water vapor) at $21.7°C$ and at a pressure of 748 mm of Hg.

After precipitation of iodide was complete, the solution was acidified, and the Br^- was removed from solution as AgBr, at a potential of 0.016 V. The volume of gas formed under the same conditions was 23.4 ml. Calculate the percent $BaBr_2$ and KI in the sample.

*3. The nitrobenzene in a 210-mg sample of an organic mixture was reduced to phenylhydroxylamine at a constant potential of -0.96 V (vs. SCE) applied to a Hg cathode:

$$C_6H_5NO_2 + 4H^+ + 4e \rightarrow C_6H_5NHOH + H_2O$$

The sample was dissolved in 100 ml of methanol; after electrolysis for 30 min, the reaction was judged complete. An electronic coulometer in series with the cell indicated that the reduction required 26.74 coulombs. Calculate the percent nitrobenzene in the sample.

4. At a potential of -1.0 V (vs. SCE), carbon tetrachloride in methanol is reduced to chloroform at a Hg cathode:

$$2CCl_4 + 2H^+ + 2e + 2Hg(l) \rightarrow 2CHCl_3 + Hg_2Cl_2(s)$$

At -1.80 V the chloroform further reacts to give methane:

$$2CHCl_3 + 6H^+ + 6e + 6Hg(l) \rightarrow 2CH_4 + 3Hg_2Cl_2(s)$$

A 0.750-g sample containing CCl_4, $CHCl_3$, and inert organic species was dissolved in methanol and electrolyzed at -1.0 V until the current approached zero. A coulometer indicated that 11.63 coulombs were used. The reduction was then continued at -1.80 V; an additional 44.24 coulombs were required to complete the reaction. Calculate the percent CCl_4 and $CHCl_3$ in the mixture.

*5. A 0.1309-g sample containing only $CHCl_3$ and CH_2Cl_2 was dissolved in methanol and electrolyzed in a cell containing a Hg cathode; the potential of the cathode was held constant at -1.80 V (vs. SCE). Both compounds were reduced to CH_4 (see Problem 4 for the reaction type). Calculate the percent $CHCl_3$ and CH_2Cl_2 if 306.7 coulombs were required to complete the reduction.

6. The iron in a 0.854-g sample of ore was converted to the $+2$ state by suitable treatment and then oxidized quantitatively at a Pt anode maintained at -1.0 V (vs. SCE). The quantity of electricity required to complete the oxidation was determined with a chemical coulometer equipped with a platinum anode

immersed in an excess of iodide ion. The iodine liberated by the passage of current required 26.3 ml of 0.0197-N sodium thiosulfate to reach a starch end point. What was the percentage of Fe_3O_4 in the sample?

*7. An apparatus similar to that shown in Figure 20-3 was employed to determine the oxygen content of a light hydrocarbon gas stream having a density of 0.00140 g/ml. A 20.0-liter sample of the gas consumed 3.13 coulombs of electricity. Calculate the parts per million O_2 in the sample on a weight basis.

8. The odorant concentration of household gas can be monitored by passage of a fraction of the gas stream through a solution containing an excess of bromide ion. Electrogenerated bromine reacts rapidly with the mercaptan odorant:

$$2RSH + Br_2 \rightarrow RSSR + 2H^+ + 2Br^-$$

Continuous analysis is made possible by means of an electrode system that signals the need for additional Br_2 to oxidize the mercaptan. Ordinarily, the current required to just react with the odorant is automatically plotted as a function of time. Calculate the average odorant concentration (as ppm C_2H_5SH) from the following information:

average density of the gas	0.00185 g/ml
rate of gas flow	9.4 liters/min
average current during a 10.0-min sampling period	1.35 mA

*9. Ascorbic acid (gfw = 176) is oxidized to dehydroascorbic acid with Br_2:

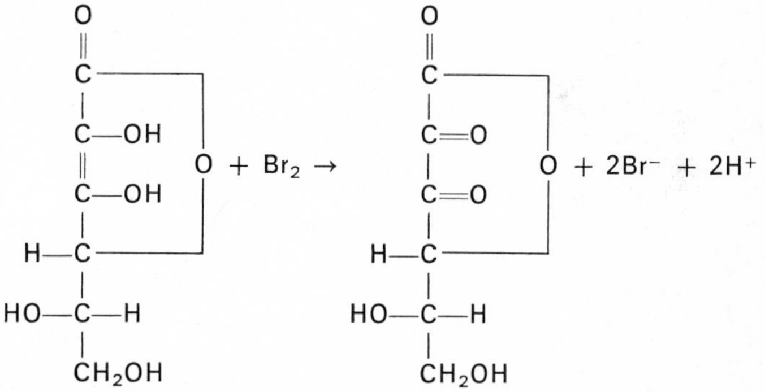

A vitamin C tablet was dissolved in sufficient water to give exactly 200 ml of solution. A 10.0-ml aliquot was then mixed with an equal volume of 0.100-F KBr. Calculate the weight of ascorbic acid in the tablet if the bromine generated by a steady current of 70.4 mA for a total of 6.51 min was required for the titration.

10. The phenol content of water downstream from a coking furnace was determined by coulometric analysis. A 100-ml sample was rendered slightly acidic and an excess of KBr was introduced. To produce Br_2 for the reaction

$$C_6H_5OH + 3Br_2 \rightarrow Br_3C_6H_2OH(s) + 3HBr$$

a steady current of 0.0313 A for 7 min and 33 sec was required. Express the results of this analysis in terms of parts phenol per million parts of water (assume that the density of water = 1.00 g/ml).

*11. The calcium content of a water sample was determined by adding an excess of an $HgNH_3Y^{2-}$ solution to a 25.0-ml sample. The anion of EDTA was then generated at a mercury cathode (see Table 20-1). A constant current of 20.1 mA was needed to reach an end point after 2 min and 56 sec. Calculate the milligrams of $CaCO_3$ per liter of sample.

12. The cyanide concentration in a 10.0-ml sample of a plating solution was determined by titration with electrogenerated hydrogen ions to a methyl orange end point. A color change occurred after 3 min and 22 sec with a current of 43.4 mA. Calculate the grams of NaCN per liter of solution.

*13. A 6.39-g sample of an ant-control preparation was decomposed by wet-ashing with H_2SO_4 and HNO_3. The arsenic in the residue was reduced to the trivalent state with hydrazine. After the excess reducing agent was removed, the arsenic(III) was oxidized with electrolytically generated I_2 in a faintly alkaline medium:

$$HAsO_3^{2-} + I_2 + 2HCO_3^- \rightarrow HAsO_4^{2-} + 2I^- + 2CO_2 + H_2O$$

The titration was complete after a constant current of 98.3 mA had been passed for 13 min and 12 sec. Express the results of this analysis in terms of the percentage As_2O_3 in the original sample.

14. The H_2S content of a water sample was assayed with electrolytically generated iodine. After 3.00 g of potassium iodide had been introduced to a 25.0-ml portion of the water, the titration required a constant current of 66.4 mA for a total of 7.25 min. Reaction:

$$H_2S + I_2 \rightarrow S(s) + 2H^+ + 2I^-$$

Express the concentration of H_2S in terms of mg/liter of sample.

*15. The equivalent weight of an organic acid was obtained by dissolving 0.0145 g of the purified compound in an alcohol-water mixture and titrating with coulometrically generated hydroxide ions. With a current of 0.0324 A, 251 sec were required to reach a phenolphthalein end point. Calculate the equivalent weight of the compound.

16. The chromium deposited on one surface of a 10.0-cm² test plate was dissolved by treatment with acid and oxidized to the $+6$ state with ammonium peroxodisulfate:

$$3S_2O_8^{2-} + 2Cr^{3+} + 7H_2O = Cr_2O_7^{2-} + 14H^+ + 6SO_4^{2-}$$

The solution was boiled to remove the excess peroxodisulfate, cooled, and then subjected to coulometric titration with Cu(I) generated from 50 ml of 0.10-F Cu^{2+}. Calculate the weight of chromium that was deposited on each square centimeter of the test plate if the titration required a steady current of 32.5 mA for a period of 7 min and 33 sec.

VOLTAMMETRY

Voltammetry comprises a group of electroanalytical procedures that are based upon the potential-current behavior of a small, easily polarized electrode in the solution being analyzed.

Historically, voltammetry developed from the discovery of *polarography* by the Czechoslovakian chemist Jaroslav Heyrovsky[1] in the early 1920s. Later in this same decade Heyrovsky and co-workers adapted the principles of polarography to the detection of end points in volumetric analyses; such methods have come to be known as *amperometric titrations*.[2]

In recent years, numerous modifications of the original polarographic method have been developed, as well as several methods that are closely related to polarography. Some of these developments are considered briefly near the end of this chapter.

[1] J. Heyrovsky, *Chem. listy*, **16,** 256 (1922). Heyrovsky was awarded the 1959 Nobel Prize in chemistry for his discovery and development of polarography.
[2] J. Heyrovsky and S. Berezicky, *Collect. Czechoslov. Chem. Commun.*, **1,** 19 (1929).

Polarography[3]

Virtually every element, in one form or another, is amenable to polarographic analysis. In addition, the method can be extended to the determination of several organic functional groups. Because the polarographic behavior of any species is unique for a given set of experimental conditions, the technique offers attractive possibilities for selective analysis.

Most polarographic analyses are performed in aqueous solution, but other solvent systems may be substituted if necessary. For quantitative analyses the optimum concentration range lies between 10^{-2} and 10^{-4} M; by suitably modifying the basic polarographic procedure, however, concentration determinations in the parts-per-billion range become possible. An analysis can be easily performed on 1 to 2 ml of solution; with a little effort, a volume as small as one drop is sufficient. The polarographic method is thus particularly useful for the determination of quantities in the milligram to microgram range.

Relative errors ranging between 2 and 3% are to be expected in routine polarographic work. This order of uncertainty is comparable to or smaller than errors affecting other methods for the analysis of small quantities.

A BRIEF DESCRIPTION OF POLAROGRAPHIC MEASUREMENTS

Polarographic data are obtained by measuring current as a function of the potential applied to a special type of electrolytic cell. A plot of the data gives current-voltage curves, called *polarograms*, which provide both qualitative and quantitative information about the composition of the solution in which the electrodes are immersed.

Polarographic Cells. A polarographic cell consists of a small, easily polarized microelectrode, a large nonpolarizable reference electrode, and the solution to be analyzed. The microelectrode, at which the analytical reaction occurs, is an inert metal surface with an area of a few square millimeters; the *dropping mercury electrode*, shown in Figure 21-1, is the most common type. Here mercury is forced by gravity through a very fine capillary to provide a continuous flow of identical droplets with a maximum diameter between 0.5 and 1 mm. The lifetime of a drop is typically 2 to 6 sec. We shall see that the dropping mercury electrode has properties that make it particularly well suited for polarographic work. Other microelectrodes, consisting of small-diameter wires or disks of platinum or other metal, can be used instead.

The reference electrode in a polarographic cell should be massive relative to the microelectrode so that its behavior remains essentially constant with the passage of small currents; that is, it should remain unpolarized during the analysis. A saturated calomel electrode and salt bridge, arranged in the manner

---

[3] The principles and applications of polarography are considered in detail in a number of monographs; see, for example, I. M. Kolthoff and J. J. Lingane, *Polarography*, 2d ed. New York: Interscience Publishers, Inc., 1952; L. Meites, *Polarographic Techniques*, 2d ed. New York: Interscience Publishers, Inc., 1965; J. Heyrovsky and J. Kůta, *Principles of Polarography*. New York: Academic Press, Inc., 1966; P. Zuman, *Organic Polarographic Analysis*. Oxford: Pergamon Press, Ltd., 1964; P. Zuman, *The Elucidation of Organic Electrode Processes*. New York: Academic Press, Inc., 1969; H. W. Nurnberg, Ed., *Electroanalytical Chemistry*, chapters 1–5. New York: John Wiley & Sons, 1974.

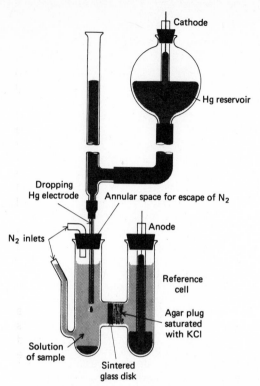

Figure 21-1 A Dropping Mercury Electrode and Cell. [From J. J. Lingane and H. A. Laitinen, *Ind. Eng. Chem., Anal. Ed.,* **11**, 504 (1939). With permission of the American Chemical Society.]

shown in Figure 21-1, is frequently employed; another common reference electrode consists simply of a large pool of mercury.

Polarograms. A polarogram is a plot of current as a function of the potential applied to a polarographic cell. The microelectrode is ordinarily connected to the negative terminal of the power supply; *by convention* the applied potential is given a negative sign under these circumstances. By convention also, the currents are designated as positive when the flow of electrons is from the power supply into the microelectrode—that is, when that electrode behaves as a cathode.

Figure 21-2 shows two polarograms. The lower one is for a solution that is 0.1 *F* in potassium chloride; the upper one is for a solution that is additionally 1×10^{-3} *F* in cadmium chloride. A step-shaped current-voltage curve, called a *polarographic wave*, is produced as a result of the reaction

$$Cd^{2+} + 2e + Hg \rightleftarrows Cd(Hg)$$

The sharp increase in current at about -2 V in both plots is associated with reduction of potassium ions to give a potassium amalgam.

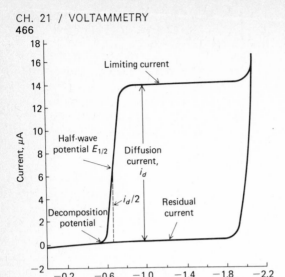

Figure 21-2 Polarogram for Cadmium Ion. The upper curve is for a solution that is $1 \times 10^{-3} F$ with respect to Cd^{2+} and 1 F with respect to KCl. The lower curve is for a solution that is 1 F in KCl only.

For reasons to be considered presently, a polarographic wave suitable for analysis is obtained only in the presence of a large excess of a *supporting electrolyte*; potassium chloride serves this function in the present example. Examination of the polarogram for the supporting electrolyte alone reveals that a small current, called the *residual current*, passes through the cell even in the absence of cadmium ions. The voltage at which the polarogram for the electrode-reactive species departs from the residual-current curve is called the *decomposition potential*.

A characteristic feature of a polarographic wave is the region in which the current, after increasing sharply, becomes essentially independent of the applied voltage; under these circumstances, it is called a *limiting current*. We shall see that the limiting current is the result of a restriction in the rate at which the participant in the electrode process can be brought to the surface of the micro-electrode; with proper control over experimental conditions, this rate is determined exclusively for all points on the wave by the velocity at which the reactant diffuses. A diffusion-controlled limiting current is given a special name, the *diffusion current*, and is assigned the symbol i_d. Ordinarily, the diffusion current is directly proportional to the concentration of the reactive constituent and is thus of prime importance from the standpoint of analysis. As shown in Figure 21-2, the diffusion current is the difference between the limiting and the residual currents.

One other important quantity, the *half-wave potential*, is the potential at which the current is equal to one-half the diffusion current. The half-wave potential is usually given the symbol $E_{1/2}$; it may permit qualitative identification of the reactant.

INTERPRETATION OF POLAROGRAPHIC WAVES

This section provides a qualitative description of the electrode phenomena that lead to the characteristic, step-shaped polarographic wave. The reduction of cadmium ion to yield cadmium amalgam at a dropping mercury electrode will be used as a specific example; the conclusions, however, are applicable to other types of electrodes, oxidation processes, and reactions that lead to other types of products.

The half-reaction

$$Cd^{2+} + Hg + 2e \rightleftarrows Cd(Hg)$$

is said to be reversible. In the context of polarography, reversibility implies that the electron transfer process is sufficiently rapid that the activities of reactants and products in the liquid film at the interface between the solution and the mercury electrode are determined by the electrode potential alone; thus, for the reversible reduction of cadmium ion, it may be assumed that at any instant the activities of reactant and product at this interface are given by

$$E_{applied} = E^0_A - \frac{0.0591}{2} \log \frac{[Cd]_0}{[Cd^{2+}]_0} - E_{ref} \qquad (21\text{-}1)$$

Here $[Cd]_0$ is the activity of metallic cadmium dissolved in the surface film of the mercury, and $[Cd^{2+}]_0$ is the activity of the ion in the aqueous solvent. Note that the subscript zero has been employed for the activity terms to emphasize that this relationship *applies to surface films of the two media only*; the activity of cadmium ion in the bulk of the solution and of elemental cadmium in the interior of the mercury drop *will ordinarily be quite different from the surface activities*. The films we are concerned with are no more than a few molecules thick. The term $E_{applied}$ is the potential applied to the dropping electrode, and E^0_A is the standard potential for the half-reaction in which a saturated cadmium amalgam is the product; the difference between E^0_A and the standard potential when the product is elemental cadmium is about +0.05 V.

Consider now what occurs when $E_{applied}$ is sufficiently negative to cause appreciable reduction of cadmium ion. Because the reaction is reversible, the activity of cadmium ion in the film surrounding the electrode decreases and the activity of cadmium in the outer layer of the mercury drop increases instantaneously to the level demanded by Equation 21-1; a momentary current results. This current would rapidly decay to zero were it not for the fact that cadmium ions are mobile in the aqueous medium and can migrate to the surface of the mercury as a consequence. Because the reduction reaction is instantaneous, *the magnitude of the current depends upon the rate at which the cadmium ions move from the bulk of the solution to the surface where reaction occurs*. That is,

$$i = k' \times v_{Cd^{2+}}$$

where i is the current at an applied potential $E_{applied}$, $v_{Cd^{2+}}$ is the rate of migration of cadmium ions, and k' is a proportionality constant.

We have noted (p. 422) that the ions or molecules in a cell migrate as a consequence of diffusion, thermal or mechanical convection, or electrostatic attraction. In polarography every effort is made to eliminate the latter two effects.

Thus vibration or stirring of the solution is avoided, and an excess of a non-reactive supporting electrolyte is employed; if the concentration of supporting electrolyte is 50 times (or more) greater than that of the reactant, the attractive (or repulsive) force between the electrode and the reactant becomes negligible.

When the forces of mechanical mixing and electrostatic attraction are eliminated, the only force responsible for the transport of cadmium ions to the electrode surface is diffusion. Since the rate of diffusion is directly proportional to the concentration (strictly activity) difference between two parts of a solution, we may write

$$v_{Cd^{2+}} = k''([Cd^{2+}] - [Cd^{2+}]_0)$$

where $[Cd^{2+}]$ is the concentration *in the bulk of the solution* from which ions are diffusing and $[Cd^{2+}]_0$ is the concentration in the aqueous film surrounding the electrode. If diffusion is the only process bringing cadmium ions to the surface, it follows that

$$i = k'v_{Cd^{2+}} = k'k''([Cd^{2+}] - [Cd^{2+}]_0)$$
$$= k([Cd^{2+}] - [Cd^{2+}]_0)$$

Note that $[Cd^{2+}]_0$ becomes smaller as $E_{applied}$ is made more negative. Thus the rate of diffusion as well as the current increases with increases in applied potential. When the applied potential becomes sufficiently negative, however, the concentration of cadmium ion in the surface film approaches zero with respect to the concentration in the bulk of the solution; under this circumstance, the rate of diffusion and thus the current becomes constant. That is, when

$$[Cd^{2+}]_0 \ll [Cd^{2+}]$$

the expression for current becomes

$$i_d = k[Cd^{2+}]$$

where i_d is the potential-independent diffusion current. Note that *the magnitude of the diffusion current is directly proportional to the concentration of the reactant in the bulk of the solution*. Quantitative polarography is based upon this fact.

When the current in a cell is limited by the rate at which a reactant can be brought to the surface of an electrode, a state of *complete concentration polarization* is said to exist. With a microelectrode, the current required to reach this condition is small—typically 3 to 10 μA (microampere) for a 10^{-3} *M* solution. Such current levels do not significantly alter the reactant concentration, as shown by the following example.

Example. The diffusion current for a 1.00×10^{-3} *M* solution of Zn^{2+} was found to be 8.4 μA. Calculate the percent decrease in the Zn^{2+} concentration after passage of this current for 8.0 min through 10.0 ml of the solution.

$$q = 8.0 \text{ min} \times 60 \frac{\text{sec}}{\text{min}} \times 8.4 \times 10^{-6} \text{ A}$$

$$= 4.03 \times 10^{-3} \text{ coulomb}$$

$$\text{no. of meq } Zn^{2+} \text{ consumed} = \frac{4.03 \times 10^{-3} \text{ coulomb}}{96,493 \text{ coulombs/faraday}} \times \frac{10^3 \text{ meq}}{\text{faraday}}$$

$$= 4.18 \times 10^{-5}$$

$$\text{no. of millimoles } Zn^{2+} \text{ consumed} = 2.09 \times 10^{-5}$$

$$\% \text{ decrease in } Zn^{2+} \text{ concentration} = \frac{2.09 \times 10^{-5}}{1.00 \times 10^{-3} \times 10} \times 100 = 0.21$$

Half-Wave Potential. An equation relating the applied potential, $E_{applied}$, and current, i, is readily derived.[4] For the reduction of cadmium ion to cadmium amalgam, the equation takes the form

$$E_{applied} = E_{1/2} - \frac{0.0591}{n} \log \frac{i}{i_d - i} \qquad (21\text{-}2)$$

where

$$E_{1/2} = E_A^0 - \frac{0.0591}{n} \log \frac{f_{Cd}k_{Cd}}{f_{Cd^{2+}}k_{Cd^{2+}}} - E_{ref} \qquad (21\text{-}3)$$

Here, f_{Cd} and $f_{Cd^{2+}}$ are activity coefficients for the metal in the amalgam and the ion in the solution; k_{Cd} and $k_{Cd^{2+}}$ are proportionality constants related to the rates at which the two species diffuse in the respective media.

Examination of Equation 21-3 reveals that the half-wave potential is a reference point on a polarographic wave; it is independent of the reactant concentration but directly related to the standard potential for the half-reaction. In practice, the half-wave potential can be a useful quantity for identification of the species responsible for a given polarographic wave.

It is important to note that the half-wave potential may vary considerably with concentration for electrode reactions that are not rapid and reversible; in addition, Equation 21-2 no longer describes such waves.

Effect of Complex Formation on Polarographic Waves. We have already seen (Chapter 14) that the potential for the oxidation or reduction of a metallic ion is greatly affected by the presence of species that form complexes with that ion. It is not surprising, therefore, that similar effects are observed with polarographic half-wave potentials. The data in Table 21-1 indicate that the half-wave potential for the reduction of a metal complex is generally more negative than that for reduction of the corresponding simple metal ion.

Lingane[5] has shown that the shift in half-wave potential as a function of the concentration of complexing agent can be employed to determine the formula and the formation constant for complexes, provided that the cation involved reacts reversibly at the dropping electrode. Thus, for the reactions

$$M^{n+} + Hg + ne \rightleftarrows M(Hg)$$

[4] I. M. Kolthoff and J. J. Lingane, *Polarography*, vol. 1, p. 190. New York: Interscience Publishers, Inc., 1952.

[5] J. J. Lingane, *Chem. Rev.*, **29**, 1 (1941).

TABLE 21-1   Effect of Complexing Agents on Polarographic Half-Wave
             Potentials at the Dropping Mercury Electrode

Ion	Noncomplexing Media	1-F KCN	1-F KCl	1-F NH$_3$, 1-F NH$_4$Cl
Cd^{2+}	-0.59	-1.18	-0.64	-0.81
Zn^{2+}	-1.00	NR[a]	-1.00	-1.35
Pb^{2+}	-0.40	-0.72	-0.44	-0.67
Ni^{2+}	—	-1.36	-1.20	-1.10
Co^{2+}	—	-1.45	-1.20	-1.29
Cu^{2+}	$+0.02$	NR[a]	$+0.04$ and -0.22	-0.24 and -0.51

[a] No reduction occurs before involvement of the supporting electrolyte.

and

$$M^{n+} + xA^- \rightleftarrows MA^{(n-x)+}$$

he derived the relationship

$$(E_{1/2})_c - E_{1/2} = -\frac{0.0591}{n} \log K_f - \frac{0.0591x}{n} \log F_A \qquad (21\text{-}4)$$

where $(E_{1/2})_c$ is the half-wave potential when the formal concentration of A is F_A while $E_{1/2}$ is the half-wave potential in the absence of complexing agent; K_f is the formation constant of the complex.

Equation 21-4 can be employed for determination of the formula of the complex. Thus, a plot of the half-wave potential against log F_A for several formal concentrations of the complexing agent gives a straight line, the slope of which is $0.0591x/n$. If n is known, the combining ratio of ligand to metal ion is thus obtained. Equation 21-4 can then be employed to calculate K_f.

Polarograms for Irreversible Reactions. Many polarographic electrode processes, particularly those associated with organic systems, are irreversible; drawn-out and less well-defined waves result. The quantitative description of such waves requires an additional term in Equation 21-3, involving the activation energy of the reaction, to account for the kinetics of the electrode process. Although half-wave potentials for irreversible reactions ordinarily show a dependence upon concentration, diffusion currents remain linearly related to this variable, and such processes are readily adapted to quantitative analysis.

THE DROPPING MERCURY ELECTRODE

Most polarographic work has been performed with a dropping mercury electrode; it is therefore of interest to consider some of the unique features of this device.

Current Variations during the Lifetime of a Drop. The current passing through a cell containing a dropping electrode undergoes periodic fluctuations corresponding in frequency to the drop rate. As a drop breaks, the current falls to zero; it then increases rapidly as the electrode area grows because of the

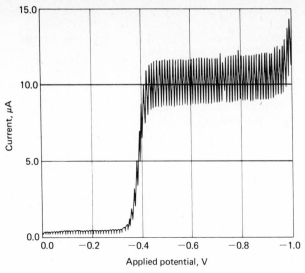

Figure 21-3 Typical Polarogram Produced by an Instrument That Continuously and Automatically Records Current-Voltage Data.

greater surface to which diffusion can occur. For convenience, a well-damped galvanometer is generally employed for current measurement. As shown in Figure 21-3, the oscillations under these circumstances are limited to a reasonable magnitude, and the average current is readily determined provided the drop rate is reproducible. Note the effect of irregular drops in the center of the limiting current region, probably caused by vibration of the apparatus.

Advantages and Limitations of the Dropping Mercury Electrode. The dropping mercury electrode offers several advantages over other types of microelectrodes. The first is the large overvoltage for the formation of hydrogen from hydrogen ions. As a consequence, the reduction of many substances from acidic solutions can be studied without interference. Second, because a new metal surface is continuously generated, the behavior of the electrode is independent of its past history. Thus, reproducible current-voltage curves are obtained regardless of how the electrode has been used previously. The third useful feature of a dropping electrode is that reproducible average currents are immediately achieved at any given applied potential.

The most serious limitation to the dropping electrode is the ease with which mercury is oxidized; this property severely restricts the use of mercury as an anode. At applied potentials much above +0.4 V (versus the saturated calomel electrode), formation of mercury(I) occurs; the resulting current masks the polarographic waves of other oxidizable species in the solution. Thus, the dropping mercury electrode can be employed only for the analysis of reducible or very easily oxidizable substances. Other disadvantages with the dropping mercury electrode are that it is cumbersome to use and tends to malfunction by clogging.

POLAROGRAPHIC DIFFUSION CURRENTS

The Ilkovic Equation. In 1934 D. Ilkovic[6] derived a fundamental equation relating the various parameters that determine the magnitude of the diffusion currents obtained with a dropping mercury electrode. He showed that at 25°C,

$$i_d = 607nD^{1/2}m^{2/3}t^{1/6}C \tag{21-5}$$

Here, i_d is the time-average diffusion current (in microamperes) during the lifetime of a drop; n is the number of faradays per mole of reactant, D is the diffusion coefficient for the reactive species expressed in units of square centimeters per second, m is the rate of mercury flow in milligrams per second, t is the drop time in seconds, and C is the concentration of the reactant in millimoles per liter. The quantity 607 represents the combination of several constants.

The Ilkovic equation contains certain assumptions that cause discrepancies of a few percent between calculated and experimental diffusion currents. Corrections to the equation have been derived that yield better correspondence;[7] for most purposes, however, the simple equation gives a satisfactory accounting of the factors that influence the current.

Capillary Characteristics. The product $m^{2/3}t^{1/6}$ in the Ilkovic equation, called the *capillary constant*, describes the influence of dropping-electrode characteristics upon the diffusion current; since both m and t are readily evaluated experimentally, comparison of diffusion currents from different capillaries is thus possible.

Two factors, other than the geometry of the capillary itself, play a part in determining the magnitude of the capillary constant. The head that forces the mercury through the capillary influences both m and t such that the diffusion current is directly proportional to the square root of the mercury column height. The drop time t of a given electrode is also affected by the applied potential since the interfacial tension between the mercury and the solution varies with the charge on the drop. Generally t passes through a maximum at about -0.4 V (versus the saturated calomel electrode) and then falls off fairly rapidly; at -2.0 V, t may be only half of its maximum value. Fortunately, the diffusion current varies only as the one-sixth power of the drop time so that, over small potential ranges, the decrease in current due to this variation is negligibly small.

Diffusion Coefficient. The Ilkovic equation indicates that the diffusion current for a particular reactant is proportional to the square root of its diffusion coefficient D. This quantity is a measure of the rate at which the species migrates through a unit concentration gradient and is dependent upon such factors as the size of the ion or molecule, its charge, if any, and the viscosity and composition of the solvent. The coefficient for a simple, hydrated metal ion is often different from that of its complexes; as a result, the diffusion current as well as the half-wave potential may be affected by the presence of a complexing reagent.

---

[6] D. Ilkovic, *Collect. Czechoslov. Chem. Commun.*, **6**, 498 (1934).
[7] See J. J. Lingane and B. A. Loveridge, *J. Amer. Chem. Soc.*, **72**, 438 (1950).

Temperature. Temperature affects several of the variables that govern the diffusion current for a given species, and its overall influence is thus complex. The most temperature-sensitive of the factors in the Ilkovic equation is the diffusion coefficient, which ordinarily can be expected to change by about 2.5% per degree. As a consequence, temperature control to a few tenths of a degree is needed for accurate polarographic analysis.

POLAROGRAMS FOR MIXTURES OF REACTANTS

Ordinarily, the reactants of a mixture will behave independently of one another at a microelectrode so that a polarogram for a mixture is simply the summation of the waves for the individual components. Figure 21-4 shows polarograms for a pair of two-component mixtures. The half-wave potentials of the two reactants differ by about 0.1 V in curve *A* and by about 0.2 V in curve *B*. Figure 21-4 suggests that a single polarogram may permit the analysis for several components in a mixture. Success depends upon the existence of a sufficient difference between succeeding half-wave potentials to permit evaluation of individual diffusion currents. Approximately 0.2 V is required if the more reducible species undergoes a two-electron reduction; a minimum between 0.2 and 0.3 V is needed if the first reduction is a one-electron process. The analysis of mixtures is considered in a later section.

ANODIC WAVES AND MIXED ANODIC-CATHODIC WAVES

Anodic waves as well as cathodic waves are encountered in polarography. The former are less common because of the relatively small range of anodic potentials that can be covered with the dropping mercury electrode before

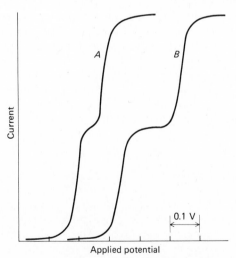

Figure 21-4 Polarograms for Two-Component Mixtures. Half-wave potentials differ by 0.1 V in curve *A* and by 0.2 V in curve *B*.

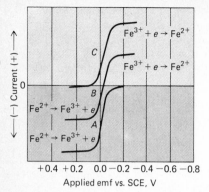

Figure 21-5 Polarographic Behavior of Iron(II) and Iron(III) in a Citrate Medium. Curve A: anodic wave for a solution in which $[Fe^{2+}] = 1 \times 10^{-3}$. Curve B: anodic-cathodic wave for a solution in which $[Fe^{2+}] = [Fe^{3+}] = 0.5 \times 10^{-3}$. Curve C: cathodic wave for a solution in which $[Fe^{3+}] = 1 \times 10^{-3}$.

oxidation of the electrode itself commences. An example of an anodic wave is illustrated in curve A of Figure 21-5, where the electrode reaction involves the oxidation of iron(II) to iron(III) in the presence of citrate ion. A diffusion current is obtained at zero volt (versus the saturated calomel electrode), which is due to the half-reaction

$$Fe^{2+} \rightleftarrows Fe^{3+} + e$$

As the potential is made more negative, a decrease in the anodic current occurs; at about -0.2 V the current becomes zero because the oxidation of iron(II) ion has ceased.

Curve C represents the polarogram for a solution of iron(III) in the same medium. Here a cathodic wave results from the reduction of the iron(III) to the divalent state. The half-wave potential is identical with that for the anodic wave, indicating that the oxidation and reduction of the two iron species are perfectly reversible at the dropping electrode.

Curve B is the polarogram of an equiformal mixture of iron(II) and iron(III). The portion of the curve below the zero-current line corresponds to the oxidation of the iron(II); this reaction ceases at an applied potential equal to the half-wave potentiai. The upper portion of the curve is due to the reduction of iron(III).

CURRENT MAXIMA

The shapes of polarograms are frequently distorted by so-called *current maxima* (see Figure 21-6). These are troublesome because they interfere with the accurate evaluation of diffusion currents and half-wave potentials. Although the cause or causes of maxima are not fully understood, considerable empirical knowledge of methods for their elimination exists. These generally involve the addition of traces of such high-molecular-weight substances as gelatin, Triton X-100 (a commercial surface-active agent), methyl red, other dyes, or carpenter's glue. The first two of these additives are particularly useful.

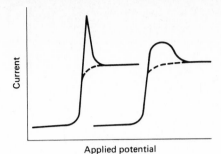

Applied potential

Figure 21-6   Typical Current Maxima.

RESIDUAL CURRENT

The typical residual current curve, such as that shown in Figure 21-2, has two sources. The first of these is the reduction of trace impurities that are almost inevitably present in the blank solution; contributors here include small amounts of dissolved oxygen, heavy-metal ions from the distilled water, and impurities present in the salt used as the supporting electrolyte. Generally, the concentrations of these contaminants are not great enough to produce a distinct wave; their presence does, however, affect the residual current. The purity of the salt employed as a supporting electrolyte is particularly important. For example, if the solution to be studied is made 1.0 F in potassium nitrate, as little as 0.001% of reducible impurity in that salt may contribute appreciably to the residual current.

A second component of the residual current is the so-called *charging*, or *condenser*, *current* resulting from a flow of electrons that charges the mercury droplets with respect to the solution; this current may be either negative or positive. At potentials more negative than -0.4 V (versus the saturated calomel electrode), an excess of electrons provides the surface of each droplet with a negative charge. These excess electrons are carried down with the drop as it breaks; since each new drop is charged as it forms, a small but steady current results. At applied potentials smaller than about -0.4 V, the mercury tends to be positive with respect to the solution; thus, as each drop is formed, electrons are repelled from the surface toward the bulk of mercury, and a negative current is the result. At about -0.4 V, the mercury surface is uncharged, and the condenser current is zero.

Ultimately, the accuracy and the sensitivity of the polarographic method depends upon the magnitude of the residual current and the accuracy with which a correction for its effect can be determined. Two procedures for correction are employed. One requires a polarogram of a blank solution that is as nearly identical as possible to the solution being analyzed except for the component of interest; the diffusion current is then taken as the arithmetic difference between the current for the sample and the blank at an identical potential. In the second procedure the correction is determined from a linear extrapolation of the polarogram of the sample in the region short of the decomposition potential (see Figure 21-9). The first method is usually the more accurate.

SUPPORTING ELECTROLYTE

An electrode process will be diffusion-controlled only if the solution contains a sufficient concentration of a supporting electrolyte. The data in Table 21-2

TABLE 21-2   Effect of Supporting Electrolyte Concentration on Polarographic Currents for Lead Ion[a,b]

Potassium Nitrate Concentration, F	Limiting Current, μA
0	17.6
0.0001	16.2
0.001	12.0
0.005	9.8
0.10	8.45
1.00	8.45

[a] Data from J. J. Lingane and I. M. Kolthoff, *J. Amer. Chem. Soc.*, **61**, 1045 (1939). With permission of the American Chemical Society.
[b] Solution: $9.5 \times 10^{-4}\ F$ in $PbCl_2$.

illustrate the effect of the supporting electrolyte concentration upon the limiting current. It is seen that the limiting current for lead ion decreases markedly with the addition of potassium nitrate and becomes constant only in the presence of high concentrations of that salt. In solutions of low electrolyte concentration, the portion of the limiting current that is due to electrostatic forces is sometimes called the *migration current*; for the first entry in Table 21-2, the migration current associated with a $9.5 \times 10^{-4}\ F$ solution of lead ion is about $(17.6 - 8.45)$ 9.2 μA.

It is of interest to note that the limiting current for the reduction of anions (such as iodate or chromate) becomes larger as the supporting electrolyte concentration increases, since the electrostatic forces are repulsive rather than attractive.

The migration current can quite generally be eliminated by the employment of a supporting electrolyte whose concentration exceeds that of the reactive species by a factor of 50 to 100. Under these circumstances, the fraction of the current carried through the solution by the species of interest is negligibly small because of the large excess of other particles of the same charge. The limiting current then becomes a diffusion current which is independent of electrolyte concentration.

OXYGEN WAVES

Dissolved oxygen is readily reduced at the dropping mercury electrode; an aqueous solution saturated with air exhibits two distinct waves attributable to this element (see Figure 21-7). The first results from the reduction of oxygen to peroxide:

$$O_2(g) + 2H^+ + 2e \rightleftarrows H_2O_2$$

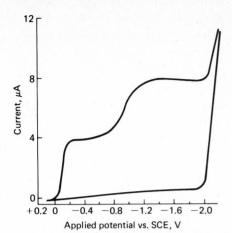

Figure 21-7 Polarogram for the Reduction of Oxygen in an Air-saturated 0.1-*F* KCl Solution. The lower curve is for 0.1-*F* KCl alone.

The second corresponds to the further reduction of the hydrogen peroxide:

$$H_2O_2 + 2H^+ + 2e \rightleftarrows 2H_2O$$

As would be expected from stoichiometric considerations, the two waves are of equal height.

While the polarographic waves are convenient for the determination of the concentration of dissolved oxygen, the presence of this element often interferes with the accurate determination of other species. Thus, oxygen removal is ordinarily the first step in a polarographic analysis. Aeration of the solution for several minutes with an inert gas accomplishes this end; a stream of the same gas, usually nitrogen, is passed over the surface during the analysis to prevent reabsorption.

KINETIC AND CATALYTIC CURRENTS

There are instances where polarographic limiting currents are controlled not only by the diffusion rate of the reactive species but also by the rate of some *chemical reaction* related to the electrode process. These currents are no longer predictable by the Ilkovic equation and are abnormally influenced by temperature, capillary characteristics, and solution composition; they are known as *kinetic currents*. The polarographic reduction of formaldehyde illustrates this behavior. In aqueous solution two forms of this compound are in equilibrium:

$$CH_2(OH)_2 \rightleftarrows HCHO + H_2O$$

The hydrated form predominates, but only the unhydrated form is reduced at a dropping electrode. Thus, when a suitable potential is applied, the concentration of the latter approaches zero at the electrode surface. This results in a shift in the equilibrium to the right and the formation of more unhydrated formaldehyde which then can react. However, the rate of the equilibrium shift is slow; thus, the supply of available reactant is controlled by a reaction rate rather than by a diffusion rate. The net effect is a smaller limiting current than would be observed if the process were totally diffusion-controlled.

A *catalytic current* represents another type of limiting current that depends upon the rate of a chemical reaction. Here the reactive species is regenerated by a chemical reaction involving some other component of the solution. The reduction of iron(III) in the presence of hydrogen peroxide provides an example of this phenomenon. The magnitude of the limiting current for iron is greatly enhanced, even when the applied potential is kept well below the value required to reduce the peroxide. This effect is readily explained by the following reaction that is thought to occur in the surface layer following the electrolytic formation of iron(II) ion:

$$2Fe^{2+} + H_2O_2 \rightarrow 2Fe^{3+} + 2OH^-$$

The current is then controlled in part by the rate of this reaction.

Both kinetic and catalytic currents can be used for analytical purposes. The latter are particularly useful for determining very small concentrations of certain species. Both are quite sensitive to the variables that influence the rates of chemical reactions.

Applications of Polarography

APPARATUS

Cells. A typical general-purpose cell for polarographic analysis is shown in Figure 21-1. The solution to be analyzed is separated from the calomel electrode by a sintered disk backed by an agar plug that has been rendered conducting by the addition of potassium chloride. Such a bridge is readily prepared and lasts for extended periods, provided a potassium chloride solution is kept in the reaction compartment when the cell is not in use. A capillary sidearm permits the passage of nitrogen or other gas through the solution. Provision is also made for blanketing the solution with the gas during the analysis.

In a simpler arrangement, a mercury pool in the bottom of the sample container can be used as the nonpolarizable electrode. Here the observed half-wave potentials differ from published values, which are based upon a saturated calomel reference electrode.

Dropping Electrodes. A dropping electrode, such as that shown in Figure 21-1, can be purchased from commercial sources.[8] A 10-cm length ordinarily has a drop time of 3 to 6 sec under a mercury head of about 50 cm. The tip of the capillary should be as nearly perpendicular to the base as possible, and care should be taken to assure a vertical mounting of the electrode; otherwise erratic and nonreproducible drop times and sizes will be observed.

With reasonable care, a capillary can be used for several months or even years. Such performance, however, requires the use of scrupulously clean

[8] For example, Sargent-Welch Scientific Co., Skokie, Ill., and Fisher Scientific Co., Pittsburgh, Pa.

mercury and the maintenance of a mercury head, no matter how slight, at all times. If solution comes in contact with the inner surface of the tip, malfunction of the electrode is to be expected. For this reason, the head of mercury should always be increased to provide a good flow before the tip is immersed in a solution.

Storage of an electrode always presents a problem. One method is to rinse the capillary thoroughly with water, dry it, and then carefully reduce the head until the flow of mercury in air just ceases. Care must be taken to avoid lowering the mercury too far. Before use, the head is increased, the tip is immersed in 1:1 nitric acid for a minute or so, and then rinsed with distilled water.

Electrical Apparatus. It is necessary to have the means for applying a voltage that can be varied continuously over the range from 0 to -2.5 V; the applied potential should usually be known to about 0.01 V. In addition, it must be possible to measure currents over the range between 0.01 and perhaps 100 μA with an accuracy of about 0.01 μA. A manual instrument that meets these requirements is easily constructed from equipment available in most laboratories. More elaborate devices for recording polarograms automatically are commercially available.

Figure 21-8 shows a circuit diagram of a simple instrument for polarographic work. Two 1.5-V batteries provide a voltage across the 100-ohm potential divider R_1, by means of which the potential applied to the cell can be varied. The magnitude of this voltage can be determined with the potentiometer when the double-pole, double-throw switch is in position 2. The current is measured by determining the potential drop across the precision 10,000-ohm resistance R_2 with the same potentiometer and the switch in position 1. The current oscillations associated with the dropping electrode require that the null-detecting galvanometer be damped with a suitable resistance.

The apparatus shown in Figure 21-8 provides satisfactory data for routine

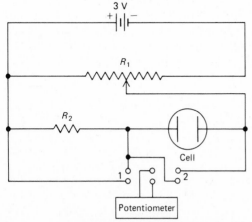

Figure 21-8 A Simple Circuit for Polarographic Measurements. [From J. J. Lingane, *Anal. Chem.*, **21**, 45 (1949). With permission of the American Chemical Society.]

quantitative work; in such applications current measurements at only two voltages are needed to define i_d (one preceding the decomposition potential and one in the limiting-current region). On the other hand, where the entire polarogram is required, a point-by-point determination of the curve with a manual instrument is tedious and time-consuming; for this type of work automatic recording is of great value.

TREATMENT OF DATA

In measuring currents obtained with the dropping electrode, it is common practice to use either the average or the maximum value of the galvanometer or recorder oscillations.

Determination of Diffusion Currents. For analytical work, limiting currents must always be corrected for the residual current. A residual-current curve is experimentally determined along with the curve for the sample. The diffusion current can be evaluated from the difference between the two at some potential in the limiting-current region. Because the residual current usually increases nearly linearly with applied voltage, it is often possible to make the correction by extrapolation; this technique is illustrated in Figure 21-9.

Analysis of Mixtures. The quantitative determination of each species in a multicomponent mixture from a single polarogram is theoretically feasible, provided the half-wave potentials of the various species are sufficiently different (see Figure 21-4). Where a major constituent is more easily reduced than a minor one, however, the accuracy with which the latter can be determined may be poor because its diffusion current can occupy only a small fraction of the current scale of the instrument. Under such circumstances, a small error in current measurement leads to a large relative error in the analysis. This problem does not arise when the minor constituent is the more easily reduced component. Here its diffusion current can be determined at high current sensitivity; then the sensitivity can be decreased to allow measurement of the major species.

Several ways exist for treating mixtures containing species in unfavorable concentration ratios. The best method is to cause the wave of the minor

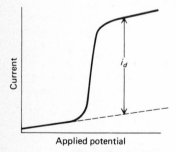

Applied potential

Figure 21-9   Determination of the Diffusion Current by Extrapolation of the Residual Current.

component to appear first through alteration of the supporting electrolyte; with the variety of complexing agents available, this technique is often feasible. Alternatively, a preliminary chemical separation can be employed. Finally, there is the so-called *compensation technique*. Here the current due to the major constituent is decreased to zero (or a very small value) by application of a counter emf in the current-measuring circuit. The current sensitivity can then be increased to give a satisfactory signal for the reduction of the minor component. Most modern polarographs are equipped with such a compensating device.

Concentration Determination. The best and most straightforward method for quantitative polarographic analysis involves preliminary calibration with a series of standard solutions; as nearly as possible, these standards should be identical with the samples being analyzed and cover a concentration range within which the unknown samples will likely fall. The linearity of the current-concentration relationship can be assessed from such data; if the relationship is nonlinear, the analysis can be based upon the calibration curve.

Another useful technique is the standard-addition method. Here the diffusion current for an accurately known volume of sample solution is measured. Then a known amount of the species of interest is introduced (ordinarily as a known volume of a standard solution), and the diffusion current is again evaluated. Provided the relationship between current and concentration is linear, the increase in wave height will permit calculation of the concentration in the original solution. The standard-addition method is particularly effective when the diffusion current is sensitive to other components of the solution that are introduced with the sample.

INORGANIC POLAROGRAPHIC ANALYSIS

The polarographic method is generally applicable to the analysis of inorganic substances. Most metallic cations, for example, are reduced at the dropping electrode to form a metal amalgam or an ion of lower oxidation state. Even the alkali metals and alkaline-earth metals are reducible, provided the supporting electrolyte does not react at the high potentials required; here the tetraalkyl ammonium halides are useful.

The successful polarographic analysis of cations frequently depends upon the supporting electrolyte that is used. To aid in this selection, tabular compilations of half-wave potential data should be consulted.[9] The judicious choice of anion often enhances the selectivity of the method. For example, with potassium chloride as a supporting electrolyte, the waves for iron(III) and copper(II) interfere with one another; in a fluoride medium, however, the half-wave potential of the former is shifted by about -0.5 V, while that for the latter is altered by only a few hundredths of a volt. The presence of fluoride thus results in the appearance of separate waves for the two ions.

The polarographic method is also applicable to the analysis of such inorganic

[9] See, for example, the references cited on page 464; another extensive source is L. Meites, Ed., *Handbook of Analytical Chemistry*. New York: McGraw-Hill Book Company, Inc., 1963.

anions as bromate, iodate, dichromate, vanadate, selenite, and nitrite. In general, polarograms for these substances are affected by the pH of the solution because the hydrogen ion is a participant in the reduction process. As a consequence, strong buffering of the solutions to some fixed pH is necessary to obtain reproducible data.

Certain inorganic anions that form complexes or precipitates with the ions of mercury are responsible for anodic waves that occur in the region of zero volt (versus the saturated calomel electrode). Here the electrode reaction involves oxidation of the electrode; for example,

$$2Hg(l) + 2Cl^- \rightleftarrows Hg_2Cl_2 + 2e$$
$$Hg(l) + 2S_2O_3{}^{2-} \rightleftarrows Hg(S_2O_3)_2{}^{2-} + 2e$$

Bromide, iodide, thiocyanate, and cyanide act similarly. The magnitude of such diffusion currents is controlled by the rate at which the anions migrate to the electrode surface; as a consequence, a linear relationship between current and anion concentration is observed.

The polarographic method can be used for the analysis of a few inorganic substances that exist as uncharged molecules in the solvent used. The determination of oxygen in gases, biological fluids, and water is a most important example of this application. Other neutral substances that react at the dropping electrode include hydrogen peroxide, hydrazine, cyanogen, elemental sulfur, and sulfur dioxide.

For further applications of polarography to inorganic analysis, the reader is referred to the monographs by Kolthoff and Lingane and by Meites.[10]

ORGANIC POLAROGRAPHIC ANALYSIS

Almost from its inception the polarographic method has been used for the study and analysis of organic compounds, and many papers have been devoted to this subject. Several common functional groups are oxidized or reduced at the dropping electrode; compounds containing these groups are thus subject to polarographic analysis.

In general, the reactions of organic compounds at a microelectrode are slower and more complex than those for inorganic cations. Thus, theoretical interpretation of the polarographic data is more difficult or even impossible; moreover, a much stricter adherence to detail is required for quantitative work. Despite these handicaps, organic polarography has proved fruitful for the determination of structure, the qualitative identification of compounds, and the quantitative analysis of mixtures.

Effect of pH on Polarograms. Organic electrode processes ordinarily involve hydrogen ions, the most common reaction being represented as

$$R + nH^+ + ne = RH_n$$

[10] I. M. Kolthoff and J. J. Lingane, *Polarography*, 2d ed. vol. 2. New York: Interscience Publishers, Inc., 1952; L. Meites, *Polarographic Techniques*. 2d ed. New York: Interscience Publishers, Inc., 1965.

where R and RH_n are the oxidized and reduced forms of the organic molecule. Half-wave potentials for organic compounds are therefore markedly pH-dependent. Furthermore, alteration of the pH often results in a change in the reaction products. For example, when benzaldehyde is reduced in a basic solution, a wave is obtained at about -1.4 V, attributable to the formation of benzyl alcohol:

$$C_6H_5CHO + 2H^+ + 2e \rightleftarrows C_6H_5CH_2OH$$

If the pH is less than 2, however, a wave occurs at about -1.0 V that is just half the size of the foregoing one. Here the reaction involves the production of hydrobenzoin:

$$2C_6H_5CHO + 2H^+ + 2e \rightleftarrows C_6H_5CHOHCHOHC_6H_5$$

At intermediate pH values two waves are observed, indicating the occurrence of both reactions.

It should be emphasized that an electrode process consuming or producing hydrogen ions tends to alter the pH of the solution *at the electrode surface*; unless the solution is well buffered, marked changes in pH can occur in the surface film during the electrolysis. These changes affect the reduction potential of the reaction and cause drawn-out and poorly defined waves. Moreover, where the electrode process is altered by the pH, nonlinearity in the diffusion current—concentration relationship must also be expected. Thus, in organic polarography good buffering is vital for the generation of reproducible half-wave potentials and diffusion currents.

Solvents for Organic Polarography. Solubility considerations frequently dictate the use of some solvent other than pure water for organic polarography; aqueous mixtures containing varying amounts of such miscible solvents as glycols, dioxane, alcohols, Cellosolve, or glacial acetic acid have been employed. Anhydrous media, such as acetic acid, formamide, and ethylene glycol have also been investigated. Supporting electrolytes are often lithium salts or tetraalkyl ammonium salts.

Irreversibility of Electrode Reactions. Few organic electrode reactions are reversible; thus, Equation 21-3 does not adequately describe their polarographic waves. Nonreversibility results in drawn-out waves; as a result, greater differences in half-wave potentials are required to permit discrimination between substances undergoing consecutive reduction.

In general, the Ilkovic equation applies to the diffusion currents for electrode reactions that are nonreversible. Thus, the quantitative aspects of organic and inorganic polarography are similar.

Reactive Functional Groups. Organic compounds containing any of the following functional groups can be expected to produce one or more polarographic waves.

1. *The carbonyl group*, including aldehydes, ketones, and quinones, produce polarographic waves. In general, aldehydes are reduced at lower potentials than ketones; conjugation of the carbonyl double bond also results in lower half-wave potentials.

2. *Certain carboxylic acids* are reduced polarographically, although simple aliphatic and aromatic monocarboxylic acids are not. Dicarboxylic acids such as fumaric, maleic, or phthalic acid, in which the carboxyl groups are conjugated with one another, give characteristic polarograms; the same is true of certain keto and aldehydo acids.

3. *Most peroxides and epoxides* yield polarograms.

4. *Nitro, nitroso, amine oxide, and azo groups* are generally reduced at the dropping electrode.

5. *Most organic halogen groups* produce a polarographic wave which results from replacement of the halogen group with an atom of hydrogen.

6. *The carbon-carbon double bond* is reduced when it is conjugated with another double bond, an aromatic ring, or an unsaturated group.

7. *Hydroquinones and mercaptans* produce anodic waves.

In addition, a number of other organic groups cause catalytic hydrogen waves that can be used for analysis. These include amines, mercaptans, acids, and heterocyclic nitrogen compounds. Numerous applications to biological systems have been reported.[11]

Amperometric Titrations

The polarographic method can be employed for the estimation of the equivalence point, provided at least one of the participants or products of the titration is oxidized or reduced at a microelectrode. Here the current passing through a polarographic cell at some fixed potential is measured as a function of the reagent volume (or of time if the reagent is generated by a constant-current coulometric process). Plots of the data on either side of the equivalence point are straight lines with differing slopes; the end point is established by extrapolation to their intersection.

The amperometric method is inherently more accurate than the polarographic method and less dependent upon the characteristics of the capillary and the supporting electrolyte. Furthermore, the temperature need not be fixed accurately, although it must be kept constant during the titration. Finally, the substance being determined need not be reactive at the electrode; a reactive reagent or product is equally satisfactory.

TITRATION CURVES

Amperometric titration curves typically take one of the forms shown in Figure 21-10. Curve (a) represents a titration in which the analyte reacts at the electrode while the reagent does not. The titration of lead with sulfate or oxalate

---

[11] M. Brezina and P. Zuman, *Polarography in Medicine, Biochemistry and Pharmacy*. New York: Interscience Publishers, Inc., 1958; P. Zuman, *Polarographic Analysis*. Oxford: Pergamon Press, Ltd., 1964.

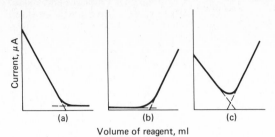

Figure 21-10 Typical Amperometric Titration Curves. (a) Analyte is reduced, reagent is not. (b) Reagent is reduced, analyte is not. (c) Both reagent and analyte are reduced.

ions may be cited as an example. The potential that is applied is sufficient to give a diffusion current for lead; a linear decrease in current is observed as lead ions are removed from the solution by precipitation. The curvature near the equivalence point reflects the incompleteness of the precipitation reaction in this region. The end point is obtained by extrapolation of the linear portions, as shown.

Curve (b) in Figure 21-10 is typical of a titration in which the reagent reacts at the microelectrode and the analyte does not; an example would be the titration of magnesium with 8-hydroxyquinoline. A diffusion current for the latter is obtained at -1.6 V (versus the saturated calomel electrode), whereas magnesium ion is inert at this potential.

Curve (c) in Figure 21-10 corresponds to the titration of lead ion with a chromate solution at an applied potential greater than -1.0 V. Both lead and chromate ions give diffusion currents, and a minimum in the curve signals the end point. This system would yield a curve resembling (b) in Figure 21-10 at zero applied potential since only chromate ions are reduced under these conditions.

Amperometric titrations in which the microelectrode serves as anode, or perhaps as anode for one reactant and cathode for another, have also been reported in the literature. The curves for such titrations are interpreted in the same way as the examples cited in the previous paragraphs. It should be recalled that an anodic current is provided with a minus sign by convention.

APPARATUS AND TECHNIQUES

Accurate amperometric titrations can be achieved with relatively simple apparatus.

Cells. Figure 21-11 shows a typical cell for an amperometric titration. A calomel half-cell is usually employed as the nonpolarizable electrode; the indicator electrode may be a dropping mercury electrode or a microwire electrode, as shown. The cell should have a capacity of 75 to 100 ml.

Volume Measurements. For linear plots, it is necessary to correct for volume changes due to the added titrant. The measured currents can be

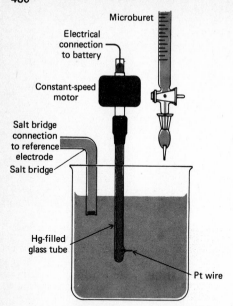

Microburet

Electrical
connection
to battery

Constant-speed
motor

Salt bridge
connection
to reference
electrode

Salt bridge

Hg-filled
glass tube

Pt wire

Figure 21-11 Typical Cell Arrangement for Amperometric Titrations with a Rotating Platinum Electrode.

multiplied by the quantity $(V + v)/V$, where V is the original volume and v is the volume of reagent; thus, all measured currents are corrected back to the original volume. An alternative, which is often satisfactory, is to use a reagent that is 20 (or more) times as concentrated as the solution being titrated. Under these circumstances, v is so small with respect to V that the correction is negligible. This approach requires the use of a microburet so that a total reagent volume of 1 or 2 ml can be measured with a suitable accuracy. The microburet should be so arranged that its tip can be touched to the surface of the solution after each addition of reagent to permit removal of the fraction of a drop that tends to remain attached.

Electrical Measurements. A simple manual polarograph is entirely adequate for amperometric titrations. An appropriate voltage is applied to the cell by means of the linear potential divider; the current is measured with a damped galvanometer (which need not be calibrated) or a low-resistance microammeter. Ordinarily, the applied voltage does not have to be known any closer than about ± 0.05 V, since it is only necessary to select a potential within the diffusion-current region of at least one of the participants in the titration.

Microelectrodes; the Rotating Platinum Electrode. Many amperometric titrations can be carried out conveniently with a dropping mercury electrode. For reactions involving oxidizing agents that attack mercury [bromine, silver ion, iron(II), among others], a rotating platinum electrode is preferable. This microelectrode consists of a short length of platinum wire sealed into the side of a glass tube; mercury within the tube provides electrical contact between the wire and the lead to the polarograph. The tube is held in the hollow chuck of a synchronous motor and rotated at a constant speed in excess of 600 rpm. Commercial models of the rotating electrode are available. A typical apparatus is shown in Figure 21-11.

Polarographic waves that are similar to those observed with the dropping electrode can be obtained with the rotating platinum electrode. Here, however, the reactive species is brought to the electrode surface not only by diffusion but also by mechanical mixing. As a consequence, the limiting currents are as much as 20 times larger than those obtained with a microelectrode that is supplied by diffusion only. With a rotating electrode, steady currents are instantaneously obtained. This behavior is in distinct contrast to the behavior of a solid micro-electrode in the absence of stirring.

Several limitations restrict the widespread application of the rotating platinum electrode to polarography. The low hydrogen overvoltage prevents its use as a cathode in acidic solutions. In addition, the high currents obtained cause the electrode to be particularly sensitive to traces of oxygen in the solution. These two factors have largely confined its employment to anodic reactions. Limiting currents from a rotating electrode are often influenced by the previous history of the electrode and are seldom as reproducible as the diffusion currents obtained with a dropping electrode. These limitations, however, do not seriously restrict the use of the rotating electrode for amperometric titrations.

APPLICATION OF AMPEROMETRIC TITRATIONS

As shown in Table 21-3, the amperometric end point has been largely confined to titrations in which a precipitate or a stable complex is the product. A notable exception is the application of the rotating platinum electrode to titrations with bromate ion in the presence of bromide and hydrogen ions (see pp. 350–353);

TABLE 21-3 Applications of Amperometric Titrations

Reagent	Reaction Product	Type Electrode[a]	Substance Determined
K_2CrO_4	Precipitate	DME	Pb^{2+}, Ba^{2+}
$Pb(NO_3)_2$	Precipitate	DME	SO_4^{2-}, MoO_4^{2-}, F^-, Cl^-
8-Hydroxyquinoline	Precipitate	DME	Mg^{2+}, Zn^{2+}, Cu^{2+}, Cd^{2+}, Al^{3+}, Bi^{3+}, Fe^{3+}
Cupferron	Precipitate	DME	Cu^{2+}, Fe^{3+}
Dimethylglyoxime	Precipitate	DME	Ni^{2+}
α-Nitroso-β-naphthol	Precipitate	DME	Co^{2+}, Cu^{2+}, Pd^{2+}
$K_4Fe(CN)_6$	Precipitate	DME	Zn^{2+}
$AgNO_3$	Precipitate	RP	Cl^-, Br^-, I^-. CN^-, RSH
EDTA	Complex	DME	Bi^{3+}, Cd^{2+}, Cu^{2+}, Ca^{2+}, etc.
$KBrO_3$, KBr	Substitution, addition, or oxidation	RP	Certain phenols, aromatic amines, olefins; N_2H_4, As(III), Sb(III)

[a] DME = dropping mercury electrode; RP = rotating platinum electrode.

beyond the equivalence point, the bromine concentration, which increases rapidly, causes a sharp increase in the current.

As seen in Table 21-3, the amperometric end point is useful for detecting end points in the titration of several metal ions with organic precipitating agents or with EDTA.

Amperometric Titrations with Two Polarized Microelectrodes

A convenient modification of the amperometric method involves the use of two identical, stationary microelectrodes immersed in a well-stirred solution of the sample. A small potential (say, 0.1 to 0.2 V) is applied between these electrodes, and such current that flows is followed as a function of the volume of added reagent. The end point is marked by a sudden current rise from zero, a decrease in the current to zero, or a minimum (at zero) in a V-shaped curve.

Although the use of two polarized electrodes for end-point detection was first proposed before 1900, almost 30 years passed before chemists came to appreciate the potentialities of the method.[12] The name *dead-stop end point* was used to describe the technique, and this term is still occasionally used. It was not until about 1950 that a clear interpretation of dead-stop titration curves was made.[13]

OXIDATION-REDUCTION TITRATIONS

Twin-polarized platinum microelectrodes are conveniently used for end-point detection for oxidation-reduction titrations. Figure 21-12 illustrates three types of titration curves that are commonly encountered. Curve (a) is observed when both reactant systems are reversible with respect to the electrodes. Curves (b) and (c) are obtained when only one of the reactants exhibits reversible behavior.

Titration Curves When Both Systems Are Reversible. The titration of iron(II) with cerium(IV) is an example of a reversible system. Interpretation of the curve requires consideration of each half-reaction system at various points in the titration.

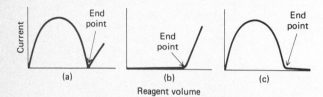

Figure 21-12 End Points for Amperometric Oxidation-Reduction Titrations with Twin Polarized Electrodes. (a) Both reactants behave reversibly at the electrode. (b) Only reagent behaves reversibly. (c) Only analyte behaves reversibly.

[12] C. W. Foulk and A. T. Bawden, *J. Amer. Chem. Soc.*, **48**, 2045 (1926).

[13] For an excellent analysis of this type of end point, see J. J. Lingane, *Electroanalytical Chemistry*, 2d ed., pp. 280–294. New York: Interscience Publishers, Inc., 1958.

When a potential of 0.1 V is applied to a pair of platinum electrodes immersed in a solution containing both iron(II) and iron(III), a current is observed, since the following electrode reactions can occur:

cathode $\qquad$ $Fe^{3+} + e \rightleftarrows Fe^{2+}$

anode $\qquad$ $Fe^{2+} \rightleftarrows Fe^{3+} + e$

These processes take place essentially instantaneously in the surface films surrounding the electrodes; that is, overvoltage effects are not observed. Since the electrodes are also small, concentration polarization will surely occur at an applied potential of 0.1 V, and the magnitude of the current will thus be determined by the rate at which the reactant in lesser concentration is brought to the electrode surface. Thus, if the concentration of iron(II) is smaller than that for iron(III), concentration polarization will occur at the anode, and the magnitude of the current will be determined by the concentration of the former. If iron(II) is in excess, however, cathodic polarization will occur, and the current will depend upon the iron(III) concentration. If the iron(III) concentration is zero (corresponding to the initial point in the titration), no current will flow in the absence of some alternative cathode process. A possible reaction might be

$$H^+ + e \rightleftarrows \tfrac{1}{2}H_2$$

The overvoltage of hydrogen on platinum is sufficiently great, however, so that this reaction does not occur at a significant rate at an applied potential of 0.1 V.

The behavior of the electrode system in a solution containing both cerium(III) and cerium(IV) is analogous to the iron system. Here the electrode processes are

cathode $\qquad$ $Ce^{4+} + e \rightarrow Ce^{3+}$

anode $\qquad$ $Ce^{3+} \rightarrow Ce^{4+} + e$

The rates of these reactions are also great, and the magnitude of the observed currents depends upon the concentration of the cerium species that is present in lesser amount.

At the equivalence point in the titration, the solution contains large amounts of iron(III) and cerium(III); the concentrations of iron(II) and cerium(IV) will be vanishingly small and can be neglected. Possible electrode reactions are then

cathode $\qquad$ $Fe^{3+} + e \rightarrow Fe^{2+}$

anode $\qquad$ $Ce^{3+} \rightarrow Ce^{4+} + e$

Let us now consider the curve for the titration of iron(II) with cerium(IV) as shown in (a) of Figure 21-12. At the outset, no current is observed because no suitable cathode reactant is available. With addition of cerium(IV), a mixture of iron(III) and iron(II) is produced, which permits the passage of current. Initially, the magnitude of the current is determined by the iron(III) concentration, but beyond the midpoint in the titration this species becomes in excess; the current is then regulated by the decreasing iron(II) concentration. At the equivalence point, the current approaches zero, since only cerium(III) and iron(III) are present in significant quantities, and the applied potential is not sufficient to cause these to react at the electrodes to any significant extent. Beyond the equivalence point, the current again rises because both cerium(III) and cerium(IV) are now present to react at the electrodes; the current in this region is limited by the cerium(IV) concentration.

Titration Curves When Only One System Is Reversible. Let us now consider the behavior of a twin-electrode system in the presence of species that are oxidized or reduced only slowly at an electrode (that is, a nonreversible system). An example is the arsenic(III)-arsenic(V) couple. Possible electrode reactions are

$$\text{cathode} \qquad H_3AsO_4 + 2H^+ + 2e \rightarrow H_3AsO_3 + H_2O$$

$$\text{anode} \qquad H_3AsO_3 + H_2O \rightarrow H_3AsO_4 + 2H^+ + 2e$$

These processes are slow at a platinum surface and occur only by applying an overpotential of several tenths of a volt. Little reaction takes place at an applied potential of 0.1 V because of this kinetic polarization; therefore, essentially no current is observed.

Upon addition of iodine to a solution of arsenious acid, a mixture of the two arsenic species and iodide ion results. While iodide can serve as an anode reactant, no cathode reactant is available because of the slow rate at which arsenic acid is reduced at a platinum surface; thus, at an applied potential of 0.1 V no current passes [see (b) of Figure 21-12]. Beyond the equivalence point, depolarization of the cell can occur at 0.1 V by virtue of the reactions

$$\text{cathode} \qquad I_2 + 2e \rightarrow 2I^-$$

$$\text{anode} \qquad 2I^- \rightarrow I_2 + 2e$$

Here the current is dependent upon the concentration of iodine.

Curve (c) of Figure 21-14 shows the titration of a dilute solution of iodine with the thiosulfate ion. In the initial stages, iodine and iodide are present, and both react reversibly at the electrodes; the current is dependent upon the concentration of the species present in lesser amount. Thiosulfate does not behave reversibly at the electrodes; therefore the current remains at zero beyond the equivalence point.

PRECIPITATION TITRATIONS

Twin silver microelectrodes have been employed for end-point detection in titrations involving silver ion. For example, in the titration of the silver ion with a standard solution of the chloride ion, currents proportional to the metal ion concentration would result from the reactions

$$\text{cathode} \qquad Ag^+ + e \rightleftarrows Ag$$

$$\text{anode} \qquad Ag \rightleftarrows Ag^+ + e$$

With the effective removal of silver ion by the analytical reaction, cathodic polarization would occur, and the current would approach zero at the end point.

APPLICATIONS

The amperometric method with two microelectrodes has been widely applied to titrations involving iodine; it is also useful with reagents such as bromine, titanium(III), and cerium(IV). An important use is in the titration of water with the Karl Fischer reagent (see Chapter 27). The technique has also been applied to end-point detection for coulometric titrations.

The principal advantage of the twin microelectrode procedure is its simplicity. One can dispense with a reference electrode; the only instrumentation needed is a simple voltage divider, powered by a dry cell, and a galvanometer or microammeter for current detection.

Modified Polarographic Methods

Classical polarography has two basic limitations that have restricted its application to certain types of routine analyses. The first is that the ordinary electrode system does not function well in nonaqueous solvents where electrical resistance is high. As a consequence, organic analyses are limited to species that are soluble in water or aqueous mixtures of polar organic solvents such as alcohol.

The second limitation is that of sensitivity, which prevents the analysis of solutions that are more dilute than about 10^{-5} M. At concentrations lower than this figure, the diffusion current is the same size as or smaller than the background currents; correction of the total current for the background then leads to large errors and ultimately limits the sensitivity of the procedure. Background currents in polarography are largely nonfaradic,[14] which cannot be reduced to less than a few hundredths of a microampere. Their main sources are the charging current, discussed on page 475, and the so-called noise currents, which are induced in the cell circuit from the instrument itself and other electrical sources as well; the effects of noise currents increase as the electrical resistance of the polarographic cell increases.

Several modifications of the classical polarographic procedure overcome these limitations to various degrees. A few of these are described briefly in this section.[15]

POLAROGRAPHY WITH POTENTIOSTATIC CONTROL

Figure 21-13 illustrates the effect of cell resistance on the polarographic wave for a reversible system. At 100 ohms the IR drop is so small that it does not influence the shape of the wave significantly. With high resistances, on the other hand, the waves become drawn out and ultimately so poorly defined as to be of little use. The cause of this effect is easily understood by reference to the equation

$$E_{applied} = E_{cathode} - E_{anode} - IR$$

[14] A faradic current is one that is carried between solution and electrodes via oxidation-reduction processes. Passage of a nonfaradic current involves no net electrochemical change; the charging current in polarography (p. 475) and in a capacitance circuit are examples.

[15] For additional details, see J. B. Flato, *Anal. Chem.*, **44** (11), 75A (1972).

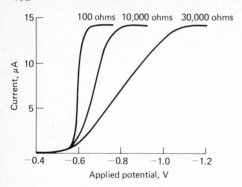

Figure 21-13   Effect of Cell Resistance on a Reversible Polarographic Wave.

Here $E_{applied}$ is varied in a regular way while E_{anode} is constant. If IR is small, the cathode potential then faithfully reflects the variation in $E_{applied}$. On the other hand, if R is large, a greater and greater fraction of $E_{applied}$ is required to overcome the IR drop as I increases. Thus, $E_{cathode}$ is no longer directly proportional to $E_{applied}$. A drawn-out wave results.

The use of potentiostatic control has permitted the extension of polarography to solvents with high electrical resistances. Here three electrodes are employed: a dropping electrode (or other microelectrode), a counter electrode, and a reference electrode. The counter electrode and the microelectrode serve the same purpose as the two electrodes in ordinary polarography. The reference electrode, on the other hand, is employed to measure and control the potential of the microelectrode. The arrangement here is analogous to that described for controlled-potential electrolysis (see Figure 19-5). The reference electrode is positioned as close as possible to the dropping electrode, and the potential between the two is measured with a high-resistance (10^{14} ohms) electrometer. This potential is then employed to control $E_{applied}$ such that the variation of $E_{cathode}$ is linear with time. Thus, the ordinate of the polarographic wave becomes $E_{cathode}$ rather than $E_{applied}$. The effect is to produce a wave form similar to that for the lowest resistance curve in Figure 21-13.

OTHER VOLTAMMETRIC METHODS

Several variants of classical polarography have been developed that increase the inherent sensitivity of the procedure and often permit better discrimination between two reductants with similar half-wave potentials. Ordinarily, the three-electrode scheme described in the previous paragraph is employed with these modifications.

Table 21-4 summarizes the characteristics of the various modifications. Two of these methods deserve brief mention.

Differential-pulsed Voltammetry. In differential-pulsed voltammetry, a dc potential, which is increased linearly with time, is applied to the polarographic cell. As in classical polarography, the rate of increase is perhaps 5 mV/sec. In

TABLE 21-4 Modified Polarographic Procedures

Name	Description	Sensitivity	$\Delta E_{1/2}$ for Resolution	Curve Form
1. Classical or linear scan	i recorded as E varied linearly at a rate of $\sim$ 5 mV/sec	10^{-5} M	0.1 V	Current steps
2. Differential pulsed voltammetry	E varied linearly as in (1); dc pulse of $\sim$ 25 mV applied for $\sim$ 60 msec; i measured for a period of $\sim$ 17 msec before pulse and at end of pulse; Δi recorded as function of linear E	10^{-8} M	0.05 V	Current maxima
3. AC or rapid scan polarography	E varied linearly as in (1); constant ac potential of a few mV superimposed on dc potential; *ac current measured* as a function of linear E	10^{-6} M	0.05 V	Current maxima
4. Rapid scan polarography	E varied linearly at rate of 100 mV/sec or greater; sweep performed during last 2–3 sec of drop lifetime; polarogram displayed on an oscilloscope	10^{-7} M	0.05 V	Current maxima
5. Stripping voltammetry	Metal deposited for 5–30 min on a single mercury drop at constant E; E then decreased linearly and i recorded	10^{-9} M	0.1 V	Anodic current maxima

contrast, however, a dc pulse of an additional 20 to 100 mV is applied at regular intervals of about 1 to 3 sec; the length of the pulse is about 60 msec and terminates with detachment of the mercury drop from the electrode. To synchronize the pulse with the drop, the latter is detached by a properly timed mechanical tap or movement of the electrode. The voltage program is shown in Figure 21-14.

Two current measurements are made alternately—one just prior to the dc pulse and one near the end of the pulse (see Figure 21-14). The *difference in current per pulse* (Δi) is recorded as a function of the linearly increasing voltage. A differential curve results which consists of a current peak for each reactive species rather than the usual current step (see Figure 21-15). The peak height is directly proportional to concentration.

One advantage of the derivative-type polarogram is that individual peak maxima can be observed for substances with half-wave potentials that differ by as little as 0.04 to 0.05 V; in contrast, classical polarography requires a potential difference of at least 0.1 V for resolution of waves. More important, however, differential pulse polarography increases the sensitivity of the polarographic method by about two orders of magnitude. This enhanced sensitivity is illustrated

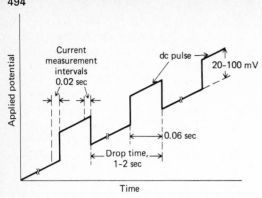

Figure 21-14 Voltage Program for Differential-pulsed Voltammetry.

in Figure 21-15. Note that a classical polarogram for a solution containing 180 ppm of the antibiotic tetracycline gives barely discernible waves; pulse polarography, in contrast, provides two well-defined peaks at a concentration level of 0.36 ppm. Note also that the current scale for Δi is in nA (nanoamperes) or 10^{-3} μA.

The greater sensitivity of pulse polarography can be attributed to two

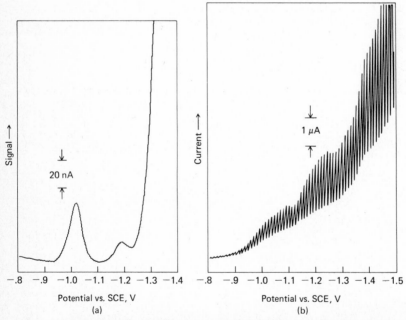

Figure 21-15 (a) Differential-pulsed Polarogram. 0.36 ppm tetracycline · HCl in 0.1-M acetate buffer, pH 4. PAR Model 174 polarographic analyzer, dropping mercury electrode, 50-mV pulse amplitude, 1-sec drop. (b) DC Polarogram. 180 ppm tetracycline · HCl in 0.1-M acetate buffer, pH 4, similar conditions. [Reproduced from J. B. Flato, *Anal. Chem.*, **44** (11), 78A (1972). With permission of the American Chemical Society.]

sources. The first is an enhancement of the faradic current, and the second is a reduction in the nonfaradic, charging current. To account for the former, let us consider the events that must occur in the surface layer around an electrode as the potential is suddenly increased by 20 to 100 mV. If a reactive species is present in this layer, a surge of current results, which reduces the concentration of the reactant to that demanded by the new potential. As the equilibrium concentration for that potential is approached, however, the current decays to a level just sufficient to counteract diffusion; that is, to the diffusion-controlled current. In classical polarography, this surge of current is not observed because the time-scale of the measurement is long relative to the lifetime of the momentary current. On the other hand, in pulse polarography, the current measurement is made before the surge has completely decayed. Thus, the current measured contains both a diffusion-controlled component and a component that has to do with reducing the surface layer to the concentration demanded by the Nernst expression; the total current is typically several times larger than the diffusion current.

When the potential pulse is first applied to the electrode, a nonfaradic current surge also occurs, which increases the charge on the drop (p. 475). This current, however, decays exponentially with time and approaches zero near the end of the life of a drop when the current is measured. Thus, the non-faradic component is virtually eliminated in pulse polarography. Enhanced sensitivity results.

Reliable instruments for pulse polarography are now available commercially at reasonable cost. The method has thus become an electroanalytical tool of considerable importance.

STRIPPING METHODS

Stripping analysis encompasses a variety of electrochemical methods having a common, characteristic initial step. In all these procedures, the analyte is first collected by electrodeposition at a mercury or a solid electrode; it is then re-dissolved (*stripped*) from the electrode to produce a more concentrated solution than originally existed. The analysis is finally based upon either electrical measurements made during the actual stripping process or some electro-analytical measurement of the more concentrated solution.

Stripping methods are of prime importance in trace work because the concentration aspects of the electrolysis permit the determination of minute amounts of an analyte with reasonable accuracy. Thus, the analysis of solutions in the 10^{-6} to 10^{-9} *M* range becomes feasible by methods that are both simple and rapid.

The stripping method that has had the most widespread application makes use of a micro mercury electrode for the deposition process, followed by anodic voltammetric measurement of the analyte; discussion in the paragraphs that follow will be largely confined to this particular application. For a more complete survey of stripping methods, the reader is referred to the review article by Shain.[16]

[16] I. Shain, *Stripping Analysis*, in I. M. Kolthoff and P. J. Elving, Eds., *Treatise on Analytical Chemistry*, part 1, vol. 4, chapter 50. New York: Interscience Publishers, Inc., 1963.

Electrodeposition Step. Ordinarily only a fraction of the analyte is deposited during the electrodeposition step; hence, to obtain quantitative results, not only must the electrode potential be controlled but care must also be taken to reproduce the electrode size, the length of deposition, and the stirring rate for both the sample and a standard solution employed for calibration.

Mercury electrodes of various forms have been most widely used for the electrolysis, although platinum or other inert metals can also be employed. Generally it is desirable to minimize the volume of the mercury to enhance the concentration of the deposited species. Several methods have been devised to produce a microelectrode of reproducible dimensions, which is essential for quantitative work. Among these is the *hanging drop* electrode, which is illustrated in Figure 21-16. Here an ordinary dropping mercury capillary is employed to transfer a reproducible quantity of mercury (usually one to three drops) to a Teflon scoop. Note that the mercury capillary does *not* serve as an electrode in this application. The hanging drop electrode is then formed by rotating the scoop and bringing the mercury in contact with a platinum wire sealed in a glass tube. The drop adheres strongly enough so that the solution can be stirred. The drop can, however, be dislodged by tapping the electrode at the completion of the electrolysis.

To use this apparatus, the drop is formed, stirring is begun, and a potential is applied that is a few tenths of a volt more negative than the half-wave

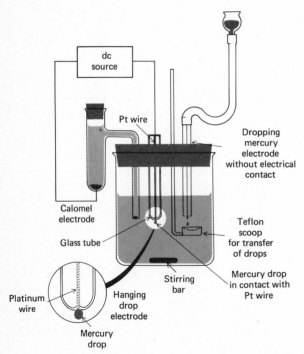

Figure 21-16 Apparatus for Stripping Analysis.

potential for the ion of interest. Deposition is allowed to occur for a carefully measured period; 5 min usually suffice for solutions that are 10^{-7} M or greater, 15 min for 10^{-8} M solutions, and 60 min for those that are 10^{-9} M. It should be emphasized that these times do not result in complete removal of the ion. The electrolysis period is determined by the sensitivity of the method ultimately employed for completion of the analysis.

Voltammetric Completion of the Analysis. The hanging drop electrode can be employed for completing the analysis by a voltammetric measurement. Here, after deposition is stopped, stirring is discontinued for perhaps 30 sec. The voltage is then decreased *at a fixed rate* from its original cathodic value toward the anodic, and the resulting anodic current is recorded as a function of the applied voltage. This technique, in which the voltage is varied at a fixed rate, is termed *linear scan voltammetry* and produces a curve of the type shown in Figure 21-17. In this experiment, cadmium was first deposited from a $1 \times 10^{-8} M$ solution by application of a potential of about -0.9 V (versus SCE), which is about 0.3 V more negative than the half-wave potential for this ion. After 15 min of electrolysis, stirring was discontinued; after an additional 30 sec, the potential was decreased at a rate of 21 mV/sec. A rapid increase in anodic current occurred at about -0.65 V as a result of the reaction

$$Cd(Hg) \rightarrow Cd^{2+} + Hg + 2e$$

After reaching a maximum, the current decayed owing to depletion of elemental cadmium in the hanging drop. The peak current, after correction for the residual current, was directly proportional to concentration of cadmium ions over a range of 10^{-6} M to 10^{-9} M and inversely proportional to deposition time. The analysis is based on calibration with standard solutions of cadmium ion. With reasonable care, an analytical precision of about 2% relative can be obtained. Analyses in the concentration range studied are not feasible by ordinary polarography. Mixtures can be resolved and analyzed by controlling the potential of deposition.

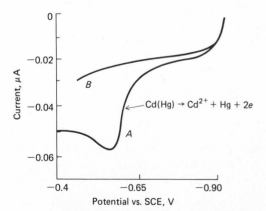

Figure 21-17 Curve *A*: Current-Voltage Curve for Anodic Stripping of Cadmium. Curve *B*: Residual Current Curve for Blank. [Adapted from R. D. DeMars and I. Shain, *Anal. Chem.*, **29**, 1825 (1957). With permission of the American Chemical Society.]

Many other variations of stripping analysis have been proposed. For example, a number of metals have been determined by electrodeposition on a platinum cathode. The quantity of electricity required to remove the deposit is then measured coulometrically. Here again the method is particularly advantageous for trace analysis.

PROBLEMS

*1. Calculate the Ni concentration (mg/liter) on the basis of the following data:

Solution	Current at -1.1 V, μA
25.0 ml of 0.2-F NaCl diluted to 50.0 ml	8.4
25.0 ml of 0.2-F NaCl plus 10.0 ml of sample diluted to 50.0 ml	46.3
25.0 ml of 0.2-F NaCl, 10.0 ml of sample, 5.00 ml of 2.30×10^{-2} M Ni^{2+} diluted to 50.0 ml	68.4

2. Calculate the concentration of Al (mg/liter) on the basis of the following data:

Solution	Current at -1.7 V, μA
20.0 ml of 0.20-N HCl plus 20.0 ml of H_2O	10.2
20.0 ml of 0.20-N HCl plus 10.0 ml of sample plus 10.0 ml of H_2O	33.3
20.0 ml of 0.20-N HCl plus 10.0 ml of sample plus 10.0 ml of 6.32×10^{-3} M Al^{3+}	52.0

3. The mercury from a dropping electrode was collected for 100 sec and found to weigh 0.196 g. The time required for 10 drops of mercury to form was 43.2 sec. When this electrode was employed with a standard 1.00×10^{-3} M solution of Pb^{2+}, a current of 8.76 μA was observed. Subsequently, an unknown lead solution produced a current of 16.31 μA with a new electrode which had a drop time of 6.13 sec and a flow rate of 3.85 mg/sec. Calculate the molar concentration of Pb^{2+} in the unknown.

*4. The following data were collected on three dropping electrodes. Complete the data for electrodes A and C.

	A	B	C
Flow rate, mg/sec	1.89	4.24	3.11
Drop time, sec	2.12	5.84	3.87
i_d/C, μA liter mmole^{-1}		4.86	

*5. Electrode C in Problem 4 was employed to study the reduction of an organic compound known to have a diffusion coefficient of 7.3×10^{-6} cm^2/sec. A 5.00×10^{-4} M solution of the compound yielded a diffusion current of 8.6 μA. Calculate n for the reaction.

6. An organic compound underwent a two-electron reduction at dropping electrode A in Problem 4. A diffusion current of 9.6 μA was produced by a 1.17×10^{-3} M solution of the compound. Calculate the diffusion coefficient for the compound.

*7. The half-wave potential for Ni(II) in a 0.10-F NaClO$_4$ solution was found to be -1.02 V (vs. SCE). In a solution that was 0.10 F in NaClO$_4$ and 0.100 F in ethylenediamine (en), the half-wave potential was -1.60 V; calculate the formation constant of the complex between the two species assuming its formula is Ni(en)$_3^{2+}$.

8. In 0.100-F KNO$_3$, the half-wave potential for the reduction of Cu^{2+} to the amalgam is $+0.021$ V. What would be the half-wave potential for the reaction in a solution which had a pH of 6.00 and was 0.200 F in EDTA?

*9. The following polarographic data were obtained for the reduction of Pb^{2+} to its amalgam from solutions that were 2.00×10^{-3} F in Pb^{2+}, 0.100 F in KNO$_3$, and that also had the following concentrations of the anion A$^-$. From the half-wave potentials, derive the formula of the complex as well as its formation constant.

Concn A$^-$, M	$E_{1/2}$ vs. SCE, V
0.0000	-0.405
0.0200	-0.473
0.0600	-0.507
0.1007	-0.516
0.300	-0.547
0.500	-0.558

10. A 1.00×10^{-3} F solution of europium(III) in 0.100-F KNO$_3$ has a polarographic wave for the reversible reduction of Eu^{3+} to the metal. The following data show the effect of increasing concentration of the anion X^{2-} on the half-wave potential.

Concn X^{2-}, M	$E_{1/2}$ vs. SCE, V
0.000	-0.692
0.0200	-1.083
0.0600	-1.113
0.100	-1.128
0.300	-1.152
0.500	-1.170

Derive the formula of the complex as well as its formation constant.

22

AN INTRODUCTION TO ABSORPTION SPECTROSCOPY

All chemical species interact with electromagnetic radiation and, in so doing, diminish the intensity or the power of the radiant beam. Absorption spectroscopy, a major branch of analytical chemistry, is based upon the measurement of this decrease in power (or *attenuation*) of the radiation brought about by the analyte.

It is convenient to characterize *absorptiometric methods* according to the type of electromagnetic radiation employed. The categories include X-ray, ultraviolet, visible, infrared, microwave, and radio-frequency radiation. We will be concerned mainly with the absorption of ultraviolet and visible radiation but will make occasional reference to the other types as well.

This chapter reviews some of the fundamental properties and concepts involving electromagnetic radiation.

Properties of Electromagnetic Radiation

Electromagnetic radiation is a type of energy that is transmitted through space at enormous velocities. It takes many forms, the most easily recognizable being light and radiant heat. Less obvious manifestations include X-ray, ultraviolet, microwave, and radio radiations.

The propagation of electromagnetic radiation through space is most conveniently described in terms of wave parameters such as velocity, frequency, wavelength, and amplitude. In contrast to other wave phenomena, such as sound, electromagnetic radiation requires no supporting medium for transmission; thus, it readily passes through a vacuum.

This wave model fails to account for phenomena associated with the absorption or the emission of radiant energy; for these processes it is necessary to view electromagnetic radiation as discrete particles of energy called *photons*. The energy of a photon is proportional to the frequency of the radiation. These dual views of radiation as particles and waves are not mutually exclusive. Indeed, the apparent duality is rationalized by wave mechanics and found to apply to other phenomena, such as the behavior of streams of electrons or other elementary particles.

WAVE PROPERTIES

For many purposes, electromagnetic radiation is conveniently treated as an electrical force field which oscillates at right angles with respect to the direction of propagation. The electrical force involved is a vector quantity; it can be represented at any instant by an arrow whose length is proportional to the magnitude of the force and whose direction is parallel to that of the force. It is seen in Figure 22-1 that a plot of the vector as a function of distance along the axis of propagation is sinusoidal.[1] This electrical force is responsible for such phenomena as the transmission, reflection, refraction, and absorption of radiation by matter.

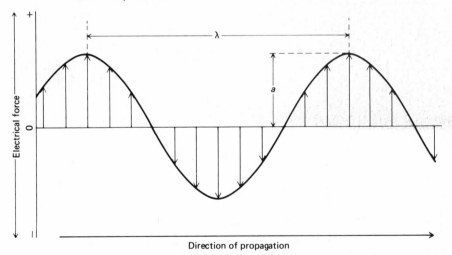

Figure 22-1 Representation of a Beam of Monochromatic Radiation of Wavelength λ and Amplitude *a*. The arrows represent the electrical vector of the radiation.

[1] Figure 22-1 is a two-dimensional representation of monochromatic (that is, a single wavelength) radiation. A better representation would be three-dimensional with a circular cross section that periodically fluctuates in radius from zero to the maximum amplitude *a*.

Wave Parameters. The time required for the passage of successive maxima past a fixed point in space is called the *period p* of the radiation. The *frequency* v is the number of oscillations of the field that occur per second[2] and is equal to $1/p$. It is of importance to realize that *the frequency is determined by the source and remains invariant* regardless of the media traversed by the radiation. In contrast, the *velocity* of propagation s_i, the rate at which the wave front moves through a medium, *is dependent* upon both the medium and the frequency; the subscript i is employed to indicate this frequency dependence. Another parameter of interest is the *wavelength* λ_i, which is the linear distance between successive maxima or minima of a wave.[3] Multiplication of the frequency (in cycles per second) by the wavelength (in centimeters per cycle) gives the velocity of the radiation (in centimeters per second); that is,

$$s_i = v\lambda_i \tag{22-1}$$

In a vacuum the velocity of radiation becomes independent of frequency and is at its maximum. This velocity, given the symbol c, has been accurately determined to be 2.99792×10^{10} cm/sec. Thus, for a vacuum,

$$c = v\lambda \cong 3 \times 10^{10} \text{ cm/sec} \tag{22-2}$$

In any other medium the rate of propagation is less because of interactions between the electromagnetic field of the radiation and the bound electrons of the medium. Since the radiant frequency is invariant and fixed by the source, the *wavelength must decrease* as radiation passes from a vacuum to a medium containing matter (Equation 22-1). It should be noted that the velocity of radiation in air differs only slightly from c (about 0.03% less); thus, Equation 22-2 is usually applicable to air as well as a vacuum.

The *wave number* σ is defined as the number of waves per centimeter and is yet another way of describing electromagnetic radiation. When the wavelength *in vacuo* is expressed in centimeters, the wave number is equal to $1/\lambda$.

Radiant Power or Intensity. The *power P* of radiation is the energy of the beam that reaches a given area per second; the *intensity I* is the power per unit solid angle. These quantities are related to the square of the amplitude a (see Figure 22-1). Although it is not strictly correct, power and intensity are often used synonymously.

PARTICLE PROPERTIES OF RADIATION

Energy of Electromagnetic Radiation. In certain interactions with matter it is necessary to consider radiation as packets of energy called *photons* or *quanta*. The energy of a photon depends upon the frequency of the radiation and is given by

$$E = hv \tag{22-3}$$

---

[2] The common unit of frequency is the *hertz* Hz, which is equal to one cycle per second.

[3] The units commonly used for describing wavelength differ considerably in the various spectral regions. For example, the ångström unit A (10^{-10} m) is convenient for X-ray, and short ultraviolet radiation; the nanometer nm or the synonymous millimicron mμ (10^{-9} m) is employed with visible and ultraviolet radiation; the micron μ (10^{-6} m) is useful for the infrared region.

where h is Planck's constant (6.63×10^{-27} erg sec). In terms of wavelength and wave number,

$$E = \frac{hc}{\lambda} = hc\sigma \qquad (22\text{-}4)$$

Note that the wave number, like the frequency, is directly proportional to energy.

THE ELECTROMAGNETIC SPECTRUM

The electromagnetic spectrum covers an immense range of wavelengths, or energies. For example, an X-ray photon ($\lambda \sim 10^{-8}$ cm) is approximately ten thousand times more energetic than a photon emitted by an incandescent tungsten wire ($\lambda \sim 10^{-4}$ cm).

Figure 22-2 depicts qualitatively the major divisions of the electromagnetic spectrum. A logarithmic scale has been employed in this representation; note that the region to which the human eye is perceptive (the *visible spectrum*) is very small. Such diverse radiations as gamma rays or radio waves differ from visible light only in the matter of frequency and hence energy.

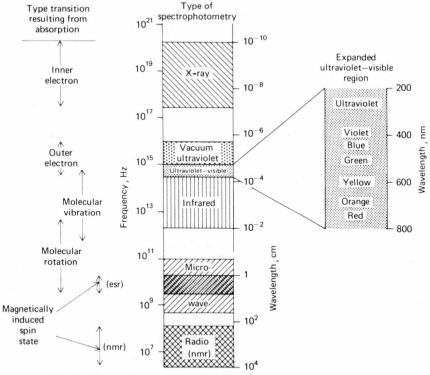

Figure 22-2 Parts of the Electromagnetic Spectrum Which Are Employed for Spectrophotometry.

Figure 22-2 also indicates the regions of the spectrum that are employed for analytical purposes and the molecular or atomic transitions responsible for absorption of radiation in each region.

Generation of Electromagnetic Radiation

Atoms or molecules have a limited number of discrete or quantized energy levels, the lowest of which is called the *ground state*. When sufficient energy is added to a system by heat from an arc or flame, exposure to a high-energy, alternating-current spark, bombardment with elementary particles such as electrons, or absorption of electromagnetic radiation, the atoms or molecules are *excited* to higher energy levels. The lifetime of an excited species is generally transitory ($\sim 10^{-8}$ sec), and relaxation to a lower energy level or to the ground state takes place with a release of energy in the form of heat or of electromagnetic radiation, or perhaps both. When radiation is the product, the energy of each emitted photon, $h\nu$, is equal to the corresponding energy difference between the excited and the lower energy level.

Radiating particles that are well separated from one another, as in the gaseous state, behave as independent bodies and often emit relatively few specific wavelengths. The resulting spectrum is thus discontinuous and termed a *line spectrum*. In contrast, a *continuous spectrum* contains all wavelengths over an appreciable range; here the individual wavelengths are so closely spaced that separation is not feasible by ordinary means. Continuous spectra result from the excitation of (1) solids or liquids, in which the atoms are so closely packed as to be incapable of independent behavior, or (2) complicated molecules that possess many closely related energy states. Continuous spectra also occur when the energy changes involve particles with unquantized kinetic energies.

Both continuous spectra and line spectra are of importance in analytical chemistry. The former are valuable as sources in methods based on the interaction of radiation with matter, such as spectrophotometry. Line spectra, on the other hand, are useful for the identification and determination of the emitting species.

Absorption of Radiation

When radiation passes through a transparent layer of a solid, liquid, or gas, certain frequencies may be selectively removed by the process of *absorption*. Here electromagnetic energy is transferred to the atoms or molecules of the sample and converts absorbing particles from their normal or ground state to an excited state.

The absorption of electromagnetic radiation by some species M can be considered to be a two-step process, the first of which can be represented by

$$M + h\nu \rightarrow M^*$$

where M^* represents the atomic or molecular particle in the excited state that results from absorption of the photon $h\nu$. The lifetime of the excited state is brief (10^{-8} to 10^{-9} sec), its existence being terminated by any of several *relaxation* processes. The most common type of relaxation involves conversion of the excitation energy to heat; that is,

$$M^* \rightarrow M + heat$$

Relaxation may also result from decomposition of M^* to form new species; such a process is called a *photochemical reaction*. Alternatively, relaxation may involve the fluorescent or phosphorescent reemission of radiation. It is important to note that the lifetime of M^* is usually so very short that its concentration at any instant is negligible under ordinary conditions. Furthermore, the amount of thermal energy created is usually not detectable. Thus, absorption measurements have the advantage of creating minimal disturbance of the system under study.

Quantitative Aspects of Absorption Measurement

The principles and laws that govern the absorption of radiation apply to all wavelengths from the X-ray region to the radio-frequency range. The absorption measurement involves the determination of the reduction in power (the *attenuation*) suffered by a beam of radiation as it passes through an absorbing medium of known dimensions.

BEER'S LAW

When monochromatic radiation passes through a sample containing an absorbing species, the radiant power of the beam is progressively decreased as more of the energy is absorbed by the particles of that species. The decrease in power depends upon the concentration of the absorber and the length of the path traversed by the beam. These relationships are expressed by *Beer's law*.

Let P_0 be the radiant power of a beam incident upon a section of solution that contains c moles of an absorbing substance per liter. Further, let P be the power of the beam after it has traversed b centimeters of the solution (Figure 22-3). As a consequence of absorption, P will be smaller than P_0. Beer's law relates these quantities as follows:

$$\log \frac{P_0}{P} = \epsilon b c = A \tag{22-5}$$

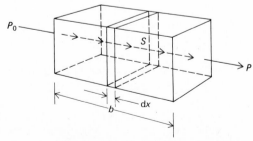

Figure 22-3 Attenuation of Radiation with Initial Power P_0 by a Solution Containing c Moles per Liter of Absorbing Solute and a Path Length of b cm. $P < P_0$.

In this equation, ϵ is a constant called the *molar absorptivity*. The logarithm (to the base 10) of the ratio between the incident power and the transmitted power is called the *absorbance* of the solution; this quantity is given the symbol A. Clearly, the absorbance increases directly with the concentration of the absorbing species and the path length traversed by the beam.

Beer's law can be rationalized as follows.[4] Consider the block of absorbing matter (solid, liquid, or gas) shown in Figure 22-3. A beam of parallel monochromatic radiation with power P_0 strikes the block perpendicular to a surface; after passing through a length b of the material, its power is decreased to P as a result of absorption. Consider now a cross section of the block having an area S and an infinitesimal thickness dx. Within this section there are dn absorbing particles (atoms, molecules, or ions); associated with each particle, we can imagine a surface at which photon capture can occur. That is, if a photon reaches one of these areas by chance, absorption will follow immediately. The total projected area of these capture surfaces within the section is designated as dS and the ratio of the capture area to the total area, dS/S. On a statistical average this ratio represents the probability for the capture of photons within the section.

The power of the beam entering the section P_x is proportional to the number of photons per square centimeter per second, and dP_x represents the quantity removed per second within the section; the fraction absorbed is then $-dP_x/P_x$ and this ratio on the average also equals the probability for capture. The term is given a minus sign to indicate that P undergoes a decrease. Thus,

$$-\frac{dP_x}{P_x} = \frac{dS}{S} \tag{22-6}$$

Recall, now, that dS is the sum of the capture areas for each particle within the section; it must therefore be proportional to the number of particles, or

$$dS = a\,dn \tag{22-7}$$

where dn is the number of particles and a is a proportionality constant that can be called the capture cross section. Combining Equations 22-6 and 22-7 and summing over the interval between zero and n, we obtain

$$-\int_{P_0}^{P} \frac{dP_x}{P_x} = \int_0^n \frac{a\,dn}{S}$$

which, upon integration, gives

$$-\ln \frac{P}{P_0} = \frac{an}{S}$$

Upon converting to base 10 logarithms and inverting the fraction to change the sign, we obtain

$$\log \frac{P_0}{P} = \frac{an}{2.303\ S} \tag{22-8}$$

[4] The discussion that follows is based on a paper by F. C. Strong, *Anal. Chem.*, **24**, 338 (1952).

where n is the total number of particles within the block shown in Figure 22-3. The cross-sectional area S can be expressed in terms of the volume of the block V and its length b. Thus,

$$S = \frac{V}{b} \text{ cm}^2$$

Substitution of this quantity into Equation 22-8 yields

$$\log \frac{P_0}{P} = \frac{anb}{2.303 \text{ V}} \tag{22-9}$$

Note that n/V has the units of concentration (that is, number of particles per cubic centimeter); we can readily convert n/V to moles per liter. Thus,

$$c = \frac{n \text{ particles}}{6.02 \times 10^{23} \text{ particles/mole}} \times \frac{1000 \text{ cm}^3/\text{liter}}{V \text{ cm}^3} = \frac{1000 \, n}{6.02 \times 10^{23} \text{ V}} \text{ mole/liter}$$

Substitution into Equation 22-9 yields

$$\log \frac{P_0}{P} = \frac{6.02 \times 10^{23} \, abc}{2.303 \times 1000}$$

Finally, the constants in this equation can be collected into a single term ϵ to give

$$\log \frac{P_0}{P} = \epsilon bc = A$$

Beer's law applies to a solution containing more than one kind of absorbing substance, provided there is no interaction among the various species. Thus, for a multicomponent system,

$$A_{\text{total}} = A_1 + A_2 + \cdots + A_n = \epsilon_1 bc_1 + \epsilon_2 bc_2 + \cdots + \epsilon_n bc_n \tag{22-10}$$

where the subscripts refer to absorbing components $1, 2, \ldots, n$.

MEASUREMENT OF ABSORPTION

Beer's law, as given by Equation 22-5, is not directly applicable to chemical analysis. Neither P nor P_0, as defined, can be conveniently measured in the laboratory because the solution to be studied must be held in some sort of container. Interaction between the radiation and the walls is inevitable, leading to a loss by reflection at each interface; moreover, significant absorption may occur within the walls. Finally, the beam may suffer a diminution in power during its passage through the solution as a result of scattering by large molecules or inhomogeneities. Reflection losses can be appreciable; for example, about 4% of a beam of visible radiation is reflected upon vertical passage across an air-to-glass or glass-to-air interface.

In order to compensate for these effects, the power of the beam transmitted through the absorbing solution is generally compared with that which passes through an identical cell containing a suitable *blank* solution. An experimental absorbance that closely approximates the true absorbance of the solution can then be calculated; that is,

$$A \cong \log \frac{P_{blank}}{P_{solution}} \cong \log \frac{P_0}{P} \qquad (22\text{-}11)$$

The term P_0, when used henceforth, refers to the power of a beam of radiation after it has passed through a cell containing the solvent for the component of interest.

TERMINOLOGY ASSOCIATED WITH ABSORPTION MEASUREMENTS

In recent years the attempt has been made to develop a standard nomenclature for the various quantities related to the absorption of radiation. The recommendations of the American Society for Testing Materials are given in Table 22-1 along with some of the alternative names and symbols frequently encountered. An important term in this table is the *transmittance T*, which is defined as

$$T = \frac{P}{P_0}$$

The transmittance is the fraction of incident radiation transmitted by the solution; it is often expressed as a percentage. The transmittance is related to the absorbance as follows:

$$-\log T = A$$

TABLE 22-1 Important Terms and Symbols Employed in Absorption Measurement

Term and Symbol[a]	Definition	Alternative Name and Symbol
Radiant power, P, P_0	Energy of radiation reaching a given area of a detector per second	Radiation intensity, I, I_0
Absorbance, A	$\log \dfrac{P_0}{P}$	Optical density, D; extinction, E
Transmittance, T	$\dfrac{P}{P_0}$	Transmission, T
Path length of radiation, in cm, b	—	l, d
Molar absorptivity,[b] ϵ	$\dfrac{A}{bc}$	Molar extinction coefficient
Absorptivity,[c] a	$\dfrac{A}{bc}$	Extinction coefficient, k

[a] Reprinted from *Anal. Chem.*, **24**, 1349 (1952). With permission of the American Chemical Society.

[b] c expressed in units of mole/liter.

[c] c may be expressed in g/liter or other specified concentration units; b may be expressed in cm or in other units of length.

LIMITATIONS TO THE APPLICABILITY OF BEER'S LAW

The linear relationship between absorbance and path length at a fixed concentration of absorbing substances is a generalization for which no exceptions are known. On the other hand, deviations from the direct proportionality between measured absorbance and concentration when b is constant are frequently encountered. Some of these deviations are fundamental and represent real limitations of the law; others occur as a consequence of the manner in which the absorbance measurements are made or as a result of chemical changes associated with concentration changes; the latter two are sometimes known, respectively, as *instrumental deviations* and *chemical deviations*.

Real Limitations to Beer's Law. Beer's law is successful in describing the absorption behavior of dilute solutions only; in this sense, it is a limiting law. At high concentrations (usually $>0.01 \; F$) the average distance between the species responsible for absorption is diminished to the point where each affects the charge distribution of its neighbors. This interaction, in turn, can alter their ability to absorb a given wavelength of radiation. Because the extent of interaction is dependent upon concentration, the occurrence of this phenomenon causes deviations from the linear relationship between absorbance and concentration.

Deviations from Beer's law also arise because ϵ is dependent upon the refractive index of the solution.[5] Thus, if concentration changes cause significant alterations in the refractive index n of a solution, departures from Beer's law are observed. A correction for this effect can be made by substitution of the quantity $\epsilon n/(n^2 + 2)^2$ for ϵ in Equation 22-5. In general, this correction is not significant at concentrations less than $0.01 \; F$.

Chemical Deviations. Apparent deviations from Beer's law are frequently encountered as a consequence of association, dissociation, or reaction of the absorbing species with the solvent. A classic example of a chemical deviation is observed with unbuffered potassium dichromate solutions, in which the following equilibria exist:

$$Cr_2O_7^{2-} + H_2O \rightleftarrows 2HCrO_4^- \rightleftarrows 2H^+ + 2CrO_4^{2-}$$

At most wavelengths, the molar absorptivities of the dichromate ion and the two chromate species are quite different. Thus, the total absorbance of any solution is dependent upon the concentration ratio between the dimeric and the monomeric forms. This ratio, however, changes markedly with dilution, causing a pronounced deviation from linearity between the absorbance and the total concentration of chromium. Nevertheless, the absorbance due to the dichromate ion remains directly proportional to its molar concentration; the same is true for chromate ions. This fact is easily demonstrated by making measurements in strongly acidic or strongly basic solutions where one or the other of these species predominates. Thus, deviations in the absorbance of

[5] G. Kortum and M. Seiler, *Angew. Chem.*, **52**, 687 (1939).

this system from Beer's law are more apparent than real, because they result from shifts in chemical equilibria. These deviations can, in fact, be readily predicted from the equilibrium constants for the reactions and the molar absorptivities of the dichromate and chromate ions.

Instrumental Deviation. Strict adherence of an absorbing system to Beer's law is observed only when the radiation employed is monochromatic. This observation is another manifestation of the limiting character of the relationship. Use of truly monochromatic radiation for absorbance measurements is seldom practical, and a polychromatic beam may cause departures from Beer's law.

Experiments show that deviations from Beer's law resulting from the use of a polychromatic beam are not appreciable, provided the radiation used does not encompass a spectral region in which the absorber exhibits large changes in absorbance as a function of wavelength. This observation is illustrated in Figure 22-4.

SPECTRAL CURVES

The relationship between absorption and the frequency or wavelength of radiation is often employed to characterize chemical species. The graphical representation of this relationship is called an *absorption spectrum*. In such plots the frequency, the wave number, or the wavelength is commonly employed for the abscissa. The ordinate is usually expressed in units of transmittance (or percent transmittance), absorbance, or the logarithm of absorbance.

Physical chemists prefer frequency or wave number for the abscissa because of the linear relationship between these functions and energy. Other chemists, on the other hand, use wavelength, expressed in nanometers or ångström units, for ultraviolet and visible absorption spectra.

Spectral data for three permanganate solutions are presented in three different ways in Figure 22-5. The concentrations stand in the ratio of 1:2:3. Note that employment of absorbance provides the greatest differentiation in the region where the absorbance is high (0.8 to 1.3) and the transmittance is low (<20%). In contrast, greater differences occur in the transmittance curves

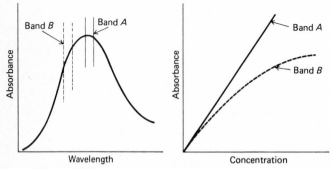

Figure 22-4 The Effect of Polychromatic Radiation upon the Beer's Law Relationship. Band *A* shows little deviation since ϵ does not change greatly throughout the band. Band *B* shows marked deviations since ϵ undergoes significant changes in this region.

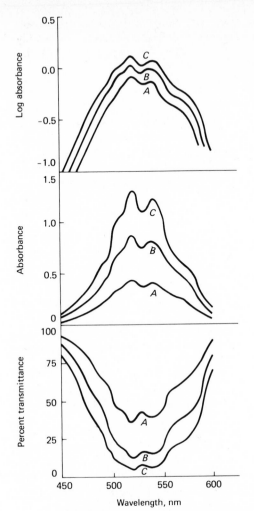

Figure 22-5 Methods for Plotting Spectral Data. Concentration relationship: $A : B : C = 1 : 2 : 3$.

when the transmittances lie in the range of 20 to 60%. With a plot of log A, spectral detail tends to be lost; on the other hand, this type of plot is particularly convenient for comparing curves for solutions that have different concentrations since the curves are displaced an equal amount along the ordinate scale regardless of wavelength.

PROBLEMS

1. Calculate the frequency in hertz and the wave number in cm⁻¹ for
    *(a) a monochromatic X-ray beam with a wavelength of 9.0 A.
    (b) the sodium D line at 589.0 nm.
    *(c) an infrared absorption peak at 12.6 μ.
    (d) a microwave beam of wavelength 200 cm.

2. Calculate the energy of each of the photons in Problem 1 in ergs per photon.
3. Express the following absorbances in terms of percent transmittance:
   *(a) 0.064.   (d) 0.209.
   *(b) 0.765.   (e) 0.437.
   *(c) 0.318.   (f) 0.413.
4. Convert the accompanying transmittance data to absorbances:
   *(a) 19.4%.   (d) 4.51%.
   *(b) 0.863.   (e) 0.100.
   *(c) 27.2%.   (f) 79.8%.
5. Use the data provided to evaluate the missing quantity.

	Absorbance, A	Molar Absorptivity, ϵ	Path Length, cm	Concentration
(a)	0.547		1.00	$3.64 \times 10^{-5}\ M$
(b)		3688	2.50	6.51 ppm (gfw = 200)
(c)	0.229	2.96×10^3		$3.86 \times 10^{-5}\ M$
(d)	0.477	6121	1.00	M
(e)	0.581	4.27×10^3	1.50	ppm (gfw = 254)

*6. Use the data provided to evaluate the missing quantity.

	Absorbance, A	Molar Absorptivity, ϵ	Path Length, cm	Concentration
(a)	0.345		2.00	$4.25 \times 10^{-4}\ M$
(b)		3.70×10^4	1.75	1.20 ppm (gfw = 325)
(c)	0.176	5.20×10^3		$2.26 \times 10^{-5}\ M$
(d)	0.982	2.75×10^4	0.980	M
(e)	0.634	2.98×10^4	2.00	ppm (gfw = 184)

7. What are the units for absorptivity when the concentration is expressed in terms of
   *(a) ppm?
   (b) w/v percent?
8. A solution containing 4.48 ppm of $KMnO_4$ has a transmittance of 0.309 when measured in a 1.00-cm cell at 520 nm. Calculate the molar absorptivity for $KMnO_4$.
*9. A solution containing 3.75 mg/100 ml of A (gfw = 220) has a transmittance of 39.6% in a 1.50-cm cell at 480 nm. Calculate the molar absorptivity for A.
10. A solution containing 10.1 ppm of B has a transmittance of 21.6% in a 1.25-cm cell. Calculate the absorptivity for B.
*11. A solution containing the complex formed between Bi(III) and thiourea has a molar absorptivity of 9.3×10^3 liter cm^{-1} mole^{-1} at 470 nm.
   (a) What will be the absorbance of a $6.2 \times 10^{-5}\ M$ solution of the complex when measured at 470 nm in a 1.00-cm cell?
   (b) What is the percent transmittance of the solution described in (a)?
   (c) What will be the concentration of the complex in a solution that has the same absorbance described in (a) when measured at 470 nm in a 5.00-cm cell?
12. At 580 nm, the wavelength of its maximum absorption, the complex $FeSCN^{2+}$ has a molar absorptivity of 7.00×10^3 liter cm^{-1} mole^{-1}. Calculate
   (a) the absorbance of a $2.50 \times 10^{-5}\ M$ solution of the complex at 580 nm when measured in a 1.00-cm cell.

    (b) the absorbance of a solution in which the concentration of the complex is twice that of (a).

    (c) the transmittance of the solutions described in (a) and (b).

    (d) the absorbance of a solution that has half the transmittance of that described in (a).

*13. A 25.0-ml aliquot of a solution that contains 3.8 ppm iron(III) is treated with an appropriate excess of KSCN and diluted to a final volume of 50.0 ml. What will be the absorbance of the resulting solution when measured at 580 nm in a 2.50-cm cell? See Problem 12 for absorptivity data.

14. Zinc(II) and the ligand L form a product that absorbs strongly at 600 nm. As long as the formal concentration of L exceeds that of zinc(II) by a factor of 5 (or more), the absorbance is dependent only on the cation concentration. Neither zinc(II) nor L absorbs at 600 nm.

    A solution that is 1.60×10^{-4} F in zinc(II) and 1.00×10^{-3} F in L has an absorbance of 0.464 when measured in a 1.00-cm cell at 600 nm. Calculate

    (a) the percent transmittance of this solution.

    (b) the percent transmittance of this solution when it is contained in a 2.50-cm cell.

    (c) the light path needed to match the absorbance of solution (a) with a 3.00-cm column that is 4.00×10^{-4} F with respect to the complex.

chapter

23

INSTRUMENTS AND METHODS FOR ABSORPTION ANALYSIS

Instruments for absorptiometric measurements with ultraviolet or visible radiation are considered in this chapter. In addition, similarities and differences among instruments and techniques that are designed for use in other spectral regions will be noted.

Components of Instruments for Absorption Measurements

Regardless of the spectral region, instruments that measure the transmittance or absorbance of solutions contain five basic components: (1) a stable source of radiant energy, (2) a device that permits employment of a restricted wave-length region, (3) transparent containers for sample and solvent, (4) a radiation detector or *transducer* that converts the radiant energy to a measurable signal (usually electrical), and (5) a signal indicator. The block diagram in Figure 23-1 shows the usual arrangement of these components.

The signal indicator for most instruments for absorption measurement is equipped with a linear scale that covers a range from 0 to 100 units. Direct percent transmittance readings can then be obtained by first adjusting the

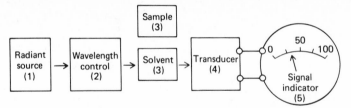

Figure 23-1 Components of Instruments for Measuring Absorption of Radiation.

indicator to read zero when radiation is blocked from the detector by a shutter. The indicator is then brought to 100 with a blank in the light path; this adjustment is accomplished by varying either the intensity of the source or the amplification of the detector signal. When the sample container is placed in the beam, the indicator gives percent transmittance directly, provided the detector responds linearly to changes in radiant power. Clearly, a logarithmic scale can be scribed on the indicator to permit direct absorbance readings as well.

The nature and the complexity of the several components of absorption instruments will depend upon the wavelength region involved and how the data are to be used; regardless of the degree of sophistication, however, the function of each component is the same. Figure 23-2 lists the common instrumental components for various wavelength regions.

Three types of instruments are employed for absorptiometric measurements in the visible region; in increasing order of sophistication these are *colorimeters*, *photometers*, and *spectrophotometers*. Most ultraviolet and infrared absorptiometric measurements are made with spectrophotometers.

Colorimeters

Colorimeters employ the human eye as a detector and the brain as a transducer and signal detector. The eye and brain, however, can only match colors; thus, they are incapable of providing numerical information about the relative power of two beams of light and therefore about absorbance. As a consequence, colorimetric methods always require the use of one or more standards for color matching with the analyte solutions.

The simplest colorimetric methods involve the comparison of the sample with a set of standards until a match is found. Flat-bottomed *Nessler tubes* are frequently employed for this purpose. These tubes are calibrated so that a uniform light path is achieved. Daylight commonly serves as a radiation source. Ordinarily, no attempt is made to restrict the portion of the spectrum employed.

A somewhat more refined colorimetric procedure involves comparison of the unknown with a single standard solution. Here the two solutions are contained in flat-bottomed tubes; the path lengths are varied by means of adjustable transparent plungers that can be moved up and down in the solutions. Balance is achieved visually, following which the path lengths are measured;

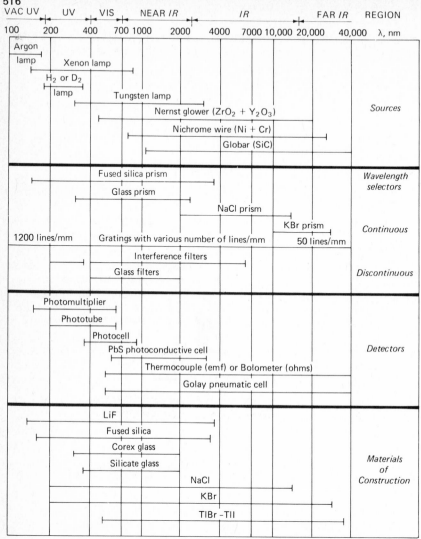

Figure 23-2 Components and Materials for Spectroscopic Instruments. (Adapted from a figure developed by Professor A. R. Armstrong, College of William and Mary. With permission.)

the concentration of the unknown may then be calculated. When the absorbances of the two solutions are identical,

$$A_x = A_s$$
$$\epsilon b_x c_x = \epsilon b_s c_s$$

or

$$c_x = c_s \frac{b_s}{b_x}$$

where *x* refers to the unknown and *s* to the standard. A *Duboscq* colorimeter embodies these principles and is equipped with an optical system that permits the ready comparison of the beams by passing these through an eyepiece with a split field.

Colorimetric methods suffer from several disadvantages. A standard or a series of standards must always be available. Furthermore, the eye usually cannot match absorbances if a second colored substance is present in the solution. Finally, the eye is not as sensitive to small differences in absorbance as a photoelectric device; as a consequence, concentration differences smaller than about 5% relative cannot be detected.

Despite their limitations, visual comparison methods find extensive application for routine analyses in which the requirements for accuracy are modest. For example, simple but useful colorimetric test kits are sold for determining the pH and the chlorine content of swimming pool water; kits are also available for the analysis of soils. Filtration plants commonly employ color comparison tests for the estimation of iron, silicon, fluorine, and chlorine in city water supplies. For such analyses a colorimetric reagent is introduced to the sample, and the resulting color is compared with permanent standard solutions or with colored glass disks. Accuracies of perhaps 10 to 50% relative are to be expected and suffice for the purposes intended.

Photometers

The photometer provides a simple, relatively inexpensive tool for the performance of absorption analyses. Convenience, ease of maintenance, and ruggedness are properties of a filter photometer that may not be found in the more sophisticated spectrophotometer. Moreover, where spectral purity is not important to a method (and often it is not), analyses can be performed as accurately with this instrument as with more complex instrumentation.

A typical photometer contains all of the components shown in Figure 23-1. Some of the properties of these components are described in the paragraphs that follow.

RADIATION SOURCE

The most common source for photometers (and spectrophotometers for the visible region as well) is the tungsten filament lamp, whose behavior approaches that of a *black-body radiator*. Sources of this kind, when heated to incandescence, emit continuous radiation that is more characteristic of the temperature of the emitting surface than of the material of which it is composed. Black-body radiation is the result of innumerable atomic and molecular oscillations (each of which is quantized) excited in the condensed solid by the thermal energy. Theoretical treatment of black-body radiation leads to the following conclusions: (1) the radiation exhibits a maximum emission at a wavelength that varies inversely with the absolute temperature; (2) the total energy emitted by a black body (per unit of time and area) varies as the fourth power of temperature; and

(3) the emissive power at a given temperature varies inversely as the fifth power of wavelength. Several laboratory sources approximate the behavior of a true black body. The wavelength distribution for a few of these is shown in Figure 23-3. Note the shift of the emission peaks to lower wavelengths with increasing temperature; clearly, very high temperatures are needed to cause a thermally excited source to emit a substantial fraction of its energy in the ultraviolet.

In most absorption instruments, the operating filament temperature of a tungsten lamp is about 2870°K; the bulk of the energy is thus emitted in the infrared region. A tungsten filament lamp is useful for the wavelength region between 320 and 2500 nm.

In the visible region the energy output of a tungsten lamp varies approximately as the fourth power of the operating voltage. As a consequence, close voltage control is required for a stable radiation source. Constant voltage transformers or electronic voltage regulators are often employed for this purpose. As an alternative, the lamp can be operated from a 6-V storage battery, which is a remarkably stable voltage source if it is maintained in good condition.

WAVELENGTH CONTROL

Both photometers and spectrophotometers employ devices that restrict the wavelength region used for an analysis. A narrow band of radiation offers three advantages. The probability of adherence of the absorbing system to Beer's law is greatly enhanced (see p. 510). In addition, a greater selectivity is assured, since substances that absorb in other wavelength regions are less likely to interfere. Finally, a greater change in absorbance per increment of concentration will be observed if only wavelengths that are strongly absorbed are employed; thus, a greater sensitivity is attained.

Devices for restricting radiation fall into two categories. *Filters*, which are

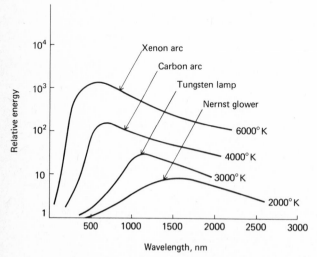

Figure 23-3   Black-Body Radiation Curves.

employed in photometers, function by absorbing large portions of the spectrum while transmitting relatively limited wavelength regions. *Monochromators*, which are used in spectrophotometers, are more sophisticated devices that permit the continuous variation of wavelength.

When illuminated with a continuous source, neither filters nor monochromators are capable of producing radiation of a single wavelength; instead, a band encompassing a narrow region of the spectrum results. The distribution of wavelengths within the region typically takes the form of the curves shown in (a) of Figure 23-4 with the peak corresponding to the wavelength setting of the monochromator or the stated wavelength of the filter. The quality of the monochromator or the filter is measured by the *effective band width*, which is the width of the band (in wavelength units) at half intensity [see curve (a), Figure 23-4].

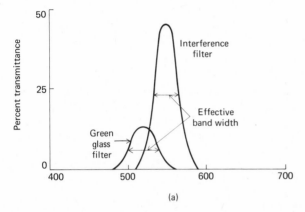

(a)

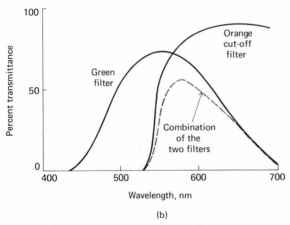

(b)

Figure 23-4 Comparison of Transmittance Characteristics of Two Filters and a Filter Combination.

Absorption Filters. Absorption filters limit radiation by absorbing certain portions of the spectrum. The most common type consists of colored glass or of a dye suspended in gelatin and sandwiched between glass plates; the former have the advantage of greater thermal stability.

Absorption filters have effective band widths ranging from perhaps 30 to 250 nm [see curve (b), Figure 23-4]. Filters that provide the narrowest band widths also absorb a significant fraction of the desired radiation and may have a transmittance that is 0.1 or less at their band peaks. Glass filters with transmittance maxima throughout the entire visible region are available commercially and are relatively inexpensive.

Cut-off filters have transmittance of nearly 100% over a portion of the visible spectrum but then rapidly decrease to zero transmittance over the remainder. A narrow spectral band can be isolated by coupling a cut-off filter with a second filter [see curve (b), Figure 23-4].

Interference Filters. As the name implies, interference filters rely on optical interference to produce relatively narrow bands of radiation [see curve (a), Figure 23-4]. An interference filter consists of a transparent dielectric (frequently calcium fluoride or magnesium fluoride) layer that occupies the space between two semitransparent metallic films coated on the inside surfaces of two glass plates. The thickness of the dielectric layer is carefully controlled and determines the wavelength of the transmitted radiation. When a perpendicular beam of collimated radiation strikes this array, a fraction passes through the first metallic layer while the remainder is reflected. The portion that is passed undergoes a similar partition upon striking the second metallic film. If the reflected portion from this second interaction is of the proper wavelength, it is partially reflected from the inner side of the first layer in phase with incoming light of the same wavelength. The result is that this particular wavelength is reinforced, while most others, being out of phase, suffer destructive interference.

Interference filters generally provide significantly narrower band widths (as low as 10 nm) and greater transmission of the desired wavelength than do absorption-type filters [see curve (b), Figure 23-4]. Interference filters that provide radiation bands from the ultraviolet up to about 6 μ in the infrared can be purchased.

SAMPLE AND SOLVENT CONTAINERS

The sample *cells* or *cuvettes* employed for photometric measurements are generally constructed of glass, although clear plastic containers have found some application.

In general, cells with windows that are perfectly normal to the direction of the beam are desirable in order to minimize reflection losses. Most instruments are provided with a pair of cells that have been matched with respect to light path and transmission characteristics to permit an accurate comparison of the power transmitted through the sample and the blank. The most common light path for visible radiation is 1 cm; matched, calibrated cells of this size are available from several commercial sources. Other path lengths from 0.1 cm (and shorter) to 10 cm can also be purchased. Transparent spacers for decreasing the path length of 1-cm cells to 0.1 cm are also available.

For reasons of economy, cylindrical cells are sometimes employed. Care must be taken to duplicate the position of such cells with respect to the beam; otherwise variations in reflection losses and path length lead to erroneous data.

The quality of absorbance data is critically dependent upon the way matched cells are used and maintained. Fingerprints, grease, or other deposits on the cell wall alter its transmission characteristics markedly. Thus, thorough cleaning before and after use is imperative; the surface of the windows must not be touched. Matched cells should never be dried by heating in an oven or over a flame, for such treatment can cause physical damage or a change in path length. They should be calibrated against each other regularly with an absorbing solution.

RADIATION DETECTORS

Photoelectric devices employed in photometers include the photovoltaic cell and the phototube. Both convert radiant energy to electrical.

To be useful, a radiation detector must respond over a broad wavelength range. It should, in addition, be sensitive to low levels of radiant power, respond rapidly to the radiation, produce an electrical signal that can be readily amplified, and have a relatively low noise level.[1] Finally, it is essential that the signal produced be directly proportional to the power of the beam striking it; that is,

$$G = k'P + k''$$
(23-1)

where G is the electrical response of the detector in units of current, resistance, or emf. The constant k' measures the sensitivity of the detector in terms of electrical response per unit of radiant power. Many detectors exhibit a small constant response, known as a *dark current k''*, when no radiation impinges on their surfaces. Instruments with detectors that have a dark-current response are ordinarily equipped with a compensating circuit that permits application of a countersignal to reduce k'' to zero. Thus, under most circumstances,

$$P = \frac{G}{k'}$$
(23-2)

and

$$P_0 = \frac{G_0}{k'}$$
(23-3)

where G and G_0 represent the electrical response of the detector to radiation passing through the solution and the blank, respectively. Substitution of Equation 23-2 and Equation 23-3 into Beer's law gives

$$\log \frac{P_0}{P} = \log \frac{k'G_0}{k'G} = \log \frac{G_0}{G} = A$$
(23-4)

---

[1] Noise in an electronic instrument refers to small random fluctuations which occur in the signal source, the detector, the amplifier, or the readout device. These fluctuations arise from such sources as vibration, pickup from 60-Hz electrical lines, temperature variations, and frequency or voltage fluctuations in the power supply.

Photovoltaic or Barrier-Layer Cells. A photovoltaic cell consists of a flat copper or iron electrode upon which is deposited a layer of semiconducting material, such as copper(I) oxide or selenium. On the surface of the semiconductor is a transparent metallic film of gold, silver, or lead, which serves as the second or collector electrode; the entire array is protected by a transparent envelope. The interface between the selenium and the metal film serves as a barrier to the passage of electrons. Irradiation with light, however, provides some electrons in the oxide layer with sufficient energy to overcome this barrier, and electrons flow from the semiconductor to the metal film. If the film is connected via an external circuit to the plate on the other side of the semiconducting layer, and if the resistance is not too great, a flow of electrons occurs. Ordinarily this current is large enough to be measured with a galvanometer or microammeter; if the resistance of the external circuit is small, the magnitude of the current is directly proportional to the power of the radiation striking the cell. Currents on the order of 10 to 100 μA are typical.

Photovoltaic cells are used primarily for the detection and measurement of radiation in the visible region. The typical cell has a maximum sensitivity at about 550 nm, and the response falls off continuously to perhaps 10% of the maximum at 250 and 750 nm. The wavelength response of a typical photovoltaic cell closely resembles that of the human eye.

The barrier-layer cell is a rugged, low-cost means for measuring radiant power. No external source of electric energy is required. On the other hand, the low internal resistance of the cell makes difficult the amplification of its output. Thus, although the barrier-layer cell provides a readily measured response at high levels of illumination, it suffers from lack of sensitivity at low levels. Finally, a photovoltaic cell exhibits fatigue, its response falling off upon prolonged illumination; proper circuit design and choice of experimental conditions largely eliminate this source of difficulty.

Phototubes. A second type of photoelectric device is the *phototube*, which consists of a semicylindrical cathode and a wire anode sealed inside an evacuated transparent envelope. The concave surface of the cathode supports a layer of photoemissive material that emits electrons upon being irradiated. When a potential is applied across the electrodes, the emitted electrons flow to the wire anode and a photocurrent results. For a given radiant intensity, the current produced is approximately one-fourth as great as that from a photovoltaic cell. In contrast, however, amplification is easily accomplished since the phototube has a very high electrical resistance. Figure 23-5 is a schematic diagram of a typical phototube arrangement.

A potential of about 90 V is impressed across the electrodes to assure proportionality between the number of electrons ejected from the photoemissive cathode and the radiant power of the beam striking it.

The photoemissive cathode surfaces of phototubes ordinarily consist of alkali metals or alkali-metal oxides, alone or combined with other metal oxides. The coating on the cathode determines the spectral response of a phototube.

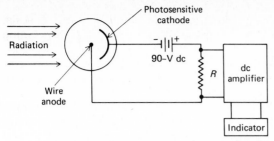

Figure 23-5 Schematic Diagram of a Phototube and Its Accessory Circuit. The current induced by the radiation causes a potential drop across the resistor R; this is amplified and measured by the indicator.

PHOTOMETER DESIGNS

Figure 23-6 presents schematic diagrams for two photometers. The first is a single-beam, direct-reading instrument consisting of a tungsten filament lamp, a lens to provide a parallel beam of light, a filter, and a photovoltaic cell. The current produced is indicated with a microammeter, the face of which is ordinarily scribed with a linear scale from 0 to 100. In some instruments, adjust-

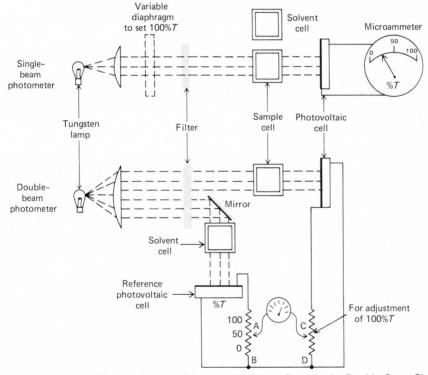

Figure 23-6 Schematic Diagram for a Single-Beam and a Double-Beam Photometer.

ment to 100% T with a blank in the light path involves changing the voltage applied to the lamp; in others, the aperture size of a diaphragm located in the light path is altered.

Also shown in Figure 23-6 is a schematic representation of a double-beam, null-type photometer. Here the light beam is split by a mirror, a part passing through the sample, and thence to a photovoltaic cell; the other half passes through the solvent to a similar detector. The currents from the two photovoltaic cells are passed through variable resistances; one of these is calibrated as a transmittance scale in linear units from 0 to 100. A sensitive galvanometer, which serves as a null indicator, is connected across the two resistances. When the potential drop across AB is equal to that across CD, no current passes through the galvanometer; under all other circumstances, a current flow is indicated. At the outset the solvent is introduced into both cells, and contact A is set at 100; contact C is then adjusted until no current is indicated. Introduction of the sample into the sample cell results in a decrease in radiant power reaching the working phototube and therefore a reduction in the potential drop across CD; this lack of balance is compensated for by moving A to a lower value. At balance, the percent transmittance is read directly from the scale.

Commercial photometers usually cost a few hundred dollars. The majority employ the double-beam principle because this design largely compensates for fluctuations in the source intensity due to voltage variations.

Filter Selection for Photometric Analysis. Photometers are generally supplied with several filters, each of which transmits a different portion of the spectrum. Selection of the proper filter for a given application is important inasmuch as the sensitivity of the measurement is directly dependent upon the filter. The color of the light absorbed is the complement of the color of the solution itself. For example, a liquid appears red because it transmits unchanged the red portion of the spectrum but absorbs the green. It is the intensity of green radiation that varies with concentration; a green filter should thus be employed. In general, the most suitable filter for a colorimetric analysis will be the color complement of the solution being analyzed. If several filters possessing the same general hue are available, the one that causes the sample to exhibit the greatest absorbance (or least transmittance) should be used.

Spectrophotometers

Spectrophotometers, in contrast to photometers, contain prism or grating monochromators which permit a continuous choice of wavelengths. Instruments exist for measurements in the ultraviolet, visible, and infrared regions. The basic design of a spectrophotometer is largely independent of wavelength.

SOURCES

The tungsten filament bulb is widely used in spectrophotometers as well as photometers. It is useful in the visible region as well as in the long ultraviolet (>320 nm) and the short infrared ($<2.5\,\mu$).

Ultraviolet Sources. The most common source of continuous radiation in the ultraviolet region is the hydrogen or deuterium lamp. Two types are encountered. The high-voltage variety employs potentials of 2000 to 6000 V to cause a discharge between aluminum electrodes; water cooling of the lamp is required if high radiation intensities are to be produced. In low-voltage lamps an arc is formed between a heated, oxide-coated filament and a metal electrode. About 40-V dc are required to maintain the arc.

Both high- and low-voltage lamps produce a continuous spectrum in the region between 180 and 375 nm. Quartz windows are required since glass absorbs strongly in this wavelength region.

Deuterium lamps produce continuous radiation of higher intensities than hydrogen lamps under the same operating conditions.

Hydrogen and deuterium lamps produce a truly continuous spectrum as a consequence of the excitation of the gaseous molecules to quantized excited states. The relaxation process involves dissociation of an excited molecule to produce a photon of ultraviolet radiation and two atoms of hydrogen in the ground state. The absorbed energy is released in two forms, namely, the kinetic energy of the two hydrogen atoms and the energy of the ultraviolet photon. Inasmuch as the kinetic energy imparted to the two atoms is not quantized, a broad spectrum of photon energies is obtained.

Infrared Sources. Continuous infrared radiation is produced by electric heating of an inert solid. A silicon carbide rod, called a *Globar*, provides radiant energy in the region of 1 to 40 μ when heated to perhaps 1500°C between a pair of electrodes. A *Nernst glower* produces radiation in the region between 0.4 and 20 μ (see Figure 23-2). This source is a rod of zirconium and yttrium oxides that is heated to about 1500°C by passage of current. A heated coil of nichrome wire is also a useful source of infrared radiation.

MONOCHROMATORS

A monochromator is a device which disperses radiation into its component wavelengths. Associated with the dispersing element is a system of lenses, mirrors, and slits that direct radiation of the desired wavelength from the monochromator toward the detector of the instrument. The construction materials for those parts of the monochromator through which the radiation is transmitted (and for the sample cells as well) determine the wavelength capability of a spectrophotometer. Figure 23-2 lists some of the most common materials and their transmittance ranges.

Prism Monochromators. Figure 23-7 is a schematic representation of a *Bunsen* monochromator, which employs a 60-deg prism for dispersion. Radiation is admitted through an entrance slit, is collimated by a lens, and then strikes the surface of the prism at an angle. Refraction occurs at both faces of the prism; the dispersed radiation is then focused on a slightly curved surface containing the exit slit. The desired wavelength can be caused to pass through this slit by rotation of the prism.

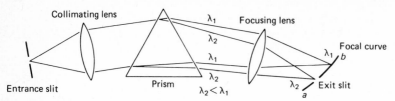

Figure 23-7 A Prism Monochromator.

The radiation emerging from a monochromator takes the form of a narrow rectangle or line which is a replica of the slit system. The distribution of wavelengths within this line is similar to that shown in curve (a), Figure 23-4. The effective band width becomes smaller as the thickness of the base of the prism is increased. To decrease bulk and still maintain a minimal band width, a *Littrow* prism is frequently employed in monochromator construction. This prism is obtained by halving the prism shown in Figure 23-7 along its vertical axis. The back face is then silvered so that radiation both enters and emerges from the same face; thus, the same dispersion is achieved as with a full prism. Figure 23-11 illustrates a monochromator that employs a Littrow prism.

Most monochromators are equipped with adjustable slits to permit a measure of control of the band width. A narrow slit width decreases the effective band width but also diminishes the intensity of the emergent beam. Thus, the minimum effective band width is often limited by the sensitivity of the detector.

For fixed slit widths, the effective band width of a prism monochromator increases continuously as the wavelength of radiation increases, because the refractive index of any prism material becomes less at longer wavelengths. For example, the quartz monochromator employed in one widely used instrument has an effective band width of 1.5 nm per millimeter of slit at 250 nm compared with 50 at 700 nm. Thus, to obtain a spectrum at a fixed band width (which is often desirable), it is necessary to decrease the slit width of a prism monochromator continuously as the wavelengths become longer.

In the visible region, glass is a more effective dispersing agent than quartz; monochromators employing glass optics are restricted to the visible region, however, because of the absorption by glass in the ultraviolet.

The distribution or dispersion of radiation from a continuous source along the line *ab* in Figure 23-7 for comparable glass and quartz prisms is shown in Figure 23-8.

For the infrared region, where glass and quartz are not transparent, prisms are constructed from materials such as sodium chloride, lithium fluoride, calcium fluoride, or potassium bromide (see Figure 23-2). All of these substances are susceptible to mechanical abrasion and attack by water vapor; consequently, their use requires special precautions.

Grating Monochromators. Dispersion of ultraviolet, visible, and infrared radiation can be brought about by passage of a beam through a *transmission*

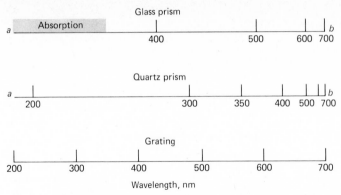

Figure 23-8   Dispersion Characteristics of Three Types of Monochromators.

grating or by reflection from a *reflection grating*. A transmission grating consists of a series of parallel grooves ruled on a piece of glass or other transparent material. A grating suitable for use in the ultraviolet and visible region has about 15,000 lines/in. It is vital that these lines be equally spaced throughout the several inches in length of the typical grating. Such gratings require elaborate apparatus for their production and are consequently expensive. Replica gratings are less costly. They are manufactured by employing a master grating as a mold for the production of numerous plastic replicas; the products of this process, while inferior in performance to an original grating, suffice for many applications.

When a transmission grating is illuminated by radiation from a slit, each groove acts as a new light source; interference among the multitude of beams results in dispersion of the radiation into its component wavelengths. If the dispersed radiation is focused on a plane surface, a spectrum consisting of images of the entrance slit is produced.

Reflection gratings are produced by ruling a polished metal surface or by evaporating a thin film of aluminum onto the surface of a replica grating. As shown in Figure 23-9, the radiation is reflected from each of the unruled portions, and interference among the reflected beams produces dispersion.

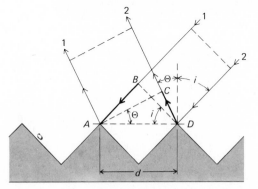

Figure 23-9   Diffraction at a Reflection Grating.

Example. Referring to Figure 23-9, the path of beam 1 differs from that of beam 2 by the amount $(\overline{AB} - \overline{CD})$; for constructive interference to occur, it is necessary that

$$n\lambda = (\overline{AB} - \overline{CD})$$

But it is readily seen that

$$\overline{AB} = d \sin i$$

where d is the spacing of the rulings and i is the angle of the incident beam to the normal. Similarly,

$$\overline{CD} = -d \sin \Theta$$

where Θ is the angle of reflection and the minus sign arises because the angle of reflection, by convention, is opposite in sign to the angle of incidence. The conditions for constructive interference then are

$$n\lambda = d(\sin i + \sin \Theta) \tag{23-5}$$

Equation 23-5 shows that several values of λ exist for a given diffraction angle Θ. Thus, if a first-order line ($n = 1$) of 800 nm is found at Θ, second-order (400 nm) and third-order (267 nm) lines also appear at this angle. Ordinarily, the first-order line is the most intense; it is possible to concentrate as much as 90% of the incident intensity in this order through suitable groove design. Higher order lines can be generally removed with suitable filters. For example, glass, which absorbs radiation below 350 nm, eliminates the high-order spectra associated with first-order radiation in most of the visible region.

A concave grating can be produced by ruling a spherical reflecting surface. Such a diffracting element serves also to focus the radiation on the exit slit and eliminates the need for a lens.

One advantage of a grating monochromator is that the dispersion along the focal plane of the exit slit is very nearly *independent of wavelength*; thus, a given slit setting will provide the same band width regardless of spectral region. Figure 23-8 contrasts the spectral dispersion of a typical grating monochromator with two prism monochromators.

Double Monochromators. Many modern monochromators contain two dispersing elements; that is, two prisms, two gratings, or a prism and a grating. This arrangement markedly reduces the amount of stray radiation[2] and also provides greater dispersion and spectral resolution. Furthermore, if one of the elements is a grating, higher order wavelengths are removed by the second element.

SAMPLE CONTAINERS

Figure 23-2 lists materials that are used to construct cells for various wavelength regions.

---

[2] Stray radiation refers to radiation reaching the exit slit of a monochromator as a result of reflections off the surfaces of the various components within the device. Stray radiation is undesirable because it usually differs in wavelength from that desired.

Detectors for the Ultraviolet and Visible Regions. Spectrophotometers for the visible and ultraviolet regions ordinarily employ a phototube (p. 522) or a photomultiplier tube rather than a photovoltaic cell. For work in the ultraviolet region, the tubes must have quartz or silica windows.

For the measurement of low radiant power, the *photomultiplier* tube offers advantages over the ordinary tube. Figure 23-10 is a schematic diagram of such a device. The cathode surface is similar in composition to that of a phototube, electrons being emitted upon exposure to radiation. The tube also contains additional electrodes (labeled 1 to 9 in Figure 23-10) called *dynodes*. Dynode 1 is maintained at a potential 90 V more positive than the cathode, and electrons are accelerated toward it as a consequence. Upon striking the dynode each photoelectron causes emission of several additional electrons; these, in turn, are accelerated toward dynode 2, which is 90 V more positive than dynode 1. Again, several electrons are emitted for each electron striking the surface. By the time this process has been repeated nine times, 10^6 to 10^7 electrons per photon will have been generated; this cascade is finally collected at the anode. The resulting enhanced current is then passed through the resistor R and can be further amplified and measured.

Photomultiplier tubes are easily damaged by exposure to strong radiation and can only be used for measurement of low radiant power. To avoid irreversible changes in their performance, the tubes must be mounted in light-tight housings, and care must be taken to avoid even momentary exposure to strong light.

Detectors for Infrared Radiation. Generally, infrared radiation is detected by measuring the temperature rise of a blackened material placed in the path of

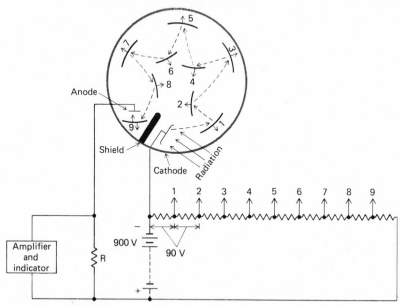

Figure 23-10   A Photomultiplier Tube.

the beam. Because the temperature changes resulting from absorption of the radiant energy are minute, close control of the ambient temperature is required if large errors are to be avoided.

One method of determining the temperature change involves use of a tiny thermocouple or a group of thermocouples called a *thermopile*. With this device, the electromotive force developed across a dissimilar metal junction is measured.

A *bolometer* is a second type of temperature detector; it consists of a resistance wire or a thermistor whose resistance varies as a function of temperature. Here it is the change in electrical resistance of the detector that is measured.

SPECTROPHOTOMETER DESIGNS

The instrumental components discussed in the previous section have been combined in various ways to produce dozens of commercial spectrophotometers that differ in complexity, performance characteristics, and cost; some of the simpler instruments can be purchased for a few hundred dollars, while the most sophisticated cost a hundred times this amount or more. No single instrument is best for all purposes; selection must be governed by the type of work for which the instrument is to be used.

Single- and Double-Beam Designs. Spectrophotometers, like photometers, are of single-beam or double-beam design. In double-beam instruments, the beam is split within the monochromator or after exiting from it by any of several means; the one beam then passes through the sample and the other through the solvent. In some instruments the power of the two beams is compared by twin detector-amplifier systems so that transmittance or absorbance data are obtained directly. In other instruments, radiation from the source is mechanically chopped so that pulses pass alternately through the sample and the blank. The resulting beams are then recombined and focused on a single detector. The pulsating electrical output from the detector is fed into an amplifier system that is programmed to compare the magnitude of the pulses and convert this information into transmittance or absorbance data.

Because comparison of the beam passing through the solvent with that passing through the sample is made simultaneously or nearly simultaneously, a split-beam instrument compensates for all but the most short-term electrical fluctuations, as well as for irregular performance in the source, the detector, and the amplifier. Therefore, the electrical components of a double-beam photometer need not be of as high quality as those for a single-beam instrument. Offsetting this advantage, however, is the greater number and complexity of components associated with double-beam instruments. Moreover, in photometers equipped with twin detectors and amplifiers, a close match between the components of the two systems is essential.

Single-beam instruments are particularly suited for the quantitative analysis that involves an absorbance measurement at a single wavelength. Here simplicity of instrumentation and the concomitant ease of maintenance offer real advantages. The greater speed and convenience of measurement, on the other hand, make the double-beam instrument particularly useful for qualitative analyses, where absorbance measurements must be made at numerous wave-lengths in order to obtain a spectrum. Furthermore, the double-beam device is readily adapted to continuous monitoring of absorbance; all modern recording spectrophotometers employ twin beams for this reason.

Both single-beam and double-beam instruments are available for ultraviolet and visible radiation. Commercial infrared spectrophotometers are always double-beam because they are ordinarily employed to scan and record a large spectral region. We shall limit our discussion of spectrophotometric design to two widely used single-beam instruments.

Single-Beam Instruments. Figure 23-11 is a schematic diagram of a high-quality, single-beam spectrophotometer—the Beckman DU-2. The first versions of this instrument appeared on the market over 30 years ago; it is one of the most widely used spectrophotometers of its kind.

The DU-2 spectrophotometer is equipped with quartz optics and can be operated in both the ultraviolet and visible regions of the spectrum. The instrument is provided with interchangeable radiation sources, consisting of a deuterium or hydrogen discharge tube for the lower wavelengths and a tungsten filament lamp for the visible and near infrared regions. A pair of mirrors reflect radiation through an adjustable slit into the monochromator compartment. After traversing the length of the instrument, the radiation is reflected into a

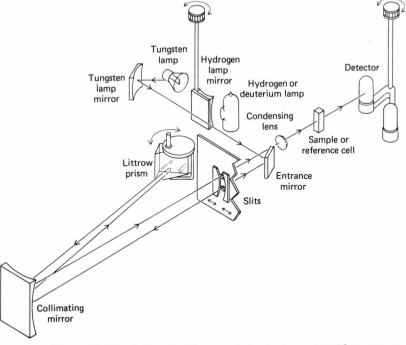

Figure 23-11   Schematic Diagram of the Beckman DU-2® Spectrophotometer. (By permission, Beckman Instruments, Inc., Fullerton, Calif.)

Littrow prism; by adjusting the position of the prism, light of the desired wavelength can be focused on the slit. The optics are so arranged that the entrance and the exit beams are displaced from one another on the vertical axis; thus, the exit beam passes above the entrance mirror as it enters the cell compartment.

Ordinarily, the cell compartment will accommodate as many as four rectangular 1-cm cells, any one of which can be positioned in the path of the beam by movement of a carriage arrangement. Compartments are also available that will hold both cylindrical cells and cells with 10-cm light paths.

The detectors are housed in a phototube compartment; control over the incoming radiation is achieved with a manually operated shutter in the path of the beam. Interchangeable detectors are provided—a red-sensitive phototube for the wavelength region beyond 625 nm and a photomultiplier tube for the range between 190 and 625 nm. The current from the phototube in the light path is passed through a fixed resistance of large magnitude; the potential drop across this resistor then gives a measure of the radiant power reaching the detector.

The DU-2 design achieves photometric accuracies as good as $\pm 0.2\%$ transmittance by employing high-quality electronic components that are operated well below their rated capacities. Narrow effective band widths (less than 0.5 nm) can be obtained throughout the spectral region by suitable adjustment of the slit. The instrument is particularly well-suited for research and quantitative analytical measurements that require absorbance data at a limited number of wavelengths.

The Bausch and Lomb Spectronic 20, shown schematically in Figure 23-12, may be considered as representative of instruments in which a degree of photometric accuracy is sacrificed in return for simplicity of operation and low cost. Its normal range is 350 to 650 nm, although this can be extended to 900 nm by the use of a red-sensitive phototube. The monochromator system consists of a reflection grating, lenses, and a pair of fixed slits. Because the grating produces a dispersion that is independent of wavelength, a constant band width of 20 nm is obtained throughout the entire operating region.

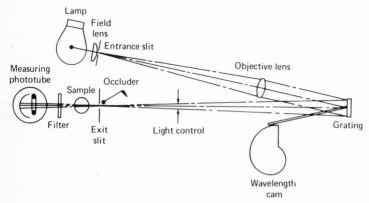

Figure 23-12 Schematic Diagram of the Bausch and Lomb Spectronic 20 Spectrophotometer. (By permission, Bausch and Lomb, Inc., Rochester, N.Y.)

The Spectronic 20 is an example of a direct-reading, single-beam spectrophotometer. The photocurrent from the detector is amplified and indicated by the position of a needle on the scale of a current-indicating meter. Since the amplified current is directly proportional to radiant power, the meter scale can be scribed to read transmittance and absorbance.

PROBLEMS

1. The following filters are available for photometric work.

Filter	Transmits All Wavelengths	Partially Transmits in Region	Absorbs All Wavelengths
A	<440 nm	440–500 nm	>500 nm
B	<500 nm	500–560 nm	>560 nm
C	<580 nm	580–640 nm	>640 nm
D	—	580–670 nm	<580 and >670 nm
E	>590 nm	540–590 nm	<540 nm
F	>560 nm	480–560 nm	<480 nm
G	>490 nm	430–490 nm	<430 nm

  *(a) What is the color of filter D?

  (b) What is the color of filter A?

  *(c) Select a filter (or combination of filters) suitable for the analysis of a blue solution.

  (d) Select a filter (or combination of filters) suitable for the analysis of a solution that absorbs strongly at 520 nm.

  *(e) Estimate the color transmitted through a combination of filters A and G.

  (f) What would be the color of solutions for which the combination of filters C and D would be appropriate?

*2. A Duboscq colorimeter was used to analyze a dilute iodine solution. A 4.54-cm light path through the unknown was found to match a 6.12-cm path through $5.00 \times 10^{-4} F$ I_2. Calculate the iodine concentration in the unknown.

3. A Duboscq colorimeter was used to analyze a dilute solution of CuA_3^-. A 7.96-cm light path through the unknown was found to match a 4.23-cm path through a standard that contained 12.3 ppm of Cu^{2+} and an excess of A^-. Calculate the copper concentration of the unknown.

24

APPLICATIONS OF MOLECULAR ABSORPTION

Molecular absorption spectroscopy has widespread application to qualitative and quantitative analysis. Infrared spectra are particularly useful for qualitative identification because they contain a large number of narrow peaks which serve to characterize the absorbing species (see, for example, Figure 24-2). In contrast, absorption in the ultraviolet and visible region usually consists of broad absorption bands (see curve *D*, Figure 24-1), which have less value for identification but great utility for quantitative analyses.

Types of Absorption Spectra

The general appearance of an absorption spectrum depends upon the complexity, the physical state, and the environment of the absorbing species. It is convenient to recognize two types of spectra, namely, those associated with atomic absorption and those resulting from molecular absorption.

ATOMIC ABSORPTION

When polychromatic ultraviolet and visible radiation is passed through a medium containing monatomic particles (such as gaseous mercury or sodium), only a relatively few well-defined frequencies are removed by absorption because the number of possible energy states for the particles is small. Excitation of atoms can occur only by an electronic transition in which an electron of the atom is raised to a higher energy level. For example, sodium has two $3p$ states which differ slightly in energy. Thus, two sharp and closely spaced peaks (589.0 and 589.6 nm), corresponding to the excitation of ground state $3s$ electrons to the two excited $3p$ states, appear in the sodium spectrum. Several other narrow absorption lines, corresponding to other permitted electronic transitions, are also observed (see, for example, curve A, Figure 24-1).

Ultraviolet and visible radiation have sufficient energy to cause transitions of the outermost or bonding electrons only. X-ray frequencies, on the other hand, are several orders of magnitude more energetic and capable of interacting with electrons closest to the nuclei of atoms. Absorption peaks corresponding to electronic transitions of the innermost electrons are thus observed in the X-ray region. Because the electrons involved do not participate in bonding, the X-ray absorption spectrum for an element tends to be independent of its chemical combination.

Regardless of the wavelength region involved, atomic absorption spectra typically consist of a limited number of discrete peaks. These peaks are useful for both qualitative identification of elements and their quantitative determination. Atomic-absorption spectroscopy is discussed in Chapter 25.

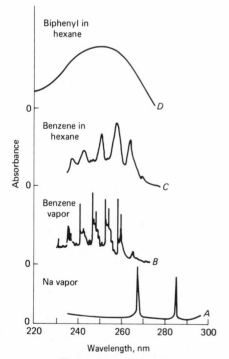

Figure 24-1 Some Typical Ultraviolet Absorption Spectra.

MOLECULAR ABSORPTION

Absorption by polyatomic molecules, particularly in the condensed state, is a considerably more complex process because the number of possible energy states is greatly increased. Here the total energy of a molecule is given by

$$E = E_{electronic} + E_{vibrational} + E_{rotational} \qquad (24\text{-}1)$$

where $E_{electronic}$ describes the energy associated with the various orbitals of the outer electrons of the molecule, while $E_{vibrational}$ refers to the energy of the molecule as a whole due to interatomic vibrations. The third term in Equation 24-1 accounts for the energy associated with the rotation of the molecule around its center of gravity.

Absorption in the Infrared and Microwave Regions. The three terms on the right in Equation 24-1 are ordered according to decreasing energy, with the average value for each term differing from the next by roughly two orders of magnitude. Pure rotational absorption spectra, which are free from electronic and vibrational transitions, can be brought about by microwave radiation which is less energetic than infrared. Spectroscopic studies of gaseous species in this region are important in gaining fundamental information concerning molecular behavior; applications of microwave absorption to analytical problems, however, have been limited.

Vibrational absorption occurs in the infrared region, where the energy of radiation is insufficient for electronic transitions. Here spectra exhibit narrow, closely spaced absorption peaks resulting from transitions among the various vibrational quantum levels (see Figure 24-2). Thus, absorption of a photon of radiant energy increases the amplitude of the vibration between atoms in the molecule; for absorption to occur, a net change in dipole must accompany the vibrational process. Variations in rotational levels may also give rise to a series of peaks for each vibrational state; with liquid or solid samples, however, rotation

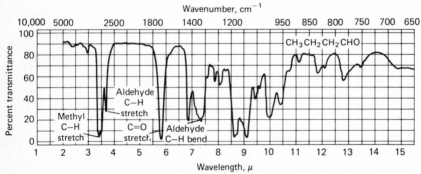

Figure 24-2 Infrared Spectrum for *n*-Butanal (*n*-Butyraldehyde). Note that transmittance rather than absorbance is plotted. [Spectrum from B. J. Zwolinski et al., "Catalog of Selected Infrared Spectral Data," Serial No. 225, Thermodynamics Research Center Data Project, Thermodynamics Research Center, Texas A & M University, College Station, Texas (loose-leaf data sheets extant, 1964).]

is often hindered or prevented, and the effects of these small energy differences are not detected. Thus, a typical infrared spectrum for a liquid such as that in Figure 24-2 consists of a series of vibrational peaks.

The number of individual ways a molecule can vibrate is largely related to the number of atoms, and thus the number of bonds, it contains. Even for a simple molecule, the number is large. Thus, n-butanal ($CH_3CH_2CH_2CHO$) has 33 vibrational modes, most differing from each other in energy. Not all of these vibrations give rise to infrared peaks; nevertheless, as shown in Figure 24-2, the spectrum for n-butanal is relatively complex.

Infrared absorption is not confined to organic molecules. Covalent metal-ligand bonds are also infrared-active in the longer wavelength region. Infrared spectrophotometric studies have thus provided much useful information about complex metal ions.

Absorption in the Ultraviolet and Visible Region. The first term in Equation 24-1 is ordinarily larger than the other two, and electronic transitions of outer electrons generally require energies corresponding to ultraviolet or visible radiation. In contrast to atomic absorption spectra, molecular spectra are often characterized by absorption bands that encompass a wide range of wavelengths (see curves B, C, and D, Figure 24-1). Here numerous vibrational and rotational energy states exist for each electronic state. Thus, for a given value of $E_{electronic}$ in Equation 24-1, values for E will exist that differ only slightly due to variations in $E_{vibrational}$ and/or $E_{rotational}$. As a consequence, the spectrum for a molecule often consists of a series of closely spaced absorption bands, such as that shown for benzene vapor in curve B of Figure 24-1. Unless a high resolution instrument is employed, individual bands may be undetected; as a result, the spectra appear as smooth curves. Furthermore, in the condensed state and in the presence of solvent molecules, the individual bands tend to broaden to give the type of spectra shown in the upper two curves in Figure 24-1.

Absorption Induced by a Magnetic Field. When electrons or the nuclei of certain elements are subjected to a strong magnetic field, additional quantized energy levels are produced; these owe their origins to magnetic properties possessed by such elementary particles. The difference in energy between the induced states is small, and transitions between them are brought about only by absorption of radiation of long wavelengths or low frequency. For nuclei, radio waves ranging from 10 to 200 MHz are generally employed; for electrons, microwaves with frequencies of 1000 to 25,000 MHz are absorbed.

Absorption by nuclei or electrons in magnetic fields is studied by *nuclear magnetic resonance* (nmr) and *electron spin resonance* (esr) techniques, respectively.

Absorbing Species

Absorption in the visible and ultraviolet regions can yield both quantitative as well as qualitative information about both inorganic and organic absorbing species.

ABSORPTION OF ULTRAVIOLET AND VISIBLE RADIATION BY ORGANIC COMPOUNDS

The electrons responsible for absorption of ultraviolet and visible radiation by organic molecules are of two types: (1) those that participate directly in bond formation and are thus associated with more than one atom; (2) nonbonding or unshared outer electrons that are largely localized about such atoms as oxygen, the halogens, sulfur, and nitrogen.

The wavelength at which an organic molecule absorbs depends upon how tightly its various electrons are bound. Thus, the shared electrons in single bonds, such as carbon-carbon or carbon-hydrogen, are firmly held and their excitations require energies corresponding to wavelengths in the vacuum ultraviolet region ($\lambda < 180$ nm). This region of the spectrum is not readily accessible because components of the atmosphere absorb here; thus absorption by single bonds is not important for analytical purposes.

Certain nonbonding electrons such as the unshared electrons in sulfur, bromine, and iodine are less energetic than the shared electrons of a saturated bond. Organic molecules containing these elements frequently exhibit useful peaks in the ultraviolet regions as a consequence.

Electrons making up double and triple bonds in organic molecules are relatively easily excited by radiation; thus, species containing unsaturated bonds generally exhibit useful absorption peaks. Unsaturated organic functional groups which absorb in the ultraviolet and visible regions are termed *chromophores*. Table 24-1 lists common chromophores and the approximate location of their absorption maxima. The data for position and peak intensity can serve only as a rough guide for identification purposes since the position of maxima is also affected by solvent and structural details. Furthermore, when two chromophores are conjugated, a shift in peak maxima to longer wavelengths usually occurs. Finally, peaks in the ultraviolet and visible regions are ordinarily broad because of vibrational effects; the precise determination of the position of the maximum is thus difficult.

ABSORPTION OF ULTRAVIOLET AND VISIBLE RADIATION BY INORGANIC SPECIES

The absorption spectra for most inorganic ions or molecules resemble those for organic compounds, with broad absorption maxima and little fine structure. In distinct contrast, the spectra for ions of the lanthanide and actinide elements consist of narrow, well-defined, and characteristic absorption peaks which are little affected by the type of ligand associated with the metal ion. In the lanthanide series, the transitions responsible for absorption involve the various energy levels of 4f electrons; in the actinide series, it is the 5f electrons that interact with radiation. In both series, these inner orbitals are largely screened from external influences by occupied orbitals with higher principal quantum numbers. As a consequence, the bands are narrow and relatively unaffected by the nature of the species bonded by the outer electrons.

With few exceptions, the ions and complexes of the 18 elements in the first two transition series are colored in one if not all of their oxidation states.

TABLE 24-1   Absorption Characteristics of Some Common Organic Chromophores

Chromophore	Example	Solvent	λ_{max} nm	ϵ_{max}
Alkene	$C_6H_{13}CH{=}CH_2$	n-Heptane	177	13,000
Conjugated alkene	$CH_2{=}CHCH{=}CH_2$	n-Heptane	217	21,000
Alkyne	$C_5H_{11}C{\equiv}C{-}CH_3$	n-Heptane	178	10,000
			196	2000
			225	160
Carbonyl	$CH_3\overset{O}{\overset{\|}{C}}CH_3$	n-Hexane	186	1000
			280	16
	$CH_3\overset{O}{\overset{\|}{C}}H$	n-Hexane	180	Large
			293	12
Carboxyl	$CH_3\overset{O}{\overset{\|}{C}}OH$	Ethanol	204	41
Amido	$CH_3\overset{O}{\overset{\|}{C}}NH_2$	Water	214	60
Azo	$CH_3N{=}NCH_3$	Ethanol	339	5
Nitro	CH_3NO_2	Isooctane	280	22
Nitroso	C_4H_9NO	Ethyl ether	300	100
			665	20
Nitrate	$C_2H_5ONO_2$	Dioxane	270	12
Aromatic	Benzene	n-Hexane	204	7900
			256	200

Here absorption involves transitions between filled and unfilled d-orbitals which differ in energy as a consequence of ligands bonded to the metal ions. The energy difference between the d-orbitals (and thus the position of the corresponding absorption peak) depends upon the oxidation state of the element, its position in the periodic table, and the kind of ligand bonded to its ion.

CHARGE-TRANSFER ABSORPTION[1]

For analytical purposes, one of the most important types of absorption by inorganic species is *charge-transfer absorption*, because the molar absorptivities of the band peaks are very large ($\epsilon_{max} > 10,000$). Thus, a highly sensitive means for detecting and determining the absorbing species is provided. Many inorganic and organic complexes exhibit charge-transfer absorption and are therefore called *charge-transfer complexes*. Common examples include the thiocyanate and phenolic complexes of iron(III), the *o*-phenanthroline complex of iron(II), the iodide complex of molecular iodine, and the ferro-ferricyanide complex responsible for the color of Prussian blue.

[1] For a brief discussion of this type of absorption, see C. N. R. Rao, *Ultra-Violet and Visible Spectroscopy, Chemical Applications*, chapter 11. New York: Plenum Press, 1967.

In order for a complex to exhibit a charge-transfer spectrum, it is necessary that one of its components have electron-donor characteristics and the other electron-acceptor properties. Absorption of radiation then involves transition of an electron of the donor group to an orbital that is largely associated with the acceptor. As a consequence, the excited state is the product of a kind of internal oxidation-reduction process.

A well-known example of charge-transfer absorption is observed in the iron(III)-thiocyanate ion complex. Absorption of a photon results in the transition of an electron from the thiocyanate ion to an orbital that is largely associated with the iron(III) ion. The product is thus an excited species involving predominantly iron(II) and the thiocyanate radical SCN. As with an intramolecular excitation, the electron, under ordinary circumstances, returns to its original state after a brief period. Occasionally, however, dissociation of an excited complex may occur to produce a photochemical oxidation-reduction process.

In most charge-transfer complexes involving a metal ion, the metal serves as the electron acceptor. An exception is observed in the o-phenanthroline complex of iron(II) or copper(I), where the ligand is the acceptor and the metal ion is the donor. Other examples of this type of complex are known.

Applications of Absorption Spectroscopy to Qualitative Analysis

Absorption spectroscopy provides a useful tool for qualitative analysis. Identification of a pure compound involves comparing the spectral details (maxima, minima, and inflection points) of the unknown with those of pure compounds; a close match is considered good evidence of chemical identity, particularly if the spectrum of the unknown contains a number of sharp and well-defined peaks. Absorption in the infrared region is particularly useful for qualitative purposes because of the wealth of fine structure that exists in the spectra of many compounds. The application of ultraviolet and visible spectrophotometry to qualitative analysis is more limited because the absorption bands tend to be broad and hence lacking in detail. Nevertheless, spectral investigations in this region frequently provide useful qualitative information concerning the presence or absence of certain functional groups (such as carbonyl, aromatic, nitro, or conjugated diene) in organic compounds. A further important application involves the detection of highly absorbing impurities in nonabsorbing media; if an absorption peak for the contaminant has a sufficiently high absorptivity, the presence of trace amounts can be readily established.

QUALITATIVE TECHNIQUES

Solvents. In choosing a solvent, consideration must be given not only to its transparency but also to its possible effects upon the absorbing system. Quite generally, polar solvents such as water, alcohols, esters, and ketones tend to obliterate spectral fine structure arising from vibrational effects; spectra that more closely approach those of the gas phase are most likely to be observed in nonpolar solvents such as hydrocarbons. In addition, the positions of absorption maxima are influenced by the nature of the solvent. Clearly, it is important to employ identical solvents when absorption spectra are being compared for identification purposes.

Table 24-2 lists common solvents for studies in the ultraviolet and visible

TABLE 24-2   Solvents for the Ultraviolet and the Visible Regions

Solvent	Lower Wavelength Limit, nm	Solvent	Lower Wavelength Limit, nm
Water	180	Carbon tetrachloride	260
Ethanol	220	Diethyl ether	210
Hexane	200	Acetone	330
Cyclohexane	200	Dioxane	320
Benzene	280	Cellosolve	320

regions and the approximate wavelengths below which they cannot be used because of absorption. These minima are strongly dependent upon the purity of the solvent.

No single solvent is transparent throughout the infrared region. Water is generally avoided, not only because it absorbs broadly but also because it attacks the infrared-transparent materials required for windows and optics. Carbon tetrachloride and carbon disulfide are the two most satisfactory solvents; the former is useful in the region up to about 7.6 μ and the latter from this wavelength to 15 μ. Many compounds are not soluble in these solvents and must be examined in the pure (or neat) form; here, very thin films of the sample are required.

Effect of Slit Width. In order to obtain an absorption spectrum useful for qualitative comparison, absorbance data should be collected with the narrowest possible band width. Otherwise significant details of the spectrum may be lost. This effect is demonstrated in Figure 24-3.

The band width of a monochromator is determined not only by the dispersion characteristics of the prism or grating but also by the width of the exit slit (Chapter 23). Most spectrophotometers are equipped with a variable slit; the operator thus has some control of the band width to be employed. The descriptive brochures supplied with most instruments give the slit setting needed to achieve a given effective band width; with prism instruments this setting must be varied as a function of wavelength if a constant band width is to be maintained.

The two factors that limit the minimum width to which the slit of a spectrophotometer may be set are the intensity of the source and the sensitivity of the detector. At wavelengths where either becomes low, wider slits and consequently wider band widths must be employed to provide a beam with sufficient power for accurate measurement.

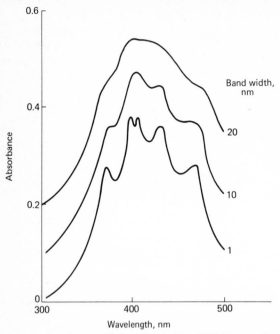

Figure 24-3 Effect of Band Width upon Spectra of Identical Solutions. Note that spectra have been displaced from one another by 0.1 in absorbance.

Effect of Scattered Radiation. The radiation emitted from the exit slit of a monochromator often contains small amounts of stray radiation comprising wavelengths that differ considerably from that to which the instrument is set. This contamination results from reflections off the various optical and housing surfaces within the monochromator. Scattered radiation amounts at most to a few tenths of a percent at all wavelength settings of high-quality instruments; under usual circumstances, its presence has no detectable effect on absorbance data. When measurements are made at the wavelength extremes of an instrument, however, scattered radiation may introduce an appreciable error. For example, the typical spectrophotometer for the visible region is equipped with glass optics and cells that begin to absorb in the region between 350 and 400 nm. Furthermore, such instruments employ phototubes or photocells that exhibit a maximum photoelectric response between 500 and 700 nm; at 350 nm, then, the response may be only 10% (or less) of the maximum. Finally, an instrument of this kind employs a tungsten filament source, the maximum energy of which is in the upper visible region; at 350 nm, its output is but a fraction of its maximum. These three factors, then, limit the low wavelength capability of the instrument; in order to obtain absorbance data in this region, it is necessary to employ maximum amplification of the detector signal, maximum intensity of the source, and a relatively wide slit width. Only under these circumstances can the instrument be made to read 100% transmittance with the blank in the light path. The effect of scattered radiation with longer wavelengths now becomes greatly magnified. First, the source generates a much greater intensity of wavelengths that can be scattered; second, the wide slit width enhances the probability of scattered radiation emerging from the monochromator; third, the scattered radiation is absorbed to a lesser extent by the cell walls; and finally, the detector is much more responsive to the wavelengths of the scattered radiation. As a consequence, the absorbance measured by such an instrument at 350 nm may be due as much to longer scattered wavelengths as to that to which the monochromator is set.

IDENTIFICATION OF COMPOUNDS

The certainty with which an identification can be made by absorption measurement is directly related to the number of similar spectral features (peaks, minima, and inflection points) that can be observed between the unknown and an authentic reference standard. For the ultraviolet and visible regions, the number of these characteristic features is often limited; identification based on a single pair of spectra may thus be ambiguous. Occasionally, the identity of unknown and standard can be confirmed by comparing their spectra in other solvents, at different pH values, or following suitable chemical treatment. Unfortunately, however, the electronic absorption peaks of many chromophores are not greatly influenced by structural features of attached nonabsorbing groups; as a consequence, only the absorbing functional groups can be identified, and other means must be used to determine the remaining features of the compound under study.

Ultraviolet and Visible Regions. Catalogs of ultraviolet and visible absorption data for organic compounds are available in several publications.[2] These compilations frequently are of help in organic qualitative work.

Infrared Region. An infrared absorption spectrum, even for relatively simple compounds, contains a bewildering array of sharp peaks and minima. It is this multiplicity of peaks, however, that imparts specificity to the spectrum; no two compounds give identical spectra. Thus, identity between the spectra of an unknown and an authentic sample is widely accepted as proof of identity of composition.

Peaks useful for the identification of certain functional groups are located in the shorter wavelength region of the infrared (below about 7.5 μ or 1300 cm^{-1}); the positions of the maxima in this region are only slightly affected by the carbon skeleton to which the groups are attached. Investigation of this portion of the spectrum thus provides considerable information regarding the overall

[2] American Petroleum Institute, *Ultraviolet Spectral Data, A.P.I. Research Project 44.* Pittsburgh: Carnegie Institute of Technology; H. M. Hershenson, *Ultraviolet and Visible Absorption Spectra Index for 1930–54.* New York: Academic Press, Inc., 1956; M. J. Kamlet and H. E. Ungnade, Eds., *Organic Electronic Spectral Data,* 2 vols. New York: Interscience Publishers, Inc., 1960; *Sadtler Ultraviolet Spectra.* Philadelphia: Sadtler Research Laboratories; R. A. Friedel and M. Orchin, *Ultraviolet Spectra of Aromatic Compounds.* New York: John Wiley & Sons, Inc., 1951; American Society for Testing Materials, Committee E-13, Philadelphia. Ultraviolet spectra are coded on IBM cards.

TABLE 24-3   Some Characteristic Infrared Absorption Peaks

Functional Group	Absorption Peaks	
	Wave Number, cm^{-1}	Wavelength, μ
O—H (alcohol)	3000–3500	2.8–3.3
—NH$_2$ (primary amine)	3300–3500	2.9–3.0
C—H (aromatic)	3000–3150	3.2–3.3
C—H (aliphatic)	2850–3000	3.3–3.5
C≡N	2200–2400	4.2–4.6
C=O (ester)	1700–1750	5.7–5.8
C=O (acid)	1670–1740	5.8–6.0
C=O (aldehyde and ketone)	1650–1750	5.8–6.1
C=C	1610–1670	6.0–6.2

constitution of the molecule under investigation. Table 24-3 gives the positions of characteristic maxima for some common functional groups.[3]

In most instances, identification of the functional groups in a molecule is not sufficient to permit positive identification of the compound, and the entire spectrum must be compared with that of known compounds. Collections of spectra are available for this purpose.[4]

Quantitative Analysis by Absorption Measurements

Absorption spectroscopy, particularly in the ultraviolet and visible regions, is one of the most useful tools available to the chemist for quantitative analysis. Important characteristics of spectrophotometric and photometric methods include:

1. *Wide applicability.* Numerous inorganic and organic species absorb in the ultraviolet and visible ranges, and are thus susceptible to quantitative determination. In addition, many nonabsorbing species can be analyzed after conversion to absorbing species by suitable chemical treatment.
2. *High sensitivity.* Molar absorptivities in the range of 10,000 to 40,000 are common, particularly for the charge-transfer complexes of inorganic species. Thus, analyses for concentrations in the range of 10^{-4} to 10^{-5} M are commonplace; the lower limit can sometimes be extended to 10^{-6} or even 10^{-7} through suitable procedural changes.
3. *Moderate to high selectivity.* Through the judicious choice of conditions, it may be possible to locate a wavelength region in which the analyte is the only absorbing component in a sample. Furthermore, where overlapping absorption bands do occur, compensations based on additional measurements at other wavelengths are sometimes possible.

---

[3] For more detailed information, see N. B. Colthup, *J. Opt. Soc. Amer.*, **40**, 397 (1950); R. M. Silverstein and G. C. Bassler, *Spectrometric Identification of Organic Compounds*, 2d ed. New York: John Wiley & Sons, Inc., 1967.

[4] American Petroleum Institute, *Infrared Spectral Data, A.P.I. Research Project 44*. Pittsburgh: Carnegie Institute of Technology; *Sadtler Standard Spectra*. Philadelphia: Sadtler Research Laboratories.

4. *Good accuracy.* For the typical spectrophotometric or photometric procedure, the relative error in concentration measurements lies in the range of 1 to 3%. With special techniques, errors can often be decreased to a few tenths of a percent.

5. *Ease and convenience.* Spectrophotometric or photometric measurements are easily and rapidly performed with modern instruments. Furthermore, the methods are often readily automated for routine analyses. As a consequence, absorption methods are widely used for chemical analysis, continuous monitoring of atmospheric and water pollutants, and control of industrial processes.

SCOPE

The applications of quantitative absorption methods are not only numerous but also touch upon every area in which quantitative chemical information is required. The reader can obtain a notion of the scope of spectrophotometry by consulting a series of review articles published periodically in *Analytical Chemistry*[5] as well as monographs on the subject.[6]

Applications to Absorbing Species. Table 24-1 lists some of the most common organic chromophoric groups. Often, when a molecule contains several of these groups in which the unsaturated bonds are conjugated, absorption is shifted to the visible region of the spectrum. Spectrophotometric analysis for any organic compound containing one or more of these groups is potentially feasible; many examples of this type of analysis are found in the literature.

A number of inorganic species also absorb and are thus susceptible to direct determination; we have already mentioned the various transition metals. In addition, other species show characteristic absorption; examples include nitrate ion, the oxides of nitrogen, the elemental halogens, and ozone.

Applications to Nonabsorbing Species. Numerous reagents react with nonabsorbing inorganic species to yield products that absorb strongly in the ultraviolet or visible regions. The successful application of such reagents to quantitative analysis usually requires that the color-forming reaction be forced near completion. It should be noted that these reagents are also frequently employed for the determination of an absorbing species, such as a transition metal ion; the molar absorptivity of the product will often be orders of magnitude greater than that of the uncombined ion.

A host of complexing agents have been employed for the determination of inorganic species. Typical inorganic reagents include thiocyanate ion for iron, cobalt, and molybdenum, the anion of hydrogen peroxide for titanium, vanadium, and chromium, and iodide ion for bismuth, palladium, and tellurium. Of even more importance are organic chelating agents which form stable, colored complexes

---

[5] For ultraviolet spectrometry, see R. A. Hummel and D. C. Kaufman, *Anal. Chem.*, **46,** 354R (1974); **44,** 535R (1972). For light-absorption spectrometry, see D. F. Boltz and M. G. Mellon, *Anal. Chem.*, **46,** 227R (1974); **44,** 300R (1972).

[6] See, for example, E. B. Sandell, *Colorimetric Determination of Traces of Metals,* 3d ed. New York: Interscience Publishers, Inc., 1959; D. F. Boltz, Ed., *Colorimetric Determination of Nonmetals.* New York: Interscience Publishers, Inc., 1958; F. D. Snell and C. T. Snell, *Colorimetric Methods of Analysis,* 3d ed. 4 vols. Princeton, N.J.: D. Van Nostrand Co., Inc., 1959.

with cations. Examples include *o*-phenanthroline for the determination of iron, dimethylglyoxime for nickel, diethyldithiocarbamate for copper, and diphenylthiocarbazone for lead.

Certain nonabsorbing organic functional groups are also determined by absorption procedures. For example, low-molecular-weight aliphatic alcohols react with cerium(IV) to produce a red 1:1 complex that can be employed for quantitative purposes.

Quantitative Infrared Methods. Quantitative measurements in the infrared region are not different in principle from similar measurements in the ultraviolet and visible region. Several practical problems, however, prevent the attainment of comparable accuracies. These include the necessity of using very narrow cell widths that are difficult to reproduce, the high probability of overlap in absorption among the components in the sample (owing to the complexity of their spectra), and the narrowness of the peaks that often leads to deviations from Beer's law.

PROCEDURAL DETAILS

Before a photometric or spectrophotometric analysis can be undertaken, it is necessary to develop a set of working conditions and prepare a calibration curve that relates concentration to absorbance.

Selection of Wavelength. In a spectrophotometric analysis, absorbance measurements are preferably made at a wavelength corresponding to an absorption peak because the change in absorbance per unit of concentration is greatest; good adherence to Beer's law can be expected (p. 510), and the measurements are less sensitive to uncertainties arising from failure to reproduce precisely the wavelength setting of the instrument. Occasionally, a wavelength other than a peak may be appropriate for a particular analysis to avoid interference from other absorbing substances. In this event the region selected should, if possible, be one in which the change in absorptivity with wavelength is not too great.

Variables That Influence Absorbance. Common variables that influence the absorption spectrum of a substance include the nature of the solvent, the pH of the solution, the temperature, high electrolyte concentrations, and the presence of interfering species. The effects of these variables must be known, and a set of analytical conditions must be chosen such that the absorbance will not be materially influenced by small, uncontrolled variations in their magnitudes.

Determination of the Relationship between Absorbance and Concentration. The calibration standards for a photometric or spectrophotometric analysis should approximate the overall composition of the actual samples and should cover a reasonable analyte concentration range. Seldom, if ever, is it safe to assume adherence to Beer's law and use only a single standard to determine the molar absorptivity. It is even less prudent to base the results of an analysis on a literature value for the molar absorptivity.

The difficulties that attend production of a set of standards whose overall composition closely resembles that of the sample can be formidable if not insurmountable. Under these circumstances, a *standard addition* approach may prove useful. Here a known quantity of standard is added to an additional aliquot of the sample. Provided Beer's law is obeyed, the analyte concentration can be calculated.

Example. A 2.00-ml urine specimen was diluted to 100 ml. Photometric analysis for the phosphate in a 25.0-ml aliquot yielded an absorbance of 0.428. Addition of 1.00 ml of a solution containing 0.050 mg of phosphate to a second 25.0-ml aliquot resulted in an absorbance of 0.517. Calculate the milligrams phosphate per milliliter of sample. We must first correct the second measurement for dilution. Thus,

$$\text{corrected absorbance} = 0.517 \times \frac{26.0}{25.0} = 0.538$$

$$\text{absorbance due to 0.050 mg phosphate} = 0.538 - 0.428 = 0.110$$

$$\text{mg phosphate in } \frac{25.0}{100} \text{ of specimen} = \frac{0.428}{0.110} \times 0.050 = 0.195$$

Thus,

$$\text{mg phosphate/ml of specimen} = \frac{100}{25.0} \times 0.195 \times \frac{1}{2.00} = 0.39$$

Analysis of Mixtures of Absorbing Substances. The total absorbance of a solution at a given wavelength is equal to the sum of the absorbances of the individual components present. This relationship makes possible the analysis of the individual components of a mixture even if an overlap in their spectra exists. Consider, for example, the spectra of M and N, shown in Figure 24-4. There is obviously no wavelength at which the absorbance of this mixture is due simply to one of the components; thus, an analysis for either M or N is

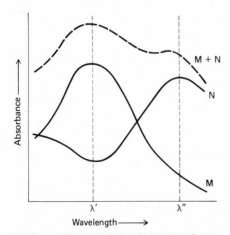

Figure 24-4 Absorption Spectrum of a Two-Component Mixture.

impossible by a single measurement. However, the absorbances of the mixture at two wavelengths λ' and λ'' may be expressed as follows:

$$A' = \epsilon'_M bc_M + \epsilon'_N bc_N$$

$$A'' = \epsilon''_M bc_M + \epsilon''_N bc_N$$

The four molar absorptivities ϵ'_M, ϵ'_N, ϵ''_M, and ϵ''_N can be evaluated from individual standard solutions of M and of N, or better, from the slopes of their Beer's law plots. The absorbances, A' and A'', for the mixture are experimentally determinable as is b, the cell thickness. Thus, from these two equations the concentration of the individual components in the mixture, c_M and c_N, can be readily calculated. These relationships are valid only if Beer's law is followed. The greatest accuracy in an analysis of this sort is attained by choosing wavelengths at which the differences in molar absorptivities are large.

Mixtures containing more than two absorbing species can be analyzed, in principle at least, if a further absorbance measurement is made for each added component. The uncertainties in the resulting data become greater, however, as the number of measurements increases.

ANALYTICAL ERRORS IN ABSORPTION ANALYSIS

In many (but certainly not all) absorption methods, the major source of indeterminate error lies in the absorbance measurement. Determination of absorbance requires evaluation of the power of a beam that has passed through the solution, as well as its power after passage through the blank. Because of the logarithmic relationship between these quantities and concentration, however, the influence of uncertainties in the power measurements on the analytical result is not immediately obvious.

The measurement of radiant power frequently involves an absolute uncertainty that is independent of the magnitude of the power. As a consequence, the ratio P/P_0 (that is, the transmittance) is affected by a constant absolute error over a considerable range. To see the effect of this constant error in transmittance upon the uncertainty of the results of an analysis, it is convenient to write Beer's law in the form

$$-\log T = \epsilon bc \tag{24-2}$$

After converting the left-hand side to a natural logarithm, the derivative of Equation 24-2 is

$$-\frac{0.434}{T} dT = \epsilon b dc \tag{24-3}$$

Dividing Equation 24-3 by Equation 24-2 and rearranging gives

$$\frac{dc}{c} = \frac{0.434}{T \log T} dT \tag{24-4}$$

and this may be written as

$$\frac{\Delta c}{c} = \frac{0.434}{T \log T} \Delta T \tag{24-5}$$

TABLE 24-4 Variations in the Percent Error in Concentration as a Function of Transmittance and Absorbance from Equation 24-5

(Error in Transmittance Measurement, ΔT, Assumed to Be ± 0.005, or 0.5%)

Transmittance, T	Absorbance, A	Percent Error in Concentration, $\frac{\Delta c}{c} \times 100$
0.95	0.022	± 10.25
0.90	0.046	± 5.27
0.80	0.097	± 2.80
0.70	0.155	± 2.00
0.60	0.222	± 1.63
0.50	0.301	± 1.44
0.40	0.399	± 1.36
0.30	0.523	± 1.38
0.20	0.699	± 1.55
0.10	1.000	± 2.17
0.030	1.523	± 4.75
0.020	1.699	± 6.39

The quantity $\Delta c/c$ is a measure of the *relative error* in a concentration measurement resulting from an absolute error ΔT in the measurement of T or P/P_0. The reader should note that the relative error in concentration varies as a function of the magnitude of T. This relationship is shown by the data in Table 24-4, in which an absolute error in T of 0.005 or 0.5% was assumed for a series of transmittance measurements;[7] the relative error in concentration passes through a minimum at an absorbance of about 0.4. (By setting the derivative of Equation 24-5 equal to zero, it can be shown that this minimum occurs at a transmittance of 0.368 or an absorbance of 0.434.) A plot of the data in Table 24-4 is shown by curve D in Figure 24-5.

The indeterminate error in transmittance measurements for commercial spectrophotometers will range between ± 0.002 and ± 0.1, with the figure ± 0.005 employed in Table 24-4 being typical. Concentration errors of about 1 to 2% relative are thus inherent in most absorption methods when absorbances measured lie in the range of 0.15 to 1.0. The higher error associated with more strongly absorbing samples can usually be avoided by suitable dilution prior to the instrumental measurement. Relative errors greater than 2% are inevitable, however, when absorbances are less than 0.1.

It is important to emphasize that these conclusions are based on the assumption that the limiting uncertainty in an analysis is a constant photometric error ΔT that is independent of transmittance. The potential for other types of errors always exists, of course, and the magnitude of these may indeed vary with transmittance. Examples include errors arising from sample preparation,

[7] C. F. Hiskey, *Anal. Chem.*, **21**, 1440 (1949), recommends that ΔT be taken as twice the average deviation in transmittance obtained by repeated measurement of T for a single solution.

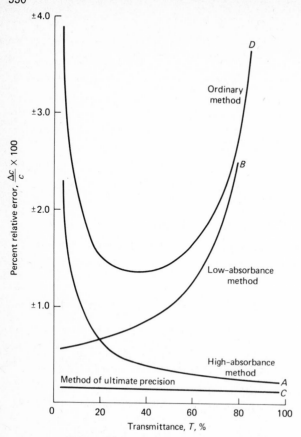

Figure 24-5 Relative Errors in Spectrophotometric Analysis. Curve *D*: ordinary method; reference solution has transmittance of 100%. Curve *A*: high-absorbance method; reference solution has transmittance of 10%. Curve *B*: low-absorbance method; reference solution has transmittance of 90%. Curve *C*: method of ultimate precision; reference solutions have transmittances of 20% and 30%, respectively. ΔT is 0.5% for each curve. See Figure 24-6 for the type of scale expansion corresponding to curves *A*, *B*, and *C*.

imperfections in the cells, uncertainties in the wavelength setting, and source fluctuations.

DIFFERENTIAL ABSORPTION METHODS[8]

By modifying the way in which absorbance is measured, it is often possible to decrease the relative concentration error associated with instrumental uncertainties to a few tenths of 1%. These methods, called *differential absorption methods*, employ solutions of the analyte for adjusting the zero and/or the 100%

[8] For a complete analysis of these methods, see C. N. Reilley and C. M. Crawford, *Anal. Chem.*, **27,** 716 (1955); C. F. Hiskey, *Anal. Chem.*, **21,** 1440 (1949).

TABLE 24-5   Comparison of Methods for Adsorption Measurements

Method	Imposed in Beam for Indicator Setting of	
	Zero	100% T
Ordinary	Shutter	Solvent
High absorbance	Shutter	Standard solution less concentrated than sample
Low absorbance	Standard solution more concentrated than sample	Solvent
Ultimate precision	Standard solution more concentrated than sample	Standard solution less concentrated than sample

transmittance setting of the photometer or spectrophotometer rather than the shutter and the solvent. The three types of differential methods are compared with the ordinary method in Table 24-5.

High-Absorbance Method. In the high-absorbance procedure, the zero adjustment is carried out in the usual way with the shutter imposed between the beam and the detector. The 100% transmittance adjustment, however, is made with a standard solution of the analyte which is less concentrated than the sample in the light path. Finally, the standard is replaced by the sample, and a relative transmittance is read directly. As shown in diagram *A* of Figure 24-6, the effect of this modification is to expand a small portion of the transmittance scale to a full 100%; thus, the transmittance of the reference standard employed is 10% *when compared with pure solvent* while the sample exhibits a 4% transmittance against the same solvent. That is, the sample transmittance is four-tenths that of the standard when compared with a common solvent. When the standard is substituted as the reference, however, its transmittance becomes 100%. The transmittance of the sample remains four-tenths that of the reference standard; here, then, it is 40%.

If the instrumental uncertainty in the measurement of absorbance is not affected by the modification, use of the standard as a reference brings the transmittance into the middle of the scale where instrumental uncertainties have a minimal effect on the relative concentration error.

It can be shown that a linear relationship exists between concentration and the relative absorbance measured by the high-absorbance technique.[9] The method is particularly useful for the analysis of samples with absorbances greater than unity.

Low-Absorbance Method. In the low-absorbance procedure, a standard solution somewhat more concentrated than the sample is employed in lieu of the shutter for zeroing the indicator scale; the full-scale adjustment is made in

[9] See D. A. Skoog and D. M. West, *Principles of Instrumental Analysis*, p. 94. New York: Holt, Rinehart and Winston, Inc., 1971.

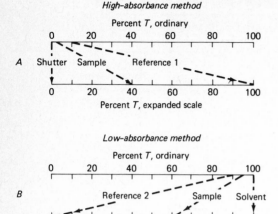

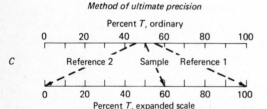

Figure 24-6   Scale Expansion by Various Methods.

the usual way with the solvent in the beam. The sample transmittance is then obtained by replacing the solvent with the sample. The effect of this procedure is shown in diagram *B* of Figure 24-6; note that, again, a small segment of the scale is expanded to 100% transmittance, and that the transmittance of the sample now lies near the center of the expanded scale.

The low-absorbance method is particularly applicable to samples having absorbances less than 0.1. It can be shown that here a nonlinear relationship exists between relative absorbance and concentration. As a consequence, a number of standard solutions must be employed to establish accurately the calibration curve for the analysis.

The Method of Ultimate Precision. The two techniques just described can be combined to give the method of ultimate precision. As shown in diagram *C* of Figure 24-6, a scale expansion is again achieved; here two reference solutions, one having a smaller and one having a greater transmittance than the sample, are employed to adjust the 100% and the 0% scale readings. A nonlinear relationship between relative absorbance and concentration is again observed.

Instrumental Requirements for Precision Methods. In order to employ the low-absorbance method, it is clearly necessary that the spectrophotometer possess a dark-current compensating circuit that is capable of offsetting larger currents than are normally produced when no radiation strikes the photoelectric detector. For the high-absorbance method, on the other hand, the instrument must have a sufficient reserve capacity to permit setting the indicator at 100% transmittance when an absorbing solvent is placed in the beam path. Here the full-scale reading may be realized either by increasing the radiation intensity (most often by widening the slits) or by increasing the amplification of the photoelectric current. The method of ultimate precision requires both of these instrumental qualities.

The capability of a spectrophotometer to be set to full scale with an absorbing solution in the radiation path will depend both upon the quality of its monochromator and the stability of its electronic circuit. Furthermore, this capacity will be wavelength-dependent since the intensity of the source and the sensitivity of the detector change as the wavelength is varied. In regions where the intensity and sensitivity are low, an increase in slit width may be necessary to realize a full-scale setting; under these circumstances, scattered radiation may lead to errors unless the quality of the monochromator is high. Alternatively, a very high current amplification may be necessary; unless the electronic stability is good, significant photometric error may result.

Precision Gain by Differential Methods. It is possible to derive relationships analogous to Equation 24-5 for the three differential methods.[10] Figure 24-5 compares the corresponding error curves with that of the ordinary method. From curve *A* it is apparent that a significant gain in precision results when the high-absorbance method is applied to samples having high absorbances or low transmittances. The plot was based upon the reference standard employed in diagram *A* of Figure 24-6; that is, for a reference having a normal transmittance of 10%. The error curve approaches that for the ordinary procedure (see curve *D*, Figure 24-5) as the percent transmittance of the reference increases and approaches 100. Error curves for the low-absorbance method (see curve *B*, Figure 24-5) and the method of ultimate precision (see curve *C*, Figure 24-5) are based on the corresponding data shown in Figure 24-6.

Photometric Titrations[11]

Photometric or spectrophotometric measurements can be employed to advantage in locating the equivalence point in a titration. The end point in a direct photometric titration is the result of a change in the concentration of a reactant or a product, or both; clearly, at least one of these species must absorb radiation at

[10] See D. A. Skoog and D. M. West, *Principles of Instrumental Analysis*, pp. 94–99. New York: Holt, Rinehart and Winston, Inc., 1971.
[11] For further information concerning this technique, see J. B. Headridge, *Photometric Titrations*. New York: Pergamon Press, Inc., 1961.

the wavelength selected. In the indirect method, the absorbance of an indicator is observed as a function of titrant volume.

TITRATION CURVES

A photometric titration curve is a plot of corrected absorbance versus the volume of titrant. If conditions are chosen properly, the curve will consist of two straight-line portions with differing slopes, one occurring at the outset of the titration and the other located well beyond the equivalence point. A marked curvature is frequently observed in the equivalence-point region; the end point is taken as the intersection of extrapolated straight-line portions. Figure 24-7 shows some typical titration curves. Titration of a nonabsorbing species with a colored titrant that is decolorized by the reaction produces a horizontal line in the initial stages followed by a rapid rise in absorbance beyond the equivalence point [curve (a), Figure 24-7]. The formation of a colored product from colorless reactants, on the other hand, initially produces a linear rise in the absorbance followed by a region in which the absorbance becomes independent of reagent volume [curve (b), Figure 24-7]. Depending upon the absorption characteristics of the reactants and the products, the other curve forms shown in Figure 24-7 are also possible.

In order to obtain a satisfactory photometric end point, it is necessary that the absorbing system(s) obey Beer's law; otherwise the titration curve will lack the linear portions needed for end-point extrapolation. Further, it is necessary to correct the absorbance for volume changes by multiplication by $(V + v)/V$, where V is the original volume of the solution and v is the volume of added titrant.

INSTRUMENTATION

Photometric titrations are ordinarily performed with a spectrophotometer or a photometer that has been modified to permit insertion of the titration vessel

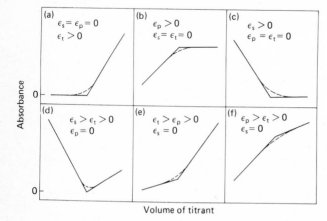

Figure 24-7 Typical Photometric Titration Curves. Molar absorptivities of the substance titrated, the product, and the titrant are given by ϵ_s, ϵ_p, ϵ_t, respectively.

in the light path.[12] After the zero adjustment of the meter scale has been made, radiation is allowed to pass through the solution of the analyte, and the instrument is adjusted by variation of the light intensity or the sensitivity of the detector until a convenient absorbance reading is obtained. Ordinarily, no attempt is made to measure the true absorbance, since relative values are perfectly adequate for the purpose of end-point detection. Data for the titration are then collected without alteration of the instrument setting.

The power of the radiation source and the response of the detector must be reasonably constant during the period required for a photometric titration. Cylindrical containers are ordinarily used, and care must be taken to avoid any movement of the vessel that might alter the length of the radiation path.

Both filter photometers and spectrophotometers have been employed for photometric titrations. The latter are preferred, however, because their narrower band widths enhance the probability of adherence to Beer's law.

APPLICATION OF PHOTOMETRIC TITRATIONS

Photometric titrations often provide more accurate results than a direct photometric analysis because the data from several measurements are pooled in determining the end point. Furthermore, the presence of other absorbing species may not interfere, since only a change in absorbance is being measured.

The photometric end point possesses the advantage over many other commonly used end points in that the experimental data are taken well away from the equivalence-point region. Thus, the titration reactions need not have as favorable equilibrium constants as those required for a titration that is dependent upon observations near the equivalence point (for example, potentiometric or indicator end points). For the same reason, more dilute solutions may be titrated.

The photometric end point has been applied to all types of reactions.[13] Most of the reagents used in oxidation-reduction titrations have characteristic absorption spectra and thus produce photometrically detectable end points. Acid-base indicators have been employed for photometric neutralization titrations. The photometric end point has also been used to great advantage in titrations with EDTA and other complexing agents. Figure 24-8 illustrates the application of the photometric end point to the successive titration of bismuth(III) and copper(II). At 745 nm, neither cation nor the reagent absorbs, nor does the more stable bismuth complex, which is formed in the first part of the titration; the copper complex, however, does absorb. Thus, the solution exhibits no absorbance until essentially all of the bismuth has been titrated. With the first formation of the copper complex, an increase in absorbance occurs. The increase continues until the copper equivalence point is reached. Further reagent additions cause no further absorbance change. Clearly, two well-defined end points result.

---

[12] Titration flasks and cells for use in a Spectronic 20 spectrophotometer are available from the Kontes Manufacturing Corp. Vineland, N.J.

[13] See, for example, the review by A. L. Underwood in C. N. Reilley, Ed., *Advances in Analytical Chemistry and Instrumentation*, vol. 3, pp. 31–104. New York: Interscience Publishers, Inc., 1964.

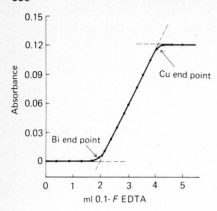

Figure 24-8 Photometric Titration Curve of 100 ml of a Solution That Was 2.0×10^{-3} F in Bi^{3+} and Cu^{2+}. Wavelength: 745 nm. [From A. L. Underwood, *Anal. Chem.*, **26**, 1322 (1954). With permission of the American Chemical Society.]

The photometric end point has also been adapted to precipitation titrations; here the suspended solid product has the effect of diminishing the radiant power by scattering; titrations are carried to a condition of constant turbidity.

Spectrophotometric Studies of Complex Ions

Spectrophotometry is one of the most useful tools for elucidating the composition of complex ions in solution and determining their formation constants. The power of the technique lies in the fact that quantitative absorption measurements can be performed without disturbing the equilibria under consideration. Although most spectrophotometric studies of complexes involve systems in which a reactant or a product absorbs, this condition is not a necessity provided that one of the components can be caused to participate in a competing equilibrium that does produce an absorbing species. For example, complexes of iron(II) with a nonabsorbing ligand might be studied by investigating the effect of this ligand on the color of the iron(II)-orthophenanthroline complex (p. 331). The formation constant and the composition of the nonabsorbing species can then be evaluated, provided the corresponding data are available for the phenanthroline complex.

Three of the most common techniques employed for complex-ion studies are (1) the method of continuous variations, (2) the mole-ratio method, and (3) the slope-ratio method. Each of these is examined briefly.

THE METHOD OF CONTINUOUS VARIATIONS[14]

In the method of continuous variations, solutions with identical formal concentrations of the cation and the ligand are mixed in such a way that the total

[14] See W. C. Vosburgh and G. R. Cooper, *J. Amer. Chem. Soc.*, **63**, 437 (1941).

volume of each mixture is the same. The absorbance of each solution is measured at a suitable wavelength and corrected for any absorbance the mixture might possess if no reaction had occurred. The corrected absorbance is then plotted against the volume fraction (which is equal to the mole fraction) of one of the reactants; that is, $V_M/(V_M + V_L)$ where V_M is the volume of the cation solution and V_L that of the ligand. A typical plot is shown in Figure 24-9. A maximum (or a minimum if the complex absorbs less than the reactants) occurs at a volume ratio V_M/V_L corresponding to the combining ratio of cation and ligand in the complex. In Figure 24-9, $V_M/(V_M + V_L)$ is 0.33, and $V_L/(V_M + V_L)$ is 0.66; thus, V_M/V_L is 0.33/0.66, which suggests that the complex has the formula ML_2.

The curvature of the experimental lines shown in Figure 24-9 is the result of incompleteness of the complex-formation reaction. By measuring the deviations from the theoretical straight lines indicated in the figure, a formation constant for the complex can be calculated.

To determine whether more than one complex forms between the reactants, the experiment is ordinarily repeated with different reactant concentrations and at several wavelengths.

THE MOLE-RATIO METHOD

In this method a series of solutions is prepared in which the formal concentration of one of the reactants (often the metal ion) is held constant while that of the other is varied. A plot of absorbance versus the mole ratio of the reactants is

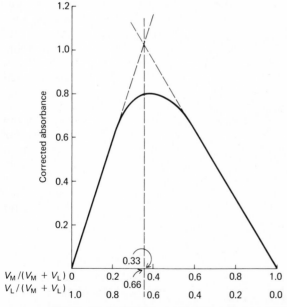

Figure 24-9 Continuous Variation Plot for the 1 : 2 Complex, ML_2.

then prepared. If the formation constant is reasonably favorable, two straight lines of different slope are obtained; the intersection occurs at a mole ratio that corresponds to the combining ratio in the complex. Typical mole-ratio plots are shown in Figure 24-10. Note that the ligand of the 1:2 complex absorbs at the wavelength selected; as a result, the slope beyond the equivalence point is greater than zero. The uncomplexed cation involved in the 1:1 complex absorbs, since the initial point has an absorbance greater than zero.

From the experimental plots, the formation constant can be obtained.

If the complex formation reaction is relatively incomplete, the mole-ratio plot appears as a continuous smooth curve with no straight-line portions that can be extrapolated to give the combining ratio. Such a reaction can sometimes be forced to completion by the addition of a several-hundredfold excess of the ligand until the absorbance becomes independent of further additions. The constant absorbance can then be employed to calculate the molar absorptivity of the complex by assuming that essentially all of the metal ion has combined with the ligand, and that the complex concentration is equal to the formal concentration of the metal. One can then successively assume that the complex has various compositions such as 1:1, 1:2, and so forth to calculate equilibrium concentrations. The composition that gives constant numerical values for the formation constant over a wide concentration range can then be assumed to be that for the complex.

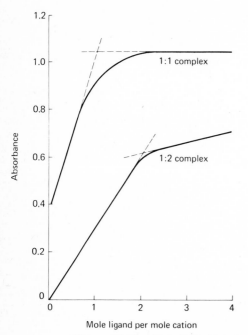

Figure 24-10 Mole-Ratio Plots for a 1:1 and a 1:2 Complex. The 1:2 complex is the more stable, as indicated by less curvature near stoichiometric ratio.

A mole-ratio plot may reveal the stepwise formation of two or more complexes as successive slope changes, provided the complexes have different molar absorptivities and provided the formation constants are sufficiently different.

THE SLOPE-RATIO METHOD

This procedure is particularly useful for weak complexes; it is applicable only to systems in which a single complex is formed. The method assumes that the complex-formation reaction can be forced to completion in the presence of a large excess of either reactant, and that Beer's law is followed under these circumstances. For the reaction

$$mM + lL \rightleftarrows M_mL_l$$

the following equation can be written, when L is present in very large excess:

$$[M_mL_l] \cong \frac{F_M}{m}$$

If Beer's law is obeyed,

$$A_m = \epsilon b[M_mL_l] = \frac{\epsilon b F_M}{m}$$

and a plot of A with respect to F_M will be linear. When M is very large with respect to L,

$$[M_mL_l] \cong \frac{F_L}{l}$$

and

$$A_l = \epsilon b[M_mL_l] = \frac{\epsilon b F_L}{l}$$

The slopes of the straight lines (A/F_m and A/F_l) are obtained under these conditions; the combining ratio between L and M is obtained from the ratio of the slopes:

$$\frac{A_m/F_m}{A_l/F_l} = \frac{S_m}{S_l} = \frac{\epsilon b/m}{\epsilon b/l} = \frac{l}{m}$$

PROBLEMS

*1. A 4.97-g petroleum specimen was decomposed by wet-ashing and subsequently diluted to 500 ml in a volumetric flask. Analysis for cobalt was performed by treating aliquots as indicated:

Volume of Diluted Sample Taken, ml	Reagent Volumes			
	Co(II), 3.00 ppm	Ligand	H_2O	Absorbance
25.00	0.00	20.00	5.00	0.398
25.00	5.00	20.00	0.00	0.510

With the added information that the Co(II)-ligand chelate obeys Beer's law, calculate the percentage of Co in the original sample.

2. A 2-tablet sample of vitamin supplement, weighing a total of 6.08 g, was wet-ashed to eliminate organic matter and then diluted to 1.00 liter in a volumetric flask. Calculate the average weight of iron in each tablet, based upon the accompanying information.

Volume of Diluted Sample, ml	Reagent Volumes			Absorbance
	Fe(III), 1.00 ppm	Ligand	H_2O	
10.00	0.00	25.00	15.00	0.492
10.00	15.00	25.00	0.00	0.571

*3. The equilibrium constant for the conjugate acid-base pair

$$HIn + H_2O \rightleftarrows H_3O^+ + In^-$$

is 8.00×10^{-3}. Additional information:

Species	Absorption Maximum, nm	Molar Absorptivity at	
		430 nm	600 nm
HIn	430	2.4×10^4	0
In$^-$	600	0	2880

Calculate the absorbance at 430 nm and at 600 nm for the following formal indicator concentrations: 3.00×10^{-4}, 2.00×10^{-4}, 1.00×10^{-4}, 0.500×10^{-4}, and 0.250×10^{-4}. Plot the data.

4. The absolute error in transmittance for a particular instrument is estimated to be 0.006. Calculate the percent relative error in concentration due to this source when
   *(a) $T = 0.015$
   *(b) $A = 0.334$
   *(c) $\%T = 64.8$
   (d) $A = 0.921$
   (e) $T = 0.804$
   (f) $\%T = 50.0$

5. The response of a photometer with respect to a particular solution yielded the accompanying transmittance data: 0.364, 0.372, 0.370, 0.362. Assume that the indeterminate error in the measurement is independent of the magnitude of T. For a solution having a transmittance of 60.0%, estimate the relative error in concentration if
   *(a) the absolute indeterminate error is assumed to be twice the average deviation from the mean of the four data (see footnote 7, p. 549).
   (b) the absolute indeterminate error is assumed to be the standard deviation for the four data.
   *(c) the absolute indeterminate error is assumed to be the 90% confidence interval for a single measurement.

*6. Ethylenediaminetetraacetic acid will abstract bismuth(III) from its thiourea complex:

$$Bi(tu)_6{}^{3+} + H_2Y^{2-} \rightarrow BiY^- + 6tu + 2H^+$$

Predict the shape of a photometric titration curve based on this process, given the further information that the Bi(III)-thiourea complex is the only species in the system that absorbs at 465 nm, the wavelength selected for the analysis.

7. The accompanying data refer to the spectrophotometric titration of a 10.00-ml aliquot of Pd(II) with 2.44×10^{-4} F Nitroso R [O. W. Rollins and M. M. Oldham, Anal. Chem., **43**, 262 (1971)].

Volume of 2.44×10^{-4} F Nitroso R, ml	Absorbance, A, 500 nm, 1.00-cm Cells
0	0
1.00	0.147
2.00	0.271
3.00	0.375
4.00	0.371
5.00	0.347
6.00	0.325
7.00	0.306
8.00	0.289

Calculate the concentration of the Pd(II) solution, given the further information that the ligand-to-cation ratio in the colored product is 2:1.

8. Solutions containing A and B are known to obey Beer's law over an extensive concentration range. Molar absorptivity data for each are as follows:

λ, nm	Molar Absorptivity, ϵ		λ, nm	Molar Absorptivity, ϵ	
	A	B		A	B
400	893	0.00	560	440	622
420	940	0.00	580	297	880
440	955	0.00	600	167	1178
460	936	24.6	620	39.7	1692
480	874	102	640	3.45	1742
500	795	185	660	0.00	1806
520	691	289	680	0.00	1809
540	574	428	700	0.00	1757

*(a) A solution containing both solutes has an absorbance of 0.360 when measured in a 1.00-cm cell at 540 nm. What is the concentration of B in this solution if it is known that $[A] = 5.00 \times 10^{-4}$?

*(b) A solution containing both species has an absorbance of 0.510 both at 440 nm and at 600 nm when measured in a 1.00-cm cell. Calculate the respective concentrations of A and B.

(c) Construct an absorption spectrum for a 5.80×10^{-4} F solution of A in a 1.00-cm cell.

(d) Construct an absorption spectrum for a 3.25×10^{-4} F solution of B in a 1.00-cm cell.

(e) Construct an absorption spectrum for a solution that is 5.80×10^{-4} F in A and 3.25×10^{-4} F in B in a 1.00-cm cell.

9. A. J. Mukhedkar and N. V. Deshpande [Anal. Chem., 35, 47 (1963)] report on a simultaneous determination for cobalt and nickel based upon absorption by their respective 8-quinolinol complexes. Molar absorptivities corresponding to absorption maxima are:

	Wavelength	
	365 nm	700 nm
ϵ_{Co}	3529	428.9
ϵ_{Ni}	3228	0.00

Calculate the concentration of nickel and cobalt in each of the following solutions based upon the accompanying data.

Absorbance, 1.00-cm Cell

Solution	365 nm	700 nm
*1	0.724	0.0710
2	0.614	0.0744
3	0.693	0.0460

10. Solutions of P and of Q individually obey Beer's law over a large concentration range. Spectral data for these species, measured in 1.00-cm cells, are tabulated below.

	Absorbance, A			Absorbance, A	
λ, nm	$8.55 \times 10^{-5}F$ P	$2.37 \times 10^{-4}F$ Q	λ, nm	$8.55 \times 10^{-5}F$ P	$2.37 \times 10^{-4}F$ Q
400	0.078	0.550	560	0.126	0.255
420	0.087	0.592	580	0.170	0.170
440	0.096	0.599	600	0.264	0.100
460	0.102	0.590	620	0.326	0.055
480	0.106	0.564	640	0.359	0.030
500	0.110	0.515	660	0.373	0.030
520	0.113	0.433	680	0.370	0.035
540	0.116	0.343	700	0.346	0.063

(a) Plot an absorption spectrum for a solution that is 8.55×10^{-5} F with respect to P and 2.37×10^{-4} F with respect to Q.

(b) Calculate the absorbance (1.00-cm cells) at 440 nm of a solution that is 4.00×10^{-5} F in P and 3.60×10^{-4} F in Q.

(c) Calculate the absorbance (1.00-cm cells) at 620 nm for a solution that is 1.61×10^{-4} F in P and 7.35×10^{-4} F in Q.

11. Use the data in the previous problem to calculate the quantities sought (1.00-cm cells are used for all measurements).

	Concentration, mole/liter		Absorbance, A	
	P	Q	440 nm	620 nm
*(a)			0.357	0.803
(b)			0.830	0.448
*(c)			0.248	0.333
(d)			0.910	0.338
*(e)			0.480	0.825
(f)			0.194	0.315

12. The indicator HIn has an acid-dissociation constant of 4.80×10^{-6} at ordinary temperatures. The accompanying absorbance data are for 8.00×10^{-5} F solutions of the indicator measured in 1.00-cm cells in strongly acidic and strongly alkaline media.

	Absorbance				Absorbance	
λ, nm	pH = 1.00	pH = 13.00		λ, nm	pH = 1.00	pH = 13.00
420	0.535	0.050		550	0.119	0.324
445	0.657	0.068		570	0.068	0.352
450	0.658	0.076		585	0.044	0.360
455	0.656	0.085		595	0.032	0.361
470	0.614	0.116		610	0.019	0.355
510	0.353	0.223		650	0.014	0.284

Estimate the wavelength at which absorption by the indicator becomes independent of the pH (that is, the isosbestic point).

13. Calculate the absorbance (1.00-cm cells) at 450 nm of a solution in which the total formal concentration of the indicator described in Problem 12 is 8.00×10^{-5} and the pH is
   *(a) 4.90.
   (b) 5.40.
   *(c) 5.90.

14. What will be the absorbance at 595 nm (1.00-cm cells) of a solution that is 1.25×10^{-4} F with respect to the indicator in Problem 12 and whose pH is
   *(a) 5.30.
   (b) 5.70.
   (c) 6.10.

15. Several buffer solutions were made 1.00×10^{-4} F with respect to the indicator in Problem 12; absorbance data (1.00-cm cells) at 450 nm and 595 nm were as follows:

	Absorbance	
Solution	450 nm	595 nm
*A	0.344	0.310
B	0.508	0.212
*C	0.653	0.136
D	0.220	0.380

Calculate the pH of each solution.

16. Construct an absorption spectrum for an 8.00×10^{-5} F solution of the indicator in Problem 12 when measurements are made with 1.00-cm cells and

   *(a) $\dfrac{[HIn]}{[In^-]} = 3$

   (b) $\dfrac{[HIn]}{[In^-]} = 1$

   (c) $\dfrac{[HIn]}{[In^-]} = \dfrac{1}{3}$

17. Absorptivity data for the cobalt and nickel complexes with 2,3-quinoxaline-dithiol at the respective absorption peaks are as follows:

	Wavelength	
	510 nm	656 nm
ϵ_{Co}	36,400	1,240
ϵ_{Ni}	5,520	17,500

A 0.425-g soil sample was dissolved and subsequently diluted to 50.0 ml. A 25.0-ml aliquot was treated to eliminate interferences; after addition of 2,3-quinoxalinedithiol, the volume was adjusted to 50.0 ml. This solution had an absorbance of 0.446 at 510 nm and 0.326 at 656 nm in a 1.00-cm cell. Calculate the respective percentages of cobalt and nickel in the soil.

*18. The chelate ZnQ_2^{2-} exhibits a maximum absorption at 480 nm. When the chelating agent is present in at least a fivefold excess, the absorbance is dependent only upon the formal concentration of Zn(II) and obeys Beer's law over a large range. Neither Zn^{2+} nor Q^{2-} absorbs at 480 nm.

A solution that is 2.30×10^{-4} F in Zn^{2+} and 8.60×10^{-3} F in Q has an absorbance of 0.690 when measured in a 1.00-cm cell at 480 nm. Under the same conditions, a solution that is 2.30×10^{-4} F in Zn^{2+} and 5.00×10^{-4} F in Q^{2-} has an absorbance of 0.540. Calculate the numerical value of K for the process.

$$Zn^{2+} + 2Q^{2-} \rightleftarrows ZnQ_2^{2-}$$

19. The sodium salt of 2-quinizarinsulfonic acid (NaQ) is reported to form a complex with aluminum(III) which absorbs strongly at 560 nm [E. G. Owens and J. H. Yoe, *Anal. Chem.*, **31**, 385 (1959)]. Use the accompanying data to establish the combining ratio between cation and ligand.

In all solutions, $C_{Al} = 3.7 \times 10^{-5}$ F; all measurements were made in 1.00-cm cells.

Formal Ligand Concentration	Absorbance, A	Formal Ligand Concentration	Absorbance, A
1.00×10^{-5}	0.131	5.00×10^{-5}	0.487
2.00×10^{-5}	0.265	6.00×10^{-5}	0.498
3.00×10^{-5}	0.396	8.00×10^{-5}	0.499
4.00×10^{-5}	0.468	1.00×10^{-4}	0.500

*20. The accompanying data were obtained in a slope-ratio investigation of the product formed between Ni(II) and 1-cyclopentene-1-dithiocarboxylic acid (CDA). The measurements were made at 530 nm in 1.00-cm cells.

$C_{CDA} = 1.00 \times 10^{-3}$ F		$C_{Ni} = 1.00 \times 10^{-3}$ F	
C_{Ni}, mfw/ml	Absorbance, A	C_{CDA}, mfw/ml	Absorbance, A
5.00×10^{-6}	0.051	9.00×10^{-6}	0.031
1.20×10^{-5}	0.123	1.50×10^{-5}	0.051
3.50×10^{-5}	0.359	2.70×10^{-5}	0.092
5.00×10^{-5}	0.514	4.00×10^{-5}	0.137
6.00×10^{-5}	0.616	6.00×10^{-5}	0.205
7.00×10^{-5}	0.719	7.00×10^{-5}	0.240

Calculate the ligand-to-cation ratio in this complex.

21. Iron(II) forms a chelate with the ligand P. Evaluate the composition of FeP_n from the accompanying data.

Fe(II) = 2.00×10^{-3} F		P = 2.00×10^{-3} F	
Formal Concentration of P	Absorbance, A, 1.00-cm Cells	Formal Concentration of Fe	Absorbance, A, 1.00-cm Cells
4.00×10^{-6}	0.025	6.00×10^{-6}	0.113
1.50×10^{-5}	0.094	1.00×10^{-5}	0.189
3.00×10^{-5}	0.189	1.75×10^{-5}	0.330
5.00×10^{-5}	0.314	3.00×10^{-5}	0.566
7.00×10^{-5}	0.440	4.00×10^{-5}	0.754

*22. The accompanying absorbance data were recorded for a continuous-variation study of the colored product formed between cadmium(II) and the complexing reagent R.

	Reactant Volumes, ml		
Solution	1.25×10^{-4} F Cd^{2+}	1.25×10^{-4} F R	Absorbance, 390 nm, 1.00-cm Cells
0	10.00	0.00	0.000
1	9.00	1.00	0.174
2	8.00	2.00	0.353
3	7.00	3.00	0.530
4	6.00	4.00	0.672
5	5.00	5.00	0.723
6	4.00	6.00	0.673
7	3.00	7.00	0.537
8	2.00	8.00	0.358
9	1.00	9.00	0.180
10	0.00	10.00	0.000

(a) Establish the ligand-to-cation ratio in the product.

(b) Determine an average value for the molar absorptivity of the complex; assume that the species in lesser amount is completely incorporated into the complex in the linear portions of the plot.

(c) Evaluate K_f for the complex, using the stoichiometric relationships that exist under conditions of maximum absorbance.

23. (a) Use the following absorbance data to evaluate the ligand-to-cation ratio in the complex formed between copper(II) and the ligand H_2B.

	Reactant Volumes, ml		
Solution	8.00×10^{-5} F Cu	8.00×10^{-5} F H_2B	Absorbance, 475 nm, 1.00-cm Cells
0	10.00	0.00	0.000
1	9.00	1.00	0.104
2	8.00	2.00	0.210
3	7.00	3.00	0.314
4	6.00	4.00	0.419
5	5.00	5.00	0.507
6	4.00	6.00	0.571
7	3.00	7.00	0.574
8	2.00	8.00	0.423
9	1.00	9.00	0.211
10	0.00	10.00	0.000

(b) Evaluate an average value for the molar absorptivity of the complex; assume that the species in lesser amount is completely incorporated into the complex in the linear portions of the plot.

(c) Evaluate K_f for the complex, using the stoichiometric relationships that exist under conditions of maximum absorption.

*24. (a) Use the following absorbance data to evaluate the ligand-to-cation ratio in the complex formed between Co(II) and the bidentate ligand Q.

| | Reactant Volumes, ml | | Absorbance, 560 nm, |
Solution	$9.50 \times 10^{-5} F$ Co(II)	$9.50 \times 10^{-5} F$ Q	1.00-cm Cells
0	10.00	0.00	0.000
1	9.00	1.00	0.094
2	8.00	2.00	0.193
3	7.00	3.00	0.291
4	6.00	4.00	0.387
5	5.00	5.00	0.484
6	4.00	6.00	0.570
7	3.00	7.00	0.646
8	2.00	8.00	0.585
9	1.00	9.00	0.295
10	0.00	10.00	0.000

(b) Determine an average value for the molar absorptivity of the complex; assume that the species in lesser amount is completely incorporated in the complex in the linear portions of the plot.

(c) Evaluate K_f for the complex, using the stoichiometric relationships that exist under conditions of maximum absorbance.

ATOMIC SPECTROSCOPY

Atomic spectroscopy, which has been extensively applied to qualitative and quantitative analysis, is based upon either absorption or emission of X-ray, ultraviolet, or visible radiation. X-ray emission or absorption involves excitation of those electrons that are closest to the nucleus and thus not involved in bonding (except for a few of the lightest elements). As a consequence, X-ray atomic spectra are produced regardless of the state of combination of the elements in a sample. Thus, the X-ray spectrum for nickel will be the same in solid nickel oxide, a solution of nickel(II) chloride, gaseous nickel carbonyl, or the elemental state.

In contrast, production of atomic spectra in the ultraviolet or visible regions requires conversion of the elements in a sample to gaseous monatomic particles (atoms or ions) because absorption or emission in these regions involves excitation of the outermost bonding electrons; only by separating an element from those to which it is bonded can pure atomic spectra be obtained.

Atomic spectra in the ultraviolet or visible region are produced by *atomization* of the sample, a process whereby the constituent molecules are decomposed and converted to gaseous elementary particles. Both the emission and the absorption spectrum of an atomized element consist of a relatively few discrete lines at wavelengths that are characteristic of the element. To the extent that molecules or complex ions are absent in the gas, band spectra are not observed, because vibrational and rotational quantum states cannot exist; thus, the lines correspond to a relatively small number of transitions only.

Table 25-1 outlines the various methods based upon atomic emission and

TABLE 25-1   Classification of Atomic Spectral Methods

	Method of Atomization	Radiation Source	Usual Sample Treatment
Emission Methods			
Common Name	Method of Atomization	Radiation Source	Usual Sample Treatment
Arc spectroscopy	Electric arc	Sample in arc	Sample placed on electrode
Spark spectroscopy	Electric spark	Sample in spark	Sample placed on electrode
Flame emission or atomic emission	Flame	Sample in flame	Sample solution aspirated into a flame
Atomic fluorescence	Flame	Discharge lamp	Sample solution aspirated into a flame
X-ray fluorescence	None required	X-ray tube	Sample exposed to X-radiation
Absorption Methods			
Flame absorption or atomic absorption	Flame	Hollow cathode lamp	Sample solution aspirated into a flame
Flameless absorption	Heated surface	Hollow cathode lamp	Sample solution pipetted onto heated surface
X-ray absorption	None required	X-ray tube	Sample held in source beam

atomic absorption. These methods offer the advantages of high specificity, wide applicability, excellent sensitivity, speed, and convenience; they are among the most selective of all analytical procedures. Perhaps 70 elements can be determined. The sensitivities are typically in the parts-per-million to parts-per-billion range. An atomic spectral analysis can frequently be completed in a few minutes.

This chapter is concerned principally with ultraviolet and visible atomic-emission and atomic-absorption spectroscopy employing flames; occasional reference will be made to some of the other methods for producing atomic spectra.

Spectroscopic Behavior of Atoms in a Flame

When an aqueous solution of inorganic salts is aspirated into the hot flame of a burner, a substantial fraction of the metallic constituent is reduced to the elemental state; to a much lesser extent, monatomic ions are also formed. Thus, within the flame there is produced a gaseous solution or *plasma* containing a significant concentration of monatomic particles.

TYPES OF FLAME SPECTROSCOPY

The temperature of a flame is sufficient to excite a small fraction of the monatomic particles to higher electronic states; their return to the ground state is accompanied by the production of emission lines, which can serve as a basis for *flame-emission spectroscopy*. Here the location of the lines provides qualitative information; their intensity permits quantitative analyses.

The much larger fraction of unexcited atoms in a flame can be employed for *atomic-absorption analysis*. Here it is only necessary to interpose the flame plasma in the beam of a spectrophotometer similar to those discussed in Chapter 23. As with flame emission, the location of absorption lines serves to identify the components of a sample; the absorbance of a line, as in solution spectrophotometry, is generally proportional to the concentration of the absorbing species.

LINE WIDTHS IN ATOMIC-ABSORPTION AND -EMISSION SPECTRA

The natural line width of an atomic-absorption or an atomic-emission peak can be shown to be about 10^{-5} nm. Two effects, however, tend to cause observed widths to range between 0.002 and 0.005 nm. *Doppler broadening* arises from the rapid motion of the absorbing or emitting particles with respect to the detector. For those atoms traveling away from the detector, the wavelength is effectively increased by the well-known Doppler effect; thus, somewhat longer wavelengths are absorbed or emitted. The reverse is true of atoms moving toward the detector. *Pressure broadening* also occurs. Here collisions among atoms cause small changes in the ground-state energy levels and a consequent broadening of peaks. Note that both of these effects are enhanced as the temperature rises. Thus, broader peaks are observed at elevated temperatures.

RELATIONSHIP BETWEEN FLAME-ABSORPTION AND FLAME-EMISSION SPECTROSCOPY

Atomic-emission spectroscopy is based upon the intensity of radiation emitted by *excited* atoms whereas the *unexcited* atoms serve as the basis for atomic-absorption measurements. The unexcited species predominate by a large margin at the temperature of a flame. For example, at $2500°K$, only about 0.02% of the sodium atoms in a flame are excited to the $3p$ state at any instant; the percentages in higher electronic states are even less. At $3000°K$ the corresponding percentage is about 0.09.

This large ratio of unexcited to excited species in a flame is important in comparing absorption and emission procedures. Because atomic-absorption methods are based upon a much larger population of particles, it might be expected to be the more sensitive procedure. This apparent advantage is offset, however, by the fact that an absorbance measurement involves a difference measurement ($\log P_0 - \log P$); when the two numbers are nearly alike, larger errors in the difference result. As a consequence, the two procedures are often complimentary in sensitivity, the one being advantageous for one group of elements and the other for a different group.

Another apparent advantage of flame-absorption spectroscopy is that the absolute number of unexcited particles is much less dependent upon temperature than is the number of excited species. For example, at about $2500°K$, a $20°K$ fluctuation causes a change of about 8% in the number of sodium atoms excited to the $3p$ state. Because so many more unexcited atoms exist at this temperature, however, their population changes by only about 0.02% for the same temperature variation.

It should be noted that temperature fluctuations do exert an indirect influence on atomic absorption by increasing the total number of atoms that are available for absorption. In addition, line broadening and a consequent decrease in peak height occurs because the atomic particles travel at greater rates and enhance the Doppler effect. Increased concentrations of gaseous atoms at higher temperatures also cause pressure broadening of the absorption lines. Because of these indirect effects, a reasonable control of the flame temperature is required for quantitative atomic-absorption measurements.

Atomic-Absorption Spectroscopy

Vaporized atoms absorb radiation with energies that exactly match those of characteristic electronic transitions. The most useful absorption lines are those involving excitation of the atom from its ground state to one of its excited states rather than a transition from one excited state to another; the latter lines are generally so weak as to go undetected because the number of excited atoms is small. Typically, then, an atomic-absorption spectrum produced with a flame consists predominantly of *resonance* lines, which are the consequence of transitions from the ground state to upper levels. Note that *the wavelength of a resonance absorption line is identical to that of the emission line that corresponds to the same transition*.

MEASUREMENT OF ATOMIC ABSORPTION

Because atomic-absorption lines are so very narrow, and because transition energies are unique for each element, analytical methods based on atomic absorption are potentially highly specific. On the other hand, the limited line widths create a measurement problem not encountered in solution absorption. Recall that although Beer's law applies only for monochromatic radiation, a linear relationship between absorbance and concentration can be expected if the band width is narrow with respect to the width of the absorption peak (p. 510). No ordinary monochromator is capable of yielding a band of radiation that is as narrow as the peak width of an atomic-absorption line (0.002 to 0.005 nm). Thus, when a continuous source is employed with a monochromator for atomic absorption, only a minute fraction of the radiation is of a wavelength that is absorbed, and the relative change in intensity of the emergent band is small in comparison to the change suffered by radiation that corresponds to the absorption peak. Under these conditions, Beer's law is not followed; in addition, the sensitivity of the method is lessened significantly.

This problem has been overcome by employing a source of radiation that emits a line of the same wavelength as the one to be used for the absorption analysis. For example, if the 589.6-nm absorption line of sodium is chosen for the analysis of that element, a sodium vapor lamp can be employed as a source. In such a lamp, gaseous sodium atoms are excited by electrical discharge; the excited atoms then emit characteristic radiation as they return to lower energy levels. The emitted line will have the same wavelength as the resonance-absorption line. With a properly designed source (one that operates at a lower temperature than a flame to minimize Doppler broadening), the emission lines, however, will have band widths that are significantly narrower than the absorption band widths. Thus, the monochromator need only have the capability of isolating a suitable emission line for the absorption measurement (see Figure 25-1). The radiation employed in the analysis is then sufficiently limited in wavelength span to permit measurements at the absorption peak. Greater sensitivity and better adherence to Beer's law result.

A separate lamp source is needed for each element being analyzed. To avoid this inconvenience, attempts have been made to employ a continuous source with a very high-resolution monchromator or, alternatively, to produce a line source by introducing a compound of the element to be determined into a high-temperature flame. Neither of these alternatives is as satisfactory as the employment of a specific lamp for each element.

Radiation Sources. The most common source for an atomic-absorption measurement is the *hollow cathode lamp*, which consists of a tungsten anode and a cylindrical cathode. The cylinder is constructed of the metal whose spectrum is desired or serves to support a layer of that metal. The electrodes are housed in a glass tube that is filled with helium or argon at 1 to 2 mm pressure.

Ionization of the gas occurs when a potential is applied across the electrodes, and a current flows as the ions migrate to the electrodes. If the potential is sufficiently large, the gaseous cations acquire enough kinetic energy to dislodge some of the metal atoms from the cathode surface and produce an atomic cloud; this process is called *sputtering*. A portion of the sputtered metal atoms are in excited states and thus emit their characteristic radiation in the usual way. Eventually, the metal atoms diffuse back to the cathode surface or the glass walls of the tube and are redeposited.

Hollow cathode lamps for various elements are sold commercially. Some have cathodes consisting of a mixture of several elements; lamps of this kind provide spectral lines for more than a single species.

In the typical atomic-absorption instrument, it is necessary to eliminate interferences caused by *emission* of radiation by the flame. Most of the emitted radiation can be removed by locating the monochromator between the flame and the detector; nevertheless, this arrangement does not remove the radiation

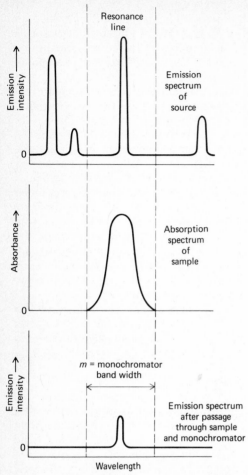

Figure 25-1   Absorption of a Resonance Line by Atoms.

corresponding to the wavelength selected for the analysis. The flame will contain such radiation since excitation and radiant emission by some atoms of the analyte can occur. This difficulty is overcome by causing the intensity of the source to fluctuate at a constant frequency; this process is called *modulation*. The detector then receives two types of signal, an alternating one from the source and a continuous one from the flame. These signals are converted to the corresponding types of electrical response. A relatively simple electronic system is then employed to respond to and amplify only the ac part of the signal and to ignore the unmodulated dc signal.

A simple and entirely satisfactory way of modulating radiation from the source is to interpose a circular disk in the beam between the source and the flame. Alternate quadrants of this disk are removed to permit passage of light. Rotation of the disk at constant speed provides a beam that is chopped to the desired frequency. As an alternative, the power supply for the source can be designed for intermittent or ac operation.

INSTRUMENTS

An instrument for atomic-absorption measurements has the same basic components as a spectrophotometer for measuring the absorption of solutions (see Figure 23-1); that is, it contains a source, a monochromator, a sample container (here a flame or a hot surface), a detector, and an amplifier-indicator. The major instrumental differences between atomic- and solution-absorption equipment are in the source and in the sample container. The features of these components require further discussion.

Both single-beam and double-beam instruments have been designed for atomic-absorption measurements; the latter are the more common.

Burners. Two types of burners are employed in atomic-absorption spectroscopy. In a *total consumption* burner, the fuel and oxidizing gases are carried through separate passages to meet and mix with the solution of the analyte at an opening at the base of the flame. In a *premix* burner, the sample is aspirated into a large chamber by a stream of the oxidant; here the fine mist of the analyte solution, the oxidant, and the fuel supply are mixed and then forced out the burner opening. Larger drops of sample collect in the bottom of the chamber and drain off. Figure 25-2 gives schematic diagrams of both burner designs.

Fuels used for flame production include natural gas, propane, butane, hydrogen, and acetylene. The last is perhaps most widely employed. The common oxidants are air, oxygen-enriched air, oxygen, and nitrous oxide. The nitrous oxide–acetylene mixture is advantageous when a hot flame is required because of its lower explosion hazard.

Low-temperature flames (natural gas-air, for example) are used to advantage for elements that are readily converted to the atomic state, such as copper, lead, zinc, and cadmium. Elements such as the alkaline earths, on the other hand, form refractory oxides which require somewhat higher temperatures for decomposition; for these, an acetylene-air mixture often produces the most sensitive results. Aluminum, beryllium, the rare earths, and certain other elements form unusually stable oxides; a reasonable concentration of their atoms can be obtained only at the high temperatures developed in an oxygen-acetylene or a nitrous oxide–acetylene flame.

Nonflame Atomizers. Recently, several nonflame atomizers have been developed, which are particularly useful for the quantitative determination of small amounts of various elements. In a nonflame atomizer, a few microliters of sample are evaporated and ashed at low temperatures on an electrically heated surface of carbon, tantalum, or other conducting material. The conductor can be a hollow tube, a strip or rod, a boat, or a trough. After ashing, a current of 100 A or more is passed through the conductor, which causes a rapid increase

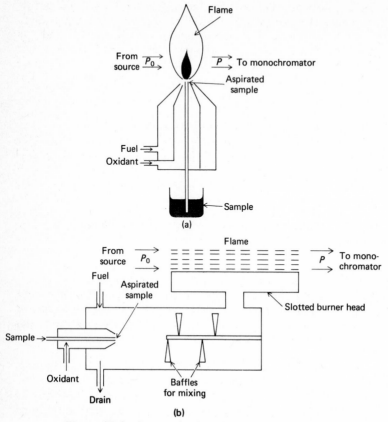

Figure 25-2 Burners for Atomic-Absorption Spectroscopy. (a) Total consumption burner. (b) Premix burner.

in temperature to perhaps 2000 to 3000°C; atomization of the sample occurs in a period of a few seconds. The atomization process is observed by means of a spectrophotometer in which the source beam is passed through the region immediately above the heated conductor. At a wavelength where absorption takes place, the absorbance is observed to rise to a maximum after a few seconds and then decay to zero, corresponding to the atomization and subsequent escape of the volatilized sample; analyses are based upon peak height.

Nonflame atomizers offer the advantage of unusually high sensitivity for small volumes of sample. Typically, sample volumes between 0.5 and 10 μl are employed; under these circumstances, absolute limits lie in the range of 10^{-10} to 10^{-13} g of analyte.

The relative precision of nonflame methods is generally in the range of 5 to 10%; 1 to 2% can be expected for flame atomization.

Monochromators or Filters. An atomic-absorption instrument must be capable of providing a sufficiently narrow band width to isolate the line chosen for the measurement from other lines that may interfere with or diminish the sensitivity of the analysis. For some of the alkali metals, which have only a few widely spaced resonance lines in the visible region, a glass filter suffices. An instrument employing readily interchangeable interference filters is available commercially. A separate filter (and light source) is used for each element. Satisfactory results for the analysis of 22 metals are claimed. Most instruments however, incorporate a good-quality, ultraviolet and visible monochromator.[1]

Detectors and Indicators. The detector-indicator components for an atomic-absorption spectrophotometer are fundamentally the same as for the typical ultraviolet and visible solution spectrophotometer (see Chapter 23). Generally, photomultiplier tubes are used to convert the radiant energy received to an electrical signal. As we have pointed out, the electronic system must be capable of discriminating between the modulated signal from the source and the continuous signal from the flame. Both null-point and direct-reading meters, calibrated in terms of absorbance or transmittance, are used; some instruments provide a digital readout.

APPLICATIONS

Atomic-absorption spectroscopy provides a sensitive means for the determination of more than 60 elements; detection limits for some are provided in Table 25-2. Details concerning the quantitative determination of these and other elements can be found in several publications.[2]

INTERFERENCES

The absorption spectrum of each element is unique to that species. With a reasonably good monochromator, it is possible to isolate an absorption peak that is free from interference by other elements. Unfortunately, however, this specificity does not extend to the chemical reactions occurring in the flame or above the heated surface of a nonflame atomizer; here interferences do occur as a consequence of interactions and chemical competitions that affect the number of atoms present in the light path under a specified set of conditions. These effects are not predictable from theory and must therefore be determined by experiment; they appear to be particularly troublesome with nonflame atomizers.

Cation Interferences. A few examples have been found in which the absorption due to one cation is affected by the presence of a second. For example, the presence of aluminum is found to cause low results in the determination of magnesium; it has been shown that this interference results from the formation

---

[1] For specifications, see W. T. Elwell and J. A. F. Gidley, *Atomic-Absorption Spectrophotometry*. New York: Pergamon Press, 1966; R. G. Martinek, *Laboratory Management*, **6**, 24 (1968); J. Ramirez-Muñoz, *Atomic-Absorption Spectroscopy*, pp. 182–202. New York: Elsevier Publishing Company, 1968.

[2] W. T. Elwell and J. A. F. Gidley, *Atomic-Absorption Spectrophotometry*. New York: Pergamon Press, 1966; J. W. Robinson, *Atomic-Absorption Spectroscopy*. New York: Marcel Dekker, Inc., 1966; J. Ramirez-Muñoz, *Atomic-Absorption Spectroscopy*. New York: Elsevier Publishing Company, 1968.

TABLE 25-2 Detection Limits for the Analysis of Selected Elements by Flame-Absorption and Flame-Emission Spectrometry

Element	Wavelength, nm	Detection Limit, µg/ml		
		Flame Emission[a]	Flame Absorption[a]	Nonflame Absorption[b]
Aluminum	396.2	0.005 (N_2O)		0.03
	309.3		0.1 (N_2O)	
Calcium	422.7	0.005 (air)	0.002 (air)	0.0003
Cadmium	326.1	2 (N_2O)		0.0001
	228.8		0.005 (air)	
Chromium	425.4	0.005 (N_2O)		0.005
	357.9		0.005 (air)	
Iron	372.0	0.05 (N_2O)		0.003
	248.3		0.005 (air)	
Lithium	670.8	0.00003 (N_2O)	0.005 (air)	0.005
Magnesium	285.2	0.005 (N_2O)	0.0003 (air)	0.00006
Potassium	766.5	0.0005 (air)	0.005 (air)	0.0009
Sodium	589.0	0.0005 (air)	0.002 (air)	0.0001

[a] Taken from data compiled by E. E. Pickett and S. R. Koirtyyohann, *Anal. Chem.*, **41** (14), 28A (1969). With permission of the American Chemical Society. Data are for an acetylene flame with the oxidant shown in parentheses.

[b] Data from J. W. Robinson and P. J. Slevin, *American Laboratory*, **4** (8), 14 (1972). With permission, International Scientific Communications, Inc.; wavelength not reported.

of a heat-stable, aluminum-magnesium compound and a consequent decrease in the concentration of magnesium atoms in the flame. Beryllium, aluminum, and magnesium are reported to have a similar effect on calcium analyses. Fortunately, this type of interference is relatively rare.

Anion Interferences. The absorption behavior for a metal may be influenced by the type and the concentration of anions present in the sample solution. This effect is not surprising inasmuch as the energy required to form atomic species from compounds must vary with the strength of the attraction between anion and cation. Such effects ordinarily become smaller with increases in flame temperature and may disappear entirely in some of the hotter flames.

Anion interference can sometimes be avoided by the addition of a complexing agent (EDTA, for example) to both the standards and the samples. In this way atom formation always results from decomposition of the complex. Alternatively, the standards can be made to approximate the anion composition of the sample to compensate for this effect.

ANALYTICAL TECHNIQUES

Both calibration curves and the standard-addition method are suitable for atomic-absorption spectroscopy.

Calibration Curves. While, in theory, absorbance should be proportional to concentration, deviations from linearity do occur. Thus, empirical calibration curves must be prepared. In addition, there are sufficient uncontrollable variables in the production of an atomic vapor to warrant measuring the absorbance of at least one standard solution each time readings are taken. Any deviation of the standard from the original calibration curve can then be employed to correct the analytical results.

Standard-Addition Method. The standard-addition method is widely used in flame-absorption spectroscopy. Here two or more aliquots of the sample are transferred to volumetric flasks. One is diluted to volume, and the absorbance of the solution is obtained. A known amount of analyte is added to the second, and its absorbance is measured after dilution to the same volume. Data for other additions may also be obtained. If a linear relationship between absorbance and concentration exists (and this should be established by several standard additions), the following relationships apply:

$$A_x = kC_x$$
$$A_T = k(C_s + C_x)$$

where C_x is the analyte concentration in the diluted sample and C_s is the contribution of the added standard to the concentration; A_x and A_T are the two measured absorbances. Combination of the two equations yields

$$C_x = C_s \frac{A_x}{(A_T - A_x)} \tag{25-1}$$

If several additions are made, A_T can be plotted against C_s. The resulting straight line can be extrapolated to $A_T = 0$. Substituting this value into Equation 25-1 reveals that at the intercept, $C_x = -C_s$.

The addition method has the advantage that it often compensates for variations caused by physical and chemical interferences in the sample solution.

Accuracy. Under usual conditions, the relative error associated with a flame absorption analysis is of the order of 1 to 2%. With special precautions, this figure can be lowered to a few tenths of 1%.

FLAME-EMISSION SPECTROSCOPY

Flame-emission spectroscopy (also called flame photometry) has found widespread application to elemental analysis. The most important applications have been to the analysis of sodium, potassium, lithium, and calcium, particularly in biological fluids and tissue. For reasons of convenience, speed, and relative freedom from interference, flame-emission spectroscopy has become the method of choice for these otherwise difficult-to-determine elements. The method has also been applied, with varying degrees of success, to the determination of perhaps half the elements in the periodic table. Thus, flame-emission spectroscopy must be considered to be one of the important tools for analysis.[3]

[3] For a more complete discussion of the theory and applications of flame-emission spectroscopy, see J. A. Dean, *Flame Photometry.* New York: McGraw-Hill Book Company, Inc., 1960; B. L. Vallee and R. E. Thiers in I. M. Kolthoff and P. J. Elving, Eds., *Treatise on Analytical Chemistry*, part I, vol. 6, chapter 65. New York: Interscience Publishers, Inc., 1965.

INSTRUMENTS FOR FLAME-EMISSION SPECTROSCOPY

Both photometers and spectrophotometers are employed in flame-emission methods. The latter find considerably more widespread use because of their greater selectivity and versatility. The typical flame spectrophotometer is similar in construction to the instruments discussed earlier in this chapter except that the flame now acts as the radiation source; the hollow cathode and chopper are not needed. Most modern instruments are adaptable to either emission or absorption analysis.

METALLIC SPECTRA IN FLAMES

An examination of the spectrum of a flame into which an aqueous solution of a metallic compound is aspirated reveals the presence of both emission lines and bands. The line spectra are characteristic of the metallic atoms present. The band spectra, on the other hand, arise from the presence of molecules; here vibrational energy states are superimposed on electronic energy states and thus produce the closely spaced lines that make up the band.

Band Spectra. When hydrogen or hydrocarbon fuels are burned, band spectra due to such species as OH radicals, CN radicals, and C_2 molecules are observed. Correction for this type of emission is necessary. In addition, some metallic elements form volatile oxides which produce analytically useful band spectra. This type of radiation is characteristic of the alkaline-earth and the rare-earth metals.

Effect of Temperature on Spectra. The temperature produced in a natural gas-air flame is so low (approximately 1800°C) that only the alkali- and alkaline-earth metals are excited. Generation of spectra for most other elements requires the use of oxygen as the oxidant and such fuels as acetylene, hydrogen, or cyanogen.

The optimum flame temperature for an analysis must be determined empirically and depends upon the excitation energy of the element, how it is combined in the sample, the sensitivity required, and what other elements are present. The high temperatures achieved with oxygen and nitrous oxide as oxidants are needed to excite many elements and may also be required for more easily excited elements when these occur as refractory compounds in the sample. Although high temperatures usually provide enhanced emission intensities, and thus higher sensitivity, there are notable instances where the use of a low-temperature flame is an advantage. For example, little enhancement in the characteristic emission by potassium is observed when the temperature is increased above 2000°C because of increases in the formation of potassium ions, which do not emit at the same wavelength as the element. At higher temperatures, moreover, there is an increased likelihood of interference from other elements. Thus, the emission analysis for potassium (and the other alkali metals as well) is best performed with a low-temperature flame.

METHODS OF QUANTITATIVE ANALYSIS

Close control of many variables is essential for the acquisition of reliable flame photometric data. Whenever possible, the standards used for calibration should closely match the overall composition of the unknown solution. Ordinarily, it is best to perform a calibration concurrently with the analysis. Even with these precautions, measurements with a filter photometer can be expected to yield good results only where the sample solution has a relatively simple composition and the element being determined is a major constituent.

Several techniques have been suggested for the performance of a flame photometric analysis.[4] The use of a calibration curve or a standard addition is equally applicable here.

APPLICATIONS

Flame photometric methods have been applied to the analysis of numerous materials, including biological fluids, vegetable matter, cements, glasses, and natural waters.[5] The most important applications are for the determination of the alkali metals and calcium.

Table 25-2 compares the detection limits of flame photometry and flame absorption for several common elements. It is apparent that the relative sensitivity of the two procedures varies from element to element.

Arc and Spark Atomic Spectroscopy

In addition to a flame, other energy sources used to produce atomic-emission spectra in the ultraviolet and visible regions include the electric arc and the electric spark. Here the sample is introduced into the hot plasma formed by the action of an arc or spark between two electrodes. The sample may be a solid or a liquid. For metallic samples, one or both of the electrodes may be formed from the sample itself. Powders can be packed into a cavity on the surface of a graphite or metal electrode. Solution samples are often evaporated in the cavity of a similar type of electrode.

ARC SOURCES

The usual arc source for a spectrochemical analysis is formed with a pair of graphite or metal electrodes spaced 1 to 20 mm apart. The arc is initiated by a low-current spark that causes formation of ions for current conduction in the gap. Once the arc is started, thermal ionization maintains the current flow. In the typical arc, currents are in the range of 1 to 30-A dc or ac. A dc source usually has an open circuit voltage of about 200 V; ac source voltages commonly range from 2200 to 4400 V.

---

[4] See J. A. Dean, *Flame Photometry*, pp. 110–122. New York: McGraw-Hill Book Company, Inc., 1960.

[5] For a summary of applications, see J. A. Dean, *Flame Photometry*. New York: McGraw-Hill Book Company, Inc., 1960.

Typically the temperature attained in an arc plasma is 4000 to 5000°C, which is sufficient to produce enough ions to carry the current once the arc is started. At these temperatures, however, the neutral species of most elements still predominate, and the spectra are, therefore, mainly those of neutral elementary particles rather than of ions. Because of the higher temperatures, an arc generally produces more excited states and therefore more emission lines than the typical flame.

SPARK EXCITATION

An electric spark is formed by impressing an ac voltage of 15,000 to 40,000 V across the air gap between a pair of electrodes containing the sample. A continuous series of oscillating discharges or sparks occurs.

The *average* current with a high-voltage spark is usually significantly less than that of the typical arc, being on the order of a few tenths of an ampere. On the other hand, during the initial pulse of the discharge the *instantaneous* current may exceed 1000 A; this current is carried by a narrow streamer that involves but a minuscule fraction of the total electrode area. The temperature within this streamer is estimated to be as great as 40,000°K. Thus, while the average electrode temperature of a spark source is much lower than that of an arc, the energy in the small volume of the streamer may be several times greater. As a consequence, ionic spectra are more pronounced in a high-voltage spark than in an arc.

SPECTROGRAPHS AND SPECTROMETERS

A *spectrograph* is an instrument that disperses radiation and then photographically records the resulting spectrum. Photoelectric recording with a *spectrometer* is also encountered in emission work, but photographic measurement is more common. It is important to differentiate the spectrograph and the spectrometer from the spectrophotometer, which was considered in Chapter 23. The spectrograph or the emission spectrometer records the intensities of all or part of the lines of an entire spectral region *simultaneously*, whereas a spectrophotometer requires a wavelength-by-wavelength scan to obtain the information. The spectrograph is particularly well suited for emission spectral analysis because it permits the detection and the determination of several elements by excitation of a single small sample.

To facilitate simultaneous intensity recording, the exit slit of a monochromator (see Figure 23-7) is replaced by a photographic plate (or film) located along the focal curve of the instrument. After exposure and development of the emulsion, the various spectral lines of the source appear as a series of black images of the entrance slit distributed along the length of the plate (see Figure 25-3). The location of the lines provides qualitative information concerning the composition of the sample; the darkness of the images can be related to line intensities and hence to concentrations.

The radiation that appears at several chosen locations along the focal curve

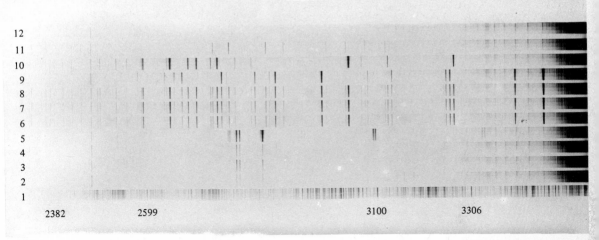

Figure 25-3 Typical Spectra Obtained with a 3.4-m Grating Spectrograph. Numbers on horizontal axis are wavelengths in Å. Spectra: (1) iron standard; (2–5) casein sample; (6–8) Cd-Ge arsenide samples; (9–11) pure Cd, Ge, and As, respectively; (12) pure graphite electrode.

of a photoelectric spectrometer is focused upon a series of photomultiplier tubes to permit the direct and simultaneous measurement of the intensities for a number of spectral lines.

APPLICATIONS

Emission spectroscopy with an electric arc or spark is a powerful tool which permits qualitative detection of some 70 elements by brief excitation of a few milligrams of sample. For qualitative analysis, this type of spectroscopy offers advantages that are virtually without parallel. The sample requires little or no preliminary treatment, the method is highly specific, and detection limits lie in the parts-per-million to parts-per-billion range.

The quantitative applications of emission spectroscopy are more limited, primarily because of difficulties encountered in reproducing radiation intensities. Only with the greatest care can relative errors be reduced to 1 to 2%; uncertainties on the order of 10 to 20% or greater are not uncommon. Where traces are being determined, errors of such magnitudes are often tolerable and not significantly greater than those associated with other methods. In the determination of a major constituent, however, emission spectroscopy suffers by comparison. Nevertheless, the method has the virtue of providing information rapidly. Frequently, several elements can be determined spectroscopically within a few minutes of receipt of the sample; situations exist (in the control of industrial processes, for example) where this speed is more important than a high level of accuracy.

THE ANALYSIS OF REAL SUBSTANCES

Thus far we have stressed the problems associated with the final step in an analysis—specifically, the measurement of some parameter that can be related to the concentration of the analyte. More often than not, this final step is relatively simple because it is performed on a solution that has been freed of interferences; consequently, the number of variables influencing the measurement are few, and our theoretical knowledge is sufficient to account for the effects of these variables. Moreover, convenient and highly refined instruments are available for performing the measurement. Indeed, if every chemical analysis consisted of determining the concentration of a single element or compound in a simple and readily soluble homogeneous mixture, analytical chemistry could profitably be entrusted to the hands of a skilled technician; assuredly, a well-trained chemist could find more useful and challenging work for his mind and his hands.

Regardless of whether a chemist is employed in academic research or in the laboratories of industry, the materials with which he works are *not*, as a rule, chemically simple. To the contrary, most substances that require analysis are complex, consisting of several or perhaps several tens of elements or compounds. Frequently these materials are far from ideal in matters of solubility, volatility, stability, and homogeneity. With such substances, several steps must precede the final measurement. As a matter of fact, this final step is often anticlimactic in a sense, being by far the easiest to perform.

To illustrate, consider the analysis for calcium, an element that occurs widely in nature and is important in many manufacturing processes. Several excellent methods exist for determining the calcium ion concentration of a simple aqueous solution, including precipitation as the oxalate, a compound that can be either titrated with a standard solution of permanganate or ignited to the carbonate for a gravimetric measurement. Calcium ion can also be titrated directly with ethylenediaminetetraacetic acid, or its concentration can be determined by flame absorption measurement. Any of these methods provides an accurate measure of the calcium content of a simple salt such as the carbonate. The chemist, however, is seldom interested in the calcium content of calcium carbonate. More likely, he needs to know the percentage of this element in a sample of animal tissue, a silicate rock, or a piece of glass. The analysis thereby acquires new complexities. For example, none of these materials is soluble in water or dilute aqueous reagents. Before the amount of calcium can be measured, therefore, the sample must be decomposed by high-temperature treatment with concentrated reagents. Unless care is taken, this step may cause losses of calcium; alternatively, the element may be introduced as a contaminant in the relatively large quantities of reagent usually required to complete the decomposition.

Even after the sample has been decomposed and the calcium put into solution, the excellent procedures mentioned previously cannot ordinarily be applied immediately to complete the analysis, for they are all based upon reactions or properties shared by several elements in addition to calcium. Thus, a sample of animal tissue, a silicate rock, or a glass would almost surely contain one or more components that would also precipitate with oxalate, react with ethylenediaminetetraacetic acid, or affect the attenuation of radiation in a flame-absorption measurement. As a consequence, steps to free the calcium from potential interferences must usually precede the final measurement; these could well involve several additional operations.

We have chosen the term *real substances* to describe materials such as those in the preceding illustration. In this context, most of the samples encountered in an elementary quantitative analysis laboratory course definitely are not real, for they are generally homogeneous, usually stable even with rough handling, readily soluble, and—above all—chemically simple. Moreover, well-established and thoroughly tested directions exist for their analysis. From the pedagogical viewpoint, there is value in introducing analytical techniques with such substances, for they do allow the student to concentrate his attention on the mechanical aspects of an analysis. Once these mechanics have been mastered, however, there is little point in the continued analysis of unreal substances; to do so creates the impression that a chemical analysis involves nothing more than the slavish adherence to a well-defined and narrow path, the end result of which is a number that is accurate to one or two parts in a thousand. All too many chemists retain this view far into their professional lives.

In truth, the pathway leading to knowledge of the composition of real substances is frequently more demanding of intellectual skills and chemical intuition than of mechanical aptitude. Furthermore, the chemist is often obliged to compromise between the time he can afford to expend in performing the analysis and the accuracy he thinks he needs. If he is realistic, he may be happy to settle for a part or two in a hundred more often than a part or two in a thousand; with very complex materials, even the former accuracy may require the expenditure of much time and effort.

The difficulties encountered in the analysis of real substances arise, of course, from their complexity as well as from differences in their composition. The chemist is frequently unable to find a clearly defined and well-tested analytical route to follow in the literature; he is thus forced to modify existing procedures to account for the composition of his material, or he must blaze a new pathway. In either case, each new component creates several new variables. Using again the analysis for calcium in calcium carbonate as an example, it is evident that the number of components is small and the variables likely to affect the results are reasonably few. Principal among the latter are the solubility of the sample in acid, the solubility of calcium oxalate as a function of pH, and the effect of the precipitation rate upon the purity and filtering character- istics of calcium oxalate. In contrast, the analysis for calcium in a real sample, such as a silicate rock, that contains a dozen or more other elements is far more complex. Here the analyst has to consider the solubility not only of the calcium oxalate but also the oxalates of the other cations present; coprecipitation of each with the calcium oxalate also becomes a concern. Furthermore, a more drastic treatment is required to dissolve the sample, and additional steps are needed to eliminate the interfering ions. Each new step creates additional variables that make a theoretical treatment of the problem difficult if not impossible.

The analysis of a real substance is thus a challenging problem requiring knowledge, intuition, and experience. The development of a procedure for such materials is not to be taken lightly even by the experienced chemist.

Choice of Method for Analysis of Complex Substances

The choice of method for the analysis of a complex substance requires good judgment and a sound knowledge of the advantages and limitations of the various available analytical tools; a familiarity with the literature on the subject is also essential. We cannot be too explicit concerning the selection of a method because there is no single best way that will apply under all circumstances. We can, however, suggest a somewhat systematic approach to the problem and present some generalities that will aid in making an intelligent decision.

DEFINITION OF THE PROBLEM

A first step, which must precede any choice of method, involves a clear definition of the analytical problem at hand. The method of approach selected by the chemist will be largely governed by his answers to the following questions:

What is the concentration range of the species to be determined?
What degree of accuracy is demanded by the use to which data will be put?
What other components are present in the sample?
What are the physical and chemical properties of the gross sample?
How many samples are to be analyzed?

The concentration range of the element or compound of interest may well limit the choice of feasible methods. If, for example, the analyst is interested in an element present to the extent of a few parts per million, he can generally eliminate gravimetric or volumetric methods and turn his attention to spectrophotometric, spectrographic, and other more sensitive procedures. He knows, moreover, that for a component in the parts-per-million range, he will have to guard against even small losses as a result of coprecipitation and volatility and must be concerned about slight contaminations of samples from reagents. On the other hand, if the analyte is a major component of the sample, these considerations become less important; furthermore, the classical analytical methods may well be preferable.

The answer to the question regarding the accuracy required is of vital importance in the choice of an analytical method and its performance. It is the height of folly to produce physical or chemical data with an accuracy significantly greater than that demanded by the use to which the data are to be put. As we have pointed out, the relationship between time expended and accuracy achieved in an analysis is not ordinarily linear; a 20-fold increase in time (and often more) may be needed to improve the reliability of an analysis from, say, 2% to 0.2%. As a consequence, a few minutes spent at the outset of an analysis in careful consideration of what degree of accuracy is really needed represents an investment that a chemist can ill afford to neglect.

The demands of accuracy will frequently dictate the procedure chosen for an analysis. For example, if the allowable error in an aluminum analysis is only a few parts in a thousand, a gravimetric procedure will probably be required. On the other hand if an error of, say, 50 ppt can be tolerated, spectroscopic or electroanalytical procedures should be considered as well. The experimental details of the method are also affected by accuracy requirements. Thus, if precipitation with ammonia were chosen for the analysis of a sample containing 20% aluminum, the presence of 0.2% iron would be of serious concern where accuracy in the parts-per-thousand range was demanded; here a preliminary separation of the two elements would be necessary. On the other hand, with a limit of error of 50 ppt, a chemist might well dispense with the separation of iron and thus appreciably shorten the analysis. Further, this tolerance would govern his performance in other aspects of the analysis. He would weigh 1-g samples to perhaps the nearest 10 mg and certainly no closer than 1 mg. In addition, he might well be less meticulous in transferring and washing the precipitate and in other time-consuming operations of the gravimetric procedure. If he chooses shortcuts intelligently, he is not being careless but instead realistic in terms of economy of time. The question of accuracy, then, must be settled in clear terms at the very outset.

The third question to be resolved early in the planning stage of an analysis is concerned with the chemical composition of the sample. An answer frequently can be reached by considering the origin of the material; in other situations, a partial or complete qualitative analysis must be undertaken. Regardless of its source, however, this information must be available before an intelligent selection of method can be made, since the various steps in any analysis are based on group reactions or group properties; that is, the analysis is based on reactions or properties shared by several elements or compounds. Thus, measurement of the concentration of a given element by a method that is simple and straightforward in the presence of one group of elements or compounds may require many tedious and time-consuming separations before it can be used in the presence of others. A solvent that is suitable for one combination of compounds may be totally unsatisfactory when applied to another. Clearly, a knowledge of the qualitative chemical composition of the sample is a prerequisite for its quantitative analysis.

The chemist must consider the physical and chemical properties of the substance closely before attempting to derive a method for its analysis. Obviously he should know whether it is a solid, liquid, or gas under ordinary conditions and whether losses by volatility are likely to be a problem. He should also try to determine whether the sample is homogeneous and, if not, what steps can be employed to bring it to this condition. It is also important to know whether or not the sample is hygroscopic or efflorescent. It is essential to know what sort of treatment is sufficient to decompose or dissolve the sample without loss of analyte. Preliminary tests of one sort or another may be needed to provide this information.

Finally, the number of samples to be analyzed is an important aspect in the choice of method. If there are many, considerable time can be expended in calibrating instruments, preparing reagents, assembling equipment, and investigating shortcuts, since the cost of these operations can be spread over the large number of analyses. On the other hand, if at most a few samples are to be analyzed, a longer and more tedious procedure involving a minimum of these preparatory operations may actually prove to be the wiser choice from the economic standpoint.

Having answered these preliminary questions, the chemist is now in a position to consider possible approaches to the problem. At this point he may have a fairly clear idea, based on his past experience, of how he wishes to proceed. He may also find it prudent to speculate on the problems likely to be encountered in the analysis and how they can be solved. He will probably have eliminated some methods from consideration and put others on the doubtful list. Ordinarily, however, he will wish to turn to the analytical literature in order to profit from the experience of others. This, then, is the next logical step in choosing an analytical method.

INVESTIGATION OF THE LITERATURE

The literature dealing with chemical analysis is extensive. For the chemist who will take advantage of them, published reports contain much of value. A list of reference books and journals concerned with various aspects of analytical chemistry appears in Appendix 1. This list is not intended to be an exhaustive catalog, but rather one that is adequate for most work. The list is divided into several categories. In many instances the division is arbitrary, since some of the works could be logically placed in more than one category.

The chemist will often begin his search of the literature by referring to one or more of the general books on analytical chemistry or to those devoted to the analysis of specific types of materials. In addition, he may find it helpful to consult a general reference work relating to the compound or element in which he is interested. From this survey he may get a clearer picture of the problem at hand—what steps are likely to be difficult, what separations must be made, what pitfalls must be avoided. Occasionally, he may find all the answers he needs or even a set of specific instructions for the analysis he wishes to perform. Alternatively, he may find journal references that will lead directly to this information. On other occasions, however, he will develop only a general notion of how to proceed; he will perhaps have several possible methods in mind, and he may also have some clear ideas of how *not* to proceed. He may then wish to consult the works concerned with specific substances and specific techniques, or he may turn directly to the analytical journals. Monographs written on methods for completing the analysis are valuable in deciding among several possible techniques.

A major problem in using the analytical journals is that of finding articles which are pertinent to the problem at hand. The various reference books are useful, since most are liberally annotated with references to the original journals. The key to a thorough search of the literature, however, is *Chemical Abstracts*. This journal contains short abstracts of all papers appearing in the major chemical publications of the world. Both yearly and cumulative indexes are provided to aid in the search; by looking under the element or compound to be determined and the type of substances to be analyzed, the chemist can make a thorough survey of the methods available. Completion of such a survey involves the expenditure of a great deal of time, however, and is often made unnecessary by consulting reliable reference works.

CHOOSING OR DERIVING A METHOD

Having defined the problem and investigated the literature for possible approaches, the chemist must next decide upon the route to be followed in the laboratory. If the choice is simple and obvious, he can proceed directly to the analysis. Frequently, however, the decision requires the exercise of considerable judgment and ingenuity; here experience, an understanding of chemical principles, and perhaps intuition all come into play.

If the substance to be analyzed occurs widely, the literature survey will probably have yielded several alternative methods for the analysis. Economic considerations may well dictate the method that will yield the desired reliability with least expenditure of time and effort. As mentioned earlier, the number of samples to be analyzed will often be a determining factor in this choice.

Investigation of the literature will not invariably reveal a method designed specifically for the type of sample in question. Ordinarily, however, the chemist will have encountered procedures for materials that are at least analogous in composition to the one in question; he will then need to decide whether the variables introduced by differences in composition are likely to have any influence on results. This judgment is often difficult and fraught with uncertainty; recourse to the laboratory may be the only way of obtaining an unequivocal answer.

If it is decided that existing procedures are not applicable, the chemist must then consider modifications that may overcome the problems imposed by the variation in composition. Again he may find that he can propose only tentative alterations, owing to the complexity of the system; he must establish in the laboratory whether these modifications will accomplish their purpose without introducing new difficulties.

After giving due consideration to existing methods and their modifications, the chemist may decide that none will fit his problem; he must then improvise his own procedure. He will need to marshal all facts that he has gathered with respect to the chemical and physical properties of the element or compound to be determined and the state in which it occurs. From this information he may be able to arrive at several possible ways of performing the desired measurement. Each of the possibilities must then be examined critically, with consideration given to the behavior of the other components in the sample as well as the reagents that must be used for solution or decomposition. At this point the chemist must try to anticipate sources of error and possible interferences arising from interactions among the components and reagents; he may well have to devise methods by which problems of this sort can be circumvented. In the end, it is to be hoped that one or more tentative methods that are worth testing will have been located. In all probability, the feasibility of some of the steps in the procedure cannot be determined on the basis of theoretical considerations alone; recourse must be made to preliminary laboratory testing of such steps. Certainly, critical evaluation of the entire procedure can come only from careful laboratory work.

TESTING THE PROCEDURE

Once a procedure for an analysis has been selected, the problem usually arises as to whether the method can be employed directly, without testing, to the problem at hand. The answer to this question is not simple and depends upon a number of considerations. If the procedure chosen has been the subject of a single, or at most a few, literature references, there may be a real point to preliminary laboratory evaluation. With experience, the chemist becomes more and more cautious about accepting claims regarding the accuracy and applicability of a new method. All too often, statements found in the literature are overly optimistic; a few hours spent in testing the procedure in the laboratory may be enlightening.

Whenever a major modification of a standard procedure is undertaken or an attempt is made to apply it to a type of sample different from that for which it was designed, a preliminary laboratory test is advisable. The effects of such alterations simply cannot be predicted with certainty, and the chemist who dispenses with such precautions is sanguine indeed.

Finally, of course, a newly devised procedure must be extensively tested before it is adapted for general use. We must now consider the means by which a new method or a modification of an existing method can be tested for reliability.

Analysis of Standard Samples. Unquestionably, the best technique for evaluating an analytical method involves the analysis of one or more standard samples whose composition with respect to the element or compound of interest is reliably known. For this technique to be of value, however, it is essential that the standards closely resemble the samples to be analyzed with respect to both the concentration range of the analyte and the overall composition. Occasionally, standards of this sort can be readily synthesized from weighed quantities of pure compounds. Others may be purchased from sources such as the National Bureau of Standards; these latter, however, are confined largely to common materials of commerce or widely distributed natural products.

As often as not, the chemist finds himself in the position of being unable to acquire a standard sample that matches closely the substance he wishes to analyze. This situation is particularly true of complex materials in which the form of the analyte is unknown or variable and quite impossible to reproduce. In these circumstances, the best the chemist can do is to prepare a solution of known concentration whose composition approximates that of the sample after it has been decomposed and dissolved. Obviously, such a standard gives no information at all concerning the fate of the substance being determined during the important decomposition and solution steps.

Analysis by Other Methods. The results of an analytical method can sometimes be evaluated by comparison with some entirely different method. Clearly, a second method must exist; in addition, it should be based on chemical principles that differ considerably from the one under examination. Comparable results from the two serve as presumptive evidence that both are yielding satisfactory results, inasmuch as it is unlikely that the same determinate errors would affect each. Such a conclusion will not apply to those aspects of the two methods that are similar.

Standard Addition to the Sample. When the foregoing approaches are inapplicable, the standard-addition method may prove useful. Here, in addition to being used to analyze the sample itself, the proposed procedure is tested against portions of the sample to which known amounts of the analyte have been added. The effectiveness of the method can then be established by evaluating the extent of recovery of the added quantity. The standard-addition method may reveal errors arising from the method of treating the sample or from the presence of the other elements or compounds.

Accuracy Obtainable in Analysis of Complex Materials

To provide a clear idea of the accuracy that can be expected when the analysis of a complex material is carried out with a reasonable amount of effort and care, data on the determination of four elements in a variety of materials are presented in the tables that follow. These data were taken from a much larger set of results collected by W. F. Hillebrand and G. E. F. Lundell of the National Bureau of Standards and published in the first edition of the excellent book on inorganic analysis.[1]

The materials analyzed were naturally occurring substances and items of commerce; they had been especially prepared to give uniform and homogeneous samples and were then distributed among chemists who were, for the most part, actively engaged in the analysis of similar materials. The analysts were allowed to use the methods they considered most reliable and best suited for the problem at hand. In most instances, special precautions were taken so that the results are better than can be expected from the average routine analysis; on the other hand, they probably do not represent the ultimate in analytical perfection.

The value in the second column of each table is a best value obtained by the most painstaking analysis for the measured quantity. It is considered to be the "true value" for calculations of the absolute and relative errors shown in the fourth and fifth columns. The fourth column was obtained by discarding results that were extremely divergent, determining the deviation of the remaining individual data from the best value (second column), and averaging these deviations. The fifth column, the percent relative error, was obtained by dividing

TABLE 26-1   Analysis of Iron in Various Materials[a]

Material	Iron Present (percent)	Number of Analysts	Average Error (absolute)	Average Error (percent relative)
Soda-lime glass	0.064 (Fe_2O_3)	13	0.01	15.6
Cast bronze	0.12	14	0.02	16.7
Chromel	0.45	6	0.03	6.7
Refractory	0.90 (Fe_2O_3)	7	0.07	7.8
Manganese bronze	1.13	12	0.02	1.8
Refractory	2.38 (Fe_2O_3)	7	0.07	2.9
Bauxite	5.66	5	0.06	1.1
Chromel	22.8	5	0.17	0.75
Iron ore	68.57	19	0.05	0.07

[a] From W. F. Hillebrand and G. E. F. Lundell, *Applied Inorganic Analysis*, p. 878. New York: John Wiley & Sons, Inc., 1929. With permission.

[1] W. F. Hillebrand and G. E. F. Lundell, *Applied Inorganic Analysis*, pp. 874–887. New York: John Wiley & Sons, Inc., 1929.

TABLE 26-2 **Analysis for Manganese in Various Materials**[a]

Material	Manganese Present (percent)	Number of Analysts	Average Error (absolute)	Average Error (percent relative)
Ferro-chromium	0.225	4	0.013	5.8
Cast iron	0.478	8	0.006	1.3
	0.897	10	0.005	0.56
Manganese bronze	1.59	12	0.02	1.3
Ferro-vanadium	3.57	12	0.06	1.7
Spiegeleisen	19.93	11	0.06	0.30
Manganese ore	58.35	3	0.06	0.10
Ferro-manganese	80.67	11	0.11	0.14

[a] From W. F. Hillebrand and G. E. F. Lundell, *Applied Inorganic Analysis*, p. 880. New York: John Wiley & Sons, Inc., 1929. With permission.

TABLE 26-3 **Analysis for Phosphorus in Various Materials**[a]

Material	Phosphorus Present (percent)	Number of Analysts	Average Error (absolute)	Average Error (percent relative)
Ferro-tungsten	0.015	9	0.003	20.
Iron ore	0.040	31	0.001	2.5
Refractory	0.069 (P_2O_5)	5	0.011	16.
Ferro-vanadium	0.243	11	0.013	5.4
Refractory	0.45	4	0.10	22.
Cast iron	0.88	7	0.01	1.1
Phosphate rock	43.77 (P_2O_5)	11	0.5	1.1
Synthetic mixtures	52.18 (P_2O_5)	11	0.14	0.27
Phosphate rock	77.56 ($Ca_3(PO_4)_2$)	30	0.85	1.1

[a] From W. F. Hillebrand and G. E. F. Lundell, *Applied Inorganic Analysis*, p. 882. New York: John Wiley & Sons, Inc., 1929. With permission.

TABLE 26-4 **Analysis for Potassium in Various Materials**[a]

Material	Potassium Oxide Present (percent)	Number of Analysts	Average Error (absolute)	Average Error (percent relative)
Soda-lime glass	0.04	8	0.02	50.
Limestone	1.15	15	0.11	9.6
Refractory	1.37	6	0.09	6.6
	2.11	6	0.04	1.9
	2.83	6	0.10	3.5
Lead-barium glass	8.38	6	0.16	1.9

[a] From W. F. Hillebrand and G. E. F. Lundell, *Applied Inorganic Analysis*, p. 883. New York: John Wiley & Sons, Inc., 1929. With permission.

the data in the fourth column by the best value found in the second column and multiplying by 100.

The results for the 4 elements shown in these tables are typical of the data for 26 elements reported in the original publication. It is to be concluded that analyses reliable to a few tenths of a percent relative are the exception when complex mixtures are analyzed by ordinary methods, and that unless the chemist is willing to invest an inordinate amount of time in the analysis, he must accept error on the order of 1 or 2%. If the sample contains less than 1% of the element of interest, even larger relative errors are to be expected.

Finally, it is clear from these data that the accuracy obtainable in the determination of an element is greatly dependent upon the nature and complexity of the substrate. Thus, the relative error for the determination of phosphorous in two phosphate rocks is 1.1%; in a synthetic mixture, it was only 0.27%. The relative error in an iron determination in a refractory was 7.8%; in a manganese bronze having about the same iron content, it was only 1.8%. Here the limiting factor in the accuracy is not in the completion step but rather in solution of the samples and the separation of interferences.

From these data it is clear that the chemist is well advised to adopt a pessimistic viewpoint regarding the accuracy of an analysis, be it his own or one performed by someone else.

27

PRELIMINARY STEPS TO AN ANALYSIS

A chemical analysis is ordinarily preceded by certain steps that are necessary if the analytical data are to have significance. These steps include (1) sampling, (2) production of a homogeneous mixture for analysis, and (3) drying the sample or, alternatively, determining its moisture content. In some instances none of these steps is important or necessary; in others, one or more is vital in determining the accuracy and significance of the analytical result. This chapter deals briefly with each of these steps.

Sampling

The use of analytical results requires the tacit assumption that the data derived from a sample are also applicable to the total mass of material from which it was taken. This supposition is justified only insofar as the chemical composition of the sample truly reflects that of the bulk of the material. The term *sampling* is used to describe the operations involved in procuring a reasonable amount of material that is representative of the whole.

Sampling is frequently the most difficult step in the entire analytical process, particularly with many of the raw materials of commerce that are sold in lots weighing hundreds of tons. Because the value of a given lot is based upon the weight of some component rather than the gross weight, the relationship between these quantities must be established by chemical analysis.

The end product of the sampling operation will be a quantity of material weighing a few grams or at most a few hundred grams. Although it may represent as little as one fifty-millionth of the entire weight of the lot, this sample must approximate closely the average composition of the total mass. Where, as with an ore or other items of commerce, the material consists of nonhomogeneous solids, the task of producing a representative sample is indeed formidable. Clearly, the reliability of the analysis cannot exceed that of the process by which the sample was acquired; the most painstaking work upon a poor sample is a waste of effort.

The literature on sampling of nonhomogeneous material is extensive.[1] We can only provide a brief outline of the methods employed. Basically, two steps are involved: (1) collection of a gross sample and (2) reduction of the gross sample to a size convenient for laboratory work.

THE GROSS SAMPLE

Ideally, the gross sample is a miniature replica of the bulk of the material to be analyzed. It corresponds to the whole, not only in chemical composition but also in particle-size distribution.

To obtain a gross sample, a certain portion of the whole must be removed in a random fashion; that is, a selection must be carried out in such a way that each portion of the whole has an equal chance of being included in the sample. The technique for obtaining a random sample will vary tremendously, depending upon the physical state of the substance, how it is contained, and its total quantity.

Size of the Gross Sample. From the standpoint of convenience and economy, it is desirable that the gross sample be no larger than absolutely necessary. Basically, sample size is determined by (1) the uncertainty that can be tolerated between the composition of the sample and the whole, (2) the degree of heterogeneity of the material being sampled, and (3) the level of particle size at which heterogeneity begins. This last point warrants amplification. In a well-mixed, homogeneous solution of a gas or a liquid, heterogeneity exists only on a molecular scale, and the size of the molecules themselves will govern the minimum size of the gross sample. A particulate solid such as an ore or a soil

[1] See, for example, F. J. Welcher, Ed., *Standard Methods of Chemical Analysis*, 6th ed., vol. 2, part A, pp. 21–52. Princeton, N.J.: D. Van Nostrand Company, Inc., 1963; *Book of Standards*. Philadelphia: American Society for Testing Materials (sampling of substances such as paints, fuels, petroleum products, and constructional materials); *Official Methods of Analysis*, 11th ed. Washington, D.C.: Association of Official Agricultural Chemists, 1970 (for soils, fertilizers, and foods); *Methods for Chemical Analysis of Metals*, 2d ed., pp. 57–72. Philadelphia: American Society for Testing Materials, 1956 (for metals and alloys); F. J. Pettijohn, *Manual of Sedimentary Petrography*. New York: Appleton-Century-Crofts, Inc., 1938 (for soils, outcroppings, and rocks). An extensive bibliography of specific sampling information has been compiled by W. W. Walton and J. I. Hoffman in I. M. Kolthoff and P. J. Elving, Eds., *Treatise on Analytical Chemistry*, part I, vol. 1, p. 93ff. New York: Interscience Publishers, Inc., 1959.

represents the opposite situation. In such materials, the individual pieces of solid can be seen to differ in composition. Here heterogeneity develops in particles that may have dimensions on the order of a centimeter or more. Intermediate between these extremes are colloidal materials and solidified metals. With the former, heterogeneity is first encountered in the particles of the dispersed phase; these typically have diameters in the range of 10^{-5} cm or less. In an alloy, heterogeneity first occurs among the crystal grains.

In order to obtain a truly representative gross sample, a certain number n of the particles referred to in (3) must be taken. The magnitude of this number is dependent upon (1) and (2) and may involve only a few particles, several millions, or even several millions of millions. Large numbers are of no great concern for homogeneous gases or liquids, since heterogeneity among particles first occurs at the molecular level; thus, even a very small weight of sample will contain the requisite number. With a particulate solid, on the other hand, the individual particles may weigh a gram or more; the gross sample may necessarily comprise several tons of material. Here sampling is a costly, time-consuming procedure at best.

Sampling Homogeneous Solutions of Liquids and Gases. For solutions of liquids or gases, the gross sample can be relatively small, since ordinarily nonhomogeneity first occurs at the molecular level, and even small volumes of sample will contain a tremendous number of particles. Whenever possible, the material to be analyzed should be well stirred prior to removal of the sample to make sure that homogeneity does indeed exist. With large volumes of solutions, mixing may be impossible; it is then best to sample several portions of the container with a "sample thief," a bottle that can be opened and filled at any desired location in the solution. This type of sampling, for example, is important in determining the constituents of liquids exposed to the atmosphere. Thus, the oxygen content of lake water may vary by as much as 1000 over a depth difference of a few feet.

Industrial gases or liquids are often sampled continuously as they flow through pipes, care being taken to ensure that the sample collected represents a constant fraction of the total flow and all portions of the stream are sampled.

Sampling Particulate Solids. The process of obtaining a random sample from a bulky particulate material is often difficult. It can be best accomplished while the material is being transferred. For example, every tenth shovelful or wheelbarrow load may be consigned to a sample pile, or portions of the material may be intermittently removed from a conveyor belt. Alternatively, the material may be forced through a riffle or a series of riffles that continuously isolate a fraction of the stream. Mechanical devices of this sort have been highly developed for handling coals and ores.

Sampling Metals and Alloys. Samples of these materials are obtained by sawing, milling, or drilling. In general, it is not safe to assume that chips of the metal removed from the surface will be representative of the entire bulk; sampling must include solid from the interior of the piece as well as from the surface. With billets or ingots of metal, a representative sample can be obtained by sawing across the piece at regularly spaced intervals and collecting the "sawdust" as the sample. Alternatively, the specimen may be drilled, again at various regularly spaced intervals, and the drillings collected as the sample; the drill should pass entirely through the block or halfway through from opposite sides. The drillings can then be broken up and mixed or melted together in a graphite crucible. In the latter case, a granular sample can often be produced by pouring the melt into distilled water.

PRODUCTION OF A LABORATORY SAMPLE

For nonhomogeneous materials, the gross sample may weigh several hundred pounds or more. Here a considerable decrease in size is desirable before the sample is brought into the laboratory, where a few pounds at most are all that can be conveniently handled. The process of reducing the sample volume by a factor of 100 or more is ordinarily multistaged, involving repeated grinding, mixing, and dividing. Diminution in particle size is essential as the weight of sample is decreased to assure that the sample composition continues to be representative of the original material.

TREATMENT OF THE LABORATORY SAMPLE[2]

When it arrives at the laboratory, the sample often requires further treatment before it can be analyzed, particularly if it is a solid. One of the objects of this pretreatment is to produce a material so homogeneous that any small portion removed for the analysis will be identical to any other fraction. Attainment of this condition usually involves decreasing the size of particles to a few tenths of a millimeter and thorough mechanical mixing. Another object of the pretreatment is to convert the substance to a form which is readily attacked by the reagents employed in the analysis; with refractory materials particularly, grinding to a very fine powder is required. Finally, the sample may have to be dried or its moisture content determined.

Crushing and Grinding of Laboratory Samples. In dealing with solid samples, a certain amount of crushing or grinding is ordinarily required to decrease the particle size. Unfortunately, these operations tend to alter the composition of the sample; for this reason, the particle size should be reduced no more than is required for homogeneity and ready attack by reagents.

Several factors may cause appreciable changes in the composition of the sample as a result of grinding. Among these is the heat that is inevitably generated, which can cause losses of volatile components in the sample. In addition, grinding increases the surface area of the solid and thus increases its suscep-

[2] For further discussion, see W. F. Hillebrand, G. E. F. Lundell, H. A. Bright, and J. I. Hoffman, *Applied Inorganic Analysis*, 2d ed., pp. 809–814. New York: John Wiley & Sons, Inc., 1953; A. A. Benedetti-Pichler, *Essentials of Quantitative Analysis*, chapter 18. New York: The Ronald Press Company, 1956.

tibility to reactions with the atmosphere. For example, it has been observed that the iron(II) content of a rock may be altered by as much as 40% during grinding— apparently a direct result of the iron being oxidized to the +3 state.

The effect of grinding on the gain or loss of water from solids is considered in a later section.

Another potential source of error in the crushing and grinding of mixtures arises from the differences in hardness of the components in a sample. Softer materials are converted to smaller particles more rapidly than are the hard ones; any loss of sample in the form of dust will thus cause an alteration in composition. Similarly, flying fragments tend to contain a higher fraction of the harder components.

Intermittent screening often increases the efficiency of grinding. In this operation, the ground sample is placed upon a wire or cloth sieve that will pass particles of the desired size. The residual particles are then returned for further grinding; the operation is repeated until the entire sample passes through the screen. This process will certainly result in segregation of the components on the basis of hardness, the toughest materials being last through the screen; it is obvious that grinding must be continued until the last particle has been passed. The need for further mixing after screening is also apparent.

A serious error can arise during grinding and crushing as a consequence of mechanical wear and abrasion of the grinding surfaces. Even though these surfaces are fabricated from hardened steel, agate, or boron carbide, contamination of the sample nevertheless is occasionally encountered. The problem is particularly acute in the analysis for minor constituents.

The so-called *Plattner diamond mortar*, shown in Figure 27-1, is used for crushing hard, brittle materials. It is constructed of hardened tool steel and consists of a base plate, a removable collar, and a pestle. The sample to be crushed is placed on the base plate inside the collar. The pestle is then fitted into place and struck several blows with a hammer. The sample is reduced to a

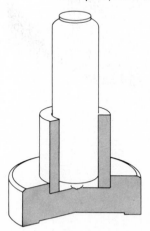

Figure 27-1 A Diamond Mortar.

fine powder; it is collected on a glazed paper after the apparatus has been disassembled.

A useful device for grinding solids that are not too hard is the *ball mill*. It consists of a porcelain crock of perhaps 2-liter capacity, which can be sealed and rotated mechanically. The container is charged with approximately equal volumes of the sample and flint or porcelain balls having a diameter of 20 to 50 mm. Upon rotation, grinding and crushing occurs as the balls tumble within the container. A finely ground and well-mixed powder can be produced in this way.

The mortar and pestle, the most ancient of man's grinding tools, still find wide use in the analytical laboratory. They now come in a variety of sizes and shapes and are fabricated from glass, porcelain, agate, mullite, and other hard materials.

Mixing Solid Laboratory Samples. It is essential that solid materials be thoroughly mixed in order to assure random distribution of the components in the analytical sample. Several methods are commonly employed in the laboratory. One of these involves rolling the sample on a sheet of glazed paper. A pile of the substance is placed in the center and mixed by lifting one corner of the paper enough to roll the particles of the sample to the opposite corner. This operation is repeated many times, the four corners of the sheet being lifted alternately.

Effective mixing of solids is also accomplished by rotating the substance for some time in a ball mill or in a twin-shell dry-blender.[3]

Moisture in Samples

The presence of water in the sample is a common and vexing problem that frequently faces the chemist. Water may be present as a contaminant from the atmosphere or the solution in which the substance was formed; it may be chemically bound within the sample. Regardless of its origin, however, water plays a part in determining the composition of the sample. Unfortunately, and with solids particularly, the water content will vary with humidity, temperature, and state of subdivision. Thus, the constitution of a sample may change significantly with environment and method of handling.

To cope with this source of variability in composition, the chemist may attempt to remove moisture from the sample prior to the weighing step; if this is not possible, he may try to bring the water content to some reproducible level that can be duplicated at a later date if necessary. A third alternative involves determination of the water content at the time the samples are weighed for analysis; in this way the results can be corrected to a dry basis. In any event, most analyses are preceded by some sort of preliminary treatment designed to take into account the presence of water.

[3] Patterson Kelley Co. Inc., Stroudsburg, Pa.

FORMS OF WATER IN SOLIDS

It is convenient to distinguish among the several ways in which water is retained by a solid. Although developed primarily with respect to minerals, the classification of Hillebrand[4] and his collaborators may be applied to other solids as well and forms the basis for the discussion that follows.

Essential Water. That water which forms an integral part of the molecular or crystal structure of a component of the solid is classed as essential water. It exists in stoichiometric quantities. Thus, the *water of crystallization* in a stable solid hydrate (for example, $CaC_2O_4 \cdot 2H_2O$, $BaCl_2 \cdot 2H_2O$) qualifies as a type of essential water.

A second form is called *water of constitution*. Here the water is not present as such in the solid but rather is formed as a product when the solid undergoes decomposition, usually as a result of heating. This is typified by the processes

$$2KHSO_4 \rightarrow K_2S_2O_7 + H_2O$$

$$Ca(OH)_2 \rightarrow CaO + H_2O$$

Nonessential Water. Nonessential water is not necessary for characterization of the chemical constitution of the sample and therefore does not occur in any sort of stoichiometric proportion. It is retained by the solid as a consequence of physical forces.

Adsorbed water is retained on the surface of solids in contact with a moist environment. The amount adsorbed is dependent upon humidity, temperature, and the specific surface area of the solid. Adsorption of water occurs to some degree with all solids.

A second type of nonessential water is called *sorbed water*, encountered with many colloidal substances such as starch, protein, charcoal, zeolite minerals, and silica gel. In contrast to adsorption, the quantity of sorbed water is often large, amounting to as much as 20% or more of the total weight of the solid. Interestingly enough, solids containing even this much water may appear as perfectly dry powders. Sorbed water is held as a condensed phase in the interstices or capillaries of the colloidal solids. The quantity contained in the solid is greatly dependent upon temperature and humidity.

A third type of nonessential moisture is *occluded water*. Here liquid water is entrapped in microscopic pockets spaced irregularly throughout the solid crystals. Such cavities often occur in minerals and rocks (also in gravimetric precipitates).

Water may also be dispersed in a solid as a *solid solution*. Here the water molecules are distributed homogeneously throughout the solid. Natural glasses may contain as much as several percent of moisture in this form.

---

[4] W. F. Hillebrand, G. E. F. Lundell, H. A. Bright, and J. I. Hoffman, *Applied Inorganic Analysis*, 2d ed., p. 815. New York: John Wiley & Sons, Inc., 1952.

EFFECT OF TEMPERATURE AND HUMIDITY ON WATER CONTENT OF SOLIDS

In general, the concentration of water contained in a solid tends to decrease with increasing temperature and decreasing humidity. The magnitude of these effects and the rate at which they manifest themselves differ considerably according to the manner in which water is retained.

Water of Crystallization. The relationship between humidity and the water content of crystalline hydrate is shown by a vapor pressure–composition diagram; Figure 27-2 illustrates such a plot for barium chloride. It was obtained by measuring the pressure of water vapor over a mixture of barium chloride and its hydrates in a closed system. The mole percentage of water is shown along the abscissa; the equilibrium vapor pressure is plotted as the ordinate. When anhydrous barium chloride is brought into equilibrium with a dry atmosphere, the partial pressure of water is zero. When water is added to the salt, however, an amount of the monohydrate is formed, and the following equilibrium is established:

$$BaCl_2 \cdot H_2O(s) \rightleftarrows BaCl_2(s) + H_2O(g)$$

The vapor pressure of water in the system will be determined by this equilibrium, for which the equilibrium constant K' is given by

$$K' = p'_{H_2O}$$

where p'_{H_2O} is the equilibrium pressure of water. As long as both the monohydrate

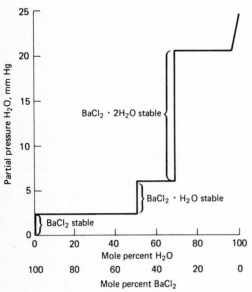

Figure 27-2 Vapor Pressure–Composition Diagram for Barium Chloride and Its Hydrates, 25°C.

and anhydrous salt are present, their activities are constant, and the partial pressure of water is *independent of the amounts* of these two compounds. This condition is shown by the horizontal line extending from just above zero mole percent water to 50 mole percent. At 50 mole percent water, anhydrous barium chloride ceases to exist, and with its disappearance the preceding equilibrium expression is no longer applicable. Increases in the amount of water result in formation of a new species, the dihydrate; the vapor pressure over the mixture is now governed by the equilibrium between this compound and the monohydrate

$$BaCl_2 \cdot 2H_2O(s) \rightleftarrows BaCl_2 \cdot H_2O(s) + H_2O(g)$$

for which

$$K'' = p''_{H_2O}$$

As shown in Figure 27-2, p''_{H_2O} is larger than p'_{H_2O}. Again, the equilibrium pressure is constant as long as both the mono- and dihydrate are present.

When the mole percent of water in the system exceeds the molar ratio of water in the dihydrate (that is, 66.7%), the monohydrate disappears completely. Higher hydrates are not formed, however; instead, the dihydrate begins to dissolve. The result is a saturated solution that is in equilibrium with the solid dihydrate; that is,

$$BaCl_2(sat'd\ soln) \rightleftarrows BaCl_2 \cdot 2H_2O(s) + H_2O(g)$$

As long as some dihydrate remains, we may write

$$K''' = p'''_{H_2O}$$

Again the equilibrium pressure of water will acquire a new value. This condition will be maintained until the solution is no longer saturated (at about 97 mole percent water). The solid dihydrate then disappears, and the vapor pressure of water increases continuously, approaching that of pure water (100 mole percent) at high dilutions.

A diagram such as Figure 27-2 is useful, for it shows clearly the stable forms of a substance at a given temperature as well as the conditions necessary to produce a given form. The behavior of hydrates under various atmospheric conditions can also be predicted. For example, Figure 27-2 indicates that barium chloride dihydrate is the stable form at 25°C when the partial pressure of water in its surroundings ranges between about 6 and 21 mm of mercury. This corresponds to a relative humidity range of 25 to 88%.[5] The relative humidity in most laboratories will be well within this range except on very dry or very damp days; thus, the dihydrate will be stable when exposed to typical laboratory conditions. Moreover, if anhydrous barium chloride was left in contact with the

---

[5] Relative humidity is the ratio of the vapor pressure of water in the atmosphere compared with the vapor pressure in air that is saturated with moisture. At 25°C the partial pressure of water in saturated air is 23.76 mm of mercury. Thus, when air contains water at a partial pressure of 6 mm, the relative humidity is

$$\frac{6.00}{23.76} = 0.253, \quad or \quad 25.3\%$$

atmosphere that had a relative humidity within this range, absorption of moisture would occur until equilibrium had been achieved—that is, until all of the anhydrous salt had been completely converted to the dihydrate. Similarly, an aqueous solution of barium chloride would lose water to the atmosphere under these conditions until finally only crystals of the equilibrium species, the dihydrate, remained. The dihydrate would, of course, lose water under some circumstances. For example, if the relative humidity dropped below 25%, as might happen during a dry winter day, equilibrium would favor formation of the monohydrate. If the dihydrate were placed in a desiccator with a reagent that kept the partial pressure of water below 2 mm of mercury, quantitative conversion to the anhydrous salt would be the ultimate result. Thus, the composition of a sample containing a hydrate or a compound capable of forming a hydrate is greatly dependent upon the relative humidity of its environment.

As we have pointed out, temperature has a marked effect on equilibrium constants. In general, the equilibrium vapor pressure of water over a hydrate increases with temperature; thus, the horizontal lines in Figure 27-2 will be displaced to higher pressures when the temperature rises. Clearly, a temperature rise favors dehydration.

Adsorbed Water. The amount of water adsorbed on the surface of a solid also increases with the moisture content of the environment. The adsorption isotherm in Figure 27-3 illustrates this effect; here the weight of water adsorbed on a typical solid is plotted against the partial pressure of water in the surrounding atmosphere. It is apparent from the diagram that the extent of adsorption is particularly sensitive to changes in water-vapor pressure at low partial pressures.

Quite generally, the amount of adsorbed water decreases with temperature increases and frequently approaches zero if the solid is dried at temperatures above 100°C.

A solid may lose or gain adsorbed moisture relatively rapidly, equilibrium often being reached after 5 or 10 min. The speed of the process is frequently

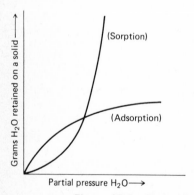

Figure 27-3 Typical Adsorption and Sorption Isotherms.

apparent during the weighing of finely divided solids that have been dehydrated by heating; a continuous increase in weight is observed unless the solid is contained in a tightly stoppered vessel.

Sorbed Water. The quantity of moisture sorbed by a colloidal solid varies tremendously with atmospheric conditions, as may be seen in Figure 27-3. In contrast to the behavior of adsorbed water, however, equilibrium may require days or even weeks for attainment, particularly at room temperatures. Furthermore, the amount of water retained by the two processes is often quite different; typically adsorption will involve quantities of water amounting to a few tenths of a percent of the solid while sorption may entail 10 or 20 percent.

The amount of water sorbed in a solid also decreases as temperature increases. However, complete removal of this type of moisture at 100°C is by no means a certainty, as is indicated by the drying curves for an organic compound shown in Figure 27-4. After drying this material for about 70 min at 105°C, constant weight was apparently reached. It is also clear, however, that additional moisture was removed by elevating the temperature. Even at 230°C dehydration was probably not entirely complete.

Occluded Water. Occluded water is not in equilibrium with the atmosphere and is therefore insensitive to changes in humidity. Heating a solid containing occluded water may cause a gradual diffusion of the moisture to the surface, where it can evaporate; temperatures substantially higher than 100°C are often required to cause this to occur at an appreciable rate. Frequently, heating is accompanied by *decrepitation*, the crystals of the solid being suddenly shattered by pressure of the steam created from moisture contained within the internal cavities.

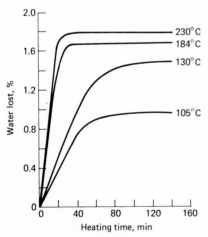

Figure 27-4 Removal of Water at Constant Temperature. [From data of C. O. Willits, *Anal. Chem.*, **23**, 1058 (1951). With permission of the American Chemical Society.]

EFFECT OF GRINDING ON MOISTURE CONTENT

The moisture content, and thus the chemical composition, of a solid is frequently altered considerably during grinding and crushing. Both increases and decreases can occur.

Decreases in water content are sometimes observed during the grinding of solids containing essential water in the form of hydrates; for example, it has been reported that the water content of gypsum, $CaSO_4 \cdot 2H_2O$, is reduced from 20 to 5% by this treatment.[6] Undoubtedly, the change is a result of localized heating during the grinding and crushing of the particles.

Losses also occur when samples containing occluded water are reduced in particle size. Here the grinding process ruptures some of the cavities and exposes the water so that it can evaporate. More commonly, perhaps, the grinding process is accompanied by an increase in moisture content, due primarily to the increase in surface exposed to the atmosphere. A corresponding increase in adsorbed water results. The magnitude of this effect is sufficient to alter the composition of a solid appreciably. For example, the water content of a piece of porcelain in the form of coarse particles was zero; after a period of grinding, it was 0.62%. Grinding a basaltic greenstone for 120 min changed its water content from 0.22 to 1.70%.[6]

From these remarks we may conclude that water determination should be made upon solids before grinding whenever possible.

DRYING THE ANALYTICAL SAMPLE

The methods for establishing the moisture content of a sample will depend upon the physical state of the substance and the information desired. Often the analytical chemist is called upon to determine the composition of a material as he receives it. Here his concern is that the moisture content of the material remain unchanged during preliminary treatment and storage. Where such changes are unavoidable or probable, it may be advantageous to determine the weight loss upon heating at some suitable temperature (say, 105°C) immediately upon receipt of the sample. Then when the analysis is to be performed, the sample can be dried again at this same temperature so that the data can be corrected back to an "as received" basis.

Analyses are also performed, and results are reported on an air-dry basis. Before samples are taken for analysis, the material is allowed to acquire a constant weight while in contact with the atmosphere. Use of an air-dry sample weight is completely satisfactory for such nonhygroscopic substances as alloys. Other particulate materials, which do not tend to adsorb moisture strongly, can also be handled conveniently in this way.

We have already noted that the moisture content of some substances is markedly changed by variations in humidity and temperature. Colloidal substances containing large amounts of sorbed moisture are particularly susceptible to the effects of these variables. For example, the moisture content of a potato

---

[6] W. F. Hillebrand, *J. Amer. Chem. Soc.*, **30**, 1120 (1908).

starch has been found to vary from 10 to 21% as a consequence of an increase in relative humidity from 20 to 70%.[7] With substances of this sort, comparable analytical data between laboratories or even within the same laboratory can be achieved only by carefully specifying a procedure for taking the moisture content into consideration; this will frequently involve drying the sample to constant weight at 105°C or at some other specified temperature. Analyses are then performed and results reported on this "dry basis." While such a procedure may not render the solid completely free of water, it will usually lower the moisture content to a reproducible level.

Frequently, the only satisfactory procedure for obtaining an analysis on a "dry basis" will require a separate determination for moisture in a set of samples taken concurrently with the samples that are to be used for the analysis.

Determination of Water[8]

DRYING PROCEDURES

Undoubtedly, the most common method for determining the water content of solids involves oven drying of a weighed sample and gravimetric determination of the evolved water, either from the loss in weight of the sample or by the gain in weight of an absorbent for water. The great virtue of this procedure is its simplicity; unfortunately, this simplicity does not necessarily extend to the interpretation of the data that the method provides, for several processes in addition to the evolution of water may also occur during the heating. Thus, one may also encounter volatilization of other components, decomposition of one or more of the constituents to give gaseous products, or perhaps air oxidation of a component in the sample. The first two of these effects will cause a decrease in sample weight; oxidation will cause an increase if the products of the reaction are nonvolatile and a decrease if they are volatile. Superimposed on these difficulties is the uncertainty with respect to the temperature required to cause complete evolution of water. Heating at 105°C will accomplish removal of adsorbed moisture and perhaps essential water as well. On the other hand, removal of sorbed and occluded water is often quite incomplete at this temperature. Many minerals, as well as such substances as alumina and silica, require temperatures of 1000°C or more.

Indirect Determination. In this method, the loss in weight of a solid during oven drying is measured; the assumption is then made that this loss equals the weight of water in the sample. The limitations to this procedure are apparent from the preceding paragraph.

In general, oven drying is carried out at as low a temperature as possible in order to minimize decomposition of the sample. The drying process is continued

[7] I. M. Kolthoff and E. B. Sandell, *Textbook of Quantitative Inorganic Analysis*, 3d ed., p. 144. New York: The Macmillan Company, 1952.

[8] For a more detailed discussion of this subject, see *Anal. Chem.*, **23**, 1058–1080 (1951).

until the weight of the sample becomes constant at the chosen temperature; attainment of this state cannot be used as an unambiguous criterion of complete dehydration, however, as is clearly shown by the data in Figure 27-4.

The rate at which drying is completed can often be accelerated by sweeping a stream of dry air over the sample during the heating. This can be done conveniently in a vacuum oven by reducing the internal pressure to a few millimeters of mercury; then, while pumping is continued, air that has been predried over a suitable desiccant is allowed to flow slowly and continuously through the drying chamber.

Direct Determination. In the direct method, the water evolved from the sample is collected on an absorbent that is specific for water; the increase in weight of the absorbent is a direct measure of the amount of water present. This procedure circumvents many of the limitations inherent in indirect drying methods; accurate results can be ordinarily expected, provided the sample is heated at a sufficier.tly high temperature to remove all water. Errors arise, however, if the sample undergoes an air oxidation that yields water as a product; this problem is often encountered with substances containing organic components.

Figure 27-5 shows a typical arrangement for direct-moisture determination. The sample is weighed into a small porcelain boat, which is then placed in the Pyrex or Vycor combustion tube. Air, which has been dried by passage through concentrated sulfuric acid and then over a desiccant such as magnesium perchlorate, is forced over the sample. Heating is accomplished by a burner or a tube furnace. The exit gases are led through a U-tube containing magnesium perchlorate or other desiccant; this tube is weighed before and after the analysis. A second guard tube containing desiccant protects the absorbent tube from becoming contaminated by diffusion of water vapor from the atmosphere.

For minerals, rocks, and many other inorganic materials, an extremely simple, direct method for moisture determination can be employed; this is sometimes called *Penfield's method.*[9] The sample is placed in the end of a hard

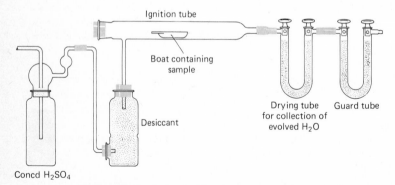

Figure labels: Ignition tube; Boat containing sample; Desiccant; Drying tube for collection of evolved H_2O; Guard tube; Concd H_2SO_4

Figure 27-5 Apparatus for the Determination of Water.

[9] S. L. Penfield, *Amer. J. Sci.* (3), **48**, 31 (1894).

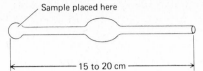

Figure 27-6 Penfield Tube for Water Determination.

glass tube such as that shown in Figure 27-6. The water, which is driven off by ignition in a Bunsen flame, collects in the center bulb of the tube, which is kept cool. After dehydration of the sample is judged to be complete, the lower end of the tube is softened, drawn off, and discarded, leaving the water in the upper end. This is weighed; the water is then removed by aspiration, and the tube is again weighed.

WATER BY DISTILLATION

Distillation, which is useful for the determination of water in materials that are readily air-oxidized, is widely employed for substances containing organic components such as fats, oils, waxes, cereals, plant materials, and foodstuffs. The sample to be analyzed is dissolved or suspended in an organic solvent that is immiscible with and has a higher boiling point than water; ordinarily, toluene or xylene is employed. Upon heating, the water in the sample is volatilized and distilled over with the organic vapors. The distillate is condensed, and the volume of the aqueous phase is measured to give the water content.

A typical distillation apparatus is illustrated in Figure 27-7. The condensed liquid is caught in a trap so constructed that the transferred water collects in the bottom while the organic liquid flows back into the distillation vessel. The trap is calibrated so that the volume of water can be determined directly.

CHROMATOGRAPHIC METHODS

Gas-liquid chromatography, which is discussed in Chapter 29, has proved useful for the determination of the water content of various liquids.[10] Water concentrations varying from a few parts per million to 1% or greater can be determined with a relative error in the 3 to 6% range in most instances.

CHEMICAL METHODS FOR WATER; THE KARL FISCHER REAGENT

A number of chemical methods for the determination of water have been devised. Unquestionably the most important of these involves the use of Karl Fischer reagent, a relatively specific reagent for water.[11]

[10] For example, see J. M. Hogan, R. A. Engel, and H. F. Stevenson, *Anal. Chem.*, **42**, 249 (1970), and references therein.
[11] For reviews on this subject, see J. Mitchell, *Anal. Chem.*, **23**, 1069 (1951); J. Mitchell and D. M. Smith, *Aquametry*. New York: Interscience Publishers, Inc., 1948.

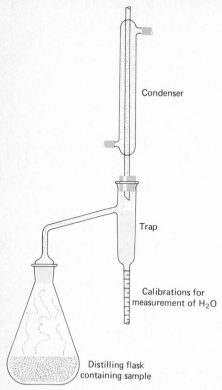

Condenser

Trap

Calibrations for
measurement of H_2O

Distilling flask
containing sample

Figure 27-7 Apparatus for the Determination of Water by Azeotropic Distillation.

Reaction and Stoichiometry. Karl Fischer reagent is composed of iodine, sulfur dioxide, pyridine, and methanol. Upon addition of this reagent to water, the following reactions occur:

$$C_5H_5N \cdot I_2 + C_5H_5N \cdot SO_2 + C_5H_5N + H_2O \rightarrow 2C_5H_5N \cdot HI + C_5H_5N \cdot SO_3$$
$$(27\text{-}3)$$

$$C_5H_5N \cdot SO_3 + CH_3OH \rightarrow C_5H_5N(H)SO_4CH_3 \qquad (27\text{-}4)$$

Only the first step, which involves the oxidation of sulfur dioxide by iodine to give sulfur trioxide and hydrogen iodide, consumes water. In the presence of a large amount of pyridine, C_5H_5N, all reactants and products exist as complexes, as indicated in the equations.

The second step in the reaction occurs when an excess of methanol is present and is important to the success of the titration, for the pyridine-sulfur trioxide complex is also capable of consuming water:

$$C_5H_5N \cdot SO_3 + H_2O \rightarrow C_5H_5NHSO_4H \qquad (27\text{-}5)$$

From the standpoint of analysis, this last reaction is undesirable because it is not as specific for water as the reaction shown by Equation 27-3; it can be prevented completely by having a large excess of methanol present.

From Equation 27-3 it is apparent that the stoichiometry of the Karl Fischer titration involves 1 mole of iodine, 1 mole of sulfur dioxide, and 3 moles of pyridine for each mole of water. In practice, excesses of both sulfur dioxide and pyridine are employed so that the combining capacity of the reagent for water is determined by its iodine content.

End-Point Detection. The end point in the Karl Fischer titration is signaled by the appearance of an excess of the pyridine-iodine complex when all water has been consumed. The color of the reagent is intense enough for a visual end point. The color change is from the yellow of the reaction products to the brown of the excess reagent. With some practice, and in the absence of other colored materials, the end point can be established with a reasonable degree of certainty (that is, to perhaps ± 0.2 ml).

Various electrometric end points are also employed for the Karl Fischer titration, the most widely used being the "dead stop" technique described on page 488.

Stability of the Reagent. As mentioned earlier, the Karl Fischer reagent is prepared so that its combining capacity for water is determined by the concentration of iodine in the solution. For typical application, the titer is about 3.5 mg of water per milliliter of reagent; a twofold excess of sulfur dioxide and a three- to fourfold excess of pyridine are provided. Stabilized Karl Fischer reagent can be purchased from commercial sources.

The titer of the Karl Fischer reagent decreases with standing. Decomposition is particularly rapid immediately after preparation; it is therefore good practice to prepare the reagent a day or two before it is to be used. Ordinarily, its titer should be established at least daily against a standard solution of water in methanol.

It is obvious that great care must be exercised to prevent contamination of the Karl Fischer reagent and the sample by atmospheric moisture. All glassware must be carefully dried before use, and the standard solution must be stored out of contact with air. It is also necessary to minimize contact between the atmosphere and the solution during the titration.

Applications. The Karl Fischer reagent has been applied to the determination of water in numerous substances.[12] The techniques employed vary considerably, depending upon the solubility of the material, the state in which the water is retained, and the physical state of the sample. If the sample can be dissolved completely in methanol, a direct and rapid titration is feasible. This method has been applied to the analysis of water in many organic acids, alcohols, esters, ethers, anhydrides, and halides. Hydrated salts of most organic acids can also be analyzed by direct titration, as well as hydrates of a number of inorganic salts that are soluble in methanol.

[12] For a complete discussion of the applications of the reagent, see J. Mitchell and D. M. Smith, *Aquametry.* New York: Interscience Publishers. Inc., 1948.

If it is impossible to produce a solution of the sample and reagent, direct titration ordinarily results in an incomplete reaction. Satisfactory results can frequently be obtained, however, by addition of an excess of reagent and back-titration with a standard solution of water in methanol after a suitable reaction time. An alternative and often effective procedure is to extract the water from the sample with an anhydrous solution of methanol or perhaps some other organic solvent; the rate of transfer of moisture is increased by refluxing the mixture of sample and solvent. The methanol can then be titrated directly with the Karl Fischer solution.

Difficulty is also encountered in the analysis of sorbed moisture and tightly bound hydrate water. For these, the preceding extraction techniques are frequently effective.

Certain substances interfere with the Fischer method. Among these are compounds that react with one of the components of the reagent to produce water. For example, carbonyl compounds combine with methanol to give acetals:

$$RCHO + 2CH_3OH \rightarrow R-CH \begin{matrix} OCH_3 \\ \diagup \\ \diagdown \\ OCH_3 \end{matrix} + H_2O$$

The result is a fading end point in the titration. Many metal oxides will react with the hydrogen iodide formed in the titration to give water:

$$MO + 2HI \rightleftarrows MI_2 + H_2O$$

Again, erroneous data result. In some instances, preliminary treatment of the sample can prevent these interferences.

Oxidizing or reducing substances frequently interfere with the Karl Fischer water titration by reoxidizing the iodide produced or reducing the iodine in the reagent.

INFRARED METHODS

Water exhibits a near infrared absorption peak (1.94 μ), which has proved to be useful for the determination of moisture in a variety of materials. A typical application is to the determination of water in various food products such as instant coffee, honey, potato chips, and flour.[13] The samples are dispersed in dimethyl sulfoxide which, after 2 to 4 hr, extracts the moisture nearly completely. Following extraction, the absorbance of the liquid is measured in a 1-cm quartz cell. A linear relationship between absorbance and water concentration exists over the range of 0.00 to 0.70 ml of water per 100 ml of solution. The accuracy of the procedure appears to be equivalent to that of the Karl Fischer method.

[13] D. M. Vomhof and J. T. Thomas, *Anal. Chem.*, **42**, 1230 (1970).

28

DECOMPOSING AND DISSOLVING THE SAMPLE

Most analyses are completed by performing measurements on a solution (usually aqueous) of the analyte. Often, converting an analyte to a soluble form requires extensive measures. For example, the determination of halogens or nitrogen in an organic compound requires vigorous treatment of the sample to rupture the strong bonds between these elements and carbon. Similarly, before the components of a siliceous mineral can be analyzed, the silicate structure must be destroyed through preliminary treatment with powerful reagents; only then can an aqueous solution of the cations be obtained.

The proper choice among the various reagents and techniques for decomposing and dissolving analytical samples can be critical to the success of an analysis, particularly where refractory substances are involved. This chapter describes some of the more common methods for obtaining aqueous solutions of samples that are difficult to decompose or dissolve.

Some General Considerations

Ideally, the reagent chosen should cause complete dissolution of the sample; attempts to leach one or more components from a mixture usually result in an incomplete separation from the unattacked residue.

611

Consideration must be given to possible interference with the final measurement by the solvent that is selected. For example, hydrochloric acid would ordinarily be avoided in a bromide determination; the similarity in behavior between the two halides would probably necessitate a separation step prior to the final measurement. In addition, impurities in the solvent may affect the outcome of an analysis. This factor is of particular importance in the determination of components that are present in low concentrations; in trace analysis, the most important consideration in choosing among possible solvents is frequently their purity.

Volatilization of important constituents in a sample may occur during the solution step unless proper precautions are taken. For example, treatment with acids can result in the loss of carbon dioxide, sulfur dioxide, hydrogen sulfide, hydrogen selenide, and hydrogen telluride. In the presence of basic reagents, loss of ammonia is common. Treatment of a sample with hydrofluoric acid will result in vaporization of silicon and boron as their fluorides, while exposure of halogen-containing substances to strong oxidizing reagents may result in the evolution of chlorine, bromine, or iodine. Reducing conditions during the preliminary treatment of a sample can cause the volatilization of such compounds as arsine, phosphine, or stibine.

A number of elements form volatile chlorides that are partially or completely lost from hot hydrochloric acid solutions. Among these are arsenic and antimony trichloride, tin(IV) and germanium tetrachloride, and mercury(II) chloride. The oxychlorides of selenium and tellurium also volatilize to some extent from hot hydrochloric acid. In the presence of chloride ion, certain other elements are lost by volatilization from hot concentrated solutions of perchloric or sulfuric acid. These include bismuth, manganese, molybdenum, thallium, vanadium, and chromium.

Boric acid, nitric acid, and the halogen acids are lost from boiling aqueous solutions, while phosphoric acid distills from hot concentrated sulfuric or perchloric acids. Volatile oxides can also be lost from hot acidic solutions; these include the tetroxides of osmium and ruthenium as well as the heptoxide of rhenium.

Liquid Reagents for Dissolving or Decomposing Samples

The most common reagents for attacking analytical samples are the mineral acids or their aqueous solutions. Solutions of sodium or potassium hydroxide also find occasional use.

HYDROCHLORIC ACID

Concentrated hydrochloric acid is an excellent solvent for many metal oxides as well as those metals that are more easily oxidized than hydrogen; in addition, it is often a better solvent for oxides than the oxidizing acids. Concentrated hydrochloric acid is about 12 F, but upon heating, hydrogen chloride is lost until a constant-boiling 6-F solution remains (boiling point about 110°C).

NITRIC ACID

Hot concentrated nitric acid will dissolve all common metals with the exception of aluminum and chromium, which become passive to the reagent as a consequence of surface oxide formation. When treated with concentrated nitric acid, tin, tungsten, and antimony form slightly soluble acids, which permit the separation of these elements from alloys by filtration immediately following the solution step.

SULFURIC ACID

Hot concentrated sulfuric acid owes part of its effectiveness to its high boiling point (about 340°C); many samples decompose and dissolve rapidly at this elevated temperature. Organic compounds are dehydrated and oxidized by hot concentrated sulfuric acid. The reagent thus serves to eliminate such components from a sample. Most metals and many alloys are attacked by the hot acid.

PERCHLORIC ACID

Hot concentrated perchloric acid, a potent oxidizing agent, attacks a number of iron alloys and stainless steels that are intractable to other mineral acids. Care must be taken in the use of the reagent, however, because of *its potentially explosive nature*. Neither the cold concentrated reagent nor the heated dilute reagent is hazardous, but violent explosions occur when hot concentrated perchloric acid comes in contact with organic material or easily oxidized inorganic substances. Because of this property, the concentrated reagent should be heated only in special hoods. Perchloric acid hoods are lined with glass or stainless steel, are seamless, and have a fog system for washing down the walls with water; their fan systems should be independent of other hoods. If proper precautions are taken,[1] perchloric acid is a safe and useful reagent.

Perchloric acid is marketed as the 60 or 72% acid. Upon heating, a constant-boiling mixture (72.4% $HClO_4$) is obtained at a temperature of 203°C.

OXIDIZING MIXTURES

More rapid solvent action can sometimes be obtained by the use of mixtures of acids or by the addition of oxidizing agents to a mineral acid. *Aqua regia*, a mixture containing three volumes of concentrated hydrochloric acid and one of nitric acid, is well known. Addition of bromine or hydrogen peroxide to mineral acids often increases their solvent action and hastens the oxidation of organic materials in the sample. Mixtures of nitric and perchloric acid are also useful for this purpose.

HYDROFLUORIC ACID

The primary use of hydrofluoric acid is for the decomposition of silicate rocks and minerals where silica is not to be determined; the silicon is evolved as the

---

[1] See H. H. Willard and H. Diehl, *Advanced Quantitative Analysis*, p. 8. Princeton, N.J.: D. Van Nostrand Company, Inc., 1942.

tetrafluoride. After decomposition is complete, the excess hydrofluoric acid is driven off by evaporation with sulfuric acid or perchloric acid. Complete removal is often essential to the success of an analysis because the fluoride complexes of several cations are extraordinarily stable; the properties of these complexes may differ markedly from those of the parent cations. Thus, for example, precipitation of aluminum with ammonia is quite incomplete if fluoride is present in a small amount. Frequently, removal of the last traces of fluoride ion from a sample is so difficult and time-consuming as to negate the attractive features of the parent acid as a solvent for silicates.[2]

Hydrofluoric acid finds occasional use in conjunction with other acids in the attack on some of the more difficultly soluble steels.

Hydrofluoric acid can cause serious damage and painful injury when brought in contact with the skin; it must be handled with great respect. A burn may not become evident until hours after exposure.

Decomposition of Samples by Fluxes

Many common substances—notably, silicates, some mineral oxides, and a few iron alloys—are attacked slowly, if at all, by the usual liquid reagents. Recourse to more potent fused-salt media, or *fluxes*, is then indicated. Fluxes will decompose most substances by virtue of the high temperature required for their use (300 to 1000°C) and the high concentration of reagent brought in contact with the sample.

Where possible, the employment of a flux is avoided, for several dangers and disadvantages attend its use. There is the possibility that significant contamination will be introduced in the rather large amount of flux (typically 10 times the sample weight) required for a successful fusion. Moreover, the aqueous solution from the fusion will have a high salt content, which may cause difficulties in the subsequent steps of the analysis. The high temperatures required for a fusion increase the danger of volatilization losses. Finally, the container in which the fusion is performed is almost inevitably attacked to some extent by the flux; again, contamination of the sample is the result.

For a sample containing only a small fraction of material that is difficultly dissolved, it is common practice to employ a liquid reagent first; the undecomposed residue is then isolated by filtration and fused with a relatively small quantity of a suitable flux. After cooling, the melt is dissolved and combined with the major portion of the sample.

METHOD OF CARRYING OUT A FUSION

Prior to the decomposition of a sample with a flux, the solid must ordinarily be ground to a very fine powder to produce a high surface area. The sample must then be thoroughly mixed with the flux; this operation is often carried out in the crucible in which the fusion is to be done by careful stirring with a glass rod.

---

[2] For methods of removal of fluoride ion, see H. H. Willard, L. M. Liggett, and H. Diehl, *Ind. Eng. Chem. Anal. Ed.*, **14**, 234 (1942).

In general, the crucible used in a fusion should never be more than half-filled. At the outset, heating must be slow and careful because the evolution of water and other gases is a common occurrence; unless care is taken, there is the danger of loss by spattering. As an added precaution, the crucible should be covered. The maximum temperature employed should be no greater than necessary; otherwise, attack upon the crucible and decomposition of the flux may occur. The time needed for fusion may range from a few minutes to a matter of hours, depending upon the nature of the sample. It is frequently difficult to decide when the heating should be discontinued. In some fusions, the production of a clear melt signals completion of the decomposition. In others, this condition is not obvious, and the analyst must base the heating time on previous experience with the type of material being analyzed. In any event, the aqueous solution from the fusion should be examined carefully for particles of unattacked sample.

When the fusion is judged complete, the mass is allowed to cool slowly; then, just before solidification occurs, the crucible is rotated to distribute the solid around the walls to produce a thin layer that can be readily detached.

TYPES OF FLUXES

With few exceptions the common fluxes used in analysis are compounds of the alkali metals. Basic fluxes employed for attack of acidic materials include carbonates, hydroxides, peroxides, and borates. The acidic fluxes are pyrosulfates, acid fluorides, as well as boric oxide. If an oxidizing flux is required, sodium peroxide can be used. As an alternative, small quantities of the alkali nitrates or chlorates can be mixed with sodium carbonate. The properties of the common fluxes are summarized in Table 28-1.

Sodium Carbonate. Silicates and certain other refractory materials can be decomposed by heating to 1000 to 1200°C with sodium carbonate. This treatment generally converts the cationic constituents of a sample to acid-soluble carbonates or oxides; the nonmetallic constituents are converted to soluble sodium salts.

The method of treatment of the fused mass depends upon the constituents of the sample and the analysis to be performed. With silicates, for example, the melt is dissolved in dilute acid; decomposition and solution of the metallic carbonates and partial separation of the silicon as hydrated silica result. Repeated dehydration completely removes the silica and leaves a solution that can be analyzed for its cationic constituents. If it is desirable to isolate the anions of the sample as soluble sodium salts, it is better to treat the fused product with water rather than with acids; with this solvent the bulk of the cations will remain undissolved as carbonates or oxides. Such treatment would be preferred, for example, in the decomposition of a sample containing the slightly soluble sulfates of barium, calcium, or lead. Treatment of the melt with water would afford a separation of the anions from cations of the sample.

Carbonate fusions are normally carried out in platinum crucibles.

TABLE 28-1   Common Fluxes

Flux	Melting Point, °C	Type of Crucible for Fusion	Type of Substance Decomposed
Na_2CO_3	851	Pt	For silicates and silica-containing samples; alumina-containing samples; sparingly soluble phosphates and sulfates
Na_2CO_3 + an oxidizing agent such as KNO_3, $KClO_3$, or Na_2O_2	—	Pt (not with Na_2O_2), Ni	For samples requiring an oxidizing environment; that is, samples containing S, As, Sb, Cr, etc.
NaOH or KOH	318 380	Au, Ag, Ni	Powerful basic fluxes for silicates, silicon carbide, and certain minerals; main limitation: purity of reagents
Na_2O_2	Decomposes	Fe, Ni	Powerful basic oxidizing flux for sulfides; acid-insoluble alloys of Fe, Ni, Cr, Mo, W, and Li; platinum alloys; Cr, Sn, Zr minerals
$K_2S_2O_7$	300	Pt, porcelain	Acid flux for slightly soluble oxides and oxide-containing samples
B_2O_3	577	Pt	Acid flux for decomposition of silicates and oxides where alkali metals are to be determined
$CaCO_3$ + NH_4Cl	—	Ni	Upon heating the flux, a mixture of CaO and $CaCl_2$ produced; used to decompose silicates for the determination of the alkali metals

Potassium Pyrosulfate. Potassium pyrosulfate provides a potent acidic flux that is particularly useful for the attack of the more intractable metal oxides. Fusions with this reagent are performed at about 400°C; at this temperature the slow evolution of the highly acidic sulfur trioxide takes place:

$$K_2S_2O_7 \rightarrow K_2SO_4 + SO_3$$

The corresponding metal sulfates are produced. Potassium pyrosulfate is not a particularly useful flux for silicates.

The effectiveness of potassium pyrosulfate decreases with prolonged heating as a consequence of sulfur trioxide evolution; it is possible, however, to regenerate the reagent by cooling and adding a few drops of concentrated sulfuric acid. Reheating must be performed with care to avoid physical losses of sample as water vapor is evolved.

Potassium pyrosulfate can be prepared by heating potassium hydrogen sulfate:

$$2KHSO_4 \rightarrow K_2S_2O_7 + H_2O$$

Direct use of the acid sulfate as a flux is occasionally recommended; great care is needed, however, at the outset of heating to avoid loss of sample by spattering and overflow as the water is evolved; it is preferable to start with the pyrosulfate.

Pyrosulfate fusions can often be carried out in ordinary porcelain crucibles or vitreous silica containers. Some attack on the porcelain may occur, but the extent is usually negligible. Platinum ware can also be used, although small amounts of the metal will be introduced to the sample.

Other Fluxes. Table 28-1 contains data for several other common fluxes. Noteworthy are boric oxide and the mixture of calcium carbonate and ammonium chloride. Both are employed to decompose silicates for the analysis of alkali metals. Boric oxide is removed after solution of the melt by evaporation to dryness with methyl alcohol; methyl borate, $B(OCH_3)_3$, distills.

When a mixture of calcium carbonate and ammonium chloride is heated, calcium oxide and calcium chloride are produced; these reagents are sufficiently active to decompose silicates. Extraction of the fused mass with water results in an aqueous solution of alkali chlorides ordinarily contaminated only with calcium ions, plus any sulfate or borate that may have been present in the original material. These contaminants are readily removed prior to the alkali-metal analysis. Many chemists prefer this procedure (known as the *J. Lawrence Smith method*[3]) to others for the determination of the alkali metals in silicates.

Decomposition of Organic Compounds

Analysis for the elemental composition of an organic sample generally requires drastic treatment to convert the elements of interest into a form susceptible to the common analytical techniques. These treatments are usually oxidative and involve conversion of carbon and hydrogen to carbon dioxide and water; occasionally, however, heating the sample with a potent reducing agent is sufficient to rupture the covalent bonds in the compound and free the element to be determined from the carbonaceous residue.

Oxidation procedures are sometimes grouped into two categories. *Wet-ashing* (or oxidation) makes use of liquid oxidizing agents such as sulfuric, nitric, or perchloric acids. *Dry-ashing* usually implies ignition of the organic compound in air or in a stream of oxygen. In addition, oxidations can be carried out in certain fused-salt media, sodium peroxide being the most common flux for this purpose.

In the sections that follow, we shall consider briefly some of the methods for decomposing organic substances prior to elemental analysis.

WET-ASHING PROCEDURES

Solutions of strong oxidizing agents will decompose organic samples. The main problem associated with the use of these reagents is the prevention of losses of the elements of interest by volatilization.

[3] J. Lawrence Smith, *Amer. J. Sci.*, (2), **50**, 269 (1871).

We have already encountered an example of wet-ashing in the Kjeldahl method for the determination of nitrogen in organic compounds (Chapter 11) where concentrated sulfuric acid is the oxidizing agent. This reagent is also frequently employed for decomposition of organic materials in which metallic constituents are to be determined. Nitric acid may be added periodically to the solution to hasten the rate at which oxidation occurs.[4] A number of elements are volatilized (at least partially) by this procedure, particularly if the sample contains chlorine; included are arsenic, boron, germanium, mercury, antimony, selenium, tin, the halogens, sulfur, and phosphorus.

An even more effective reagent than sulfuric–nitric acid mixtures is perchloric acid mixed with nitric acid. *Great care must be exercised in using this reagent*, however, because of the tendency of hot anhydrous perchloric acid to react explosively with organic material. Explosions can be avoided by starting with a solution in which the perchloric acid is diluted with sufficient nitric acid to ensure complete oxidation of the organic substance present. As the solution is heated, the nitric acid attacks the easily oxidized materials. With continued heating, water and nitric acid are lost by decomposition and evaporation, and the solution becomes progressively a stronger oxidant. If the solution becomes too concentrated in perchloric acid before most of the oxidation is complete, it will darken in color or blacken. If this occurs, the mixture should be *immediately* removed from heat and diluted with water and nitric acid. The heating can then be continued. As we mentioned earlier (p. 613), perchloric acid oxidations should be carried out only in a special hood. If properly performed, oxidations with a mixture of nitric and perchloric acids are rapid, and losses of metallic ions are negligible.[5] It cannot be too strongly emphasized that proper precautions must be taken in the use of hot concentrated perchloric acid to prevent violent explosions.

DRY-ASHING PROCEDURES

The simplest method for decomposing an organic sample is by heating with a flame in an open dish or crucible until all carbonaceous material has been oxidized to carbon dioxide. Red heat is often required to complete the oxidation. Analysis of the nonvolatile components follows solution of the residual solid. Unfortunately, a great deal of uncertainty always exists with respect to the completeness of recovery of supposedly nonvolatile elements from a dry-ashed sample. Some losses probably result from the mechanical entrainment of finely divided particulate matter in the convection currents around the crucible. In addition, volatile metallic compounds may be formed during the ignition. For example, copper, iron, and vanadium are appreciably volatilized when samples containing porphyrin compounds are heated.[6]

---

[4] *Official Methods of Analysis*, 11th ed., p. 400. Washington, D.C.: Association of Official Analytical Chemists, 1970.

[5] T. T. Gorsuch, *Analyst*, **84**, 135 (1959); G. F. Smith, *Anal. Chim. Acta*, **8**, 397 (1953); G. F. Smith, *The Wet Chemical Oxidation of Organic Compositions Employing Perchloric Acid*. Columbus, Ohio: G. F. Smith Chemical Co., 1965.

[6] See T. T. Gorsuch, *Analyst*, **84**, 135 (1959); R. E. Thiers in D. Glick, Ed., *Methods of Biochemical Analysis*, vol. 5. New York: Interscience Publishers, Inc., 1957.

In summary, although dry-ashing is the simplest of all methods for decomposing organic compounds, it is often the least reliable; it should not be employed unless tests have been performed to demonstrate its applicability to a given type of sample.

COMBUSTION-TUBE METHODS[7]

Several common and important elemental components of organic compounds are converted to gaseous products when the material is oxidized. With suitable apparatus, it is possible to trap these volatile compounds quantitatively, thus making them available for the analysis of the element of interest. The oxidation is commonly performed in a glass or quartz combustion tube through which a stream of carrier gas is passed. The stream serves to transport the volatile products to parts of the apparatus where they can be separated and retained for measurement; the gas may also serve as the oxidizing agent. Elements susceptible to this type of treatment are carbon, hydrogen, oxygen, nitrogen, the halogens, and sulfur.

Figure 28-1 shows a classical combustion train, which has been used for nearly a century for the analysis of carbon and hydrogen in an organic substance. Here dry oxygen or air is forced through the tube to oxidize the sample as well as to carry the products to the absorption section of the train. The sample, contained in a small platinum or porcelain boat, is ignited by slowly raising the temperature with a burner or furnace. Partial combustion as well as thermal decomposition occurs; the products are then carried over an oxidation catalyst consisting of a platinum gauze packing maintained at a temperature of 700 to 800°C. Downstream from the platinum catalyst is a packing of copper(II) oxide that completes the oxidation of the sample to carbon dioxide and water.

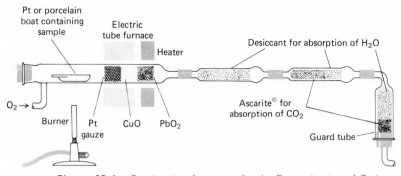

Figure 28-1 Combustion Apparatus for the Determination of Carbon and Hydrogen.

[7] For a more detailed discussion, see C. R. N. Strouts, J. H. Gilfillan, and H. N. Wilson, *Analytical Chemistry*, vol. 1, chapter 14. New York: The Oxford University Press, 1955; J. B. Niederl and V. Niederl, *Micromethods of Quantitative Organic Analysis*, 2d ed. New York: John Wiley & Sons, Inc., 1942.

Additional packing is often included to remove compounds that will interfere with the determination of the carbon dioxide and water in the exit stream. Lead chromate and silver serve to remove halogen and sulfur compounds; lead dioxide can be employed to eliminate interference from the oxides of nitrogen.

The exit gases from the combustion tube are first passed through a weighing tube packed with a desiccant that removes the water from the stream. The increase in weight of this tube gives a measure of the hydrogen content of the sample. The carbon dioxide in the gas stream is then removed in the second weighing tube, which is packed with Ascarite (sodium hydroxide held on asbestos). Because the absorption of carbon dioxide is accompanied by the formation of water, additional desiccant follows the Ascarite in this tube. Finally, the gases are passed through a *guard tube*, which protects the two weighing tubes from contamination by the atmosphere.

Table 28-2 lists other applications of the combustion-tube method. A substance containing a halogen will yield the free element upon oxidation; reduction to the corresponding halide frequently precedes the analytical step. Sulfur is converted to sulfuric acid, which can be estimated by precipitation with barium ion or by alkalimetric titration.

The *Dumas method*, which is suitable for the analysis of nitrogen in all types of organic compounds, involves mixing the sample with powdered copper(II) oxide and igniting in a stream of carbon dioxide in a combustion tube. At elevated temperatures the organic substance is oxidized to carbon dioxide and water by the copper(II) oxide. Any nitrogen in the compound is converted primarily to the elemental state, although nitrogen oxides may also be formed. These oxides are reduced to elemental nitrogen by passing the gas stream over a bed of hot copper. The products of the ignition are then swept into a gas buret filled with highly concentrated potassium hydroxide; this solution completely absorbs the carbon dioxide, water, and other products of the combustion such as sulfur dioxide and hydrochloric acid. The elemental nitrogen remains undissolved in the caustic, however, and its volume is directly measured. Potassium hydroxide is preferred over sodium hydroxide as an absorbent for carbon dioxide, owing to the appreciably greater solubility of potassium carbonate over that of sodium carbonate.

Recently, automated combustion-tube analyzers have become available which make possible the determination of carbon, hydrogen, and nitrogen in a single sample. The apparatus requires essentially no attention by the operator and completes the analysis in less than 15 min. In one analyzer, oxidation is carried out in a mixture of oxygen and helium over a cobalt oxide catalyst; halogens and sulfur are removed with a packing of silver salts. At the end of the combustion train is a packing of hot copper, which removes oxygen and converts nitrogen oxides to nitrogen. The exit gas, consisting of a mixture of water, carbon dioxide, nitrogen, and helium, is finally collected in a glass bulb. The analysis of this mixture is accomplished with three thermal conductivity measurements (see p. 657). The first is made on the intact mixture, the second on the mixture after water has been removed by passage of the gases through a dehydrating agent, and the third is on the mixture after removal of carbon dioxide with Ascarite. The relationship between thermal conductivity readings

TABLE 28-2 Combustion-Tube Methods for the Elemental Analysis of Organic Substances

Element	Name of Method	Method of Oxidation	Method of Completion of Analysis
Halogens	Pregl	Sample burned in a stream of oxygen over a red-hot platinum catalyst; halogens converted primarily to HX and X_2	Gas stream passed through a carbonate solution containing SO_3^{2-} (to reduce halogens and oxyhalogens to halides); halide ion, X^-, then determined by usual procedures
	Grote	Sample burned in a stream of air over a hot silica catalyst; products are HX and X_2	Same as above
Sulfur	Pregl	Similar to halogen determination; combustion products are SO_2 and SO_3	Gas stream passed through aqueous H_2O_2 to convert sulfur oxides to H_2SO_4, which can then be titrated with standard base
	Grote	Similar to halogen determination; products are SO_2 and SO_3	Similar to above
Nitrogen	Dumas	Sample oxidized by hot CuO to give CO_2, H_2O, and N_2	Gas stream passed through concentrated KOH solution leaving only N_2, which is measured volumetrically
Carbon and hydrogen	Pregl	Similar to halogen analysis; products are CO_2 and H_2O	H_2O adsorbed on a desiccant and CO_2 on Ascarite; determined gravimetrically
Oxygen	Unterzaucher	Sample pyrolyzed over carbon; oxygen converted to CO; H_2 used as carrier gas	Gas stream passed over I_2O_5 $[5CO + I_2O_5(s) \rightarrow 5CO_2 + I_2(g)]$; liberated I_2 titrated

and concentration is linear, and the slope for each constituent is established by calibration with a pure compound such as acetanilide.

COMBUSTION WITH OXYGEN IN SEALED CONTAINERS

A relatively straightforward method for the decomposition of many organic substances involves combustion with oxygen in a sealed container. The reaction products are absorbed in a suitable solvent before the reaction vessel is opened and are subsequently analyzed by ordinary methods.

A remarkably simple apparatus for performing such oxidations has been suggested by Schöniger (see Figure 28-2).[8] A heavy-walled flask of 300- to 1000-ml capacity fitted with a ground-glass stopper is employed. Attached to the stopper is a platinum-gauze basket that holds from 2 to 200 mg of sample. If the substance to be analyzed is a solid, it is wrapped in a piece of low-ash filter paper cut in the shape shown in Figure 28-2. Liquid samples can be weighed into gelatin capsules, which are then wrapped in a similar fashion. A tail is left on the paper and serves as an ignition point.

A small volume of an absorbing solution is placed in the flask, and the air in the container is then displaced by oxygen. The tail of the paper is ignited, and the stopper is quickly fitted into the flask; the container is then inverted, as shown in Figure 28-2, to prevent the escape of the volatile oxidation products. The reaction ordinarily proceeds rapidly, being catalyzed by the platinum gauze surrounding the sample. During the combustion the flask is shielded to minimize the damage in case of explosion.

After cooling, the flask is shaken thoroughly, disassembled, and the inner surfaces are carefully rinsed. The analysis is then performed on the resulting solution. This procedure has been applied to the determination of halogens,[8] sulfur,[8] phosphorus,[9] fluorine,[10] and various metals[11] in organic compounds.

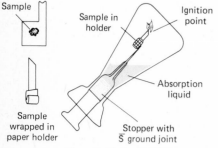

Figure 28-2 Schöniger Combustion Apparatus. (Courtesy Arthur H. Thomas Company, Philadelphia, Pa.)

[8] W. Schöniger, *Mikrochim. Acta*, **1955**, 123; **1956**, 869. See also review article by A. M. G. MacDonald in C. E. Reilley, Ed., *Advances in Analytical Chemistry and Instrumentation*, vol. 4, p. 75. New York: Interscience Publishers, Inc., 1965.
[9] R. Belcher and A. M. G. MacDonald, *Talanta*, **1**, 185 (1958).
[10] B. Z. Senkowski, E. G. Wollish, and E. G. E. Shafter, *Anal. Chem.*, **31**, 1574 (1959).
[11] R. Belcher, A. M. G. MacDonald, and T. S. West, *Talanta*, **1**, 408 (1958).

PEROXIDE FUSION[12]

Sodium peroxide, a strong oxidizing reagent in the fused state, reacts rapidly and often violently with organic matter, converting carbon to carbonate, sulfur in organic compounds to sulfate, phosphorus to phosphate, and iodine and bromine to iodate and bromate. Under suitable conditions, the oxidation is complete, and analysis for the various elements may be performed upon an aqueous solution of the fused mass.

DECOMPOSITION WITH METALLIC SODIUM OR POTASSIUM

Metallic sodium and potassium are powerful reagents for the reductive decomposition of organic compounds. Under proper conditions they will abstract the halogens (including fluorine), sulfur, nitrogen (as cyanide), and other nonmetals from the organic matrix and convert these into water-soluble sodium or potassium salts. The reaction with sodium represents a preliminary step in the common procedure for qualitative identification of the elemental composition of organic compounds.

Alkali-metal decompositions have proved useful for the quantitative determination of the halogens in general and fluorine in particular. Compounds of fluorine are often unusually refractory toward the ordinary oxidative reagents; reduction with an alkali metal provides a simple means for obtaining aqueous fluoride solutions. Several variants of the procedure have been proposed. One involves heating of the sample with the molten sodium at 400°C in a sealed glass vessel for 15 min.[13] After cooling, the excess metal is decomposed by addition of ethanol; the entire mass is then extracted with water. After filtration, the analysis is completed on the aqueous solution. Other procedures call for extended refluxing of the organic sample with a solvent containing metallic sodium. Various alcohols, ethanolamine, dioxane, and combinations of these have been proposed. Unfortunately, not all substances are completely dehalogenated by such treatments.

[12] See C. R. N. Strouts, J. H. Gilfillan, and H. N. Wilson, *Analytical Chemistry*, vol. 1, p. 301. New York: The Oxford University Press, 1955.
[13] P. J. Elving and W. B. Ligett, *Ind. Eng. Chem., Anal. Ed.*, **14**, 449 (1942).

ANALYTICAL SEPARATIONS

The physical and chemical properties upon which analytical methods are based are seldom, if ever, entirely specific. Instead, these properties are shared by numerous species; as a consequence, the elimination of interferences is more often the rule than the exception in a quantitative analysis.

Two general methods are available for coping with substances that interfere in an analytical measurement. The first involves alteration of the system to immobilize the potential interference and thereby prevent its participation in the measurement step; clearly, the alteration must not affect the species being determined. Immobilization is frequently accomplished by introducing a complexing agent that reacts selectively with the interfering substance. For example, in the iodometric determination of copper, iron(III) can be rendered unreactive toward iodide by complexation with fluoride or phosphate ion; neither anion inhibits the oxidation of iodide by copper(II). The introduction of a reagent to eliminate an interference is called *masking*. Numerous masking reagents have been discussed in earlier chapters.

The second method involves the physical separation of an interference from the analyte. Various methods for performing such separations are considered in this chapter.

Nature of the Separation Process

All separation procedures have in common the distribution of the components in a mixture between two phases which subsequently can be separated mechanically. If the ratio between the amount of a particular component in each phase (the *distribution ratio*) differs significantly from that of another, a separation of the two is potentially feasible. To be sure, the complexity of the separation process depends upon the difference between the distribution ratios for the two components. Where the difference is extreme, a single-stage process suffices. For example, a single precipitation with silver ion is adequate for the isolation of chloride from many other anions. Here the ratio of the chloride ion in the solid phase to that in equilibrium in the aqueous phase is immense, while comparable ratios for, say, nitrate or perchlorate ions approach zero.

A somewhat more complex situation is encountered when the distribution ratio for one component is essentially zero, as in the foregoing example, but the ratio for the other is not very large. Here a multistage process is required. For example, uranium(VI) can be extracted into ether from an aqueous nitric acid solution. Although the distribution ratio in the two phases is only about unity for a single extraction, uranium(VI) can nevertheless be isolated by repeated or *exhaustive* extraction of the aqueous solution with fresh portions of ether.

The most complex procedures are required when the distribution ratios of the species to be separated are greater than zero and approach one another in magnitude; here multistage *fractionation* techniques are necessary. These techniques do not differ in principle from their simpler counterparts; both are based upon differences in the distribution ratios of solutes between two phases. However, two factors account for the gain in separation efficiency associated with fractionation. First, the number of times that partitioning occurs between phases is increased enormously; second, distribution occurs between fresh portions of both phases. An exhaustive extraction differs from a fractionation in this latter respect; although many contacts between phases are provided during the former, fresh portions of only one of the phases are involved.

ERRORS RESULTING FROM THE SEPARATION PROCESS

In general, separations are based upon equilibrium processes; as a consequence, complete separation of an interference from the species of interest is never possible. At best, the separation lowers the concentration of the interference to a tolerable level. An additional requirement is that losses of the analyte during the separation must be smaller than the allowable error in the analysis. Thus, the two factors to be considered in any separation are (1) the completeness of the recovery of the analyte and (2) the degree of separation from the unwanted constituent. These factors can be expressed algebraically in terms of *recovery ratios Q*. For example, if x is the amount of the analyte X that is recovered in a separation and x_0 is the amount of X in the original sample, the recovery ratio is given by

$$Q_X = \frac{x}{x_0} \qquad (29\text{-}1)$$

Clearly, this ratio should be as close as possible to one. A similar recovery ratio can be written for the unwanted component Y

$$Q_Y = \frac{y}{y_o} \qquad (29\text{-}2)$$

where y_o and y represent the initial and final amounts. As Q_Y becomes smaller, the better the separation process becomes.

The incomplete recovery of X always results in a negative error. The error associated with the incomplete removal of Y will be positive if this species contributes to the analytically measured quantity and negative if it decreases the magnitude of this measurement. To examine the nature of these errors, let us assume that the analysis is based upon the measurement of some quantity M that is proportional to the amount of x and the amount of y in the solution following the separation process (M may represent mass, volume, absorbance, diffusion current, and so forth). Thus, we may write that

$$M_X = k_X x \qquad (29\text{-}3)$$
$$M_Y = k_Y y \qquad (29\text{-}4)$$

where k_X and k_Y are constants that are proportional to the sensitivity of the measurement for each constituent (k_Y will be negative when Y interferes by lowering the sensitivity of the measurement for X).

If both X and Y are present, the measured value of M represents the sum of their contributions:

$$M = M_X + M_Y \qquad (29\text{-}5)$$

Let us define M_o as the value of the measured quantity that would have been obtained if the sample contained no Y; then

$$M_o = k_X x_o \qquad (29\text{-}6)$$

Thus, when both X and Y are present, the relative error associated with the separation process can be expressed as

$$\text{relative error due to separation} = \frac{(M - M_o)}{M_o} \qquad (29\text{-}7)$$

Substituting Equations 29-3, 29-4, 29-5, and 29-6 into Equation 29-7 gives

$$\text{relative error} = \frac{k_X x + k_Y y - k_X x_o}{k_X x_o}$$

The further substitution of Equations 29-1 and 29-2 yields

$$\text{relative error} = \frac{k_X Q_X x_o + k_Y Q_Y y_o - k_X x_o}{k_X x_o}$$

This expression may be rearranged to

$$\text{relative error} = (Q_X - 1) + \frac{k_Y y_o}{k_X x_o} Q_Y \qquad (29\text{-}8)$$

The first term in Equation 29-8 represents the error that is associated with the losses of X during the separation process. Thus, if 99% of X is recovered during the separation ($x/x_O = Q_X = 0.99$), a relative error of -0.01 or -1% results from this source.

The second term in Equation 29-8 takes account of the error resulting from the incomplete removal of Y by the separatory operation. The magnitude of this error is not only related to the recovery ratio Q_Y, *but is also dependent upon the ratio of y_O to x_O in the sample*. Thus, a highly favorable recovery ratio Q_X is required for the separation of a wanted minor constituent X from a major component Y. This situation is often encountered in trace analysis where the ratio of y_O to x_O at the outset may be as great as 10^6 or 10^7.

The error incurred in a separation is also seen to depend upon the relative sensitivity of the measurement M for the two constituents (k_Y/k_X). If this measurement is not greatly affected by the presence of Y (that is, if k_Y is small), then a relatively incomplete separation may be adequate for an analysis. On the other hand, if the measurement system is equally sensitive to both X and Y, a more complete separation will be required.

Separation by Precipitation

The fundamental basis for all precipitation separations is the solubility difference between the analyte and the undesired components. Solubility-product considerations will generally provide guidance as to whether or not a given separation is theoretically feasible and will define the conditions required to achieve the separation. Unfortunately, other variables are influential in determining the success or failure of a gravimetric separation, and some of these are not susceptible to theoretical treatment within our present state of knowledge. We have seen, for example, that various coprecipitation phenomena may cause extensive contamination of a precipitate by an unwanted component, even though the solubility product of the contaminant has not been exceeded (Chapter 6). Likewise, the rate of an otherwise feasible precipitation process may be so slow that it becomes useless as the basis for a separation. Finally, when the precipitate forms as a colloidal suspension, coagulation may be a difficult and slow process. These latter problems are particularly formidable when the isolation of a small quantity of a solid phase is attempted.

Many precipitating agents are employed for quantitative inorganic separations; we shall limit this discussion to those that have the most general applicability.

SEPARATIONS BASED ON CONTROL OF ACIDITY

Enormous differences exist among the solubilities of the hydroxides, hydrous oxides, and acids of various elements. Moreover, the concentration of hydrogen or hydroxide ions in a solution can be varied by a factor of 10^{15} or more and can be readily controlled by the use of buffers. As a consequence, many separation procedures based on pH control are, in theory, available to the chemist. In practice, these separations can be grouped in three categories: (1) those made in relatively concentrated solutions of strong acids, (2) those made in buffered solutions at intermediate pH values, and (3) those made in concentrated solutions of sodium or potassium hydroxide.

Separations with Solutions of Strong Acids. Several elements are precipitated as slightly soluble acidic oxides from concentrated solutions of mineral acids. These solids are often formed during the solution of the sample and are thus removed at the outset of the analysis. Tungsten(VI), tantalum(V), niobium(V), and silicon(IV) precipitate as oxides in the presence of concentrated perchloric, sulfuric, hydrochloric, or nitric acids. Tin and antimony form acidic oxides only in the presence of hot concentrated perchloric or nitric acid.

Manganese can be separated as the dioxide by heating a perchloric or nitric acid solution of manganese(II) ion with an oxidizing agent such as potassium chlorate.

Precipitation of Basic Oxides from Buffered Solutions. Table 29-1 shows the pH at which precipitation of several hydrous oxides is initiated from aqueous solutions; somewhat higher values are required for complete precipitation. It is apparent from these data that many useful separations should be possible by proper control of pH with buffer mixtures that maintain the hydrogen ion concentration at a suitable predetermined level.

Many of the separations suggested by Table 29-1 are unsatisfactory. Hydrous oxide precipitates are usually gelatinous colloids that are appreciably contaminated by adsorbed impurities, and several reprecipitations may be required to achieve a clean separation. In addition, precipitation of more soluble constituents of the sample may occur, owing to local excesses of hydroxide ion during the addition of base; frequently these do not redissolve readily or completely and thus limit the effectiveness of the separation. Precipitation from homogeneous solution (Chapter 6) may eliminate many of these problems.

An ammonia–ammonium chloride buffer is perhaps the most common medium employed for the separation of iron, chromium, aluminum, and titanium from manganese(II) and the alkaline-earth hydroxides. Copper, zinc, nickel, and cobalt remain in solution as stable ammine complexes. The precipitates formed in this environment are frequently gelatinous and difficult to manipulate. Moreover, as a result of surface adsorption, they tend to coprecipitate substantial amounts of foreign ions.

TABLE 29-1   pH at Which Certain Hydrous Oxides Are Precipitated[a]

pH	Metal Ion
11	Mg(II)
9	Ag(I), Mn(II), La, Hg(II)
8	Ce(III), Co(II), Ni(II), Cd, Pr, Nd, Y
7	Sm, Fe(II), Pb
6	Zn, Be, Cu, Cr(III)
5	Al
4	U(VI), Th
3	Sn(II), Zr, Fe(III)

[a] H. T. S. Britton, *J. Chem. Soc.*, **127**, 2157 (1925). With permission.

Precipitation of the hydrous oxides of iron, aluminum, and chromium from a somewhat acidic medium provides a more satisfactory separation from the common dipositive ions. The *basic acetate* method, in which the pH is maintained by an acetic acid–ammonium acetate buffer, is widely used for this purpose. Other acidic buffers that have also been recommended include benzoic acid–benzoate, formic acid–formate, and succinic acid–succinate mixtures.

Separations by Solutions of Strong Base. In strongly alkaline solutions and in the presence of an oxidizing agent such as sodium peroxide, several amphoteric elements are quite soluble and can be separated from ions that form precipitates under these circumstances. Soluble species include zinc, chromium, vanadium, and uranium; precipitates are formed by iron, cobalt, nickel, and the rare earths.

SULFIDE SEPARATIONS

With the exception of the alkali and alkaline-earth metals, most cations form sparingly soluble sulfides. Their solubilities differ greatly; since it is a relatively easy matter to control the sulfide ion concentration of an aqueous solution by adjustment of pH, separations based on formation of sulfides have found extensive use. Sulfides can be conveniently precipitated from homogeneous solution, the anion being generated by hydrolysis of thioacetamide (see Table 6-2).[1]

A theoretical treatment of the ionic equilibria influencing the solubility of sulfide precipitates was considered in Chapter 5. Often, however, such treatment does not provide realistic conclusions regarding the feasibility of separations because of coprecipitation and the slow rate at which some sulfides form. As a consequence, resort must be made to empirical observations.

Table 29-2 shows some common separations that can be accomplished with hydrogen sulfide through control of the pH.

TABLE 29-2   Precipitation of Sulfides

Elements	Conditions for Precipitation[a]	Conditions for No Precipitation[a]
Hg(II), Cu(II), Ag(I)	1, 2, 3, 4	
As(V), As(III), Sb(V), Sb(III)	1, 2, 3	4
Bi(III), Cd(II), Pb(II), Sn(II)	2, 3, 4	1
Sn(IV)	2, 3	1, 4
Zn(II), Co(II), Ni(II)	3, 4	1, 2
Fe(II), Mn(II)	4	1, 2, 3

[a] Conditions include:

1. 3-F HCl.
2. 0.3-F HCl.
3. Buffered to pH 6 with acetate.
4. Buffered to pH 9 with NH_3, $(NH_4)_2S$.

[1] See E. H. Swift and F. C. Anson in C. E. Reilley, Ed., *Advances in Analytical Chemistry and Instrumentation*, vol. 1, pp. 293–345. New York: Interscience Publishers, Inc., 1960.

OTHER INORGANIC PRECIPITANTS

No other inorganic ions are as generally useful for separations as those just discussed. Phosphate, carbonate, and oxalate ions are often employed as precipitants for cations; their behavior is nonselective, so preliminary separations must generally precede their use.

Chloride and sulfate ions are useful because of their relatively specific behavior. The former can be used to separate silver from most other metals, while the latter is frequently employed to separate a group of metals that includes lead, barium, and strontium.

ORGANIC PRECIPITANTS

Selected organic reagents used for the separation of various inorganic ions were discussed in Chapter 6. Some organic precipitants such as dimethylglyoxime are useful because of their remarkable selectivity in forming precipitates with very few ions. Others, such as 8-hydroxyquinoline, form slightly soluble compounds with a host of cations. The solubilities of the hydroxyquinolates differ greatly; by control of reagent concentration, useful separations can be achieved. As with sulfide ion, the concentration of the precipitating reagent is readily controlled by adjustment of pH.

SEPARATION OF CONSTITUENTS PRESENT IN TRACE AMOUNTS

A problem often encountered in trace analysis is that of isolating the minor constituent, which may be present in microgram quantities, from the major components of the sample. Although the separation is sometimes based on a precipitation process, the techniques required differ from those used when the analyte is present in generous amounts.

Several problems attend the quantitative separation of a trace element by precipitation even when solubility losses are not important. Supersaturation may delay formation of the precipitate, and coagulation of small amounts of a colloidally dispersed substance is often difficult. In addition, it is likely that an appreciable fraction of the solid will be lost during transfer and filtration. To minimize these difficulties, a small quantity of some other ion that also forms a precipitate with the reagent can be added to the solution. The precipitate formed by the added ion is called a *collector* and serves to carry the desired minor species out of solution. For example, in isolating manganese as the sparingly soluble manganese dioxide, a small amount of iron(III) is frequently added. The basic iron(III) oxide carries down even the smallest amounts of the dioxide. A few micrograms of titanium can be removed from a large volume of solution by addition of aluminum ion and ammonia. Here a hydrous aluminum oxide serves as the collector. Copper sulfide is often employed to collect traces of zinc and lead ions. Many other uses of collectors are described by Sandell.[2]

---

[2] E. B. Sandell, *Colorimetric Determination of Traces of Metals*, 3d ed. New York: Interscience Publishers, Inc., 1959.

The mechanism by which a collector functions undoubtedly differs in various cases. Sometimes it simply involves the carrying down of the trace precipitate by physical entrainment. At other times the process must involve coprecipitation in which the minor component is adsorbed or incorporated in the collector precipitate as the result of mixed crystal formation.

Clearly, the collector must not interfere with the method selected for the subsequent analysis of the trace component.

SEPARATION BY ELECTROLYTIC PRECIPITATION

Electrolytic precipitation constitutes a highly useful method for accomplishing separations. In this process, the more easily reduced species, be it the wanted or the unwanted component of the mixture, is isolated as a second phase. The method becomes particularly effective when the potential of the working electrode is controlled at a predetermined level (p. 431).

The mercury cathode (Figure 19-8) has found wide application for the removal of many metal ions prior to the analysis of the residual solution. In general, metals more easily reduced than zinc are conveniently deposited in the mercury, leaving such ions as aluminum, beryllium, the alkaline earths, and the alkali metals in solution. The potential required to reduce the concentration of a metal ion to any desired level is readily calculated from polarographic data.

Extraction Methods

The distribution of a solute between two immiscible phases is an equilibrium process that can be treated by the law of mass action. Equilibrium constants for this process vary enormously among solutes, thus making possible many useful separations based on extraction. The extraction technique has been widely used to separate the components of organic systems. For example, carboxylic acids are readily separated from phenolic compounds by extracting a nonaqueous solution of the sample with dilute aqueous sodium bicarbonate. The carboxylic acids are almost completely transferred to the aqueous phase, while the phenolic constituents remain in the organic phase.

Extraction methods have also found useful application to inorganic materials. For example, the nitrates, chlorides, and thiocyanates of a surprisingly large number of cations are readily extracted into organic solvents, making feasible the separation of these cations from salts that are not extracted. In addition, organic chelating agents render many inorganic cations readily extractable by organic solvents. Owing to the selective nature of both the chelation and the extraction processes, many useful separations are thus possible.

THEORY

The partition of a solute between two immiscible solvents is governed by the *distribution law*. If we assume that the solute species A distributes itself between an aqueous and an organic phase, the resulting equilibrium may be written

$$A_{aq} \rightleftarrows A_{org}$$

where the subscripts "aq" and "org" refer to the aqueous and organic phases, respectively. Ideally, the ratio of activities of A in the two phases will be constant and independent of the total quantity of A; that is, at any given temperature,

$$K = \frac{[A_{org}]}{[A_{aq}]} \qquad (29\text{-}9)$$

where the equilibrium constant K is the *partition coefficient* or *distribution coefficient*. The terms in brackets are strictly the activities of A in the two solvents, but molar concentrations can frequently be substituted without serious error. Often, K is approximately equal to the ratio of the solubility of A in each solvent.

In some systems the solute may exist in different states of aggregation in the two solvents; then the equilibrium becomes

$$x(A_y)_{aq} \rightleftarrows y(A_x)_{org}$$

and the partition coefficient takes the form

$$K = \frac{[(A_x)_{org}]^y}{[(A_y)_{aq}]^x}$$

Expressions such as these are useful in describing the behavior of simple systems where the solute exists as a single species in either solvent phase. Often, however, analytical extractions are complicated by association or dissociation processes that occur in one or both of the solvents. For example, in the extraction of a metal-organic chelate, it may be necessary to account for dissociation of the chelate in the aqueous phase. Thus, if A were the extractable chelate, and M and X represented the metal ion and the chelating agent, respectively, the following equilibrium would have to be considered as well:

$$A_{aq} \rightleftarrows M_{aq} + X_{aq}$$

for which

$$K_{inst} = \frac{1}{K_f} = \frac{[M_{aq}][X_{aq}]}{[A_{aq}]}$$

Here the concentration of A in the organic layer becomes additionally dependent upon the formation constant of the complex in water and the concentration of the chelating agent.

Partition coefficients enable the chemist to determine the experimental conditions required to transfer a solute from one solvent to another. Consider, for example, a simple system that is adequately described by Equation 29-9.[3] Assume, further, that we have V_{aq} ml of an aqueous solution containing a mmole of A and that this solution is to be extracted with V_{org} ml of an immiscible organic solvent. At equilibrium, x_1 mmole of A will remain in the aqueous layer; thus

$$[A_{aq}] = \frac{x_1}{V_{aq}}$$

---

[3] Suitable modification of this treatment can be made to take into account other equilibria; see H. A. Laitinen, *Chemical Analysis*, pp. 258–269. New York: McGraw-Hill Book Company, Inc., 1960.

It follows, then, that

$$[A_{org}] = \frac{(a - x_1)}{V_{org}}$$

Substituting into Equation 29-9 and rearranging gives

$$x_1 = \left(\frac{V_{aq}}{V_{org} K + V_{aq}}\right) a \qquad (29\text{-}10)$$

The number of millimoles, x_2, remaining after a second extraction of the water with an identical volume of solvent will, by the same reasoning, be

$$x_2 = \left(\frac{V_{aq}}{V_{org} K + V_{aq}}\right) x_1$$

When this expression is substituted into Equation 29-10, we obtain

$$x_2 = \left(\frac{V_{aq}}{V_{org} K + V_{aq}}\right)^2 a$$

After n extractions, the number of millimoles remaining is given by the expression

$$x_n = \left(\frac{V_{aq}}{V_{org} K + V_{aq}}\right)^n a \qquad (29\text{-}11)$$

Example. The distribution coefficient for iodine between CCl_4 and water is 85. Calculate the number of millimoles of I_2 remaining in 100 ml of an aqueous solution that was originally $1.00 \times 10^{-3} M$ after extraction with two 50-ml portions of CCl_4.

$$a = 100 \text{ ml} \times 1.00 \times 10^{-3} \frac{\text{mmole}}{\text{ml}} = 0.100 \text{ mmole}$$

$$x_2 = \left(\frac{100 \text{ ml}}{50 \text{ ml} \times 85 + 100 \text{ ml}}\right)^2 0.100 \text{ mmole}$$

$$= 5.28 \times 10^{-5} \text{ mmole}$$

Thus, the two extractions should diminish the number of millimoles of I_2 in the aqueous solution from 0.1 to 5.28×10^{-5}.

The exponential nature of the foregoing relationship shows that a more efficient extraction is achieved with several small volumes of solvent than with a single large one. This effect is illustrated by comparing the results from the following calculation with the previous one.

Example. Calculate the number of millimoles of I_2 remaining if the aqueous solution in the preceding example had been extracted with a single 100-ml portion of CCl_4 rather than two 50-ml portions.

$$x_1 = \frac{100}{100 \times 85 + 100} \times 0.1$$

$$= 1.16 \times 10^{-2} \times 0.1 = 1.16 \times 10^{-3} \text{ mmole}$$

Here the single extraction with the larger volume of CCl_4 leaves more than 20 times as much I_2 in the aqueous layer as two extractions with half-volumes.

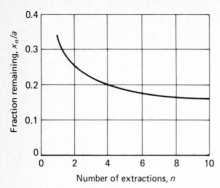

Figure 29-1   Plot of Equation 29-11, Assuming $K = 2$ and $V_{aq} = 100$. The total volume of the organic solvent was also assumed to be 100, so that $V_{org} = 100/n$.

Figure 29-1 shows that the improved efficiency brought about by multiple extraction falls off rapidly as the number of subdivisions is increased; clearly, little is to be gained by dividing the extracting solvent into more than five or six portions.

TYPES OF EXTRACTION PROCEDURES

Simple Extraction. A separation by extraction can be a simple process, provided the partition coefficient for one species is one reasonably favorable (of the order of 10 or greater) while that for the other is unfavorable. Just how different the two coefficients must be will depend on the variables discussed earlier in this section. Typically, however, a useful separation is possible if the second coefficient is in the range of 0.1 to 0.001 (or smaller). Under these circumstances, the extraction can be carried out in a separatory funnel, the original solution being extracted successively with up to five or six portions of fresh solvent.

Exhaustive or Continuous Extraction. The arrangement shown in Figure 29-2 permits the continuous extraction of a solution with a less dense, immiscible solvent. This apparatus is useful for the removal of extractable components from those with partition ratios that approach zero. Even components with relatively unfavorable partition ratios will be separated after an extraction period of several hours.

Countercurrent Fractionation.[4] Automated devices have been developed to permit 200 or more successive extractions to be performed mechanically. Here a *countercurrent* scheme causes distribution to occur at each step between

[4] For an extended discussion of this topic, see L. C. Craig and D. Craig in A. Weissberger, Ed., *Technique of Organic Chemistry*, 2d ed., vol. III, part I, pp. 149–332. New York: Interscience Publishers, Inc., 1956.

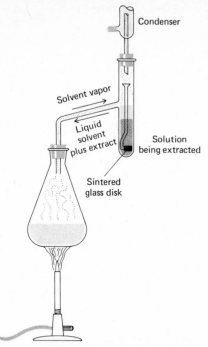

Condenser

Solvent vapor

Liquid solvent plus extract

Solution being extracted

Sintered glass disk

Figure 29-2   Apparatus for the Continuous Extraction of a Sample. A solvent that is less dense than the solution being extracted is required.

fresh portions of both phases. An exhaustive extraction differs from the counter-current technique in that fresh portions of only one phase are introduced in the former.

The countercurrent method permits separation of components with nearly identical partition ratios. For example, Craig[5] has demonstrated that 10 amino acids can be separated by countercurrent extraction even though their partition coefficients differ by less than 0.1.

Applications of Extraction Procedures

For separating inorganic species, an extraction is often more attractive than a classical precipitation method. The process of equilibration and separation of phases in a separatory funnel is inherently less tedious and time-consuming than precipitation, filtration, and washing. In addition, problems of copre-cipitation and postprecipitation are avoided. Finally, and in contrast to the precipitation process, extraction procedures are ideally suited for the isolation of trace quantities of a species.

[5] L. C. Craig, *Anal. Chem.*, **22**, 1346 (1950).

TABLE 29-3 Ethyl Ether Extractions of Various Chlorides from
6-F Hydrochloric Acid[a]

Percent Extracted	Elements and Oxidation State
90–100	Fe(III), 99%; Sb(V), 99%[b]; Ga(III), 97%; Ti(III), 95%[b]; Au(III), 95%
50–90	Mo(VI), 80–90%; As(III), 80%[b,c]; Ge(IV), 40–60%
1–50	Te(IV), 34%; Sn(II), 15–30%; Sn(IV), 17%; Ir(IV), 5%; Sb(III), 2.5%[a]
< 1 > 0	As(V),[a] Cu(II), In(III), Hg(II), Pt(IV), Se(IV), V(V), V(IV), Zn(II)
0	Al(III), Bi(III), Cd(II), Cr(III), Co(II), Be(II), Fe(II), Pb(II), Mn(II), Ni(II), Os(VIII), Pd(II), Rh(III), Ag(I), Th(IV), Ti(IV), W(VI), Zr(IV)

[a] Data from Ernest H. Swift, *Introductory Quantitative Analysis*, p. 431. Englewood Cliffs, N.J.: Prentice-Hall, Inc., 1950. With permission.
[b] Isopropyl ether employed rather than ethyl ether.
[c] 8-F HCl rather than 6-F.

SOME EXAMPLES OF INORGANIC SEPARATIONS

Ether Extractions of Metal Chlorides. The data in Table 29-3 indicate that a substantial number of metal chlorides can be extracted into ether from 6-F hydrochloric acid solution; equally important, a large number of metal ions are either unaffected or extracted only slightly under these conditions. Thus, many useful separations are possible. One of the most important of these is the separation of iron(III) (99% extracted) from a host of other cations. The greater part of iron from steel or iron ore samples can be removed by extraction prior to analysis for such trace elements as chromium, aluminum, titanium, or nickel. The species extracted has been shown to be $HFeCl_4$. It has also been found that the percentage of iron transferred to the organic phase is dependent upon the hydrochloric acid content of the aqueous phase (little is removed from solutions that are below 3-F and above 9-F HCl) and to some extent upon the iron content. Unless special precautions are taken, extraction of the last traces of iron is incomplete.[6]

Extraction of Nitrates. Certain nitrate salts are selectively extracted by ether as well as other organic solvents. For example, uranium is conveniently separated from such elements as lead and thorium by ether extraction of an aqueous solution that is saturated with ammonium nitrate and has a nitric acid concentration of about 1.5 F; the uranium must be in the +6 oxidation state. Bismuth and iron(III) nitrates are also extracted to some extent under these conditions.

Extraction of Chelate Compounds. Many of the organic reagents mentioned in Chapter 6, as well as others, form chelates with various metal ions; these chelates are frequently soluble in such solvents as chloroform, carbon tetrachloride, benzene, and ether. Thus, quantitative transfer of the metallic ions to the organic phase is possible.

[6] See S. E. Q. Ashley and W. M. Murray, *Ind. Eng. Chem., Anal. Ed.*, **10**, 367 (1938).

A reagent that has widespread application for extraction separations is 8-hydroxyquinoline (Chapter 6). Most of its metal chelates are soluble in chloroform as well as other organic solvents. The reaction which occurs when an aqueous solution of a divalent metal ion M^{2+} is extracted with an organic solvent containing 8-hydroxyquinoline (symbolized as HQ) can be formulated as

$$2(HQ)_{org} + (M^{2+})_{aq} \rightleftarrows (MQ_2)_{org} + 2(H^+)_{aq}$$

where the subscript indicates the phase. The equilibrium is clearly pH-dependent; thus, by controlling the pH of the aqueous solvent, separations among metals that have different formation constants with the ligand are possible. The method has proved particularly useful for separation of traces of metals.[7]

Another useful reagent for separating minute quantities of metal ions is dithizone[8] (diphenylthiocarbazone). Its reaction with a divalent metallic ion can be written as

dithizone

Both dithizone and its metal chelates are soluble in chloroform or carbon tetrachloride. As with 8-hydroxyquinoline, the equilibrium between the metal ion and the reagent is pH-dependent; thus, by controlling the pH of the aqueous phase, various separations of metallic ions are possible.

The dithizone complexes of many metal ions are intensely colored. Spectrophotometric measurement of the organic extract often serves to complete the analysis after the separation has been effected.

Information concerning the use of other organic chelating agents for separations by extraction can be found in several reference works.[9]

Separations by Distillation

Distillation permits the separation of components in a mixture whose partition coefficients between solution and vapor phases differ significantly. If one species has a partition coefficient that is large compared with the other components of

[7] See E. B. Sandell, *Colorimetric Determination of Traces of Metals*, 3d ed., pp. 179–188. New York: Interscience Publishers, Inc., 1959.

[8] See E. B. Sandell, *Colorimetric Determination of Traces of Metals*, 3d ed., pp. 151–154. New York: Interscience Publishers, Inc., 1959.

[9] G. H. Morrison and H. Freiser, *Solvent Extraction in Analytical Chemistry*. New York: John Wiley & Sons, Inc., 1957; A. K. De, S. M. Khophar, and R. A. Chalmers, *Solvent Extraction of Metals*. New York: D. Van Nostrand, 1970; E. B. Sandell, *Colorimetric Determination of Traces of Metals*, 3d ed. New York: Interscience Publishers, Inc., 1959.

the mixture, the separation process is simple. For example, ammonium ion is readily separated from other cations in an aqueous solution after conversion to ammonia by the addition of base. The distribution coefficient for ammonia between the gas phase and the solution is large, particularly if the solution temperature is elevated. By passing an inert gas through the mixture, the ammonia in the gas phase can be continuously removed and subsequently collected; the process is quite analogous to the exhaustive extraction procedure discussed earlier.

In common with extraction procedures, separations by distillation can be applied to those systems in which the differences in partition coefficients are small; a fractionation procedure is required, however. Fractional distillation is used extensively for separations involving organic systems. The technique is not as generally applied to inorganic species; nevertheless, some inorganic separations can be performed by simple distillation, as may be seen in Table 29-4.

Halogens can be distilled from aqueous solution in their elemental form or as the hydrohalides. Separation is sometimes accomplished by the selective oxidation of a halide followed by distillation of the element. Thus, iodide can be separated from chloride and bromide as volatile iodine by treatment with nitrous acid. Bromides can be separated from chlorides by oxidation with telluric acid or with potassium hydrogen iodate followed by distillation of bromine.

Arsenic, *antimony*, and *tin* can be isolated from most other elements as well as from one another by distillation. Trivalent arsenic is quantitatively distilled at $110°C$ as the chloride from an aqueous solution containing sulfuric and hydrochloric acids. With the exception of germanium, no other element is volatilized under these conditions. After removal of the arsenic, water is distilled off until the boiling point of the mixture reaches 155 to $165°C$, whereupon quantitative removal of antimony trichloride occurs; concentrated hydrochloric acid must be added to the solution during the distillation. Phos-

TABLE 29-4 **Separation of Some Inorganic Species by Distillation**

Analyte	Sample Treatment	Volatile Species	Method of Collection
CO_3^{2-}	Acidification	CO_2	$Ba(OH)_2(aq) + CO_2(g) \rightarrow BaCO_3(s) + H_2O$ or on Ascarite
SO_3^{2-}	Acidification	SO_2	$SO_2(g) + H_2O_2(aq) \rightarrow H_2SO_4(aq)$
S^{2-}	Acidification	H_2S	$Cd^{2+} + H_2S(g) \rightarrow CdS(s) + 2H^+$
F^-	Addition of SiO_2 and acidification	H_2SiF_6	Basic solution
Si	Addition of HF	SiF_4	Basic solution
H_3BO_3	Addition of H_2SO_4 and methanol	$B(OCH_3)_3$	Basic solution
$Cr_2O_7^{2-}$	Addition of concd HCl	CrO_2Cl_2	Basic solution
NH_4^+	Addition of HCl	NH_3	Acidic solution

phoric acid is added to complex with tin(IV) and prevent the partial distillation of its chloride. After arsenic and antimony have been removed, tin can be separated as the tetrabromide at 140°C by the addition of hydrobromic acid to the residual solution.

Chromatographic Separations[10]

Chromatography encompasses a diverse group of separation methods (see Table 29-5) that are of great importance to the analytical chemist, for they often enable him to separate, isolate, and identify components of mixtures that might otherwise be resolved with great difficulty if at all. The term "chromatography" is difficult to define rigorously, owing to the variety of systems and techniques to which it has been applied. In its broadest sense, however, chromatography refers to processes that are based on differences in rates at which the individual components of a mixture migrate through a stationary medium under the influence of a moving phase.

TYPES OF STATIONARY AND MOBILE PHASES

In chromatography the components of a mixture are carried through a porous *stationary phase* by means of a moving liquid or a gas called the *mobile phase*. In some applications, the stationary phase is a finely divided solid held in a narrow glass or metal tube. The mobile phase then percolates through the solid under the influence of gravity or as a result of pumping. In other methods, the stationary phase may be a porous paper or a finely ground solid that has been

TABLE 29-5   Classification of Chromatographic Separations

Name	Type Mobile Phase	Type Stationary Phase	Method of Fixing the Stationary Phase
Gas-liquid	Gas	Liquid	Adsorbed on a porous solid held in a tube or adsorbed on the inner surface of a capillary tube
Gas-solid	Gas	Solid	Held in a tubular column
Partition	Liquid	Liquid	Adsorbed on a porous solid held in a tubular column
Paper	Liquid	Liquid	Held in the pores of a thick paper
Thin layer	Liquid	Liquid or solid	Finely divided solid held on a glass plate; liquid may be adsorbed on particles
Gel	Liquid	Liquid	Held in the interstices of a polymeric solid
Ion exchange	Liquid	Solid	Finely divided ion-exchange resin held in a tubular column

[10] For discussions of chromatography, see E. Heftmann, *Chromatography*, 2d ed. New York: Reinhold Publishing Corp., 1967; E. Lederer and M. Lederer, *Chromatography*, 2d ed. New York: American Elsevier Publishing Company, Inc., 1957; H. G. Cassidy, *Fundamentals of Chromatography*. New York: Interscience Publishers, Inc., 1957.

spread on a glass plate; here the mobile phase moves through the solid either by capillary action or under the influence of gravity. The stationary phase can also be an immobilized liquid which is immiscible with the mobile phase. Several procedures are employed to fix the stationary liquid in place. For example, a finely divided solid, coated with a thin layer of liquid, may be held in a glass or metal tube through which the mobile phase percolates. Ordinarily, the solid plays no direct part in the separation, functioning only to hold the stationary liquid phase in place. Alternatively, the inner walls of a capillary tube can be coated with a thin layer of liquid; a gaseous mobile phase is then caused to flow through the tube. A liquid phase can also be held in place on the fibers of paper or on the surface of finely ground particles held on a glass plate.

In Table 29-5 the common chromatographic procedures are classified on the basis of the nature of the fixed and mobile phases.

THE PARTITION RATIO

All chromatographic separations are based upon differences in the extent to which solutes are partitioned between the mobile and the stationary phase. The equilibria involved can be described quantitatively by means of a temperature-dependent constant, the *partition ratio* or *partition coefficient K*

$$K = \frac{C_s}{C_m} \tag{29-12}$$

where C_s is the total analytical concentration of a solute in the stationary phase and C_m is its concentration in the mobile phase.

The assumption that K in Equation 29-12 is constant is generally satisfactory, provided the solute concentration is low. Marked variations are often observed if concentrations are high, and conclusions that are based on the constancy of K must be modified accordingly. Fortunately, the concentrations in a chromatographic column are ordinarily low. Chromatography carried out under conditions such that K is constant is called *linear chromatography*. We shall treat only this type.

A GENERAL DESCRIPTION OF THE CHROMATOGRAPHIC PROCESS

Chromatographic methods are grouped into three categories: (1) *elution analysis*, (2) *frontal analysis*, and (3) *displacement analysis*. We shall consider only elution analysis, which is the most widely encountered of the three.

In elution chromatography, a single portion of the sample, dissolved in the mobile phase, is introduced at the head of the column (see Figure 29-3), whereupon the components of the sample distribute themselves between the two phases. The introduction of additional mobile phase forces the solvent containing a part of the sample down the column where further partition between the mobile phase and fresh portions of the stationary phase can occur. Simultaneously, partitioning between the fresh solvent and the stationary

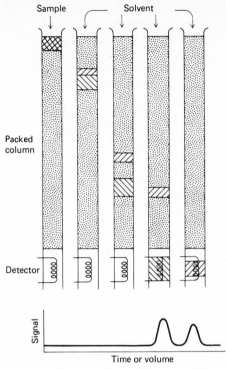

Figure 29-3 Schematic Diagram of an Elution Chromatographic Separation of a Two-Component Mixture.

phase occurs at the site of the original sample. With continued addition of solvent, solute molecules are carried down the column in a continuous series of transitions between the mobile and the stationary phase. Because solute movement can only occur in the mobile phase, however, the average *rate* at which a solute migrates depends upon the fraction of time it spends in that phase. This fraction is small for solutes with partition ratios that favor retention in the stationary phase and large where retention in the mobile phase is more likely. Ideally, the resulting differences in rates cause the components in a mixture to separate into bands located along the length of the column (see Figures 29-3 and 29-4). Isolation can then be accomplished by passing a sufficient quantity of mobile phase through the column to cause these various bands to pass out the end where they can be collected. Alternatively, the column packing can be removed and divided into portions containing the various components of the mixture.

The process whereby the solute is washed through the column by addition of fresh solvent is called *elution*. If a detector that responds to solute concentration is placed at the end of the column, and its signal is plotted as a function of time (or of volume of the added mobile phase), a series of symmetric peaks is obtained, as shown in the lower part of Figure 29-3. Such a plot, called a *chromatogram*, is useful for both qualitative and quantitative analysis. The positions of peaks may serve to identify the components of the sample; the areas under the peaks can be related to concentration.

Theories of Elution Chromatography

Figure 29-4 shows concentration profiles for solutes A and B on a chromatographic column at an early and a late state of elution. The partition ratio of A is the larger of the two; thus, B lags behind during the migration process. It is apparent that movement down the column increases the distance between the two peaks. At the same time, however, broadening of both bands takes place, which lowers the efficiency of the column as a separating device. Zone broadening is unavoidable; fortunately, however, it occurs more slowly than zone separation. Thus, a clean separation of species is possible provided the column is sufficiently long.

Two theories have been developed to account for the rate at which a solute zone migrates and its concentration profile or shape. One is the *plate theory*; the other is the *rate* or *kinetic theory*. The latter is the more useful theory and therefore the one that will be treated in detail here.

The rate theory can be applied, with slight modification, to all types of chromatography. In this discussion, we have chosen to focus on the most common types in which the stationary phase is a liquid that is immobilized on a solid surface.

PLATE THEORY

The plate theory of chromatography, which was originally developed by Martin and Synge,[11] envisages a chromatographic column as being composed of a series of discrete but contiguous, narrow, horizontal layers called *theoretical plates*. At each plate, equilibration of the solute between the mobile and the stationary phase is assumed to take place. Movement of the solute and solvent is then viewed as a series of stepwise transfers from one plate to the next.

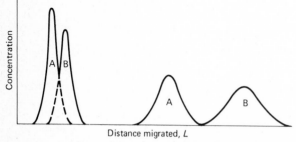

Figure 29-4 Concentration Profiles for Solutes A and B at Different Points in Their Migration down a Column.

[11] A. J. P. Martin and R. L. M. Synge, *Biochem. J.*, **35**, 1358 (1941).

The efficiency of a chromatographic column as a separation device increases as the number of equilibrations increases—that is, as the number of theoretical plates increases. Thus, the *number of theoretical plates N* is used as a measure of column efficiency. A second term, the *height equivalent of a theoretical plate H*, also serves this purpose. The relationship between these two parameters is

$$N = \frac{L}{H} \tag{29-13}$$

where L is the length of the column packing. Note that H decreases as the efficiency of a column becomes greater. That is, as H becomes smaller, the number of equilibrations that occur in a given length of column becomes larger.

Although the plate theory yields equations that satisfactorily describe the rate of solute migration and gives approximate equations that account for the shape of elution bands, it fails to provide a rationale for the effects of such variables as flow rate and packing characteristics on the width of a solute band and therefore upon H and N; nor does it permit theoretical derivation of these parameters from the properties of the two phases. These weaknesses have led to abandonment of the plate theory in favor of the more useful kinetic or rate theory, which approaches the problem from an entirely different viewpoint. It is important to note, however, *that H and N are retained as efficiency parameters in the rate theory and that Equation 29-13 continues to apply. No physical significance can be attached to these quantities, however*; they should be viewed simply as criteria for column efficiency.

THE RATE THEORY OF CHROMATOGRAPHY[12]

Rate theory successfully describes the effects of variables which affect the width of an elution band as well as its time of appearance at the end of a column.

Migration Rate of Solutes. If we define the *retention time t* as the time required for a solute peak to appear at the end of a column of length L, the rate of migration is given by L/t. Similarly, let t_m be the time required for passage of solute that is not retained at all by the stationary phase (t_m, then, is also the time required for the average mobile-phase molecule to pass through the column); the rate of movement for this type of species is thus L/t_m. A useful parameter in chromatography is the *retardation factor* or the *retention ratio R*, which is the rate of movement of a solute molecule *relative* to the rate of movement of the solvent. That is,

$$R = \frac{L/t}{L/t_m} = \frac{t_m}{t} \tag{29-14}$$

Another useful term in chromatography is the *retention volume V_R*, which is related to the retention time by

$$V_R = tF$$

[12] For a detailed presentation of the rate theory, see J. C. Giddings, *J. Chem. Educ.*, **44**, 704 (1967); J. C. Giddings, *Dynamics of Chromatography*, part I. New York: Marcel Dekker, Inc., 1965.

where F is the flow rate of the mobile phase. For a component that is not retained by the stationary phase, the retention volume V_m is given by

$$V_m = t_m F$$

Substitution of these relationships into Equation 29-14 yields

$$R = \frac{V_m}{V_R}$$

It is important to note that R can be defined in other terms as well. First, it corresponds to the fraction of time that an average solute molecule resides in the mobile phase because it can only migrate when it is in that phase; the quantity $(1 - R)$ is then the fraction of time in which the solute is fixed in the stationary phase. The retardation factor also measures the *probability* for a particular solute molecule to be in the mobile phase at any instant. Finally, to the extent that chemical equilibrium exists between the two phases (as it usually does at the center of a band), R represents the fraction of the solute which is present in the mobile phases and is thus dependent upon the partition coefficient K.

If it is now assumed that dynamic equilibrium exists in a small cross section of a solute zone, the ratio of fractional time spent in each of the two phases will equal the ratio of the quantity of solutes in the two phases. That is,

$$\frac{R}{(1 - R)} = \frac{C_m V_m}{C_s V_s} \tag{29-15}$$

where C_m and C_s are molar concentrations of solutes in the mobile and the liquid stationary phase, respectively, and V_m and V_s are the corresponding volumes in the cross section chosen. Substituting Equation 29-12 (p. 640) into Equation 29-15 and rearranging, we obtain

$$R = \frac{V_m}{V_m + K V_s} \tag{29-16}$$

Clearly, the relative rate of migration of a solute decreases as its partition ratio increases.

Equation 29-16 relates the fraction of solute in the mobile phase to K and the volume of the two phases. By suitable rearrangement of Equation 29-15, the fraction in the stationary phase $(1 - R)$ can also be calculated. That is,

$$(1 - R) = 1 - \frac{V_m}{V_m + K V_s} = \frac{K V_s}{V_m + K V_s} \tag{29-17}$$

Zone Shapes. Examination of a typical chromatogram (Figure 29-3) or the concentration profile for a solute on a chromatographic column (Figure 29-4) reveals a similarity to the normal error or Gaussian curves shown in Chapter 4. Recall that the latter plots could be rationalized by assuming that the distribution of replicate results around the mean of an experimental measurement was the consequence of the additive combination of a very large number of small and random uncertainties. In a similar way, the typical Gaussian shape of a chromatogram can be attributed to the additive combination of the random motions of the millions of solute particles in the chromatographic band or zone.

Let us first consider the behavior of an individual solute particle, which, during migration, undergoes many thousands of transfers between the stationary and mobile phases. The time it spends in either phase after a transfer is highly irregular and depends upon its accidentally gaining sufficient thermal energy from its environment to accomplish a reverse transfer. Thus, in some instances, the residence time in a given phase may be transitory; in others, the period may be relatively long. Recall that the particle can move *only during residence in the mobile phase*; as a result, its migration down the column is also highly irregular. Because of variability in the residence time, the average rate at which individual particles move relative to the mobile phase varies considerably. Certain individuals travel rapidly by virtue of their accidental inclusion in the mobile phase for a majority of the time. Others, in contrast, may lag because they happen to have been incorporated in the stationary phase for a greater-than-average time. The consequence of these random individual processes is a symmetric spread of velocities around the mean value, which represents the behavior of the average and most common particle.

The breadth of a zone increases as it moves down the column because more time is allowed for migration to occur. Thus, the zone breadth is directly related to residence time in the column and inversely related to the velocity at which the mobile phase flows.

Recall that the breadth of a Gaussian curve is conveniently related to a single parameter called the standard deviation σ (p. 57), and that approximately 96% of the area under such a curve lies within plus or minus two standard deviations ($\pm 2\sigma$) of its maximum. Thus, σ derived from a chromatogram serves as a convenient quantitative measure of zone broadening. Figure 29-5 illustrates

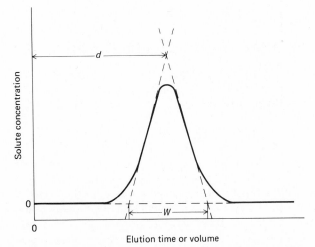

Figure 29-5 Theoretical Plate Determination from a Chromatogram. $N = 16(d/W)^2$.

a simple means for approximating σ from an experimentally derived chromatogram. Tangents to the two sides of the Gaussian curve are extended to form a triangle with the abscissa. The intercepts then occur at approximately $\pm 2\sigma$ from the maximum; that is, $W \cong 4\sigma$. Alternatively, the width of the curve at half-peak $W_{1/2}$ can be employed as a measure of broadening; here, $W_{1/2} = 2.35\sigma$.

Column Efficiency. In the rate theory, the column efficiency is expressed in terms of the height equivalent of a theoretical plate H, which is *defined* as

$$H = \frac{\sigma^2}{L} \qquad (29\text{-}18)$$

where L is the distance a zone maximum has migrated down a column and σ is the standard deviation in units of column height. Thus, H represents the broadening expressed in terms of σ^2 (the variance) per unit length of column. Note that H becomes smaller as the column becomes more efficient (that is, with decreasing zone broadening).

The efficiency of a column can also be expressed in terms of its number of theoretical plates N by substituting Equation 29-13 into 29-18; that is,

$$N = \frac{\sigma^2}{H^2} \qquad (29\text{-}19)$$

Clearly, a large number of plates leads to high efficiency.

Ordinarily in elution chromatography, the time required for a zone peak to pass through a column of fixed length L is recorded. The standard deviation is then more conveniently expressed in units of time τ rather than distance. It is possible to relate τ and σ by considering that the time required for a solute zone to move a distance of σ (or $W/4$ in Figure 29-5) is given by

$$\tau = \frac{\sigma}{L/t}$$

where L/t is the zone velocity (p. 643); that is, τ is the time required for 1σ of the zone to emerge from the column.

Substituting into Equation 29-18, we obtain

$$H = \frac{(\tau L/t)^2}{L}$$

which rearranges to

$$H = \frac{L\tau^2}{t^2} \qquad (29\text{-}20)$$

Here t is the variable normally employed as the abscissa of a chromatogram. Substitution of Equation 29-13 into Equation 29-20 relates N to time parameters. Thus,

$$N = \frac{t^2}{\tau^2} \qquad (29\text{-}21)$$

Experimental Evaluation of N and H. A method for evaluation of N is shown in Figure 29-5. Ordinarily, d and W will be measured in units of time; thus,

$$W = 4\tau$$

$$t = d$$

Substitution of these quantities into Equation 29-21 yields

$$N = \frac{d^2}{(W/4)^2} = 16 \left(\frac{d}{W}\right)^2 \tag{29-22}$$

If $W_{1/2}$ is used in place of W,

$$N = 5.5 \left(\frac{d}{W_{1/2}}\right)^2 \tag{29-23}$$

Equation 29-13 can then be employed to calculate H; here the column height L would have to be measured.

Zone Broadening. Chromatographic peaks are generally broader than would be expected from the random nature of the migration process alone because of three processes, *eddy diffusion, longitudinal diffusion,* and *mass transfer.* The magnitudes of these effects are determined by controllable variables such as flow rate, particle size of packing, and thickness of the stationary layer on the packing.

Zone broadening from eddy diffusion arises from the multitude of pathways by which a molecule can find its way through a packed column. As shown in Figure 29-6, the lengths of these pathways differ; thus, the residence times in the column for molecules of the same species are also variable. As a result, solute molecules reach the end of the column over a time interval that tends to broaden the elution band. Eddy diffusion is dependent upon the homogeneity of the packing and the particle size; it is *independent* of flow rate. With carefully packed columns, broadening due to eddy diffusion is believed to be minimal.

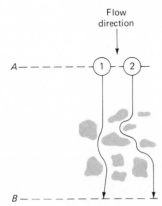

Figure 29-6 Typical Pathways of Two Solute Molecules during Elution. Note that distance traveled by molecule 2 is greater than that traveled by molecule 1. Thus, molecule 2 would arrive at *B* later than molecule 1.

Longitudinal diffusion results from the tendency of molecules to migrate from the concentrated center part of a band toward regions on either side where solute concentrations are lower. This type of diffusion, which can occur in both the mobile and the stationary phase, causes further band broadening. Longitudinal diffusion is most important when the mobile phase is a gas, because diffusion rates in the gas phase are several orders of magnitude greater than those in liquids. The amount of diffusion increases with time; thus, the extent of broadening increases as the flow rate decreases.

Chromatographic bands are also broadened because the flow of the mobile phase is ordinarily so rapid that true equilibrium between phases cannot be realized. For example, at the front of the zone, where the mobile phase encounters fresh stationary phase, equilibrium is not instantly achieved, and solute is therefore carried farther down the column than would be expected under true equilibrium conditions. Similarly, at the end of the zone, solutes in the stationary phase encounter fresh mobile phase. The rate of transfer of solute molecules is not instantaneous, however; thus, the tail of the zone is more drawn out than it would be if time existed for equilibration. The net effect here is a broadening at both ends of the solute band.

The effects of nonequilibrium mass transfer become smaller as the flow rate is decreased because more time is allowed for equilibrium to be approached. Furthermore, a closer approach to true equilibrium is to be expected if the channels through which the mobile phase flows are narrow so that solute molecules do not have far to diffuse in order to reach the stationary phase. For the same reason, the layers of immobilized liquid on a stationary phase should be as thin as possible.

Effect of Column Variables on Zone Broadening. From theoretical considerations, a number of equations have been developed that relate column efficiency in terms of plate height to controllable variables such as flow rate, particle size, and other packing characteristics. The simplest of these, which provides only an approximate relationship between the flow rate v and plate height, is known as the *van Deemter* equation

$$H = A + \frac{B}{v} + Cv \qquad (29\text{-}24)$$

In this equation, A, B, and C are constants which have been derived from consideration of the kinetics of eddy diffusion, longitudinal diffusion, and mass transfer, respectively.

The quantity A can be related to the particle size, geometry, and tightness with which the stationary phase is packed. As a first approximation, A is independent of flow rate; more sophisticated treatments, however, reveal that A is in fact dependent in a complex way upon this variable. Band broadening due to eddy diffusion is minimized by careful packing of a column with small, spherical particles. Particular care is needed to avoid open channels.

The second term (the longitudinal diffusion term) in Equation 29-24 is seen to be inversely proportional to flow rate. The constant B is related to the diffusion coefficient of molecules in the mobile phase.

For a system in which the stationary phase is an immobilized liquid, the third term describes the effect of mass transfer kinetics on H. The constant C is dependent upon the thickness of the immobilized film of liquid, the diffusion coefficient of the solute in this liquid, and the volume of liquid compared with the volume of the mobile phase. The most important variable appears to be the thickness of the immobilized liquid layer. A marked improvement in column efficiency is obtained with very thin layers of liquid phase. Equilibrium is also more closely approached at high temperatures and with low solvent viscosities. Note that the mass transfer term is directly proportional to flow rate.

Figure 29-7 shows the contribution of each term in the van Deemter equation as a function of mobile-phase velocity (broken line) as well as their net effect (solid line) on H. Clearly, the optimum efficiency is realized at a flow rate corresponding to the minimum in the solid curve. From experimental curves of this type, A, B, and C are readily evaluated for any column. Such data provide hints as to how the performance of a given type of packing can be improved.

COLUMN RESOLUTION

The ability of a column to resolve two solutes is of prime interest in chromatography. A quantitative term for expressing the resolving power of a column for solutes X and Y is the *resolution R_s*, which is defined as

$$R_s = \frac{\Delta Z}{W} \tag{29-25}$$

As shown in Figure 29-8, ΔZ is the separation (usually in units of time) when the second solute is exiting from the column. The zone front, peak, or tail, can be used as the reference point; in Figure 29-8 the peak has been chosen, as is most common.

The width of two adjacent peaks will ordinarily be approximately the same. That is, referring again to the figure,

$$W_X \cong W_Y = W$$

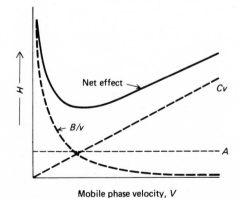

Mobile phase velocity, V

Figure 29-7 Effect of Variables in Equation 29-24 on Plate Height.

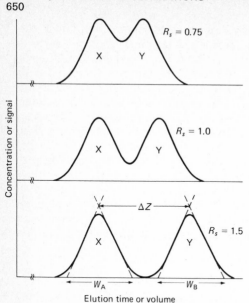

Figure 29-8 Separations at Three Resolutions and Method for Determining Resolution. Here $R_s = 2\Delta Z/(W_X + W_Y)$.

If desired, however, the average of W_X and W_Y can be employed for W.

Figure 29-8 indicates that a resolution (R_s) of 1.5 gives an essentially complete separation of X and Y, whereas a resolution of 0.75 does not. At a resolution of unity, zone X contains about 4% of Y, and conversely; at a resolution of 1.5, the overlap is about 0.3%. For the same kind of packing, the resolution can be improved by lengthening the column and thus increasing the number of theoretical plates.

Retardation Factor and Resolution. If Equation 29-14 is applied to the solutes X and Y in Figure 29-8, it is found that

$$t_X = \frac{t_m}{R_X} \qquad \text{and} \qquad t_Y = \frac{t_m}{R_Y}$$

where t_X and t_Y are the times at which the maxima appear at the end of the column of length L. From the definition of ΔZ, we can write

$$\Delta Z = t_Y - t_X$$

or

$$\Delta Z = t_m \left(\frac{1}{R_Y} - \frac{1}{R_X} \right) = t_m\, \Delta \left(\frac{1}{R} \right) \tag{29-26}$$

where

$$\Delta \left(\frac{1}{R} \right) = \frac{1}{R_Y} - \frac{1}{R_X} \tag{29-27}$$

In order to express W in terms of R, Equation 29-21 is first rearranged to

$$\tau = \frac{t_Y}{\sqrt{N}}$$

followed by substitution for t_Y with Equation 29-14; thus,

$$\tau = \frac{t_m}{R_Y\sqrt{N}}$$

Recall, however, that $4\tau \cong W$ (p. 647); thus,

$$W = \frac{4t_m}{R_Y\sqrt{N}}$$

Note that we have expressed R and thus W in terms of the second solute to leave the column. When this relationship and Equation 29-26 are substituted into Equation 29-25, it is found upon rearrangement that

$$R_s = \frac{\sqrt{N}}{4} R_Y \,\Delta\left(\frac{1}{R}\right) \tag{29-28}$$

or

$$N = \left(\frac{4R_s}{R_Y \,\Delta(1/R)}\right)^2 \tag{29-29}$$

This last equation permits calculation of the number of plates and thus the column length required to achieve a resolution of a desired quality.

Partition Ratio and Resolution. It is convenient to rewrite Equation 29-16 in the form

$$\frac{1}{R} = 1 + K\frac{V_s}{V_m}$$

Applying this relationship to solutes X and Y in Equation 29-27 yields

$$\Delta\left(\frac{1}{R}\right) = (K_Y - K_X)\frac{V_s}{V_m} = \Delta K\frac{V_s}{V_m} \tag{29-30}$$

where $\Delta K = (K_Y - K_X)$. Substitution of this relationship into Equation 29-28 gives

$$R_s = \frac{\sqrt{N}}{4} R_Y \,\Delta K\frac{V_s}{V_m} \tag{29-31}$$

If Equation 29-16 is employed to remove R_Y from the foregoing expression, it is found that

$$R_s = \frac{\sqrt{N}}{4}\frac{\Delta K\,V_s}{(V_m + K_Y V_s)} \tag{29-32}$$

Equation 29-32 provides a convenient relationship between R_s and N in terms of the two measurable equilibrium constants and the volumes of the stationary and mobile solvent; the latter two quantities are readily obtained experimentally.

Another useful expression is obtained by multiplication of the numerator and denominator of Equation 29-32 by K_Y. Thus,

$$R_s = \frac{\sqrt{N}}{4} \frac{\Delta K\, K_Y V_s}{(V_m + K_Y V_s) K_Y}$$

But $K_Y V_s / (V_m + K_Y V_s)$ is equal to $(1 - R_Y)$ (Equation 29-17). Therefore,

$$R_s = \frac{\sqrt{N}}{4} \frac{\Delta K (1 - R_Y)}{K_Y}$$

Frequently, R_Y will be small with respect to unity. Thus, when $R_Y \ll 1$, this equation further simplifies to

$$R_s \cong \frac{\sqrt{N}}{4} \frac{\Delta K}{K_Y} \tag{29-33}$$

Equation 29-33 gives a simple means of approximating the required number of plates from the partition ratios for the two solutes.

It is useful to substitute the definition for ΔK into Equation 29-33. Thus,

$$R_s = \frac{(\sqrt{N}/4)(K_Y - K_X)}{K_Y} = \frac{\sqrt{N}}{4}\left(1 - \frac{K_X}{K_Y}\right) \tag{29-34}$$

We then define the *separation factor a* as

$$a = \frac{K_Y}{K_X}$$

Substitution into Equation 29-34 yields

$$R_s = \frac{\sqrt{N}}{4}\left(1 - \frac{1}{a}\right) \tag{29-35}$$

or upon rearrangement

$$N = \left(\frac{4 R_s a}{a - 1}\right)^2 \tag{29-36}$$

Figure 29-9 shows a plot which gives the number of plates required to resolve two solutes as a function of a and of the desired resolution.

Some Common Types of Chromatography

The more common types of chromatography and their applications to analytical problems are described in this section.

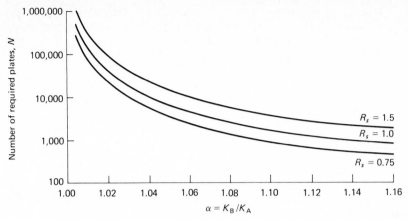

Figure 29-9   Number of Theoretical Plates Required to Resolve Solutes A and B as a Function of α.

GAS-LIQUID CHROMATOGRAPHY[13]

In gas-liquid chromatography, the mobile phase is an inert gas while the stationary phase is a liquid held on a solid surface. The sample is introduced as a vapor at the head of the column; those components that have a finite solubility in the stationary liquid phase distribute themselves between this phase and the gas according to the equilibrium law. Elution is then accomplished by forcing a gas, such as nitrogen or helium, through the column. The rate at which the various components migrate depends upon their tendency to dissolve in the stationary liquid phase. A favorable partition coefficient results in a low rate. On the other hand, components with low solubility in the liquid phase migrate rapidly. Qualitative identification of a component is based upon the time required for its peak to appear at the end of the column; quantitative data are obtained from evaluation of peak areas. A typical gas chromatogram is shown in Figure 29-10.

APPARATUS

The apparatus for gas-liquid chromatography can be relatively simple and inexpensive; more than two dozen models are offered by various instrument manufacturers. The essential components are illustrated in Figure 29-11. A brief description of each component follows.

   Carrier-Gas Supply. Helium, nitrogen, carbon dioxide, and hydrogen from tank sources constitute the most commonly used carrier gases. Hydrogen

[13] For detailed discussions of this technique, see R. A. Jones, *An Introduction to Gas-Liquid Chromatography*. New York: Academic Press, 1970; A. B. Littlewood, *Gas Chromatography*, 2d ed. New York: Academic Press, 1970; S. Dal Nogare and R. S. Juret, Jr., *Gas-Liquid Chromatography*. New York: Interscience Publishers, Inc., 1962; J. H. Purnell, *Gas Chromatography*. New York: John Wiley & Sons, Inc., 1962.

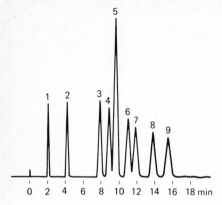

Figure 29-10 A Typical Gas Chromatogram. Column: 5.0% SP-1200/1. 75% Bentone 34 on 100/120 mesh Supelcoport, 6 ft × $\frac{1}{8}$ in. stainless steel; column temperature: 75°C; flow rate: 20 ml/min N_2; sample size: 0.10 µl. (1) Benzene (2) toluene, (3) ethylbenzene, (4) p-xylene, (5) m-xylene, (6) o-xylene, (7) isopropylbenzene, (8) styrene, (9) n-propylbenzene. [Reproduced from D. M. Otterstein, D. A. Bartley, and W. R. Supina, *Anal. Chem.*, **46**, 2225 (1974). With permission of the American Chemical Society.]

has the obvious disadvantage of explosion danger. Suitable flow-regulating valves are required, and some means of reproducing the flow rate is desirable.

Sampling System. Good column efficiency requires that the sample be of suitable size and be introduced as a "plug" of vapor; slow injection and over-sized samples cause band spreading and poor resolution. Liquid samples are introduced by injection through a silicone rubber diaphragm with a hypodermic syringe. To prevent overloading of the column, the volumes must be small (between 1 and 20 µl). The injection port ordinarily is heated to a temperature above the boiling point of the sample so that vaporization is rapid. Special apparatus has been developed for introduction of solids and gases.

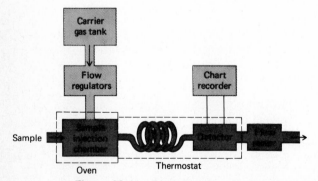

Figure 29-11 Block Diagram of a Gas-Chromatographic Apparatus.

Columns. Two types of columns are employed for gas-liquid chromatography. The capillary type is fabricated from small-diameter tubing, the bore of which is coated with a very thin film of the liquid phase. A capillary column has a very low pressure drop and can thus be of great length; columns of several hundred thousand theoretical plates have been described. These columns have the disadvantage of very low sample capacities.

More convenient is the packed column, which consists of a glass or metal tube of roughly 2.5-mm diameter that ranges in length from 1 to 100 m or more. The tubes are ordinarily folded or coiled so that they can be conveniently fitted into a thermostat. Typically, columns contain 100 to 1000 theoretical plates per foot. The best packed columns have a total of 30,000 to 60,000 theoretical plates.

Solid Support. The ideal solid support would consist of small (20 to 40 μ), uniform, spherical particles with good mechanical strength. In addition, this material would be inert at elevated temperatures and be readily wetted by the liquid phase to give a uniform coating. No substance that meets all of these criteria perfectly is yet available.

The most widely used supports are made from diatomaceous earth and marketed under such trade names as Celite, Dicalite, Chromosorb, C-22 Firebrick, and Sterchamol. Other supports have been fashioned from powdered Teflon, alumina, Carborundum, and micro glass beads.

Liquid Phase. Desirable properties for the liquid phase in a gas-liquid chromatographic column include (1) *low volatility* (ideally the boiling point should be at least 200°C higher than the maximum operating temperature for the column), (2) *thermal stability*, (3) *chemical inertness*, and (4) *solvent characteristics* such that K values for the solutes to be resolved fall within a suitable range.

No single liquid meets all of these requirements, the last in particular. As a consequence, it is common practice to have available several interchangeable columns, each with a different stationary phase. Although some qualitative guidelines exist to aid in the choice of the liquid phase, selection remains largely a matter of trial and error.

The retention time for a solute depends directly upon its partition coefficient which, in turn, is related to the properties of the liquid serving as the stationary phase. Clearly, to be useful for gas-liquid chromatography, the liquid must possess differing affinities for the solutions to be separated; in addition, however, the partition coefficients of these solutes in the liquid must be neither extremely large nor extremely small. This latter requirement is particularly important; if K values are too small, the solutes pass through the column so rapidly that no significant separation occurs. If, on the other hand, the ratios are large, the time required to remove solutes from the columns becomes inordinate.

In order to have a reasonable residence time in the column, a solute must show at least some degree of compatibility (solubility) with the solvent; thus, their polarities should be at least somewhat alike. A stationary liquid such as squalane (a high-molecular-weight saturated hydrocarbon) or dinonylphthalate might be chosen to separate members of a nonpolar homologous series such as hydrocarbons, ethers, or esters. On the other hand, a more polar liquid such as polyethyleneglycol would probably be more effective for the separation of alcohols or amines. For aromatic hydrocarbons, benzyldiphenyl might prove appropriate.

Among solutes of similar polarity, the elution order usually follows the order of boiling points; where these differ sufficiently, clean separations are feasible. Solutes with nearly identical boiling points but different polarities frequently require a liquid phase that will selectively retain one (or more) of the components by dipole interaction or by adduct formation. An important interaction that often enhances selectivity is hydrogen bond formation. For this effect to operate, the solute must have a polar hydrogen atom, and the solvent must have an electronegative group (oxygen, fluorine, or nitrogen), or conversely.

Column Preparation. The support material is first screened to limit the particle size range. It is then made into a slurry with a volatile solvent that contains an amount of the stationary liquid calculated to produce a thin coating on all of the particles. After solvent evaporation, the particles appear dry and are free flowing.

Columns are fabricated from glass, stainless steel, copper, or aluminum. They are filled by slowly pouring the coated support into the straight tube with gentle tapping or shaking to provide a uniform packing. Care must be taken to avoid channeling. After it has been packed, a column is bent or coiled in an appropriate shape to fit the thermostat.

A properly prepared column may be employed for several hundred determinations.

Column Thermostatting. Column temperature is an important variable that must be controlled to a few tenths of a degree for precise work. Control systems that have been employed include circulating air baths, electrically heated metal blocks, and jackets fed with vapor from a constant boiling liquid.

The optimum column temperature depends upon the boiling point of the sample and the degree of separation required. Roughly, a temperature equal to or slightly above the average boiling point of a sample results in a reasonable elution period (10 to 30 min). For samples with a broad boiling range, it is often desirable to increase the column temperature, either continuously or in steps during the separation.

In general, optimum resolution is associated with minimal temperature; the cost of lowered temperature, however, is an increase in band broadening, elution time, and, therefore, the time required to complete an analysis. Figure 29-12 illustrates this principle.

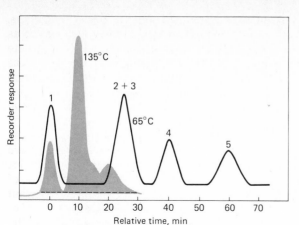

Figure 29-12   Effect of Temperature on the Separation of Hexane Isomers: (1) 2,2-Dimethyl-butane; (2) 2,3-Dimethylbutane; (3) 2-Methylpentane; (4) 3-Methylpentane; (5) *n*-Hexane. (From C. E. Bennett, S. Dal Nogare, and L. W. Safranski in I. M. Kolthoff and P. J. Elving, Eds., *Treatise on Analytical Chemistry*, part I, vol. 3, p. 1690. New York: Interscience Publishers, Inc., 1961. With permission.)

DETECTION SYSTEMS

Detection devices for a gas-liquid chromatograph must respond rapidly and reproducibly to the low concentrations of the solutes eluted from the column. The solute concentration in a carrier gas at any instant is only a few parts in a thousand at most; frequently the detector is called upon to respond to concentrations that are smaller than this figure by one or two orders of magnitude (or perhaps more). In addition, the interval during which a peak passes the detector is usually a second or less; the detector must thus be capable of exhibiting its full response during this brief period.

Other desirable properties of the detector include linear response, good stability over extended periods, and uniform response for a wide variety of compounds. No single detector meets all of these requirements; altogether, more than a dozen different types have been proposed. We shall describe the most widely used of these.

Thermal Conductivity Detectors. A relatively simple and broadly applicable detection system is based upon changes in the thermal conductivity of the gas stream; an instrument employed for this purpose is sometimes called a *katharometer*. The sensing element of this device is an electrically heated source whose temperature at constant electrical power depends upon the thermal conductivity of the surrounding gas. The heated element may consist of a fine platinum or tungsten wire or, alternatively, a semiconducting thermistor. The resistance of the wire or thermistor gives a measure of the thermal conductivity of the gas; in contrast to the wire detector, the thermistor has a negative temperature coefficient.

In chromatographic applications, a double detector is always employed, one element being placed in the gas stream *ahead* of the sample injection chamber and the other immediately beyond the column. In this way the thermal conductivity of the carrier gas is canceled, and the effects of variations in column temperature, pressure, and electrical power are minimized. The resistances of the twin detectors are usually compared by incorporating them into two arms of a simple bridge circuit such as that shown in Figure 29-13.

The thermal conductivities of hydrogen and helium are roughly 6 to 10 times greater than most organic compounds. Thus, the presence of even small amounts of organic materials causes a relatively large decrease in thermal conductivity of the column effluent; the detector undergoes a marked rise in temperature as a result. The conductivities of nitrogen and carbon dioxide more closely resemble those of organic constituents; thus, detection by thermal conductivity is less sensitive when these gases are used as carriers.

Thermal conductivity detectors are simple, rugged, inexpensive, non-selective, accurate, and nondestructive of the sample. They are not as sensitive, however, as some of the other devices to be described.

Flame Ionization Detectors. Most organic compounds, when pyrolyzed at the temperature of a hydrogen-air flame, produce ionic intermediates which provide a mechanism by which current can be carried through the flame. With a device such as that shown in Figure 29-14, these ions can be collected and

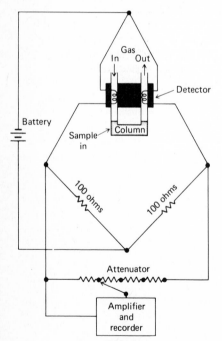

Figure 29-13 Schematic Diagram of a Thermal Conductivity Detector-Recorder System.

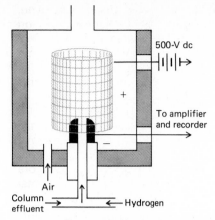

Figure 29-14 Hydrogen Flame Ionization Detector.

the resulting ion current measured. The electrical resistance of a flame is very high (perhaps 10^{12} ohms), and the resulting currents are minuscule. As can be imagined, moderately expensive electronic circuitry is needed to measure these tiny currents.

The hydrogen flame detector currently is one of the most popular and most sensitive detectors. It is more complicated and more expensive than the thermal conductivity detector but has the advantage of higher sensitivity. In addition, it has a wide range of linear response. It is, of course, destructive of the sample.

β-Ray Detectors. In a *β*-ray detector, the effluent from the column is passed over a *β*-emitter such as strontium-90 or tritium; argon, which serves as the carrier gas, is excited to a metastable state by the radiation. The excited argon atoms cause ionization of the sample molecules by collision, and the resulting ion current can be measured in essentially the same way as in the hydrogen flame detector.

β-Ray detectors are highly sensitive and possess the advantage of not altering the sample significantly (in contrast to the flame detector). The requirement of argon as the carrier is a disadvantage.

APPLICATIONS OF GAS-LIQUID CHROMATOGRAPHY

In evaluating the importance of gas-liquid chromatography, it is important to distinguish between the two roles the method plays in its application to chemical problems. The first is as a tool for performing separations; in this capacity it is unsurpassed when applied to complex organic, metal-organic, and biochemical systems. The second and distinctly different function is that of providing the means for completion of an analysis. Here retention times or volumes are employed for qualitative identification, while peak heights or areas provide quantitative information. In its analytical role, gas-liquid chromatography is considerably less versatile than some of the other methods considered in earlier chapters. An important current trend in gas-liquid chromatography is in the direction of combining the remarkable fractionating qualities of the method with the superior analytical properties of other instruments.

Qualitative Analysis. Gas chromatograms are widely used as criteria of purity for organic compounds. Contaminants, if present, are revealed by the appearance of additional peaks; the areas under these peaks provide rough estimates of the extent of contamination. The technique is also useful for evaluating the effectiveness of purification procedures.

In theory, retention-time data should be useful for the identification of components in mixtures. In fact, however, the applicability of the data is limited by the number of variables that must be controlled in order to obtain reproducible results. Nevertheless, gas-liquid chromatography provides an excellent means of confirming the presence of a suspected compound in a mixture, provided an authentic sample of the substance is available. No new peaks in the chromatogram of the mixture should appear upon addition of the pure compound, and enhancement of an existing peak should be observed. The evidence is particularly convincing if the effect can be duplicated on different columns and at different temperatures.

We have seen (p. 652) that the separation factor a for compounds X and Y is given by the relationship

$$a = \frac{K_y}{K_x} \cong \frac{t_y}{t_x}$$

If a standard substance is chosen as compound Y, then a can provide an index for identification of compound X that is largely independent of column variables other than temperature; that is, numerical tabulations of separation factors for pure compounds relative to a common standard can be prepared and then used for the characterization of samples on any column. The amount of such data available in the literature is presently limited.

Quantitative Analysis. The detector signal from a gas-liquid chromatographic column has had wide use for quantitative and semiquantitative analyses. Under carefully controlled conditions, an accuracy of about 1% relative is attainable. As with most analytical tools, reliability is directly related to the amount of effort spent in calibration and control of variables as well as to the nature of the sample.

Quantitative analysis can be based either on peak heights or on peak areas. Peak heights, while more convenient to measure, are less satisfactory because of their greater dependence upon experimental variables. Flow rate is particularly critical, since peaks tend to become broader and lower as the residence time of the solute on the column increases. In addition, peak heights are more strongly affected by column temperature, porosity of the packing, and column length. Despite these limitations, peak height is often a better analytical parameter than peak area for solutes with low retention times. The peaks for such solutes are narrow and tall; accurate determination of their areas is thus difficult.

Peak areas generally increase linearly with the reciprocal of flow rate, and a satisfactory correction for change in this variable can be applied. In contrast to peak heights, the *relative* areas of the peaks for two compounds remain constant and independent of flow rate, a valuable characteristic that permits use of internal standards. Temperature fluctuations have only a small effect on absolute peak areas and none whatsoever on relative ones.

Area Measurement. The area A of a chromatographic peak can be evaluated in any of several ways.

1. Cut out the peak; compare its mass with that for a portion of the recorder paper of known area.
2. Measure the area with a planimeter.
3. Multiply the height of the peak by its width at half-height; insofar as the curve has Gaussian characteristics, this product is equal to 0.84 A.
4. Form a triangle by drawing tangents through the inflection points on either side of the peak, and connect these with a third line along the recorder base line; the resulting triangle will have an area that is approximately 0.96 A.
5. Use a mechanical or electronic integrator; such devices are included as accessories to many recorders.

Methods 4 and 5 appear to be most widely used.

Calibration curves can be based upon weight, volume, or mole percent. A separate curve is needed for each solute to be determined, since no detector responds in exactly the same way for each compound. As always, the accuracy of the analysis depends upon how closely the standard samples approximate the unknown in composition, the care with which they are prepared, and the closeness with which the operational variables of the columns are controlled. The effect of column variables can be partially offset by adding an internal standard in known amount to both calibration standards and samples. The internal standard should have a retention time similar to the solute of interest; its peak, however, must be separated from all other components of the sample—a requirement that is often difficult to realize.

All of the detection systems for qualitative and quantitative analysis that we have considered thus far are nonselective in the sense that they respond in more or less the same way to the solutes emerging from a column (ideally, the nonselective detector would respond identically to every compound). The employment of selective or specific detectors makes it possible in principle to combine the tremendous separatory power of gas-liquid chromatography with the superior analytical qualities of the instruments that have been considered in earlier chapters.

Two general approaches are possible. In the first, the solute vapors are collected as separate fractions in a cold trap, a nondestructive and nonselective detector being employed to indicate their appearance. The fractions are then identified by infrared, mass, or other spectroscopic techniques. The main limitation to this approach lies in the very small (usually micromolar) quantities of solute contained in a fraction; nonetheless, the general procedure has proved useful in the qualitative analysis of complex mixtures.

A second general method involves the use of a selective detector to monitor the column effluent continuously. Some examples will illustrate this approach.

In their first publication of gas chromatography, James and Martin employed an automatic titration cell as a detector for volatile fatty acids; the same device was later applied to mixtures of aromatic and aliphatic amines. The effluent from the column was led directly into a cell containing an acid-base indicator solution. The absorbance of the solution was monitored photometrically, the output from the photocell being employed to control continuously the addition of titrant; the volume of the latter was recorded on a chart to produce an integral curve of volume versus time. Both qualitative and quantitative information could be retrieved from the recorded curve.

A coulometric method has been widely employed as the detector in the analysis of the various chlorinated compounds contained in pesticide residues. After the halogenated compounds have been separated on the column, the effluent is mixed with oxygen and burned over a catalyst. The hydrochloric acid formed is absorbed in an automatic coulometric cell which generates silver ions on demand. An integral curve that relates quantity of electricity to time is obtained.

Rapid-scan techniques have also been employed to yield the entire spectrum of a fraction during the brief period of its appearance at the end of a column; both infrared and mass spectral instruments that produce a spectrum in a few seconds have been designed for this purpose.

LIQUID PARTITION CHROMATOGRAPHY[14]

Liquid–liquid partition chromatography was developed by Martin and Synge, who showed that the height equivalent of a theoretical plate of a properly prepared column can be as small as 0.002 cm.[15] Thus, a 10-cm column of this type may contain as many as 5000 plates; high separation efficiencies are to be expected even with relatively short columns.

Stationary Phase. The most widely used solid support for partition chromatography has been silicic acid or silica gel. This material adsorbs water strongly; the stationary phase is thus aqueous. For some separations the inclusion of a buffer or a strong acid (or base) in the water film has proved helpful. Polar solvents, such as aliphatic alcohols, glycols, or nitromethane, have also been employed as the stationary phase on silica gel. Other support media include diatomaceous earth, starch, cellulose, and powdered glass; water and a variety of organic liquids have been used to coat these solids.

Mobile Phase. The mobile phase may be a pure solvent or a mixture of solvents which are at least partially immiscible with the stationary phase. Better

[14] For detailed discussion of liquid partition chromatography, see J. J. Kirkland, Ed., *Modern Practice of Liquid Chromatography*. New York: Wiley-Interscience, 1971; P. R. Brown, *High Pressure Liquid Chromatography*. New York: Academic Press, 1973.

[15] A. J. P. Martin and R. L. M. Synge, *Biochem. J.*, **35**, 1358 (1941).

separations sometimes are realized if the composition of a mixed solvent is changed continuously as elution progresses (*gradient elution*). Other separations are improved by elution with a series of different solvents. The choice of mobile phase is largely empirical.

Modern liquid partition apparatus is often equipped with a pumping device to hasten the flow of liquid through the column.

Analysis of Fractions. No single method has found general use for the detection of bands emerging from a partition chromatographic column. The composition of the eluent can be monitored continuously by comparing its refractive index, absorbance, conductivity, or other property with that of the pure solvent. Alternatively, fractions can be collected and analyzed by physical or chemical methods.

Applications. Partition chromatography has become a powerful tool for the separation of substances that resemble one another chemically. Typical examples include the resolution of the numerous amino acids formed in the hydrolysis of a protein, the separation and analysis of closely related aliphatic alcohols, and the separation of sugar derivatives. Figure 29-15 illustrates an application of the procedure to the separation of carboxylic acids. Here silica gel was employed as the support with 0.5-*M* sulfuric acid as the stationary phase. The column was eluted with a series of chloroform-butanol mixtures that varied from 15 to 50% in the alcohol. The eluted fractions were analyzed by titration with a standard solution of base. Where separation was incomplete, further treatment of the appropriate fractions on a longer column gave satisfactory isolation of the various components.

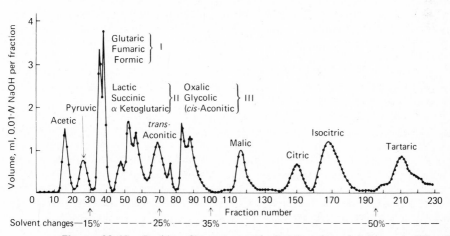

Figure 29-15 Partition Chromatographic Fractionation of Acids on a Silica Gel Column. Elution solvent is chloroform containing the indicated percentages of *n*-butyl alcohol. [From W. A. Bulen, J. E. Varner, and R. C. Burrell, *Anal. Chem.*, **24**, 187 (1952). With permission of the American Chemical Society.]

ADSORPTION CHROMATOGRAPHY[16]

Historically, the first chromatographic technique was based on adsorption. Its discoverer was a Russian botanist, M. Tswett, who in 1903 developed the technique to separate the various colored components of a plant extract. A quarter of a century passed before the significance of his discovery was fully appreciated by investigators engaged in the separation of biological and organic materials.

Separations based on adsorption chromatography rely on the equilibria that govern the distribution of the various solute species between the solvent and the surface of the solid. Large differences exist in the tendencies of compounds to be adsorbed. For example, a positive correlation can be discerned between adsorption properties and the number of hydroxyl groups in a molecule; a similar correlation exists for double bonds. Compounds containing certain functional groups are more strongly held than others. The tendency to be adsorbed decreases in the following order: acid > alcohol > carbonyl > ester > hydrocarbon. The nature of the adsorbent is also influential in determining the order of adsorption. Much of the available knowledge in this field is empirical; the choice of adsorbent and solvent for a given separation frequently must be made on a trial-and-error basis.

Adsorption chromatography has been used primarily for separation of organic compounds; several monographs concerned with applications have been published.

ION-EXCHANGE CHROMATOGRAPHY[17]

Ion exchange is a process in which an interchange of ions of like sign occurs between a solution and an essentially insoluble solid in contact with the solution. Many substances, both natural and synthetic, act as ion exchangers. Among the former are clays and zeolites; the ion-exchange properties of these materials have been recognized and studied for over a century. Synthetic ion-exchange resins were first produced in 1935 and have since found widespread laboratory and industrial application for water softening, water deionization, solution purification, and ion separation.

Synthetic ion-exchange resins are high-molecular-weight, polymeric materials containing a large number of ionic functional groups per molecule. For cation exchange there is a choice between strong acid resins containing sulfonic acid groups (RSO_3H) or weak acid resins containing carboxylic acid ($RCOOH$) groups. The former have wider application. Anion exchange resins contain amine functional groups attached to the polymer molecule. Strong base exchangers are quaternary amines ($RN(CH_3)_3{}^+OH^-$); weak base types contain secondary and tertiary amines.

[16] For a detailed discussion, see L. R. Snyder, *Principles of Adsorption Chromatography*. New York: Marcel Dekker, Inc., 1969.

[17] O. Samuelson, *Ion Exchangers in Analytical Chemistry*. New York: John Wiley & Sons, Inc., 1953; H. L. Rothbart in B. L. Karger, L. R. Snyder, and C. Horvath, *An Introduction to Separation Science*, p. 337. New York: John Wiley & Sons, 1973.

Ion-Exchange Equilibria. A cation-exchange process is illustrated by the equilibrium

$$x\text{RSO}_3^-\text{H}^+ + \text{M}^{x+} \rightleftarrows (\text{RSO}_3^-)_x\,\text{M}^{x+} + x\text{H}^+$$
$$\quad\text{solid}\qquad\quad\text{solution}\qquad\quad\text{solid}\qquad\qquad\text{solution}$$

where M^{x+} represents a cation and R represents *a part* of a resin molecule. The analogous process involving a typical anion-exchange resin can be written as

$$x\text{RN}(\text{CH}_3)_3^+\text{OH}^- + \text{A}^{x-} \rightleftarrows (\text{RN}(\text{CH}_3)_3^+)_x\,\text{A}^{x-} + x\text{OH}^-$$
$$\quad\text{solid}\qquad\qquad\text{solution}\qquad\qquad\text{solid}\qquad\qquad\text{solution}$$

where A^{x-} is an anion.

The identity of the mobile ion attached to an ion-exchange resin is determined by the electrolyte content of the solution with which the resin is in contact. For example, if a sulfonic acid resin is treated with a solution of sodium chloride, the following exchange reaction takes place:

$$\text{RSO}_3^-\text{H}^+ + \text{Na}^+ \rightleftarrows \text{RSO}_3^-\text{Na}^+ + \text{H}^+$$

where R again represents a part of the resin molecule. If the salt concentration is high, essentially all of the resin is converted to the sodium form. Similarly, if an anion-exchange resin is treated with strong sodium chloride, the reaction

$$\text{RN}(\text{CH}_3)_3^+\text{OH}^- + \text{Cl}^- \rightleftarrows \text{RN}(\text{CH}_3)_3^+\text{Cl}^- + \text{OH}^-$$

is strongly favored, and the chloride form of the resin predominates. More than one functional unit of the resin is involved in exchange reactions with polyvalent ions. Thus, when a sulfonic acid resin is treated with calcium ions, the equilibrium can be expressed as

$$(\text{RSO}_3^-\text{H}^+)_2 + \text{Ca}^{2+} \rightleftarrows (\text{RSO}_3^-)_2\text{Ca}^{2+} + 2\text{H}^+$$

Ion-Exchange Equilibrium Constants. The simplest approach to the theoretical treatment of ion-exchange equilibria is to employ the mass-action law. In general, the reaction can be expressed as

$$\text{A} + \text{BR} \rightleftarrows \text{B} + \text{AR}$$

where A and B are ions in the aqueous phase, and AR and BR represent the corresponding ions bound on the resin. The equilibrium constant for the system can be written as

$$K = \frac{a_\text{B}a_\text{AR}}{a_\text{A}a_\text{BR}} \tag{29-37}$$

where a_B and a_A represent activities of the two ions in the aqueous solution, and a_AR and a_BR represent their activities on the solid-resin phase. The latter two terms can be replaced by products of activity coefficients and mole-fraction terms. That is,

$$K = \frac{a_\text{B}X_\text{AR}f_\text{AR}}{a_\text{A}X_\text{BR}f_\text{BR}} \tag{29-38}$$

where X_{AR} represents the mole fraction of the resin that is in the form of AR, and f_{AR} is an activity coefficient; X_{BR} and f_{BR} are analogously defined.

Equation 29-38 can be rearranged to give

$$\frac{a_B X_{AR}}{a_A X_{BR}} = K \frac{f_{BR}}{f_{AR}} = K_P \qquad (29\text{-}39)$$

where K_P is called the *apparent*, or *practical equilibrium quotient*.

It is important to note that the activity ratio f_{BR}/f_{AR} cannot be directly measured by independent means, and the value of K is thus unknown. On the other hand, K_P can be determined experimentally by measuring the concentration of ions on a resin and their activities in the solution in equilibrium with the resin. Such experiments show that K_P is not entirely independent of changes in ionic concentrations on the resin or the activities of ions in the solution, particularly when the ions bear different charges. It has been found, however, that if one of the ions is in large excess, *both in the eluting solution and on the resin*, K_P is relatively constant and independent of changes in concentration of the other ion; that is, when $a_B \gg a_A$, $X_{BR} \gg X_{AR}$, and a_B is fixed, Equation 29-39 becomes

$$\frac{X_{AR}}{a_A} = \frac{X_{BR} K_P}{a_B} \cong K_D \qquad (29\text{-}40)$$

where K_D is a partition ratio analogous to K in Equation 29-12. Thus, to the extent that K_D is constant, the general theory developed for chromatography also can be applied to column separations involving ion-exchange resins.

Note that values of K_D represent the affinity of a given resin for some ion A *relative* to some other ion B. Where K_D is large, there is a strong tendency for the solid phase to retain ion A; where K_D is small, the reverse obtains. By selecting a common reference ion B, distribution ratios for different ions on a given type of resin can be compared. Such experiments reveal that polyvalent ions are much more strongly held than singly charged species. Within a given charge group, differences appear that are related to the size of the hydrated ion and other properties. Thus, for a typical sulfonated cation-exchange resin, values for K_D decrease in the order $Cs^+ > Rb^+ > K^+ > NH_4^+ > Na^+ > H^+ > Li^+$; for divalent cations, the order is $Ba^{2+} > Pb^{2+} > Sr^{2+} > Ca^{2+} > Cd^{2+} > Cu^{2+} > Zn^{2+} > Mg^{2+}$.

Application of Ion-Exchange Chromatography. Techniques for the fractionation of ions with K_D values that are relatively close to one another are analogous to those already discussed. For example, Beukenkamp and Rieman[18] report the separation of potassium and sodium ions by introducing the sample at the top of a column containing a sulfonic acid resin in its acidic form. The column is then eluted with a solution of hydrochloric acid. Sodium ions, being the less strongly held, move down the column more rapidly and can be collected before potassium ions appear in the effluent.

A number of important separations are based upon ion exchange. Among

[18] J. Beukenkamp and W. Rieman, *Anal. Chem.*, **22**, 582 (1950).

these is the separation of the rare earths, primarily for preparative purposes. Here fractionation is enhanced by elution with complexing reagents which form complexes of differing stabilities with the various cations.[19] Ion-exchange resins have also been widely used to resolve mixtures of biological importance. For example, Figure 29-16 shows a partial chromatogram of a synthetic mixture which simulates the composition of a protein hydrolysate. The amino acids, histidine, lysine, and arginine appear with higher volumes of eluent. Note that elutions were performed under varying conditions for different parts of the chromatogram. Peak locations are reported to be reproducible to better than 5%.

Separation of Interfering Ions of Opposite Charge. Ion-exchange resins are useful for removal of interfering ions, particularly where these ions have a charge opposite to the species being determined. For example, iron(III), aluminum(III), and other cations cause difficulty in the determination of sulfate by virtue of their tendency to coprecipitate with barium sulfate. Passage of a solution to be analyzed through a column containing a cation-exchange resin results in retention of all cations and the liberation of a corresponding number of protons. The sulfate ion, on the other hand, passes freely through the column, and the analysis can be performed on the effluent. In a similar manner, phosphate ion, which interferes in the analysis of barium or calcium ions, can be removed by passing the sample through an anion-exchange resin.

Concentration of Traces of an Electrolyte. A useful application of ion exchangers involves the concentration of traces of an ion from a very dilute solution. Cation-exchange resins, for example, have been employed to collect traces of metallic elements from large volumes of natural waters. The ions are then liberated from the resin by treatment with acid; the result is a considerably more concentrated solution for analysis.

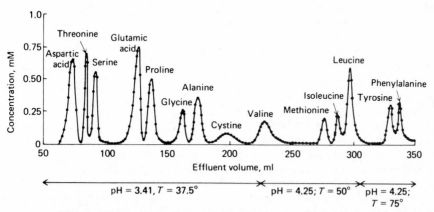

Figure 29-16   Separation of Amino Acids on a Cation Exchanger. (Dowex 50, Na form: 100-cm column.) Total sample ∼ 6 mg. [Taken from S. Moore and W. H. Stein, *J. Biol. Chem.*, **192**, 663 (1951). With permission of the copyright owners.]

[19] See a series of papers in *J. Amer. Chem. Soc.*, **69**, 2769–2881 (1947).

Conversion of Salts to Acids or Bases. An interesting application of ion-exchange resins is the determination of the total salt content of a sample. This analysis can be accomplished by passing the sample through the acid form of a cation-exchange resin; absorption of the cations causes the release of an equivalent quantity of hydrogen ion, which can be collected in the washings from the column and then titrated. Similarly, a standard acid solution can be prepared from a salt; for example, a cation-exchange column in the acid form can be treated with a weighed quantity of sodium chloride. The salt liberates an equivalent quantity of hydrochloric acid, which is then collected in the washings and diluted to a known volume.

Treatment of an anion-exchange resin with salts liberates hydroxide ion by an analogous mechanism.

PAPER CHROMATOGRAPHY[20]

Paper chromatography is a remarkably simple version of the chromatographic method which has become an extremely valuable analytical tool to the organic chemist and the biochemist. In this application a coarse paper serves in lieu of a packed column. A drop of solution containing the sample is introduced at some point on the paper; migration then occurs as a result of flow by a mobile phase called the *developer*. Movement of the developer is caused by capillary forces. In some applications the flow is in a downward direction (*descending development*), in which case gravity also contributes to the motion. In *ascending development* the motion of the mobile phase is upward; in *radial development* it is outward from a central spot.

Techniques. The paper used in paper chromatography is ordinarily 15 to 30 cm long and one to several centimeters in width. A drop of the sample solution is placed a short distance from one end of the paper, and its position is marked with a pencil. The original solvent is allowed to evaporate. The end of the paper nearest the sample is then brought in contact with the developer. Both paper and developer are sealed in a container to prevent evaporation losses. After the solvent has traversed the length of the strip, the paper is removed and dried. The positions of the various components are then determined. Reagents that form colored compounds in the presence of certain functional groups are useful for this purpose. For example, a ninhydrin solution causes the development of blue or purple stains that reveal the location of amines and amino acids on the paper. Detection has also been based upon the absorption of ultraviolet radiation, fluorescence, radioactivity, and so forth.

Development can be carried out in two directions (*two-dimensional paper chromatography*). Here the sample is placed on one corner of a square sheet of paper. Development along one axis is then performed in the usual way. After evaporation of the solvent, the paper is rotated 90 degrees and is again developed but with a different solvent. Solutes that cannot be resolved by a single solvent

---

[20] For a more complete discussion, see H. J. Pazdera and W. H. McMullen in I. M. Kolthoff and P. J. Elving, Eds., *Treatise on Analytical Chemistry*, part I, vol. 3, chapter 36. New York: Interscience Publishers, Inc., 1961.

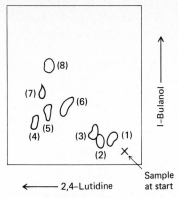

Figure 29-17 Two-Dimensional Chromatogram of an Amino Acid Mixture. (1) Serine, (2) glycine, (3) alanine, (4) tryptophan, (5) methionine, (6) valine, (7) phenylalanine, and (8) leucine.

are frequently separated by this technique. Figure 29-17 shows a typical two-dimensional chromatogram.

Applications. Paper chromatography has been widely employed for the qualitative analysis of inorganic, organic, and biochemical mixtures. Its greatest impact has been felt in the biochemical field where complicated mixtures of closely related compounds are so frequently encountered. One of the most important advantages of the method is its sensitivity; ordinarily, a few micrograms of sample are sufficient for an analysis.

THIN-LAYER CHROMATOGRAPHY[21]

The techniques of thin-layer chromatography closely resemble those of paper chromatography. Here, however, partition occurs on a layer of finely divided adsorbent which is supported upon a glass plate.

Adsorbents. Adsorbent materials used for thin-layer chromatography include silica gel, alumina, diatomaceous earth, and powdered cellulose. A thin-layer plate is prepared by spreading an aqueous slurry of the finely ground adsorbent over the surface of a glass plate or a microscope slide. The plate is then allowed to stand until the layer has set up. For many purposes, it may be activated by heating in an oven for several hours.

The type of equilibrium involved in thin-layer chromatography depends upon the composition of the layer and the way it has been prepared. Thus, if a silica gel film is deposited from aqueous solution and allowed to set up at room temperature, the particles undoubtedly remain coated with a thin film of water. If the sample is then distributed with an organic solvent, resolution will most

[21] See R. Maier and H. K. Mangold in C. N. Reilley, Ed., *Advances in Analytical Chemistry and Instrumentation*, vol. 3, p. 369. New York: Interscience Publishers, Inc., 1964; K. Randerroth, *Thin-Layer Chromatography*. New York: Academic Press, 1970.

likely be the result of a liquid-liquid partition. On the other hand, if the silica film has been dried by heating, partition may involve solid-liquid adsorption equilibria.

Development and Detection. The sample is spotted at one end of the plate and then developed by the ascending technique described for paper chromatography. Development is carried out in a closed container saturated with developer vapor. The plate is then dried and sprayed with a reagent for detection of the components or, more commonly, exposed to iodine vapor. Brown spots indicate solute positions; identification is based upon retention values.

Applications. Thin-layer chromatography is generally faster than paper chromatography and gives more reproducible R values. It is widely used for the identification of components in drugs, biochemical preparations, and natural products.

PROBLEMS

*1. The partition coefficient for X between chloroform and water is 9.6. Calculate the concentration of X remaining in the aqueous phase after 50.0 ml of 0.150-M X is treated by extraction with the following quantities of chloroform: (a) one 40.0-ml portion, (b) two 20.0-ml portions, (c) four 10.0-ml portions, and (d) eight 5.00-ml portions.

2. The partition coefficient for Z between n-hexane and water is 6.25. Calculate the percent of Z remaining in 25.0 ml of water, which was originally 0.0600 M in Z, after extraction with the following volumes of n-hexane: (a) one 25.0-ml portion, (b) two 12.5-ml portions, (c) five 5.00-ml portions, and (d) ten 2.50-ml portions.

*3. What total volume of $CHCl_3$ would be required to decrease the concentration of X (see Problem 1) to 1.00×10^{-4} M if 25.0 ml of 0.0500-M X were extracted with (a) 25.0-ml portions of solvent, (b) 10.0-ml portions of solvent, and (c) 2.0-ml portions of solvent?

4. What volume of n-hexane would be required to decrease the concentration of Z (Problem 2) to 1.00×10^{-5} M if 40.0 ml of 0.0200-M Z were to be extracted with (a) 50.0-ml portions of the solvent, (b) 25.0-ml portions of the solvent, and (c) 10.0-ml portions of the solvent?

*5. What would be the minimum partition coefficient that would permit removal of 99% of a solute from 50.0 ml of water with (a) two 25.0-ml extractions with benzene and (b) five 10.0-ml extractions with benzene?

6. If 30.0 ml of water which is 0.0500 M in Q are to be extracted with four 10.0-ml portions of an immiscible organic solvent, what would be the minimum partition ratio that would allow transfer of all but the following percentages of the solute to the organic layer:
   *(a) 1.00×10^{-1}, (b) 1.00×10^{-2}, and (c) 1.00×10^{-3}?

*7. A 0.200-F aqueous solution of the weak organic acid HA was prepared from the pure compound, and three 50.0-ml aliquots were transferred to 100-ml volumetric flasks. Solution 1 was diluted to 100 ml with 1.0-N $HClO_4$ while solution 2 was diluted to the mark with 1.0-N NaOH. Solution 3 was diluted to the mark with water. Aliquots of 25.0 ml each were extracted with 25.0 ml of n-hexane. The extract from the basic solution contained no detectable trace of A-containing species, which indicated that A^- is not soluble in the organic solvent. The extract from solution 1 contained no ClO_4^- or $HClO_4$ but was found to be 0.0737 M in HA (by extraction with standard NaOH and back-titration with standard HCl). The extract of solution 3 was found to be 0.0320 M in HA. Assume HA does not associate or dissociate in the organic solvent, and calculate

(a) the partition ratio for HA between the two solvents.
(b) the concentration of the *species* HA and A^- in the aqueous solution 3 after extraction.
(c) the dissociation constant for HA in water.

8. To determine the equilibrium constant for the reaction

$$I_2 + 2SCN^- \rightleftarrows I(SCN)_2^- + I^-$$

a 0.0100-F solution of an aqueous I_2 solution was prepared; 25.0 ml of the solution were extracted with 10.0 ml of CCl_4. After extraction, spectrophotometric measurements revealed that the I_2 concentration *of the aqueous layer* was 1.12 $\times$ 10^{-4} M. An aqueous solution was then prepared that was 0.0100 F in I_2 and 0.100 F in KSCN. After extraction of 25.0 ml of the aqueous solute with 10.0 ml of CCl_4, the concentration of I_2 *in the CCl_4* was found from spectrophotometric measurement to be 1.02 $\times$ 10^{-3} M.

(a) What is the partition coefficient for I_2 between CCl_4 and H_2O ?
(b) What is the formation constant for I_2SCN^- ?

*9. The chromatographic peak for compound X was found 15.0 min after sample injection (while the peak for a species Y, which was not retained by the packing, appeared after 1.32 min). The peak for X was Gaussian in form and had a base width of 24.2 sec. The column length was 40.2 cm.

(a) Calculate the number of theoretical plates in the column.
(b) Calculate the H for the column.
(c) Calculate τ and σ for the column.
(d) Calculate the retardation factor for X.
(e) From its method of preparation, the volume of liquid held on the surface of the packing was known to be 9.9 ml. The volume of the mobile phase was 12.3 ml. Calculate the partition coefficient for X.
(f) What would be the resolution of X and compound Z if the separation factor for Z and X was 1.011 ?
(g) How long would the column need to be to obtain a resolution of 1.50 for X and Z?

10. A chromatogram exhibited three peaks for compounds A, B, and C at 0.84 min, 10.60 min, and 11.08 min, respectively. Compound A was not retained by the liquid stationary phase. The peaks appeared to be Gaussian, and the base widths for the peaks for B and C were 0.56 and 0.59 min. The column was 28.3 cm in length.

(a) Calculate an average value for N and H from the peaks for B and C.
(b) Calculate average values for τ and σ.
(c) Calculate retardation factors for B and C.
(d) The volume of liquid held on the surface of the packing was 12.3 ml, and the volume of the mobile phase was 17.6 ml. Calculate partition coefficients for B and C.
(e) Calculate the separation factor for the two compounds B and C.
(f) What would be the column resolution for B and C ?
(g) What length of column with the same packing would be required to give a resolution of 1.5 ?

CHEMICALS, APPARATUS, AND UNIT OPERATIONS FOR ANALYTICAL CHEMISTRY

This chapter is concerned with the practical aspects of the unit operations that are used in most analyses as well as the description of the apparatus and chemicals required for these operations.

Choosing and Handling Chemicals and Reagents

Of constant concern in analytical chemistry is the quality of reagents, inasmuch as the accuracy of an analysis is often affected by this factor.

CLASSIFICATION OF COMMERCIAL CHEMICALS

Technical or Commercial Grade. Chemicals labeled technical or commercial grade are of indeterminate quality and should be used only where high purity is not of paramount importance. Thus, the potassium dichromate and the sulfuric acid used in the preparation of cleaning solution can be of this grade. In general, however, technical- or commercial-grade chemicals are not used in analytical work.

Chemically Pure, or CP Grade. The term *chemically pure* has little meaning. A chemical that is given this label is usually more refined than technical grade, but specifications to indicate the nature and extent of impurities are not provided. In general, then, the quality of a CP reagent is too uncertain for analytical work. Any such reagent that must be used should be tested for contaminants that might affect the analysis; the performance of frequent reagent blanks may also be required.

USP Grade. USP chemicals have been found to conform to tolerances set forth in the *United States Pharmacopoeia.*[1] The specifications are designed to limit contaminants that are dangerous to health; thus, chemicals passing USP tests may still be quite heavily contaminated with impurities that are not physiological hazards.

Reagent Grade. Reagent-grade chemicals conform to the minimum specifications of the Reagent Chemical Committee of the American Chemical Society;[2] these are used, wherever possible, in analytical work. Some suppliers label their products with the maximum limits of impurity allowed by these specifications; others print the actual results of analyses for the various impurities.

Primary-Standard Grade. Primary standards are substances that are obtainable in extraordinarily pure form (see Chapter 7). Primary-standard grade reagents, which are available commercially, have been carefully analyzed, and the assay value is printed on the label. An excellent source for primary-standard chemicals is the National Bureau of Standards. This agency also supplies *reference standards*—complex mixtures that have been exhaustively analyzed.[3]

HANDLING REAGENTS AND SOLUTIONS

The availability of reagents and solutions with established purity is of prime importance for successful analytical work. A freshly opened bottle of a reagent-grade chemical can be used with confidence in most applications; whether the same confidence is justified when this bottle is half-full depends entirely upon the way it has been handled after being opened. The rules that are given here will be successful in preventing contamination of reagents only if they are conscientiously followed.

1. Select the best available grade of chemical for analytical work. If there is a choice, pick the smallest bottle that will supply the quantity that is needed.
2. Replace the top of every container immediately after removal of reagent. Do not rely on someone else to do this.
3. Hold stoppers between the fingers; stoppers should never be set on the desk top.

---

[1] U.S. Pharmacopoeial Convention, *Pharmacopoeia of the United States of America*, 18th rev. Easton, Pa.: Mack Publishing Co., 1970.

[2] Committee on Analytical Reagents, *Reagent Chemicals, American Chemical Society Specifications*, 5th ed. Washington, D.C.: American Chemical Society, 1974.

[3] United States Department of Commerce, *Catalog and Price List of Standard Materials Issued by the National Bureau of Standards*, NBS Misc. Publ. 260 and its semiannual supplements. Washington, D.C.: Government Printing Office, 1968–

4. Unless specifically directed to the contrary, *never return any excess reagent or solution to a bottle*. The minor saving represented by the return of an excess is indeed a false economy compared to the risk of contaminating the entire bottle.
5. Again, unless specifically instructed otherwise, do not insert spoons, spatulas, or knives into a bottle containing a reagent chemical. Instead, shake the capped bottle vigorously to dislodge the contents. Then pour out the desired quantity.
6. Keep the reagent shelf and the laboratory balances clean. Be sure to clean up any spilled chemicals immediately.

Cleaning and Marking of Laboratory Ware

Beakers and crucibles should be marked to permit identification. The etched area on the sides of beakers and flasks can be marked semipermanently with a pencil. Special marking inks are available for porcelain surfaces; the marking is baked permanently into the glaze. A saturated solution of iron(III) chloride can also be used, although it is not as satisfactory as the commercial preparations.

Care must be taken to ensure that glass and porcelain ware is thoroughly clean before being used. Apparatus should be washed with detergent, rinsed first with copious amounts of tap water, and finally with several small portions of distilled water. A properly cleaned object will be coated with a uniform and unbroken film of water. It is seldom necessary to dry glassware before use; in fact, this practice should be discouraged on the grounds of being wasteful of time as well as a potential source for contamination.

If a grease film persists after thorough cleaning with detergent, a cleaning solution consisting of sodium or potassium dichromate in concentrated sulfuric acid may be used. Extensive rinsing is required after the use of this solution in order to remove the last traces of dichromate ions, which adhere strongly to glass or porcelain surfaces. Cleaning solution is most effective when warmed to about 70°C; at this temperature it rapidly attacks plant and animal matter and is thus a potentially dangerous preparation. Any spillages should be diluted promptly with copious volumes of water.

PROCEDURE

Preparation of Cleaning Solution. In a 500-ml, heat-resistant conical flask, mix 10 to 15 g of potassium dichromate with about 15 ml of water. Add concentrated sulfuric acid slowly; swirl the flask thoroughly between increments. The contents of the flask will become a semisolid red mass; add just enough sulfuric acid to dissolve this mass. Allow to cool somewhat before attempting to transfer to a storage bottle. The solution may be reused until it acquires the green color of chromium(III) ion, at which time it must be discarded.

PLATINUM WARE

Platinum requires special mention, not only because of its high cost but also because of the possibilities for its misuse. The chemically valuable properties of this soft, dense metal include its resistance to attack by most mineral acids (including hydrofluoric acid), its inertness with respect to many molten salts, its resistance to oxidation (even at elevated temperatures), and its very high melting point.

With respect to limitations, platinum is readily dissolved on contact with aqua regia as well as with mixtures of chlorides and oxidizing agents. At elevated temperatures, it is also dissolved by fused alkali oxides, peroxides, and to some extent hydroxides. When heated strongly, it readily alloys with such metals as gold, silver, copper, bismuth, lead, and zinc; because of this predilection toward alloy formation, contact between heated platinum and other metals or their readily reduced oxides must be avoided. Slow solution of platinum accompanies contact with fused nitrates, cyanides, alkali, and alkaline-earth chlorides at temperatures above 1000°C; hydrogen sulfates attack the metal slightly at temperatures above 700°C. Surface changes result from contact with ammonia, chlorine and volatile chlorides, sulfur dioxide, and gases possessing a high percentage of carbon. At red heat, platinum is readily attacked by arsenic, antimony, and phosphorous, the metal being embrittled as a consequence. A similar effect occurs as a result of high-temperature contact with selenium, tellurium, and, to a lesser extent, sulfur and carbon. Finally, when heated in air for prolonged periods at temperatures greater than 1500°C, a significant loss in weight due to volatilization of the metal must be expected.

Rules Governing the Use of Platinum Ware.

1. Use platinum equipment only in those applications that will not affect the metal. Where the nature of the system is in doubt, demonstrate the absence of potentially damaging components before committing platinum ware to use.
2. Avoid violent changes in temperature; deformation of a platinum container can result if its contents expand upon cooling.
3. Supports made of clean, unglazed ceramic materials, fused silica, or platinum itself may be safely used in contact with incandescent platinum; tongs of nichrome or stainless steel may be employed only after the platinum has cooled below the point of incandescence.
4. Clean platinum ware with an appropriate chemical agent immediately following use; recommended cleaning agents are hot chromic acid solution for removal of organic materials, boiling hydrochloric acid for removal of carbonates and basic oxides, and fused potassium hydrogen sulfate for the removal of silica, metals, and their oxides. A bright surface should be maintained by burnishing with sea sand.
5. Avoid heating platinum under reducing conditions, particularly in the presence of carbon. Specifically, (a) do not allow the reducing portion of a burner flame to contact a platinum surface, and (b) char filter papers under the mildest possible heating conditions and with free access of air.
6. Precautions relating to the electroanalytical uses of platinum are discussed in Chapter 19.

Evaporation of Liquids

In the course of an analysis, the chemist often finds it necessary to decrease the volume of a solution without loss of a nonvolatile solute. An arrangement such

Figure 30-1   Arrangement for the Evaporation of Liquids.

as that illustrated in Figure 30-1 is generally satisfactory for most evaporations. A ribbed cover glass permits vapors to escape and protects the solution from accidental contamination. Less satisfactory is the use of glass hooks to provide space between the lip of the container and a conventional watch glass.

The evaporation process is occasionally difficult to control, owing to the tendency of some solutions to superheat locally. The bumping that results, if sufficiently violent, can cause a partial loss of the sample. This danger is minimized by careful and gentle heating. The introduction of glass beads, where permissible, is also helpful.

Unwanted constituents of a solution can frequently be eliminated by evaporation. For example, chloride and nitrate ions can be removed by adding sulfuric acid and evaporating until copious white fumes of sulfur trioxide are observed. Nitrate ion and nitrogen oxides can be eliminated by adding urea to an acidic solution, evaporating to dryness, and gently igniting the residue. If large quantities of ammonium chloride must be removed, it is better to add concentrated nitric acid after first evaporating the solution to a small volume. Rapid oxidation of the ammonium ion occurs upon heating; the solution is then evaporated to dryness.

Unwanted organic substances are frequently eliminated by adding sulfuric acid and evaporating the solution until sulfur trioxide fumes are observed. Nitric acid may be added at this point to hasten oxidation of the last traces of organic matter.

The Measurement of Mass

Before the final measurement in every analysis, an accurate measurement of mass or volume (or perhaps both) will have been required. The practical problems associated with the precise determination of mass will be considered first; the measurement of volume is discussed in a later section.

THE DISTINCTION BETWEEN MASS AND WEIGHT

The reader should clearly recognize the difference between the concepts of mass and weight. The more fundamental of these is *mass*—an invariant measure of the quantity of matter in an object. The *weight* of an object, on the other hand, is the force of attraction exerted between the object and its surroundings, principally the earth. Since gravitational attraction is subject to slight geographical variation with altitude and latitude, the weight of an object is likewise a somewhat variable quantity. For example, the weight of a crucible would be less in Denver than in Atlantic City, since the attractive force between it and the earth is less at the higher altitude. Similarly, it would weigh more in Seattle than in Panama, since the earth is somewhat flattened at the poles and the force of attraction increases appreciably with latitude. The mass of this crucible, on the other hand, remains constant regardless of the location in which it is measured.

Weight and mass are simply related to each other through the familiar expression

$$W = Mg$$

where the weight, W, is given by the product of the mass, M, of the object and the acceleration due to gravity, g.

Chemical analyses are always based on mass in order to free the results from dependence on locality. The mass of an object is readily determined by comparison with a known mass. A *balance* is used for this purpose; since g affects both known and unknown to exactly the same extent, a fair comparison of masses is obtained.

STRUCTURAL FEATURES OF THE ANALYTICAL BALANCE

Beams and Pans. Figure 30-2 illustrates the two principal types of analytical balance. Common to both is a *beam* which acts, in essence, as a first-class lever as it pivots on a prism-shaped *knife edge* (1). In an *equal-arm balance* this central knife edge is situated at the exact center of the beam so that the lengths of the two lever arms L_1 and L_2 are identical. In a *single-pan balance* the knife edge that acts as fulcrum is offset from the center of the beam.

Linkage between beam and pan is provided by a *stirrup* (2) that also pivots on a knife edge. A single-pan balance thus has one less bearing surface than an equal-arm balance.

The performance of an analytical balance is in part limited by the mechanical perfection of the knife edges and the surfaces upon which these edges bear. Both parts are fabricated from agate or synthetic sapphire and are precision ground to minimize friction. Because the area of contact between knife edges and bearing surfaces is so small, stresses at these points are severe; the loading on a central knife edge may be on the order of hundreds of kilograms per square centimeter. Thus, even though they are constructed of very hard materials, these parts can be damaged by sudden shocks. To prevent such damage, analytical balances are equipped with auxiliary mechanisms that support the weight of the beam when the balance is not actually in use and when changes are being made in the loading.

Dampers. The beam acts as the sensing element for the balance, since its equilibrium position is determined by the loading on the two sides of the fulcrum. *Dampers* are devices that shorten the time required for the beam to come to

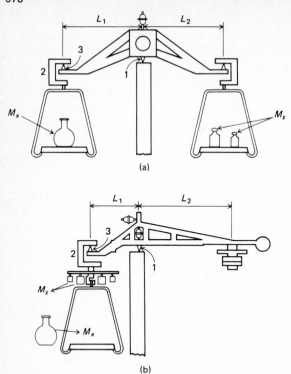

Figure 30-2 Schematic Diagram Illustrating the Operating Principles of (a) Equal-Arm and (b) Single-Pan Balances. (Diagram courtesy of Mettler Instrument Corporation, Princeton, N.J.)

rest. A single-pan balance is ordinarily equipped with an *air damper*, consisting of a piston attached to the beam which moves within a concentric cylinder mounted to the balance case [see (3) and (6), Figure 30-4]. When the beam is set in motion, the enclosed air undergoes slight expansions and contractions because of the close spacing between piston and cylinder; the beam comes rapidly to rest as a result of this opposition to its motion. An equal-arm balance is likely to be equipped with a *magnetic damper*, which consists of a metal plate (generally aluminum) secured to the end of the beam and positioned between the poles of a permanent magnet. Motion in the beam causes the plate to move through the field of the magnet. The forces induced act to oppose further motion; the beam rapidly acquires its rest position as a result.

Weights. Every analytical balance is supplied with a set of weights that will accommodate loadings to the designed maximum.

THEORY OF OPERATION OF A BALANCE

When the beam of a balance is horizontal, the force moments acting on the two arms are identical; that is,

$$F_1 = F_2$$

This relationship may also be expressed as

$$M_1L_1 = M_2L_2 \tag{30-1}$$

where M_1 is the mass acting at a distance L_1 to the left of the fulcrum, while M_2 and L_2 are the analogous quantities to the right.

Single-Pan Balances. As may be seen from (b) in Figure 30-2, L_1 and L_2 are ordinarily unequal; M_1 and M_2 must necessarily be different as well. In this balance M_1 includes the masses of the pan, stirrup, the arm of length L_1, and a full complement of weights; M_2 is the combined mass of the arm L_2, the damper, and a counterweight. Note that the product M_2L_2 is a constant throughout.

When an unknown mass M_x is placed on the pan, the beam undergoes a deflection. This tendency is offset by *removing* a mass M_s from the standards mounted on the beam above the pan. In terms of Equation 30-1

$$(M_1 + M_x - M_s)L_1 = M_2L_2$$

The beam will be restored to its horizontal state when $M_s = M_x$.

The process by which balance is achieved through removal of weights is called *weighing by substitution*. An important feature of this process is that the total load on the beam in its horizontal position is the same regardless of the loading on the pan. As will be shown presently, substitution weighing is a design requirement for single-pan balances.

Equal-Arm Balances. In the construction of an equal-arm balance, great care is taken to assure that L_1 and L_2 are identical (any difference between M_1 and M_2 of the unloaded balance is compensated for by adjustment of small counterweights on the beam). Thus, placement of an unknown mass M_x on the left pan would generate a counterclockwise force moment if it were not balanced by the clockwise moment of the known masses M_s located on the right pan. Here Equation 30-1 becomes

$$(M_1 + M_x)L_1 = (M_2 + M_s)L_2$$

Again, the beam will return to horizontal when $M_s = M_x$.

Stability. When set in motion, the beam of an undamped analytical balance will oscillate reproducibly about the equilibrium point at which it will eventually come to rest. This behavior occurs because the center of gravity for the beam is below the fulcrum; under these circumstances, the combined mass of the moving parts acts as a restoring force to return the beam to its stable equilibrium position.

Sensitivity. Regardless of balance type, the weighing process involves ascertaining the known mass M_s needed to return the beam to its original rest position. Much time can be saved if the difference between the rest position of the beam under zero load and under a particular load can be translated into the increment of mass needed to bring the two rest positions into coincidence. The *sensitivity* is a measure of this difference; it is defined[4] as the ratio of the change in response to the quantity measured (that is, mass). Sensitivity can be expressed as a function of the angular deflection of the beam that results from a small imbalance—say, 1.00 mg; alternatively, it can be described as the linear displacement of the indicating element caused by a unit difference in load. The indicating element for a single-pan balance is an optical lever that is deflected by one scale division per milligram. The corresponding device on an equal-arm balance is a pointer attached to the beam; the scale divisions against which displacement is measured are arbitrarily spaced for convenient observation.

The sensitivity of an analytical balance is directly proportional to the mass causing the deflection and the length of the beam; it is inversely proportional to the total mass of the moving parts (beam, stirrups, pans), as well as the loading, and to the distance between the fulcrum and the center of gravity of the moving parts when the beam is horizontal. Figure 30-3 illustrates how the sensitivity should vary with total load. Clearly, if the optical lever for a single-pan balance is to read directly in milligrams under any loading, the sensitivity must remain constant; it is for this reason that these instruments are designed for substitution weighing.

Evaluation of the sensitivity greatly shortens the time required to complete a weighing with an undamped equal-arm balance (p. 694).

SINGLE-PAN BALANCES

Operational Features. The essential parts of a single-pan balance are shown in Figure 30-4. The pan (1) and a set of nonmagnetic, stainless steel weights (2) for loads that are larger than 100 mg are located on one side of the fulcrum; a fixed weight (3), which also serves as the piston of an air damper (6), is located on the other side. Restoration of the beam to its original position after an object

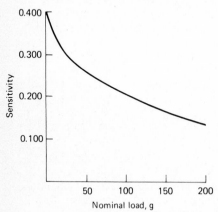

Figure 30-3 Theoretical Variation of Sensitivity with Loading of an Analytical Balance. The sensitivity function plotted is the tangent of the angle caused by a 1.00-mg difference in loading on the beam.

[4] 1954 Report and Recommendations of the Committee on Balances and Weights, Division of Analytical Chemistry, American Chemical Society [*Anal. Chem.*, **26**, 1190 (1954)].

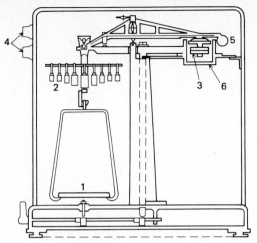

Figure 30-4   Schematic Diagram of a Single-Pan Balance. (Diagram courtesy of Mettler Instrument Corporation, Princeton, N.J.)

has been placed on the pan is accomplished by removing weights in an amount equal to the mass of the object. In actual operation, sufficient weights are removed to leave the system with a residual imbalance of 100 mg or less. Under these conditions, the deflection of the beam with respect to its original position is directly proportional to the difference between the actual mass of the object and the weights that have been removed. An optical level (5) translates this deflection into units of mass and displays the fractional weight to the operator as an image on a frosted-glass surface.

Weighing. The weight of an object is very simply obtained with a modern single-pan balance.

PROCEDURE

To zero the empty balance, rotate the arrest knob counterclockwise to the full release position. Then manipulate the zero-adjusting control until the illuminated scale indicates a reading of zero. Next arrest the balance, and place the object to be weighed on the pan. Again turn the arrest knob, this time clockwise to its partial release position. Rotate the dial controlling the heaviest likely weight for the object until the illuminated scale changes position or the notation "remove weight" appears; then turn back the knob one stop. Repeat this procedure with the other dials, working successively through the lighter weights. Then turn the arrest knob to its counterclockwise position, and allow the balance to achieve equilibrium. The weight of the object is found by taking the sum of the weights indicated on the dials and that which appears on the illuminated scale. A vernier is helpful in reading this scale to the nearest tenth of a milligram.

NOTE

  Directions for operation of single-pan balances vary with make and model. Consult with the instructor for any modifications to this procedure that may be required.

EQUAL-ARM BALANCES

Weights. The weights for an equal-arm balance are manufactured from corrosion-resistant materials. Brass, plated with platinum, gold, or rhodium is preferred for denominations larger than 1 g, although highly polished, non-magnetic stainless steel and lacquered brass weights also give good service. These integral weights are either carefully machined from a single piece of metal or built in two pieces; with the latter, the final adjustment of mass is accomplished by inserting small compensating weights in the space between the two parts.

Fractional weights (that is, weights smaller than 1 g) are fabricated from sheets of platinum, tantalum, or aluminum. The smallest weights are introduced by means of a rider or a chain; with the former, the beam is calibrated in units of 0.1 mg. Movement of a small wire rider along the beam has the effect of altering the loading of the balance from 0 to 5 (or 10) mg. The position of the rider is adjusted with a rodlike control that extends through the balance case.

Chain weights greatly facilitate the addition of fractional weights to an equal-arm balance. As illustrated in Figure 30-5, a small gold chain is attached some distance to the right of the fulcrum and also to an externally operated crank. The amount of weight supported by the beam can be varied by lengthening or shortening the chain. A scale arrangement and vernier makes possible the direct reading of the weight that has been added. Since the chain generally allows introduction of any weight within the range of 0 to 100 mg, the beam of such a balance will be ordinarily notched for a rider that permits addition of 100-mg increments from 0 to 1 g. Thus, no weight smaller than 1 g need be manually introduced. A chain balance fitted with magnetic damping can provide accurate results with a minimum expenditure of time.

Notwithstanding painstaking care in manufacture, some variation must be expected between the nominal and actual value of a weight; any correction that must be applied can be determined through calibration of the weight against a known standard. Clearly, the production cost for a set of weights

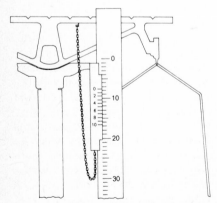

Figure 30-5 The Chain Balance. Note arrangement of chain and vernier. Indicated loading is 7.6 mg. Note also that beam is notched to accommodate the rider.

increases as the acceptable tolerances are made smaller. As a corollary, it should also be clear that the routine use of extremely precise weights is unnecessary; a well-calibrated set will serve as well.

The National Bureau of Standards has established classes of weights based upon the type of work for which they are intended, acceptable tolerance limits for weights within each class, and specifications with respect to the materials and construction that are acceptable for each class. The interested reader should consult the *National Bureau of Standards Circular* 547, section 1 (1962), for further information on this subject.

Weighing. Several techniques exist for determining the weight of an object with an equal-arm analytical balance. Regardless of method, however, two terms must be defined. The *zero point* refers to the equilibrium position of the pointer for the empty balance. A *rest point* is the equilibrium position assumed by the pointer when a load has been imposed on the balance. The weighing operation, then, consists of determining the weight necessary to produce a rest point that is identical with the zero point. The beam of an undamped balance comes to rest very slowly; thus, its equilibrium position is estimated by observing the swings of its pointer.

The deflection scale of the typical analytical balance generally does not have numerical values assigned to the division lines, these being left to the discretion of the operator. We shall employ the convention that assigns a value of 10 to the midline of the scale. The lines to the left, then, represent integers less than 10, while those to the right have values greater than 10.

We shall describe the method of long swings as well as the method of short swings. The former is included because of its established reliability for the most exacting work. The latter, being the quicker method, is recommended for all routine weighing. The principal difference between these schemes is the number of observations made upon the pointer; the long-swing method involves an odd number of these observations (either five or three), while the short-swing method uses only two observations under somewhat restricted conditions.

PROCEDURE

The Method of Long Swings. To obtain an estimate of the zero point, close the balance case, lower the beam support gently, and then disengage the pan arrests (this order is always followed in freeing the beam; the opposite order is used when arresting the beam). Allow the beam to swing freely through several cycles. Motion can be imparted, if necessary, by temporarily displacing the rider (or chain) from its zero position. Observe and record the maximum positions attained by the pointer, starting with a right-hand swing; make estimates to the nearest tenth of a division. Continue until observations of three or five consecutive maxima have been made. Calculate the average value for the swings in each direction; take the mean of these two quantities as the zero point of the balance. The calculations should not be carried beyond the first decimal place.

Example. Calculate the zero point for a balance if the following consecutive readings were observed: 12.4, 6.9, 12.2, 7.1, 11.9.

readings	6.9	12.4
	7.1	12.2
	14.0	11.9
		36.5
averages	7.0	12.2

$$\text{zero point} = \frac{7.0 + 12.2}{2} = 9.6$$

It should be noted in passing that the zero point will undergo slight changes with time, owing to changes in temperature and humidity. As a consequence, its value should be determined for each series of weighings. Pronounced shifts in the zero point are indicative of trouble; in such event, the instructor should be consulted.

After ascertaining that the pan arrests and beam supports have been reengaged, place the object to be weighed on the left pan of the balance. Make a rough estimate of its mass, and, using forceps, place on the right pan the single weight that is believed to have a slightly greater mass. Gently and partially disengage the beam support, and note the behavior of the pointer; it will swing away from the pan that has the heavier load. If no movement is seen, the loadings are within 1 to 2 g of each other. Under such circumstances, cautiously disengage the pan arrests to determine which pan has the heavier load. Continue a systematic trial-and-error loading of weights until the closest approach to balance has been achieved where the object is still heavy. Then use the rider in a similar manner; the balance case should be closed henceforth. When balance has apparently been achieved, obtain a rest point in the manner described for determining the zero point, and compare with the zero point. Subsequent corrections in the position of the rider can be checked in the same fashion until balance is attained.

Time can be saved by employing the balance sensitivity to calculate the number in the final decimal place. Here the weight of the object is determined to the nearest whole milligram, and a rest point (RP_1) is obtained. Then 1 mg is either added to or subtracted from the load on the balance, and a second rest point (RP_2) is taken. The difference between these rest points yields the sensitivity of the balance for the particular loading involved. The correction needed to shift either rest point into coincidence with the zero point can then be readily calculated.

Example.

zero point (ZP)	10.2
rest point with 10.422 g (RP_1)	11.1
rest point with 10.423 g (RP_2)	8.7
sensitivity (11.1 − 8.7)	2.4 divisions/mg
difference between RP_1 and ZP	0.9 division

Since

$$\frac{1 \text{ mg}}{2.4 \text{ div}} = \frac{x \text{ mg}}{0.9 \text{ div}}$$

then

$$x = 0.4 \text{ mg}$$

This correction should be *added* to 10.422 g because the rest point corresponding to this loading (RP_1) lies to the *right* of the zero point. Thus, the weight of the object is 10.422 + 0.0004 = 10.4224 g. Note that the same result is obtained from RP_2 data:

difference between ZP and RP_2 = 1.5 div

Since

$$\frac{1 \text{ mg}}{2.4 \text{ div}} = \frac{x \text{ mg}}{1.5 \text{ div}}$$

then

$$x = 0.6 \text{ mg}$$

This correction should be *subtracted* from 10.423 g because the rest point corresponding to this loading lies to the *left* of the zero point. Thus, the weight is again 10.423 − 0.0006 = 10.4224 g.

The Method of Short Swings. The difference between consecutive swings is negligible when the amplitude is small (from one to three divisions on either side of the mean). Under these circumstances, the arithmetic mean of a consecutive left and right swing of the pointer gives a close approximation of the equilibrium point. Therefore, only two observations are made, and the average is taken for use. This technique represents a considerable saving of time. Whether the weight is directly determined to the nearest 0.1 mg or the number in the final decimal place is calculated from the sensitivity is a matter of personal preference; with a knowledge of the approximate sensitivity of the balance, the direct method is quite rapid.

Weighing with a Damped Balance. A damped balance, particularly if also equipped with a chain, is well adapted for the direct determination of the weight of an object. Here the zero point is established by releasing the beam support and pan arrest and allowing time for the pointer to cease moving. Weighing is accomplished as before, the rider and chain being adjusted to give a rest point that is identical with the zero point.

SUMMARY OF RULES GOVERNING THE USE OF AN ANALYTICAL BALANCE

Continued good performance from an analytical balance depends entirely upon the treatment accorded it. Similarly, reliable data are obtained only when careful attention is paid to the details of the weighing operation. Since weighing data are required for virtually every quantitative analysis, it is desirable to summarize the rules and precautions that relate to their acquisition.

To Avoid Damage or Minimize Wear to the Balance and Weights.

1. Be certain that arresting mechanisms for the beam are engaged whenever the loading on the balance is being changed and when the balance is not in use.
2. Insofar as possible, center the load on the pan.
3. Protect the balance from corrosion. Only vitreous materials and nonreactive metal or plastic objects should be placed directly on the pans.
4. The weighing of volatile materials requires special precautions (p. 693).
5. Do not attempt to adjust the balance without the prior consent of the instructor.
6. Balance and case should be kept scrupulously clean. A camel's-hair brush is useful for removal of spilled material or dust.
7. Free the beam of an equal-arm balance by releasing the beam support slowly, followed by the pan arrests. Reverse this order when arresting the beam; engage the pan arrests as the pointer passes through the center of the deflection scale.
8. Handle weights for an equal-arm balance gently and with special forceps. Weights should never be touched, as moisture from one's hands can initiate corrosion. Keep weights enclosed in their box when they are not actually in use.
9. If possible, avoid bringing weights for an equal-arm balance into the laboratory.

To Obtain Reliable Weighing Data.

10. Do not attempt to weigh an object until it has returned to room temperature.
11. Do not touch a dried object with bare hands; handle it with tongs or use finger pads to protect against the uptake of moisture.
12. Place the object on the left pan of an equal-arm balance, the weights on the right.
13. Recalibrate a weight that has been mishandled before using it again.
14. At the completion of the weighing, sum the weights at least twice to avoid arithmetic and reading errors. Some analysts first count the weights on the pan and then the empty spots in the weight box.
15. Perform all weighings for an analysis on the same balance, using the same set of weights.

OTHER TYPES OF BALANCES

A typical analytical balance can accommodate loads of 200 g with a sensitivity that ranges from 1.5 to 4 scale divisions per milligram; the reciprocal sensitivity will thus be on the order of 0.7 to 0.25 mg per scale division. By way of contrast, *semimicrobalances* usually have capacities no greater than 100 g and reciprocal sensitivities on the order of 0.05 mg per division. *Microbalances* accommodate only 20 to 50 g on each pan but have reciprocal sensitivities in the range of 0.01 to 0.001 mg per division. Yet another order of diminution is possible through the use of quartz fibers, either in the form of beams or helices. Such weighing devices have very low capacities but are sensitive to as little as 0.005 μg $(1\ \mu g = 1 \times 10^{-6}\ g)$.

Since approximate weights are frequently needed, the analytical laboratory is also supplied with balances that are less sensitive but more rugged than the analytical balance. It is necessary to develop the ability to distinguish the weighing operations in an analysis that must be performed with great reliability from those that do not require such care, and to select the balance to be used on this basis.

ERRORS IN THE WEIGHING OPERATION

In common with all other types of physical measurement, weighing data are subject to uncertainty. To be sure, the weighing operation is ordinarily the most reliable step in a chemical analysis; uncertainties attributable to this source typically amount to fractions of a milligram. Among the sources of error in weighing, the effects of buoyancy and temperature are particularly important.

Buoyancy. An object surrounded completely by a medium will have displaced a volume of the medium equal to its own. The *apparent mass* of the object will be the difference between its true mass and that of the medium it has displaced. A weighing in air thus involves comparison of apparent masses. The condition of balance for an equal-arm instrument is described by the equation

$$(M_1 - m_1)L_1 = (M_2 - m_2)L_2 \tag{30-2}$$

where m_1 and m_2 represent the masses of air displaced by the object M_1 and the weights M_2, respectively. Since L_1 and L_2 are identical, Equation 30-2 can be simplified and rearranged to

$$M_1 = M_2 + (m_1 - m_2)$$

The volumes of air displaced by the object and the weights are equal to V_1 and V_2, respectively; m_1 and m_2 can be expressed in terms of V_1 and V_2 through use of the density of air $(d = m/V)$. Thus,

$$M_1 = M_2 + d_{air}(V_1 - V_2)$$

The volumes can also be expressed in terms of the densities of the object and the weights. Thus,

$$V_1 = \frac{M_1}{d_1} \qquad \text{and} \qquad V_2 = \frac{M_2}{d_2}$$

where d_1 and d_2 are the respective densities of object and weights. Then

$$M_1 = M_2 + d_{air} \left(\frac{M_1}{d_1} - \frac{M_2}{d_2} \right)$$

Finally, since M_1 and M_2 are very nearly identical, this equation rearranges to

$$M_1 = M_2 + M_2 \left(\frac{d_{air}}{d_1} - \frac{d_{air}}{d_2} \right) \tag{30-3}$$

An analogous equation can be generated for the effect of buoyancy upon weighings performed with a single-pan balance.

The densities of materials commonly employed in the construction of analytical weights are compiled in Table 30-1. The density of air can be taken as 0.0012 g/ml.

TABLE 30-1 Density of Metals Used in the Manufacture of Weights

Metal	Density, g/ml
Aluminum	2.7
Brass	8.4
Gold	19.3
Platinum	21.4
Stainless steel	7.8
Tantalum	16.6

Example. Stainless steel weights (d = 7.8 g/ml) were used to determine the mass of a benzene (d = 0.878 g/ml) sample. Correct this weighing for the effects of buoyancy, given the accompanying data:

$$\text{wt bottle plus sample, g} = 12.4137$$

$$\text{wt bottle, empty, g} = 10.9873$$

Only the buoyant effect upon 1.4264 g of benzene (12.4137 − 10.9873) needs to be considered, because the same force acted on the container during both weighings. Substitution of numerical values into Equation 30-3 yields

$$\text{corrected wt} = 1.4264 + 1.4264 \left(\frac{0.0012}{0.878} - \frac{0.0012}{7.8} \right)$$

$$= 1.4264 + 0.0017 = 1.4281 \text{ g}$$

The consequences of Equation 30-3 are shown in Figure 30-6, in which the relative error due to buoyancy is correlated with the density of objects weighed in air against brass or stainless steel weights. It is seen that this error becomes appreciable only where objects of low density are being weighed. In routine analytical work, correction of weighing data for buoyancy is limited to the calibration of glassware.

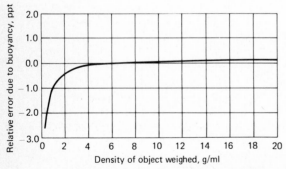

Figure 30-6 Effect of Buoyancy upon Weighing Data. The relative error incurred (with brass weights) is shown as a function of the density of the object weighed.

Effects of Temperature. Weighing data are significantly influenced by temperature differences that may exist between the object to be weighed and the balance; these effects are more pronounced in equal-arm than in single-pan instruments. If, as is usually the situation, the object is warmer than its surroundings, its weight will tend to be low, owing to bouyant convection currents caused by the temperature difference within the case of an equal-arm balance. If, in addition, the object is a covered container, the enclosed air will weigh less than an equal volume of ambient air; again, a low apparent weight will be given by both types of balance.

Errors in Weights. A common source of weighing error resides in the weights of an equal-arm balance. Although fabricated from corrosion-resistant metals, weights can be significantly in error as the result of hard use, or more seriously, misuse. Periodic calibration against a standard set is recommended.

The weights for a single-pan balance are not as vulnerable to change through use as those for an equal-arm balance. Methods have been developed to permit calibration of built-in weights without the need to remove them from the balance.[5] The reliability of the optical scale of a single-pan balance should be checked regularly, particularly under loading conditions that require essentially the full range of the scale; a 100-mg weight is useful for this check.

OTHER SOURCES OF ERROR IN THE WEIGHING OPERATION

Many weighing errors are directly attributable to the analyst himself. The misreading of weights is common.

Occasionally, a porcelain or glass object will acquire a static charge that is sufficient to cause the balance to perform erratically; this problem is particularly serious when the humidity is low. Spontaneous discharge frequently occurs after a short period. The use of a faintly damp chamois to wipe the object is a recommended preventative.

Equipment and Manipulations Associated with Weighing

Most solids absorb atmospheric moisture and, as a consequence, change in composition (Chapter 27). This effect assumes appreciable proportions when a large surface area is exposed, as with a sample or a reagent chemical that has been finely ground. It is ordinarily necessary to dry such solids before weighing in order to free the results from dependence upon the humidity of the surrounding atmosphere.

[5] ASTM Designation 319-67T, *Tentative Methods for Testing the Performance of Laboratory Balances,* 1967.

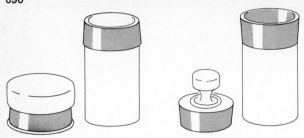

Figure 30-7   Typical Weighing Bottles.

EQUIPMENT FOR DRYING AND WEIGHING

Weighing Bottles. Solids are conveniently dried and stored in weighing bottles, two common varieties of which are illustrated in Figure 30-7. Ground-glass contacting surfaces ensure a snug fit between container and lid. In the newer design, the lid acts as a cap; with this style of weighing bottle, there is less possibility of the sample being entrained on, and subsequently lost from, a ground-glass surface. Weighing bottles usually have numbers from 1 to 100 etched on their sides for purposes of identification. Plastic weighing bottles are available; ruggedness is the principal advantage possessed by these bottles over their glass counterparts.

Desiccator, Desiccants. Oven drying is, by all odds, the most convenient method for removing adsorbed moisture from a sample. The technique, of course, is not appropriate for samples that undergo decomposition at the ambient temperature of the oven. Furthermore, with some solids, the temperatures attainable in ordinary drying ovens are insufficient to effect complete removal of the bound water.

While they cool, dried materials are stored in desiccators; these provide a measure of protection from the uptake of moisture. As illustrated in Figure 30-8, the base section of a desiccator contains a quantity of a chemical drying agent. Samples are placed on a perforated plate that is supported by a constriction in the wall. Lightly greased ground-glass surfaces provide a tight seal between lid and base.

Table 30-2 lists the characteristics of common drying agents.

MANIPULATIONS ASSOCIATED WITH DRYING AND WEIGHING

Use of the Desiccator. Whether it is being replaced or removed, the lid of the desiccator is properly moved by a sliding, rather than a lifting, motion. An airtight seal is achieved by slight rotation and direct downward pressure upon the positioned lid.

When a heated object is placed in a desiccator, the increased pressure of the enclosed air is often sufficient to break the seal between lid and base. If heating has caused the grease on the ground-glass surfaces to soften, there is the further danger that the lid may slide off and break. Upon cooling, the

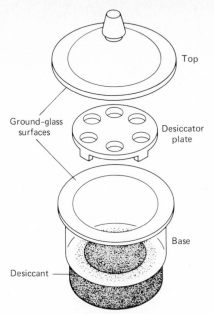

Figure 30-8   Components of a Typical Desiccator.

opposite effect is likely to occur, the interior of the desiccator now being under a partial vacuum. Both of these conditions can cause the contents of the desiccator to be physically lost or contaminated. While it defeats the purpose of the desiccator somewhat, it is wise to allow some cooling to occur before finally sealing the lid. It also helps to break the seal several times during cooling in order to relieve any vacuum that may have developed. Finally, it is prudent to lock the lid in place with one's thumbs while moving the desiccator to ensure against accidental breakage.

Very hygroscopic materials should be stored in containers equipped with snugly fitting covers, the covers remaining in place during storage in the desiccator. Other substances may be stored with container covers removed.

Manipulation of Weighing Bottles. The moisture bound on the surface of many solid materials can be removed by heating at 105 to 110°C for about an hour. Figure 30-9 depicts the arrangement recommended for the drying of a sample in a weighing bottle. The bottle is contained in a beaker, which in turn is covered by a watch glass supported on glass hooks (or by a ribbed watch glass). The sample is protected from accidental contamination while the free access of air is maintained. This arrangement also satisfactorily accommodates crucibles containing precipitates that can be freed of moisture by simple drying. The beaker should be marked to permit identification.

Weighing data can be significantly affected by moisture picked up as a consequence of handling a dried weighing bottle with one's fingers. For this reason, the bottle should be manipulated with tongs or with strips of clean paper.

TABLE 30-2 Common Drying Agents

Agent	Capacity	Deliquescent	Condition for Regeneration	Nature of Reaction with Water	mg Water Remaining in 1 Liter of Dried Air[a]
Phosphorous pentoxide	Low	Yes	Difficult	Acidic	2.6×10^{-4}
Barium oxide	Moderate	No	Difficult	Alkaline	6.5×10^{-4}
Aluminum oxide	Low	No	175°C	Neutral	1×10^{-3}
Anhydrous magnesium perchlorate (Anhydrone[b] or Dehydrite[c])	High	Yes	240°C Vacuum	Neutral	2×10^{-3}
Calcium oxide	Moderate	No	500°C	Alkaline	3×10^{-3}
Calcium sulfate (Drierite[d])	Moderate	No	275°C	Neutral	5×10^{-3}
Silica gel	Low	No	120°C	Neutral	6×10^{-3}
Potassium hydroxide, stick	Moderate	Yes	Difficult	Alkaline	1.4×10^{-2}
Anhydrous calcium chloride	High	Yes	Difficult	Neutral	1.4

[a] J. H. Bower, J. Research Nat. Bur. Standards, **12**, 241 (1934); **33**, 199 (1944).
[b] ™ J. T. Baker Chemical Company, Phillipsburg, N.J.
[c] ® Arthur H. Thomas Company, Philadelphia, Pa.
[d] ® W. A. Hammond Drierite Co., Xenia, Ohio.

Figure 30-9   Arrangement for Drying of Samples.

The latter technique is illustrated in Figure 30-10, which shows the weighing of a sample by difference. The weight of the bottle and its contents is first obtained. Then the sample is transferred from the bottle to the container, the utmost care being taken to avoid losses during transfer. Gentle tapping of the weighing bottle with its top provides adequate control over the process; slight rotation of the bottle is also helpful. Finally, the bottle is again weighed.

Weighing of Hygroscopic Substances. Many substances are hygroscopic to a relatively limited extent. These often equilibrate rapidly, taking up moisture almost to capacity in a short time. Where this effect is pronounced, the approximate amount of each sample should be introduced to individual weighing bottles. After drying and cooling, the exact weight is determined by difference, care being taken to remove the sample and replace the top of the weighing bottle as rapidly as possible.

Weighing of Liquids. The weight of a liquid is always obtained by difference. Samples that are noncorrosive and relatively nonvolatile can be weighed into containers fitted with snugly fitted covers, such as weighing bottles; the mass of the container is subtracted from the total weight. If the sample is volatile or corrosive, it should be sealed in a weighed glass ampoule. The bulb of the

Figure 30-10   Method for Transferring Sample.

ampoule is first heated. Then the neck is immersed in the sample. As cooling occurs, the liquid is drawn into the bulb. The neck is then sealed off with a small flame. After cooling, the bulb and contents are weighed, with any glass removed during the sealing also included in this weighing. The ampoule is then broken in the vessel where the sample is desired. A small correction for the volume of the glass in the ampoule is required for the most precise work.

Equipment and Manipulations for Filtration and Ignition

EQUIPMENT

Simple Crucibles. Simple crucibles serve as containers only and are of two general types. The weight of the more common variety remains constant within the limits of experimental error while in use. These crucibles are fabricated from porcelain, aluminum oxide, silica, or platinum and are employed to convert precipitates into suitable weighing forms. Nickel, iron, silver, or gold crucibles serve as containers for high-temperature fusions of difficultly soluble samples. Crucibles of this type may be appreciably attacked by the atmosphere and by their contents. Both effects will cause a change in weight; the latter will also contaminate the contents with species derived from the crucible. The analyst selects the crucible whose components will offer the least interference in subsequent steps of the analysis.

Filtering Crucibles. Filtering crucibles are also of two types. For some, the filtering medium is an integral part of the crucible. In addition, there is the *Gooch crucible*, in which a filter mat (usually asbestos, but not always) is supported upon a perforated bottom.

A partial vacuum is used to draw the supernatant liquid through a filtering crucible; this procedure frequently shortens the time needed for filtration. Connection is made between the crucible and a heavy-walled filtering flask with a rubber adapter (see Figure 30-11). A diagram for the complete filtration train is shown in Figure 30-16.

Asbestos is commonly formed as a mat over the small holes in the bottom of a Gooch crucible. This mineral is of variable composition and structure, some forms being appreciably water soluble. Thus, especially selected and prepared grades are required for filtration. Long-fibered, acid-washed asbestos can be obtained from chemical suppliers.

Since asbestos is hygroscopic, care must be taken to bring the crucible and mat to constant weight under exactly the same conditions as will be required to convert the precipitate to a form suitable for weighing.

Small circles of glass matting are available commercially and can be used instead of asbestos in a Gooch crucible; they are used in pairs to protect against accidental disintegration while liquid is added. Glass mats can tolerate temperatures in excess of 500°C and are substantially less hygroscopic than asbestos.

Glass also finds application as a filtering medium in the form of fritted

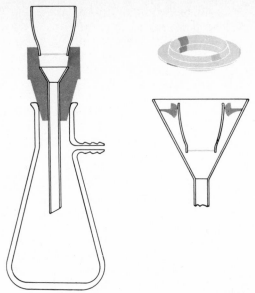

Figure 30-11   Adapters for Filtering Crucibles.

disks sealed permanently into filtering crucibles. These are known as *sintered glass crucibles* and are available in various porosities. With extreme care, their use can be extended to temperatures as high as 500°C; normally, however, 150 to 200°C represents the upper practical temperature limit.

Filtering crucibles made entirely of fused quartz may be taken to high temperatures and can be cooled rapidly without damage.

Unglazed porcelain is also used as a filtering medium. Crucibles of this type have the versatility of temperature range possessed by the Gooch crucible and are not as costly as fused quartz. They require no preparation comparable to that for the Gooch crucible and are far less hygroscopic. Filtering crucibles made of aluminum oxide offer similar advantages.

Filter Paper. Paper is an important filtering medium. Because it is appreciably hygroscopic, a paper filter is always destroyed by ignition if its contents are to be weighed. Ashless paper is manufactured from fibers that have been washed with hydrochloric and hydrofluoric acids and neutralized with ammonia. The residual ammonium salts in many papers are sufficient to affect analyses for amine nitrogen. Typically, 9- or 11-cm circles of such paper will leave an ash weighing less than 0.1 mg, an amount that is ordinarily negligible. Ashless paper is manufactured in various grades of porosity.

Gelatinous precipitates, such as hydrous iron(III) oxide, present special problems, owing to their tendency to clog the pores of the paper upon which they are being retained. This problem can be minimized by mixing a dispersion of ashless filter paper pulp with the precipitate prior to filtration. The pulp may be prepared by briefly treating a piece of ashless paper with concentrated

TABLE 30-3 Comparison of Filtering Media for Gravimetric Analysis

Characteristic	Paper	Asbestos (Gooch)	Glass Crucible	Porcelain Crucible	Aluminum Oxide Crucible
Speed of filtration	Slow	Rapid	Rapid	Rapid	Rapid
Convenience and ease of preparation	Somewhat troublesome and inconvenient	Somewhat troublesome and inconvenient	Convenient	Convenient	Convenient
Maximum ignition temperature	None	1200°C	200–500°C	1100°C	1450°C
Chemical reactivity	Carbon from paper has reducing properties	Inert	Inert	Inert	Inert
Control of porosity	Many porosities available	Little control possible	Several porosities available	Several porosities available	Several porosities available
Convenience with gelatinous precipitates	Satisfactory	Unsuitable; filter tends to clog	Unsuitable; filter tends to clog	Unsuitable; filter tends to clog	Unsuitable; filter tends to clog
Cost	Low	Low	High	High	High

hydrochloric acid and washing the disintegrated mass free of acid. Tablets of pulp are also commercially available.

Table 30-3 summarizes the characteristics of common filtering media.

Heating Equipment. Many precipitates may be weighed directly after the low-temperature removal of moisture; drying ovens, which are heated electrically and capable of maintaining uniform temperatures to within a degree or less, are convenient for this purpose. The maximum attainable temperature will range from 140 to 260°C, depending upon make and model. For many precipitates, 110°C is a satisfactory drying temperature.

The efficiency of a drying oven is increased significantly by the forced circulation of air. Further refinement is achieved by predrying the air to be circulated and by using vacuum ovens through which a small flow of predried air is maintained.

Ordinary heat lamps are also useful for laboratory drying and provide temperatures capable of charring filter paper. A convenient method of treating precipitates collected on paper involves the use of heat lamps for initial drying and charring, followed by ignition at elevated temperatures in a muffle furnace.

Burners represent convenient sources of intense heat. The maximum temperature attainable from a burner depends upon its design as well as the combustion properties of the gas used. Of the three common laboratory burners, the Meker provides the greatest heat, followed by the Tirrill and Bunsen types in that order.

Heavy-duty electric furnaces are capable of maintaining temperatures of 1100°C or higher with a control superior to that obtainable with a burner. Special long-handled tongs and asbestos gloves are frequently required for protection while transferring objects to and from such furnaces.

A rough judgment of the temperature of an object can be gained from observation of its color; Table 30-4 will serve as a guide.

MANIPULATIONS ASSOCIATED WITH THE FILTRATION AND IGNITION PROCESSES

Preparation of Crucibles. A crucible used to convert a precipitate into a form suitable for weighing must maintain a substantially constant weight throughout the drying or ignition process. To demonstrate this property, the

TABLE 30-4   Estimation of Temperature by Color[a]

Temperature, °C	Approximate Color of Object at This Temperature
700	Dull red
900	Cherry red
1100	Orange
1300	White

[a] Adapted from T. B. Smith, *Analytical Processes*, 2d ed., p. 431. London: Edward Arnold (Publishers) Ltd., 1940.

crucible first is cleaned thoroughly (filtering crucibles are conveniently cleaned by backwashing with suction) and then brought to constant weight, using the same heating and cooling cycle as will be required for the precipitate. For most purposes, agreement within 0.2 mg between consecutive cycles can be considered as constant weight.

Filtration and Washing of Precipitates. The actual filtration process consists of the three operations: decantation, washing, and transfer. Decantation involves gently pouring off the bulk of liquid phase while leaving the precipitated solid essentially undisturbed. The pores of any filtering medium become clogged with precipitate; hence, the longer the transfer of solid can be delayed, the more rapid will be the overall filtration process. A stirring rod is employed to direct the flow of the decantate (see Figure 30-12). Wash liquid is then added to the

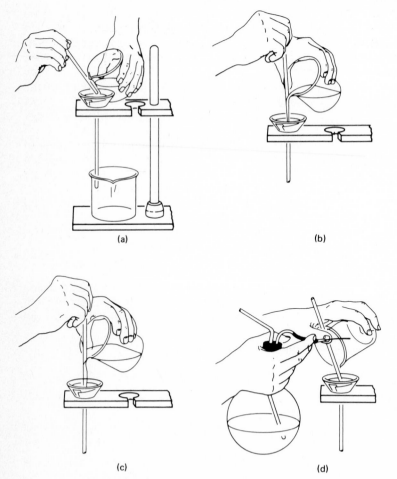

(a)

(b)

(c)

(d)

Figure 30-12 The Filtering Operation. Techniques for decantation and transfer of precipitates are illustrated.

beaker and thoroughly mixed with the precipitate; after the solid has again settled, this liquid is also decanted through the filter. It can be seen that the principal washing of the precipitate is carried out *before* the solid is transferred; a more thoroughly washed precipitate and a more rapid filtration are the result.

The decantation and transfer processes are illustrated by (d) in Figure 30-12. The bulk of the precipitate is moved from beaker to filter by suitably directed streams of liquid. As always, a stirring rod is used to provide direction for the flow of liquid to the filtering medium.

The last traces of precipitate that cling to the walls of the beaker are dislodged with a *rubber policeman,* a small section of rubber tubing which has been crimped shut at one end; this device is fitted on the end of a stirring rod. A rubber policeman should be wetted with wash liquid before use. Any solid collected is combined with the main portion on the filter. The last traces of hydrous oxide precipitates are conveniently collected with small pieces of ashless filter paper; clean forceps are useful here.

Many precipitates have the exasperating property of *creeping* or spreading over wetted surfaces against the force of gravity. Filters are never filled to more than three-quarters of their capacity, owing to the possibility for losses of solid as a result of creeping.

DIRECTIONS FOR FILTRATION AND IGNITION WITH ASHLESS FILTER PAPER

Preparation of a Filter Paper. Figure 30-13 illustrates the sequence followed in folding a filter paper and seating it in a 58-deg or 60-deg funnel. The paper is

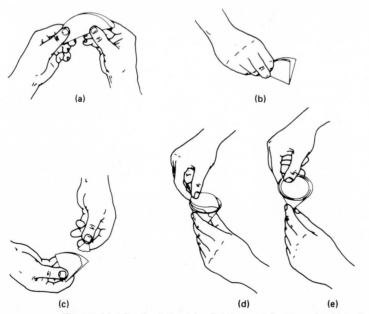

(a)

(b)

(c)

(d)

(e)

Figure 30-13 Technique for Folding and Seating of a Filter Paper.

first folded exactly in half (a), firmly creased, and next loosely folded into quarters (b). A triangular portion is torn from one of the two single corners (c) parallel to the fold. The paper is then opened out so that a cone is formed (d); after fitting in the funnel, the second fold is creased. Seating (e) is completed by dampening the cone with water from a wash bottle and gentle patting with a finger. When the cone is properly seated, there will be no leakage of air between paper and funnel, and the stem of the funnel will be filled with an unbroken column of liquid.

A gelatinous precipitate should not be allowed to dry out before the washing cycle is complete, because the dried mass shrinks and develops cracks. Any liquid subsequently added merely passes through these cracks and accomplishes little or no washing.

Transfer of Paper and Precipitate to Crucible. Upon completion of the filtration and washing steps, the filter paper and its contents must be transferred from the funnel to a weighed crucible. Ashless paper has very low wet strength; considerable care must be exercised in performing this operation. The danger of tearing can be reduced considerably if the paper is allowed to dry partially prior to removal from the funnel.

Figure 30-14 illustrates the preferred method of transfer. First, the triple-folded portion is drawn across the filter (a) to flatten the cone along its upper edge; the corners then are folded inward (b). Next, the top is folded over (c). Finally, the paper and contents are eased into the crucible (d) so that the bulk of the precipitate is near the bottom.

Ashing of a Filter Paper. If a heat lamp is to be used, the crucible is placed on a clean, nonreactive surface; an asbestos pad covered with a layer of aluminum foil is satisfactory. The lamp is then positioned about one-half inch from the top of the crucible and turned on. Charring of the paper will take place without further intervention; the process is considerably accelerated if the paper can

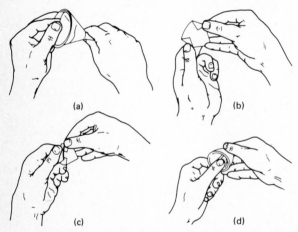

(a)   (b)

(c)   (d)

Figure 30-14 Method for Transfer of a Filter Paper and Precipitate to a Crucible.

be moistened with no more than one drop of strong ammonium nitrate solution. Removal of the remaining carbon is accomplished with a burner, as described in the following paragraphs.

Considerably more attention is required when a burner is employed to ash a filter paper. Since the burner can produce much higher temperatures, there exists the danger of expelling moisture so rapidly in the initial stages of heating that mechanical loss of the precipitate occurs. A similar possibility arises if the paper is allowed to flame. Finally, as long as there is carbon present, there is also the possibility that the precipitate will be partially reduced; this is a serious problem where reoxidation following ashing of the paper is not convenient.

In order to minimize these difficulties, the crucible is placed as illustrated in Figure 30-15; the tilted position allows for the ready access of air. A clean crucible cover should be located nearby, ready for use if necessary. Heating is then commenced with a small burner flame. The temperature is gradually increased as moisture is evolved and the paper begins to char. The smoke that is given off serves as a guide with respect to the intensity of heating that can be safely tolerated. Normally, this will appear in thin wisps. An increase in the volume of smoke emitted indicates that the paper is about to flash; heating should be temporarily discontinued if this condition is observed. If, despite precautions, a flame does appear, it should be immediately snuffed out with the crucible cover. (The cover may become discolored, owing to the condensation of carbonaceous products. These must be ultimately removed by ignition to confirm the absence of entrained particles of precipitate.) Finally, when no further smoking can be detected, the residual carbon is removed by gradually increasing the flame. Strong heating, as necessary, can then be undertaken. Care must be exercised to avoid heating the crucible in the reducing portion of the flame.

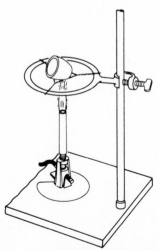

Figure 30-15 Ignition of the Precipitate. Arrangement of the crucible for the preliminary charring of the paper is illustrated.

The foregoing sequence will ordinarily precede final ignition of the sample in a muffle furnace, in which a reducing atmosphere is equally undesirable.

DIRECTIONS FOR THE USE OF FILTERING CRUCIBLES

Preparation of a Gooch Crucible. A mat is prepared by arranging the Gooch crucible in a vacuum filtration train (see Figure 30-16). A few milliliters of the well-mixed and diluted suspension are poured into the crucible and allowed to stand for about a minute. In this interval the heavier filaments tend to settle and form a support for the finer fibers. Most of the liquid also drains off. Next, suction is applied to set the mat. The mat is then washed until no further loss of asbestos can be detected in the washings; several hundred milliliters may be required. Finally, the crucible is dried and ignited to constant weight.

A proper mat is one through which the pattern of holes in the crucible can be barely discerned when viewed against a strong light.

A vacuum must always be applied before attempting to transfer liquid to a Gooch crucible. Care must also be taken to avoid destroying the mat as liquids are introduced; the flow is always directed down the wall of the crucible with a stirring rod. The mat can be further protected with a small perforated disk called a *Witt plate*.

Ignitions with a Filtering Crucible. The use of a filtering crucible tends to shorten the time required for the filtration of a precipitate. The porous bottom of such a crucible, however, greatly increases the danger of reducing the precipitate with the burner flame. This difficulty is readily circumvented by placing the filtering crucible inside an ordinary crucible during the ignition step.

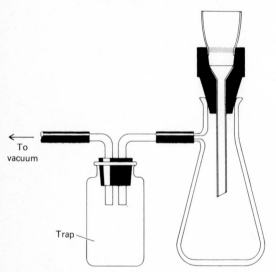

To vacuum

Trap

Figure 30-16   Train for Vacuum Filtration.

The reader is again cautioned against subjecting crucibles to unnecessarily abrupt changes in temperature.

RULES FOR THE MANIPULATION OF HEATED OBJECTS

1. A crucible that has been subjected to the full flame of a burner or to a muffle furnace should be allowed to cool momentarily on an asbestos plate before being moved to the desiccator.
2. Hot objects should not be placed directly on the desk top but should be set on an asbestos pad or a clean wire gauze.
3. Manipulations should be practiced first to assure that adequate control can be maintained with the implements to be used.
4. The tongs and forceps employed in handling heated objects should be kept scrupulously clean. The tips should not be allowed to come in contact with the desk top.

The Measurement of Volume

UNITS OF VOLUME

The fundamental unit of volume is the *liter*, defined as the volume occupied by one kilogram of water at the temperature of maximum density (3.98°C) and at one atmosphere of pressure.

The *milliliter* is defined as one one-thousandth of a liter and is widely used when the liter represents an inconveniently large volume unit.

Another unit of volume is the *cubic centimeter*. While this and the milliliter can be used interchangeably without effect in most situations, the two units are not strictly identical, the milliliter being equal to 1.000028 cm³. It was originally intended that 1 kg of water should occupy exactly 1 dm³. Owing to inadequacies of early experimental measurements, however, this relationship was not realized, and the small difference in units was the result. For volumetric analysis, the liter or milliliter is used.

EFFECT OF TEMPERATURE UPON VOLUME MEASUREMENTS

The volume occupied by a given mass of liquid varies with temperature. So also does the volume of the container that holds the liquid. The accurate measurement of volume may require taking both of these temperature effects into account.

Most volumetric measuring devices are constructed of glass, which, fortunately, has a small temperature coefficient. Thus, for example, a soft glass vessel will change in volume by about 0.003% per degree; with heat-resistant glass, the change is about one-third of this value. Clearly, variations in the volume of a container due to changes in temperature need be considered only for the most exacting work.

The coefficient of expansion for dilute aqueous solutions is approximately 0.025% per degree. The magnitude of this figure is such that a temperature variation of about 5 degrees will measurably affect the precision of ordinary volumetric measurements.

Example. A 40.00-ml sample is taken from a liquid refrigerated at 5°C. Calculate the volume this sample will occupy at 20°C.

$$V_{20°} = V_{5°} + 0.00025(20 - 5)(40.00) \qquad (30\text{-}4)$$
$$= 40.00 + 0.15$$
$$= 40.15 \text{ ml}$$

Volumetric measurements must be referred to some standard temperature; to minimize the need for calculations such as these, 20.0°C (the average room temperature) has been chosen for this reference point. Since in most laboratories the ambient temperature is close to 20°C, the need seldom arises for a temperature correction in ordinary analytical work. The coefficient of cubic expansion for many organic liquids, however, is considerably greater than that for water or dilute aqueous solutions. Good precision in the measurement of these liquids may require corrections for temperature variations of a degree or less.

APPARATUS FOR THE PRECISE MEASUREMENT OF VOLUME

The reliable measurement of volume is performed with the *pipet*, the *buret*, and the *volumetric flask*. These can be calibrated either to *deliver* or, alternatively, to *contain* a specified volume. Volumetric equipment is marked by the manufacturer to indicate not only the manner of calibration (usually with a TD for "to deliver" or a TC for "to contain") but also the temperature for which the calibration strictly refers. Ordinarily, pipets and burets are designed and calibrated to deliver specified volumes, while volumetric flasks are calibrated on a to-contain basis.

Pipets. All pipets are designed to transfer known volumes of liquid from one container to another. Some, called *volumetric* or *transfer pipets*, deliver a single, fixed volume. Others, known as *measuring pipets*, are calibrated in convenient units so that any volume up to the maximum capacity can be delivered.

In situations where a particular volume must be repeatedly dispensed, *syringe pipets* are valuable. These are available with capacities ranging from 0.005 to 25 ml.

All pipets are filled to an initial calibration mark at the outset; the manner in which the transfer is completed depends upon the particular type. Because of the attraction between most liquids and glass, a drop tends to remain in the tip of a drained transfer pipet. This drop is blown from some pipets but not from others. Table 30-5 and Figure 30-17 summarize the several varieties most likely to be encountered in an analytical laboratory.

Burets. Burets, like measuring pipets, enable the analyst to deliver any volume up to the maximum capacity. The precision attainable with a buret is appreciably better than that with a measuring pipet.

A buret consists of a calibrated tube containing the liquid and a valve arrangement by which flow from a tip can be controlled. Principal differences among burets are to be found in the type of valve employed. The simplest

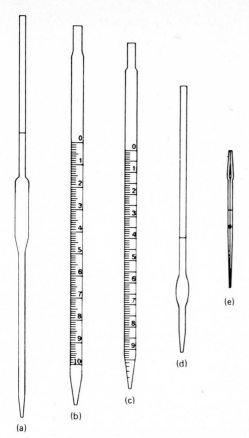

Figure 30-17 Typical Pipets: (a) Volumetric, (b) Mohr, (c) Serological, (d) Ostwald-Folin, (e) Lambda.

consists of a closely fitting glass bead within a short length of rubber tubing. Only when the tubing is deformed can liquid flow past the bead.

Burets equipped with glass stopcocks rely upon a lubricant between the ground-glass surfaces of stopcock and barrel for a liquid-tight seal. Some solutions, notably bases, will cause a stopcock to freeze upon long contact; thorough cleaning is necessary after each use.

Valves made of Teflon are commonly encountered; these are inert to attack by most common reagents and require no lubricant.

More elaborate burets are designed so that they are filled and zeroed automatically; these are of particular value in routine analysis.

Volumetric Flasks. Volumetric flasks are manufactured with capacities ranging from 5 ml to 5 liters and are usually calibrated to contain a specified volume when filled to the line etched on the neck. They are used in the preparation of standard solutions and the dilution of samples to known volumes

TABLE 30-5 Pipets

Name	Type of Calibration	Function	Available Capacities, ml	Type of Drainage
Volumetric	TD	Delivery of a fixed volume	1–200	Free drainage
Mohr	TD	Delivery of a variable volume	1–25	Drain to lower calibration line
Serological	TD	Same	0.1–10	Blow out last drop[a]
Serological	TD	Same	0.1–10	Drain to lower calibration line
Ostwald-Folin	TD	Delivery of a fixed volume	0.5–10	Blow out last drop[a]
Lambda	TC	To contain a fixed volume	0.001–2	Wash out with suitable solvent
Lambda	TD	Delivery of a fixed volume	0.001–2	Blow out last drop[a]

[a] A frosted ring near the top of recently manufactured pipets indicates that the last drop is to be blown out.

prior to taking aliquot portions with a pipet. Some are also calibrated on a "to deliver" basis. These are readily distinguishable by two reference lines; if delivery of the stated volume is desired, the flask is filled to the upper of the two lines.

MANIPULATIONS ASSOCIATED WITH THE MEASUREMENT OF VOLUME

Only clean glass surfaces will support a uniform film of liquid; the presence of dirt or oil will tend to cause breaks in this film. The appearance of water breaks is a certain indication of an unclean surface. Volumetric glass ware is carefully cleansed by the manufacturer before being supplied with calibrations; if these calibrations are to have meaning, the equipment must be kept equally clean when in use.

As a general rule, the heating of calibrated glass equipment should be avoided. Rapid cooling can permanently distort the glass and cause a change in volume.

When a liquid is confined in a narrow tube such as a buret or a pipet, the surface exhibits a marked curvature, called a *meniscus*. It is common practice to use the bottom of the meniscus as the point of reference in calibrating and using volumetric ware. This minimum can often be established more exactly if an opaque card or piece of paper is held behind the graduations (see Figure 30-18).

In judging volumes, the eye must be level with the liquid; otherwise, the reading will be in error due to *parallax*. Thus, if one's eye level is above that of the liquid, it will appear that a smaller volume has been taken than is actually the case. An error in the opposite direction can be expected if the point of observation is too low (see Figure 30-18).

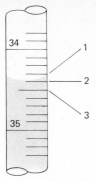

Figure 30-18   The Meniscus. Liquid level in the buret is 34.38 ml.

Directions for the Use of a Pipet. The following instructions pertain specifically to the manipulation of transfer pipets, but with minor modifications they may be used for other types as well. Liquids are usually drawn into pipets through the application of a slight vacuum. The mouth should not be used for suction because of the danger of accidentally ingesting liquids. Use of a rubber suction bulb or a rubber tube connected to an aspirator pump is strongly recommended (see Figure 30-19).

PROCEDURE

Cleaning. Pipets may be cleaned with a warm solution of detergent or with tepid cleaning solution (see p. 674). Draw in sufficient liquid to fill the bulb to about one-third of its capacity. While holding it nearly horizontal, carefully rotate the pipet so that all interior surfaces are wetted. Drain and then rinse thoroughly with distilled water. Inspect for water breaks, and repeat the cleaning cycle if necessary.

Measurement of an Aliquot. As in cleaning, draw in a small quantity of the liquid to be sampled, and thoroughly rinse the interior surfaces. Repeat with at least two more portions. Then carefully fill the pipet somewhat past the graduation mark. Quickly place a *forefinger* over the upper end of the pipet to arrest the outflow of liquid. Make certain that there are no bubbles in the bulk of the liquid or foam at the surface. Tilt the pipet slightly from the vertical, and wipe the exterior free of adhering liquid. Touch the tip of the pipet to the wall of a glass vessel (not the actual receiving vessel), and slowly allow the liquid level to drop by partially releasing the forefinger. Halt further flow as the bottom of the meniscus coincides exactly with the graduation mark. Then place the tip of the pipet well into the receiving vessel, and allow the sample to drain. When free flow ceases, rest the tip against an inner wall for a full 10 sec. Finally, withdraw the pipet with a rotating motion to remove any droplet still adhering to the tip. *The small volume remaining inside the tip is not to be blown or rinsed into the receiving vessel.*

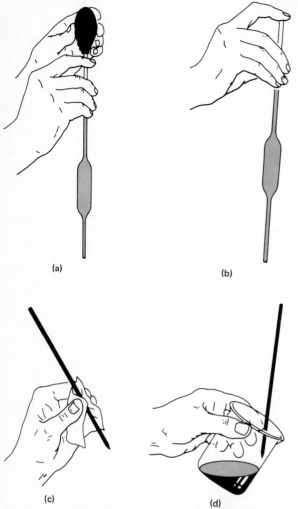

(a)

(b)

(c)

(d)

Figure 30-19 Technique for Use of a Volumetric Pipet. (a) Draw liquid past graduation mark. (b) Use forefinger to maintain liquid level above the graduation mark. (c) Tilt pipet slightly, and wipe away any drop on outside surface. (d) Allow pipet to drain freely.

This sequence is illustrated in Figure 30-19.

NOTES
1.   The liquid can best be held at a constant level in the pipet if one's forefinger is slightly moist; too much moisture, however, makes control difficult.
2.   It is good practice to avoid handling a pipet by the bulb.
3.   Pipets should be thoroughly rinsed with distilled water after use.

Directions for the Use of a Buret. Before being placed in service, a buret must be scrupulously clean. In addition, it must be established that the valve is liquid-tight.

PROCEDURE

Cleaning. Thoroughly clean the tube with detergent and a long brush. If water breaks persist after rinsing, clamp the buret in an inverted position with the end dipped in a beaker of cleaning solution. Connect a hose from the buret tip to a vacuum line. Gently pull the cleaning solution into the buret, stopping well short of the stopcock (Note 1). Allow to stand for at least 15 min and then drain. Rinse thoroughly with distilled water, and again inspect for water breaks. Repeat the treatment if necessary.

Lubrication of a Stopcock Buret. Carefully remove all old grease from the stopcock and barrel with a paper towel, and dry both parts completely. Lightly grease the stopcock, taking care to avoid the area near the hole. Insert the stopcock into the barrel, and rotate it vigorously. When the proper amount of lubricant has been used, the area of contact between stopcock and barrel appears nearly transparent, the seal is liquid-tight, and no grease has worked its way in to the tip.

NOTES
1. Cleaning solution often disperses more stopcock lubricant than it oxidizes and leaves a buret with a heavier grease film than before treatment. For this reason, cleaning solution should *not* be allowed to come in contact with lubricated stopcock assemblies.
2. Grease films unaffected by cleaning solution may yield to treatment with such organic solvents as acetone or benzene. Thorough washing with detergent should follow such treatment. The use of silicone lubricants is not recommended; contamination by such preparations is very difficult to remove.
3. As long as the flow of liquid is not impeded, fouling of the buret tip with lubricant is not a serious matter. Removal is best accomplished with organic solvents. A stoppage in the middle of a titration can be freed by *gently* warming the tip with a lighted match.
4. Before returning a buret to service after reassembly, it is advisable to test for leakage. Simply fill the buret with water, and establish that the volume reading does not change with time.

Filling. Make certain that the stopcock is closed. Add 5 to 10 ml of solution, and carefully rotate the buret to wet the interior completely. Allow the liquid to drain through the tip. Repeat this procedure two more times. Then fill the buret well above the zero mark. Free the tip of air bubbles by rapidly rotating the stopcock and allowing small quantities of solution to pass. Finally, lower the level of the solution to, or somewhat below, the zero marking. After allowing about a minute for drainage, take an initial volume reading.

Titration. Figure 30-20 illustrates the preferred method for manipulation of a stopcock. Any tendency for lateral movement of the stopcock will be in the direction of firmer seatings.

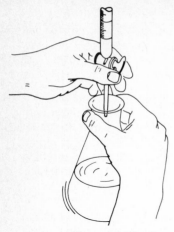

Figure 30-20   Recommended Technique for Manipulation of a Buret Stopcock.

With the tip well within the titration vessel, introduce solution from the buret in increments of 1 ml or so. Swirl (or stir) the sample constantly to assure efficient mixing. Decrease the volume of the additions as the titration progresses. In the immediate vicinity of the end point, add reagent a drop at a time. When it is judged that only a few more drops are needed, rinse down the walls of the titration vessel before completing the titration. Allow 30 sec to elapse between the last addition of reagent and the reading of the buret.

NOTES

1.   Many analysts, when unfamiliar with a particular titration, prepare an extra sample. No care is lavished on its titration since its functions are to reveal the nature of the end point and provide a rough estimate of titrant requirements. This deliberate sacrifice of one sample often results in an overall saving of time.

2.   Instead of being rinsed near the end of the titration, a flask can be carefully tipped and rotated so that the bulk of the liquid picks up any droplets adhering to the walls.

3.   Volume increments smaller than a normal drop may be taken by allowing a small volume of liquid to form on the tip of the buret and then touching the tip to the wall of the flask. This droplet is then combined with the bulk of the solution as in Note 2.

Directions for the Use of a Volumetric Flask. Before use, volumetric flasks should be washed with detergent and, if necessary, cleaning solution. Only rarely need they be dried. If required, however, drying is best accomplished by clamping the flasks in an inverted position. Insertion of a glass tube connected to a vacuum line will hasten the process.

PROCEDURE

Weighing Directly into a Volumetric Flask. Direct preparation of a standard solution requires that a known weight of solute be introduced into a volumetric flask. To minimize the possibility of loss during transfer, insert a powder funnel into the neck of the flask. The funnel is subsequently washed free of solid.

Dilution to the Mark. After transferring the solute, fill the flask about half-full, and swirl the contents to achieve solution. Add more solvent, and again mix well. Bring the liquid level almost to the mark, and allow time for drainage. Then use a medicine dropper to make such final additions of solvent as are necessary. Firmly stopper the flask, and invert repeatedly to assure uniform mixing. Finally, transfer the solution to a dry storage bottle.

NOTE

If, as sometimes happens, the liquid level accidently exceeds the calibration mark, the solution can be saved by correcting for the excess volume. Use a gummed label to mark the actual position of the meniscus. After the flask has been emptied, carefully refill to the graduation mark with water. Then, with a buret, determine the additional volume needed to duplicate the actual volume of the solution. This volume, of course, should be added to the nominal value for the flask when calculating the concentration of the solution.

CALIBRATION OF VOLUMETRIC WARE

The reliability of a volumetric analysis depends upon agreement between the volumes actually and purportedly contained (or delivered) by the apparatus. Calibration simply verifies this agreement if such already exists or provides the means for attaining agreement if it is lacking. The latter involves either the assignment of corrections to the existing volume markings or the striking of new markings that agree more closely with the nominal values.

In general, a calibration consists of determining the mass of a liquid of known density contained (or delivered) by the apparatus. Although this appears to be a straightforward process, a number of important variables must be controlled. Principal among these is the temperature, which influences a calibration in two ways. First and most important, the volume occupied by a given mass of liquid varies with temperature. Second, the volume of the apparatus itself is variable, owing to the tendency of the glass to expand or contract with changes in temperature.

We noted earlier (p. 688) that the effect of buoyancy upon weighing data is most pronounced when the density of the object is significantly less than that of the weights. As a general rule, a buoyancy correction must be applied to data where water is the calibration fluid.

Finally, the liquid employed for calibration requires consideration. Water is the liquid of choice for most work. Mercury is also useful, particularly where small volumes are involved. Because mercury does not wet glass surfaces, the volume contained by the apparatus will be identical with that which is delivered. In addition, the convex meniscus of mercury gives rise to a small correction that must be applied to give the corresponding volume for a liquid forming a concave meniscus. The magnitude of this correction is dependent upon the diameter of the apparatus at the graduation mark.

The calculations associated with calibrations, while not difficult, are somewhat involved. First, the raw weighing data are corrected for buoyancy, using Equation 30-3. Next, the volume of the apparatus at the temperature (t) of calibration is obtained by dividing the density of the liquid at that temperature into the corrected weight. Finally, this volume is corrected to the standard temperature of 20°C by means of Equation 30-4.

TABLE 30-6   **Volume Occupied by 1.000 g of Water Weighed in Air against Stainless Steel Weights[a]**

Temperature, t, °C	Volume, ml	
	At Temperature, t	Corrected to 20°C
10	1.0013	1.0016
11	1.0014	1.0016
12	1.0015	1.0017
13	1.0016	1.0018
14	1.0018	1.0019
15	1.0019	1.0020
16	1.0021	1.0022
17	1.0022	1.0023
18	1.0024	1.0025
19	1.0026	1.0026
20	1.0028	1.0028
21	1.0030	1.0030
22	1.0033	1.0032
23	1.0035	1.0034
24	1.0037	1.0036
25	1.0040	1.0037
26	1.0043	1.0041
27	1.0045	1.0043
28	1.0048	1.0046
29	1.0051	1.0048
30	1.0054	1.0052

[a] Corrections for buoyancy (stainless steel weights) and change in volume of the container have been applied.

Table 30-6 is provided to ease the computational burden of calibration. Corrections for buoyancy with respect to stainless steel and brass weights and for volume change of the water, as well as its glass container, have been incorporated into these data. Multiplication by the appropriate factor from the table converts the mass of water measured at some other temperature to the volume it would occupy at 20°C.

Example. A 25-ml pipet was found to deliver 24.976 g of water when calibrated against stainless steel weights at 25°C. Use the data in Table 30-6 to calculate the volume delivered by the pipet at this temperature and at 20°C.

At 25°C $V = 24.976 \times 1.0040 = 25.08$ ml
At 20°C $V = 24.976 \times 1.0037 = 25.07$ ml

General Directions for Calibration Work. All volumetric apparatus should be painstakingly freed of water breaks before being tested. Burets and pipets need not be dried. Volumetric flasks should be thoroughly drained.

The water used for calibration should be drawn well in advance of use to permit it to reach thermal equilibrium with its surroundings. This condition is best assured by noting the temperature of the water at frequent intervals and waiting until no further changes are observed.

An analytical balance can be used for calibrations involving 50 ml or less. Weighings need never be more reliable than the nearest milligram; this order of reproducibility is within the capabilities of most modern top-loading, single-pan balances. Weighing bottles or small, well-stoppered conical flasks are convenient receivers for small volumes.

PROCEDURE

Calibration of a Volumetric Pipet. Determine the empty weight of the receiver. Transfer a volume of water to the receiver with the pipet (p. 707), weigh the receiver and contents to the nearest milligram, and calculate the weight of water delivered from the difference in these weights. Repeat the calibration several times.

Calibration of a Buret. Fill the buret, and make certain that no bubbles are entrapped in the tip. Withdraw water until the level is at, or just below, the zero mark. Touch the tip to the wall of a beaker to remove any adhering drop. After allowing time for drainage, take an initial reading of the meniscus, estimating the volume to the nearest 0.01 ml. Allow the buret to stand for 5 min, and recheck the reading; if the stopcock is tight, there should be no noticeable change. During this interval, weigh (to the nearest milligram) a 125-ml conical flask fitted with a rubber stopper.

Once tightness of the stopcock has been established, run approximately 10 ml into the flask at about 10 ml/min. Touch the tip to the wall of the flask. Wait 1 min, record the volume, and refill the buret. Weigh the flask and its contents to the nearest milligram; the difference between this and the initial weight gives the mass of water actually delivered. Convert this mass into volume delivered, using Table 30-6. Compute the correction in this interval by subtracting the apparent volume from the true volume. The difference is the correction that must be applied to the *apparent volume* to give the *true volume*.

Starting again from the zero mark, repeat the calibration, using about 20 ml. Test the buret at 10-ml intervals over its entire length. Prepare a plot of the correction to be applied as a function of the volume delivered.

NOTES
1. Any correction larger than 0.10 ml should be verified by duplicate determinations before being accepted.
2. Corrections associated with any interval may be determined from the plot.

Calibration of a Volumetric Flask. Weigh the clean, dry flask, placing it on the right-hand pan of a Harvard trip balance. Set a beaker on the left pan, and add lead shot until balance is achieved. Remove the flask, and in its place substitute known weights until the same point of balance is reached (Note 2). Carefully fill the flask with water of known temperature until the meniscus coincides with the graduation mark. Return the flask to the right-hand pan and the beaker to the left. Repeat the process of counterweighing with lead shot, followed by substituting weights for the flask. The difference between the two weighings gives the mass of water contained by the apparatus. Calculate the corresponding volume with the aid of Table 30-6.

NOTES

1. A glass tube with one end drawn to a tip is useful in making final adjustments of the liquid level.
2. The substitution method described here is employed to eliminate any error arising from unequal arm lengths of the balance. If a top-loading, single-pan balance is available, the substitution technique is unnecessary.

Calibration of a Volumetric Flask Relative to a Pipet. The calibration of a flask relative to a pipet provides an excellent method for partitioning a sample into aliquots. The following directions pertain specifically to a 50-ml pipet and a 500-ml flask; other combinations are equally convenient.

With a 50-ml pipet, carefully transfer 10 volumes to a 500-ml volumetric flask. Mark the location of the meniscus with a gummed label. Coat the label with paraffin to assure permanence. When a sample is diluted to this mark, the same 50-ml pipet will deliver a one-tenth aliquot of the total sample.

The Laboratory Notebook

RULES FOR KEEPING THE LABORATORY NOTEBOOK

1. The notebook should be permanently bound with consecutively numbered pages.
2. Entries should be legible and well spaced from one another. Most notebooks have more than ample room, so crowding of data is unnecessary.
3. The first few pages of the notebook should be reserved for a table of contents, which should be conscientiously kept up to date.
4. *All data should be entered directly into the notebook, and in ink.*
    (a) Entries should be liberally identified with labels. If a series of weights refers to a set of empty crucibles, they should be labeled "Empty Crucible Weight" or something similar. The significance of an entry is obvious when it is recorded but may become unclear with the passage of time.
    (b) Each notebook page should be dated as it is used.
    (c) An erroneous entry should not be erased, nor should it be obliterated. Instead, it should be crossed out with a *single* horizontal line with the corrected entry located as nearby as possible. Numbers should never be written over; in time it may be difficult to decide what the correct number is.
    (d) Pages should not be removed from the notebook. It is sufficient to draw a single line diagonally across a page that is to be disregarded. A brief notation of the reason for striking out the page is useful.

SUGGESTED FORM

A satisfactory format involves the consecutive use of all pages for the recording of data. Upon completion of the analysis, the next pair of facing pages is used to summarize the results. The right-hand page should contain the following:

1. The title of the experiment—for example, *The Gravimetric Determination of Chloride.*

Gravimetric Determination of Chloride

The chloride in a soluble sample was precipitated as AgCl and weighed as such

Sample weights	1	2	3
Wt. bottle plus sample, g	27.6115	27.2185	26.8105
- less sample, g	27.2185	26.8105	26.4517
wt. sample, g	0.3930	0.4080	0.3588
Crucible weights, empty	~~20.7925~~	~~22.8311~~	~~21.2488~~
	20.7926	22.8311	~~21.2482~~
			21.2483
Crucible weights, with AgCl, g	~~21.4294~~	~~23.4920~~	~~21.8324~~
	~~21.4297~~	~~23.4914~~	21.8323
	21.4296	23.4915	
Weight of AgCl, g	0.6370	0.6604	0.5840
Percent Cl⁻	40.10	40.04	40.27
Average percent Cl⁻		40.12	
Relative standard deviation		3.0 parts per thousand	

Date Started   1-7-76
Date Completed 1-14-76

Figure 30-21   Sample Summary Page.

2. A brief statement of the principles upon which the analysis is based.
3. A summary of the data collected and the result calculated for each sample in the set.
4. A report of the best value for the set and a statement of the precision attained in the analysis.

A sample summary is shown in Figure 30-21. The left-hand page should show:

1. Equations for the principal reactions in the analysis.
2. An equation that shows the calculation employed in computing the results.
3. The calculations themselves.

SELECTED METHODS
OF ANALYSIS

This chapter contains specific instructions for the performance of selected chemical analyses. These are grouped according to the nature of the final measurement; Experiments 1–6 are gravimetric determinations, Experiments 7–32 are volumetric analyses, and Experiments 33–41 require the use of various instruments. Extensive references to discussions in earlier chapters have been provided. The student is urged to read these citations before starting laboratory work.

Gravimetric Methods of Analysis

EXPERIMENT 1:   Determination of Water in
Barium Chloride Dihydrate

The water in a crystalline hydrate, such as $BaCl_2 \cdot 2H_2O$, is readily determined gravimetrically. A sample is heated at a suitable temperature, and its water content is taken as the difference in its weight before and after heating. Less frequently, the evolved water is collected and weighed. Gravimetric methods based upon volatilization are briefly discussed in Chapters 6 and 27.

PROCEDURE

Throughout this experiment, perform all weighings to the nearest 0.1 mg.

Carefully clean two weighing bottles. Dry them for about 1 hr at 105 to 110°C; use a covered beaker to prevent accidental contamination (see Figure 30-9). After they have cooled to room temperature in a desiccator, determine the weight of each bottle. Repeat this cycle of heating, cooling, and weighing until successive weighings agree within 0.2 mg. Next introduce a quantity of unknown into each bottle, and reweigh. Heat the samples for about 2 hr at 105 to 110°C; then cool and weigh as before. Repeat the heating cycle until constant weights for the bottles and their contents have been attained. Report the percentage of water in the sample.

NOTES
1. If the unknown consists of the pure dihydrate, $BaCl_2 \cdot 2H_2O$, take samples weighing approximately 1 g. If the unknown is a mixture of the dihydrate and some anhydrous diluent, obtain the proper sample size from the instructor.
2. Barium chloride can be heated to elevated temperatures without danger of decomposition. If desired, the analysis can be performed in crucibles with a Bunsen flame as the source of heat.
3. Magnesium sulfate heptahydrate can be used instead of barium chloride; heating to 140°C is needed to eliminate the last traces of moisture.

EXPERIMENT 2: Determination of Chloride in a Soluble Sample

The chloride content of a soluble salt can be determined by precipitation as silver chloride:

$$Ag^+ + Cl^- \rightarrow AgCl(s)$$

The precipitate is collected in a filtering crucible, washed, and brought to constant weight by drying at 105 to 110°C. Precipitation is carried out in acid solution to eliminate potential interference from anions of weak acids (for example, CO_3^{2-}) which form precipitates with silver in neutral media. A moderate excess of silver ion is required to diminish the solubility of the precipitate; a large excess will lead to serious coprecipitation and should be avoided.

Silver chloride precipitates first as a colloid; it is coagulated with heat in the presence of a relatively high electrolyte concentration. A small quantity of nitric acid is added to the wash liquid to maintain the electrolyte concentration and prevent peptization during washing; the acid is volatilized during the subsequent heat treatment. See pages 127–131 for further information concerning the characteristics of colloidal precipitates.

In common with other silver halides, silver chloride is susceptible to photodecomposition, the reaction being

$$AgCl(s) \rightarrow Ag(s) + \tfrac{1}{2}Cl_2(aq)$$

The precipitate acquires a violet color due to the accumulation of finely divided silver. If photochemical decomposition occurs in the presence of excess silver ion, the additional reaction,

$$3Cl_2(aq) + 3H_2O + 5Ag^+ \rightarrow 5AgCl(s) + ClO_3^- + 6H^+$$

will cause the analytical results to be high. In the absence of silver ion, the results will be low. Dry silver chloride is virtually unaffected by exposure to light.

Unless elaborate precautions are taken, some photochemical decomposition of silver chloride is unavoidable; however, with reasonable care this effect will not induce an appreciable error in the analysis.

Iodide, bromide, and thiocyanate, if present in the sample, will be precipitated along with silver chloride and cause high results. In addition, the chlorides of tin and antimony are likely to hydrolyze and precipitate under the conditions of the analysis.

PROCEDURE

Clean (Note 1) and dry three fritted glass or porcelain filtering crucibles at 105 to 110°C; supply each with identifying marks, and bring them to constant weight during periods of waiting in the analysis.

Dry the sample at 105 to 110°C for 1 to 2 hr in a weighing bottle (see Figure 30-9). Store in a desiccator while cooling. Weigh (to the nearest 0.1 mg) individual 0.4-g samples into 400-ml beakers (p. 693). To each, add about 200 ml of distilled water and 3 to 5 ml of 6-F nitric acid. Slowly, and with good stirring, add 5% silver nitrate to the cold solution until the precipitate is observed to coagulate (Note 2); then add an additional 3 to 5 ml. Heat almost to boiling, and digest the precipitate at this temperature for about 10 min. Check for completeness of precipitation by adding a few drops of silver nitrate to the supernatant liquid; should additional silver chloride appear, continue the addition of silver nitrate until precipitation is complete. Store in a dark place for at least 1 to 2 hr, preferably until the next laboratory period. Then decant (p. 695) the supernatant through a weighed filtering crucible. Wash the precipitate several times (while it is still in the beaker) with a cold solution consisting of 2 to 5 ml of 6-F nitric acid per liter of distilled water; decant these washings through the filter also. Finally, transfer the bulk of the precipitate to the crucible, using a rubber policeman to dislodge any particles that adhere to the walls of the beaker. Continue washing until the filtrate is found to be substantially free of silver ion (Note 3). Dry the precipitates at 105 to 110°C for about 1 hr. Store crucibles in a desiccator until they have cooled to room temperature. Determine the weight of the crucibles and their contents. Repeat the cycle of heating, cooling, and weighing until constant weight has been achieved. Report the percentage of chloride in the sample.

NOTES

1. Residual silver chloride can be removed from filtering crucibles by soaking in a strong solution of sodium thiosulfate.

2. To determine the approximate amount of silver nitrate needed, calculate the volume that would be required if the sample were pure sodium chloride.

3. Washings are readily tested for their silver content by collecting a few milliliters in a test tube and treating with a few drops of hydrochloric acid. Washing is judged complete when little or no turbidity is observed with this test.

EXPERIMENT 3:   Analysis of a Soluble Sulfate

The analysis of a soluble sulfate is based upon precipitation with barium ion

$$Ba^{2+} + SO_4^{2-} \rightarrow BaSO_4(s)$$

The barium sulfate is collected on a suitable filter, washed with water, and strongly ignited. See Chapter 6 for further discussion concerning the characteristics of crystalline precipitates.

Superficially, this method appears straightforward. In fact, however, it is subject to numerous interferences, due chiefly to the tendency of barium sulfate to occlude foreign anions and cations. Table 31-1 summarizes the more common interferences affecting this analysis. Purification by reprecipitation is not feasible because there is no practical solvent for barium sulfate. It is therefore necessary to eliminate the principal interferences by preliminary treatment of the sample and then to precipitate the barium sulfate from hot, dilute solutions. Even so, the excellent agreement often observed between theoretical and experimental results is due in considerable measure to a cancellation of errors.

PROCEDURE

Dry the unknown (see Figure 30-9) for at least 1 hr at 105 to 110°C; allow it to cool to room temperature in a desiccator. Weigh (to the nearest 0.1 mg) individual 0.5- to 0.7-g samples into 400-ml beakers (p. 693). Dissolve each in 200 ml of distilled water to which 4 ml of 6-F hydrochloric acid have been added.

TABLE 31-1   Interferences Attending the Gravimetric Determination of Sulfate as $BaSO_4$

Effect upon Analysis		Nature of Interference
Low results	1.	Excessive amounts of mineral acid present. (Solubility of $BaSO_4$ is appreciably greater in strong acid media.)
	2.	Coprecipitation of sulfuric acid. (Note that this is a source of error in a gravimetric determination of sulfate but not of barium, since this H_2SO_4 is driven off during ignition.)
	3.	Coprecipitation of alkali metal and calcium ions. (Sulfates of these ions weigh less than the equivalent amount of $BaSO_4$, which should have formed.)
	4.	Coprecipitation of ammonium ion. (Ammonium sulfate is volatilized upon ignition of the precipitate.)
	5.	Coprecipitation of iron as a basic iron(III) sulfate.
	6.	Partial reduction of $BaSO_4$ to BaS if filter paper is charred too rapidly.
	7.	Presence of trivalent chromium. [May not achieve complete precipitation of $BaSO_4$ owing to formation of soluble complex sulfates of chromium(III).]
High results	1.	Absence of mineral acid. (The slightly soluble carbonate or phosphate of barium can precipitate under these conditions.)
	2.	Coprecipitation of barium chloride.
	3.	Coprecipitation of anions, particularly nitrate and chlorate as barium salts.

For each sample, dissolve 1.3 g of barium chloride dihydrate in 100 ml of distilled water, and filter if necessary. Heat nearly to boiling before quickly adding, with vigorous stirring, to hot solutions of the sample.

Digest the precipitated barium sulfate for 1 to 2 hr (see Note 1). Decant the hot supernatant through a fine ashless paper (p. 698 and Note 2). Wash the precipitate three times with hot water, decanting the washings through the filter. Finally, transfer the precipitate to the paper (p. 700). Place papers and contents in marked porcelain crucibles that have been ignited to constant weight; gently char off the papers (p. 700). Ignite to constant weight at 900°C. Report the percentage of sulfate in the sample.

NOTES
1. The digested precipitates can be allowed to stand until the following laboratory period without harm.
2. Use of Schleicher and Schuell No. 589 Blue Ribbon or Whatman No. 42 paper is recommended. If desired, the precipitate can be collected in a Gooch crucible or a porcelain filtering crucible. Complete removal of $BaSO_4$ from the latter type crucible is difficult, however.

EXPERIMENT 4: Analysis of Iron in a Soluble Sample

The analysis of iron in a soluble sample is based upon the precipitation of iron(III) as a hydrous oxide, followed by ignition in Fe_2O_3:

$$2Fe^{3+} + 6NH_3 + (x + 3)H_2O \rightarrow Fe_2O_3 \cdot xH_2O(s) + 6NH_4^+$$

$$Fe_2O_3 \cdot xH_2O(s) \rightarrow Fe_2O_3(s) + xH_2O(g)$$

The hydrous oxide forms as a gelatinous mass that rapidly clogs the pores of most filtering media; for this reason, a very coarse grade of ashless paper is employed. Even here, however, it is best to delay transfer of the precipitate as long as possible and to wash by decantation.

The use of Fe_2O_3 as a weighing form requires that all iron present be in the +3 state, a condition that is readily achieved by treating the sample with nitric acid before precipitating the hydrous oxide:

$$3Fe^{2+} + NO_3^- + 4H^+ \rightarrow 3Fe^{3+} + NO + 2H_2O$$

The complex $FeSO_4 \cdot NO$ occasionally imparts a very dark color to the solution; it decomposes upon further heating.

Since it has negligible solubility, the precipitate can be safely washed with a hot solution of ammonium nitrate.

PROCEDURE

Clean and supply identifying markings to three porcelain crucibles and covers; bring these to constant weight during periods of waiting throughout the analysis.

1. Double-Precipitation Method. Unless directed otherwise, do not dry the sample. Consult the instructor for the proper sample size to be weighed into 600-ml beakers (p. 693). Dissolve in 20 to 30 ml of distilled water to which about 5 ml of concentrated hydrochloric acid have been added. Then add 1 to

2 ml of concentrated nitric acid; heat gently to complete the oxidation and remove any oxides of nitrogen. Dilute to 350 or 400 ml, and slowly add, with good stirring, freshly filtered aqueous ammonia until precipitation is complete (Notes 1 and 2). Digest briefly, and then check for completeness of precipitation with a few additional drops of ammonia. Decant the warm, clear supernatant liquid through ashless filter paper (Note 3), and wash the precipitate in the beaker with two 30-ml portions of hot 1% ammonium nitrate solution (see pp. 698–699 for information concerning the filtration and washing of precipitates).

Return the filter papers to their appropriate beakers. Add about 5 ml of concentrated hydrochloric acid to each, and macerate the paper thoroughly with a stirring rod. Dilute to about 300 ml with distilled water, and reprecipitate the hydrous iron(III) oxide as before. Again decant the filtrate through ashless filter paper, and wash the precipitate repeatedly with hot 1% ammonium nitrate. When the filtrate gives little or no test for chloride ion (Note 4), transfer the precipitate quantitatively to the filter cone. Remove the last traces of precipitate adhering to the walls of the beakers by scrubbing with a small piece of ashless paper. Allow the precipitate to drain overnight, if possible. Then transfer the paper and contents (p. 700) to a porcelain crucible that has previously been ignited to constant weight. Char the paper at low temperature, taking care to allow free access of air (Note 5). Gradually increase the temperature until all carbon has been burned away. Then ignite to constant weight at 900 to 1000°C.

NOTES

1.   Aqueous ammonia solutions attack the glass of their containers upon prolonged contact and thereby become contaminated with particles of silica. It is a wise precaution to filter the ammonia prior to use.
2.   Aqueous ammonia should be added until its odor is unmistakable over the solution. The precipitate should appear reddish brown. If it is black (or nearly so), the presence of iron(II) is indicated; the sample is best discarded.
3.   A porous grade of ashless paper is required for this gelatinous precipitate. Schleicher and Schuell No. 589 Black Ribbon or Whatman No. 41 is a satisfactory paper. The time required for filtration is shortened by performing the principal washing of the precipitate while it is still in the beaker.
4.   Before making a test for chloride in the washings, the sample of filtrate taken must be first acidified with dilute nitric acid.
5.   Heat lamps are convenient for the initial charring of the filter papers.

2. *Single-Precipitation Method.* Proceed as directed in the first paragraph of the double-precipitation method, making the following changes:

a. Instead of washing with only two 30-ml portions of 1% ammonium nitrate, wash repeatedly until the decantate is essentially free of chloride.

b. Then transfer the precipitate quantitatively to the filter and complete the analysis as before.

EXPERIMENT 5:   Precipitation of Aluminum from
                Homogeneous Solution

Aluminum may be determined by precipitation as a hydrous oxide followed by ignition to Al_2O_3. Iron(III), if present, reacts analogously and causes the results of the analysis to be high. An effective scheme for separating these cations is based upon differences in their behavior with respect to thiocyanate ion. Iron(III) forms stable anionic complexes in the presence of high concentrations of thiocyanate while aluminum(III) does not. After being treated with a relatively large quantity of thiocyanate, the sample is passed through a column containing an anion-exchange resin (Chapter 29). Aluminum ion passes through while iron(III) is retained on the column as a thiocyanate complex.

The accompanying directions were adapted from data reported by Gordon, Salutsky, and Willard.[1] After separation from iron(III), aluminum is precipitated as a basic succinate with hydroxide ions homogeneously generated by the decomposition of urea (Chapter 6).

PROCEDURE

To receive the sample, submit a clean, labeled, 500-ml volumetric flask. Dilute the sample to volume (p. 711).

Mark three crucibles and covers. Bring these to constant weight by ignition at 1000°C.

Prepare 1 liter of 0.30-F NH$_4$SCN wash solution, adjusting this to pH 1.0 $\pm$ 0.2 with dilute HCl or NH$_3$ (Note 3). *Do not continue past this point unless there is time to complete the steps in the next two paragraphs during the same day.*

Withdraw three 50-ml aliquots (p. 707) of the sample solution, and transfer to individual 250-ml beakers. Add 5 g of NH$_4$SCN to each sample, and adjust the pH to 1.0 with dilute HCl or NH$_3$, as needed (Note 3).

Wash the resin in an anion-exchange column with 50 ml of 0.30-F NH$_4$SCN (Note 2). Adjust the flow rate to 8 to 10 ml per minute (no faster than one drop per second); discard this wash liquid. Then pass the aliquot, prepared as in the previous paragraph, through the column at the same rate, collecting the eluate in a 600-ml beaker. Wash the column with 250 ml of 0.30-F NH$_4$SCN solution, which should be added in increments of about 15 ml, and collect the eluate in the same beaker as the sample. *Never allow the solution level to fall below the top level of the resin.* Treat the other two samples identically. When finished with the column, leave the solution level at least 3 cm above the resin level.

For each sample, dissolve 5 g of succinic acid, 10 g of NH$_4$Cl, and 4 g of urea in a minimum volume of water; filter, if necessary, to remove any undissolved solids. Then add this solution to the sample, bring the volume to about 400 ml with distilled water, and adjust the pH to 3.1 to 3.4 with dilute HCl or NH$_3$ (Notes 3 and 4).

Provide each beaker with a special stirring rod (Note 5), and heat the solution to boiling over a burner. Locate the flame directly beneath the stirring rod so that a stream of bubbles emanates from the end of the rod. Use considerable care while heating, for the solution may tend to bump when the precipitate forms. Boil the solution gently for 2 hr *after* appearance of the first

[1] L. Gordon, M. L. Salutsky, and H. H. Willard, *Precipitation from Homogeneous Solution.* New York: John Wiley & Sons, Inc., 1959.

opalescence that develops 15 to 30 min after heating has commenced. Keep the solution level constant during boiling by the periodic addition of distilled water.

Prepare 1 liter of 1% succinic acid, adding sufficient NH_3 to make this solution just alkaline to methyl red. The solution should be heated almost to boiling before use in washing the precipitate.

After precipitation is complete, decant the supernatant liquid through ashless paper suitable for collection of fine, crystalline solids (Note 6). Transfer the precipitate quantitatively to the paper, and wash with hot 1% succinic acid until essentially free of chloride ion.

To remove the last traces of precipitate from beaker and stirring rod, add about 10 ml of dilute HCl, cover the beaker with a watch glass, and gently reflux for about 5 min. Add NH_3 until the solution is faintly alkaline to methyl red, warm for a few minutes, and then filter through a fresh cone of ashless paper. Wash with hot 1% succinic acid, as before.

Place the combined precipitates for each sample into individual porcelain crucibles that have been brought, with their covers, to constant weight. Gently dry, and then burn off the papers. Finally ignite at 1000°C. Weigh the ignited Al_2O_3 in the covered crucible. Repeat the ignition step until constant weight has been attained (Note 7).

Report the number of milligrams of Al(III) contained in the original sample.

NOTES

1.   A stock solution for unknowns may be prepared by dissolving 50.00 g of reagent-grade aluminum and 0.15 to 0.2 g of $FeCl_3$ in 6-*F* HCl; dilute to 1.000 liter. Unknowns, consisting of 16 to 25 ml of this solution, are issued from a buret into 500-ml volumetric flasks. The stock solution is somewhat viscous. Several minutes should be allowed for drainage between delivery of the sample and reading of the buret.

2.   A 30-cm column of 20- to 50-mesh Amberlite IRA 400 is satisfactory for removal of iron(III). The column will give satisfactory service until it is completely discolored with the iron(III)-thiocyanate complex. The resin can be regenerated with 2-*F* HCl; a substantial volume of acid is required.

3.   A pH meter is convenient for adjustment of the acidity. When using this instrument, remember that (a) the glass electrode is very fragile and (b) a few drops adhere to the electrodes when they are removed from solution. These drops should be washed into the proper beaker with distilled water.

4.   When adjusting samples to pH 3.1 to 3.4, add NH_3 slowly and with good stirring. Be sure the solution is free of precipitate when the proper pH has been achieved; if any traces remain, dissolve with dilute HCl, and readjust the pH.

Adjustment of the pH can be achieved by adding NH_3 until the first permanent traces of precipitate appear and then discharging this turbidity with a minimum volume of 1-*F* HCl.

5.   In addition to a cover glass, each beaker should be supplied with a special stirring rod to aid in the prevention of bumping. The end of an ordinary stirring rod is heated to the softening point and then firmly pressed upon the point of a thumbtack. The indentation produces a stream of bubbles which significantly reduces the danger of loss due to local overheating.

6.   Whatman No. 40 or Schleicher and Schuell No. 589 White Ribbon is satisfactory for this filtration. See page 700 for folding instructions. The filtering process is greatly hastened through use of hot solutions.

7.   Ignite the precipitates in the full flame of a Meker burner (or in a furnace at 1000°C) for at least 40 min. Subsequent heatings need not exceed 20 to 30 min.

After each heating, allow the crucibles to cool on an asbestos mat for 3 to 4 min before transferring them to a desiccator. Allow a uniform cooling time (25 to 30 min) in the desiccator before each weighing.

Crucible covers should be in place during all cooling and weighing operations.

EXPERIMENT 6:   Determination of Nickel in Steel

The nickel in a steel sample can be precipitated from a slightly alkaline medium with an alcoholic solution of dimethylglyoxime. Tartaric acid is introduced to prevent interference from iron. The organic nickel compound serves as a convenient weighing form.

Owing to the bulky character of the precipitate, only a small quantity of nickel can be conveniently handled. The sample weight taken is governed by this consideration. The excess of precipitating agent must be controlled, not only because its solubility in water is low but also because the nickel compound becomes appreciably more soluble as the alcohol content of precipitating medium is increased.

PROCEDURE

Weigh individual samples containing between 30 and 35 mg of nickel into 400-ml beakers, and dissolve by warming with about 50 ml of 6-F hydrochloric acid. Carefully introduce about 15 ml of 6-F nitric acid, and boil gently to expel the oxides of nitrogen. Dilute the resulting solution to 200 ml, and heat nearly to boiling. Introduce 5 to 6 g of tartaric acid (Note 1), and neutralize with aqueous ammonia until a faint odor of ammonia can be detected in the vapors over the solution; add 1 to 2 ml in excess. If the solution is not clear at this stage, proceed as directed in Note 2. Make the solution slightly acidic with hydrochloric acid, heat to 60 to 80°C, and add 20 ml of a 1% alcoholic solution of dimethyl-glyoxime. Then, with good stirring, introduce sufficient dilute ammonia until a slight excess is present as indicated by the odor, plus an additional 1 to 2 ml. Digest for 30 to 60 min at about 60°C, cool for at least 1 hr, and filter through filtering crucibles that have been previously brought to constant weight. Wash with water until free of chloride. Finally, bring crucibles and contents to constant weight by drying at 110 to 120°C. Report the percentage of nickel in the sample. The precipitate, $NiC_8H_{14}O_4N_4$, contains 20.31% nickel.

NOTES

1.   Tartaric acid is conveniently introduced as a concentrated solution (25 g diluted to 100 ml). If this solution is not clear, it should be filtered prior to use.
2.   If a precipitate is formed upon the addition of base, the solution should be acidified, treated with additional tartaric acid, and again made alkaline. Alternatively, the precipitate can be removed by filtration. Thorough washing of the entire filter paper with a hot, dilute NH_3/NH_4Cl solution is required; the washings should be combined with the remainder of the sample.
3.   Gooch crucibles, or filtering crucibles of porcelain or fritted glass, may be used for collection of the precipitate.

Volumetric Precipitation Analysis with Silver Ion

Argentometric methods of analysis are discussed in Chapter 8.

Specific directions follow for the argentometric determination of chloride ion by the Mohr, the Volhard, and the Fajans methods. These may be used with little or no modifications for the titration of other anions; see Table 8-2.

SILVER NITRATE AND ITS SOLUTIONS

Silver nitrate may be obtained in primary-standard purity. It has a high equivalent weight and is readily soluble in water. Both the solid and the aqueous solution must be scrupulously protected from dust and other organic materials and from sunlight. Metallic silver is produced by chemical reduction in the former instance and photodecomposition in the latter. The reagent is expensive.

Silver nitrate crystals may be freed of surface moisture by drying at 110°C for about 1 hr. Some discoloration of the solid may result, but the amount of decomposition occurring in this time is ordinarily negligible.

EXPERIMENT 7: Preparation of Standard 0.1-F Silver Nitrate

Use a laboratory balance to weigh approximately 17 g of silver nitrate into a clean, dry weighing bottle. Heat bottle and contents (see Figure 30-9) for 1 hr, but not much longer, at 110°C; store in a desiccator. When cooled, determine the weight (to the nearest milligram) with an analytical balance. Carefully transfer the bulk of the solid into a powder funnel held in the neck of a 1-liter volumetric flask. Reweigh the bottle to obtain the weight of solid by difference. Wash the silver nitrate into the volumetric flask with copious amounts of distilled water. Remove and rinse the funnel; fill the flask half-full of water, and swirl until the crystals are completely dissolved. Dilute to the mark, and *mix the solution thoroughly* by upending the stoppered flask several times. Compute the normality of the solution.

NOTES

1. Weighing silver nitrate to the nearest milligram will incur a maximum error of only 1 part in 17,000 in the value of the normality; since uncertainties in the subsequent analyses commonly exceed this figure, more accurate weighing is of no value.
2. If desired, silver nitrate solutions prepared to approximately the desired strength can be standardized against sodium chloride.
3. Once prepared, and when not actually in use, $AgNO_3$ solutions should be stored in a dark place.

EXPERIMENT 8: Determination of Chloride by the Mohr Method

The Mohr method is discussed on page 179.

Dry the unknown for 1 hr at 100 to 110°C. Carefully weigh 0.25- to 0.35-g samples to the nearest 0.1 mg; dissolve each in about 100 ml of water. Add a pinch of calcium carbonate or sodium hydrogen carbonate, making further additions, if necessary, until effervescence ceases. Then introduce 1 to 2 ml of 5% potassium chromate, and titrate with standard silver nitrate solution to the first permanent appearance of a buff color due to silver chromate. Determine an indicator blank by suspending a small quantity of $CaCO_3$ in about 100 ml of water containing 1 to 2 ml of 5% K_2CrO_4; use the color developed in this determination as a standard for judgment of the end point in the actual titrations.

NOTE

Mohr titrations should be performed at room temperature. Elevated temperatures significantly increase the solubility of silver chromate; its sensitivity as an indicator for this titration undergoes a corresponding decrease.

EXPERIMENT 9:   Determination of Chloride by the Volhard Method

The Volhard method is discussed on page 180.

As applied to the analysis of chloride ion, the Volhard method involves the addition of a measured excess of standard silver nitrate to the sample and back-titration of the excess with a standard potassium thiocyanate solution.

1. Preparation and Standardization of 0.1-F Potassium Thiocyanate. Dissolve approximately 9.8 g of KSCN in about 1 liter of water. *Mix well.*

Measure 25- to 30-ml samples (to the nearest 0.01 ml) of standard $AgNO_3$ solution into conical flasks, and dilute to approximately 100 ml. Add about 5 ml of 6-*F* HNO_3, followed by 5 ml of iron(III) ammonium sulfate indicator (Note 3). Titrate with the KSCN solution, swirling the flask vigorously, until the red-brown color of $FeSCN^{2+}$ is permanent for 1 min. Calculate the concentration of the KSCN solution. Results from duplicate standardizations should show agreement within 2 to 3 ppt; if this precision has not been attained, perform further standardization titrations.

NOTES

1.   Potassium thiocyanate is usually somewhat moist; the direct preparation of a standard solution is not ordinarily attempted. However, Kolthoff and Lingane[2] report that gentle fusion followed by storage over calcium chloride yields a product that can be used for the direct preparation of standard solutions.

2.   A potassium thiocyanate solution retains its titer over extended periods of time.

3.   The indicator is readily prepared by dissolving about 10 g of iron(III) ammonium sulfate, $NH_4Fe(SO_4)_2 \cdot 12H_2O$, in 100 ml of freshly boiled 6-*F* nitric acid. The acid prevents hydrolysis of the iron(III) ion.

2. The Analysis of Chloride. Dry the unknown at 100 to 110°C for 1 hr. Weigh several 0.25- to 0.35-g samples to the nearest 0.1 mg into numbered 250-ml conical flasks. Dissolve each sample in 100 ml of distilled water, and acidify with 5 ml of 6-*F* HNO_3. Introduce an excess of standard silver nitrate, and be sure to note the volume taken. Add 5 ml of iron(III) indicator and 5 ml of chloride-free nitrobenzene. Shake vigorously. Titrate the excess silver with standard thiocyanate until the color of $FeSCN^{2+}$ is permanent for 1 min.

NOTES

1.   With the concurrence of the instructor, a larger quantity of unknown can be weighed into a volumetric flask and diluted to known volume. The determination can then be made upon aliquot portions of this solution.

2.   To obtain an approximation of the volume of standard $AgNO_3$ that constitutes an excess, calculate the amount that would be required for one of the samples, assuming that it is 100% NaCl. When actually adding silver solution to the sample, swirl the flask vigorously, and add 3 or 4 ml in excess of the volume required to cause the AgCl to coagulate.

3.   Nitric acid is introduced to improve observation of the end point. Since the lower oxides of nitrogen tend to attack thiocyanate, the acid should be freshly boiled.

4.   Nitrobenzene is a hazardous chemical and must be handled with respect. Poisoning can result not only from prolonged breathing of its vapors but also from absorption of the liquid through the skin. In the event of spillage on one's person, the affected areas should be promptly and thoroughly washed with soap and warm water. Clothing soaked with the liquid should be removed and laundered.

[2] I. M. Kolthoff and J. J. Lingane, *J. Amer. Chem. Soc.,* **57,** 2126 (1935).

5.   At the outset of the back-titration, an appreciable quantity of silver ion is adsorbed on the surface of the precipitate. As a result, there is a tendency for a premature appearance of the end-point color. Since success of the method depends upon an accounting for all excess silver ions, thorough and vigorous agitation is essential to bring about desorption of this ion from the precipitate. A magnetic stirrer is helpful for this purpose.

OTHER APPLICATIONS AND LIMITATIONS OF THE VOLHARD METHOD

As indicated in Table 8-2, the Volhard method may be employed for the determination of numerous substances. Where the solubility of the silver salt formed is increased by the presence of strong acid, a filtration is mandatory prior to back-titration. The use of nitrobenzene (or filtration) can be omitted when the salts produced are less soluble than silver thiocyanate in acid media; silver bromide and silver iodide are the only common examples.

The Volhard method cannot be employed in the presence of oxidizing agents because of the susceptibility of thiocyanate to attack. It is often possible to eliminate this source of interference by prior treatment of the sample with a reducing agent. The method also fails in the presence of cations that form slightly soluble thiocyanates, notably palladium and mercury. Again, preliminary treatment may eliminate these sources of interference.

EXPERIMENT 10:   Determination of Chloride by the Fajans Method

The Fajans method employs a direct titration with dichlorofluorescein as the indicator; dextrin is added to maintain the silver chloride in the colloidal state. Only a standard silver nitrate solution is required.

Dry the unknown for 1 hr at 100 to 110°C. Weigh individual 0.20- to 0.29-g samples (to the nearest 0.1 mg) into 500-ml flasks. Dissolve in 175 to 200 ml of distilled water. Introduce 10 drops of dichlorofluorescein solution and about 0.1 g of dextrin; immediately titrate with standard silver nitrate to the first permanent appearance of the pink color of the indicator.

NOTES

1.   Kolthoff[3] states that the chloride concentration should be within the range of 0.025 to 0.005 *M*. These directions yield solutions approaching this maximum concentration only when the sample taken is pure sodium chloride at the upper weight limit. If the approximate percentage of chloride in the sample is known, a corresponding decrease in volume or increase in sample size can be tolerated.
2.   Silver chloride is particularly sensitive to photodecomposition in the presence of the indicator; the titration will fail if attempted in direct sunlight. Where this problem exists, the approximate equivalence point should first be ascertained by a trial titration, this value being used to calculate the volume of silver nitrate required for the other samples. The addition of indicator and dextrin should be delayed until the bulk of the silver nitrate has been added to subsequent samples, after which the titration should be completed without delay.
3.   The indicator solution may be prepared as a 0.1% solution of dichlorofluorescein in 70% alcohol or as a 0.1% aqueous solution of the sodium salt.

[3] I. M. Kolthoff, W. M. Lauer, and C. J. Sunde, *J. Amer. Chem. Soc.*, **51**, 3273 (1929).

Neutralization Titrations in Aqueous Solution

The theory of neutralization titrations is discussed in Chapters 9 and 10; practical aspects and typical applications are considered in Chapter 11.

EXPERIMENT 11: Preparation and Standardization of Reagents for Neutralization Titrations

PREPARATION OF INDICATOR SOLUTIONS

Acid-base indicators exist for virtually any pH range between 1 and 13.[4] The theory of indicator behavior is discussed in Chapter 9. Directions follow for the preparation of indicator solutions that will permit the performance of most common analyses.

1. Stock solutions generally contain 0.5 to 1.0 g of indicator per liter of solution.

Methyl orange and methyl red. Dissolve the sodium salt directly in distilled water.

Phenolphthalein and thymolphthalein. Dissolve the solid indicator in a solution that is 80% ethyl alcohol by volume.

Sulfonphthaleins. Dissolve the sulfonphthaleins in water by adding sufficient NaOH to react with the sulfonic acid group of the indicator. To prepare stock solutions, triturate 100 mg of the solid indicator with the specified volume of 0.1-N NaOH; then dilute to 100 ml with distilled water. The volumes of base (ml) required are as follows: *bromocresol green,* 1.45; *bromothymol blue,* 1.6; *bromophenol blue,* 1.5; *thymol blue,* 2.15; *cresol red,* 2.65; *phenol red,* 2.85. The sodium salts of several sulfonphthaleins are available commercially. These substances can be dissolved directly in water.

PREPARATION OF SOLUTIONS FOR NEUTRALIZATION TITRATIONS

The reagents that are commonly used for volumetric neutralization titrations are discussed in Chapter 11. Directions follow for the preparation of standard hydrochloric acid solutions by two different methods, the one requiring a standardization, the other involving dilution of a weighed quantity of the constant-boiling reagent to a known volume. Directions for preparation of carbonate-free base solutions are also provided.

2. Preparation of Approximately 0.1-N HCl. Add about 8 ml of concentrated HCl to approximately 1 liter of distilled water. Mix thoroughly; store in a glass-stoppered bottle.

NOTE

For very dilute HCl solutions, it is advisable to eliminate CO_2 from the water by a preliminary boiling.

[4] See, for example, L. Meites, *Handbook of Analytical Chemistry*, p. **3**-35. New York: McGraw-Hill Book Company, Inc., 1963.

3. Preparation of Constant-Boiling HCl. Prepare a stock hydrochloric acid solution by diluting approximately 600 ml of the concentrated acid to 1 liter. Place the acid in a distillation flask, and distill at a rate of 3 to 4 ml/min. Reject the first three-fourths of the distillate; collect the remainder in a clean dry container until only 50 to 60 ml remain in the distilling flask. Note the barometric pressure. Transfer the distillate to a clean, dry bottle fitted with a tight stopper. The composition of this acid can be determined from published tables[5] or by means of the equation given on page 239.

4. Preparation of 0.1-N HCl from Constant-Boiling HCl. Clean and dry a 50-ml glass-stoppered flask, and weigh to the nearest 5 mg. Transfer approximately 18 g of the constant-boiling mixture to the flask, being careful to avoid wetting the ground-glass surfaces. Stopper the flask tightly and reweigh. Immediately transfer the contents quantitatively to a calibrated 1-liter volumetric flask. Dilute to volume with distilled water; mix the contents thoroughly. Calculate the normality of the acid.

5. Preparation of Carbonate-Free 0.1-N NaOH. If directed by the instructor, prepare a bottle for protected storage as in Figure 11-1. Boil approximately 1 liter of distilled water; after it has cooled, transfer the water to the storage bottle. Decant 4 to 5 ml of commercial 50% NaOH, add to the water, and *mix thoroughly*. Protect the solution from unnecessary contact with the atmosphere.

6. Determination of Acid-Base Ratio. Establishment of the volume ratio between an acid and a base makes it possible to calculate the normality of one solution from that of the other. In addition, knowledge of this ratio permits use of a back-titration, should such become necessary.

Rinse the burets with three or four portions of the solutions they are to contain (p. 709). Then fill each buret, and remove the air bubble in the tip by opening the stopcock briefly. Cover the top of the buret that contains the base with a test tube. Record the initial buret readings. Deliver a 35- to 40-ml portion of the acid into a 250-ml conical flask. Touch the tip of the buret to the inside wall of the flask, and rinse down with a little distilled water. Add 2 drops of phenolphthalein; then introduce NaOH until the solution is definitely pink. Now add HCl dropwise until the solution is again colorless, rinse the inside surface of the flask with water, and again add NaOH until the solution acquires a light pink hue that persists for 30 sec. Drops that are smaller than normal can be delivered by touching the buret tip to the wall of the flask and rinsing down with water. Record the final buret readings. Repeat the experiment, and calculate the volume ratio between the acid and the base. Duplicate titrations should yield values that lie within 1 to 2 ppt of the mean.

NOTES
1. The end point slowly fades as CO_2 is absorbed from the atmosphere.
2. The volume ratio can also be obtained with an indicator that has an acidic transition range, such as bromocresol green. With this particular indicator, titrate the solution to the faintest tinge of green. If appreciable carbonate is present, the ratios obtained with phenolphthalein will differ significantly from those with bromocresol green.

[5] See, for example, L. Meites, *Handbook of Analytical Chemistry*. p. **3**-32. New York: McGraw-Hill Book Company, Inc., 1963.

STANDARDIZATION OF ACID OR BASE SOLUTIONS

Sodium carbonate is an excellent primary standard for acids, while potassium hydrogen phthalate is equally useful for the standardization of bases. Directions follow for each of these titrations.

7. Standardization of 0.1-N HCl. Dry a quantity of primary-standard sodium carbonate for 2 hr at 110°C, and cool in a desiccator. Weigh 0.2- to 0.25-g samples (to the nearest 0.1 mg) into 250-ml conical flasks, and dissolve in about 50 ml of distilled water. Introduce 3 drops of bromocresol green, and titrate until the solution just changes from blue to green. Boil the solution for 2 to 3 min, cool to room temperature, and complete the titration.

Determine an indicator correction by titrating approximately 100 ml of 0.05-*F* sodium chloride and 3 drops of indicator. Subtract the volume of the blank from the titration data.

NOTE

The indicator should change from green to blue as a result of the heating step. If it does not, an excess of acid was added originally. This excess can be back-titrated with base, provided its combining ratio with the acid has been established; otherwise the sample must be discarded.

8. Standardization of 0.1-N NaOH. Dry a quantity of primary-standard potassium hydrogen phthalate for 2 hr at 110°C, and cool in a desiccator.

Weigh 0.7- to 0.9-g samples (to the nearest 0.5 mg) into 250-ml conical flasks, and dissolve in 50 to 75 ml of freshly boiled and cooled distilled water. Add 2 drops of phenolphthalein, and titrate with base until the pink color of the indicator persists for 30 sec.

Applications of Neutralization Titrations

Directions for typical neutralization titrations in aqueous solutions are given in Experiments 12 and 13. Chapter 11 should be consulted for discussion of these analyses. Experiments 14 through 16 involve nonaqueous neutralization titrations; Chapter 12 is a source for further information.

Acid-base indicators are specified for these titrations. Other methods of end-point detection exist; the most important of these involves measurement of the potential of a pH-sensitive electrode immersed in the solution being titrated. Potentiometric methods are discussed in Chapter 17; the instructor may choose to substitute a potentiometric end point for the indicator suggested for these titrations.

EXPERIMENT 12:   Analysis of Weak Acids

All of the experiments in this section involve the titration of weak acids with a standard solution of a strong base. Thus, an indicator with a basic transition range is required. Although it is not absolutely necessary, a standard hydrochloric acid solution improves the reliability in determining the end point and permits an accidentally overtitrated sample to be saved. The acid-base ratio must be established as described on page 730.

1. Determination of Potassium Hydrogen Phthalate in an Impure Sample. The sample consists of potassium hydrogen phthalate mixed with a neutral salt. The percentage of the former is to be determined.

PROCEDURE

Dry the sample for 2 hr at 110°C. Weigh samples of suitable size into 250-ml conical flasks, and dissolve in 50 to 75 ml of freshly boiled and cooled distilled water. Introduce 2 drops of phenolphthalein, and titrate with standard 0.10-*N* NaOH. Consider the first pink that persists for 30 sec as the end point.

2. Determination of the Equivalent Weight for a Weak Acid. The equivalent weight of an acid is useful in establishing its identity. This quantity is readily determined by titrating a weighed quantity of the purified acid with standard sodium hydroxide.

PROCEDURE

Weigh 0.3-g samples (to the nearest 0.1 mg) of the purified acid into 250-ml conical flasks, and dissolve in 50 to 75 ml of freshly boiled water (Note). Add 2 drops of phenolphthalein, and titrate with standard base to the first persistent pink color. Calculate the equivalent weight of the acid, assuming 100.0% purity for the sample.

NOTE

If the acid is sparingly soluble in water, use ethanol or an ethanol-water mixture. Alternatively, dissolve the acid in a known volume of standard base, and determine the excess by back-titration with standard HCl.

3. Analysis of Vinegar for Its Acid Content. The total acid content of vinegar is conveniently determined by titration with standard base. Even though other acids are also present, the results of the analysis are customarily reported in terms of acetic acid, the principal acidic constituent. Vinegars assay approximately 5% acid, expressed as acetic acid.

Pipet 25 ml of vinegar into a 250-ml volumetric flask (p. 707), and dilute to the mark with boiled and cooled distilled water. Mix thoroughly and pipet 50 ml-aliquots into 250-ml flasks. Add 50 ml of water, 2 drops of phenolphthalein, and titrate with standard 0.1-*N* NaOH to the first permanent pink color.

Calculate total acidity as grams of acetic acid per 100 ml of sample.

EXPERIMENT 13: Analysis of Bases

The analysis of bases is analogous to that for acids. A standard acid solution is used as titrant. For the analysis of weak bases, an indicator that changes color in the acid range is ordinarily required. Strong bases, the conjugate bases of very weak acids, ammonia, and organic derivatives of ammonia can be titrated. Typical examples of these titrations are discussed in Chapter 11; directions follow for several specific analyses.

THE ANALYSIS OF CARBONATE AND CARBONATE MIXTURES

The determination of carbonate, alone as well as in mixtures with other bases, is discussed on page 247. The following directions are suitable for samples containing sodium carbonate and various neutral salts; the analysis is conveniently performed at the same time the acid is standardized.

1. Determination of Sodium Carbonate in an Impure Sample. Dry the sample for 2 hr at 110°C, and then cool in a desiccator. Weigh samples of the proper size (see instructor) into 250-ml flasks. Dissolve in 50 to 75 ml of boiled water, and add 2 drops of bromocresol green. Titrate with standard 0.1-N HCl until the indicator just begins to turn green. Boil the solution for 2 to 3 min, cool, and complete the titration. If additional acid is not required after boiling, either discard the sample or back-titrate the excess acid with standard base.

Calculate the percentage of sodium carbonate in the sample.

2. Determination of a Sodium Carbonate and Sodium Hydrogen Carbonate Mixture. For a discussion of this titration, see page 247.

If the sample is a solid, weigh dried portions into 250-ml volumetric flasks, dissolve in 50 to 75 ml of water, and dilute to the mark with boiled and cooled water. If the sample is a solution, pipet suitable aliquots into 250-ml volumetric flasks, and dilute to volume. Mix thoroughly.

To determine the total number of milliequivalents of the two components, transfer 25-ml aliquots of each solution to 250-ml conical flasks, and titrate with standard 0.1-N HCl, following the directions given in the preceding section for a sodium carbonate sample.

To determine the bicarbonate content, pipet additional 25-ml portions into flasks; *treat each aliquot individually from here.* Add a carefully measured excess of standard 0.1-N NaOH to an aliquot (conveniently 50.00 ml). Immediately add 10 ml of 10% $BaCl_2$ and 2 drops of phenolphthalein. Titrate the excess NaOH at once with standard 0.1-N HCl to the disappearance of the pink color.

Titrate a blank consisting of 25 ml of water, 10 ml of the $BaCl_2$ solution, and *exactly* the same volume of NaOH as used with the samples. The difference between the volume of HCl needed for the blank and the sample corresponds to the $NaHCO_3$ present.

Calculate the percentage of $NaHCO_3$ and of Na_2CO_3 in the sample.

3. Analysis of a Sodium Carbonate and Sodium Hydroxide Mixture. This analysis is discussed on page 248.

The sample is an aqueous solution containing sodium carbonate and sodium hydroxide.

PROCEDURE

Transfer the sample to a 250-ml volumetric flask immediately upon receipt, and dilute to the mark with boiled and cooled distilled water. Mix thoroughly. Keep the flask tightly stoppered to avoid absorption of CO_2.

Pipet 25-ml aliquots of the sample solution into 250-ml conical flasks, add 2 drops of bromocresol green, and titrate with 0.1-N HCl, following the procedure in section 1 for the analysis of carbonate. This titration will give the total number of milliequivalents of NaOH and Na_2CO_3.

To determine the quantity of NaOH present, transfer a 25-ml aliquot of the sample solution into a 250-ml conical flask. Slowly add 10 ml of a neutral 10% $BaCl_2$ solution and 2 drops of phenolphthalein. Titrate immediately with standard 0.1-N HCl. To avoid contamination of the samples by atmospheric CO_2, complete this part of the analysis as rapidly as possible. Repeat with additional aliquots of the sample.

Calculate the weight of NaOH and Na_2CO_3 in the original sample.

THE ANALYSIS OF NITROGEN-CONTAINING COMPOUNDS

Sections 4 and 5 of Experiment 13 make possible the analysis for ammonium ion and organic amine nitrogen, respectively. These methods are discussed on pages 244–247.

4. Determination of Ammonium Ion. Transfer the sample, which should contain 2 to 4 meq of ammonium salt, to a 500-ml Kjeldahl flask, and add enough water to give a total volume of about 200 ml.

Arrange a distillation apparatus similar to that shown in Figure 11-2. Use a buret or pipet to measure precisely 50 ml of standard 0.1-N HCl into the receiver flask. Clamp the flask so that the tip of the adapter extends just below the surface of the standard acid. Circulate the cooling water through the jacket of the condenser.

For each sample, prepare a solution containing approximately 45 g of NaOH in about 75 ml of water. Cool this solution to room temperature before use. With the Kjeldahl flask tilted, slowly pour the caustic down the side of the container so that little mixing occurs with the solution in the flask (Note 1). Add several pieces of granulated zinc (Note 2) and a small piece of litmus paper. *Immediately* connect the flask to the spray trap. Very *cautiously* mix the solution by gentle swirling. After mixing is complete, the litmus paper should indicate that the solution is basic.

Immediately bring the solution to a boil, and distill at a steady rate until one-half to one-third of the original solution remains. Control the rate of heating during this period to prevent the receiver acid from being drawn back into the distillation flask. After the distillation is judged complete, lower the receiver flask until the tip of the adapter is well clear of the standard acid. Then discontinue heating, disconnect the apparatus, and rinse the inside of the condenser with a small amount of water. Disconnect the adapter, and rinse it thoroughly. Add 2 drops of bromocresol green or methyl red, and titrate the distillate with standard 0.1-N NaOH to the color change of the indicator.

A modification of this procedure makes use of about 50 ml of 4% boric acid solution in place of the standard HCl in the receiver flask. The distillation is then carried out in an identical fashion. When complete, the ammonium borate produced is titrated with a standard 0.1-N HCl solution, using 2 to 3 drops of bromocresol green as indicator.

Calculate the percent nitrogen in the sample.

NOTES

1.   The dense caustic solution should form a second layer on the bottom of the flask. Mixing must be avoided at this point in order to prevent loss of the volatile ammonia. Manipulations should be carried out as rapidly as possible.

2.   Granulated zinc is added to minimize bumping during the distillation. It reacts slowly with the alkali to give small bubbles of hydrogen that minimize superheating of the liquid.

5. The Kjeldahl Analysis. These directions are suitable for the analysis of protein in samples such as blood meal, wheat flour, macaroni, dry cereals, and dry pet foods. Prereduction is not required for such materials. A simple modification will permit analysis of samples that contain more highly oxidized forms of nitrogen.[6] The Kjeldahl analysis is discussed on page 244.

Weigh three samples of 0.25 to 2.5 g, depending upon the nitrogen content, and wrap each in a 9-cm filter paper. Drop each into a 500-ml Kjeldahl flask (the paper wrapping will prevent the sample from clinging to the neck of the flask). Add 25 ml of concentrated sulfuric acid and 10 g of powdered K_2SO_4. Add catalyst (Note 1), and clamp the flask in an inclined position in a hood. Heat the mixture carefully until the H_2SO_4 is boiling. Continue the heating until the solution becomes colorless or light yellow; this may take as long as 2 to 3 hr. If necessary, cautiously replace the acid lost by evaporation.

Remove the flame, and allow the flask to cool. Swirl the flask if the liquid begins to solidify. Cautiously dilute with 250 ml of water, and cool to room temperature under a water tap. If mercury was used as the catalyst, introduce 25 ml of a 4% sodium sulfide solution. Then complete the analysis as described in the third and fourth paragraphs of the preceding directions for the analysis of an inorganic ammonium salt.

Calculate the percent nitrogen in the samples.

NOTES

1.   As a catalyst, one may use any of the following: a drop of mercury, 0.5 g of HgO, a crystal of $CuSO_4$, 0.1 g of Se, or 0.2 g $CuSeO_3$. Alternatively, the catalyst may be omitted.

2.   It is recommended that a blank, which includes the filter paper, be carried through all steps of the analysis.

Neutralization Titrations in Glacial Acetic Acid

It was pointed out in Chapter 12 that the substitution of a nonaqueous solvent system may permit the analysis of species that are insufficiently strong as acids or as bases for titration in an aqueous environment. The experiments that follow illustrate the use of glacial acetic acid as a solvent;[7] further information is to be found on pages 263–266.

[6] See *Official Methods of Analysis*, 11th ed., p. 16. Washington, D.C.: Association of Official Analytical Chemists, 1970.

[7] These directions are based on J. S. Fritz, *Acid-Base Titrations in Nonaqueous Solvent*. Columbus, Ohio: The G. F. Smith Chemical Company, 1952.

EXPERIMENT 14: Preparation and Standardization of Reagents

1. Preparation of 0.1-N HClO₄. Mix 8.5 ml of 72% perchloric acid with about 300 ml of glacial acetic acid, and add 20 ml of acetic anhydride. Dilute to approximately 1 liter with glacial acetic acid. Allow the solution to stand overnight or longer before use in order to permit the reaction between acetic anhydride and water to become complete.

2. Preparation of 0.1-N Sodium Acetate. Dissolve 8.2 g anhydrous sodium acetate in glacial acetic acid, and dilute to about 1 liter with the acid.

3. Preparation of Methyl Violet Indicator. Dissolve 0.2 g of methyl violet in 100 ml of chlorobenzene.

NOTE

If any of these reagents comes in contact with the skin, wash the affected area with water immediately.

4. Determination of Acid-Base Ratio. Fill burets with the perchloric acid and the sodium acetate solutions, and measure about 30 ml of the former into a flask. Add 2 drops of the methyl violet indicator, and titrate with the sodium acetate solution until the indicator color is a faint violet. Practice the end point by making further small additions of acid and base to the solution.

Repeat the determination with a fresh portion of the perchloric acid solution. Calculate the volume ratio of acid to base.

5. Standardization of Solutions. Dry potassium hydrogen phthalate, $KHC_8H_4O_4$, for 1 to 2 hr at 110°C and cool. Weigh 0.5- to 0.6-g samples (to the nearest 0.5 mg) into flasks, and add 60 ml of glacial acetic acid to each. Cautiously heat the mixtures until all solid is dissolved; then cool to room temperature.

Measure and record the temperature of the perchloric acid reagent. Then fill burets with the perchloric acid and the sodium acetate solutions. Add 2 drops of the methyl violet solution to each of the samples, and titrate to the same end point as in the determination of the acid-base ratio.

Calculate the normality of the perchloric acid and the sodium acetate solutions.

EXPERIMENT 15: Analysis of Amines

PROCEDURE

Dissolve a sample containing 2 to 4 meq of the amine in 50 ml of glacial acetic acid. Measure and record the temperature of the perchloric acid solution. Add 2 drops of methyl violet to the sample, and titrate to the disappearance of the violet color. Back-titrate with standard sodium acetate until the first faint violet color reappears.

Correct the volumes of the two standard solutions to the standardization temperature with the aid of Equation 12-12. Then calculate the number of milliequivalents of amine in the sample.

EXPERIMENT 16: Analysis of Amino Acids

The behavior of amino acids in aqueous solution was considered in Chapter 10. When acetic acid is substituted as solvent, the acid function of these solutes is completely repressed; they behave simply as weak bases. Owing to their limited solubility in glacial acetic acid, amino acid samples are ordinarily dissolved in a known excess of perchloric acid, and the excess is determined by back-titration with standard sodium acetate.

PROCEDURE

Weigh samples containing 2 to 4 meq of the amino acid into flasks. Measure and record the temperature of the standard perchloric acid solution; then pipet exactly 50.00 ml of this reagent into each sample. After solution is complete, add 2 drops of the methyl violet indicator, and titrate with the standard sodium acetate to the first violet color.

Titrations Based on Chelate Formation

The principles upon which titrations with ethylenediaminetetraacetic acid are based are considered in Chapter 13. Directions are provided here for the preparation and standardization of EDTA solutions and for representative titrations involving this reagent.

EXPERIMENT 17: Preparation and Standardization of Reagents

1. Preparation of 0.01-F EDTA Solution. Dry the purified dihydrate ($Na_2H_2Y \cdot 2H_2O$) at 80°C to remove superficial moisture. After cooling, weigh about 3.8 g (to the nearest milligram) into a 1-liter volumetric flask; dilute to the mark with distilled water.

NOTES
1. W. J. Blaedel and H. T. Knight [*Anal. Chem.*, **26**, 741 (1954)] give specific instructions for the purification of commercial preparations of the disodium salt.
2. Direct preparation of standard solutions requires the total exclusion of polyvalent cations. If any doubt exists regarding the quality of the distilled water, pretreatment by passage through a cation-exchange resin is recommended.
3. If desired, the anhydrous salt may be employed instead of the dihydrate. The weight taken should be adjusted accordingly.
4. If desired, an EDTA solution that is approximately 0.01 F can be prepared and standardized against a Mg^{2+} solution of known strength.

2. Preparation of the Magnesium Complex of EDTA, 0.1-F Solution. To 37.2 g of $Na_2H_2Y \cdot 2H_2O$ in 500 ml of distilled water, add an equivalent quantity (24.65 g) of $MgSO_4 \cdot 7H_2O$. Introduce a few drops of phenolphthalein followed by sufficient sodium hydroxide to turn the solution faintly pink. Dilute the solution to 1 liter. When properly prepared, portions of this solution should assume a dull violet color when treated with pH-10 buffer and a few drops of Eriochrome black T (Erio T) indicator. Furthermore, a single drop of 0.01-*F*

Na_2H_2Y should cause a color change to blue, while an equal quantity of 0.01-F Mg^{2+} should cause a change to red. The composition of the solution should be adjusted by addition of Mg^{2+} or Na_2H_2Y until these criteria are met.

3. Preparation of Eriochrome Black T Solution. Dissolve 200 mg of the solid in a solution consisting of 15 ml of triethanolamine and 5 ml of absolute ethanol. Solutions should be freshly prepared every two weeks; refrigeration slows their deterioration.

4. Preparation of Buffer, pH 10. Dilute 570 ml of aqueous NH_3 (sp gr 0.90) and 70 g of NH_4Cl to approximately 1 liter.

5. Determination of Magnesium by Direct Titration. The sample will be issued as an aqueous solution; transfer it to a clean 500-ml volumetric flask, and dilute to the mark. Mix thoroughly. Take 50.00-ml aliquots, adding 1 to 2 ml of pH-10 buffer and 2 to 4 drops of Erio ,T indicator solution to each. Titrate with 0.01-F Na_2H_2Y to a color change from red to pure blue. Express the results of the analysis in terms of milligrams of Mg^{2+} per liter of solution.

NOTES
1. The color change of the indicator is slow in the vicinity of the end point. Care must be taken to avoid overtitration.
2. Other alkaline earths, if present, will also be titrated along with magnesium and should be removed prior to the analysis; $(NH_4)_2CO_3$ is a suitable reagent. Most polyvalent cations also interfere and should be precipitated as hydroxides.
3. Since this reaction may be expressed as
$$Mg^{2+} + H_2Y^{2-} \rightleftarrows MgY^{2-} + 2H^+$$
the normality of the chelating solution is twice its formal concentration. For a 40-ml titration, each aliquot should contain about 10 mg of magnesium ion.

6. Determination of Calcium by Substitution Titration. Weigh the sample into a 500-ml volumetric flask, and dissolve in a minimum quantity of dilute HCl. Neutralize the solution with sodium hydroxide (Note 2), and dilute to the mark. Take 50.00-ml aliquots for titration, treating each as follows: add approximately 2 ml of pH-10 buffer, 1 ml of the magnesium chelate solution, and 2 to 4 drops of Erio T indicator. Titrate with standard 0.01-F Na_2H_2Y to a color change from red to blue. Report the percentage of calcium oxide in the sample.

NOTES
1. The sample taken should contain about 150 to 160 mg of Ca^{2+}.
2. To neutralize the solution, introduce a few drops of methyl red, and add base until the red color is discharged.
3. The amount of MgY^{2-} solution added is not critical. Its presence is required because the indicator does not exhibit a satisfactory color change when Ca^{2+} is present alone. Since the calcium chelate is more stable, the reaction
$$Ca^{2+} + MgY^{2-} \rightleftarrows CaY^{2-} + Mg^{2+}$$
takes place. Introduction of titrant leads to the preferential formation of the calcium chelate for the same reason. At the equivalence point, the reaction is actually between H_2Y^{2-} and Mg^{2+}, for which Erio T is an excellent indicator.
4. Interferences with this method are substantially the same as with the direct determination of magnesium and are eliminated in the same way.

7. Determination of Calcium and Magnesium in Hard Water. Acidify 100-ml aliquots with a few drops of HCl, and boil gently for a few minutes to remove CO_2. Cool, add a few drops of methyl red, and neutralize the solution with NaOH. Introduce 2 ml of pH-10 buffer, 2 to 4 drops of Erio T indicator, and titrate with standard 0.01-*F* Na_2H_2Y to a color change from red to pure blue. Report results of the analysis in terms of milligrams of $CaCO_3$ per liter of water

NOTE

If the color change of the indicator is sluggish, the absence of magnesium is indicated. In this event, 1 to 2 ml of standard 0.1-*F* MgY^{2-} solution should be added.

8. Determination of Calcium by Back-Titration. Prepare the sample as directed for the substitution analysis of calcium. To each 50.00-ml aliquot, add about 2 ml of pH-10 buffer and 2 to 4 drops of Erio T indicator. Run in an excess of 0.01-*F* Na_2H_2Y solution from a buret, and record the volume taken. Titrate the excess chelating agent to a color change from blue to red with standard 0.01-*F* $MgSO_4$ solution.

NOTE

The magnesium sulfate solution is conveniently prepared by dissolving about 2.5 g of the heptahydrate in 1 liter of distilled water. The equivalence of this solution with respect to the Na_2H_2Y solution should be determined by a direct titration.

Titrations Based on Oxidation-Reduction Reactions

The concept of equilibrium as it relates to oxidation-reduction is developed in Chapter 14 and applied to volumetric analysis in Chapter 15. Discussion of specific reagents and methods is found in Chapter 16. The experiments given here are organized according to the titrant that is used.

EXPERIMENT 18:   Preparation and Use of a Jones Reductor

It is essential that the analyte exist in a single and known oxidation state prior to the start of an oxidation-reduction titration. For many species, passage through a Jones reductor (p. 339) assures this condition.

1. Preparation of a Jones Reductor. The reductor (Figure 16-1) should be thoroughly cleaned and fitted with a porcelain disk to support a mat of glass wool or asbestos at its lower end. This mat must be sufficiently thick to prevent passage of zinc granules when the reductor is in use.

The zinc should be 20 to 30 mesh and free of impurities such as iron. Cover a suitable quantity of the metal with 1-*F* HCl, and let stand for about 1 min. Decant the liquid, and cover the zinc with a 0.25-*F* solution of $Hg(NO_3)_2$ or $HgCl_2$. Stir the mixture vigorously for 3 min, decant, and wash two or three times with distilled water. Fill the reductor tube with water, and slowly add the amalgam until a column of about 30 cm is achieved. Wash with 500 ml of water, being sure that the packing is always covered by liquid. Keep the reductor filled with water during storage.

2. *Use of the Jones Reductor.* Solutions that are 0.5 to 5 N in HCl or H_2SO_4 may be used in the reductor. Warm solutions may be passed through the column without harm. Wash the column with several 20- to 30-ml portions of a solution that has about the same acid concentration as the sample. After each addition of wash liquid, drain the column to about 1 cm above the zinc. Then add the next portion. Leave the last washing on the column.

Attach a receiver to the column, and add the sample to the reservoir at the top. The sample can be pulled through the column at a rate of 75 to 100 ml/min. When the sample solution has been drawn to within 1 cm of the top of the packing, add 25 ml of a dilute acid as a wash. Follow this with two additional acid washes and then with 100 to 200 ml of water. Leave the column covered with water, and place a beaker over its top during storage.

EXPERIMENT 19: Preparation and Standardization of
Permanganate Solutions

The precautions that must be observed in the preparation and storage of permanganate solutions are discussed on pages 339–346.

1. *Preparation of Approximately 0.1-N KMnO₄.* Dissolve 3.2 g of $KMnO_4$ in about 1 liter of distilled water. Heat to boiling, and keep hot for about 1 hr. Cover and let stand overnight. Filter the solution through a fine-porosity sintered glass crucible or through a Gooch crucible with an asbestos mat (p. 702). Store the solution in a clean, glass-stoppered bottle, and keep in the dark when not in use.

STANDARDIZATION AGAINST SODIUM OXALATE

Directions are provided for standardization by the McBride method and the method of Fowler and Bright; see page 342 for discussion of these methods as well as other primary standards for permanganate solutions.

2. *Method of McBride.* Weigh samples of dried $Na_2C_2O_4$ into 400-ml beakers, and dissolve in approximately 250 ml of 0.72-F H_2SO_4. Heat to 80 to 90°C, and titrate with the $KMnO_4$, stirring vigorously with a thermometer. The reagent should be introduced slowly so that the pink color is discharged before further additions are made (Note 4). If the solution temperature drops below 60°C, heat. The end point is the first persistent pink color. Correct the titration for an end-point blank determined by titrating an equal volume of the water and sulfuric acid.

3. *Method of Fowler and Bright.* Dry primary-standard grade $Na_2C_2O_4$ for 1 hr at 110 to 120°C. Cool in a desiccator, and accurately weigh suitable portions (0.2 to 0.3 g for 0.1-N $KMnO_4$) into 400-ml beakers. Add 250 ml of 0.90-F sulfuric acid that has been boiled for 10 to 15 min and cooled to room temperature. Stir until dissolved. A thermometer is convenient for this purpose since the temperature must be measured later. Introduce from a buret sufficient

permanganate to consume 90 to 95% of the oxalate (about 40 ml for 0.1-N $KMnO_4$ and a 0.3-g sample; a preliminary titration by the McBride method will provide the approximate volume required). Let stand until the solution is decolorized. Then warm to 55 to 60°C, and complete the titration, taking the first pale pink color that persists for 30 sec as the end point. Determine an end-point correction by titrating 250 ml of 0.90-F sulfuric acid at this same temperature. Correct for the blank and calculate the normality.

NOTES

1. To measure the volume of $KMnO_4$, take the surface of the liquid as a point of reference. Alternatively, provide sufficient backlighting with a flashlight or match to permit reading of the meniscus in the conventional manner.

2. Permanganate solutions should not be allowed to stand in burets any longer than necessary, as decomposition to MnO_2 may occur. Freshly formed MnO_2 can be removed from burets and glassware with a solution of 1-F H_2SO_4 containing a small amount of 3% H_2O_2.

3. Any $KMnO_4$ spattered on the sides of the titration vessel should be washed down immediately with a stream of water.

4. If the addition of $KMnO_4$ is too rapid, some MnO_2 will be produced in addition to Mn^{2+}; evidence for this is a faint brown discoloration of the solution. This is not a serious problem as long as sufficient oxalate remains to reduce the MnO_2 to Mn^{2+}; the titration is temporarily discontinued until the solution clears. The solution must be free of MnO_2 at the equivalence point.

EXPERIMENT 20: Determination of Iron in an Ore

Success of the method depends upon having all iron in the $+2$ state at the outset of the titration. Directions are provided for prereduction with a Jones reductor and with tin(II) chloride.

1. Prereduction with a Jones Reductor. The preparation and use of the Jones reductor are described on page 789.

a. *Sample preparation.* Dry the sample, and weigh portions of suitable size into 250-ml beakers. Add 10 ml of concentrated HCl, and heat until the sample is decomposed. If solution of the sample is incomplete, filtration and fusion of the dark residue with Na_2CO_3 (Note 1) may be necessary.

Heat a blank consisting of 10 ml of HCl for the same time as the samples.

Add 15.0 ml of 1:1 H_2SO_4 to the solution, and heat until fumes of SO_3 are observed (Hood). Cool and slowly add 100 ml of water. Swirl the solution to dissolve any iron salts that may form on the side of the beaker. The first few milliliters of water should be added a drop at a time down the side of the beaker.

b. *Reduction of iron.* Drain the solution from the Jones reductor to within 3 cm of the top of the packing (Note). Pass 200 ml of 1-F H_2SO_4 through the column, again draining to about 3 cm from the top of the zinc. Disconnect the receiver, discard its contents, and rinse it with distilled water. Again attach the receiver, and pass 50 ml of 1-F H_2SO_4, followed by the sample, through the reductor at a rate of about 50 to 100 ml/min. Wash the beaker with five 10-ml portions of 1-F H_2SO_4, passing each through the reductor. Finally, pass 100 ml of water through the column. Leave the reductor filled with water. Throughout the entire process, the amalgam should be covered with liquid and never exposed to the atmosphere.

c. *Titration.* Disconnect the receiver at once, rinse the tip of the reductor into the flask, and add 5 ml of 85% H_3PO_4. Titrate with standard 0.10-N $KMnO_4$.

The blank should be treated in the same way as the sample, the volume of $KMnO_4$ consumed being subtracted from that required for the sample.

NOTES

1. If dark particles persist after treatment with acid, filter the solution through ashless paper, wash with 5 to 10 ml of 6-F HCl, and retain the filtrate and washings. Place the paper in a small platinum crucible and ignite. Mix 0.5 to 0.7 g of finely ground anhydrous Na_2CO_3 with the residue, and heat until a clear liquid melt is obtained. Cool, add 5 ml of water, followed by the cautious addition of an equal volume of 6-F HCl. Warm the crucible, and combine the contents with the original filtrate. Evaporate to about 15 ml, and proceed with the reduction step.
2. If the reductor has not been in constant use, it may be partially clogged with various oxides. The first acid passed through may consume some oxidant. Therefore, test the reductor before use by passage of about 200 ml of 1-F H_2SO_4 and 100 ml of water. The emergent solution should require no more than 0.06 ml of 0.1-N $KMnO_4$. If necessary, repeat the operation until a blank of the required size is obtained.

2. Prereduction with Tin(II) Chloride. The chemistry of the tin(II) reduction for iron(III) is discussed on page 345. It is necessary to use the Zimmermann-Reinhardt reagent (p. 345), owing to the presence of chloride ion in the titration mixture.

a. *Special solutions.* Approximately 100 titrations can be performed with the following solutions:

(1) Tin(II) chloride, 0.5 N. Dissolve 60 g of iron-free $SnCl_2 \cdot 2H_2O$ in 100 ml of concentrated HCl. Heat if necessary. After solution is complete, dilute to about 1 liter, and store in a well-stoppered bottle.

(2) Mercury(II) chloride. Dissolve 50 g of $HgCl_2$ in 1000 ml of water.

(3) Zimmermann-Reinhardt reagent. Dissolve 210 g of $MnSO_4 \cdot 4H_2O$ in about 1 liter of water. Cautiously add 375 ml of concentrated H_2SO_4 and 375 ml of 85% phosphoric acid. Dilute to about 3 liters.

b. *Sample preparation.* Dry the ore at 105 to 110°C, and weigh individual samples into 500-ml conical flasks. A sample of optimum size will require 25 to 40 ml of the standard $KMnO_4$. Add 10 ml of concentrated HCl, 3 ml of $SnCl_2$ solution, and heat at just below boiling until the sample is decomposed; this is indicated by the disappearance of all of the dark particles (see Note 1 of the preceding directions). A pure white residue may remain. A blank consisting of 10 ml of HCl and 3 ml of $SnCl_2$ should be heated for the same length of time. If the solutions become yellow during the heating, add another milliliter or two of $SnCl_2$. After the decomposition is complete, add approximately 0.2-F $KMnO_4$ dropwise until the solution is just yellow. Dilute to about 15 ml. Add $KMnO_4$ to the blank until the solution just turns pink. Then just decolorize with $SnCl_2$. *Carry samples individually through subsequent steps.*

c. *Reduction of iron.* Heat the solution containing the sample nearly to boiling, and add $SnCl_2$ drop by drop until the yellow color disappears. Add 2 drops in excess (Note). Cool to room temperature, and *rapidly* add 10 ml of the $HgCl_2$ solution. A small quantity of white precipitate should appear. If no precipitate forms or if the precipitate is gray, the sample should be discarded. The blank solution should also be treated with 10 ml of the $HgCl_2$ solution.

    d. *Titration.* After 2 to 3 min, add 25 ml of Zimmermann-Reinhardt reagent and 300 ml of water. Titrate *immediately* with the $KMnO_4$ to the first faint pink that persists for 15 to 20 sec. Do not titrate rapidly at any time. Correct the volume of $KMnO_4$ for the blank titration.

NOTE

The solution may not become entirely colorless but may instead acquire a pale yellow-green hue. Further additions of $SnCl_2$ will not alter this color. If too much $SnCl_2$ is inadvertently introduced, add 0.2-F $KMnO_4$ until the yellow color is restored, and repeat the reduction.

EXPERIMENT 21: Determination of Calcium

The precipitation of calcium as the oxalate, followed by titration of the oxalic acid liberated when the solid is dissolved, is discussed on pages 345–346 and is the basis for the directions that follow.

    1. Analysis for Calcium in Impure Calcium Carbonate. Dry the unknown at 110°C. Weigh samples large enough to contain about 100 mg of calcium into 600-ml beakers. Cover each with a watch glass. Add 10 ml of water, followed by 10 ml of 6-F HCl from a pipet; the acid should be added slowly to avoid losses by spattering. Heat the solution to drive off the CO_2. Wash the watch glass and sides of the beaker with water, and dilute to about 150 ml. Heat to 60 to 80°C, and add 50 ml of a warm solution containing about 3 g of $(NH_4)_2C_2O_4 \cdot H_2O$ [if the $(NH_4)_2C_2O_4$ solution is not clear, filter before use]. Add 3 to 4 drops of methyl red. Then introduce 1:1 NH_3 dropwise from a pipet until the color changes from red to yellow. Allow the solution to stand for about 30 min (but no longer than 1 hr if magnesium is present) without further heating.

    Filter the solution through a medium-porosity glass crucible or a Gooch crucible fitted with a mat of asbestos (p. 702) or glass fiber. Wash the beaker, and precipitate with 10- to 20-ml portions of chilled distilled water until the washings show only a faint cloudiness when tested with an acidified $AgNO_3$ solution (Note). A quantitative transfer of the precipitate is unnecessary.

    Rinse the outside of the crucible with water, and return it to the beaker in which the precipitate was formed. Add 150 ml of water and 50 ml of 3-F H_2SO_4. Heat to 80 to 90°C to dissolve the precipitate. Titrate with 0.1-N $KMnO_4$ with the crucible still in the beaker. The temperature of the solution should not be allowed to drop below 60°C. Take as an end point the first pink color that persists for 15 to 20 sec.

    Report the percentage of $CaCO_3$ in the sample.

NOTE

If the sample contains large concentrations of sodium or magnesium ions, more accurate results can be obtained by reprecipitation of the calcium oxalate. To do this, filter the precipitate through paper and wash four or five times with 0.1% $(NH_4)_2C_2O_4$ solution. Pour 50 ml of hot 1:4 HCl (p. 15) through the paper, collecting the washings in the beaker in which the precipitation was made. Wash the paper several times with hot 1:100 HCl, and dilute all of the washings to about 200 ml. Reprecipitate the calcium oxalate as before, this time collecting the precipitate in a Gooch or glass crucible and washing with cold water. Proceed with the analysis as before.

2. Determination of Calcium in Limestone. Limestones are composed principally of calcium carbonate. Dolomitic limestones contain large concentrations of magnesium carbonate in addition. Also present in smaller amounts are calcium and magnesium silicates as well as carbonates and silicates of iron, aluminum, manganese, titanium, the alkalies, and other metals.

Hydrochloric acid will often decompose limestones completely; only silica remains undissolved. Some limestones are more readily decomposed if first ignited; a few will yield only to a carbonate fusion.

The following method is remarkably effective for the analysis of calcium in most limestones. Iron and aluminum, in amounts equivalent to the calcium, do not interfere. Small amounts of titanium and manganese can be tolerated.[8]

PROCEDURE

Dry the sample for 1 to 2 hr at 110°C. If the material can be readily decomposed with acid, weigh 0.25- to 0.3-g samples into 250-ml beakers, and cover each beaker with a watch glass. Add 10 ml of water and 10 ml of concentrated HCl, taking care to avoid loss by spattering. Proceed with the analysis as described in the next paragraph. If the limestone is not completely decomposed by acid, weigh the sample into a small porcelain crucible, and ignite. Raise the temperature slowly to 800 to 900°C, and maintain this temperature for 30 min. After cooling, place the crucible in a 250-ml beaker, add 5 ml of water, and cover with a watch glass. Carefully add 10 ml of concentrated HCl, and heat to boiling. Remove the crucible with a stirring rod, rinsing thoroughly with water.

Add 5 drops of saturated bromine water to oxidize any iron present, and boil for 5 min to remove the excess bromine. Dilute to 50 ml, heat to boiling, and add 100 ml of hot, filtered 5% $(NH_4)_2C_2O_4$ solution. Add 3 to 4 drops of methyl red, and precipitate the calcium oxalate by the dropwise addition of 1:1 NH_3. The rate of addition should be 1 drop every 3 or 4 sec until the solution turns to the intermediate orange-yellow color of the indicator (pH 4.5 to 5.5). Allow the solution to stand for 30 min but no longer and filter (Note). A Gooch crucible or a medium-porosity filtering crucible can be used. Wash the precipitate with several 10-ml portions of cold water. Rinse the outside of the crucible, and return it to the beaker in which the calcium oxalate was originally formed. Add 150 ml of water and 50 ml of 3-*F* H_2SO_4.

Heat the solution to 80 to 90°C. If a Gooch crucible was used, stir to break up the asbestos pad. Then titrate with 0.1-*N* $KMnO_4$. The solution should be kept above 60°C throughout the titration.

Report the percentage of CaO in the sample.

NOTE
If Mg^{2+} is absent, the period of standing need not be limited to 30 min.

EXPERIMENT 22:   Preparation and Standardization of
                                 Cerium(IV) Solutions

The properties and the standardization of cerium(IV) solutions are discussed on pages 346–349.

[8] For further details of the method, see J. J. Lingane, *Ind. Eng. Chem., Anal. Ed.*, **17**, 39 (1945).

1. Preparation of Approximately 0.1-F Cerium(IV) Solutions. Carefully add 50 ml of concentrated H_2SO_4 to 500 ml of water, and then add 63 g of $Ce(SO_4)_2 \cdot 2(NH_4)_2SO_4 \cdot 2H_2O$ with continual stirring. Cool, filter if the solution is not clear, and dilute to about 1 liter.

If $Ce(NO_3)_4 \cdot 2NH_4NO_3$ is used, weigh about 55 g into a 1-liter beaker. Add about 60 ml of 95% H_2SO_4, and stir for 2 min. *Cautiously* introduce 100 ml of water, and again stir for 2 min. Repeat the operations of adding water and stirring until all of the salt has dissolved. Then dilute to about 1 liter.

2. Standardization of Cerium(IV) Solutions against Arsenious Oxide. Dry primary-standard As_2O_3 for 1 hr at 110°C, and weigh 0.2-g portions (to the nearest 0.1 mg) into 250-ml flasks. Dissolve samples in 15 ml of 2-*F* NaOH. After solution is complete, promptly acidify with 25 ml of 3-*F* H_2SO_4 (Note 1). Dilute to about 100 ml, add 3 drops of 0.01-*F* osmium tetroxide (Note 2) and 1 drop of iron(II) orthophenanthroline indicator (Note 3). Titrate to a color change from red to very pale blue or colorless.

NOTES

1.   Arsenic(III) will undergo air oxidation in an alkaline environment. Arsenite solutions should be neutralized as soon as possible.
2.   The catalyst solution may be purchased from the G. Frederick Smith Chemical Co., Columbus, Ohio; it should be about 0.01 *F* in OsO_4 and 0.1 *F* in H_2SO_4.
3.   The orthophenanthroline indicator is prepared by dissolving 1.485 g of the compound in 100 ml of water that contains 0.695 g of $FeSO_4 \cdot 7H_2O$. Indicator solutions are available commercially.

3. Standardization of Cerium(IV) Solutions against Sodium Oxalate. Weigh 0.25 to 0.3-g portions of dried sodium oxalate (to the nearest 0.1 mg) into 250-ml beakers, and dissolve in 75 ml of water. Add 20 ml of concentrated HCl and 1.5 ml of 0.017-*F* ICl (Note 1). Heat to 50°C, and add 2 to 3 drops of iron(II) orthophenanthroline indicator (Note 3 above). Titrate with cerium(IV) until the solution turns pale blue or colorless and the pink does not return within 1 min. The temperature should be between 45 and 50°C throughout (Note 2).

NOTES

1.   Iodine monochloride can be prepared by mixing 25 ml of 0.04-*F* KI, 40 ml of concentrated HCl, and 20 ml of 0.025-*F* KIO_3. Add 5 to 10 ml of CCl_4, and shake thoroughly. Titrate with either KI or KIO_3 until the CCl_4 layer is barely pink after shaking. The former should be added if the CCl_4 is colorless and the latter if it is too pink.
2.   Do not permit the temperature to exceed 50°C, as the indicator may be destroyed.

EXPERIMENT 23:   Determination of Iron in an Ore

Quadrivalent cerium oxidizes iron(II) smoothly and rapidly at room temperature. Orthophenanthroline is an excellent indicator for the titration. In contrast to the analysis based upon oxidation with permanganate, consumption of the reagent by chloride ion is of no concern.

The problems associated with solution of the sample and prereduction of the iron are the same as in the permanganate method (p. 741), the only major difference being that here there is no need for the Zimmermann-Reinhardt solution.

1. Special Solutions. (a) Tin(II) chloride. (b) Mercury(II) chloride (p. 742). (c) Orthophenanthroline indicator (0.025 *F*) (Note 3, p. 745). (d) Standard quadrivalent cerium solution.

2. Preparation of the Sample. Use the directions on page 741 if a Jones reductor is to be used. See page 742 if tin(II) chloride reduction is to be employed.

3. Titration. If the tin(II) chloride is used, allow about 2 to 3 min after addition of the $HgCl_2$. Then add 300 ml of 1-*F* HCl, a drop or two of ortho-phenanthroline, and titrate to the color change of the indicator. For accurate work a blank should be carried through the entire procedure.

If the Jones reductor is used for prereduction, complete the analysis by titration with the cerium(IV) solution, using 1 drop of orthophenanthroline indicator.

EXPERIMENT 24: Preparation and Standardization of Dichromate Solutions

See pages 349–350 for further information concerning the properties of potassium dichromate as an oxidizing reagent.

Preparation of a 0.1-N Potassium Dichromate Solution. Dry primary-standard $K_2Cr_2O_7$ for 2 hr at 150 to 200°C. After cooling, weigh 4.9 g of the solid (to the nearest mg) into a 1-liter volumetric flask, and dilute to the mark with distilled water.

If the purity of the salt is suspect, recrystallize three times from water before drying. Alternatively, prepare a solution of approximate normality by dissolving about 5 g of $K_2Cr_2O_7$ in a liter of water, and standardize the solution against weighed samples of electrolytic iron wire employing the procedure that follows. Dissolve 0.20- to 0.25-g wire samples (weighed to the nearest 0.1 mg) in a minimum amount of dilute HCl, prereduce the iron with Sn(II) (p. 742) or a Jones reductor (p. 741), and proceed as in Experiment 25.

EXPERIMENT 25: Determination of Iron in an Ore

To the prereduced solution, introduce about 10 ml of concentrated H_2SO_4 and 15 ml of syrupy H_3PO_4. Add water, if necessary, to bring the volume to about 250 ml. Cool, add 8 drops of diphenylamine sulfonate indicator (Note), and titrate with dichromate to the violet-blue end point.

NOTE
The indicator solution should contain 0.2 g of sodium diphenylamine sulfonate in 100 ml of water.

Iodimetric Methods of Analysis

The oxidizing properties of iodine, the composition and stability of triiodide solutions, and the application of this reagent to volumetric analysis are discussed in Chapter 16.

EXPERIMENT 26: Preparation and Standardization of
Iodine Solutions

Directions follow for the generation of an iodine solution and its standardization against trivalent arsenic; starch is used to indicate the end point.

1. Preparation of Approximately 0.1-N Iodine Solution. Weigh about 40 g of KI into a 100-ml beaker. Add 12.7 g of I_2 and 10 ml of water. Stir until solution is complete. Filter through a Gooch or a sintered glass crucible; then dilute to about 1 liter. Store in a glass-stoppered bottle. If possible, allow the solution to stand for two to three days before standardizing.

2. Preparation of Starch Indicator. Make a paste by rubbing about 2 g of soluble starch and 10 mg of HgI_2 in about 30 ml of water. Pour this into 1 liter of boiling water, and heat until a clear solution results. Cool, and store in stoppered bottles. For most titrations, 3 to 5 ml of this solution should be sufficient.

3. Standardization of Iodine against Arsenic(III). Dry a quantity of primary-standard-grade As_2O_3 for about 1 hr at 110°C. For standardization of 0.1-N I_2 solutions, weigh 0.2-g samples (to the nearest 0.1 mg). Dissolve in 10 ml of 1-F NaOH. When solution is complete, dilute with about 75 ml of water, and add 2 drops of phenolphthalein. Promptly introduce 6-F HCl until the red color just disappears. Then add about 1 ml of acid in excess. Carefully add 3 to 4 g of solid $NaHCO_3$, in small portions at first to avoid losses of solution due to effervescence of the CO_2. Add 5 ml of starch indicator, and titrate to the first faint purple or blue color that persists for at least 30 sec.

4. Alternatively, prepare a standard arsenic(III) solution by dissolving a known weight of dried, primary-standard As_2O_3 in 1-F NaOH (about 6 ml of base for each 100 ml of final solution). Neutralize with 1-F HCl; a small piece of litmus paper in the solution is an adequate indicator. Transfer the solution quantitatively to a volumetric flask, and dilute to the mark.
To standardize an iodine solution, withdraw convenient aliquots of the arsenite solution, acidify each with 1 ml of 6-F HCl, and proceed as in section 3.

EXPERIMENT 27: Determination of Antimony in Stibnite

See page 362 for additional information concerning this experiment.
Dry the sample for 1 hr at 110°C. After cooling, weigh sufficient quantities

of the ore to consume 25 to 35 ml of 0.1-N I_2 into 500-ml conical flasks. Add about 0.3 g of KCl and 10 ml of concentrated HCl. Heat the mixture to just below boiling, and maintain at this temperature until only a white or slightly gray residue of silica remains.

Add 3 g of solid tartaric acid to the solution, and heat for another 10 to 15 min. While swirling the solution, slowly add water from a pipet until the volume is about 100 ml. The addition of water should be slow enough to prevent formation of white SbOCl. If reddish Sb_2S_3 forms, stop the addition of water, and heat further to remove H_2S, adding more acid if necessary.

Add 3 drops of phenolphthalein to the solution and 6-F NaOH until the first pink color is observed. Add 6-F HCl dropwise until the solution is decolorized and then 1 ml in excess. Add 4 to 5 g of $NaHCO_3$, taking care to avoid losses of solution during the addition. Add 5 ml of starch, and titrate to the first blue color that persists for 30 sec or longer.

Iodometric Methods of Analysis

Numerous methods are based upon the reducing properties of iodide; the reaction product, iodine, is ordinarily titrated with a standard thiosulfate solution. Iodometric methods are discussed on pages 362–369.

EXPERIMENT 28: Preparation and Standardization of
Sodium Thiosulfate Solutions

The precautions that must be observed with thiosulfate solutions are described on pages 363–365.

1. Preparation of Approximately 0.1-N Sodium Thiosulfate Solution. Boil about 1 liter of distilled water for at least 5 min. Cool and add about 25 g of $Na_2S_2O_3 \cdot 5H_2O$ and 0.1 g of Na_2CO_3. Stir until solution is complete, then transfer to a clean stoppered bottle (glass or plastic). Store in the dark.

STANDARDIZATION OF THIOSULFATE SOLUTIONS

2. Against Potassium Iodate. Weigh (to the nearest 0.1 mg) about 0.12-g samples of dried, primary-standard KIO_3 into 250-ml conical flasks. Dissolve in 75 ml of water, and add about 2 g of iodate-free KI. *From here, treat each sample individually.* After solution is complete, add 10 ml of 1.0-N HCl, and titrate immediately with the thiosulfate solution until the color of the solution becomes pale yellow. Add 5 ml of starch, and titrate to the disappearance of the blue color.

To minimize the weighing error, this procedure can be modified by taking a 0.6-g sample, dissolving in water, and diluting to exactly 250 ml in a volumetric flask. A 50-ml aliquot of this solution can then be used for each standardization.

3. Against Potassium Dichromate. Dry primary-standard $K_2Cr_2O_7$ for 1 to 2 hr at 150 to 200°C, and weigh (to the nearest 0.1 mg) 0.20- to 0.23-g portions into 500-ml flasks. Dissolve in 50 ml of water. Then add a freshly prepared solution consisting of 3 g of KI, 5 ml of 6-N HCl, and 50 ml of water. Swirl gently, cover the flask with a watch glass, and let it stand in a dark place for 5 min. Wash down the sides of the flask, add 200 ml of water, and titrate with the thiosulfate solution. When the yellow color of the iodine becomes faint, add 5 ml of starch. Continue the titration until a color change from the blue starch-iodine complex to the green color of the Cr(III) ion is observed.

EXPERIMENT 29: Iodometric Methods

The analysis for copper is typical of iodometric methods. The directions that follow will permit the determination of copper in samples of ore and brass; additional information concerning the analysis is to be found on pages 366–369.

Directions are also supplied for the determination of dissolved oxygen by the Winkler method.

1. Determination of Copper in an Ore. Weigh appropriate-sized samples of the finely ground and dried ore (about 1 g for a sample containing 10 to 30% Cu) into 150-ml beakers, and add 20 ml of concentrated HNO_3. Heat until all of the Cu is in solution. If the volume becomes less than 5 ml, add more HNO_3. Continue the heating until only a white or slightly gray siliceous residue remains (Note 1). Evaporate to about 5 ml.

Add 25 ml of distilled water, and boil to bring all soluble salts into solution. If the residue is small and nearly colorless, no filtration is necessary. Otherwise, filter the suspension, collecting the filtrate in a 250-ml conical flask. Wash the paper with several small portions of hot 1:100 HNO_3. Evaporate the filtrate and washings to about 25 ml, cool, and slowly add concentrated NH_3 to the first appearance of the deep blue tetraammine copper(II) complex. A faint odor of NH_3 should be detectable over the solution. If it is not, add another drop of NH_3 and repeat the test. Avoid an excess (Note 2).

From this point on, treat each sample individually. Add 2.0 ± 0.1 g NH_4HF_2 (Caution! Note 3), and swirl until completely dissolved. Then add 3 g of KI, and titrate immediately with 0.1-N $Na_2S_2O_3$. When the color of the iodine is nearly discharged, add 2 g of KSCN and 3 ml of starch. Continue the titration until the blue starch-iodine color is decolorized and does not return for several minutes.

NOTES
1.   If the ore is not readily decomposed by the HNO_3, add 5 ml of concentrated HCl, and heat until only a small white or gray residue remains. Do not evaporate to dryness. Cool, add 10 ml of concentrated H_2SO_4, and evaporate until copious white fumes of SO_3 are observed (Hood). Cool and carefully add 15 ml of water and 10 ml of saturated bromine water. Boil the solution vigorously in a hood until all of the bromine has been removed. Cool and proceed as indicated in the second paragraph.
2.   If too much NH_3 is added, neutralize the excess with 3-F H_2SO_4.
3.   Ammonium hydrogen fluoride is a toxic and corrosive chemical. Avoid contact with the skin. If exposure does occur, immediately rinse the affected area with copious amounts of water.

2. Determination of Copper in Brass. Weigh (to the nearest 0.1 mg) about 0.3-g samples of the clean, dry metal into 250-ml conical flasks, and add 5 ml of 6-*F* HNO_3. Warm the solution (Hood) until decomposition is complete. Then add 10 ml of concentrated H_2SO_4, and evaporate to copious white fumes of SO_3 (Hood). Allow the mixture to cool. Then carefully add 20 ml of water. Boil for 1 to 2 min and cool.

With good mixing, add concentrated NH_3 dropwise until the first dark-blue color of the tetraammine copper(II) complex appears. The solution should smell faintly of NH_3. Add 3-*F* H_2SO_4 dropwise until the color of the complex just disappears. Then add 2.0 ml of syrupy phosphoric acid. Cool to room temperature.

From this point on, treat each sample individually. Dissolve 4.0 g of KI in 10 ml of water, and add this to the sample. Titrate immediately with standard $Na_2S_2O_3$ until the iodine color is no longer distinct. Add 5 ml of starch solution, and titrate until the blue begins to fade. Add 2 g of KSCN, and complete the titration.

3. The Winkler Method for Dissolved Oxygen. In an alkaline solution, dissolved oxygen will oxidize manganese(II) to the trivalent state

$$8OH^- + 4Mn^{2+} + O_2 + 2H_2O = 4Mn(OH)_3(s)$$

The analysis is completed by titrating the iodine produced from potassium iodide by manganese(III); sodium thiosulfate is used as the titrant.

Success of the method is critically dependent upon the manner in which the sample is manipulated; at all stages, every effort must be made to assure that oxygen is neither introduced to nor lost from the sample. Biological Oxygen Demand (BOD) bottles are designed to minimize the entrapment of air.

The sample should be free of any solutes that will oxidize iodide or reduce iodine. Numerous modifications have been developed to permit use of the Winkler method in the presence of such species.

4. Special Solutions. a. Manganese(II) sulfate. Dissolve 48 g of $MnSO_4 \cdot 4H_2O$ in sufficient water to give 100 ml of solution.

b. Potassium iodide–sodium hydroxide. Dissolve 15 g of KI in about 25 ml of water, add 66 ml of 50% NaOH, and dilute to 100 ml.

c. Starch. See page 747.

d. Sodium thiosulfate, 0.025 *N*. See Experiment 28 for the preparation and standardization of thiosulfate solutions.

Transfer the sample to a BOD bottle, taking care to minimize exposure to air. Use a tube to introduce the water to the bottom of the bottle; remove the tube slowly while the bottle is overflowing.

Add 1 ml of $MnSO_4$ with a dropper; discharge the reagent well below the surface (some overflow will occur). Similarly, introduce 1 ml of the KI-NaOH solution. Place the stopper in the bottle; be sure that no air becomes entrapped. Invert the bottle to distribute the precipitate uniformly.

When the precipitate has settled at least 3 cm below the stopper, introduce 1 ml of concentrated (18 *F*) H_2SO_4 well below the surface (Note 1). Replace the stopper and mix until the precipitate dissolves (Note 2). Measure 200 ml of the acidified sample into a 500-ml conical flask. Titrate with 0.025-N $Na_2S_2O_3$ until the iodine color becomes faint. Then introduce 5 ml of starch indicator, and complete the titration.

Report the milliliters of O_2 (STP) dissolved in each liter of sample.

NOTES
1. Care should be taken to avoid exposure to the overflow, as the solution is quite alkaline.
2. A magnetic stirrer is effective in bringing about solution of the precipitate.

EXPERIMENT 30:   Determination of Iodine and Iodide in a Mixture

The oxidizing properties and applications of iodate ion as a volumetric reagent are considered on pages 353–354. The accompanying directions permit the assay of iodine plus iodide by iodate titration; the results of this analysis, in conjunction with a thiosulfate (or arsenite) titration of the iodine only, are used to establish the concentration of each species in a mixture.

1. Special Solutions, Reagents. a. Potassium iodate, approximately 0.05 F. Weigh (to the nearest milligram) about 2.67 g of dried, primary-standard KIO_3 into a 250-ml volumetric flask, and dilute to the mark. Mix thoroughly.
b. Chloroform, $CHCl_3$.
c. Standard solution of sodium thiosulfate (p. 748) or sodium arsenite (p. 747).

2. Determination of I_2 and I^-. Obtain the unknown in a clean, 500-ml volumetric flask. Dilute to the mark, and mix thoroughly.
Fill one buret with the diluted I_2-I^- unknown and the other with standard KIO_3. Run about 25 ml of sample into a 250-ml conical flask. Add about 40 ml of concentrated HCl and 5 ml of $CHCl_3$. Titrate with KIO_3 until the purple color disappears from the chloroform layer. As the end point is approached, make dropwise additions of reagent, and swirl vigorously between additions. To eliminate the likelihood of overtitration, introduce enough of the sample solution to reestablish the iodine color in the nonaqueous phase. Titrate, as before, with iodate. Repeat as necessary to establish the end point.

3. Determination of I_2 Only. Titrate 50-ml aliquots of the unknown with a standard solution of arsenite or thiosulfate.
Report the grams of I_2 and of KI in the sample.

EXPERIMENT 31:   The Periodate Oxidation of Glycerol

Glycerol is cleanly oxidized by periodate, the products of the reaction being 2 moles of formaldehyde and 1 mole of formic acid. The oxidation requires about 30 min at room temperature.

A general description of the Malaprade reaction is given on pages 354–357. The procedure given here is a modification of the method published by Voris,

Ellis, and Maynard;[9] it involves reduction of the excess periodate to iodate with a known volume of arsenite solution, followed by titration of the excess arsenic(III) with standard iodine.

1. Special Solutions. a. Periodate solution, approximately 6.9 g of KIO_3/liter (0.03 *F*). The solution should be approximately 0.1 *N* with respect to H_2SO_4; it is conveniently standardized concurrently with the analysis.

b. Standard 0.1-*N* arsenious acid solution (p. 748).

c. Iodine solution, approximately 0.1 *N*, the volume ratio of which is known with respect to the arsenite solution.

d. $MgSO_4$ solution, approximately 15 g/100 ml.

e. Starch suspension (p. 747).

2. Oxidation and Titration. To a 250-ml iodine flask introduce a 10-ml aliquot of a solution that is 0.02 to 0.03 *M* in glycerol and a 25-ml aliquot of periodate solution; allow a 30-min reaction period at room temperature. Then add 3 drops of $MgSO_4$ solution, followed by the dropwise introduction of NaOH until the first faint turbidity of $Mg(OH)_2$ is observed (see Note). Add dilute H_2SO_4 until the turbidity is just discharged. Saturate the solution with $NaHCO_3$, and introduce a measured excess of standard arsenite solution. After 5 min, add starch, and titrate the excess arsenite with iodine solution. Calculate the weight of glycerol in the sample.

NOTE

The $MgSO_4$ solution serves as a rough indicator of the acidity. The concentration of NaOH used to adjust the pH may be 3 to 6 *F*, while that of the H_2SO_4 should be somewhat less. Neither concentration is critical.

EXPERIMENT 32:   Determination of Phenol by Bromine Substitution

Applications of potassium bromate solutions to organic analysis are described on pages 350–353. Typical of analyses based on substitution is the bromination of phenol, directions for which follow.

1. Special Solutions. a. Potassium bromate, 0.12 *N*. Dry reagent-grade $KBrO_3$ for 1 hr at 100 to 110°C and cool. Weigh approximately 3.3 g (to the nearest milligram) into a 1-liter volumetric flask, dilute to the mark, and mix thoroughly.

b. Starch indicator (p. 747).

c. Sodium thiosulfate, 0.1 *N*. Prepare 1 liter of approximately 0.1-*N* $Na_2S_2O_3$, following the instructions on page 748. This solution is conveniently standardized against $KBrO_3$ solution by the following procedure at the time of its use for the phenol analysis.

2. Standardization of Thiosulfate. Pipet 25 ml of the $KBrO_3$ solution into a flask, and add about 2 to 3 g of KI and 5 ml of 6-*N* H_2SO_4. Titrate the liberated

[9] L. Voris, G. Ellis, and L. Maynard, *J. Biol. Chem.,* **133**, 491 (1940).

I_2 until the solution becomes faintly yellow. Add 5 ml of starch solution, and continue the titration to the disappearance of the blue color.

3. *Titration of Phenol*. Measure a sample containing between 3 and 4 mmole of phenol, and dilute to exactly 250 ml in a volumetric flask. Pipet 25-ml aliquots of this solution into 250-ml glass-stoppered conical flasks, and add exactly 25 ml of the standard bromate reagent. Add about 0.5 g of KBr to each flask and about 5 ml of 3-F H_2SO_4. Stopper each flask *immediately* after addition of the acid to prevent loss of Br_2. Mix and let stand for at least 10 min. Weigh 2 to 3 g of KI (not accurately) for each analysis. Add the KI rapidly to each flask, restoppering immediately after the addition. Swirl the solution until the KI is dissolved. Then titrate with the standard $Na_2S_2O_3$ until the solution is faintly yellow. Add 5 ml of starch, and complete the titration.

See the instructor for the method of reporting the results.

Potentiometric Methods

The uses of potential measurements in analytical chemistry are discussed in Chapter 17; directions for several typical applications follow.

EXPERIMENT 33: Potentiometric Titrations

1. *General Instructions*. These instructions are applicable to most potentiometric titrations. Minor modifications may be required in specific instances.

a. Dissolve the sample in 50 to 250 ml of water. Rinse the electrodes with distilled water, and then immerse them in the sample solution. Make provision for magnetic (or mechanical) stirring. Position the buret so that reagent can be delivered without splashing.

b. Connect the electrodes to the potentiometer, commence stirring, measure, and record the initial potential.

c. Measure and record the potential after each addition of reagent. Introduce fairly large volumes (0 to 5 ml) at the outset. Withhold each succeeding addition until the potential remains constant within 1 to 2 mV (0.05 pH unit) for 30 sec. A stirring motor will occasionally cause erratic potential readings; it may be advisable to turn off the motor during the actual measuring process. Judge the volume of reagent to be added by calculating an approximate value of $\Delta E/\Delta V$ after each addition. In the immediate vicinity of the equivalence point, introduce the reagent in exact 0.1-ml increments. Continue the titration 2 to 3 ml past the equivalence point. Increase the volumes added as $\Delta E/\Delta V$ once again acquires small values.

d. Locate the end point by one of the methods described on pages 408–410.

2. *Titration of a Chloride-Iodide Mixture*. The potentiometric titration of halide mixtures is discussed on page 411. The indicator electrode can be simply a polished silver wire or a commercial billet-type variety. A calomel electrode can be used as reference, although diffusion of Cl^- from the salt bridge will

cause results to be slightly high. An alternative is to place the calomel electrode in a saturated KNO_3 solution and make contact with the solution to be titrated with a KNO_3 bridge. If desired, the solution can be made acidic with several drops of nitric acid; a glass electrode can then be employed as the reference, since the pH of the solution will remain substantially constant during the titration.

PROCEDURE

1. Prepare and standardize a 0.100-F $AgNO_3$ solution as instructed on page 726.

2. The sample will be issued as a solution. Take aliquots containing a total of 2 to 4 mfw of Cl^- and I^-; dilute to 100 $\pm$ 10 ml with water. Acidify with HNO_3, and titrate with the silver nitrate as instructed in section 1; use small increments of titrant near the two end points. Plot the data, and determine the end point for each ion. Plot $\Delta E/\Delta V$ versus V to ascertain the end points. Plot a theoretical titration curve, assuming the measured concentrations of the two constituents to be correct. Report the number of milligrams of I^- and Cl^- in the sample or as instructed.

Directions for several neutralization titrations follow. The theoretical basis for these titrations is to be found in Chapters 9 and 10. A glass electrode and a saturated calomel electrode serve as indicator and reference, respectively.

3. *Determination of the pH of Water.* To demonstrate the problems associated with the measurement of the pH of unbuffered and nearly neutral solutions, determine the pH of tap water; continue to take samples until successive specimens give readings that agree within 0.02 pH unit of one another.
Repeat the experiment, this time with distilled water.

4. *Titration of a Weak Acid.* A standard solution of carbonate-free NaOH (p. 730) is required. Prepare a solution containing between 1 and 4 meq of the acid in about 100 ml of water. A larger volume may be required to dissolve the less soluble organic acids. Add 2 drops of phenolphthalein to the solution.
Titrate as directed in the general instructions. Some samples will contain more than one replaceable hydrogen; be alert for more than one break in the titration curve. Note the volume at which the indicator changes color.
Plot the titration data, and determine the end point or points. Compare these with the phenolphthalein end point. The derivative method for end-point determination may also be used.
Calculate the number of milliequivalents of H^+ present in the sample. Evaluate the dissociation constant(s) for the species titrated.

5. *Titration of a Carbonate-Bicarbonate Mixture.* A standard solution of HCl is employed as the titrant. Dissolve the sample, which contains a total of 2 to 3 mfw of the two solutes in about 200 ml of water. Add 2 drops of

phenolphthalein, and titrate as directed in the general instructions. After the phenolphthalein indicator has become colorless, add 2 drops of methyl orange. Carry the titration 3 to 5 ml beyond the second end point. Note the titrant volumes where the two indicators change color.

Plot the data, and determine the end points. Compare the potentiometric and indicator end points. Estimate the two dissociation constants for carbonic acid from the curve, and compare these with literature values.

Calculate the percent Na_2CO_3 and $NaHCO_3$ in the sample.

6. Titration of a Phosphate Mixture. Standard solutions of carbonate-free NaOH and of HCl are required.

The sample will contain one or two of the following components: HCl, H_3PO_4, NaH_2PO_4, Na_2HPO_4, Na_3PO_4, and NaOH. The analytical problem is to determine which compatible components are present as well as their amounts.

Depending upon its composition, the sample may require a titration with only standard HCl or with standard NaOH. For other compositions, separate aliquots may have to be titrated, the one with HCl and the other with NaOH. Use the initial pH of the solution and the titration curve for H_3PO_4 (p. 230) to select the appropriate titrant(s).

The sample will be issued as a solution, which should be diluted to volume in a volumetric flask. Mix well.

Pipet exactly 50.0 ml of the sample into a beaker, and determine its pH. Titrate with either standard HCl or NaOH until one or two end points have been passed; follow the general titration instructions. If necessary, titrate a second aliquot with the reagent not employed in the first titration. On the basis of the titration curves, choose acid-base indicators that would be suitable for the titrations, and perform the duplicate analyses with these.

Identify and report the total number of millimoles of the component(s) in the unknown solution. Also, calculate and report approximate values for any dissociation constants for H_3PO_4 that can be obtained from the titration data.

7. Titration of Organic Salts in Glacial Acetic Acid. Standard solutions of perchloric acid and sodium acetate are required (p. 736). Establish the acid-base ratio between these solutions, using a glass-calomel electrode system in place of the indicator. Rinse the electrodes well with water and dry with a tissue immediately before use. Follow the general instructions for a potentiometric titration; set the pH meter to read in millivolts.

Standardize the perchloric acid against potassium hydrogen phthalate (p. 736); again determine the end point potentiometrically.

The sample will be the sodium, potassium, or ammonium salt of an organic acid. Weigh 0.1- to 0.2-g portions (to the nearest 0.1 mg) into beakers, and dissolve in 25 ml of glacial acetic acid. Heat, if necessary, until solution is complete; then cool to room temperature. Titrate immediately with standard acid, using the glass-calomel electrode system and the millivolt scale of the pH meter. If the sample is not readily soluble, add an accurately measured volume of the standard $HClO_4$ solution (about 20 ml) to the suspension of the salt in acetic acid, and stir for about 10 min. The presence of the $HClO_4$ should bring the salt into solution. Back-titrate the excess $HClO_4$ with the standard solution of sodium acetate, determining the end point potentiometrically as in the acid-base ratio determination.

Repeat the experiment, and report the average equivalent weight of the sample.

8. Analysis of Amines and Amino Acids by Titration in Glacial Acetic Acid. Follow the procedure on page 737, but employ the glass-calomel system to detect the end point.

9. Titration of Iron(II) with Quadrivalent Cerium. The iron in a sample of matter can be determined conveniently by a potentiometric titration with cerium(IV). Prereduction of any iron(III) is accomplished with tin(II) chloride (p. 742).

a. Prepare a platinum-calomel electrode system, and connect the leads to a pH meter or a potentiometer.

Prepare and standardize a 0.100-N solution of quadrivalent cerium in sulfuric acid as directed in Experiment 22. Weigh, dissolve, and reduce the iron sample as directed on page 742. Dilute the solution to about 50 ml, and titrate immediately with the cerium(IV).

Plot the voltage data against the volume of reagent, and determine the end point. Calculate the percent iron in the sample.

b. Procedure employing a differential end point (p. 414). In this experiment a pair of platinum electrodes is used; one is housed in a glass tube such as that shown in Figure 17-19 (Note).

Prepare the reagent and sample as directed in the preceding section. Add about 1 ml of the reagent, stir, and measure the potential. Then homogenize the solution around the shielded electrode by squeezing the rubber bulb, and add more reagent. Again record the potential and volume of reagent before mixing the solution around the shielded electrode with the bulk of the solution. Continue this process, decreasing the volume increments to 0.1 ml in the vicinity of the end point. Carry the titration 2 to 5 ml beyond the end point.

Plot the data, and calculate the percent iron.

NOTE

A small hole in the rubber bulb will minimize the danger of drawing solutions into the bulb. The hole is covered with a finger to push liquid from the tube and is uncovered at all other times.

Electrogravimetric Analysis

Phenomena associated with the passage of current through electrochemical cells are discussed in Chapter 18; analytical applications of electrodeposition are considered in Chapter 19. Directions follow for the electrogravimetric analysis of copper in solution and for the consecutive analysis of copper and nickel in an alloy.

EXPERIMENT 34: Electrodeposition of Copper and Nickel

1. Determination of Copper in an Aqueous Solution. The sample to be analyzed should be free of chloride ion (Note 1) and contain from 0.2 to 0.3 g of Cu in 100 ml of water. Transfer the solution to a 150-ml electrolytic beaker. Add 3 ml of concentrated H_2SO_4 and 2 ml of freshly boiled and cooled 6-*F* HNO_3.

Prepare the electrodes by immersion in hot 6-*F* HNO_3 that contains about 1 g of KNO_2 (Note 2). Wash thoroughly with distilled water, rinse several times with small portions of ethyl alcohol or acetone, and dry in an oven at 110°C for 2 to 3 min. Cool and weigh the cathode carefully on an analytical balance (Note 3).

Attach the cathode to the negative terminal of the electrolytic apparatus and the anode to the positive terminal. Elevate the beaker containing the solution so that all but a few millimeters of the cathode is covered. Start the stirring motor, and adjust the potential so that a current of about 2 A passes through the cell (Note 4). When the blue copper color has entirely disappeared from the solution, add sufficient water to raise the liquid level by a detectable amount, and continue the electrolysis with a current of about 0.5 A. If no further copper deposit appears on the newly covered portion of the cathode within 15 min, the electrolysis is complete. If additional copper does deposit, continue testing for completeness of deposition from time to time as directed above.

When no more copper is deposited after a 15-min period, stop the stirring, and slowly lower the beaker while continuously playing a stream of wash water on the electrodes. Rinse the electrodes thoroughly with a fine stream of water. *Maintain the applied potential until rinsing is complete* (Note 5). Disconnect the cathode, and immerse it in a beaker of distilled water. Then rinse with several portions of alcohol or acetone. Dry the cathode in an oven for 2 to 3 min at 110°C, cool, and weigh.

NOTES
1.   The presence of chloride results in attack on the platinum anode. This is not only destructive but will also lead to errors, since the dissolved platinum will codeposit with the copper at the cathode.
2.   Grease and organic material can be removed by bringing the electrode to red heat in a flame.
3.   The cathode surface should not be touched with the fingers, because grease and oil cause nonadherent deposits.
4.   Alternatively, the electrolysis may be carried out without stirring. Here the current should be kept below 0.5 A. Several hours will be required to complete the analysis.
5.   It is important to maintain the application of a potential until the electrodes have been removed from the solution and washed free of acid. If this precaution is not observed, solution of some copper may occur.

2. Separation and Determination of Copper and Nickel in Alloys. Owing to the preferential reduction of hydrogen and nitrate ions, nickel ion is not reduced in the nitric–sulfuric acid environment suitable for the deposition of copper. Once the copper has been removed, however, nickel can be caused to deposit after the solution has been rendered somewhat ammoniacal. Nitrate remains a source of interference and must first be removed by volatilization.

Metals such as iron and aluminum interfere with the nickel analysis by forming sparingly soluble hydrous oxides in the basic medium required for the electrolysis. Unless they are removed by filtration, these precipitates will occlude with the electrolytic deposit and cause high results. Double precipitation is advised if hydrous oxides are formed in large amounts; otherwise, loss of nickel may occur as a result of adsorption.

PROCEDURE

Into 150- to 200-ml electrolytic beakers weigh samples that will give 0.1- to 0.3-g deposits of the two metals; dissolve in a mixture of 12 ml of 3-F H_2SO_4, and 5 ml of 3-F HNO_3. Boil the solution, cool, and dilute to about 100 ml.

Deposit the copper as directed in section 1; retain the solution and washings for the nickel analysis. Keep the cathode in a desiccator with the copper deposit intact.

Evaporate the solution and washings from the copper electrolysis until fumes of SO_3 are observed. Cool; then carefully add 25 ml of water. Add 1:1 NH_3 until the solution is basic to litmus; remove any hydrous oxide precipitate that appears by filtration through a small paper. Collect the filtrate in an electrolysis beaker. Wash the precipitate with several small portions of water. If precipitation is extensive, redissolve the solid by pouring a little warm 3-F H_2SO_4 through the paper and collecting the solution in a fresh beaker. Wash the paper with water; reprecipitate the hydrous oxide with ammonia. Filter through the same paper, and collect the filtrate in the electrolysis beaker containing the original solution. Wash the precipitate with water. Adjust the volume of the solution to about 100 ml, and add 15 ml of concentrated NH_3. Deposit the nickel on the cathode used for the copper. Follow the instructions for the electrolysis of the copper solution. Weigh the combined precipitates of copper and nickel.

Coulometric Analysis

The equivalence between chemical change and quantity of electricity is the basis for coulometric analysis. The topic is discussed at length in Chapter 20. Typical of constant-current coulometry is the electrolytic generation of iodine. This reagent is used to titrate arsenic(III) in the accompanying experiment.

The apparatus required is a constant-current supply (5 to 30 mA/sec), a stopclock, and a potentiometer (see Figure 20-4). The potentiometer serves to determine the current by measurement of the potential drop across a precision standard resistor (50 to 100 ohms). The titration vessel can be a 150- to 250-ml beaker provided with magnetic stirring. The electrode system consists of a platinum wire cathode and a platinum foil anode (at least 2 cm^2 in area). Iodine is generated at the anode and hydrogen at the cathode. The latter does not interfere with the analysis. Reaction conditions are similar to those described in Experiment 26 for the iodimetric determination of As(III).

EXPERIMENT 35:   Titration of Trivalent Arsenic with Iodine

Weigh a sample (to the nearest 0.1 mg) that will contain about 0.1 g As_2O_3, and dissolve in 30 ml of 1-F NaOH. Neutralize with 1-F HCl (litmus paper); add 1 or 2 drops excess HCl, and dilute to exactly 1 liter in a volumetric flask.

See page 747 for the preparation of a starch suspension.

Turn on the constant-current generator. Set and measure the current with the potentiometer. Place about 50 ml of water containing 3 to 4 g of $NaHCO_3$ in the titration vessel. Add about 5 ml of 1-F KI and 5 ml of the starch indicator.

Immerse the electrodes in the solution, and start the stirrer. Switch on the electrolysis and timer unit, and titrate until the solution develops a permanent faint blue color. Stop the electrolysis, and reset the timer to zero. Pipet 10 ml of the sample into the cell, and titrate to the same faint blue color. Record the elapsed time.

Additional titrations can be performed by adding fresh aliquots of sample to the solution just titrated.

Calculate the percentage of As_2O_3 in the sample.

Voltammetry

This section contains directions for the performance of selected polarographic analyses and amperometric titrations. Chapter 21 should be consulted for a discussion of voltammetry.

EXPERIMENT 36:   Polarographic Analysis

Enormous variation exists among equipment available for polarographic work; the reader should consult instructional literature for the particular instrument that is available. The characteristics of many commercial instruments have been summarized by Meites.[10]

1. Polarographic Behavior of Cadmium(II). Into 50-ml volumetric flasks transfer 5.0 ml of 0.1% gelatin, 10.0 ml of 0.5-F KNO_3, and, respectively, 1.00, 3.00, 10.0, and 30.0 ml of 5.0 × 10^{-3} F Cd^{2+}; dilute each solution to the mark and mix well. Prepare an additional solution with 10.0 ml of Cd^{2+} and 10.0 ml of KNO_3, but no gelatin.

Rinse the polarographic cell with the solution containing no Cd^{2+} ion, and equilibrate in a constant temperature waterbath (about 25°C). Obtain a polarogram from 0 to −2.0 V. Bubble nitrogen through the solution for 15 min, and again obtain the polarogram.

Obtain polarograms at 25°C for each of the Cd^{2+}-containing solutions after removal of O_2. If a manual polarograph is used, establish complete curves

[10] L. Meites, *Polarographic Techniques*, 2d ed., pp. 36–55. New York: Interscience Publishers, Inc., 1965.

for the sample containing 10.0 ml of Cd^{2+}; take only enough points to determine i_d for the remainder.

    a. Compare the experimentally determined oxygen wave with Figure 21-7.

    b. Compare the cadmium waves obtained with and without the addition of gelatin.

    c. Calculate $E_{1/2}$ for each determination; compare the average value with published data.

    d. Plot E versus $\log i/(i_d - i)$ for the solution containing 10.0 ml of Cd^{2+} solution and gelatin; see Equation 21-6. Evaluate n from this plot.

    2. Polarogram of a Mixture of Two Ions. Prepare 50 ml of a solution that contains 5 ml of the gelatin, 10 ml of 0.5-F KNO_3, 10 ml of the standard cadmium solution, and 10 ml of 5×10^{-3} F Zn^{2+}. After removal of oxygen, obtain the polarogram.

    Determine i_d/C and $E_{1/2}$ for each wave.

    3. Anodic and Cathodic Polarograms. Prepare a saturated solution of oxalic acid. Transfer exactly 25 ml (or some other carefully measured volume) of this solution to a polarographic cell, add 1 drop of methyl red indicator, and deaerate. Record the polarogram from $+0.2$ to -0.6 V (versus SCE). Add 1 ml of 0.05-F solution of iron(III) sulfate, deaerate again, and record the polarogram over the same voltage range.

    Transfer another 25-ml aliquot of the oxalic acid to the cell, add 1 drop of methyl red, and remove the oxygen. Quickly add 1 ml of freshly prepared 0.05-F iron(II) ammonium sulfate in 0.05-F H_2SO_4 and deaerate briefly. Record the polarogram as before while nitrogen is passed over the surface of the solution.

    Finally, add 1 ml of the iron(III) sulfate to the solution in the cell, and after a brief aeration again obtain a polarogram.

    Calculate the wave heights and half-wave potentials for the three solutions. Interpret the results.

EXPERIMENT 37: Amperometric Titrations

Amperometric titrations are discussed in Chapter 21. Directions follow to illustrate use of the dropping mercury electrode (1, 2), the rotating platinum electrode (3), and twin polarized electrodes (4).

    1. Titration of Lead. Prepare 0.100-F $Pb(NO_3)_2$ and 0.050-F $K_2Cr_2O_7$ solutions and a pH-4.2 acetate buffer (add sufficient 0.1-F NaOAc to a solution that is 0.2 F in HOAc and 0.2 F in KNO_3 until the desired pH is obtained).

    Transfer 5.00 ml of the $Pb(NO_3)_2$ to the titration vessel; add 5 ml of 0.1% gelatin and 40 ml of the acetate buffer. Insert a dropping electrode and a saturated calomel electrode. Bubble nitrogen through the solution for 15 min; then measure the current at -1.0 V after the gas flow has been stopped. Add exactly 1.00 ml of $K_2Cr_2O_7$, bubble with nitrogen, and again determine the current. Continue to add the $K_2Cr_2O_7$ in 1-ml increments until a total of 10 ml has been added. Correct the currents for dilution, and plot the titration curve. Determine the end point by extrapolation.

Repeat the titration at zero applied potential. For this titration, oxygen removal is unnecessary. Contrast the curves, and explain the differences.

2. Titration of Sulfate Ion. Transfer 25.0 ml of 0.0200-F K_2SO_4 to the titration vessel, and add 25 ml of a 40% (by volume) ethyl alcohol solution containing a drop of methyl red. Deaerate the solution, and titrate with the 0.1-F $Pb(NO_3)_2$, following the directions given in the preceding section. Determine the end point graphically.

3. Titration of Arsenic(III) with Bromate. A rotating platinum electrode, a saturated calomel electrode, and a 100-ml titration vessel are required.

Prepare a 0.0075-F arsenious acid solution (p. 747) and a 0.0100-F potassium bromate solution (p. 752).

Transfer 10 ml of the arsenious acid solution to the titration vessel, and add 40 ml of a solution that is about 2 F in HCl and 0.1 F in KBr. Fill a 10-ml buret with the bromate solution. Adjust the electrode so that it rotates at about 600 rpm. Apply +0.2 V to the rotating electrode, and titrate as described in the foregoing section. Locate the end points graphically.

4. Titration with Twin-Polarized Electrodes. Assemble a cell consisting of a 100-ml beaker, a good stirrer, and a pair of small platinum wire electrodes. Place 5 ml of 0.01-N iodine in the cell, add 0.1 g of KI, and dilute to about 50 ml. Apply a potential of 0.1 V to the electrodes, start the stirrer, and titrate the solution with 0.01-N $Na_2S_2O_3$. Add the reagent in 0.5-ml increments, recording the current after each addition. Plot the data and determine the end point.

Methods Based on the Absorption of Radiation

The analytical uses to which absorption measurements have been put are reviewed in Chapter 24; further information is to be found in Chapters 22 and 23. Directions follow for a determination of iron that requires generation of a calibration curve, a method for the analysis of manganese in steel that makes use of a standard addition, and an experiment involving the resolution of a mixture that contains two species with mutually overlapping absorption spectra.

EXPERIMENT 38:   Determination of Iron in Water

An excellent and sensitive method for the determination of iron is based upon the formation of the orange-red iron(II)-orthophenanthroline complex, the structure of which is shown on page 331. Orthophenanthroline is a weak base; in acidic solution the principal species is the phenanthrolium ion PhH^+. Thus, the complex-formation reaction is best described by the equation

$$Fe^{2+} + 3PhH^+ \rightleftarrows Fe(Ph)_3^{2+} + 3H^+$$

The equilibrium constant for this reaction is 2.5×10^6 at 25°C. Quantitative formation of the complex is observed in the pH region between 2 to 9. Ordinarily, a pH of about 3.5 is recommended to prevent precipitation of various iron salts such as phosphates. Careful control of pH is not required.

In using orthophenanthroline for the analysis of iron, an excess of reducing agent is added to the solution to keep the iron in the $+2$ state; either hydroquinone or hydroxylamine hydrochloride is convenient for this purpose. Once formed, the color of the complex is stable for long periods of time.

Certain ions interfere with the analysis for iron and must therefore be absent. These include colored ions in general; silver and bismuth, which form precipitates with the reagent; and cadmium, mercury, and zinc, which form colorless soluble complexes with the reagent and lower color intensity. Under certain conditions, molybdenum, tungsten, copper, cobalt, nickel, and tin may also interfere.[11]

The experiment can be performed with a spectrophotometer at 508 nm or with a photometer equipped with a green filter. Consult the manufacturer's instructions for operating details of the instrument to be used.

1. Special Solutions. a. Hydroxylamine hydrochloride, 10%. Dissolve 10 g of $H_2NOH \cdot HCl$ in about 100 ml of water. Introduce sufficient sodium citrate to bring the pH to 4.5.

b. Sodium citrate solution, 250 g per liter of solution.

c. *o*-Phenanthroline, 0.3% of the monohydrate in water. Store in a dark place. Discard the reagent when it becomes colored.

d. Standard iron solution, 0.1 mg Fe/ml. Dissolve 0.702 g of analytical reagent grade $FeSO_4 \cdot (NH_4)_2SO_4 \cdot 6H_2O$ in 50 ml of water containing 1 ml of concentrated H_2SO_4. Transfer to a volumetric flask, and dilute to exactly 1 liter.

Alternatively, dissolve 0.1000 g of electrolytic iron wire in 6 to 10 ml of 6-*F* HCl, and dilute to exactly 1 liter.

2. Preparation of Calibration Curve. Transfer a 5-ml aliquot of the standard iron solution into a beaker, and add a drop of bromophenol blue indicator. Introduce sodium citrate solution from a pipet until the intermediate color of the indicator is achieved. Note the volume of citrate required, and discard the solution. Now measure a second 5-ml aliquot of the iron standard into a 100-ml volumetric flask, and add 1 ml of the hydroxylamine and 3 ml of the orthophenanthroline solution. Introduce the same quantity of citrate solution as was required for the preliminary titration, and allow the mixture to stand 5 min. Dilute to the mark.

Clean the cells for the instrument, rinse one of these with the solution of the complex, and then fill. Rinse and fill the second cell with a blank containing all reagents except the iron solution. Carefully wipe the windows of the cells with tissue, and place them in the instrument. Measure the absorbance of the standard against the blank.

[11] See E. B. Sandell, *Colorimetric Determination of Traces of Metals*, 3d ed., p. 537. New York: Interscience Publishers, Inc., 1959.

Prepare at least three other standards so that an absorbance range between about 0.1 to 1.0 will be covered. Construct a calibration curve for the instrument.

3. Analysis of Sample. Transfer a 5-ml aliquot of the sample into a beaker, and add 1 drop of bromophenol blue. Adjust the pH of the solution with sodium citrate solution or 0.1-N H_2SO_4 from a pipet until the intermediate color of the indicator is achieved. Note the volume of reagent required, and then discard the solution. Transfer a fresh 5-ml aliquot to a 100-ml volumetric flask; add 1 ml of the hydroxylamine and 3 ml of the orthophenanthroline solution. Introduce the same quantity of citrate solution or sulfuric acid as was required to adjust the pH. After 5 min, dilute to the mark, and measure the absorbance. Repeat the analysis if necessary, using quantities of sample that will give an absorbance in the range of the calibration curve.

Calculate the milligrams of iron per liter of sample solution.

EXPERIMENT 39:   Determination of Manganese in Steel

Small quantities of manganese are readily determined colorimetrically by oxidation to the highly colored permanganate ion. Potassium periodate is an effective oxidizing reagent:

$$5IO_4^- + 2Mn^{2+} + 3H_2O \rightarrow 2MnO_4^- + 5IO_3^- + 6H^+$$

Permanganate solutions containing an excess of periodate are relatively stable.

Interferences to this procedure are few. The presence of colored ions can be compensated by employing as a blank an aliquot of the sample that has not been oxidized by the periodate. This method of correction is not effective in the presence of appreciable quantities of cerium(III) or chromium(III) ions, for both are oxidized by the periodate to a greater or lesser extent, and their reaction products absorb in the region commonly employed for the permanganate.

The accompanying method is applicable to most steels except those containing large amounts of chromium. The sample is dissolved in nitric acid; any carbon present is removed by oxidation with peroxodisulfate. Phosphoric acid is added to complex the iron(III) and prevent the color of this species from interfering with the analysis. The standard-addition method (p. 547) is used to establish the relationship between absorbance and concentration.

A spectrophotometer set at 525 nm or a photometer with a green filter may be used for absorbance measurements. Consult the instruction manual of the instrument to be used for specific operating instructions.

1. Preparation of Standard Manganese(II) Solution. Dissolve approximately 0.100 g of Mn (weighed to the nearest 0.1 mg) in about 10 ml of HNO_3. Boil the solution gently to eliminate oxides of nitrogen; cool, then transfer the solution quantitatively to a 1-liter volumetric flask, and dilute to the mark.

2. Analysis of Steel. Weigh duplicate 1.0-g samples of the steel (to the nearest mg), and dissolve in 50 ml of 4-F HNO_3 with gentle boiling; heating for 5 min should suffice. Cautiously add about 1 g of ammonium peroxodisulfate, and boil gently for 10 to 15 min. If the solution is pink or contains a brown oxide of manganese, add approximately 0.1 g of sodium hydrogen sulfite or ammonium hydrogen sulfite, and heat for another 5 min. Cool and dilute the solution to exactly 100 ml in a volumetric flask.

Pipet three 20.0-ml aliquots of each sample into small beakers. Treat as follows:

	H_3PO_4, ml	Std Mn Soln, ml	KIO_4, g
Aliquot 1	5	0.00	0.4
Aliquot 2	5	5.00	0.4
Aliquot 3	5	0.00	0.0

Boil each solution gently for 5 min; then cool and dilute to 50.0 ml in volumetric flasks. Determine the absorbance of Aliquots 1 and 2, with Aliquot 3 serving as blank.

Report the average percentage of manganese in the sample.

EXPERIMENT 40: Spectrophotometric Investigation of a
Two-Color System

The concentrations of absorbing species in a two-color system can be obtained from absorbance measurements of the solution and individual absorptivity data for each component at two wavelengths; see page 547.

1. *Preparation of a Stock Methyl Red Solution.* Dissolve approximately 40 mg (weighed to the nearest 0.1 mg) of methyl red in a minimum of dilute NaOH, and dilute to 1 liter in a volumetric flask.

2. *Determination of Individual Absorption Spectra.* Transfer 25.0-ml aliquots of stock methyl red solution to each of two 100-ml volumetric flasks. To one, add 25.0 ml of 0.4-*F* HCl; to the other, add 25.0 ml of 0.4-*F* NaOH. Dilute to the mark; mix thoroughly. Obtain the absorption spectra for the acid and conjugate-base forms of the indicator between 400 and 600 nm, using water as a blank; record absorbance values at 10-nm intervals routinely and at smaller intervals as necessary to define the curves. Evaluate the absorptivity for each species at wavelengths that correspond to their respective absorption maxima.

3. *Determination of the pH of an Unknown Buffer.* Transfer a 25.0-ml aliquot of the stock methyl red solution to a 100-ml volumetric flask. Add 50.0 ml of unknown buffer, dilute to the mark, and measure the absorbance of the diluted solution at the wavelengths for which absorptivity data were calculated. Report the pH of the buffer.

Flame Photometry

A discussion of flame emission photometry is presented in Chapter 25. The accompanying directions are suitable for the analysis of sodium, potassium, and calcium in water samples; radiation buffers are used to minimize the effect of each ion upon the emission intensity of the others.

EXPERIMENT 41: Determination of Sodium, Potassium, and
Calcium in Mineral Waters[12]

1. Special Solutions. a. Prepare stock solutions from reagent-grade chemicals. For 1-liter quantities of 500-ppm standards, take 1.2708 g of NaCl, 0.9535 g of KCl, or 1.2486 g of $CaCO_3$. In preparing the last, make a preliminary addition of about 300 ml of water, followed by about 10 ml of concentrated HCl. Dilute to volume after solution is complete and the evolution of CO_2 has ceased.

b. Radiation buffer for sodium determination. Saturate a solution with reagent-grade $CaCl_2$, KCl, and $MgCl_2$, in that order.

c. Radiation buffer for potassium determination. Saturate a solution with reagent-grade NaCl, $CaCl_2$, and $MgCl_2$, in that order.

d. Radiation buffer for calcium determination. Saturate a solution with reagent-grade NaCl, KCl, and $MgCl_2$, in that order.

2. Preparation of Working Curves. Introduce 5.00 ml of the appropriate radiation buffer to each of a series of 100-ml volumetric flasks. Then add volumes of standard in the amounts needed to give coverage to the concentration range from 0 to 100 ppm. Dilute to 100 ml with distilled water and mix well. Measure the emission intensity for these samples; take at least three readings for each. Between each set of measurements, aspirate distilled water through the burner. Correct the average values for background luminosity, and prepare a working curve from these data.

3. Analysis of a Water Sample. Prepare aliquot portions of the sample as directed above for standards. If necessary, use a standard to calibrate the response of the spectrophotometer to the working curve. Then measure the emission intensity for the unknown. After correcting the data for background, determine the concentration by comparison with the working curve.

[12] The method has been adapted from P. W. West, P. Folse, and D. Montgomery, *Anal. Chem.* **22**, 667 (1950).

ANSWERS TO PROBLEMS

Chapter 2

1. (a) 0.119 fw or 119 mfw,
   (b) 2.40×10^{-6} fw or 2.40×10^{-3} mfw,
   (c) 0.277 fw or 277 mfw,
   (d) 2.00×10^{-5} fw or 2.00×10^{-2} mfw,
   (e) 2.23×10^{-5} fw or 2.23×10^{-2} mfw.
3. (a) 88.0 g,   (b) 0.144 g,   (c) 1600 g,
   (d) 0.258 g,   (e) 1480 g.
5. (a) 0.0127 F,   (b) 0.0254 M,
   (c) 0.121%,   (d) 0.120%,
   (e) 0.318 mmole.

7.
Species	F	Species	F
Cl	5.49×10^{-1}	Ca	1.02×10^{-2}
Na	4.72×10^{-1}	K	9.95×10^{-3}
Mg	5.35×10^{-2}	Br	8.34×10^{-4}
S	2.82×10^{-2}	C	2.39×10^{-3}
		Sr	1.52×10^{-4}

9. 3.04 F
11. $2.50 \times 10^{-4} F$
13. (a) Dissolve 20 g glucose in H_2O, and dilute to 200 ml.

(b) Dissolve 20 g glucose in 180 g H_2O.
(c) Dissolve 20 ml of ethanol in H_2O, and dilute to 200 ml.

15. (a) Dissolve 51.2 g $BaCl_2 \cdot 2H_2O$ in H_2O, and dilute to 525 ml.
    (b) Dissolve 40.0 g K_2SO_4 in H_2O, and dilute to 2.30 liters.
    (c) Dilute 22.7 ml of the $AgNO_3$ to 100 ml.
    (d) Dilute 26.1 ml of the $KMnO_4$ to 500 ml.
    (e) Dilute 960 ml of the K_2SO_4 to 3.0 liters.

17. (a) Dissolve 37.9 g $K_2Cr_2O_7$ in H_2O, and dilute to 750 ml.
    (b) Dissolve 710 g of Na_2SO_4 in H_2O, and dilute to 50.0 liters.
    (c) Dissolve 1.75 g NaCl in H_2O, and dilute to 2.00 liters.
    (d) Dilute 835 ml of the Na_2SO_4 to 20.0 liters.
    (e) Dissolve 21.3 g Na_2SO_4 in H_2O, and dilute to 2.00 liters.

19. (a) 490 g HNO_3/500 ml,
    (b) dilute about 10.3 ml concentrated HNO_3 to about 800 ml.

21. Dilute about 164 ml of concentrated H_3PO_4 to 400 ml.

23. (a) 4.25 g KIO_3,   (b) 1284 g KIO_3,
   (c) 1.09 g $La(NO_3)_3$,   (d) 3.10 g $La(IO_3)_3$,
   (e) 0.159 g $La(IO_3)_3$.

25. (a) 2.02 g KOH,   (b) 66.7 ml KOH,
   (c) 351 ml HCl,   (d) 134 ml $AgNO_3$,
   (e) 553 ml $Ba(OH)_2$.

27. 78.1 g $BaCl_2 \cdot 2H_2O$

29. (a) 0.953 g $CaCO_3$,   (b) 6.88 g loss.

Chapter 3

1. (a) $K_{sp} = [Tl^+][Cl^-]$,
   (c) $K_{sp} = [Ag^+]^2[SO_4^{2-}]$,
   (e) $K_{sp} = [Pr^{3+}][OH^-]^3$,
   (g) $K_{sp} = [Mg^{2+}][NH_4^+][PO_4^{3-}]$.

2. (a) $s = (K_{sp})^{1/2}$,   (c) $s = (K_{sp}/4)^{1/3}$,
   (e) $s = (K_{sp}/27)^{1/4}$,   (g) $s = (K_{sp})^{1/3}$.

3. (a) $C_2H_5NH_2 + H_2O \rightleftarrows C_2H_5NH_3^+ + OH^-$
    $K_b = 4.28 \times 10^{-4}$
    $= [C_2H_5NH_3^+][OH^-]/[C_2H_5NH_2]$,
   (c) $C_5H_5NH^+ + H_2O \rightleftarrows C_5H_5N + H_3O^+$
$$K_a = \frac{K_w}{K_b} = \frac{1.00 \times 10^{-14}}{1.7 \times 10^{-9}} = 5.9 \times 10^{-6}$$
    $= [C_5H_5N][H_3O^+]/[C_5H_5NH^+]$,
   (e) $AgCl(s) + Cl^- \rightleftarrows AgCl_2^-$
    $K_{s2} = [AgCl_2^-]/[Cl^-] = 2.0 \times 10^{-5}$,
   (g) $Cd^{2+} + 2NH_3 \rightleftarrows Cd(NH_3)_2^{2+}$
$$\beta_2 = K_1 K_2 = \frac{[Cd(NH_3)_2^{2+}]}{[Cd^{2+}][NH_3]^2}$$
    $= 3.2 \times 10^2 \times 9.1 \times 10^1$
    $= 2.9 \times 10^4$,
   (i) $H_3AsO_4 + 3H_2O \rightleftarrows AsO_4^{3-} + 3H_3O^+$
    $K_1 K_2 K_3 = [H_3O^+]^3[AsO_4^{3-}]/[H_3AsO_4]$
    $= 6.0 \times 10^{-3} \times 1.05 \times 10^{-7}$
    $\times 3.0 \times 10^{-12}$
    $= 1.9 \times 10^{-21}$.

4. (a) 4.0×10^{-16},   (c) 2.0×10^{-2},
   (e) 7.1×10^{-9},   (g) 3.5×10^{-10}.

5. (a) 5.5×10^{-8},   (c) 3.4×10^{-7}.

6. (a) 0.056 g/100 ml,
   (b) 8.2×10^{-4} g/100 ml,
   (c) 0.0137 g/100 ml.

8. (a) $2.2 \times 10^{-7} M$,   (b) 0.98 M.

10. (a) $3.3 \times 10^{-4} M$

11. (a) 0.025 M,   (b) $1.9 \times 10^{-3} F$,
   (c) $7.6 \times 10^{-7} M$,   (d) 0.0167.

13. (a) $[H_3O^+] = 2.4 \times 10^{-5}$
    $[OH^-] = 4.1 \times 10^{-10}$,
   (c) $[H_3O^+] = 1.04 \times 10^{-12}$
    $[OH^-] = 9.6 \times 10^{-3}$,
   (e) $[H_3O^+] = 5.0 \times 10^{-11}$
    $[OH^-] = 2.0 \times 10^{-4}$.

15. (a) $1.10 \times 10^{-2} M$,   (b) $1.17 \times 10^{-8} M$,
   (e) $1.47 \times 10^{-4} M$.

16. (a) $2.38 \times 10^{-6} M$

17.

	[SCN$^-$]	[Hg^{2+}]
(a)	2.4×10^{-9}	0.0050
(b)	1.0×10^{-6}	5.2×10^{-7}
(c)	0.010	5.6×10^{-16}

18. (a) $8 \times 10^{-7} M$,   (b) $8 \times 10^{-6} M$,
   (c) $8 \times 10^{-5} M$.

Chapter 4

1. (a) $\bar{x} = 396$ mg,   (b) median $= 397$ mg,
   (c) $w = 9$ mg,
   (d) abs dev from mean $= 2.7$ mg, rel dev $= 7$ ppt,
   (e) abs error $= -2$ mg, rel error $= -5$ ppt,
   (f) abs error $= -4$ mg, rel error $= -9$ ppt.

3.

		Student A	Student B
(a)	abs dev from mean	0.01% H_2O	0.03% H_2O
	rel dev from mean	9 ppt	3 ppt
(b)	abs error	-0.03% H_2O	-0.15% H_2O
	rel error	-25 ppt	-17 ppt

5. (a) -200 ppt,   (b) -40 ppt,
   (c) -20 ppt,   (d) -4 ppt,
   (e) -2 ppt.

7. (a) $s = 3$ mg Ca, $s_r = 9$ ppt,
   (b) $s = 4$ mg Ca, $s_r = 9$ ppt.

9.

		Student A	Student B
(a)	s	0.015% H_2O	0.036% H_2O
	s_r	12 ppt	4.2 ppt
(b)	s	0.018% H_2O	0.041% H_2O
	s_r	14 ppt	4.8 ppt

11. $s = 0.29\%$

13. $s = 2.6$ ppb

15.

	80% C.I.	95% C.I.
(a)	18.5 ± 3.1	18.5 ± 4.7
(b)	18.5 ± 2.2	18.5 ± 3.3
(c)	18.5 ± 1.5	18.5 ± 2.4

17. For 95% level, $N = 6$; for 99% level, $N = 10$.

19. (a) $s = 0.25$, 90% C.I. $= 7.24 \pm 0.42$,
   (b) 90% CI $= 7.24 \pm 0.26$.

21. $s = 0.031$, 95% C.I. $= 0.72 \pm 0.077$.

23. (a) $s = 0.061$, 95% C.I. $= 3.22 \pm 0.15$,
   (b) 95% C.I. $= 3.22 \pm 0.06$.

24. (a) $s = 0.13, 0.18, 0.21, 0.15,$ and 0.11, respectively,
   (b) pooled $s = 0.15$ meq K$^+$/ml,
   (c) yes,
   (d) 6.26 ± 0.12 and 6.26 ± 0.10,
   (e) 5.0 ± 1.6 and 5.02 ± 0.21.

26. (a) 4,   (c) 3.

27. 90% limit $= \pm 0.25$, 95% limit $= \pm 0.29$.

28. (a) 3

30.

	$\bar{x} - \mu$	$ts/\sqrt{N}$	Error
(a)	$-0.01\%\,V$	0.004%	yes
(b)	$-0.03\%\,Ni$	0.017%	yes
(c)	$-0.08\%\,Cu$	0.021%	yes

32. (a) $\bar{x} - \mu = 0.11\%$ CaO, $z\sigma/\sqrt{N} = 0.092$, determinate error suggested,

   (b) $\bar{x} - \mu = 0.11\%$ CaO, $ts/\sqrt{N} = 0.14\%$, determinate error not demonstrated.

34. (a) $\bar{x} - \mu = 0.06\%$ Cu, $z\sigma/\sqrt{N} = 0.039\%$, difference suggested,

   (b) $\bar{x} - \mu = -0.6$ ppm, $ts/\sqrt{N} = 0.087$ ppm, difference not demonstrated.

36. (a) Significant difference not demonstrated,
   (b) significant difference indicated,
   (c) significant difference indicated.

38. Significant difference in results for La and Th. Therefore, there is reasonable doubt as to the identity of the two samples.

40. (a) 3, (b) 4, (c) 4, (d) 7, (e) 4, (f) 5.

42.

	s_y	y
(a)	0.03	$0.57(\pm 0.03)$
(b)	0.09	$21.26(\pm 0.09)$
(c)	0.1×10^{-16}	$6.9(\pm 0.1) \times 10^{-16}$
(d)	0.01×10^5	$1.84(\pm 0.01) \times 10^5$
(e)	0.5	$6.0(\pm 0.5)$
(f)	0.1×10^{-3}	$8.1(\pm 0.1) \times 10^{-3}$
(g)	55	$8.8(\pm 0.6) \times 10^2$
(h)	0.06×10^{-6}	$4.69(\pm 0.06) \times 10^{-6}$

44.

	s_y	y
(a)	0.008	$0.238(\pm 0.008)$
(b)	0.04	$-1.36(\pm 0.04)$
(c)	0.001	$23.800(\pm 0.001)$
(d)	0.4×10^{-4}	$3.4(\pm 0.4) \times 10^{-4}$
(e)	6×10^5	$5(\pm 6) \times 10^5$
(f)	1	10 ± 1

46. (a) Retain, (b) reject.

Chapter 5

1. (a) $PbI_2(1.2 \times 10^{-3}F) > TlI(2.6 \times 10^{-4}F)$
   $> BiI_3(1.3 \times 10^{-5}F)$
   $> AgI(9.1 \times 10^{-9}F)$,

   (b) $PbI_2(7.1 \times 10^{-7}F)$
   $> TlI(6.5 \times 10^{-7}F)$
   $> AgI(8.3 \times 10^{-16}F)$
   $> BiI_3(8.1 \times 10^{-16}F)$,

(c) $PbI_2(1.3 \times 10^{-4}F)$
   $> BiI_3(6.7 \times 10^{-7}F)$
   $> TlI(6.5 \times 10^{-7}F)$
   $> AgI(8.3 \times 10^{-16}F)$.

3. (a) $UO_2(OH)_2$,   (b) $3.3 \times 10^{-8}\,M$,
   (c) yes, 8.0×10^{-7} to $3.3 \times 10^{-8}\,M$.

4. (b) Feasible, 1.1×10^{-6} to $1.2 \times 10^{-5}\,M$,
   (d) not feasible,
   (f) feasible, 1.3×10^{-9} to $3.4 \times 10^{-11}\,F$.

5. (a) Feasible, 3.4×10^{-4} to $0.18\,M$,
   (c) feasible, 4.8×10^{-9} to 1.6×10^{-10},
   (e) not feasible.

6. (a) 2.8×10^{-10},   (b) 5.1×10^{-19},
   (c) 6.4×10^{-10},   (d) 1.6×10^{-31}.

8. (a) 1.4×10^{-6} and $1.0 \times 10^{-6}\,F$,
   (b) 2.0×10^{-3} and $1.2 \times 10^{-3}\,F$,
   (c) 2.8×10^{-8} and 3.9×10^{-9}
   (d) 1.4×10^{-5} and $2.0 \times 10^{-6}\,F$.

10. (a) $1.1 \times 10^{-5}\,F$,   (b) $3.5 \times 10^{-5}\,F$.

11. (a) $2.1 \times 10^{-4}\,F$,   (b) $2.9 \times 10^{-4}\,F$,
    (c) $1.2 \times 10^{-3}\,F$.

14. (a) $7.8 \times 10^{-4}\,F$

15. (a) $2 \times 10^{-18}\,F$ and $3.7 \times 10^{-14}\,F$,
    (c) $5 \times 10^{-7}\,F$ and $1.4 \times 10^{-4}\,F$.

16. $7.1 \times 10^{-5}\,F$ and $1.3 \times 10^{-4}\,F$

18. (a) $4 \times 10^{-17}\,F$,   (c) $4.0 \times 10^{-6}\,F$.

19. (a) $1.2 \times 10^{-4}\,F$,   (b) $1.5 \times 10^{-5}\,F$,
    (c) $1.1 \times 10^{-6}\,F$ (d) $3.7 \times 10^{-7}\,F$.

21. 14.1 ml

23. $3.1 \times 10^{-3}\,F$

Chapter 6

1. (a) $\dfrac{gfw\ Na_2B_4O_7}{4 \times gfw\ KBF_4}$,   (b) $\dfrac{2\ gfw\ Nb}{gfw\ Nb_2O_5}$,

   (c) $\dfrac{gfw\ Nb_2O_3}{gfw\ Nb_2O_5}$,

   (d) $\dfrac{gfw\ SiF_4}{gfw\ SiO_2 \cdot 12MoO_3}$,

   (e) $\dfrac{12\ gfw\ Mo}{gfw\ SiO_2 \cdot 12MoO_3}$,

   (f) $\dfrac{2\ gfw\ P}{gfw\ Zn_2P_2O_7}$.

3. If $Y = gfw\ Fe_3Al_2Si_3O_{12}$,

   (a) $\dfrac{2Y}{3\ gfw\ Fe_2O_3}$,   (b) $\dfrac{Y}{3\ gfw\ SiO_2}$,

   (c) $\dfrac{Y}{gfw\ Al_2O_3}$,

   (d) $\dfrac{Y}{2\ gfw\ Al(C_9H_6ON)_3}$.

5. If $Z = gfw\ CuCrO_4 \cdot 2CuO \cdot 2H_2O$,

   (a) $\dfrac{Z}{3\ gfw\ Cu}$,   (b) $\dfrac{2Z}{gfw\ Ag_2Cr_2O_7}$,

   (c) $\dfrac{Z}{gfw\ PbCrO_4}$,   (d) $\dfrac{Z}{2\ gfw\ H_2O}$.

(e) $\dfrac{2Z}{3\,\text{gfw Cu}_2(\text{SCN})_2}$, (f) $\dfrac{Z}{3\,\text{gfw CuO}}$.

6. Largest is $Cu_2(SCN)_2$ and smallest is H_2O.
8. 1.46%
10. 0.0393 g
12. 18.8 ml
14. (a) 16.1 ml,   (b) 14.5 ml.
16. 25.1%
18. 66.9%
20. (a) 53.2%,   (b) 853 lb.
22. 1.42%
24. 10.7%
27. 0.0203 F
29. Dry basis, 26.2% Si and 75.8% Ca; as received, 24.9% Si and 72.0% Ca.
31. 20.7% $MgSO_4$ and 12.5% $Al_2(SO_4)_3$
33. 40.8% $Al_2(SO_4)_3$ and 59.2% $BeSO_4$
35. 30.6% NaBr and 5.86% $NaBrO_3$
37. 0.087 g
39. (a) 8.8×10^{-6} g,   (b) 7.6×10^{-3} g,
    (c) 76 ppt.

Chapter 7

1. (a) pNa = pBr = 2.00,
   (c) pBa = 2.46, pOH = 2.16,
   (e) pCa = 2.28, pBa = 2.44, pCl = 1.75.
2. (a) pI = 0.89,
   (c) pPb = 1.90, pI = 3.12,
   (e) pPb = −0.26.
3. (a) 2.1×10^{-9} M,   (c) 0.92 M,
   (e) 2.09 M,   (g) 0.99 M.
4. (a) Neutralization, $\dfrac{\text{gfw H}_2\text{SO}_4}{2}$, $\dfrac{\text{gfw Ca(OH)}_2}{2}$,

   $\dfrac{\text{gfw Ca}_2\text{P}_2\text{O}_7}{4}$,

   (b) oxidation-reduction, $\dfrac{\text{gfw I}_2}{2}$, $\dfrac{\text{gfw KI}}{1}$,

   $\dfrac{\text{gfw SnCl}_4}{2}$, $\dfrac{\text{gfw Sn}_2\text{O(NO}_3)_2}{4}$,

   (c) precipitation, $\dfrac{\text{gfw Ce}^{3+}}{3}$, $\dfrac{\text{gfw Na}_2\text{C}_2\text{O}_4}{2}$,

   $\dfrac{\text{gfw C}}{1}$, $\dfrac{\text{gfw NaHC}_2\text{O}_4}{2}$,

   (d) complex formation, $\dfrac{\text{gfw Al(OH)}_2\text{Cl}}{3}$,

   2 gfw NaF, gfw BaF_2, $\dfrac{2\,\text{gfw C}_6\text{H}_3\text{F}_3}{3}$.

6. (a) Complex formation, $\dfrac{\text{gfw Ag}_2\text{SO}_4}{2}$, gfw Ag,

   2 gfw $Na_2S_2O_3$,

   (b) oxidation-reduction, $\dfrac{\text{gfw I}_2}{2}$,

gfw $Na_2S_2O_3$, $\dfrac{\text{gfw Na}_2\text{S}_4\text{O}_6}{2}$,

(c) precipitation, $\dfrac{\text{gfw LaCl}_3}{3}$, gfw $NaIO_3$,

$\dfrac{\text{gfw AlOH(IO}_3)_2}{2}$, $\dfrac{\text{gfw I}_2\text{O}_5}{2}$,

(d) neutralization, $\dfrac{\text{gfw H}_3\text{PO}_4}{2}$, $\dfrac{\text{gfw P}_2\text{O}_5}{4}$,

$\dfrac{\text{gfw Ba(OH)}_2}{2}$.

8. (a) 5.98×10^{-3} F,   (b) 0.0120 N,
   (c) 0.0120 N,   (d) 0.0120 N,
   (e) 0.336 mg/ml.
10. (a) 0.200 N,   (b) 1.00 N,
    (c) 0.600 N,   (d) 0.800 N.
11. (a) 0.0108 F,   (b) 0.0108 N,
    (c) 0.0324 N,   (d) 2.47 mg/ml.
13. (a) 2.00 meq, (b) 4.00 meq, (c) 2.00 meq,
    (d) 2.00 meq, (e) 1.00 meq, (f) 12.0 meq.
15. (a) 0.0834 g, (b) 11.0 g,   (c) 1.24 g,
    (d) 9.13 g,   (e) 2.14 g,   (f) 4.28 g.
17. Dilute each of the following quantities to 3.00 liters:
    (a) 40.0 ml,   (b) 90.5 g,
    (c) 1.3 liters,   (d) 6.7 ml.
19. Dilute each of the following to 500.0 ml:
    (a) 3.73 g,   (b) 4.95 g,
    (c) 13.6 g,   (d) 3.51 g.
21. 0.0914 N
23. 0.0462 N
25. 0.297 N $HClO_4$ and 0.326 N NaOH
27. 29.1%
31. 7.47 g/100 ml

Chapter 8

1. (a) Dissolve 24.5 g and dilute to 2.00 liters.
   (c) Dissolve 11.8 g and dilute to 750 ml.
2. (a) 0.0220 N,   (b) 2.21 mg/ml.
4. 32.4% F and 67.9% Na_2SeF_6
6. 21.5%
8. 25.9 ppm
10. 14.1 mg/ml
12. 32.4% $BaCl_2$ and 67.6% KBr

13. Reagent Volume

Reagent Volume from End Point	p-Cations		
	pAg (a)	pAg (c)	pBa (e)
−10.0	2.40	3.70	2.12
−1.00	3.47	4.61	3.17
−0.100	4.48	4.84	4.18
0.000	5.98	4.87	4.94
+0.100	7.48	4.90	5.71
+1.00	8.47	5.12	6.70
+10.0	9.42	5.90	7.66

Chapter 9

1. (a) 2.12,   (b) 2.48,   (c) 11.40,
   (d) 11.52,   (e) 2.12,   (f) 7.00.

3. 0.16

5. (a) 9.91,   (b) 2.30,   (c) 12.00,
   (d) 4.70,   (e) 5.13,   (f) 4.1,
   (g) 4.52.

7. (a) 2.13 (exact), 2.10 (approx.),
   (b) 2.87 (exact), 2.83 (approx.),
   (c) 7.83,   (d) 12.60,   (e) 4.26,
   (f) 2.77 (exact), 2.74 (approx.),
   (g) 0.60.

9. (a) 4.41,   (b) 4.81,   (c) 12.72.

10. (a) 1.98,   (b) 4.59,   (c) 8.15.

11. (a) 9.22,   (b) 10.16,   (c) 0.79.

13. (a) 3.58,   (b) 0.13,
    (c) 0.075,   (d) 5.3×10^{-3}.

15. (a) 9.55,   (b) 9.79.

17. (a) 1.21,   (b) 0.43,   (c) 0.09.

19. (a) -0.41,   (b) -0.41,   (c) -0.04.

21. (a) 0.96,   (b) 0.48,   (c) 0.04.

23. 0.294 g

25. Mix 26.6 ml HA with 223.4 ml NaA.

27. Mix 122 ml $HClO_4$ with 178 ml pyridine.

29.

ml Reagent	(a)	(b)	(c)	(d)
0.00	11.07	11.98	9.47	11.45
5.00	10.09	11.72	6.87	10.85
20.00	9.25	11.08	6.03	10.00
35.00	8.40	10.25	5.18	9.16
39.00	7.65	9.51	4.44	8.41
40.00	5.30	6.23	3.69	5.68
41.00	2.96	2.96	2.96	2.96
45.00	2.28	2.28	2.28	2.28
50.00	2.00	2.00	2.00	2.00

pH (column header above (a)–(d))

Chapter 10

1. (a) 4.23,   (b) 1.16,   (c) 12.78,   (d) 11.35.

3. (a) 9.92,   (b) 2.95,   (c) 4.56,   (d) 8.39.

5. (a) 1.55,   (b) 2.00,   (c) 12.07,   (d) 12.00.

7. (a) $[HAsO_4^{2-}]/[H_2AsO_4^-] = 0.105$,
   (b) $[Cit^{3-}]/[HCit^{2-}] = 0.402$,
   (c) $[M^{2-}]/[HM^-] = 2.01$,
   (d) $[T^{2-}]/[HT^-] = 43.1$.

9. (a) 2.06,   (b) 7.20,   (c) 10.63,
   (d) 2.09,   (e) 2.01.

11. (a) 2.11,   (b) 7.37.

13. Mix 558 ml HCl with 442 ml Na_2CO_3.

16. 44.0 g

18. (a)

ml Reagent	pH
0.00	11.66
12.50	10.33
20.00	9.73
24.00	8.95
25.00	8.34
26.00	7.74
37.50	6.35
45.00	5.75
49.00	4.97
50.00	3.83
51.00	2.70
60.00	1.74

(c)

ml Reagent	pH
0.00	0.96
12.50	1.32
20.00	1.52
24.00	1.62
25.00	1.64
26.00	1.67
37.50	2.14
45.00	2.65
49.00	3.40
50.00	7.31
51.00	11.30
60.00	12.26

19. (a)

pH	a_0	a_1
2.00	0.899	0.101
4.00	0.079	0.886
6.00	1.8×10^{-4}	0.204
8.00	2.3×10^{-8}	2.6×10^{-3}
10.00	2.3×10^{-12}	2.6×10^{-5}
12.00	2.3×10^{-16}	2.6×10^{-7}

pH	a_2
2.00	3.9×10^{-5}
4.00	0.035
6.00	0.796
8.00	0.997
10.00	1.000
12.00	1.000

(c)

pH	a_0	a_1
2.00	0.931	0.069
4.00	0.103	0.765
6.00	5.3×10^{-5}	0.039
8.00	1.9×10^{-10}	1.4×10^{-5}
10.00	1.9×10^{-16}	1.4×10^{-9}
12.00	1.9×10^{-22}	1.4×10^{-13}

pH	a_2	a_3
2.00	1.2×10^{-4}	4.8×10^{-9}
4.00	0.132	5.3×10^{-4}
6.00	0.685	0.275
8.00	0.024	0.976
10.00	2.5×10^{-4}	0.999
12.00	2.5×10^{-6}	1.000

Chapter 11

1. Dissolve the following quantities, and dilute to 2.00 liters:
   (a) 17 g,   (b) 4.7 g,
   (c) 120 ml,   (d) 54.0 g.

3. (a) 0.0996,   (b) 4.6×10^{-4}

5. (a) 0.056 to 0.072 g,   (c) 0.85 to 1.1 g,
   (e) 0.044 to 0.056 g.
6. 0.119 N
8. 0.121 g/100 ml
10. 76.8 g/equiv
12. 14.1%
14. 19.0%
16. 1200 ppm
18. 0.0274 N
20. 14.0%
22. 0.524% acetaldehyde and 0.651% ethyl acetate
24. 6.38%
26. 69.5% KOH, 22% K_2CO_3, 8.5% H_2O.
28. (a) 18.1 ml,   (b) 45.3 ml,
    (c) 38.2 ml   (d) 12.2 ml.
30. The answers are in units of mg/ml.
    (a) NaOH, 4.31,
    (b) Na_2CO_3, 7.97 and $NaHCO_3$, 4.35,
    (c) Na_2CO_3, 3.45 and NaOH, 4.39,
    (d) Na_2CO_3, 8.20,   (e) $NaHCO_3$, 13.4.
32. The answers are in units of mg/ml.
    (a) H_3PO_4, 2.30 and HCl, 2.42,
    (b) HCl 1.39,
    (c) NaH_2PO_4, 9.66,
    (d) H_3PO_4, 6.17 and NaH_2PO_4, 5.16,
    (e) H_3PO_4, 6.27.

Chapter 12

1. (a) $2H_2O \rightleftarrows H_3O^+ + OH^-$
       $K_w = [H_3O^+][OH^-]$,
   (c) $2H_2SO_4 \rightleftarrows H_3SO_4^+ + HSO_4^-$
       $K_s = [H_3SO_4^+][HSO_4^-]$,
   (e) $2HCOOH \rightleftarrows HCOOH_2^+ + HCOO^-$
       $K_s = [HCOOH_2^+][HCOO^-]$.
2. (a) 7.00,   (c) 1.93,   (e) 3.11.

3. (a)

Volume	pH	Volume	pH
0.00	1.30	25.0	9.55
12.5	1.70	25.1	15.2
24.0	2.87	26.0	16.2
24.9	3.87	30.0	16.9

   (b) ΔpH for ethanol = 11.3, for water = 6.25.

5.

	pH in Ethanol	pH in Water
(a)	5.78	3.03
(b)	10.25	4.76
(c)	14.0	8.72

7.

	pH in Ethanol	pH in Water
(a)	11.4	8.30
(b)	6.0	4.90
(c)	3.8	3.30

9.

ml Base	pH Ethanol	pH Water	ml Base	pH Ethanol	pH Water
0	3.3	2.80	49.9	8.4	7.29
10.0	5.1	3.99	50.0	11.8	8.65
25.0	5.7	4.60	50.1	15.1	10.00
40.0	6.3	5.20	51.0	16.1	11.00
49.0	7.4	6.28	60.0	17.1	11.96

Chapter 13

1. (a) 0.0452 F,   (b) 0.0226 N,
   (c) 0.0226 N,   (d) 2.44 mg/ml,
   (e) 0.807 mg/ml.
3. (a) 0.0192 F,   (b) 0.774 mg/ml,
   (c) 1.54 mg/ml.
5. 0.0103 F
7. 0.01005 F
9. 322 mg Ca^{2+} and 257 mg Mg^{2+}, both normal.
11. 93.9%
13. 31.4% NaBr and 48.3% $NaBrO_3$
15. 0.0839 F
17. 7.51% Pb, 9.44% Zn, and 4.29% Mg.
19. 8.52% Pb, 24.8% Zn, 64.1% Cu, and 2.6% Sn.
20. (a) 1.4×10^9,   (b) 3.3×10^{11},
    (c) 2.2×10^{13}.
22. (a) 8.1×10^{12},   (b) 1.3×10^{14},
    (c) 5.0×10^9,   (d) 8.1×10^{10}.

24.

ml	pSr	ml	pSr
0.00	2.00	25.00	5.37
10.00	2.30	25.10	6.17
24.00	3.57	26.00	7.17
24.90	4.57	30.00	7.88

26.

ml	pCo	ml	pCo
0.00	2.48	20.00	8.87
5.00	2.68	22.00	14.02
10.00	2.92	30.00	14.72
18.00	3.71		

Chapter 14

1. (a) $Tl^{3+} + 2Ag(s) + 3Br^- \rightleftarrows$
            $TlBr(s) + 2AgBr(s)$,
   (b) $2Fe^{2+} + UO_2^{2+} + 4H^+ \rightleftarrows$
            $2Fe^{3+} + U^{4+} + 2H_2O$,
   (c) $N_2(g) + 2H_2(g) + H^+ \rightleftarrows N_2H_5^+$,
   (d) $Cr_2O_7^{2-} + 9I^- + 14H^+ \rightleftarrows$
            $2Cr^{3+} + 3I_3^- + 7H_2O$,
   (e) $2Ce^{3+} + O_2(g) + 2H^+ \rightleftarrows 2Ce^{4+} + H_2O_2$,
   (f) $IO_3^- + 5I^- + 6H^+ \rightleftarrows 3I_2(s) + 3H_2O$.

3. Oxidizing Agent (Underlined)

(a) $Tl^{3+} + Br^- + 2e \rightleftarrows TlBr(s)$
(b) $\underline{UO_2^{2+}} + 4H^+ + 2e \rightleftarrows U^{4+} + 2H_2O$
(c) $\underline{N_2(g)} + 5H^+ + 4e \rightleftarrows N_2H_5^+$
(d) $\underline{Cr_2O_7^{2-}} + 14H^+ + 6e \rightleftarrows 2Cr^{3+} + 7H_2O$
(e) $\underline{O_2(g)} + 2H^+ + 2e \rightleftarrows H_2O_2$
(f) $\underline{2IO_3^-} + 12H^+ + 10e \rightleftarrows I_2(s) + 6H_2O$

Reducing Agent (Underlined)

(a) $AgBr(s) + e \rightleftarrows \underline{Ag}(s) + Br^-$
(b) $Fe^{3+} + e \rightleftarrows \underline{Fe^{2+}}$
(c) $2H^+ + 2e \rightleftarrows \underline{H_2}(g)$
(d) $I_3^- + 2e \rightleftarrows \underline{3I^-}$
(e) $Ce^{4+} + e \rightleftarrows \underline{Ce^{3+}}$
(f) $I_2(s) + 2e \rightleftarrows \underline{2I^-}$

5. (a) 3×10^{45}, (b) 2×10^{-15},
(c) 3×10^{-16}, (d) 4×10^{80},
(e) 2×10^{-26} in 1-F H_2SO_4 or 3×10^{-35} in 1-F $HClO_4$, (f) 6×10^{55}.
7. (a) 0.813 V, (b) 0.748 V,
(c) 0.351 V, (d) 0.664 V.
9. (a) 0.794 V, (b) −0.689 V,
(c) 0.600 V, (d) 0.39 V,
(e) 0.437 V, (f) 0.171 V.
11. (a) anode, −0.235 V, (b) cathode, 0.163 V,
(c) anode, −0.014 V, (d) anode, −0.01 V,
(e) cathode, 0.331 V, (f) cathode, 0.119 V.
13. (a) right, (b) left, (c) left,
(d) right, (e) left, (f) right.
15. (a) 0.496 V, galvanic,
(b) −0.117 V, electrolytic,
(c) −0.077 V, electrolytic,
(d) 0.804 V, galvanic,
(e) −0.589 V, electrolytic,
(f) 1.23 V, galvanic.
17. 0.391 V
19. 1.3×10^{-6}
21. −1.25 V
23. 1.5×10^{-14}
25. 2.7×10^8
27. 4.5×10^{14}
29. 9.47×10^{-6}

Chapter 15

1. (a) −0.02 V, (b) 1.17 V, (c) 1.25 V.
3. (a) 0.07 = 0.1 V
4. (a) 5.4×10^{17},
(b) 6.5×10^9 (in 1-F H_2SO_4),
(c) 10^{96}.
6. (a) 6×10^{-8} M, (b) 1.4×10^{-6} M,
(c) 1.0×10^{-16} M.

8.

ml	E, V (a)	(c)	(e)
10.00	−0.291	0.32	0.32
25.00	−0.255	0.36	0.33
49.00	−0.155	0.46	0.38
49.90	−0.096	0.52	0.41
50.00	0.018	0.95	1.17
50.10	0.074	1.17	1.48
51.00	0.104	1.20	1.49
60.00	0.133	1.23	1.50

Chapter 16

1. (a) $2Mn^{2+} + 5S_2O_8^{2-} + 8H_2O \rightleftarrows$
$2MnO_4^- + 10SO_4^{2-} + 16H^+$,
(b) $2Ce^{3+} + BiO_3^- + 6H^+ \rightleftarrows$
$2Ce^{4+} + Bi^{3+} + 3H_2O$,
(c) $U^{4+} + H_2O_2 \rightleftarrows UO_2^{2+} + 2H^+$,
(d) $V(OH)_4^+ + Ag + Cl^- + 2H^+ \rightleftarrows$
$AgCl + VO^{2+} + 3H_2O$,
(e) $2SCN^- + 3IO_3^- + 6Cl^- + 4H^+ \rightleftarrows$
$2SO_4^{2-} + 2HCN + 3ICl_2^- + H_2O$,
(f) $3H_2C_2O_4 + BrO_3^- \rightleftarrows$
$6CO_2 + Br^- + 3H_2O$,
(g) $5H_2O_2 + 2MnO_4^- + 6H^+ \rightleftarrows$
$5O_2 + 2Mn^{2+} + 8H_2O$,
(h) $ClO_3^- + 6I^- + 6H^+ \rightleftarrows$
$3I_2 + Cl^- + 3H_2O$.
3. (a) 2 moles H_2CO and 4 moles HCOOH,
(b) 2 moles HCOOH,
(c) 1 mole HCOOH and 1 mole CH_3CHO.
5. Dissolve 12.2 g and dilute to 2.50 liters.
7. Dissolve 3.13 g and dilute to 750 ml.
9. 8.00 mg/ml
11. 2.35 mg/ml
13. 0.0668 N
15. 35.9%
17. (a) 15.3%, (b) 21.4%.
19. 0.700%
21. 0.552 g (eq wt Tl = 2 gfw/3)
23. (a) 15.7%, (b) 7.28%.
25. (a) $4Mn^{2+} + MnO_4^- + 20F^- + 8H^+ \rightleftarrows$
$5MnF_4^- + 4H_2O$,
(b) 0.0936, (c) 40.8%.
27. 63.4% Fe and 15.5% Cr
29. 26.2%
31. 1.76%
33. 0.365 H_2S mg/liter and 0.311 mg SO_2/liter
35. 26.4%
37. 0.0397 mg/ml
39. 7.64
41. (a) $SCN^- + 3I_2 + 4H_2O \rightleftarrows$
$SO_4^{2-} + HCN + 7H^+ + 6I^-$,
(b) 15.4%.
43. 84.0%

Chapter 17

1. (a) -0.32 V,
   (b) Cu|CuSCN (sat'd), SCN$^-$(xM)$||$ SCE,
   (c) $E_{cell} = E_{SCE} - (E^0_{CuSCN}$
   $- 0.0591 \log [SCN^-])$
   $-\log [SCN^-] = pSCN$
   $= \dfrac{E_{SCE} - E^0_{CuSCN} - E_{cell}}{0.0591}$
   $pSCN = (0.56 - E_{cell})/0.0591,$
   (d) 8.2.
3. (a) SCE $||$ Hg$_2$Cl$_2$(sat'd), Cl$^-$(xM)|Hg
   $pCl = (E_{cell} - 0.026)/0.0591,$
   (b) SCE $||$ Ag$_2$CO$_3$(sat'd), CO$_3{}^{2-}$(xM)|Ag
   $pCO_3 = (E_{cell} - 0.229) \times 2/0.0591,$
   (c) SCE$||$Sn^{4+}(xM), Sn^{2+}(1.00 $\times$ 10^{-4}M)|Pt
   $pSn(IV) = (0.030 - E_{cell}) \times 2/0.0591.$
5. 6.70
7. 0.606 V
9. (a) 0.104 V,   (b) 0.045 V,   (c) -0.014 V.
11. (a) -0.238 V,   (b) -0.247 V,
    (c) -0.259 V.
13. (a) 5.558,   (b) 1.665.
14. (a) 0.2% and -2%,   (b) 0.5% and -2%.
16. 9.3 $\times$ 10^{-16}

18. (a)

Volume	$E_{vs.\ SCE}$, V	Volume	$E_{vs.\ SCE}$, V
5.00	0.67	49.0	0.75
10.0	0.68	50.0	0.86
15.0	0.69	51.0	1.10
25.0	0.70	55.0	1.14
40.0	0.72	60.0	1.16

20. (a) 0.35 V,   (b) 0.097 V.

Chapter 19

1. (a) -0.92 V,   (b) 0.26 V,
   (c) 0.85 V,   (d) $E = -2.03$ V,
   (e) -2.04 V.
3. (a) Feasible,
   (b) -0.04 to -0.40 V,
   (c) -0.545 V.
5. (a) Not feasible,   (b) feasible,
   (c) 0.038 to -0.063 V.
7. (a) 0.435 V,   (b) 0.279 V,
   (c) -0.125 V.
8. (a) 1.18 V,   (b) 0.157 A,   (c) 0.96 V,
   (d) 0.128 A; unlikely that current would be this high because concentration polarization would occur at the cathode.

Chapter 20

1. 13.3% Cd and 5.43% Zn
3. 4.06%
5. 53.9% CH$_2$Cl$_2$ and 46.1% CHCl$_3$

7. 9.27 ppm
9. 0.502 g
11. 73.4 mg/liter
13. 0.625%
15. 172 g/equiv

Chapter 21

1. 1.16 $\times$ 10^3 mg/liter
4. For A, 2.39; for C, 3.68.
5. 4
7. 4 $\times$ 10^{22}
9. 7.2 $\times$ 10^5

Chapter 22

1. (a) $\nu = 3.3 \times 10^{17}$ Hz, $\sigma = 1.11 \times 10^7$ cm^{-1},
   (c) $\nu = 2.38 \times 10^{13}$ Hz, $\sigma = 794$ cm^{-1}.
3. (a) 86.3%,   (c) 48.1%,   (e) 36.6%.
4. (a) 0.712,   (c) 0.565,   (e) 1.00.
6. (a) 406,   (b) 0.239,
   (c) 1.50 cm,   (d) 3.64 $\times$ 10^{-5} M,
   (e) 1.96 ppm.
7. (a) cm^{-1} ppm^{-1} or cm^{-1} liter mg^{-1}
9. 1.57 $\times$ 10^3
11. (a) 0.577,   (b) 26.5%,
    (c) 1.24 $\times$ 10^{-5} M.
13. 0.595

Chapter 23

1. (a) Orange,   (c) $A + G$,   (e) blue.
2. 6.74 $\times$ 10^{-4} F

Chapter 24

1. 0.0214%

3.

F_{HIn}	[In$^-$]	[HIn]
3.00 $\times$ 10^{-4}	2.90 $\times$ 10^{-4}	1.05 $\times$ 10^{-6}
2.00 $\times$ 10^{-4}	1.95 $\times$ 10^{-4}	4.7 $\times$ 10^{-5}
1.00 $\times$ 10^{-4}	9.88 $\times$ 10^{-5}	1.2 $\times$ 10^{-6}
0.500 $\times$ 10^{-4}	4.97 $\times$ 10^{-5}	3.1 $\times$ 10^{-7}
0.250 $\times$ 10^{-4}	2.49 $\times$ 10^{-5}	7.8 $\times$ 10^{-8}

F_{HIn}	A_{430}	A_{600}
3.00 $\times$ 10^{-4}	0.25	0.835
2.00 $\times$ 10^{-4}	0.11	0.562
1.00 $\times$ 10^{-4}	0.029	0.284
0.500 $\times$ 10^{-4}	0.0074	0.143
0.250 $\times$ 10^{-4}	0.0019	0.072

4. (a) $\pm 9.5\%$,   (b) $\pm 1.7\%$,   (c) $\pm 2.1\%$.
5. (a) $\pm 2.6\%$,   (c) $\pm 3.6\%$.
6. Absorbance decreases linearly and then levels off to zero at the equivalence point.

8. (a) $1.71 \times 10^{-4}\,M$,   (b) $3.57 \times 10^{-4}\,M$.
9. (a) Co, $1.66 \times 10^{-4}\,F$ and Ni, $4.28 \times 10^{-5}\,F$
11. (a) concn P $= 2.08 \times 10^{-4}\,M$
    concn Q $= 4.90 \times 10^{-5}\,M$,
  (c) concn P $= 8.36 \times 10^{-5}\,M$
    concn Q $= 6.10 \times 10^{-5}\,M$,
  (e) concn P $= 2.10 \times 10^{-4}\,M$
    concn Q $= 9.64 \times 10^{-5}\,M$.
13. (a) 0.497,   (c) 0.197.
14. (a) 0.301
15. (a) 5.60,   (c) 4.80.
16. (a)

λ	A	λ	A	λ	A
420	0.414	470	0.490	585	0.123
445	0.510	510	0.320	595	0.114
450	0.512	550	0.170	610	0.103
455	0.513	570	0.139	650	0.082

18. 1.8×10^8
20. Ni(CDA)$_3$

22. (a) CdR^{2+},   (b) 1.42×10^4,
  (c) 3.8×10^5.
24. (a) CoQ_3^{2+},   (b) 3.05×10^4,
  (c) 1×10^{17}.

Chapter 29

1. (a) $1.73 \times 10^{-2}\,M$,   (b) $6.40 \times 10^{-3}\,M$,
  (c) $2.06 \times 10^{-3}\,M$,   (d) $6.89 \times 10^{-4}\,M$.
3. (a) 75.0 ml,   (b) 40.0 ml,   (c) 22.0 ml.
5. (a) 18.0,   (b) 7.56.
6. (a) 13.9
7. (a) 2.80,
  (b) $[HA]_{eq} = 0.0114\,M$ and $[A^-] = 0.0566\,M$,
  (c) $K_a = 0.28$.
9. (a) 2.22×10^4 plates,
  (b) 1.81×10^{-3} cm,
  (c) $\sigma = 0.270$ cm, $\tau = 0.101$ min,
  (d) 0.0880, (e) 12.9, (f) 0.40, (g) 550 cm.

APPENDIXES

APPENDIXES

APPENDIX 1
SELECTED REFERENCES TO THE LITERATURE OF ANALYTICAL CHEMISTRY

The references and journals in the accompanying list will augment the titles cited throughout this text and will serve as a point of departure in a search for specific information concerned with analytical chemistry.

Series Devoted to Topics in Analytical Chemistry

American Society for Testing and Materials, *Annual Book of Standards*. Philadelphia: American Society for Testing and Materials, 1974. In 47 volumes, many of which are concerned with chemical methods.

R. Belcher and L. Gordon, *International Series of Monographs on Analytical Chemistry*. New York: Pergamon Press, Inc., 1961–. A multivolume series.

P. J. Elving and I. M. Kolthoff, Eds., *Chemical Analysis*. New York: Interscience Publishers, Inc., 1948–1973. In 39 volumes.

N. H. Furman and F. J. Welcher, Eds., *Scott's Standard Methods of Chemical Analysis*, 6th ed. New York: D. Van Nostrand Company, Inc., 1962–1966. In 5 volumes; devoted to specific applications.

I. M. Kolthoff and P. J. Elving, Eds., *Treatise on Analytical Chemistry*. New York: John Wiley & Sons, Inc., 1959–1972. A multivolume series: Part I is concerned with theoretical aspects, Part II with methods for the analysis of organic and inorganic compounds, and Part III with industrial analytical chemistry.

C. N. Reilley, Ed., *Advances in Analytical Chemistry and Instrumentation*. New York: Interscience Publishers, Inc., 1960–1974. A series of monographs on topics of analytical interest.

General References

American Public Health Association, *Standard Methods for the Examination of Water and Wastewater*, 13th ed. New York: American Public Health Association, 1971.

American Society for Testing and Materials, *Chemical Analysis of Metals and Metal-Bearing Ores*. Philadelphia: American Society for Testing and Materials, 1974. This is Volume 12 of *Annual Book of ASTM Standards*.

Association of Official Analytical Chemists, *Methods*, 11th ed. Washington, D.C.: Association of Official Analytical Chemists, 1970.

W. G. Berl, Ed., *Physical Methods in Chemical Analysis*, 2d ed. New York: Academic Press, Inc., 1960–1961. In 4 volumes.

N. W. Hanson, *Official, Standardised and Recommended Methods of Analysis*, 2d ed. London: Society for Analytical Chemistry, 1973.

W. F. Hillebrand, G. E. F. Lundell, H. A. Bright, and J. I. Hoffman, *Applied Inorganic Analysis*, 2d ed. New York: John Wiley & Sons, Inc., 1953.

J. J. Lingane, *Analytical Chemistry of Selected Metallic Elements*. New York: Reinhold Publishing Corporation, 1966.

L. Meites, Ed., *Handbook of Analytical Chemistry*. New York: McGraw-Hill Book Company, Inc., 1963.

J. Mitchell, Jr., I. M. Kolthoff, E. S. Proskauer, and A. Weissberger, Eds., *Organic Analysis*. New York: Interscience Publishers, Inc., 1953–1960. In 4 volumes.

G. H. Morrison, Ed., *Trace Analysis: Physical Methods*. New York: John Wiley & Sons, Inc., 1965.

D. D. Perrin, *Masking and Demasking of Chemical Reactions*. New York: John Wiley & Sons, Inc., 1970.

A. Weissberger, Ed., *Techniques of Organic Chemistry*, 3d ed. New York: Interscience Publishers, Inc., 1960. Volume 1, Parts III and IV.

C. L. Wilson and D. W. Wilson, Eds., *Comprehensive Analytical Chemistry*. New York: American Elsevier Publishing Company, Inc., 1959–1964.

Titrimetric Methods

M. R. F. Ashworth, *Titrimetric Organic Analysis*. New York: Interscience Publishers, Inc., 1963. In 2 volumes.

W. Huber, *Titrations in Nonaqueous Solvents*. New York: Academic Press, 1967.

I. M. Kolthoff, V. A. Stenger, and R. Belcher, *Volumetric Analysis*. New York: Interscience Publishers, Inc., 1942–1957. In 3 volumes.

W. Wagner and C. J. Hull, *Inorganic Titrimetric Analysis*. New York: Marcel Dekker, Inc., 1971.

Organic Analysis

S. Siggia, *Quantitative Organic Analysis Via Functional Groups*, 3d ed. New York: John Wiley & Sons, Inc., 1963.

S. Siggia, Ed., *Instrumental Methods of Organic Functional Group Analysis*. New York: John Wiley & Sons, Inc., 1972.

S. Siggia and H. J. Stolten, *An Introduction to Modern Organic Analysis*. New York: Interscience Publishers, Inc., 1956.

Electrochemical Methods

I. M. Kolthoff and J. J. Lingane, *Polarography*, 2d ed. New York: Interscience Publishers, Inc., 1952. In 2 volumes.

J. J. Lingane, *Electroanalytical Chemistry*, 2d ed. New York: Interscience Publishers, Inc., 1954.

L. Meites, *Polarographic Techniques*. New York: Interscience Publishers, Inc., 1965.

H. W. Nurnberg, Ed., *Electroanalytical Chemistry*. New York: John Wiley & Sons, Inc., 1974.

Spectrometric Methods

W. T. Elwell and J. A. F. Gidley, *Atomic Absorption Spectrophotometry*, 2d ed. New York: Pergamon Press, Inc., 1966.

E. B. Sandell, *Colorimetric Determination of Traces of Metals*, 3d ed. New York: Interscience Publishers, Inc., 1959.

E. I. Stearns, *The Practice of Absorption Spectrophotometry*. New York: John Wiley & Sons, Inc., 1969.

Analytical Separations

B. L. Karger, L. R. Snyder, and C. Horvath, *An Introduction to Separation Science*. New York: John Wiley & Sons, Inc., 1973.

G. H. Morrison and H. Freiser, *Solvent Extraction in Analytical Chemistry*. New York: John Wiley & Sons, Inc., 1957.

J. H. Purnell, Ed., *Progress in Gas Chromatography*. New York: Interscience Publishers, Inc., 1968.

O. Samuelson, *Ion Exchange Separations in Analytical Chemistry*. New York: Interscience Publishers, Inc., 1968.

Journals

Analytica Chimica Acta
Analytical Abstracts
Analytical Chemistry
Analytical Letters
Analyst
Chimie analitique
Collection of Czechoslovak Chemical Communications
Current Chemical Papers
Journal of the Association of Official Analytical Chemists
Microchemical Journal
Mikrochimica Acta
Talanta
Zeitschrift für analytische Chemie

APPENDIX 2
SOME STANDARD AND FORMAL ELECTRODE POTENTIALS[a]

Half-Reaction	E^0, V	Formal Potential, V
$F_2(g) + 2H^+ + 2e \rightleftarrows 2HF(aq)$	3.06	
$O_3(g) + 2H^+ + 2e \rightleftarrows O_2(g) + H_2O$	2.07	
$S_2O_8^{2-} + 2e \rightleftarrows 2SO_4^{2-}$	2.01	
$Co^{3+} + e \rightleftarrows Co^{2+}$	1.842	
$H_2O_2 + 2H^+ + 2e \rightleftarrows 2H_2O$	1.776	
$MnO_4^- + 4H^+ + 3e \rightleftarrows MnO_2(s) + 2H_2O$	1.695	
$Ce^{4+} + e \rightleftarrows Ce^{3+}$		1.70, 1-F HClO$_4$; 1.61, 1-F HNO$_3$; 1.44, 1-F H$_2$SO$_4$
$HClO + H^+ + e \rightleftarrows \frac{1}{2}Cl_2(g) + H_2O$	1.63	
$H_5IO_6 + H^+ + 2e \rightleftarrows IO_3^- + 3H_2O$	1.60	
$BrO_3^- + 6H^+ + 5e \rightleftarrows \frac{1}{2}Br_2(l) + 3H_2O$	1.52	
$MnO_4^- + 8H^+ + 5e \rightleftarrows Mn^{2+} + 4H_2O$	1.51	
$Mn^{3+} + e \rightleftarrows Mn^{2+}$		1.51, 7.5-F H$_2$SO$_4$
$ClO_3^- + 6H^+ + 5e \rightleftarrows \frac{1}{2}Cl_2(g) + 3H_2O$	1.47	
$PbO_2(s) + 4H^+ + 2e \rightleftarrows Pb^{2+} + 2H_2O$	1.455	
$Cl_2(g) + 2e \rightleftarrows 2Cl^-$	1.359	
$Cr_2O_7^{2-} + 14H^+ + 6e \rightleftarrows 2Cr^{3+} + 7H_2O$	1.33	
$Tl^{3+} + 2e \rightleftarrows Tl^+$	1.25	0.77, 1-F HCl
$IO_3^- + 2Cl^- + 6H^+ + 4e \rightleftarrows ICl_2^- + 3H_2O$	1.24	
$MnO_2(s) + 4H^+ + 2e \rightleftarrows Mn^{2+} + 2H_2O$	1.23	1.24, 1-F HClO$_4$
$O_2(g) + 4H^+ + 4e \rightleftarrows 2H_2O$	1.229	

APPENDIX 2 (CONTINUED)
SOME STANDARD AND FORMAL ELECTRODE POTENTIALS[a]

Half-Reaction	E^0, V	Formal Potential, V
$IO_3^- + 6H^+ + 5e \rightleftarrows \frac{1}{2}I_2(s) + 3H_2O$	1.195	
$IO_3^- + 6H^+ + 5e \rightleftarrows \frac{1}{2}I_2(aq) + 3H_2O$	1.178[b]	
$SeO_4^{2-} + 4H^+ + 2e \rightleftarrows H_2SeO_3 + H_2O$	1.15	
$Br_2(l) + 2e \rightleftarrows 2Br^-$	1.065	1.05, 4-F HCl
$Br_2(aq) + 2e \rightleftarrows 2Br^-$	1.087[b]	
$ICl_2^- + e \rightleftarrows \frac{1}{2}I_2(s) + 2Cl^-$	1.06	
$V(OH)_4^+ + 2H^+ + e \rightleftarrows VO^{2+} + 3H_2O$	1.00	1.02, 1-F HCl, HClO$_4$
$HNO_2 + H^+ + e \rightleftarrows NO(g) + H_2O$	1.00	
$Pd^{2+} + 2e \rightleftarrows Pd(s)$	0.987	
$NO_3^- + 3H^+ + 2e \rightleftarrows HNO_2 + H_2O$	0.94	0.92, 1-F HNO$_3$
$2Hg^{2+} + 2e \rightleftarrows Hg_2^{2+}$	0.920	0.907, 1-F HClO$_4$
$HO_2^- + H_2O + 2e \rightleftarrows 3OH^-$	0.88	
$Cu^{2+} + I^- + e \rightleftarrows CuI(s)$	0.86	
$Hg^{2+} + 2e \rightleftarrows Hg(l)$	0.854	
$Ag^+ + e \rightleftarrows Ag(s)$	0.799	0.228, 1-F HCl; 0.792, 1-F HClO$_4$; 0.77, 1-F H$_2$SO$_4$
$Hg_2^{2+} + 2e \rightleftarrows 2Hg(l)$	0.789	0.274, 1-F HCl; 0.776, 1-F HClO$_4$; 0.674, 1-F H$_2$SO$_4$
$Fe^{3+} + e \rightleftarrows Fe^{2+}$	0.771	0.700, 1-F HCl; 0.732, 1-F HClO$_4$; 0.68, 1-F H$_2$SO$_4$
$H_2SeO_3 + 4H^+ + 4e \rightleftarrows Se(s) + 3H_2O$	0.740	
$PtCl_4^{2-} + 2e \rightleftarrows Pt(s) + 4Cl^-$	0.73	
$C_6H_4O_2(quinone) + 2H^+ + 2e \rightleftarrows C_6H_4(OH)_2$	0.699	0.696, 1-F HCl, H$_2$SO$_4$, HClO$_4$
$O_2(g) + 2H^+ + 2e \rightleftarrows H_2O_2$	0.682	
$PtCl_6^{2-} + 2e \rightleftarrows PtCl_4^{2-} + 2Cl^-$	0.68	
$Hg_2SO_4(s) + 2e \rightleftarrows 2Hg(l) + SO_4^{2-}$	0.615	
$Sb_2O_5(s) + 6H^+ + 4e \rightleftarrows 2SbO^+ + 3H_2O$	0.581	
$MnO_4^- + e \rightleftarrows MnO_4^{2-}$	0.564	
$H_3AsO_4 + 2H^+ + 2e \rightleftarrows H_3AsO_3 + H_2O$	0.559	0.577, 1-F HCl, HClO$_4$
$I_3^- + 2e \rightleftarrows 3I^-$	0.536	
$I_2(s) + 2e \rightleftarrows 2I^-$	0.5355	
$I_2(aq) + 2e \rightleftarrows 2I^-$	0.620[b]	
$Cu^+ + e \rightleftarrows Cu(s)$	0.521	
$H_2SO_3 + 4H^+ + 4e \rightleftarrows S(s) + 3H_2O$	0.45	
$Ag_2CrO_4(s) + 2e \rightleftarrows 2Ag(s) + CrO_4^{2-}$	0.446	
$VO^{2+} + 2H^+ + e \rightleftarrows V^{3+} + H_2O$	0.361	
$Fe(CN)_6^{3-} + e \rightleftarrows Fe(CN)_6^{4-}$	0.36	0.71, 1-F HCl; 0.72, 1-F HClO$_4$, H$_2$SO$_4$
$Cu^{2+} + 2e \rightleftarrows Cu(s)$	0.337	
$UO_2^{2+} + 4H^+ + 2e \rightleftarrows U^{4+} + 2H_2O$	0.334	
$BiO^+ + 2H^+ + 3e \rightleftarrows Bi(s) + H_2O$	0.32	
$Hg_2Cl_2(s) + 2e \rightleftarrows 2Hg(l) + 2Cl^-$	0.268	0.242, sat'd KCl; 0.282, 1-F KCl
$AgCl(s) + e \rightleftarrows Ag(s) + Cl^-$	0.222	0.228, 1-F KCl
$SO_4^{2-} + 4H^+ + 2e \rightleftarrows H_2SO_3 + H_2O$	0.17	
$BiCl_4^- + 3e \rightleftarrows Bi(s) + 4Cl^-$	0.16	
$Sn^{4+} + 2e \rightleftarrows Sn^{2+}$	0.154	0.14, 1-F HCl
$Cu^{2+} + e \rightleftarrows Cu^+$	0.153	
$S(s) + 2H^+ + 2e \rightleftarrows H_2S(g)$	0.141	
$TiO^{2+} + 2H^+ + e \rightleftarrows Ti^{3+} + H_2O$	0.1	0.04, 1-F H$_2$SO$_4$
$AgBr(s) + e \rightleftarrows Ag(s) + Br^-$	0.095	
$S_4O_6^{2-} + 2e \rightleftarrows 2S_2O_3^{2-}$	0.08	

APPENDIX 2 (CONTINUED)
SOME STANDARD AND FORMAL ELECTRODE POTENTIALS[a]

Half-Reaction	E^0, V	Formal Potential, V
$Ag(S_2O_3)_2{}^{3-} + e \rightleftarrows Ag(s) + 2S_2O_3{}^{2-}$	0.01	
$2H^+ + 2e \rightleftarrows H_2(g)$	0.000	-0.005, 1-F HCl, HClO$_4$
$Pb^{2+} + 2e \rightleftarrows Pb(s)$	-0.126	-0.14, 1-F HClO$_4$;
		-0.29, 1-F H$_2$SO$_4$
$Sn^{2+} + 2e \rightleftarrows Sn(s)$	-0.136	-0.16, 1-F HClO$_4$
$AgI(s) + e \rightleftarrows Ag(s) + I^-$	-0.151	
$CuI(s) + e \rightleftarrows Cu(s) + I^-$	-0.185	
$N_2(g) + 5H^+ + 4e \rightleftarrows N_2H_5{}^+$	-0.23	
$Ni^{2+} + 2e \rightleftarrows Ni(s)$	-0.250	
$V^{3+} + e \rightleftarrows V^{2+}$	-0.255	-0.21, 1-F HClO$_4$
$Co^{2+} + 2e \rightleftarrows Co(s)$	-0.277	
$Ag(CN)_2{}^- + e \rightleftarrows Ag(s) + 2CN^-$	-0.31	
$Tl^+ + e \rightleftarrows Tl(s)$	-0.336	-0.551, 1-F HCl;
		-0.33, 1-F HClO$_4$, H$_2$SO$_4$
$PbSO_4(s) + 2e \rightleftarrows Pb(s) + SO_4{}^{2-}$	-0.356	
$Ti^{3+} + e \rightleftarrows Ti^{2+}$	-0.37	
$Cd^{2+} + 2e \rightleftarrows Cd(s)$	-0.403	
$Cr^{3+} + e \rightleftarrows Cr^{2+}$	-0.41	
$Fe^{2+} + 2e \rightleftarrows Fe(s)$	-0.440	
$2CO_2(g) + 2H^+ + 2e \rightleftarrows H_2C_2O_4$	-0.49	
$Cr^{3+} + 3e \rightleftarrows Cr(s)$	-0.74	
$Zn^{2+} + 2e \rightleftarrows Zn(s)$	-0.763	
$Mn^{2+} + 2e \rightleftarrows Mn(s)$	-1.18	
$Al^{3+} + 3e \rightleftarrows Al(s)$	-1.66	
$Mg^{2+} + 2e \rightleftarrows Mg(s)$	-2.37	
$Na^+ + e \rightleftarrows Na(s)$	-2.714	
$Ca^{2+} + 2e \rightleftarrows Ca(s)$	-2.87	
$Ba^{2+} + 2e \rightleftarrows Ba(s)$	-2.90	
$K^+ + e \rightleftarrows K(s)$	-2.925	
$Li^+ + e \rightleftarrows Li(s)$	-3.045	

[a] Sources for E^0 values: W. M. Latimer, *The Oxidation States of the Elements and Their Potentials in Aqueous Solutions* 2d ed. Englewood Cliffs, N.J.: Prentice Hall, Inc., 1952; A. J. deBethune and N. A. S. Loud, *Standard Aqueous Electrode Potentials and Temperature Coefficients at 25°C*. Skokie, Ill.: Clifford A. Hampel, 1964. Source of formal potentials: E. H. Swift and E. A. Butler, *Quantitative Measurements and Chemical Equilibria*. W. H. Freeman and Company. Copyright © 1972.

[b] These potentials are hypothetical since they correspond to solutions that are 1.00 M in Br$_2$ or I$_2$. The solubilities of these two compounds at 25°C are 0.21 M and 0.0133 M, respectively. In saturated solutions containing an excess of Br$_2$(l) or I$_2$(s), the standard potentials for the half-reactions Br$_2$(l) + $2e \rightleftarrows 2Br^-$ or I$_2$(s) + $2e \rightleftarrows 2I^-$ should be used. On the other hand, at Br$_2$ and I$_2$ concentrations less than saturation, these hypothetical electrode potentials should be employed.

APPENDIX 3
SOLUBILITY PRODUCT CONSTANTS[a]

Substance	Formula	K_{sp}
Aluminum hydroxide	$Al(OH)_3$	2×10^{-32}
Barium carbonate	$BaCO_3$	5.1×10^{-9}
Barium chromate	$BaCrO_4$	1.2×10^{-10}
Barium iodate	$Ba(IO_3)_2$	1.57×10^{-9}

APPENDIX 3 (CONTINUED)
SOLUBILITY PRODUCT CONSTANTS[a]

Substance	Formula	K_{sp}
Barium manganate	$BaMnO_4$	2.5×10^{-10}
Barium oxalate	BaC_2O_4	2.3×10^{-8}
Barium sulfate	$BaSO_4$	1.3×10^{-10}
Bismuth oxide chloride	$BiOCl$	7×10^{-9}
Bismuth oxide hydroxide	$BiOOH$	4×10^{-10}
Cadmium carbonate	$CdCO_3$	2.5×10^{-14}
Cadmium hydroxide	$Cd(OH)_2$	5.9×10^{-15}
Cadmium oxalate	CdC_2O_4	9×10^{-8}
Cadmium sulfide	CdS	2×10^{-28}
Calcium carbonate	$CaCO_3$	4.8×10^{-9}
Calcium fluoride	CaF_2	4.9×10^{-11}
Calcium oxalate	CaC_2O_4	2.3×10^{-9}
Calcium sulfate	$CaSO_4$	1.2×10^{-6}
Copper(I) bromide	$CuBr$	5.2×10^{-9}
Copper(I) chloride	$CuCl$	1.2×10^{-6}
Copper(I) iodide	CuI	1.1×10^{-12}
Copper(I) thiocyanate	$CuSCN$	4.8×10^{-15}
Copper(II) hydroxide	$Cu(OH)_2$	1.6×10^{-19}
Copper(II) sulfide	CuS	6×10^{-36}
Iron(II) hydroxide	$Fe(OH)_2$	8×10^{-16}
Iron(Ii) sulfide	FeS	6×10^{-18}
Iron(III) hydroxide	$Fe(OH)_3$	4×10^{-38}
Lanthanum iodate	$La(IO_3)_3$	6.2×10^{-12}
Lead carbonate	$PbCO_3$	3.3×10^{-14}
Lead chloride	$PbCl_2$	1.6×10^{-5}
Lead chromate	$PbCrO_4$	1.8×10^{-14}
Lead hydroxide	$Pb(OH)_2$	2.5×10^{-16}
Lead iodide	PbI_2	7.1×10^{-9}
Lead oxalate	PbC_2O_4	4.8×10^{-10}
Lead sulfate	$PbSO_4$	1.6×10^{-8}
Lead sulfide	PbS	7×10^{-28}
Magnesium ammonium phosphate	$MgNH_4PO_4$	3×10^{-13}
Magnesium carbonate	$MgCO_3$	1×10^{-5}
Magnesium hydroxide	$Mg(OH)_2$	1.8×10^{-11}
Magnesium oxalate	MgC_2O_4	8.6×10^{-5}
Manganese(II) hydroxide	$Mn(OH)_2$	1.9×10^{-13}
Manganese(II) sulfide	MnS	3×10^{-13}
Mercury(I) bromide	Hg_2Br_2	5.8×10^{-23}
Mercury(I) chloride	Hg_2Cl_2	1.3×10^{-18}
Mercury(I) iodide	Hg_2I_2	4.5×10^{-29}
Silver arsenate	Ag_3AsO_4	1×10^{-22}
Silver bromide	$AgBr$	5.2×10^{-13}
Silver carbonate	Ag_2CO_3	8.1×10^{-12}
Silver chloride	$AgCl$	1.82×10^{-10}
Silver chromate	Ag_2CrO_4	1.1×10^{-12}
Silver cyanide	$AgCN$	7.2×10^{-11}
Silver iodate	$AgIO_3$	3.0×10^{-8}
Silver iodide	AgI	8.3×10^{-17}
Silver oxalate	$Ag_2C_2O_4$	3.5×10^{-11}
Silver sulfide	Ag_2S	6×10^{-50}
Silver thiocyanate	$AgSCN$	1.1×10^{-12}
Strontium oxalate	SrC_2O_4	5.6×10^{-8}
Strontium sulfate	$SrSO_4$	3.2×10^{-7}

APPENDIX 3 (CONTINUED)
SOLUBILITY PRODUCT CONSTANTS[a]

Substance	Formula	K_{sp}
Thallium(I) chloride	TlCl	1.7×10^{-4}
Thallium(I) sulfide	Tl_2S	1×10^{-22}
Zinc hydroxide	$Zn(OH)_2$	1.2×10^{-17}
Zinc oxalate	ZnC_2O_4	7.5×10^{-9}
Zinc sulfide	ZnS	4.5×10^{-24}

[a] Taken from L. Meites, *Handbook of Analytical Chemistry*, pp. 1–13. New York: McGraw-Hill Book Company, Inc., 1963.

APPENDIX 4
DISSOCIATION CONSTANTS FOR ACIDS[a]

Name	Formula	Dissociation Constant, 25°C		
		K_1	K_2	K_3
Acetic	CH_3COOH	1.75×10^{-5}		
Arsenic	H_3AsO_4	6.0×10^{-3}	1.05×10^{-7}	3.0×10^{-12}
Arsenious	H_3AsO_3	6.0×10^{-10}	3.0×10^{-14}	
Benzoic	C_6H_5COOH	6.14×10^{-5}		
Boric	H_3BO_3	5.83×10^{-10}		
1-Butanoic	$CH_3CH_2CH_2COOH$	1.51×10^{-5}		
Carbonic	H_2CO_3	4.45×10^{-7}	4.7×10^{-11}	
Chloroacetic	$ClCH_2COOH$	1.36×10^{-3}		
Citric	$HOOC(OH)C(CH_2COOH)_2$	7.45×10^{-4}	1.73×10^{-5}	4.02×10^{-7}
Ethylenediamine- tetraacetic	H_4Y	1.0×10^{-2}	2.1×10^{-2} $K_4 = 5.5 \times 10^{-11}$	6.9×10^{-7}
Formic	HCOOH	1.77×10^{-4}		
Fumaric	*trans*-HOOCCH : CHCOOH	9.6×10^{-4}	4.1×10^{-5}	
Glycolic	$HOCH_2COOH$	1.48×10^{-4}		
Hydrazoic	HN_3	1.9×10^{-5}		
Hydrogen cyanide	HCN	2.1×10^{-9}		
Hydrogen fluoride	H_2F_2	7.2×10^{-4}		
Hydrogen peroxide	H_2O_2	2.7×10^{-12}		
Hydrogen sulfide	H_2S	5.7×10^{-8}	1.2×10^{-15}	
Hypochlorous	HOCl	3.0×10^{-8}		
Iodic	HIO_3	1.7×10^{-1}		
Lactic	$CH_3CHOHCOOH$	1.37×10^{-4}		
Maleic	*cis*-HOOCCH : CHCOOH	1.20×10^{-2}	5.96×10^{-7}	
Malic	$HOOCCHOHCH_2COOH$	4.0×10^{-4}	8.9×10^{-6}	
Malonic	$HOOCCH_2COOH$	1.40×10^{-3}	2.01×10^{-6}	
Mandelic	$C_6H_5CHOHCOOH$	3.88×10^{-4}		
Nitrous	HNO_2	5.1×10^{-4}		
Oxalic	HOOCCOOH	5.36×10^{-2}	5.42×10^{-5}	
Periodic	H_5IO_6	2.4×10^{-2}	5.0×10^{-9}	
Phenol	C_6H_5OH	1.00×10^{-10}		

APPENDIX 4 (CONTINUED)
DISSOCIATION CONSTANTS FOR ACIDS[a]

Name	Formula	Dissociation Constant, 25°C		
		K_1	K_2	K_3
Phosphoric	H_3PO_4	7.11×10^{-3}	6.34×10^{-8}	4.2×10^{-13}
Phosphorous	H_3PO_3	1.00×10^{-2}	2.6×10^{-7}	
o-Phthalic	$C_6H_4(COOH)_2$	1.12×10^{-3}	3.91×10^{-6}	
Picric	$(NO_2)_3C_6H_2OH$	5.1×10^{-1}		
Propanoic	CH_3CH_2COOH	1.34×10^{-5}		
Pyruvic	$CH_3COCOOH$	3.24×10^{-3}		
Salicylic	$C_6H_4(OH)COOH$	1.05×10^{-3}		
Sulfamic	H_2NSO_3H	1.03×10^{-1}		
Sulfuric	H_2SO_4	strong	1.20×10^{-2}	
Sulfurous	H_2SO_3	1.72×10^{-2}	6.43×10^{-8}	
Succinic	$HOOCCH_2CH_2COOH$	6.21×10^{-5}	2.32×10^{-6}	
Tartaric	$HOOC(CHOH)_2COOH$	9.20×10^{-4}	4.31×10^{-5}	
Trichloroacetic	Cl_3CCOOH	1.29×10^{-1}		

[a] Taken from L. Meites, *Handbook of Analytical Chemistry*, pp. 1–21. New York: McGraw-Hill Book Company, Inc., 1963.

APPENDIX 5
DISSOCIATION CONSTANTS FOR BASES[a]

Name	Formula	Dissociation Constant, K, 25°C
Ammonia	NH_3	1.76×10^{-5}
Aniline	$C_6H_5NH_2$	3.94×10^{-10}
1-Butylamine	$CH_3(CH_2)_2CH_2NH_2$	4.0×10^{-4}
Dimethylamine	$(CH_3)_2NH$	5.9×10^{-4}
Ethanolamine	$HOC_2H_4NH_2$	3.18×10^{-5}
Ethylamine	$CH_3CH_2NH_2$	4.28×10^{-4}
Ethylenediamine	$NH_2C_2H_4NH_2$	$K_1 = 8.5 \times 10^{-5}$
		$K_2 = 7.1 \times 10^{-8}$
Hydrazine	H_2NNH_2	1.3×10^{-6}
Hydroxylamine	$HONH_2$	1.07×10^{-8}
Methylamine	CH_3NH_2	4.8×10^{-4}
Piperidine	$C_5H_{11}N$	1.3×10^{-3}
Pyridine	C_5H_5N	1.7×10^{-9}
Trimethylamine	$(CH_3)_3N$	6.25×10^{-5}

[a] Taken from L. Meites, *Handbook of Analytical Chemistry*, pp. 1–21. New York: McGraw-Hill Book Company, Inc., 1963.

APPENDIX 6
STEPWISE FORMATION CONSTANTS[a]

Name and Formula, Ligand	Metal Ion	Ionic Strength	K_1	K_2	K_3	K_4
Acetate, CH₃COO⁻	Ag⁺	0	5.4	0.8		
	Cd²⁺	0	5.0×10^1	1.0×10^1	1.4	0.38
	Cu²⁺	3	2.0×10^1	1.1×10^1		
	Hg²⁺	0	1.4×10^2	$K_1K_2 = 2.7 \times 10^8$		
	Pb²⁺	0	2.7×10^2	3.3×10^1		
Ammonia, NH₃	Ag⁺	0	2.0×10^3	6.9×10^3	2.0×10^1	6.2
	Cd²⁺	0	3.2×10^2	9.1×10^1	8.5	4.4
	Co²⁺	0	9.8×10^1	3.2×10^1	$K_6 = 0.18$	$K_5 = 1.1$
	Cu²⁺	0	9.8×10^3	2.2×10^3	5.4×10^2	9.3×10^1
	Hg²⁺	2	6×10^8	5×10^8	1.0×10^1	6
	Ni²⁺	0	4.7×10^2	1.3×10^2	4.1×10^1	1.2×10^1
				$K_5 = 0.43$	$K_6 = 0.11$	
Bromide, Br⁻	Zn²⁺	0	1.9×10^2	2.2×10^2	2.5×10^2	1.1×10^2
	Ag⁺	0	$AgBr(s) + Br^- \rightleftharpoons AgBr_2^-$ $K_{S2} = 2.0 \times 10^{-5}$	$AgBr_2^- + Br^- \rightleftharpoons AgBr_3^{2-}$ $K_3 = 4.6$		
	Hg²⁺	0.5	1.1×10^9	1.9×10^8	2.6×10^2	1.8×10^1
	Pb²⁺	0	1.7×10^1			
Chloride, Cl⁻	Ag⁺	0	$AgCl(s) + Cl^- \rightleftharpoons AgCl_2^-$ $K_{S2} = 2.0 \times 10^{-5}$	$AgCl_2^- + Cl^- \rightleftharpoons AgCl_3^{2-}$ $K_3 = 1$		
	Bi³⁺	2	1.5×10^2	3.6×10^1	1.3×10^1	1.1×10^2
	Cd²⁺	0	1.0×10^2	5.0	0.26	
	Cu⁺	2	$Cu^+ + 2Cl^- \rightleftharpoons CuCl_2^-$		$\beta_2 = 8.7 \times 10^4$	
	Fe²⁺		2.3	1.1		
	Fe³⁺		3.0×10^1	4.5	0.1	
	Hg²⁺	0.5	5.5×10^6	3.0×10^6	8.9	1.1×10^1
	Pb²⁺	0	4.0×10^1	$Pb^{2+} + 3Cl^- \rightleftharpoons PbCl_3^-$	$\beta_3 = 4.8 \times 10^1$	
	Sn²⁺	0	3.2×10^1	5.4	0.62	0.28
Cyanide, CN⁻	Ag⁺	0	$Ag^+ + 2CN^- \rightleftharpoons Ag(CN)_2^-$		$\beta_2 = 1 \times 10^{20}$	
	Cd²⁺	3	5.0×10^5	1.3×10^5	4.3×10^4	3.5×10^3
	Hg²⁺	0.1	1.0×10^{18}	5×10^{16}	7×10^3	1×10^3
	Ni²⁺	0	$Ni^{2+} + 4CN^- \rightleftharpoons Ni(CN)_4^{2-}$		$\beta_4 = 1 \times 10^{22}$	
Ethylenediamine-tetraacetate, Y⁴⁻	See Table 13-2					

Ligand	μ	Ion	K_1	K_2	K_3	K_4	Other
Fluoride, F^-	0.5	Al^{3+}	1.3×10^6	1.0×10^5	7×10^3	5×10^2	$K_5 = 4 \times 10^1$; $K_6 = 5$
Hydroxide, OH^-	0.5	Fe^{3+}	1.8×10^5	1.0×10^4			
	0	Al^{3+}	6×10^9		1.0×10^3		$Al(OH)_3(s) + OH^- \rightleftarrows Al(OH)_4^-$, $K_{S4} \sim 10$
	0.1	Cd^{2+}	$\sim 10^4$				
	0	Cu^{2+}	$\sim 10^6$				
	0	Fe^{2+}	$\sim 10^6$				
	0	Fe^{3+}	1×10^{11}	5×10^{10}			
	0	Hg^{2+}	3×10^{11}				
	0	Ni^{2+}	$\sim 10^4$				
	0	Pb^{2+}	7×10^7				$Pb(OH)_2(s) + OH^- \rightleftarrows Pb(OH)_3^-$, $K_{S3} = 5 \times 10^{-2}$
	0	Zn^{2+}	1×10^4				$Zn(OH)_2(s) + 2OH^- \rightleftarrows Zn(OH)_4^{2-}$, $K_{S4} = 0.13$
	0	Cd^{2+}	1.9×10^2	4.4×10^1	1.2×10^1	1.3×10^1	
Iodide, I^-	0	Cu^+					$CuI(s) + I^- \rightleftarrows CuI_2^-$, $K_{S2} = 8 \times 10^{-4}$
	0.5	Hg^{2+}	7.4×10^{12}	9.0×10^{10}	6.0×10^3	2.1×10^2	
	0	Pb^{2+}	1×10^2				$PbI_2(s) + I^- \rightleftarrows PbI_3^-$, $K_{S3} = 2.2 \times 10^{-5}$; $PbI_3^- + I^- \rightleftarrows PbI_4^{2-}$, $K_4 = 1.4 \times 10^{-4}$
Oxalate, $C_2O_4^{2-}$	0	Al^{3+}	$K_1K_2 = 1 \times 10^{13}$		2×10^3		
	0	Fe^{3+}	2.5×10^9	6×10^6	1×10^4		
	0	Mg^{2+}	6.6×10^3	2.7×10^1			
	2	Mn^{3+}	1.0×10^{10}	3.9×10^6	7×10^2		
	0	Pb^{2+}					$Pb^{2+} + 2C_2O_4^{2-} \rightleftarrows Pb(C_2O_4)_2^{2-}$, 8×10^1
Sulfate, SO_4^{2-}	0	Al^{3+}	1.6×10^3				
	0	Cd^{2+}	2×10^2				
	0	Cu^{2+}	1.6×10^2				
	0	Fe^{3+}	1.1×10^4				
Thiocyanate, SCN^-	0	Ag^+		2×10^1			$AgSCN(s) + SCN^- \rightleftarrows Ag(SCN)_2^-$, $K_{S2} = 6 \times 10^{-8}$
	5	Cu^+		2×10^1			$CuSCN(s) + SCN^- \rightleftarrows Cu(SCN)_2^-$, $K_{S2} = 4 \times 10^{-4}$
	0	Fe^{3+}	1.4×10^2	2×10^1			
	0	Hg^{2+}	$K_1K_2 = 1.8 \times 10^{17}$	5.1×10^2		6×10^1	$\beta_2 = 3.5 \times 10^6$

ᵃ Taken from L. G. Sillén and A. E. Martell, *Stability Constants of Metal-Ion Complexes.* London: The Chemical Society, 1964. With permission.

APPENDIX 7
DESIGNATIONS AND POROSITIES FOR FILTERING CRUCIBLES[a]

Type	Coarse		Medium	Fine	
Glass, Pyrex ®[b]		C(60)	M(15)	F(5.5)	
Glass, Kimax ®[c]	EC(170–220)	C(40–60)	M(10–15)	F(4–55)	VF(2–2.5)
Porcelain, Coors U.S.A. ®[d]		Medium(15)		Fine(5)	Very fine(1.2)
Porcelain, Selas[e]	XF(100)	XFF(40)	#10(8.8)	Extra fine #01(6)	
Aluminum oxide ALUNDUM ®[f]	Extra coarse(30)	Coarse(20)	Medium(5)	Fine(0.1)	

[a] Nominal maximum pore diameter in microns is given in parentheses.
[b] Corning Glass Works, Corning, N.Y.
[c] Owens-Illinois, Toledo, Ohio.
[d] Coors Porcelain Company, Golden, Colo.
[e] Selas Corporation of America, Dresher, Pa.
[f] Norton Company, Worcester, Mass.

APPENDIX 8
DESIGNATIONS CARRIED BY ASHLESS FILTER PAPERS[a]

Manufacturer	Fine Crystals	Moderately Fine Crystals	Coarse Crystals		Gelatinous Precipitates	
Schleicher and Schuell[b]	507, 590 589 blue ribbon	589 white ribbon	589 green ribbon	589 black ribbon	589 black ribbon	589-1H
Munktell[c]	OOH	OK OO	OOR		OOR	
Whatman[d]	42	44, 40	41		41	41H
Eaton-Dikeman[e]	90	80	60		50	

[a] Manufacturers' literature should be consulted for more complete specifications. Tabulated are manufacturer designations of papers suitable for filtration of the indicated type of precipitate.
[b] Schleicher and Schuell, Inc., Keene, N.H.
[c] E. H. Sargent and Company, Chicago, Ill., agents.
[d] H. Reeve Angel and Company, Inc., Clifton, N.J., agents.
[e] Eaton-Dikeman Company, Mount Holly Springs, Pa.

INDEX

Boldface entries refer to specific laboratory directions.

Absolute error(s), 45; in sums and differences, 72

Absorbance, *A*, 506, 508*t*; variables affecting, 546

Absorption of electromagnetic radiation, 504 (*see also* Atomic-absorption spectrometry); applications, 540 (*see also* Absorption spectrometry); by complex ions, 556–559; by inorganic species, 538; instruments for measurement of, 514–533; measurement of, 507 (*see also* Beer's law); by mixtures, 507, 547, **764**; nomenclature relating to, 508*t*; by organic compounds, 538

Absorption filter(s), for photometric analysis, 520

Absorption spectra, 510, 534–537; effect of slit width on, 541

Absorption spectrometry, 500; applications, 540–559, **761–764**; effect of scattered radiation on, 542; errors in, 548–553

Absorptivity, *a*, 508*t* (*see also* Molar absorptivity)

Accuracy, 45; of gravimetric methods, 137; of photometric methods, 548 (*see also* Photometric error)

Acetic acid, apparent dissociation constant for, 109*t*; autoprotolysis constant for, 257*t*; determination of, in vinegar, **732**; neutralization titrations in, 263, **735**, **755**; as a solvent, 9, 261

Acid(s), coulometric titration of, 458*t*; definitions of, 7; dissociation constants for, Appendix 4 (*see also* Dissociation constant); pH calculation for solutions of, 30 (*see also* pH calculation); potentiometric titration of, 412, **754**; preparation and standardization of, 239, **729**, **736**; primary standard, 243; primary standards for, 239; titration of, **731**; titration curves for, 193 (*see also* Titration curve)

Acid-base concepts, 7 (*see also* Brønsted-Lowry concepts)

Acid-base equilibria, effect of ionic strength upon, 200; quantitative treatment of, 30, 187–234

Acid-base indicator(s), preparation of solutions, **729**; selection of, 195, 210, 264; theory of behavior of, 188; transition pH range for, 190*t*

Acid-base ratio, **730**, **736**

Acid-base titration(s), applications of, 238, 731

FOUR-PLACE LOGARITHMS OF NUMBERS

n	0	1	2	3	4	5	6	7	8	9
10	0000	0043	0086	0128	0170	0212	0253	0294	0334	0374
11	0414	0453	0492	0531	0569	0607	0645	0682	0719	0755
12	0792	0828	0864	0899	0934	0969	1004	1038	1072	1106
13	1139	1173	1206	1239	1271	1303	1335	1367	1399	1430
14	1461	1492	1523	1553	1584	1614	1644	1673	1703	1732
15	1761	1790	1818	1847	1875	1903	1931	1959	1987	2014
16	2041	2068	2095	2122	2148	2175	2201	2227	2253	2279
17	2304	2330	2355	2380	2405	2430	2455	2480	2504	2529
18	2553	2577	2601	2625	2648	2672	2695	2718	2742	2765
19	2788	2810	2833	2856	2878	2900	2923	2945	2967	2989
20	3010	3032	3054	3075	3096	3118	3139	3160	3181	3201
21	3222	3243	3263	3284	3304	3324	3345	3365	3385	3404
22	3424	3444	3464	3483	3502	3522	3541	3560	3579	3598
23	3617	3636	3655	3674	3692	3711	3729	3747	3766	3784
24	3802	3820	3838	3856	3874	3892	3909	3927	3945	3962
25	3979	3997	4014	4031	4048	4065	4082	4099	4116	4133
26	4150	4166	4183	4200	4216	4232	4249	4265	4281	4298
27	4314	4330	4346	4362	4378	4393	4409	4425	4440	4456
28	4472	4487	4502	4518	4533	4548	4564	4579	4594	4609
29	4624	4639	4654	4669	4683	4698	4713	4728	4742	4757
30	4771	4786	4800	4814	4829	4843	4857	4871	4886	4900
31	4914	4928	4942	4955	4969	4983	4997	5011	5024	5038
32	5051	5065	5079	5092	5105	5119	5132	5145	5159	5172
33	5185	5198	5211	5224	5237	5250	5263	5276	5289	5302
34	5315	5328	5340	5353	5366	5378	5391	5403	5416	5428
35	5441	5453	5465	5478	5490	5502	5514	5527	5539	5551
36	5563	5575	5587	5599	5611	5623	5635	5647	5658	5670
37	5682	5694	5705	5717	5729	5740	5752	5763	5775	5786
38	5798	5809	5821	5832	5843	5855	5866	5877	5888	5899
39	5911	5922	5933	5944	5955	5966	5977	5988	5999	6010
40	6021	6031	6042	6053	6064	6075	6085	6096	6107	6117
41	6128	6138	6149	6160	6170	6180	6191	6201	6212	6222
42	6232	6243	6253	6263	6274	6284	6294	6304	6314	6325
43	6335	6345	6355	6365	6375	6385	6395	6405	6415	6425
44	6435	6444	6454	6464	6474	6484	6493	6503	6513	6522
45	6532	6542	6551	6561	6571	6580	6590	6599	6609	6618
46	6628	6637	6646	6656	6665	6675	6684	6693	6702	6712
47	6721	6730	6739	6749	6758	6767	6776	6785	6794	6803
48	6812	6821	6830	6839	6848	6857	6866	6875	6884	6893
49	6902	6911	6920	6928	6937	6946	6955	6964	6972	6981
50	6990	6998	7007	7016	7024	7033	7042	7050	7059	7067
51	7076	7084	7093	7101	7110	7118	7126	7135	7143	7152
52	7160	7168	7177	7185	7193	7202	7210	7218	7226	7235
53	7243	7251	7259	7267	7275	7284	7292	7300	7308	7316
54	7324	7332	7340	7348	7356	7364	7372	7380	7388	7396